Operator Theory: Advances and Applications
Vol. 151

Aspects of Boundary Problems in Analysis and Geometry

Juan Gil
Thomas Krainer
Ingo Witt
Editors

Advances in Partial Differential Equations

Springer Basel AG

Editors:

Juan Gil
Department of Mathematics
Penn State Altoona
3000 Ivyside Park
Altoona, PA 16601-3760
USA
e-mail: jgil@psu.edu

Thomas Krainer
Ingo Witt
Institute of Mathematics
University of Potsdam
P.O. Box 60 15 53
14415 Potsdam
Germany
e-mail: krainer@math.uni-potsdam.de
ingo@math.uni-potsdam.de

2000 Mathematics Subject Classification 58Jxx, 35Jxx, 53Cxx, 19Kxx, 19Lxx, 35Sxx

A CIP catalogue record for this book is available from the
Library of Congress, Washington D.C., USA

Bibliographic information published by Die Deutsche Bibliothek
Die Deutsche Bibliothek lists this publication in the Deutsche Nationalbibliografie; detailed bibliographic data is available in the Internet at <http://dnb.ddb.de>

ISBN 978-3-0348-9595-8 ISBN 978-3-0348-7850-0 (eBook)
DOI 10.1007/978-3-0348-7850-0

Originally published by Birkhäuser Verlag, Basel - Boston - Berlin in 2004
Softcover reprint of the hardcover 1st edition 2004

Printed on acid-free paper produced from chlorine-free pulp. TCF ∞
Cover design: Heinz Hiltbrunner, Basel

ISBN 978-3-0348-9595-8

9 8 7 6 5 4 3 2 1

www.birkhauser-science.

Contents

Index defects in the theory of spectral boundary value problems
by ANTON SAVIN AND BORIS STERNIN

Part II. Elliptic Boundary Value Problems

Toeplitz operators, and ellipticity of boundary value problems with global projection conditions

by B.-W. Schulze

On the tangential oblique derivative problem - methods, results, open problems
by Peter Popivanov

A note on boundary value problems on manifolds with cylindrical ends
by Marius Mitrea and Victor Nistor

Relative elliptic theory
by Vladimir Nazaikinskii and Boris Sternin

Preface

The present volume consists of invited contributions by outstanding experts in the field of boundary problems in analysis and geometry. This subject is in the center of current research activities and encompasses a variety of connections and interactions. Despite of their differences in scope and methods, boundary problems constitute an essential field of common interest. The intention of this volume is to highlight several analytic and geometric aspects of boundary problems with special emphasis on their interplay. It includes surveys on classical topics presented from a modern perspective as well as reports on current research.

The collection of articles in this book roughly splits into two related groups: The first group concerning the analysis and geometry of geometric operators and their index theory; the second one dealing with elliptic theory of boundary value problems and the Shapiro-Lopatinsky condition. There is a total of nine papers.

The paper by Bleecker and Booss-Bavnbek is devoted to classical aspects of the index of Dirac operators on manifolds with boundary, and gives an extensive introduction to general index theory. In particular, the Atiyah-Patodi-Singer boundary condition and the η-invariant are studied in detail.

In the second paper, Loya discusses the heat equation method for Dirac operators on manifolds with boundary having metrics that degenerate near the boundary. Generalized η-invariants are studied as they appear as global correction terms to the topological index.

Starting from the classical index problem on manifolds with boundary, Savin and Sternin develop a general framework for the index defect arising in the investigation of topological and analytical indices. Index defects are considered in several examples.

In the paper by Benameur, Brodzki, and Nistor, the authors study index theory by adopting ideas from noncommutative geometry in terms of algebras of pseudodifferential operators and associated homologies and cohomologies.

Bunke and Ma introduce some aspects of index and secondary index theory. Diverse K- and L-functors are considered, including a thorough discussion of their various interrelations, and their main properties are derived building upon the machinery of η-forms and adiabatic limits.

The article by Schulze deals with the parametrix construction for elliptic boundary value problems that do not satisfy the Shapiro-Lopatinsky condition, as it is the case for the Atiyah-Patodi-Singer problem. This involves the consideration of projections defining subspaces of admissible boundary data, and gives rise to the study of Toeplitz operators.

In the next article, Popivanov gives a survey on the tangential oblique derivative problem. Historically, this problem laid the ground for various developments in analysis that were later particularly applied to solve many geometric problems. The author demonstrates the use of several analytic techniques and their interplay for the problem at hand, emphasizing their methodical background.

The paper by Mitrea and Nistor is concerned with the Laplace equation on a manifold with cylindrical ends under Dirichlet and Neumann boundary conditions. The authors illustrate the use of the layer potential method from harmonic analysis combined with pseudodifferential techniques on the boundary.

The final paper by Nazaikinskii and Sternin generalizes concepts from elliptic boundary value problems to the case of surfaces of higher codimension. This includes, in particular, the construction of a related algebra of pseudodifferential operators. A generalization of the Shapiro-Lopatinsky condition is considered, and the index problem is investigated in K-theoretical terms.

The initiative for editing this volume emerged during the workshop "Ellipticity and Parabolicity in Analysis and Geometry" held at the University of Potsdam, Germany, in August 2001. We wish to express our gratitude to the participants of the workshop for many valuable discussions and for suggesting to edit this collection. We would also like to thank Professor B.-W. Schulze for his invaluable support and for promoting this project. Finally, we are grateful to the Deutsche Forschungsgemeinschaft (DFG) and the EU Research and Training Network "Geometric Analysis" for funding the workshop, and to Birkhäuser Verlag for publishing the volume.

Juan Gil, Thomas Krainer, and Ingo Witt.

Operator Theory:
Advances and Applications, Vol. 151, 1–130

Spectral invariants of operators of Dirac type on partitioned manifolds

David Bleecker and Bernhelm Booss–Bavnbek

ABSTRACT. We review the concepts of the *index* of a Fredholm operator, the *spectral flow* of a curve of self-adjoint Fredholm operators, the *Maslov index* of a curve of Lagrangian subspaces in a symplectic Hilbert space, and the *eta invariant* of operators of Dirac type on closed manifolds and manifolds with boundary. We emphasize various (occasionally overlooked) aspects of rigorous definitions and explain the quite different stability properties. Moreover, we utilize the heat equation approach in various settings and show how these topological and spectral invariants are mutually related in the study of additivity and nonadditivity properties on partitioned manifolds.

Introduction

For many decades now, workers in differential geometry and mathematical physics have been increasingly concerned with differential operators (exterior differentiation, connections, Laplacians, Dirac operators, etc.) associated to underlying Riemannian or space-time manifolds. Of particular interest is the interplay between the spectral decomposition of such operators and the geometry/topology of the underlying manifold. This has become a large, diverse field involving index theory, the distribution of eigenvalues, zero sets of eigenfunctions, Green functions, heat and wave kernels, families of elliptic operators and their determinants, canonical sections, etc.. Moreover, Donaldson's analysis of moduli of solutions of the nonlinear Yang-Mills equations and Seiberg-Witten theory have led to profound insights into the classification of four-manifolds, which were not accessible by techniques that are effective in higher dimensions.

We shall touch upon many of these topics, but we focus on three invariants characterizing the asymmetry of the spectrum of operators of Dirac type: the *index* which gives the chiral asymmetry of the kernel or null space (i.e., the difference in the number of independent left and right zero modes) of a total Dirac operator; the

2000 *Mathematics Subject Classification.* 53C21, 58G30.

Key words and phrases. Determinant, Dirac operators, eta invariant, heat equation, index, Maslov index, partitioned manifolds, pasting formulas, spectral flow, symplectic analysis.

spectral flow of a curve of Dirac operators that counts the net number of eigenvalues moving from the negative half line over to the positive; and the *eta invariant* which describes the overall asymmetry of the spectrum of a Dirac operator. The three invariants appear both for Dirac operators and curves of Dirac operators on a closed manifold or on a smooth compact manifold with boundary subject to suitable boundary conditions.

Index and spectral flow can be described in general functional analytic terms, namely for bounded and unbounded, closed Fredholm operators and *curves* of bounded and unbounded self-adjoint Fredholm operators. Correspondingly, stability properties of index and spectral flow and their topological and geometric meaning are relatively well understood. There is, however, not an easily identifiable operator class for eta invariant, and its behavior under perturbations is rather delicate.

Most significant differences between the behavior of these three invariants are met when we address splitting properties on partitioned manifolds (with product metrics assumed near the partitioning hypersurface). It is well known that the index can be described in local terms. This is one aspect of the Atiyah-Singer Index Theorem. So, splitting formulas for the index are relatively easily obtainable, once one specifies and understands boundary and transmission conditions for the gluing of the parts of the underlying manifold. In this process, precise additivity is obtained only for a few invariants, Euler characteristic and signature, which actually can be characterized by cutting and pasting invariance. In general, an error term appears which can be expressed either by spectral flow or, equivalently, by the index of a boundary value problem on a cylinder over the separating hypersurface or, alternatively, by the index of a suitable Fredholm pair.

Surprisingly, simple splitting formulas can be obtained also for the spectral flow and eta invariant. This is particularly surprising for the eta invariant where we only have a local formula for the first derivative. Now the integer error terms are expressed by the Maslov index for curves of Cauchy data spaces. These formulas relate the symmetric category of self-adjoint Dirac operators over closed partitioned manifolds (and self-adjoint boundary value problems over compact manifolds with boundary) to the symplectic analysis of Lagrangian subspaces (the Cauchy data spaces).

One message of this review is, that index, spectral flow, and eta invariant, in spite of their quite different appearance, share various features which become most visible on partitioned manifolds. Roughly speaking, one reason for that is that the spectral flow of a path of Dirac type operators (say, on a closed manifold) with unitarily equivalent ends $A_1 = gA_0g^{-1}$ equals the index of the induced suspension operator $\partial_t + A_t$ on the underlying mapping torus (see [**BoWo93**, Theorems 17.3, 17.17, and Proposition 25.1]). Another reason is that the integer part of the derivative of the eta invariant along a path of Dirac operators or boundary problems can be expressed by the spectral flow (see, e.g., [**DoWo91**], [**LeWo96**, p. 39], and [**KiLe00**, Section 3]).

We also summarize recent discussion on additivity and non-additivity of the zeta regularized determinant. It should be mentioned that the various splitting formulas all depend decisively on the well-established unique continuation property (UCP) for operators of Dirac type. We give a full proof of weak UCP below in Section 1.4. Moreover (modulo some technicalities for the computation of the index density form, to appear in our forthcoming book [**BlBo04**]), we provide proofs of the Atiyah-Singer and Atiyah-Patodi-Singer Index Theorems in important special cases using heat equation methods. That we explain concrete calculations only for the case of the Atiyah-Patodi-Singer (spectral – "APS") boundary condition is no big loss of generality since each admissible boundary condition for an operator of Dirac type $\mathcal{D}$ with tangential part $\mathcal{B}$ can be written as the APS projection of a perturbed operator $\mathcal{B}'$ (see below Lemma 2.29, following a recent result, Gerd Grubb [**Gr02**]).

Acknowledgments. The second author wants to thank K. P. Wojciechowski (Indiana University – Purdue University Indianapolis) for many discussions about the subject(s) of this review and the Erwin Schrödinger International Institute for Mathematical Physics at Vienna for generous hospitality during the finalizing phase of this review. We both thank the referees for their corrections, thoughtful comments, and helpful suggestions which led to many improvements. They clearly went beyond the call of duty, and we are in their debt.

1. Basic notations and results

1.1. Index of Fredholm operators and spectral flow of curves of self-adjoint Fredholm operators.

1.1.1. *Notation.* Let H be a separable complex Hilbert space. First let us introduce some notation for various spaces of operators in H:

$$\begin{aligned}
\mathcal{C}(H) &:= \text{closed, densely defined operators on } H,\\
\mathcal{B}(H) &:= \text{bounded linear operators } H \to H,\\
\mathcal{U}(H) &:= \text{unitary operators } H \to H,\\
\mathcal{K}(H) &:= \text{compact linear operators } H \to H,\\
\mathcal{F}(H) &:= \text{bounded Fredholm operators } H \to H,\\
\mathcal{CF}(H) &:= \text{closed, densely defined Fredholm operators on } H.
\end{aligned}$$

If no confusion is possible we will omit "(H)" and write $\mathcal{C}$, $\mathcal{B}, \mathcal{K}$, etc.. By $\mathcal{C}^{\mathrm{sa}}, \mathcal{B}^{\mathrm{sa}}$, etc., we denote the set of self-adjoint elements in $\mathcal{C}$, $\mathcal{B}$, etc..

1.1.2. *Operators with index – Fredholm operators.* The topology of the operator spaces $\mathcal{U}(H)$, $\mathcal{F}(H)$, and $\mathcal{F}^{\mathrm{sa}}(H)$ is quite well understood. The key results are:

(1) the *Kuiper Theorem* which states that $\mathcal{U}(H)$ is contractible;
(2) the *Atiyah-Jänich Theorem* which states that $\mathcal{F}(H)$ is a classifying space for the functor K. Explicitly, the construction of the index bundle of

a continuous family of bounded Fredholm operators parametrized over a compact topological space X yields a homomorphism of semigroups, namely index : $[X, \mathcal{F}] \to K(X)$. In particular, the index is a homotopy invariant, and it provides a one-to-one correspondence of the connected components $[point, \mathcal{F}]$ of $\mathcal{F}$ with $K(point) = \mathbb{Z}$.

(3) the corresponding observation that $\mathcal{F}^{\mathrm{sa}}(H)$ consists of three connected components, the contractible subsets $\mathcal{F}^{\mathrm{sa}}(H)_+$ and $\mathcal{F}^{\mathrm{sa}}(H)_-$ of essentially positive, respectively essentially negative, Fredholm operators, and the topologically nontrivial component $\mathcal{F}^{\mathrm{sa}}_*(H)$ which is a classifying space for the functor K^{-1}. In particular, the spectral flow gives an isomorphism of the fundamental group $\pi_1(\mathcal{F}^{\mathrm{sa}}_*(H))$ onto the integers.

Full proofs can be found in [**BlBo04**] for items 1 and 2, and in [**AtSi69**], [**BoWo93**], and [**Ph96**] for item 3.

So much for the bounded case. Of course, the Dirac operators of interest to us are not bounded in L^2. On closed manifolds, however, they can be considered as bounded operators from the first Sobolev space H^1 into L^2, and by identifying these two Hilbert spaces we can consider a Dirac operator as a bounded operator of a Hilbert space in itself. The same philosophy can be applied to Dirac operators on compact manifolds with boundary when we consider the domain not as dense subspace in L^2 but as Hilbert space (with the graph inner product) and then identify.

Strictly speaking, the concept of unbounded operator is dispensable here. However, for varying boundary conditions it is necessary to keep the distinction and to follow the variation of the domain as a variation of subspaces in L^2.

Following [**CorLab**], we generalize the concept of *Fredholm operators* to the unbounded case.

DEFINITION 1.1. Let H be a complex separable Hilbert space. A linear (not necessarily bounded) operator F with domain $\mathrm{Dom}(F)$, null-space $\mathrm{Ker}(F)$, and range $\mathrm{Im}(F)$ is called *Fredholm* if the following conditions are satisfied.

(i) $\mathrm{Dom}(F)$ is dense in H.
(ii) F is closed.
(iii) The range $\mathrm{Im}(F)$ of F is a closed subspace of H.
(iv) Both $\dim \mathrm{Ker}(F)$ and $\operatorname{codim} \mathrm{Im}(F) = \dim \mathrm{Im}(F)^\perp$ are finite. The difference of the dimensions is called $\mathrm{index}(F)$.

So, a closed operator F is characterized as a Fredholm operator by the same properties as in the bounded case. Moreover, as in the bounded case, F is Fredholm if and only if F^* is Fredholm (proving the closedness of $\mathrm{Im}(F^*)$ is delicate: see [**CorLab**, Lemma 1.4]) and (clearly) we have

$$\mathrm{index}\, F = \dim \mathrm{Ker}\, F - \dim \mathrm{Ker}\, F^* = -\,\mathrm{index}\, F^*.$$

In particular, $\mathrm{index}\, F = 0$ in case F is self-adjoint.

The composition of (not necessarily bounded) Fredholm operators yields again a Fredholm operator. More precisely, we have the following composition rule. For the proof, which is considerably more involved than that in the bounded case, we refer to [**GoKr57**], [**CorLab**, Lemma 2.3 and Theorem 2.1].

THEOREM 1.2 (Gohberg, Krein). *If F and G are (not necessarily bounded) Fredholm operators then their product GF is densely defined with*

$$\operatorname{Dom}(GF) = \operatorname{Dom}(F) \cap F^{-1} \operatorname{Dom}(G)$$

and is a Fredholm operator. Moreover,

$$\operatorname{index} GF = \operatorname{index} F + \operatorname{index} G .$$

1.1.3. *Metrics on the space of closed operators.* For S, $T \in \mathcal{C}(H)$ the orthogonal projections $P_{\mathfrak{G}(S)}$, $P_{\mathfrak{G}(T)}$ onto the graphs of S, T in $H \oplus H$ are bounded operators and

$$\gamma(S,T) := \| P_{\mathfrak{G}(T)} - P_{\mathfrak{G}(S)} \|$$

defines a metric for $\mathcal{C}(H)$, the *projection metric.*

It is also called the *gap metric* and it is (uniformly) equivalent with the metric given by measuring the distance between the (closed) graphs, namely

$$d(\mathfrak{G}(S), \mathfrak{G}(T)) := \sup_{\{x \in \mathfrak{G}(S) : \|x\|=1\}} d\left(x, \mathfrak{G}(T)\right) + \sup_{\{x \in \mathfrak{G}(T) : \|x\|=1\}} d\left(x, \mathfrak{G}(S)\right) .$$

For details and the proof of the following lemma and theorem, we refer to [**CorLab**, Section 3].

LEMMA 1.3. *For $T \in \mathcal{C}(H)$ the orthogonal projection onto the graph of T in $H \oplus H$ can be written (where $R_T := (I + T^*T)^{-1}$) as*

$$P_{\mathfrak{G}(T)} = \begin{pmatrix} R_T & R_T T^* \\ T R_T & T R_T T^* \end{pmatrix} = \begin{pmatrix} R_T & T^* R_{T^*} \\ T R_T & T T^* R_{T^*} \end{pmatrix} = \begin{pmatrix} R_T & T^* R_{T^*} \\ T R_T & I - R_{T^*} \end{pmatrix} .$$

THEOREM 1.4 (Cordes, Labrousse).

(a) *The space $\mathcal{B}(H)$ of bounded operators is dense in the space $\mathcal{C}(H)$ of all closed operators in H. The topology induced by the projection ($\cong$ gap) metric on $\mathcal{B}(H)$ is equivalent to that given by the operator norm.*

(b) *Let $\mathcal{CF}(H)$ denote the space of closed (not necessarily bounded) Fredholm operators. Then the index is constant on the connected components of $\mathcal{CF}(H)$ and yields a bijection between the integers and the connected components.*

EXAMPLE 1.5. Let H be a Hilbert space with $e_1, e_2, \ldots$ a complete orthonormal system (i.e., an orthonormal basis for H). Consider the multiplication operator M_{id}, given by the domain

$$D := \operatorname{Dom}(M_{\text{id}}) := \left\{ \sum_{j=1}^{\infty} c_j e_j \mid \sum_{j=1}^{\infty} j^2 |c_j|^2 < +\infty \right\}$$

and the operation

$$D \ni u = \sum c_j e_j \quad \mapsto \quad M_{\text{id}}(u) := \sum j c_j e_j .$$

It is a densely defined closed operator which is injective and surjective. Let P_n denote the orthogonal projection of H onto the linear span of the n-th orthonormal basis element e_n. Clearly the sequence (P_n) does not converge in $\mathcal{B}(H)$ in the operator norm. However, the sequence $M_{\mathrm{id}} - 2nP_n$ of self-adjoint Fredholm operators converges in $\mathcal{C}(H)$ with the projection metric to M_{id}.

This can be seen by the following argument: On the subset of self-adjoint (not necessarily bounded) operators in the space $\mathcal{C}^{\mathrm{sa}}(H)$ the projection metric is uniformly equivalent to the metric γ given by

$$\gamma(A_1, A_2) := \|(A_1 + i)^{-1} - (A_2 + i)^{-1}\|\,,$$

(see below Theorem 1.10). Then for $T_n := M_{\mathrm{id}} - 2nP_n$,

$$\left\| (T_n + i)^{-1} - (M_{\mathrm{id}} + i)^{-1} \right\| = \left\| (-n+i)^{-1} e_n - (n+i)^{-1} e_n \right\| = \tfrac{2n}{n^2+1} \to 0.$$

REMARK 1.6. The results by Heinz Cordes and Jean-Philippe Labrousse may appear to be rather counter-intuitive. For (a), it is worth mentioning that the operator-norm distance and the projection metric on the set of bounded operators are equivalent but not uniformly equivalent since the operator norm is complete while the projection metric is not complete on the set of bounded operators. Actually, this is the point of the first part of (a); see also the preceding example.

Assertion (b) says two things: (i) that the index is a homotopy invariant, i.e., two Fredholm operators have the same index if they can be connected by a continuous curve in $\mathcal{CF}(H)$; (ii) that two Fredholm operators having the same index always can be connected by a continuous curve in $\mathcal{CF}(H)$. Note that the topological results are not as far reaching as for bounded Fredholm operators.

1.1.4. *Self-adjoint Fredholm operators and spectral flow.* We investigate the topology of the subspace of self-adjoint (not necessarily bounded) Fredholm operators. Many users of the notion of spectral flow feel that the definition and basic properties are too trivial to bother with. However, there are some difficulties both with extending the definition of spectral flow from loops to paths and from curves of bounded self-adjoint Fredholm operators to curves of not necessarily bounded self-adjoint Fredholm operators.

To overcome the second difficulty, the usual way is to apply the *Riesz transformation* which yields a bijection

$$\mathcal{R} : \mathcal{C}^{\mathrm{sa}} \to \{S \in \mathcal{B}^{\mathrm{sa}} \mid \|S\| \leq 1 \text{ and } S \pm I \text{ both injective}\}, \text{ where} \tag{1.1}$$
$$\mathcal{R}\,(T) := T(I + T^2)^{-1/2}.$$

In [**BoFu98**] the following theorem was proved:

THEOREM 1.7. *Let S be a self-adjoint operator with compact resolvent in a real separable Hilbert space $\mathcal{H}$ and let C be a bounded self-adjoint operator. Then the sum $S + C$ also has compact resolvent and is a closed Fredholm operator. We have*

$$\|\mathcal{R}(S + C) - \mathcal{R}(S)\| \leq c\|C\|\,,$$

where the constant c does not depend on S or on C.

The preceding theorem is applied in the following form:

COROLLARY 1.8. *Curves of self-adjoint (unbounded) Fredholm operators in a separable real Hilbert space of the form $\{S+C_t\}_{t\in I}$ are mapped into continuous curves in $\mathcal{F}^{\mathrm{sa}}$ by the transformation $\mathcal{R}$ when S is a self-adjoint operator with compact resolvent and $\{C_t\}_{t\in I}$ is a continuous curve of bounded self-adjoint operators.*

REMARK 1.9. Define

$$\mathcal{T}_S : \mathcal{B}^{\mathrm{sa}} \to \mathcal{CF}^{\mathrm{sa}} \text{ by } \mathcal{T}_S(C) := S + C$$

This is translation by S, mapping bounded self-adjoint operators on $\mathcal{H}$ into self-adjoint Fredholm operators in $\mathcal{H}$. On $\mathcal{CF}^{\mathrm{sa}}$, the gap topology is defined by the metric

$$g(A_1, A_2) := \sqrt{\|R_{A_1} - R_{A_2}\|^2 + \|A_1 R_{A_1} - A_2 R_{A_2}\|^2}\,,$$

where $R_{A_j} := (I + A_j^2)^{-1}$ as before (see Cordes and Labrousse, [**CorLab**] and also Kato, [**Kat**]). Theorem 1.7 says that the composition $\mathcal{R}\circ\mathcal{T}_S$ is continuous.

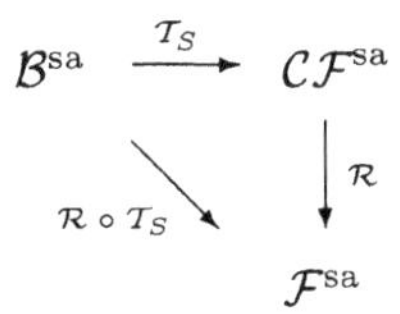

Further, we can prove that the translation operator $\mathcal{T}_S$ is a continuous operator from $\mathcal{B}^{\mathrm{sa}}$ onto the subspace $\mathcal{B}^{\mathrm{sa}} + S \subset \mathcal{CF}^{\mathrm{sa}}$.

The preceding arguments permit to treat continuous curves of Dirac operators in the same way as continuous curves of self-adjoint bounded Fredholm operators under the precondition that the domain is fixed and the perturbation is only by bounded self-adjoint operators. That precondition is satisfied when we have a curve of Dirac operators on a closed manifold which differ only by the underlying connection. It is also satisfied for curves of Dirac operators on a manifold with boundary as long the perturbation is bounded. In particular, this demands that the domain remains fixed.

A closer look at Example 1.5 shows that the preceding argument cannot be generalized: The sequence of the Riesz transforms $\mathcal{R}(T_n)$ of $T_n := M_{\mathrm{id}} - 2nP_n$ does not converge, since $\mathcal{R}(T_n)$ is not a Cauchy sequence:

$$\|\mathcal{R}(T_n) - \mathcal{R}(T_{n+1})\| \geq \|\mathcal{R}(T_n)e_n - \mathcal{R}(M_{\mathrm{id}})\, e_n\|$$
$$= \left\| \tfrac{n-2n}{\sqrt{1+(n-2n)^2}} e_n - \tfrac{n}{\sqrt{1+n^2}} e_n \right\| = \tfrac{2n}{\sqrt{1+n^2}} \to 2 \ \text{ as } n \to \infty.$$

In particular, the Riesz transformation is not continuous on the whole space $\mathcal{C}^{\mathrm{sa}}$, nor on $\mathcal{CF}^{\mathrm{sa}}$, neither on the whole space of self-adjoint operators with compact resolvent. Other methods are needed for working with varying domains.

Of course, we can define a different metric in $\mathcal{C}^{\text{sa}}$; e.g. the metric which makes the Riesz transformation a homeomorphism. That approach was chosen by L. Nicolaescu in [**Ni00**]. He shows that quite a large class of naturally arising curves of Dirac operators with varying domain are continuous under this 'Riesz metric', as opposed to the aforementioned example.

Here, we choose a different approach and adopt the gap metric. Note that continuity in the gap metric is much easier to establish than continuity in the Riesz metric. We follow [**BoLePh01**] where the proofs can be found. A feature of this approach is the use of the Cayley Transform:

THEOREM 1.10.

(a) *On $\mathcal{C}^{\text{sa}}$ the gap metric is (uniformly) equivalent to the metric γ given by*

$$\gamma(T_1,T_2) = \|(T_1+i)^{-1} - (T_2+i)^{-1}\|.$$

(b) *Let $\kappa : \mathbb{R} \to S^1 \setminus \{1\}$, $x \mapsto \frac{x-i}{x+i}$ denote the Cayley transform. Then κ induces a homeomorphism*

$$\begin{aligned} \kappa : \mathcal{C}^{\text{sa}}(H) &\to \{U \in \mathcal{U}(H) \mid U - I \text{ is injective } \} =: \mathcal{U}_{\text{inj}} \\ T &\mapsto \kappa(T) = (T-i)(T+i)^{-1}. \end{aligned} \tag{1.2}$$

More precisely, the gap metric is (uniformly) equivalent to the metric $\tilde{\delta}$ defined by $\tilde{\delta}(T_1,T_2) = \|\kappa(T_1) - \kappa(T_2)\|$.

We note some immediate consequences of the Cayley picture:

COROLLARY 1.11.

(a) *With respect to the gap metric the set $\mathcal{B}^{\text{sa}}(H)$ is dense in $\mathcal{C}^{\text{sa}}(H)$.*

(b) *For $\lambda \in \mathbb{R}$ the sets*

$$\{T \in \mathcal{C}^{\text{sa}}(H) \mid \lambda \notin \operatorname{spec} T\} \quad \textit{and} \quad \{T \in \mathcal{C}^{\text{sa}}(H) \mid \lambda \notin \operatorname{spec}_{\text{ess}} T\}$$

are open in the gap topology.

(c) *The set of (not necessarily bounded) self-adjoint Fredholm operators*

$$\mathcal{CF}^{\text{sa}} = \{T \in \mathcal{C}^{\text{sa}} \mid 0 \notin \operatorname{spec}_{\text{ess}} T\} = \kappa^{-1}({}_{\mathcal{F}}\mathcal{U}), \ \textit{where}$$

${}_{\mathcal{F}}\mathcal{U} \coloneqq \{U \in \mathcal{U} \mid -1 \notin \operatorname{spec}_{\text{ess}} U\} = \{U \in \mathcal{U} \mid U + I \textit{ Fredholm operator}\}$, *is open in $\mathcal{C}^{\text{sa}}$.*

The preceding corollary implies that the set $\mathcal{F}^{\text{sa}}$ is dense in $\mathcal{CF}^{\text{sa}}$ with respect to the gap metric.

Contrary to the bounded case, we have the following somewhat surprising result in the unbounded case: In particular, it shows that not every gap continuous path in $\mathcal{CF}^{\text{sa}}$ with endpoints in $\mathcal{F}^{\text{sa}}$ can be continuously deformed into an operator norm continuous path in $\mathcal{F}^{\text{sa}}$, in spite of the density of $\mathcal{F}^{\text{sa}}$ in $\mathcal{CF}^{\text{sa}}$.

THEOREM 1.12.

(a) *$\mathcal{CF}^{\text{sa}}$ is path connected with respect to the gap metric.*

(b) *Moreover, its Cayley image*

$$ {}_{\mathcal{F}}\mathcal{U}_{\text{inj}} := \{U \in \mathcal{U} \mid U + I \text{ Fredholm and } U - I \text{ injective}\} = \kappa(\mathcal{CF}^{\text{sa}}) $$

is dense in ${}_{\mathcal{F}}\mathcal{U}$.

PROOF. (a) Once again we look at the Cayley transform picture. Note that so far we have introduced three different subsets of unitary operators

$$ \begin{aligned} \mathcal{U}_{\text{inj}} &:= \{U \in \mathcal{U} \mid U - I \ \text{ injective}\} = \kappa(\mathcal{C}^{\text{sa}}) \\ {}_{\mathcal{F}}\mathcal{U} &:= \{U \in \mathcal{U} \mid U + I \ \text{ Fredholm}\}, \text{ and} \\ {}_{\mathcal{F}}\mathcal{U}_{\text{inj}} &:= {}_{\mathcal{F}}\mathcal{U} \cap \mathcal{U}_{\text{inj}} = \kappa(\mathcal{CF}^{\text{sa}}). \end{aligned} $$

Let $U \in {}_{\mathcal{F}}\mathcal{U}_{\text{inj}}$. Then H is the direct sum of the spectral subspaces $H_\pm$ of U corresponding to the intervals $[0,\pi)$ and $[\pi,2\pi]$, respectively, and we may decompose $U = U_+ \oplus U_-$. More precisely, we have

$$ \operatorname{spec}(U_+) \subset \{e^{it} \mid t \in [0,\pi)\} \quad \text{and} \quad \operatorname{spec}(U_-) \subset \{e^{it} \mid t \in [\pi,2\pi]\}. $$

Note that there is no intersection between the spectral spaces in the endpoints: if -1 belongs to $\operatorname{spec}(U)$, it is an isolated eigenvalue by our assumption and hence belongs only to $\operatorname{spec}(U_-)$; if 1 belongs to $\operatorname{spec}(U)$, it can belong both to $\operatorname{spec}(U_+)$ and $\operatorname{spec}(U_-)$, but in any case, it does not contribute to the decomposition of U since, by our assumption, 1 is not an eigenvalue at all.

By spectral deformation ("squeezing the spectrum down to $+i$ and $-i$") we contract U_+ to iI_+ and U_- to $-iI_-$, where $I_\pm$ denotes the identity on $H_\pm$. We do this on the upper half arc and the lower half arc, respectively, in such a way that 1 does not become an eigenvalue under the course of the deformation: actually it will no longer belong to the spectrum; neither will -1 belong to the spectrum. That is, we have connected U and $iI_+ \oplus -iI_-$ within $\kappa(\mathcal{CF}^{\text{sa}})$.

We distinguish two cases: If H_- is *finite-dimensional*, we now rotate $-iI_-$ up through -1 into iI_-: More precisely, we consider $\{iI_+ \oplus e^{i(\pi/2+(1-t)\pi)}I_-\}_{t\in[0,1]}$. This proves that we can connect U with $iI_+ \oplus iI_- = iI$ within $\kappa(\mathcal{CF}^{\text{sa}})$ in this first case.

If H_- is *infinite-dimensional*, we "dilate" $-iI_-$ in such a way that no eigenvalues remain. To do this, we identify H_- with $L^2([0,1])$. Now multiplication by $-i$ on $L^2([0,1])$ can be connected to multiplication by a function whose values are a short arc centered on $-i$ and so that the resulting operator V_- on H_- has no eigenvalues. This will at no time introduce spectrum near $+1$ or -1. We then rotate this arc up through $+1$ (which keeps us in the right space) until it is centered on $+i$. Then we contract the spectrum on H_- to be $+i$. That is, also in this case we have connected our original operator U to $+iI$. To sum up this second case (see also Figure 1):

$$ \begin{aligned} U \sim iI_+ \oplus -iI_- \sim iI_+ \oplus V_- &\sim iI_+ \oplus e^{it\pi}V_- \ \text{ for } t \in [0,1], \\ &\sim iI_+ \oplus -V_- \sim iI_+ \oplus -(-iI_-) \sim iI. \end{aligned} $$

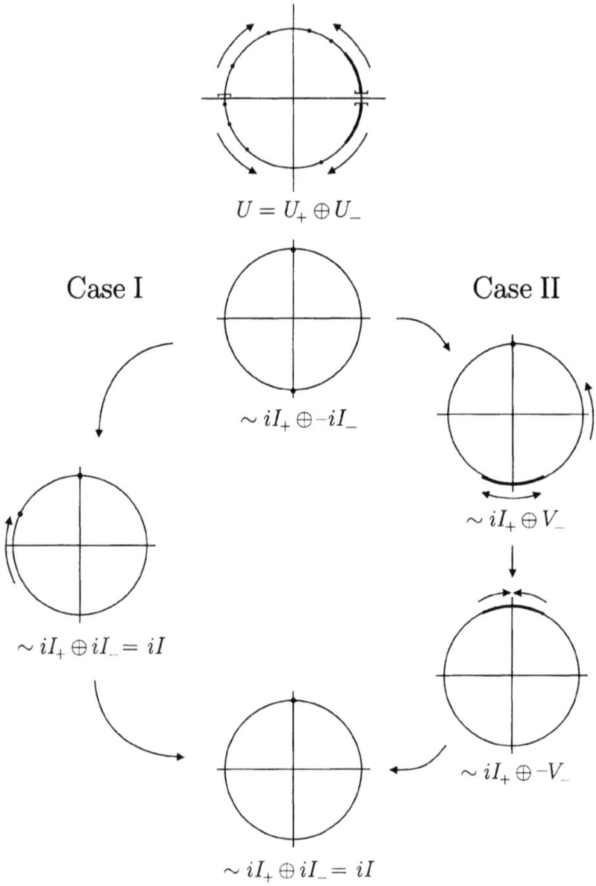

FIGURE 1. Connecting a fixed U in ${}_{\mathcal{F}}\mathcal{U}_{\text{inj}}$ to iI.
Case I (finite rank U_-) and Case II (infinite rank U_-)

To prove (b), we just decompose any $V \in {}_{\mathcal{F}}\mathcal{U}$ into $V = U \oplus I_1$, where $U \in {}_{\mathcal{F}}\mathcal{U}_{\text{inj}}(H_0)$ and I_1 denotes the identity on the 1-eigenspace $H_1 = \operatorname{Ker}(V - I)$ of V with $H = H_0 \oplus H_1$ an orthogonal decomposition. Then for $\varepsilon > 0$, $U \oplus e^{i\varepsilon} I_1 \in {}_{\mathcal{F}}\mathcal{U}_{\text{inj}}$ approaches U for $\varepsilon \to 0$. □

REMARK 1.13. The preceding proof shows also that the two subsets of $\mathcal{CF}^{\text{sa}}$

$$\mathcal{CF}^{\text{sa}}_{\pm} = \{T \in \mathcal{CF}^{\text{sa}} \mid \operatorname{spec}_{\text{ess}}(T) \subset \mathbb{C}_{\mp}\},$$

the spaces of all essentially positive, resp. all essentially negative, self-adjoint Fredholm operators, are no longer open. The third of the three complementary subsets

$$\mathcal{CF}^{\mathrm{sa}}_{*} = \mathcal{CF}^{\mathrm{sa}} \setminus \left(\mathcal{CF}^{\mathrm{sa}}_{+} \cup \mathcal{CF}^{\mathrm{sa}}_{-}\right)$$

is also not open. We do not know whether the two "trivial" components are contractible as in the bounded case nor whether the whole space is a classifying space for K^1 as the nontrivial component in the bounded case. Independently of Example 1.5, the connectedness of $\mathcal{CF}^{\mathrm{sa}}$ and the disconnectedness of $\mathcal{F}^{\mathrm{sa}}$ show that the Riesz map is not continuous on $\mathcal{CF}^{\mathrm{sa}}$ in the gap topology.

In analogy to [**Ph96**], we can give an explicit description of the winding number (spectral flow across -1) $\operatorname{wind}(f)$ of a curve f in ${}_{\mathcal{F}}\mathcal{U}$. Alternatively, it can be used as a definition of wind:

PROPOSITION 1.14. *Let $f : [0,1] \to {}_{\mathcal{F}}\mathcal{U}$ be a continuous path.*

(a) *There is a partition $\{0 = t_0 < t_1 < \cdots < t_n = 1\}$, and positive real numbers $0 < \varepsilon_j < \pi$, $j = 1, \dots, n$, such that $\operatorname{Ker}(f(t) - e^{i(\pi \pm \varepsilon_j)}) = \{0\}$ for $t_{j-1} \le t \le t_j$.*

(b) *Then*

$$\operatorname{wind}(f) = \sum_{j=1}^{n} k(t_j, \varepsilon_j) - k(t_{j-1}, \varepsilon_j),$$

where

$$k(t, \varepsilon_j) := \sum_{0 \le \theta < \varepsilon_j} \dim \operatorname{Ker}(f(t) - e^{i(\pi+\theta)}).$$

(c) *In particular, this calculation of $\operatorname{wind}(f)$ is independent of the choice of the partition of the interval and of the choice of the barriers.*

PROOF. In (a) we use the continuity of f and the fact that $f(t) \in {}_{\mathcal{F}}\mathcal{U}$. (b) follows from the path additivity of wind. (c) is immediate from (b). □

The idea of a *spectral flow across* -1 was introduced first in [**BoFu98**, Sec. 1.3], where it was used to give a definition of the Maslov index in an infinite dimensional context (see also Definition 1.24 further below).

After these explanations, the definition of spectral flow for paths in $\mathcal{CF}^{\mathrm{sa}}$ is straightforward:

DEFINITION 1.15. Let $f : [0,1] \to \mathcal{CF}^{\mathrm{sa}}(H)$ be a continuous path. Then the *spectral flow* $\operatorname{SF}(f)$ is defined by

$$\operatorname{SF}(f) := \operatorname{wind}(\kappa \circ f).$$

From the properties of κ and of the winding number, we infer immediately:

PROPOSITION 1.16. SF *is path additive and homotopy invariant in the following sense: Let $f_1, f_2 : [0,1] \to \mathcal{CF}^{\mathrm{sa}}$ be continuous paths and let $h : [0,1] \times [0,1] \to \mathcal{CF}^{\mathrm{sa}}$ be a homotopy such that $h(0,t) = f_1(t)$, $h(1,t) = f_2(t)$, and such that* $\dim \operatorname{Ker} h(s,0)$, $\dim \operatorname{Ker} h(s,1)$ *are independent of s. Then* $\operatorname{SF}(f_1) = \operatorname{SF}(f_2)$. *In particular,* SF *is invariant under homotopies leaving the endpoints fixed.*

From Proposition 1.14 we get

PROPOSITION 1.17. *For a continuous path* $f : [0,1] \to \mathcal{F}^{\mathrm{sa}}$ *our definition of spectral flow coincides with the definition in* [**Ph96**].

Note that also the conventions coincide for $0 \in \operatorname{spec} f(0)$ or $0 \in \operatorname{spec} f(1)$.

COROLLARY 1.18. *For any* $S \in \mathcal{CF}^{\mathrm{sa}}$ *with compact resolvent and any continuous path* $C : [0,1] \to \mathcal{B}^{\mathrm{sa}}$ *we have* $\mathrm{SF}(S+C) = \mathrm{SF}(\mathcal{R}(S+C))$, *where* $\mathcal{R}$ *denotes the Riesz transformation of* (1.1).

Note that the curve $S+C$ is in $\mathcal{CF}^{\mathrm{sa}}$, so that $\mathrm{SF}(S+C)$ is defined via Cayley transformation, whereas the curve $\mathcal{R}(S+C)$ of the Riesz transforms is in $\mathcal{F}^{\mathrm{sa}}$.

REMARK 1.19. The spectral flow induces a surjection of $\pi_1(\mathcal{CF}^{\mathrm{sa}})$ onto $\mathbb{Z}$. Because $\mathbb{Z}$ is free, there is a right inverse of SF and a normal subgroup G of $\pi_1(\mathcal{CF}^{\mathrm{sa}})$ such that we have a split short exact sequence

$$0 \to G \to \pi_1(\mathcal{CF}^{\mathrm{sa}}) \to \mathbb{Z} \to 0.$$

For now, an open question is whether G is trivial: Does the spectral flow distinguish the homotopy classes? That is, the question is whether each loop with spectral flow 0 can be contracted to a constant point, or equivalently, whether two continuous paths in $\mathcal{CF}^{\mathrm{sa}}$ with same endpoints and with same spectral flow can be deformed into each other. Or is $\pi_1(\mathcal{CF}^{\mathrm{sa}}) \cong \mathbb{Z} \times\!| \; G$ the semi-direct product of a nontrivial factor G with $\mathbb{Z}$? In that case, homotopy invariants of a curve in $\mathcal{CF}^{\mathrm{sa}}$ are not solely determined by the spectral flow (contrary to the folklore behind parts of the topology and physics literature). For now, we can only speculate about the existence of an additional invariant and its possible definition. For example, one can try to define a *spectral flow at infinity*. Then, continuity of the Riesz transformation $\mathcal{R}$ on a subclass $\mathcal{S} \subset \mathcal{CF}^{\mathrm{sa}}$ would imply vanishing spectral flow at infinity. Non-vanishing spectral flow at infinity will typically appear with families of the type discussed in Example 1.5 (and after Remark 1.9). One may also expect it with curves of differential operators of second order. However, the results of [**Ni00**], though only obtained under quite restrictive conditions, may indicate that perhaps spectral flow at infinity will not be exhibited for continuous curves of Dirac operators. If this is true, it will also explain why the mentioned unfounded folklore has not yet led to clear contradictions.

1.2. Symmetric operators and symplectic analysis. In an interview with Victor M. Buchstaber in the Newsletter of the European Mathematical Society [**No01**, p. 20], Sergej P. Novikov recalls his idea of the late 1960's and the early 1970's, which were radically different of the main stream in topology at that time, "that the explanation of higher signatures and of other deep properties of multiply connected manifolds had a symplectic origin (...) In 1971 I. Gelfand went into my algebraic ideas: They impressed him greatly. In particular, he told me of his observation that the so-called von Neumann theory of self-adjoint extensions of symmetric operators is simply the choice of a Lagrangian subspace in a Hilbert space with symplectic structure." Closely following [**BoFu98**], [**BoFu99**], and [**BoFuOt01**] (see also the Krein-Vishik-Birman theory summarized in [**AlSi80**] and [**LaSnTu75**], for first pointing to symplectic aspects), we will elaborate on that thought.

1.2.1. *Symplectic Hilbert space, Fredholm Lagrangian Grassmannian, and Maslov index.* We fix the following notation. Let $(\mathcal{H}, \langle\cdot,\cdot\rangle)$ be a separable real Hilbert space with a fixed symplectic form ω, i.e., a skew-symmetric bounded bilinear form on $\mathcal{H}\times\mathcal{H}$ which is nondegenerate. We assume that ω is compatible with $\langle\cdot,\cdot\rangle$ in the sense that there is a corresponding almost complex structure $J:\mathcal{H}\to\mathcal{H}$ defined by

$$\omega(x,y) = \langle Jx, y\rangle \tag{1.3}$$

with $J^2=-I$, ${}^tJ=-J$, and $\langle Jx,Jy\rangle = \langle x,y\rangle$. Here ${}^tJ$ denotes the transpose of J with regard to the (real) inner product $\langle x,y\rangle$. Let $\mathcal{L}=\mathcal{L}(\mathcal{H})$ denote the set of all Lagrangian subspaces λ of $\mathcal{H}$ (i.e., $\lambda=(J\lambda)^\perp$, or equivalently, let λ coincide with its annihilator λ^0 with respect to ω). The topology of $\mathcal{L}$ is defined by the operator norm of the orthogonal projections onto the Lagrangian subspaces.

Let $\lambda_0\in\mathcal{L}$ be fixed. Then any $\mu\in\mathcal{L}$ can be obtained as the image of $\lambda_0^\perp$ under a suitable unitary transformation

$$\mu = U(\lambda_0^\perp)$$

(see also Figure 2a). Here we consider the real symplectic Hilbert space $\mathcal{H}$ as a complex Hilbert space via J. The group $\mathcal{U}(\mathcal{H})$ of unitary operators of $\mathcal{H}$ acts transitively on $\mathcal{L}$; i.e., the mapping

$$\begin{aligned} \rho:\mathcal{U}(\mathcal{H}) &\longrightarrow \mathcal{L},\\ U &\mapsto U({\lambda_0}^\perp) \end{aligned} \tag{1.4}$$

is surjective and defines a principal fibre bundle with the group of orthogonal operators $\mathcal{O}(\lambda_0)$ as structure group.

EXAMPLE 1.20.

(a) In finite dimensions one considers the space $\mathcal{H}:=\mathbb{R}^n\oplus\mathbb{R}^n$ with the symplectic form

$$\omega\left((x,\xi),(y,\eta)\right) := -\langle x,\eta\rangle + \langle \xi, y\rangle \quad \text{for } (x,\xi),\,(y,\eta)\in\mathcal{H}.$$

To emphasize the finiteness of the dimension we write $\operatorname{Lag}(\mathbb{R}^{2n}) := \mathcal{L}(\mathcal{H})$. For linear subspaces of $\mathbb{R}^{2n}$ one has

$$l\in\operatorname{Lag}(\mathbb{R}^{2n}) \iff \dim l = n \text{ and } l\subset l^0 := \omega\text{-annihilator of } l;$$

i.e., Lagrangian subspaces are true half-spaces which are maximally isotropic ('isotropic' means $l\subset l^0$).

Note that $\mathbb{R}^{2n} = \mathbb{R}^n \otimes \mathbb{C}$ with the Hermitian product

$$((x,\xi),(y,\eta))_{\mathbb{C}} = (x+i\xi)(y-i\eta) \\ := \langle x,y\rangle + \langle \xi,\eta\rangle + i\langle \xi,y\rangle - i\langle x,\eta\rangle .$$

Then every $U \in \mathrm{U}(n)$ can be written in the form

$$U = A + iB = \begin{pmatrix} A & -B \\ B & A \end{pmatrix}$$

with ${}^tAB = {}^tBA,\ A^tB = B^tA$, and $A^tA + B^tB = I$, and

$$\mathrm{O}(n) \ni A \to \begin{pmatrix} A & 0 \\ 0 & A \end{pmatrix} = A + 0i \in \mathrm{U}(n)$$

gives the embedding of $\mathrm{O}(n)$ in $\mathrm{U}(n)$. One finds $\mathrm{Lag}(\mathbb{R}^{2n}) \cong \mathrm{U}(n)/\mathrm{O}(n)$ with the fundamental group

$$\pi_1(\mathrm{Lag}(\mathbb{R}^{2n})), \lambda_0) \cong \mathbb{Z}.$$

The mapping is given by the 'Maslov index' of loops of Lagrangian subspaces which can be described as an intersection index with the 'Maslov cycle'. There is a rich literature on the subject, see e.g. the seminal paper [**Ar67**], the systematic review [**CaLeMi94**], or the cohomological presentation [**Go97**].

(b) Let $\{\varphi_k\}_{k\in\mathbb{Z}\setminus\{0\}}$ be a complete orthonormal system for $\mathcal{H}$. We define an almost complex structure, and so by (1.3) a symplectic form, by

$$J\varphi_k := \mathrm{sign}(k)\, \varphi_{-k}\, .$$

Then the spaces $\mathcal{H}_- := \mathrm{span}\{\varphi_k\}_{k<0}$ and $\mathcal{H}_+ := \mathrm{span}\{\varphi_k\}_{k>0}$ are complementary Lagrangian subspaces of $\mathcal{H}$.

In infinite dimensions, the space $\mathcal{L}$ is contractible due to Kuiper's Theorem (see [**BlBo04**], Part I) and therefore topologically not interesting. Also, we need some restrictions to avoid infinite-dimensional intersection spaces when counting intersection indices. Therefore we replace $\mathcal{L}$ by a smaller space. This problem can be solved, as first suggested in Swanson [**Sw78**], by relating symplectic functional analysis with the space $\mathrm{Fred}(\mathcal{H})$ of Fredholm operators, we obtain finite dimensions for suitable intersection spaces and at the same time topologically highly nontrivial objects.

DEFINITION 1.21.

(a) The space of *Fredholm pairs* of closed infinite-dimensional subspaces of $\mathcal{H}$ is defined by

$$\mathrm{Fred}^2(\mathcal{H}) := \{(\lambda,\mu) \mid \dim(\lambda\cap\mu) < \infty, \text{ and} \\ \lambda+\mu \subset \mathcal{H} \text{ closed and } \dim \mathcal{H}/(\lambda+\mu) < \infty\}.$$

Then

$$\mathrm{index}(\lambda,\mu) := \dim(\lambda\cap\mu) - \dim \mathcal{H}/(\lambda+\mu)$$

is called the *Fredholm intersection index of* (λ,μ).

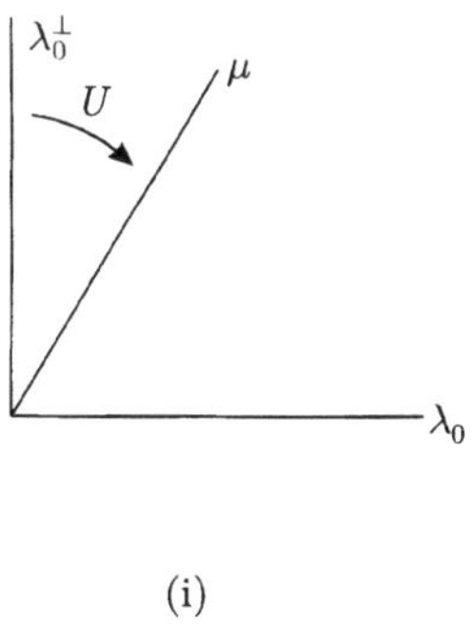

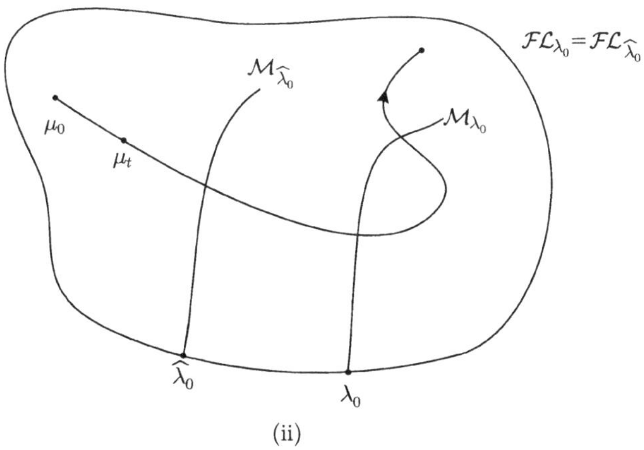

FIGURE 2. (i) The generation of $\mathcal{L} = \{\mu = U(\lambda_0^\perp) \mid U \in \mathcal{U}(\mathcal{H})\}$
(ii) One curve and two Maslov cycles in $\mathcal{FL}\lambda_0 = \mathcal{FL}\widehat{\lambda}_0$

(b) The *Fredholm-Lagrangian Grassmannian* of a real symplectic Hilbert space $\mathcal{H}$ at a fixed Lagrangian subspace λ_0 is defined as

$$\mathcal{FL}_{\lambda_0} := \{\mu \in \mathcal{L} \mid (\mu, \lambda_0) \in \operatorname{Fred}^2(\mathcal{H})\}.$$

(c) The *Maslov cycle* of λ_0 in $\mathcal{H}$ is defined as

$$\mathcal{M}_{\lambda_0} := \mathcal{FL}_{\lambda_0} \setminus \mathcal{FL}_{\lambda_0}^{(0)},$$

where $\mathcal{FL}_{\lambda_0}^{(0)}$ denotes the subset of Lagrangians intersecting λ_0 transversally; i.e., $\mu \cap \lambda_0 = \{0\}$.

Recall the following algebraic and topological characterization of general and Lagrangian Fredholm pairs (see [**BoFu98**] and [**BoFu99**], inspired by [**BoWo85**], Part 2, Lemma 2.6).

PROPOSITION 1.22.

(a) *Let $\lambda, \mu \in \mathcal{L}$ and let π_λ, π_μ denote the orthogonal projections of $\mathcal{H}$ onto λ respectively μ. Then*

$$(\lambda, \mu) \in \operatorname{Fred}^2(\mathcal{H}) \iff \pi_\lambda + \pi_\mu \in \operatorname{Fred}(\mathcal{H}) \iff \pi_\lambda - \pi_\mu \in \operatorname{Fred}(\mathcal{H}).$$

(b) *The fundamental group of $\mathcal{FL}_{\lambda_0}$ is $\mathbb{Z}$, and the mapping of the loops in $\mathcal{FL}_{\lambda_0}(\mathcal{H})$ onto $\mathbb{Z}$ is given by the Maslov index*

$$\mathbf{mas} : \pi_1(\mathcal{FL}_{\lambda_0}(\mathcal{H})) \xrightarrow{\cong} \mathbb{Z}.$$

To define the Maslov index, one needs a systematic way of counting, adding and subtracting the dimensions of the intersections $\mu_s \cap \lambda_0$ of the curve $\{\mu_s\}$ with the Maslov cycle $\mathcal{M}_{\lambda_0}$. We recall from [**BoFu98**], inspired by [**Ph96**], a functional analytical definition for continuous curves without additional assumptions.

First we recall from (1.4) that any $\mu \in \mathcal{FL}_{\lambda_0}(\mathcal{H})$ can be obtained as the image of $\lambda_0^\perp$ under a suitable unitary transformation $\mu = U(\lambda_0^\perp)$. As noted before, the operator U is not uniquely determined by μ. Actually, from $\beta \cong \lambda_0 \otimes \mathbb{C}$ we obtain a complex conjugation so that we can define the transpose by the following formula

$$ {}^T U := \overline{U^*} $$

and obtain a unitary operator $W(\mu) := U\,{}^T U$ which can be defined invariantly as the *complex generator* of the Lagrangian space μ relative to λ_0.

PROPOSITION 1.23. *The composed mapping W*

$$ \mathcal{FL}_{\lambda_0}(\mathcal{H}) \ni \mu \mapsto U \mapsto W(\mu) := U\,{}^T U \in {}_{\mathcal{F}}\mathcal{U}\,(\mathcal{H}) $$

is well defined. In particular, we have

$$ \operatorname{Ker}(I+W) = (\mu \cap \lambda_0) \otimes \mathbb{C} = (\mu \cap \lambda_0) + J\,(\mu \cap \lambda_0) $$

PROOF. The operator $I + W$ is a Fredholm operator because (μ, λ_0) is a Fredholm pair. □

DEFINITION 1.24. Let $\mathcal{H}$ be a symplectic Hilbert space and $\lambda_0 \in \mathcal{L}(\mathcal{H})$. Let

$$ [0,1] \ni s \mapsto \mu_s \in \mathcal{FL}_{\lambda_0}(\mathcal{H}) $$

be a continuous curve. Then $W \circ \mu$ is a continuous curve in ${}_{\mathcal{F}}\mathcal{U}\,(\mathcal{H})$, and the Maslov index can be defined by

$$ \mathbf{mas}\,(\mu, \lambda_0) := \operatorname{wind}\,(W \circ \mu)\,, $$

where wind is defined as in Proposition 1.14.

REMARK 1.25. To define the Maslov (intersection) index $\mathbf{mas}(\{\mu_s\}, \lambda_0)$, we count the change of the eigenvalues of W_s near -1 little by little. For example, between $s = 0$ and $s = s'$ we plot the spectrum of the complex generator W_s close to $e^{i\pi}$. In general, there will be no parametrization available of the spectrum near -1. For sufficiently small s', however, we can find barriers $e^{i(\pi+\theta)}$ and $e^{i(\pi-\theta)}$ such that no eigenvalues are lost through the barriers on the interval $[0, s']$. Then we count the number of eigenvalues (with multiplicity) of W_s between $e^{i\pi}$ and $e^{i(\pi+\theta)}$ at the right and left end of the interval $[0, s']$ and subtract. Repeating this procedure over the length of the whole s-interval $[0, 1]$ gives the Maslov intersection index $\mathbf{mas}(\{\mu_s\}, \lambda_0)$ without any assumptions about smoothness of the curve, 'normal crossings', or non-invertible endpoints.

It is worth mentioning that the construction can be simplified for a *complex* symplectic Hilbert space $\mathcal{H}$. Then each Lagrangian subspace of $\mathcal{H}$ is the graph of a uniquely determined unitary operator from $\operatorname{Ker}(J - iI)$ to $\operatorname{Ker}(J + iI)$. Moreover, a pair (λ, μ) of Lagrangian subspaces is a Fredholm pair if and only if $U^{-1}V$ is Fredholm, where $\lambda = \mathfrak{G}\,(U)$ and $\mu = \mathfrak{G}\,(V)$. We have $\mathbf{mas}\,(\{\mu_t\}, \lambda_0) = \operatorname{SF}\left(U_t^{-1}V, 1\right)$ where the 1 indicates that the spectral flow is taken at the eigenvalue 1.

REMARK 1.26.

(a) By identifying $\mathcal{H} \cong \lambda_0 \otimes \mathbb{C} \cong \lambda_0 \oplus \sqrt{-1}\,\lambda_0$, we split in [**BoFuOt01**] any $U \in \mathcal{U}(\mathcal{H})$ into a real and imaginary part

$$U = X + \sqrt{-1}\,Y$$

with $X, Y : \lambda_0 \to \lambda_0$. Let $\mathcal{U}(\mathcal{H})^{\mathrm{Fred}}$ denote the subspace of unitary operators which have a Fredholm operator as real part. This is the total space of a principal fibre bundle over the Fredholm Lagrangian Grassmannian $\mathcal{FL}_{\lambda_0}$ as base space and with the orthogonal group $\mathcal{O}(\lambda_0)$ as structure group. The projection is given by the restriction of the trivial bundle $\rho : \mathcal{U}(\mathcal{H}) \to \mathcal{L}$ of (1.4). This bundle

$$\mathcal{U}(\mathcal{H})^{\mathrm{Fred}} \xrightarrow{\ \rho\ } \mathcal{FL}_{\lambda_0}$$

may be considered as the infinite-dimensional generalization of the familiar bundle $\mathrm{U}(n) \to \mathrm{Lag}(\mathbb{R}^{2n})$ for finite n and provides an alternative proof of the homotopy type of $\mathcal{FL}_{\lambda_0}$.

(b) The Maslov index for curves depends on the specified Maslov cycle $\mathcal{M}_{\lambda_0}$. It is worth emphasizing that two *equivalent* Lagrangian subspaces λ_0 and $\widehat{\lambda}_0$ (i.e., $\dim \lambda_0/(\lambda_0 \cap \widehat{\lambda}_0) < +\infty$) always define the same Fredholm Lagrangian Grassmannian $\mathcal{FL}_{\lambda_0} = \mathcal{FL}_{\widehat{\lambda}_0}$ but may define different Maslov cycles $\mathcal{M}_{\lambda_0} \neq \mathcal{M}_{\widehat{\lambda}_0}$. The induced Maslov indices may also become different; in general,

$$\mathbf{mas}(\{\mu_s\}_{s\in[0,1]}, \lambda_0) - \mathbf{mas}(\{\mu_s\}_{s\in[0,1]}, \widehat{\lambda}_0) \neq 0 \tag{1.5}$$

(see [**BoFu99**], Proposition 3.1 and Section 5). However, if the curve is a loop, then the Maslov index does not depend on the choice of the Maslov cycle. From this property it follows that the difference in (1.5), beyond the dependence on λ_0 and $\widehat{\lambda}_0$, depends only on the initial and end points of the path $\{\mu_s\}$ and may be considered as the infinite-dimensional generalization $\sigma_{\mathbf{H\ddot{o}r}}(\mu_0, \mu_1; \lambda_0, \widehat{\lambda}_0)$ of the Hörmander index. It plays a part as the transition function of the universal covering of the Fredholm Lagrangian Grassmannian (see also Figure 2b).

1.2.2. *Symmetric operators and symplectic analysis.* Let H be a real separable Hilbert space and A an (unbounded) closed symmetric operator defined on the domain $D_{\min}$ which is supposed to be dense in H. Let A^* denote its adjoint operator with domain $D_{\max}$. We have that $A^*|_{D_{\min}} = A$ and that A^* is the maximal closed extension of A in H. Note that $D_{\max}$ is a Hilbert space with the graph scalar product

$$(x, y)_{\mathcal{G}} := (x, y) + (A^*x, A^*y),$$

and $D_{\min}$ is a closed subspace of this $D_{\max}$ since A is closed (each sequence in $D_{\min}$ which is Cauchy relative to the graph norm defines a sequence in the graph $\mathfrak{G}(A)$ which is Cauchy relative to the simple norm in $H \times H$).

We form the space β of *natural boundary values* with the *natural trace map* γ in the following way:

$$\begin{aligned} D_{\max} &\xrightarrow{\gamma} D_{\max}/D_{\min} =: \beta \\ x &\mapsto \gamma(x) = [x] := x + D_{\min}. \end{aligned}$$

The space β becomes a symplectic Hilbert space with the induced scalar product and the symplectic form given by Green's form

$$\omega([x],[y]) := (A^* x, y) - (x, A^* y) \quad \text{for } [x],[y] \in \beta.$$

We define the *natural Cauchy data space* $\Lambda := \gamma(\operatorname{Ker} A^*)$. It is a Lagrangian subspace of β under the assumption that A admits at least one self-adjoint Fredholm extension A_D. Actually, we shall assume that A has a self-adjoint extension A_D with compact resolvent. Then $(\Lambda, \gamma(D))$ is a Fredholm pair of subspaces of β; i.e., $\Lambda \in \mathcal{FL}_{\gamma(D)}(\beta)$.

We consider a continuous curve $\{C_s\}_{s\in[0,1]}$ in the space of bounded self-adjoint operators on H. We assume that the operators $A^* + C_s - r$ have no 'inner solutions'; i.e., they satisfy the *weak inner unique continuation property* (UCP)

$$\operatorname{Ker}(A^* + C_s - r) \cap D_{\min} = \{0\} \tag{1.6}$$

for $s \in [0,1]$ and $|r| < \varepsilon_0$ with $\varepsilon_0 > 0$. For a discussion of UCP see Section 1.4 below.

Clearly, the domains $D_{\max}$ and $D_{\min}$ are unchanged by the perturbation C_s for any s. So, β does not depend on the parameter s. Moreover, the symplectic form ω is invariantly defined on β and so also independent of s. It follows (see [**BoFu98**, Theorem 3.9]) that the curve $\{\Lambda_s := \gamma(\operatorname{Ker}(A^* + C_s))\}$ is continuous in $\mathcal{FL}_{\gamma(D)}(\beta)$.

We summarize the basic findings:

PROPOSITION 1.27.

(a) *Assume that there exists a self-adjoint Fredholm extension A_D of A with domain D. Then the Cauchy data space $\Lambda(A)$ is a closed Lagrangian subspace of β and belongs to the Fredholm Lagrangian Grassmannian $\mathcal{FL}_{\gamma(D)}(\beta)$.*

(b) *For arbitrary domains D with $D_{\min} \subset D \subset D_{\max}$ and $\gamma(D)$ Lagrangian, the extension $A_D := A_{\max}|_D$ is self-adjoint. It becomes a Fredholm operator, if and only if the pair $(\gamma(D), \Lambda(A))$ of Lagrangian subspaces of β becomes a Fredholm pair.*

(c) *Let $\{C_t\}_{t\in[0,1]}$ be a continuous family (with respect to the operator norm) of bounded self-adjoint operators. Assume the weak inner UCP for all operators $A^* + C_t$. Then the spaces $\gamma(\operatorname{Ker}(A^* + C_t, 0))$ of Cauchy data vary continuously in β.*

Given this, the family $\{A_D + C_s\}$ can be considered at the same time in the spectral theory of self-adjoint Fredholm operators, defining a spectral flow, and in the symplectic category, defining a Maslov index. Under the preceding assumptions, the main result obtainable at that level is the following general spectral flow formula (proved in [**BoFu98**, Theorem 5.1] and inviting to generalizations for varying domains D_s instead of a fixed domain D):

THEOREM 1.28. *Let A_D be a self-adjoint extension of A with compact resolvent and let $\{A_D + C_s\}_{s\in[0,1]}$ be a family satisfying the weak inner UCP assumption. Let Λ_s denote the Cauchy data space $\gamma(\operatorname{Ker}(A^* + C_s))$ of $A^* + C_s$. Then*

$$\operatorname{SF}\{A_D + C_s\} = \mathbf{mas}(\{\Lambda_s\}, \gamma(D)).$$

1.3. Operators of Dirac type and their ellipticity. There are different notions of operators of Dirac type. We shall not discuss the original hyperbolic Dirac operator (in the Minkowski metric), but restrict ourselves to the elliptic case related to Riemannian metrics.

Recall that, if (M,g) is a compact smooth Riemannian manifold (with or without boundary) with $\dim M = m$, we denote by $\mathfrak{Cl}(M) = \{\mathfrak{Cl}(TM_x, g_x)\}_{x\in M}$ the bundle of Clifford algebras of the tangent spaces. For $S \to M$, a smooth complex vector bundle of Clifford modules, the *Clifford multiplication* is a bundle map $\mathbf{c} : \mathfrak{Cl}(M) \to \operatorname{Hom}(S,S)$ which yields a representation $\mathbf{c} : \mathfrak{Cl}(TM_x, g_x) \to \operatorname{Hom}_{\mathbb{C}}(S_x, S_x)$ in each fiber. We may assume that the bundle S is equipped with a Hermitian metric which makes the Clifford multiplication skew-symmetric

$$\langle \mathbf{c}(v)s, s'\rangle = -\langle s, \mathbf{c}(v)s'\rangle \quad \text{for } v \in TM_x \text{ and } s \in S_x.$$

We note that it is not necessary to assume that (M,g) admits a spin structure in order that $\mathfrak{Cl}(M)$ and S exist. Indeed, one special case is obtained by taking $S = \Lambda^*(TM)$ and letting $\mathbf{c}$ be the extension of $\mathbf{c}(v)(\alpha) = v \wedge \alpha - v \llcorner \alpha$ for $v \in TM_x \subset \mathfrak{Cl}(TM_x, g_x)$ and $\alpha \in \Lambda^*(TM_x)$, where "$\llcorner$" denotes the interior product (i.e., the dual of the exterior product $\wedge$). The extension is guaranteed by the fact that $\mathbf{c}(v)^2 = -g_x(v,v)\, I$. However, we do need a spin structure in the case where S is a bundle of spinors.

Any choice of a smooth connection

$$\nabla : \mathrm{C}^\infty(M;S) \to \mathrm{C}^\infty(M;T^*M \otimes S)$$

defines an *operator of Dirac type* $\mathcal{D} := \mathbf{c} \circ \nabla$ under the Riemannian identification of the bundles TM and T^*M. In local coordinates we have $\mathcal{D} := \sum_{j=1}^m \mathbf{c}(e_j)\nabla_{e_j}$ for any orthonormal base $\{e_1, \dots, e_m\}$ of TM_x. Actually, we may choose a local frame in such a way that

$$\nabla_{e_j} = \frac{\partial}{\partial x_j} + \text{zero order terms}$$

for all $1 \le j \le m$. So, locally, we have

$$\mathcal{D} := \sum_{j=1}^m \mathbf{c}(e_j)\frac{\partial}{\partial x_j} + \text{zero order terms.} \tag{1.7}$$

It follows at once that the principal symbol $\sigma_1(\mathcal{D})(x,\xi)$ is given by Clifford multiplication with $i\xi$, so that any operator of Dirac type is elliptic with symmetric principal symbol. If the connection ∇ is *compatible* with Clifford multiplication (i.e.

$\nabla \mathbf{c} = 0$), then the operator $\mathcal{D}$ itself becomes symmetric. We shall, however, admit incompatible metrics. Moreover, the *Dirac Laplacian* $\mathcal{D}^2$ has principal symbol $\sigma_2(\mathcal{D}^2)(x,\xi)$ given by scalar multiplication by $\|\xi\|^2$ using the Riemannian metric. So, the principal symbol of $\mathcal{D}^2$ is a real multiple of the identity, and $\mathcal{D}^2$ is elliptic. In the special case above where $S = \Lambda^*(TM)$ and we identify $\Lambda^*(TM)$ with $\Lambda^*(T^*M)$ by means of the metric g, $\mathcal{D}$ becomes $d+\delta : \Omega^*(M,\mathbb{C}) \to \Omega^*(M,\mathbb{C})$ and $\mathcal{D}^2 = d\delta + \delta d$ is the Hodge Laplacian on the space $\Omega^*(M,\mathbb{C})$ of complex-valued forms on M.

On a closed manifold M, a key result for any operator $\mathcal{D}$ of Dirac type (actually, for any linear elliptic operator of first order) is the *a priori* estimate (*Gårding's inequality*)

$$\|\psi\|_1 \leq C\left(\|\mathcal{D}\psi\|_0 + \|\psi\|_0\right) \text{ for all } \psi \in H^1(M;S). \tag{1.8}$$

Here $\|\cdot\|_0$ denotes the L^2 norm and $H^1(M;S)$ denotes the first Sobolev space with the norm $\|\cdot\|_1$. Note that the same symbol $\mathcal{D}$ is used for the original operator (defined on smooth sections) and its closed L^2 extension with domain $H^1(M;S)$. Combined with the simple continuity relation $\|\mathcal{D}\psi\|_0 \leq C'\|\psi\|_1$, inequality (1.8) shows that the first Sobolev norm $\|\cdot\|_1$ and the graph norm coincide on $H^1(M;S)$.

1.4. Weak unique continuation property. A linear or non-linear operator $\mathfrak{D}$, acting on functions or sections of a bundle over a compact or non-compact manifold M has the *weak Unique Continuation Property* (*UCP*) if any solution ψ of the equation $\mathfrak{D}\psi = 0$ has the following property: If ψ vanishes on a nonempty open subset Ω of M, then it vanishes on the whole connected component of M containing Ω. Note that weak inner UCP, as defined in (1.6), follows from weak UCP, but not vice versa.

There is also a notion of *strong UCP*, where, instead of assuming that a solution ψ vanishes on an open subset, one assumes only that ψ vanishes 'of high order' at a point. The concepts of weak and strong UCP extend a fundamental property of analytic functions to *some* elliptic equations other than the Cauchy-Riemann equation.

Up to now, (almost) all work on UCP goes back to two seminal papers [**Ca33**], [**Ca39**] by Torsten Carleman, establishing an inequality of Carleman type (see our inequality 1.13 below). In this approach, the difference between weak and strong UCP and the possible presence of more delicate nonlinear perturbations are related to different choices of the weight function in the inequality, and to whether L^2 estimates suffice or L^p and L^q estimates are required.

The weak UCP is one of the basic properties of an operator of Dirac type $\mathcal{D}$. Contrary to common belief, UCP is *not* a general fact of life for elliptic operators. See [**Pl61**] where counter-examples are given with smooth coefficients. Lack of UCP invalidates the continuity of the Cauchy data spaces and of the Calderón projection (Proposition 1.27c and Theorem 2.28b) and of the main continuity lemma (Lemma 2.27). It corrupts the invertible double construction (Section 2.3.4) and threatens Bojarski type theorems (like Proposition 2.33 and Theorem 2.37).

For partitioned manifolds $M = M_1 \cup_\Sigma M_2$ (see Section 2.3), it guarantees that there are no *ghost* solutions of $\mathcal{D}\psi = 0$; that is, there are no solutions which vanish on M_1 and have nontrivial support in the interior of M_2. This property is also called UCP *from open subsets* or *across any hypersurface*. For Euclidean (classical) Dirac operators (i.e., Dirac operators on $\mathbb{R}^m$ with constant coefficients and without perturbation), the property follows by squaring directly from the well-established UCP for the classical (constant coefficients and no potential) Laplacian.

From [**BoWo93**, Chapter 8] we recall a very simple proof of the weak UCP for operators of Dirac type, inspired by [**Ni73**, Sections 6-7, in particular the proof of inequality (7.11)] and [**Tr80**, Section II.3]. We refer to [**Boo00**] for a further slight simplification and a broader perspective, and to [**BoMaWa02**] for perturbed equations.

The proof does not use advanced arguments of the Aronszajn/Cordes type (see [**Ar57**] and [**Co56**]) regarding the diagonal and real form of the principal symbol of the Dirac Laplacian nor any other reduction to operators of second order (like [**We82**]), but only the following product property of Dirac type operators (besides Gårding's inequality).

LEMMA 1.29. *Let Σ be a closed hypersurface of M with orientable normal bundle. Let u denote a normal variable with fixed orientation such that a bicollar neighborhood N of Σ is parameterized by $[-\varepsilon, +\varepsilon] \times \Sigma$. Then any operator of Dirac type can be rewritten in the form*

$$\mathcal{D}|_N = \mathbf{c}(du)\left(\frac{\partial}{\partial u} + B_u + C_u\right), \tag{1.9}$$

where B_u is a self-adjoint elliptic operator on the parallel hypersurface Σ_u, and $C_u : S|_{\Sigma_u} \to S|_{\Sigma_u}$ is a skew-symmetric operator of order zero, actually a skew-symmetric bundle homomorphism.

PROOF. Let (u, y) denote the coordinates in a tubular neighborhood of Σ. Locally, we have $y = (y_1, \ldots, y_{m-1})$. Let $\mathbf{c}_u, \mathbf{c}_1, \ldots, \mathbf{c}_{m-1}$ denote Clifford multiplication by the unit tangent vectors in normal, respectively tangential, directions. By (1.7), we have

$$\begin{aligned}\mathcal{D} &= \mathbf{c}_u \frac{\partial}{\partial u} + \sum_{k=1}^{m-1} \mathbf{c}_k \frac{\partial}{\partial y_k} + \text{ zero order terms}\\ &= \mathbf{c}_u \left(\frac{\partial}{\partial u} + \mathcal{B}_u\right),\end{aligned}$$

where

$$\mathcal{B}_u := \sum_{k=1}^{m-1} -\mathbf{c}_u \mathbf{c}_k \frac{\partial}{\partial y_k} + \text{ zero order terms.}$$

We shall call $\mathcal{B}_u$ the *tangential* operator component of the operator A. Clearly it is an elliptic differential operator of first order over Σ_u. From the skew-hermicity of $\mathbf{c}_u$ and $\mathbf{c}_k$, we have

$$\begin{aligned}\left(\mathbf{c}_u\mathbf{c}_k\frac{\partial}{\partial y_k}\right)^* &= \left(-\frac{\partial}{\partial y_k}\right)(-\mathbf{c}_k)(-\mathbf{c}_u)\\ &= -\mathbf{c}_k\,\mathbf{c}_u\frac{\partial}{\partial y_k} + \text{ zero order terms}\\ &= \mathbf{c}_u\mathbf{c}_k\,\frac{\partial}{\partial y_k} + \text{ zero order terms.}\end{aligned}$$

So,

$$\mathcal{B}_u^* = \mathcal{B}_u + \text{ zero order terms.}$$

Hence, the principal symbol of $\mathcal{B}_u$ is self-adjoint. Then the assertion of the lemma is proved by setting

$$B_u := \frac{1}{2}\left(\mathcal{B}_u + \mathcal{B}_u^*\right) \quad \text{and} \quad C_u := \frac{1}{2}\left(\mathcal{B}_u - \mathcal{B}_u^*\right). \tag{1.10}$$

□

REMARK 1.30.

(a) It is worth mentioning that the product form (1.9) is invariant under perturbation by a bundle homomorphism. More precisely: Let $\mathfrak{D}$ be an operator on M which can be written in the form (1.9) close to any closed hypersurface Σ, with B_u and C_u as explained in the preceding lemma. Let R be a bundle homomorphism. Then

$$(\mathfrak{D} + R)\,|_N = \mathbf{c}(du)\left(\frac{\partial}{\partial u} + B_u + C_u\right) + \mathbf{c}(du)S|_N$$

with $T|_N := \mathbf{c}(du)^* R|_N$. Splitting $T = \frac{1}{2}(T+T^*)+\frac{1}{2}(T-T^*)$ into a symmetric and a skew-symmetric part and adding these parts to B_u and C_u, respectively, yields the desired form of $(\mathfrak{D} + R)\,|_N$.

(b) For operators of Dirac type, it is well known that a perturbation by a bundle homomorphism is equivalent to modifying the underlying connection of the operator. This gives an alternative argument for the invariance of the form (1.9) for operators of Dirac type under perturbation by a bundle homomorphism.

(c) By the preceding arguments (a), respectively (b), establishing weak UCP for sections belonging to the kernel of a Dirac type operator, respectively an operator which can be written in the form (1.9), implies weak UCP for all eigensections. Warning: For general linear elliptic differential operators, weak UCP for "zero-modes" does not imply weak UCP for all eigensections.

To prove the weak UCP, in combination with the preceding lemma, the standard lines of the UCP literature can be radically simplified, namely with regard to the weight functions and the integration order of estimates. These simplifications make it also very easy to generalize the weak UCP to the perturbed case.

We replace the equation $\mathcal{D}\psi = 0$ by

$$\widetilde{\mathcal{D}}\psi := \mathcal{D}\psi + \mathfrak{P}_A(\psi) = 0\,, \tag{1.11}$$

where $\mathfrak{P}_A$ is an *admissible* perturbation, in the following sense.

DEFINITION 1.31. A perturbation is *admissible* if it satisfies the following estimate:

$$|\mathfrak{P}_A(\psi)|_x| \le P(\psi, x)|\psi(x)| \quad \text{for } x \in M, \tag{1.12}$$

for a real-valued function $P(\psi, \cdot)$ which is locally bounded on M for each fixed ψ.

EXAMPLE 1.32. Some typical examples of perturbations satisfying the admissibility condition of Definition 1.31 are:

1. Consider a nonlinear perturbation

$$\mathfrak{P}_A(\psi)|_x := \omega(\psi(x)) \cdot \psi(x),$$

where $\omega(\psi(x))|_{x\in M}$ is a (bounded) function which depends continuously on $\psi(x)$, for instance, for a fixed (bounded) spinor section $a(x)$ we can take

$$\omega(\psi(x)) := \langle \psi(x), a(x) \rangle$$

with $\langle \cdot, \cdot \rangle$ denoting the Hermitian product in the fiber of the spinor bundle over the base point $x \in M$. This satisfies (1.12).

2. Another interesting example is provided by (linear) nonlocal perturbations with

$$\omega(\psi, x) = \left| \int k(x, z)\psi(z)dz \right|$$

with suitable integration domain and integrability of the kernel k. These also satisfy (1.12).

3. Clearly, an unbounded perturbation may be both nonlinear and global at the same time. This will, in fact, be the case in our main application. In all these cases the only requirement is the estimate (1.12) with bounded $\omega(\psi(\cdot))$.

We now show that (admissible) perturbed Dirac operators always satisfy the weak Unique Continuation Property. In particular, we show that this is true for unperturbed Dirac operators.

THEOREM 1.33. *Let $\mathcal{D}$ be an operator of Dirac type and $\mathfrak{P}_A$ an admissible perturbation. Then any solution ψ of the perturbed equation* (1.11) *vanishes identically on any connected component of the underlying manifold M if it vanishes on a nonempty open subset of the connected component.*

PROOF. Without loss of generality, we assume that M is connected. Let ψ be a solution of the perturbed (or, in particular) unperturbed equation which vanishes on an open, nonempty set Ω. First we localize and convexify the situation and we introduce spherical coordinates (see Figure 3). Without loss of generality we may assume that Ω is maximal, namely the union of all open subsets on which ψ vanishes; i.e., $\Omega = M \setminus \operatorname{supp} \psi$.

Since M is connected, to prove that $\Omega = M$ it suffices to show that Ω is closed (i.e., $\overline{\Omega} = \Omega$). If $\overline{\Omega} \neq \Omega$, then let $y_0 \in \partial\Omega := \overline{\Omega} \setminus \Omega$, and let B be an open, normal

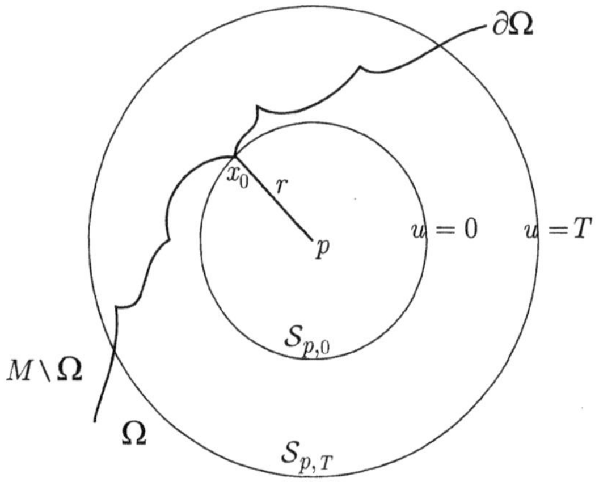

FIGURE 3. Local specification for the Carleman estimate

coordinate ball about y_0. Let $p \in \Omega \cap B$ and let $x_0 \in \operatorname{supp} \psi$ be a point of the non-empty, compact set $\overline{B} \setminus \Omega = \overline{B} \cap \operatorname{supp} \psi$ which is closest to p. Let

$$r := d(p, x_0) = \min_{x \in \overline{B} \setminus \Omega} d(p, x).$$

For $z \in B$, let $u(z) := d(p, z) - r$ be a "radial coordinate". Note that $u = 0$ defines a sphere, say $\mathcal{S}_{p,0}$, of radius r about x_0. We have larger hyperspheres $\mathcal{S}_{p,u} \subset M$ for $0 \leq u \leq T$ with $T > 0$ sufficiently *small*. In such a way we have parameterized an annular region $N_T := \{\mathcal{S}_{p,u}\}_{u \in [0,T]}$ around p of width T and inner radius r, ranging from the hypersphere $\mathcal{S}_{p,0}$ which is contained in $\overline{\Omega}$, to the hypersphere $\mathcal{S}_{p,T}$. Note that N_T contains some points where $\psi \neq 0$, for otherwise $x_0 \in \Omega$. Let y denote a variable point in $\mathcal{S}_{p,0}$ and note that points in N_T may be identified with $(u, y) \in [0, T] \times \mathcal{S}_{p,0}$. Next, we replace the solution $\psi | N_T$ by a cutoff

$$v(u, y) := \varphi(u)\psi(u, y)$$

with a smooth bump function φ with $\varphi(u) = 1$ for $u \leq 0.8T$ and $\varphi(u) = 0$ for $u \geq 0.9T$. Then $\operatorname{supp} v$ is contained in N_T. More precisely, it is contained in the annular region $N_{0.9T}$. Now our proof goes in two steps: first we establish a Carleman inequality for any spinor section v in the domain of $\mathcal{D}$ which satisfies $\operatorname{supp}(v) \subset N_T$. More precisely, we are going to show that for T sufficiently small there exists a constant C, such that

$$(1.13) \qquad R \int_{u=0}^{T} \int_{\mathcal{S}_{p,u}} e^{R(T-u)^2} |v(u, y)|^2 \, dy \, du \leq C \int_{u=0}^{T} \int_{\mathcal{S}_{p,u}} e^{R(T-u)^2} |\mathcal{D}v(u, y)|^2 \, dy \, du$$

holds for any real R sufficiently large. In the second step we apply (1.13) to our cutoff section v and conclude that then ψ is equal 0 on $N_{T/2}$.

Step 1. First consider a few technical points. The Dirac operator $\mathcal{D}$ has the form $G(u)(\partial_u+\mathcal{B}_u)$ on the annular region $[0,T]\times\mathcal{S}_{p,0}$, and it is obvious that we may consider the operator $(\partial_u+\mathcal{B}_u)$ instead of $\mathcal{D}$. Moreover, we have by Lemma 1.29 that $\mathcal{B}_u = B_u + C_u$ with a self-adjoint elliptic differential operator B_u and an anti-symmetric operator C_u of order zero, both on $\mathcal{S}_{p,u}$. Note that the metric structures depend on the normal variable u.

Now make the substitution $v =: e^{-R(T-u)^2/2}v_0$ which replaces (1.13) by

$$\text{(1.14)}\qquad R\int_0^T\int_{\mathcal{S}_{p,u}} |v_0(u,y)|^2\,dy\,du \le C\int_0^T\int_{\mathcal{S}_{p,u}}\left|\frac{\partial v_0}{\partial u}+\mathcal{B}_u v_0 + R(T-u)v_0\right|^2 dy\,du.$$

We denote the integral on the left-hand side by J_0 and the integral on the right-hand side by J_1. Now we prove (1.14). Decompose $\dfrac{\partial}{\partial u}+\mathcal{B}_u+R(T-u)$ into its symmetric part $B_u+R(T-u)$ and anti-symmetric part $\partial_u + C_u$. This gives

$$\begin{aligned}
J_1 &= \int_0^T\int_{\mathcal{S}_{p,u}}\left|\frac{\partial v_0}{\partial u}+\mathcal{B}_u v_0+R(T-u)v_0\right|^2 dy\,du\\
&= \int_0^T\int_{\mathcal{S}_{p,u}}\left|\frac{\partial v_0}{\partial u}+C_u v_0\right|^2 dy\,du + \int_0^T\int_{\mathcal{S}_{p,u}}\left|(B_u+R(T-u))\,v_0\right|^2 dy\,du\\
&\quad + 2\Re\int_0^T\int_{\mathcal{S}_{p,u}}\left\langle\frac{\partial v_0}{\partial u}+C_u v_0, B_u v_0 + R(T-u)v_0\right\rangle dy\,du\,.
\end{aligned}$$

Integrate by parts and use the identity $\Re\langle f, Pf\rangle = \frac{1}{2}\langle f,(P+P^*)f\rangle$ in order to investigate the last and critical term which will be denoted by J_2. This yields (where we drop domains of integration)

$$\begin{aligned}
J_2 &= 2\Re\iint\left\langle\frac{\partial v_0}{\partial u}+C_u v_0,\ B_u v_0+R(T-u)v_0\right\rangle dy\,du\\
&= 2\Re\iint\left\langle\frac{\partial v_0}{\partial u},\ B_u v_0+R(T-u)v_0\right\rangle dy\,du + 2\Re\iint\langle C_u v_0,\ B_u v_0\rangle\,dy\,du\\
&= -2\Re\iint\left\langle v_0,\left\{\frac{\partial}{\partial u}(B_u+R(T-u))\right\}v_0\right\rangle dy\,du - 2\Re\iint\langle v_0, C_u B_u v_0\rangle\,dy\,du\\
&= 2\iint\left\langle v_0, -\frac{\partial B_u}{\partial u}v_0+Rv_0\right\rangle dy\,du + \iint\left(v_0,[B_u,C_u]\,v_0\right)dy\,du\\
&= 2R\int_0^T\|v_0\|_0^2\,du + \iint\left\langle v_0, -2\frac{\partial B_u}{\partial u}v_0+[B_u,C_u]\,v_0\right\rangle dy\,du = 2RJ_0+J_3,
\end{aligned}$$

where $\|\cdot\|_m$ denotes the m-th Sobolev norm on $E|_{\mathcal{S}_{p,u}}$ and J_3 requires a careful analysis. It follows from the preceding decompositions of J_1 and J_2 that the proof of (1.14) will be completed with $C = \frac{1}{2}$ when $J_3 \geq 0$. If $J_3 < 0$ and $C = \frac{1}{2}$, it suffices to show that

$$|J_3| \leq \frac{1}{2}\left(R\int_0^T \|v_0\|_0^2\, du + \int_0^T \|(B_u + R(T-u))v_0\|^2\, du\right). \tag{1.15}$$

Since the operators B_u are elliptic of order 1, Gårding's inequality (1.8) yields

$$\|f\|_1 \leq c(\|f\|_0 + \|B_u f\|_0)$$

for any section f of E on $\mathcal{S}_{p,u}$ (and $0 \leq u \leq T$). Then, also using the fact that $-2\frac{\partial B_u}{\partial u} + [B_u, C_u]$ is a *first-order* differential operator on $E|\mathcal{S}_{p,u}$, we obtain

$$\begin{aligned}
|J_3| &\leq \int_0^T \|v_0\|_0 \left\|-2\tfrac{\partial B_u}{\partial u} v_0 + [B_u, C_u]v_0\right\|_0 du \leq c_1 \int_0^T \|v_0\|_0 \|v_0\|_1\, du \\
&\leq c_1 c \int_0^T \|v_0\|_0 \left(\|B_u v_0\|_0 + \|v_0\|_0\right) du \\
&\leq c_1 c \int_0^T \|v_0\|_0 \left\{\|(B_u + R(T-u))v_0\|_0 + (R(T-u)+1)\|v_0\|_0\right\} du \\
&\leq c_1 c(RT+1)\int_0^T \|v_0\|_0^2\, du + c_1 c \int_0^T \|(B_u + R(T-u))v_0\|_0 \|v_0\|_0\, du\,.
\end{aligned}$$

The integrand of the second summand can be estimated by

$$\begin{aligned}
&\frac{\|(B_u + R(T-u))v_0\|_0}{\sqrt{c_1 c}} \left(\sqrt{c_1 c}\|v_0\|_0\right) \\
&\qquad \leq \frac{1}{2}\left\{\frac{1}{c_1 c}\|(B_u + R(T-u))v_0\|_0^2 + c_1 c\|v_0\|_0^2\right\}
\end{aligned} \tag{1.16}$$

due to the inequality $ab \leq \frac{1}{2}(a^2+b^2)$. By inserting (1.16) in the preceding inequality for $|J_3|$, we obtain

$$\begin{aligned}
|J_3| &\leq c_1 c(RT+1)\int_0^T \|v_0\|_0^2\, du \\
&\qquad + c_1 c \int_0^T \left(\tfrac{1}{2}\left\{\tfrac{1}{c_1 c}\|(B_u + R(T-u))v_0\|_0^2 + c_1 c\|v_0\|_0^2\right\}\right) du \\
&= c_1 c(RT+1)\int_0^T \|v_0\|_0^2\, du + c_1 c \int_0^T \tfrac{1}{2} c_1 c\|v_0\|_0^2\, du \\
&\qquad + \int_0^T \left(\tfrac{1}{2}\|(B_u + R(T-u))v_0\|_0^2\right) du \\
&= \int_0^T \tfrac{1}{2}\|(B_u + R(T-u))v_0\|_0^2\, du + c_1 c\left((RT+1) + \tfrac{1}{2}c_1 c\right)\int_0^T \|v_0\|_0^2\, du \\
&= \tfrac{1}{2}\int_0^T \|(B_u + R(T-u))v_0\|_0^2\, du + Rc_1 c\left(T + \frac{c_1 c + 2}{2R}\right)\int_0^T \|v_0\|_0^2\, du.
\end{aligned}$$

So (1.15) holds for T and $\frac{1}{R}$ sufficiently small, and we then have the Carleman inequality (1.13) for $C = \frac{1}{2}$.

Step 2. To begin with, we have

$$(1.17)\qquad e^{RT^2/4} \int_0^{\frac{T}{2}} \int_{\mathcal{S}_{p,u}} |\psi(u,y)|^2 \, dy \, du \\ \leq \int_0^T \int_{\mathcal{S}_{p,u}} e^{R(T-u)^2} \, |(\varphi\psi)(u,y)|^2 \, dy \, du.$$

We apply our Carleman type inequality (1.13):

$$(1.18)\qquad I := \int_0^T \int_{\mathcal{S}_{p,u}} e^{R(T-u)^2} \, |(\varphi\psi)(u,y)|^2 \, dy \, du \\ \leq \tfrac{C}{R} \int_0^T \int_{\mathcal{S}_{p,u}} e^{R(T-u)^2} \, |\mathcal{D}(\varphi\psi)(u,y)|^2 \, dy \, du.$$

Assuming that ψ is a solution of the perturbed equation $\mathcal{D}\psi + \mathfrak{P}_A(\psi) = 0$, we get

$$\mathcal{D}(\varphi\psi) = \varphi\mathcal{D}\psi + \mathbf{c}(du)\varphi'\psi = -\varphi\mathfrak{P}_A(\psi) + \mathbf{c}(du)\varphi'\psi.$$

Using this in (1.18) and noting that $(a+b)^2 \leq 2\left(a^2 + b^2\right)$, yields

$$I \leq \tfrac{2C}{R} \int_0^T \int_{\mathcal{S}_{p,u}} e^{R(T-u)^2} \left(|\varphi(u)\mathfrak{P}_A(\psi)(u,y)|^2 + |\mathbf{c}(du)\varphi'(u)\psi(u,y)|^2\right) dy \, du.$$

Now we exploit our assumption

$$(1.19)\qquad |\mathfrak{P}_A(\psi)(x)| \leq P(\psi,x)|\psi(x)| \quad \text{for } x \in M,$$

about the perturbation with locally bounded $P(\psi,\cdot)$, say

$$|P(\psi,(u,y))| \leq C_0 := \max_{x\in K} |P(\psi,x)| \ \text{ for all } y \in \mathcal{S}_{p,u},\ u \in [0,T],$$

where K is a suitable compact set. We obtain at once

$$\left(1 - \tfrac{2CC_0}{R}\right) I \leq \tfrac{2C}{R} \int_0^T \int_{\mathcal{S}_{p,u}} e^{R(T-u)^2} \, |\mathbf{c}(du)\varphi'(u)\psi(u,y)|^2 \, dy \, du \\ \leq \tfrac{2C}{R} e^{RT^2/25} \int_0^T \int_{\mathcal{S}_{p,u}} |\mathbf{c}(du)\varphi'(u)\psi(u,y)|^2 \, dy \, du.$$

Here we use that $\varphi'(u) = 0$ for $0 \leq u \leq 0.8T$ so that we can estimate the exponential and pull it in front of the integral. Using (1.17),

$$\int_0^{\frac{T}{2}} \int_{\mathcal{S}_{p,u}} |\psi(u,y)|^2 \, dy \, du \leq e^{-RT^2/4} I \\ \leq \left(\tfrac{2C}{R-2CC_0}\right) e^{RT^2\left(\frac{1}{25}-\frac{1}{4}\right)} \int_0^T \int_{\mathcal{S}_{p,u}} |\mathbf{c}(du)\varphi'(u)\psi(u,y)|^2 \, dy \, du.$$

As $R \to \infty$, we get

$$\int_0^{\frac{T}{2}} \int_{S_{p,u}} |\psi(u,y)|^2 \, dy \, du = 0$$

which contradicts $x_0 \in \operatorname{supp} \psi$. □

2. The index of elliptic operators on partitioned manifolds

2.1. Examples and the Hellwig-Vekua index theorem.

We begin with some elementary examples.

EXAMPLE 2.1. Consider the (trivially elliptic) ordinary differential operator on the unit interval $I = [0,1]$ defined by

$$P : C^\infty(I) \times C^\infty(I) \to C^\infty(I) \times C^\infty(I),$$
$$(f,g) \mapsto (f', -g'),$$

with the boundary conditions $C^\infty(I) \times C^\infty(I) \to C^\infty(\partial I) \cong \mathbb{C} \times \mathbb{C}$ given by

$$\text{(i)} \quad R_1 : (f,g) \mapsto (f-g)\,|_{\partial I},$$
$$\text{(ii)} \quad R_2 : (f,g) \mapsto f|_{\partial I},$$
$$\text{(iii)} \quad R_3 : (f,g) \mapsto (f+g')|_{\partial I},$$

where $\partial I := \{0,1\}$. We determine the index of the operators

$$(P, R_i) : C^\infty(I) \times C^\infty(I) \to C^\infty(I) \times C^\infty(I) \times C^\infty(\partial I).$$

for $i = 1,2,3$. Clearly, dim $\operatorname{Ker}(P, R_i) = 1$. To determine the cokernel, one writes $P(f,g) = (F,G)$ and $R_i(f,g) = h$, with $F, G \in C^\infty(I)$ and $h = (h_0, h_1) \in \mathbb{C} \times \mathbb{C}$ obtaining,

$$f(t) = \int_0^t F(\tau)d\tau + c_1, \quad g(t) = -\int_0^t G(\tau)d\tau + c_2$$

and two more equations for the boundary condition. The dimension of $\operatorname{Coker}(P, R_i)$ is then the number of linearly independent conditions on F, G, and h which must be imposed in order to eliminate the constants of integration. For each $i \in \{1,2,3\}$, there is only one condition, namely,

$$h_0 = h_1 - \int_0^1 F(\tau)d\tau - \int_0^1 G(\tau)d\tau; \qquad (i = 1)$$

$$h_0 = h_1 - \int_0^1 F(\tau)d\tau; \qquad (i = 2)$$

$$h_0 = h_1 - \int_0^1 F(\tau)d\tau - G(0) + G(1). \quad (i = 3)$$

So, the index vanishes in all three cases.

For a more comprehensive treatment of the existence and uniqueness of boundary value problems for ordinary differential equations (including systems), we refer to [**CodLev**] and [**Har64**, p. 322-403].

We now consider the Laplace operator $\Delta := \partial^2/\partial x^2 + \partial^2/\partial y^2$, as a linear elliptic differential operator from $C^\infty(X)$ to $C^\infty(X)$, where X is the unit disk $\{z = x + iy \mid |z| \leq 1\} \subset \mathbb{C}$, so $\partial X = \{z \in \mathbb{C} \mid |z| = 1\}$.

EXAMPLE 2.2. For the Dirichlet boundary condition

$$R : C^\infty(X) \to C^\infty(\partial X) \text{ with } R(u) = u|_{\partial X},$$

we show that

$$\text{(a) } \operatorname{Ker}(\Delta, R) = \{0\} \quad \text{and} \quad \text{(b) } \operatorname{Im}(\Delta, R)^\perp = \{0\},$$

where $\perp$ is orthogonal complement in $L^2(X) \times L^2(\partial X)$.[1] In particular, it follows that $\operatorname{index}(\Delta, R) = 0$.

For (a): $\operatorname{Ker}(\Delta, R)$ consists of functions of the form $u + iv$, where u and v are real-valued. Since the coefficients of the operators Δ and R are real, we may assume $v = 0$ without loss of generality. Thus, consider a real solution u with $\Delta u = 0$ in X and $u = 0$ on ∂X. Then (where $\nabla u := (u_x, u_y) := (\frac{\partial u}{\partial x}, \frac{\partial u}{\partial y})$)

$$0 = -\int_X u \Delta u \, dx dy = \int_X |\nabla u|^2 \, dx dy, \tag{2.1}$$

whence $\nabla u = 0$. Thus, u is constant, and indeed zero since $u = 0$ on ∂X. The trick lies in the equality (2.1), which follows from Stokes' formula, namely $\int_X d\omega = \int_{\partial X} \omega$, where ω is a 1-form. Indeed, setting $\omega := u \wedge *du$, where $*$ is the Hodge star operator ($*du = *(u_x dx + u_y dy) := u_x dy - u_y dx$), we obtain

$$d\omega = du \wedge *du + u \wedge d * du = |\nabla u|^2 \, dx \wedge dy + (u\Delta u) \, dx \wedge dy.$$

Using Stokes' formula and $u|_{\partial X} = 0$, we have

$$\int_X |\nabla u|^2 \, dx dy + \int_X (u\Delta u) \, dx dy = \int_X d\omega = \int_{\partial X} \omega = \int_{\partial X} u \wedge *du = 0.$$

From this and $\Delta u = 0$, we conclude that $\nabla u = 0$ and u is constant.

For (b): Choose $L \in C^\infty(X)$ and $l \in C^\infty(\partial X)$ with (L, l) orthogonal to $\operatorname{Im}(\Delta, R)$, whence (relative to the usual measures on X and ∂X)

$$\int_X (\Delta u) L + \int_{\partial X} ul = 0 \text{ for all } u \in C^\infty(X). \tag{2.2}$$

[1] Here, consider that the intersection of the orthogonal complement of the range $\operatorname{Im}(\Delta, R)$ relative to the usual inner product in $L^2(X) \times L^2(\partial X)$ with the space $C^\infty(X) \times C^\infty(\partial X)$ is isomorphic to $\operatorname{Coker}(\Delta, R)$. This is true, since it turns out that the image of the natural Sobolev extension of (Δ, R) is closed in the L^2-norm, and its L^2-orthogonal complement is contained in $C^\infty(X) \times C^\infty(\partial X)$.

Using a 2-fold integration by parts (in the exterior calculus), we obtain

$$
\begin{aligned}
\int_X u\Delta L - \int_X (\Delta u)\, L &= \int_X (u(d * dL) - d\,(*du)\, L) \\
&= \int_{\partial X} (u * dL - L * du)\,.
\end{aligned} \tag{2.3}
$$

First we consider u with support $\operatorname{supp}(u) :=$ the closure of $\{z \in X \mid u(z) \neq 0\}$, contained in the interior of X. Then

$$
\int_X u\Delta L = \int_X (\Delta u)\, L = -\int_{\partial X} ul = 0,
$$

and so $\Delta L = 0$. Now for $u \in C^\infty(X)$ we apply (2.2) and (2.3) to deduce that

$$
\begin{aligned}
\int_{\partial X} ul = -\int_X (\Delta u)\, L &= \int_{\partial X} (u * dL - L * du) \\
&= \int_{\partial X} (u\,(xL_x + yL_y) - L\,(xu_x + yu_y))\,.
\end{aligned}
$$

Thus, $l = xL_x + yL_y$ and $L|_{\partial X} = 0$, and we finally apply (a). Details can be found in [**Ho63**, p. 264].

The preceding result $\operatorname{index}(\Delta, R) = 0$ (for $Ru = u|_{\partial X}$) can also be obtained by proving the symmetry of Δ and that the L^2 extension on the domain defined by $Ru = 0$ is a self-adjoint Fredholm extension.

We now consider a C^∞ vector field $\nu : \partial X \to \mathbb{C}$ on the boundary $\partial X = \{z \in \mathbb{C} \mid |z| = 1\}$. For $u \in C^\infty(X)$, $z \in \partial X$, and $\nu(z) = \alpha(z) + i\beta(z)$, the "directional derivative" of the function u relative to ν at the point z is

$$
\frac{\partial u}{\partial \nu}(z) := \alpha(z)\frac{\partial u}{\partial x}(z) + \beta(z)\frac{\partial u}{\partial y}(z)\,.
$$

From the standpoint of differential geometry it is better, either to denote the vector field by $\frac{\partial}{\partial \nu}$ or to write the directional derivative as simply as $\nu\,[u]\,(z)$, since $\frac{\partial}{\partial x}$ and $\frac{\partial}{\partial y}$ can be regarded as vector fields. The pair $(\Delta, \frac{\partial}{\partial \nu})$ defines a linear operator

$$
\left(\Delta, \frac{\partial}{\partial \nu}\right) : C^\infty(X) \to C^\infty(X) \oplus C^\infty(\partial X) \text{ given by}
$$

$$
u \mapsto \left(\Delta u, \frac{\partial u}{\partial \nu}\right).
$$

THEOREM 2.3 (I. N. Vekua 1952). *For $p \in \mathbb{Z}$ and $\nu(z) := z^p$, we have that $(\Delta, \frac{\partial}{\partial \nu})$ is an operator with finite-dimensional kernel and cokernel, and*

$$
\operatorname{index}\left(\Delta, \frac{\partial}{\partial \nu}\right) = 2\,(1 - p)\,.
$$

REMARK 2.4. The theorem of Vekua remains true, if we replace z^p by any nonvanishing "vector field" $\nu : \partial X \to \mathbb{C} \setminus \{0\}$ with "winding number" p.

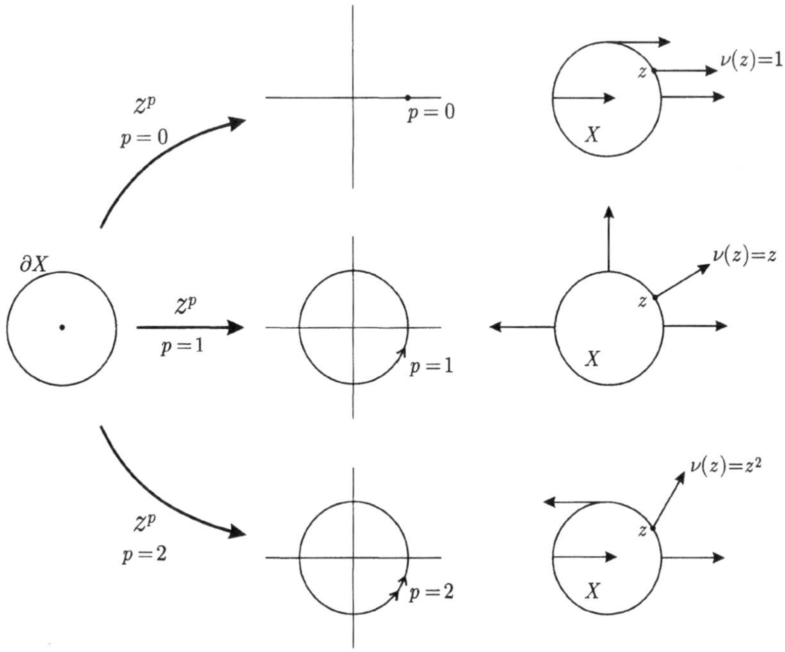

FIGURE 4. The vector field $\nu : \partial X \to \mathbb{C}$ with winding number $p = 0, 1, 2$

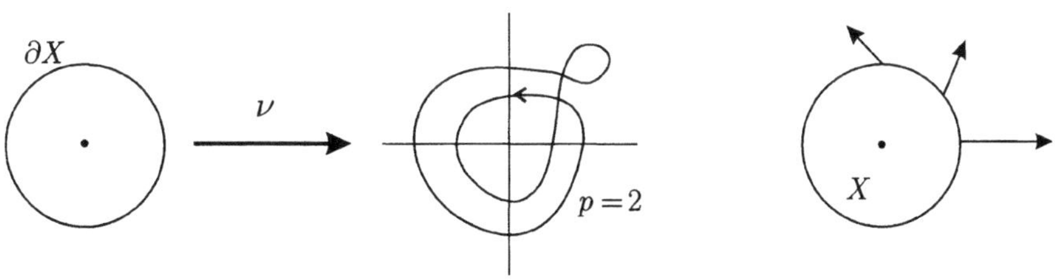

FIGURE 5. Another vector field ν with winding number 2

Moreover, in place of the disk, we can take X to be any simply-connected domain in $\mathbb{C}$ with a "smooth" boundary ∂X. The reason is the homotopy invariance of the index.

REMARK 2.5. In the theory of Riemann surfaces (e.g., in the Riemann-Roch Theorem), one also encounters the number $2(1-p)$ as the Euler characteristic of a closed surface of genus p. This is no accident, but rather it is connected with the relation between elliptic boundary value problems and elliptic operators on closed manifolds. Specifically, there is a relation between the index of $(\Delta, \frac{\partial}{\partial\nu})$ and the index of the Cauchy-Riemann operator for complex line bundles over $S^2 = \mathbb{P}^1(\mathbb{C})$ with Chern number $1-p$ (see Example 2.19 for a start).

REMARK 2.6. Motivated by the method of replacing a differential equation by difference equations, David Hilbert and Richard Courant expected "linear problems of mathematical physics which are correctly posed to behave like a system of N linear algebraic equations in N unknowns (...) If for a correctly posed problem in linear differential equations the corresponding homogeneous problem possesses only the trivial solution zero, then a uniquely determined solution of the general inhomogeneous system exists. However, if the homogeneous problem has a nontrivial solution, the solvability of the inhomogeneous system requires the fulfillment of certain additional conditions." This is the "heuristic principle" which Hilbert and Courant saw in the Fredholm Alternative [**CoHi**]. Günter Hellwig [**He52**] (nicely explained in [**Ha52**]) in the real setting and Ilya Nestorovich Vekua [**Ve56**] in complex setting disproved it with their independently found example where the principle fails for $p \neq 1$.
We remark that in addition to these "oblique-angle" boundary value problems, "coupled" oscillation equations, as well as restrictions of boundary value problems, even with vanishing index, to suitable half-spaces, furnish further more or less elementary examples for index $\neq 0$. The simplest example of a system of first order differential operators on the disc is provided in Example 2.7 below. A world of more advanced, and for differential geometry much more meaningful examples, is approached by the Atiyah-Patodi-Singer Index Theorem, see Section 2.4 below.

PROOF OF THEOREM 2.3. (After [**Ho63**, p. 266 f.]) Since the coefficients of the differential operators $(\Delta, \frac{\partial}{\partial\nu})$ are real, we may restrict ourselves to real functions. Thus, $u \in C^\infty(X)$ denotes a single real-valued function, rather than a complex-valued function.

$\operatorname{Ker}(\Delta, \frac{\partial}{\partial\nu})$: It is well-known that $\operatorname{Ker}(\Delta)$ consists of real (or imaginary) parts of holomorphic functions on X (e.g., see [**Ah53**, p. 175 ff.]). Hence, $u \in \operatorname{Ker}(\Delta)$, exactly when $u = \Re(f)$ where $f = u + iv$ is holomorphic; i.e., the Cauchy-Riemann equation $\frac{\partial f}{\partial \bar z} = 0$ holds, where $\frac{\partial}{\partial \bar z} := \frac{1}{2}(\frac{\partial}{\partial x} + i\frac{\partial}{\partial y})$. Explicitly,

$$0 = \frac{\partial f}{\partial \bar z} = \frac{1}{2}\left(\frac{\partial}{\partial x} + i\frac{\partial}{\partial y}\right)(u+iv) = \frac{1}{2}\left(\frac{\partial u}{\partial x} - \frac{\partial v}{\partial y}\right) + \frac{i}{2}\left(\frac{\partial u}{\partial y} + \frac{\partial v}{\partial x}\right).$$

Every holomorphic (= complex differentiable) function f is twice complex differentiable and its derivative is given by

$$\begin{aligned}\frac{\partial f}{\partial z} &:= \frac{1}{2}\left(\frac{\partial}{\partial x} - i\frac{\partial}{\partial y}\right)(u+iv)\\ &= \frac{1}{2}\left(\frac{\partial u}{\partial x} + \frac{\partial v}{\partial y}\right) + \frac{i}{2}\left(-\frac{\partial u}{\partial y} + \frac{\partial v}{\partial x}\right) = \frac{\partial u}{\partial x} - i\frac{\partial u}{\partial y}.\end{aligned}$$

In this way we have a holomorphic function $\phi := f'$ for each $u \in \operatorname{Ker}(\Delta)$. Since

$$\frac{\partial u}{\partial \nu} = \Re(z^p)\frac{\partial u}{\partial x} + \Im(z^p)\frac{\partial u}{\partial y} = \Re\left(\left(\frac{\partial u}{\partial x} - i\frac{\partial u}{\partial y}\right)z^p\right) = \Re(\phi(z)\,z^p),$$

the boundary condition $\frac{\partial u}{\partial \nu} = 0$ $(\nu = z^p)$ then means that the real part $\Re(\phi(z)z^p)$ vanishes for $|z| = 1$. For $p \geq 0$, $\phi(z)z^p$ is holomorphic as well as ϕ, and hence for $\phi(z) := \frac{\partial u}{\partial x} - i\frac{\partial u}{\partial y}$, we have

$$\begin{aligned} &u \in \operatorname{Ker}\left(\Delta, \frac{\partial}{\partial \nu}\right) \text{ with } \nu = z^p, \quad p \geq 0, \\ &\Rightarrow \Re\left(\phi(z)z^p\right) \in \operatorname{Ker}\left(\Delta, R\right) \text{ where } R\left(\cdot\right) = \left(\cdot\right)|_{\partial X}. \end{aligned}$$

Thus, we succeed in associating with the "oblique-angle" boundary value problem for u a Dirichlet boundary value problem for $\Re\left(\phi(z)z^p\right)$, which has only the trivial solution by Example 2.2a. Since $\phi(z)z^p$ is holomorphic with $\Re\left(\phi(z)z^p\right) = 0$, the partial derivatives of the imaginary part vanish, and so there is a constant $C \in \mathbb{R}$ such that $\phi(z)z^p = iC$ for all $z \in X$. If $p > 0$, then we have $C = 0$ (set $z = 0$). Hence $\phi = 0$, and (by the definition of ϕ) the function u is constant (i.e., $\dim \operatorname{Ker}\left(\Delta, \frac{\partial}{\partial \nu}\right) = 1$). If $p = 0$, then $\frac{\partial u}{\partial x} - i\frac{\partial u}{\partial y} = \phi(z) = iC$, and so $u(x,y) = -Cy + \widetilde{C}$, whence $\dim \operatorname{Ker}\left(\Delta, \frac{\partial}{\partial \nu}\right) = 2$ in this case.

We now come to the case $p < 0$, which curiously is not immediately reducible to the case $q > 0$ where $q := -p$. One might try to look for a solution by simply turning $\frac{\partial}{\partial \nu}$ around to $-\frac{\partial}{\partial \nu}$, but this is futile since the winding numbers of ν and $-\nu$ about 0 are the same. Besides, if $\nu_p(z) = z^p$, we do *not* have $\frac{\partial}{\partial \nu_{-p}} = -\frac{\partial}{\partial \nu_p}$. In order to reduce the boundary value problem with $p < 0$ to the elementary Dirichlet problem, we must go through a more careful argument. Note that $\phi(z)z^p$ can have a pole at $z = 0$, whence $\Re\left(\phi(z)z^p\right)$ is not necessarily harmonic. We write the holomorphic function $\phi(z)$ as

$$\phi(z) = \sum_{j=0}^{q} a_j z^j + g(z) z^{q+1}$$

where $q := -p$, and g is holomorphic. We define a holomorphic function ψ by

$$\psi(z) := g(z) z + \sum_{j=0}^{q-1} \overline{a}_j z^{q-j},$$

with $\psi(0) = 0$. Then one can write

$$\phi(z)z^p = a_q + \sum_{j=0}^{q-1} \left(a_j z^{j-q} - \overline{a}_j z^{q-j}\right) + \psi(z)$$

The boundary condition $\frac{\partial u}{\partial \nu} = 0$ implies $\Re\left(\phi(z)z^p\right) = 0$ for $|z| = 1$. By the above equation, we have $0 = \Re\left(\phi(z)z^p\right) = \Re(\psi(z) + a_q)$ for $|z| = 1$ since then $z^{-1} = \bar{z}$. Since ψ is holomorphic, we have again arrived at a Dirichlet boundary value problem; this time for the function $\Re(\psi(z) + a_q)$. From Example 2.2a, it follows again that $\psi(z) + a_q$ is an imaginary constant, whence $\psi(z) = \psi(0) = 0$ and a_q is pure imaginary. We have

$$\phi(z) = \phi(z) z^p z^q = a_q z^q + \sum_{j=0}^{q-1} \left(a_j z^j - \overline{a}_j z^{2q-j} \right)$$

for arbitrary $a_0, a_1, ..., a_{q-1} \in \mathbb{C}$, and $a_q \in i\mathbb{R}$. As a vector space over $\mathbb{R}$, the set

$$\left\{ \tfrac{\partial u}{\partial x} - i \tfrac{\partial u}{\partial y} \mid u \in \operatorname{Ker}\left(\Delta, \tfrac{\partial}{\partial \nu}\right) \right\}$$

has dimension $2q+1$; here we have restricted ourselves to real u, according to our convention above. Since u is uniquely determined by ϕ up to an additive constant, it follows that for $\nu = z^p$ and $p < 0$,

$$\dim \operatorname{Ker}\left(\Delta, \tfrac{\partial}{\partial \nu}\right) = 2q + 2 = 2 - 2p$$

Coker $\left(\Delta, \frac{\partial}{\partial \nu}\right)$: As Example 2.2b shows the equation $\Delta u = F$ has a solution for each $F \in C^\infty(X)$. In view of this we can show

$$\begin{aligned} \operatorname{Coker}\left(\Delta, \tfrac{\partial}{\partial \nu}\right) &= \left(C^\infty(X) \times C^\infty(\partial X) \right) / \operatorname{Im}\left(\Delta, \tfrac{\partial}{\partial \nu}\right) \\ &\cong C^\infty(\partial X) / \tfrac{\partial}{\partial \nu} \left(\operatorname{Ker} \Delta \right) \end{aligned}$$

as follows. We assign to each representative pair $(F, h) \in C^\infty(X) \times C^\infty(\partial X)$ the class of $h - \frac{\partial u}{\partial \nu} \in C^\infty(\partial X)$, where u is chosen so that $\Delta u = F$. This map is clearly well defined on the quotient space of pairs, and the inverse map is given by $h \mapsto (0, h)$. Hence, we have found a representation for Coker$(\Delta, \frac{\partial}{\partial \nu})$ in terms of the "boundary functions" $\left\{ \frac{\partial u}{\partial \nu} \mid u \in \operatorname{Ker} \Delta \right\}$, rather than the cumbersome pairs in Im $\left(\Delta, \frac{\partial}{\partial \nu}\right)$. (This trick can always be applied for the boundary value problems (P, R), when the operator P is surjective.)

We therefore investigate the existence of solutions of the equation $\Delta u = 0$ with the "inhomogeneous" boundary condition $\frac{\partial u}{\partial \nu} = h$, where h is a given C^∞ function on ∂X. According to the trick introduced in the first part of our proof, it is equivalent to ask for the existence of a holomorphic function ϕ with the boundary condition $\Re\left(\phi(z) z^p\right) = h$, $|z| = 1$, i.e., for a solution of a Dirichlet problem for $\Re\left(\phi(z) z^p\right)$. By Example 2.2b, there is a unique harmonic function which restricts to h on the boundary ∂X; hence, we have a (unique up to an additive imaginary constant) holomorphic function θ with $\Re\theta(z) = h$ for $|z| = 1$.

In the case $p < 0$, the boundary problem for ϕ is always solvable; namely, set $\phi(z) := z^{-p} \theta(z)$. Hence, we have

$$\dim \operatorname{Coker}\left(\Delta, \tfrac{\partial}{\partial \nu}\right) = 0 \text{ for } \nu(z) = z^p \text{ and } p \leq 0.$$

For $p > 0$, we can construct a solution of the boundary value problem in for ϕ from θ, if and only if there is a constant $C \in \mathbb{R}$, such that $(\theta(z) - iC)/z^p$ is holomorphic (i.e., the holomorphic function $\theta(z) - iC$ has a zero of order at least p at $z = 0$. Using the Cauchy Integral Formula, these conditions on the derivatives of θ at $z = 0$ correspond to conditions on line integrals around ∂X. In this way, we have $2p - 1$ linear (real) equations that h must satisfy in order that the boundary value problem have a solution. We summarize our results in the following table ($\nu(z) = z^p$) and Figure 6:

p	$\dim \operatorname{Ker} (\Delta, \frac{\partial}{\partial\nu})$	$\dim \operatorname{Coker} (\Delta, \frac{\partial}{\partial\nu})$	$\operatorname{index} (\Delta, \frac{\partial}{\partial\nu})$
> 0	1	$2p - 1$	$2 - 2p$
≤ 0	$2 - 2p$	0	$2 - 2p$

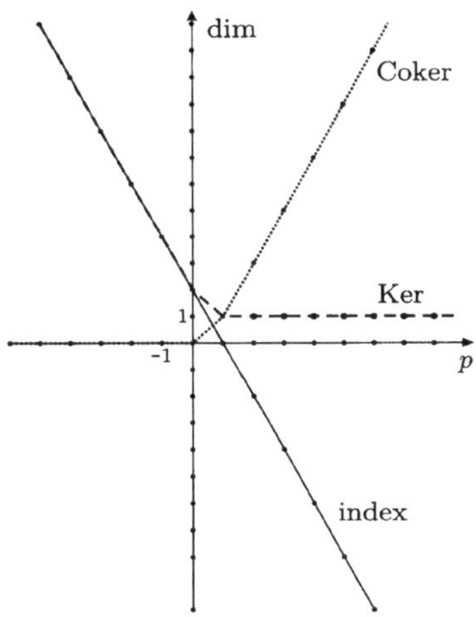

FIGURE 6. The dimensions of kernel, cokernel and the index of the Laplacian with boundary condition given by $\nu(z) := z^p$ for varying p.

□

Note 1: We already noted in the proof the peculiarity that the case $p < 0$ cannot simply be played back to the case $p > 0$. This is reflected here in the asymmetry of the dimensions of kernel and cokernel and the index. It simply reflects the fact that there are "more" rational functions with prescribed poles than there are polynomials with "corresponding" zeros.

Note 2: In contrast to the Dirichlet Problem, which we could solve via integration by parts (i.e., via Stokes' Theorem), the above proof is function-theoretic in nature and cannot be used in higher dimensions. This is no loss in our special case, since the index of the "oblique-angle" boundary value problem must vanish anyhow in higher dimensions for topological reasons; see [**Ho63**, p. 265 f.]. The actual mathematical challenge of the function-theoretic proof arises less from the restriction $\dim X = 2$ than from a certain arbitrariness, namely the tricks and devices of the definitions of the auxiliary functions ϕ, ψ, θ, by means of which the oblique-angle problem is reduced to the Dirichlet problem.

Note 3: The theory of ordinary differential equations easily conveys the impression that partial differential equations also possess a "general solution" in the form of a functional relation between the unknown function ("quantity") u, the independent variables x and some arbitrary constants or functions, and that every "particular solution" is obtained by substituting certain constants or functions f, h, etc. for the arbitrary constants and functions. (Corresponding to the higher degree of freedom in partial differential equations, we deal not only with constants of integration but with arbitrary functions.) The preceding calculations, regarding the boundary value problem of the Laplace operator, clearly indicate how limited this notion is which was conceived in the 18th century on the basis of geometric intuition and physical considerations. The classical recipe of first searching for general solutions and only at the end determining the arbitrary constants and functions fails. For example, the specific form of boundary conditions must enter the analysis to begin with.

EXAMPLE 2.7. Let $X := \{z = x + iy \mid |z| < 1\}$ be the unit disk, and define an operator $T : C^\infty(X) \times C^\infty(X) \to C^\infty(X) \oplus C^\infty(X) \oplus C^\infty(\partial X)$ by

$$T(u,v) := \Big(\frac{\partial u}{\partial \bar z}, \frac{\partial v}{\partial z}, (u-v)\,|_{\partial X}\Big),$$

where $\frac{\partial}{\partial z} = \frac{1}{2}(\frac{\partial}{\partial x} - i\frac{\partial}{\partial y})$ is complex differentiation, and $\frac{\partial}{\partial \bar z} = \frac{1}{2}(\frac{\partial}{\partial x} + i\frac{\partial}{\partial y})$ is the Cauchy-Riemann differential operator "formally adjoint" to $\frac{\partial}{\partial z}$. We show that $\dim(\operatorname{Ker} T) = 1$ and $\operatorname{Coker}(T) = \{0\}$, and hence that $\operatorname{index}(T) = 1$.

Suppose that $(u,v) \in \operatorname{Ker} T$. Then $\frac{\partial u}{\partial \bar z} = 0$ and $\frac{\partial v}{\partial z} = 0$ in which case u and v are harmonic. Then, since $(u-v)\,|_{\partial X} = 0$, we have $u = v$ on X. Now $\frac{\partial v}{\partial \bar z} = \frac{\partial u}{\partial \bar z} = 0$ implies that v is holomorphic, and $v'(z) = \frac{\partial v}{\partial z} = 0$ implies that v $(= u)$ is constant. Thus, $\dim(\operatorname{Ker} T) = 1$. To show that $\operatorname{Coker}(T) = \{0\}$, or more precisely $(\operatorname{Im} T)^\perp = \{0\}$ (see the footnote to Example 2.2), we choose arbitrary functions $f, g \in C^\infty(X)$, and $h \in C^\infty(\partial X)$, and prove that f, g, and h must identically vanish, if

$$\int_X \Big(\frac{\partial u}{\partial \bar z} f + \frac{\partial v}{\partial z} g\Big) + \int_{\partial X} (u-v)h = 0 \tag{2.4}$$

for all $u, v \in C^\infty(X)$. Note that for $P, Q \in C^\infty(X)$ we have

$$\begin{aligned} d(Pdz + Qd\bar z) &= \frac{\partial P}{\partial \bar z} d\bar z \wedge dz + \frac{\partial Q}{\partial z} dz \wedge d\bar z = \left(\frac{\partial P}{\partial \bar z} - \frac{\partial Q}{\partial z}\right) d\bar z \wedge dz \\ &= \left(\frac{\partial P}{\partial \bar z} - \frac{\partial Q}{\partial z}\right)(dx - idy) \wedge (dx + idy) \\ &= 2i \left(\frac{\partial P}{\partial \bar z} - \frac{\partial Q}{\partial z}\right) dx \wedge dy. \end{aligned}$$

Thus we obtain the complex version of Stokes' Theorem,

$$\int_{\partial X} P dz + Q d\bar{z} = \int_X d\,(P dz + Q d\bar{z}) = 2i \int_X \left(\frac{\partial P}{\partial \bar{z}} - \frac{\partial Q}{\partial z} \right).$$

From this, we get

$$\int_X \frac{\partial u}{\partial \bar{z}} f = \int_X \frac{\partial}{\partial \bar{z}} (uf) - \int_X u \frac{\partial f}{\partial \bar{z}} = \frac{1}{2i} \int_{\partial X} uf\, dz - \int_X u \frac{\partial f}{\partial \bar{z}},$$
$$\int_X \frac{\partial v}{\partial z} g = \int_X \frac{\partial}{\partial z} (vg) - \int_X v \frac{\partial g}{\partial z} = -\frac{1}{2i} \int_{\partial X} vg\, d\bar{z} - \int_X v \frac{\partial g}{\partial z}.$$

Hence,

$$\int_X \left(\frac{\partial u}{\partial \bar{z}} f + \frac{\partial v}{\partial z} g \right) = -\int_X \left(u \frac{\partial f}{\partial \bar{z}} + v \frac{\partial g}{\partial z} \right) + \frac{1}{2i} \int_{\partial X} (uf\, dz - vg\, d\bar{z}).$$

Assuming (2.4), we have

$$0 = \int_X \left(\frac{\partial u}{\partial \bar{z}} f + \frac{\partial v}{\partial z} g \right) + \int_{\partial X} (u - v) h$$
$$= -\int_X \left(u \frac{\partial f}{\partial \bar{z}} + v \frac{\partial g}{\partial z} \right) + \frac{1}{2i} \int_{\partial X} (uf\, dz - vg\, d\bar{z}) + \int_{\partial X} (u - v) h.$$

By considering u and v with compact support inside the open disk, we deduce that $\frac{\partial f}{\partial \bar{z}} = 0$ and $\frac{\partial g}{\partial z} = 0$ (i.e., f and g are analytic and conjugate analytic, respectively). Thus, (2.4) implies

$$0 = \frac{1}{2i} \int_{\partial X} (uf\, dz - vg\, d\bar{z}) + \int_{\partial X} (u - v) h,$$

for all $u, v \in C^\infty(X)$. Choosing $v = u$, we have

$$0 = \frac{1}{2i} \int_{\partial X} u\,(f\, dz - g\, d\bar{z}) \text{ for all } u \Rightarrow f\, dz = g\, d\bar{z} \text{ on } \partial X$$
$$\Rightarrow f\left(e^{i\theta}\right) ie^{i\theta} d\theta = -g\left(e^{i\theta}\right) ie^{-i\theta} d\theta \Rightarrow f\left(e^{i\theta}\right) e^{i\theta} = -g\left(e^{i\theta}\right) e^{-i\theta}.$$

However, since f is analytic, the Fourier series of $f\left(e^{i\theta}\right) e^{i\theta}$ has a nonzero coefficient for $e^{im\theta}$ only when $m > 0$, and since g is conjugate analytic, $g\left(e^{i\theta}\right) e^{-i\theta}$ has a nonzero coefficient for $e^{im\theta}$ only when $m < 0$. Thus, $f = g = 0$. Choosing $v = -u$, the equation (2.4) then yields

$$0 = \int_{\partial X} (u - v) h = 2 \int_{\partial X} uh$$

for all $u \in C^\infty(X)$, which implies $h = 0$.

REMARK 2.8. In engineering, a system of separate differential equations

$$Pu = f,$$
$$Qv = g,$$

related by a "transfer condition" $R(u, v) = h$, is called a "coupling problem". When the domains of u and v are different but have a common boundary (or boundary part) on which the transfer condition is defined, then we have a "transmission problem"; see e.g. [**Boo72**, p.7 ff]. Thus, we may think of T as an operator for a problem on the spherical surface $X \cup_{\partial X} X$ with different behavior on the upper and lower hemispheres, but with a fixed coupling along the equator.

2.2. The index of twisted Dirac operators on closed manifolds.

Recall that the index of a Fredholm operator is a measure of its asymmetry: It is defined by the difference between the dimension of the kernel (the null space) of the operator and the dimension of the kernel of the adjoint operator (= the codimension of the range). So, the index vanishes for self-adjoint Fredholm operators. For an elliptic differential or pseudo-differential operator on a closed manifold M, the index is finite and depends only on the homotopy class of the principal symbol σ of the operator over the cotangent sphere bundle S^*M. It follows (see [**LaMi**, p. 257]) that the index for elliptic *differential* operators always vanishes on *closed* odd-dimensional manifolds. On even-dimensional manifolds one has the Atiyah-Singer Index Theorem which expresses the index in explicit topological terms, involving the Todd class defined by the Riemannian structure of M and the Chern class defined by gluing two copies of a bundle over S^*M by σ.

The original approach in proving the Index Theorem in the work of Atiyah and Singer [**AtSi69**] is based on the following clever strategy. The invariance of the index under homotopy implies that the index (say, the *analytic index*) of an elliptic operator is stable under rather dramatic, but continuous, changes in its principal symbol while maintaining ellipticity. Moreover, the index is functorial with respect to certain operations, such as addition and composition. Thus, the indices of elliptic operators transform predictably under various global operations (or relations) such as direct sums, embedding and cobordism. Using K-theory, a topological invariant (say, the *topological index*) with the same transformation properties under these global operation is built from the symbol of the elliptic operator. It turns out that the global operations are sufficient to construct enough vector bundles and elliptic operators to deduce the Atiyah-Singer Index Theorem (i.e., *analytic index* = *topological index*). With the aid of Bott periodicity, it suffices to check that the two indices are the same in the trivial case where the base manifold is just a single point. A particularly nice exposition of this approach is found in E. Guentner's article [**Gu93**] following an argument of P. Baum.

Not long after this first proof (given in quite different variants), there emerged a fundamentally different means of proving the Atiyah-Singer Index Theorem, namely the *heat kernel method.* This is outlined here in the important case of the chiral half $\mathcal{D}^+$ of a twisted Dirac operator $\mathcal{D}$. (In the index theory of closed manifolds, one usually studies the index of a chiral half $\mathcal{D}^+$ instead of the total Dirac operator $\mathcal{D}$, since $\mathcal{D}$ is symmetric for compatible connections and then index $\mathcal{D} = 0$.) The heat kernel method had its origins in the late 1960s (e.g., in [**McK-Si**]) and was pioneered in the works [**Pa71**], [**Gi73**], [**ABP73**], etc.. In the final analysis, it is debatable as to whether this method is really much shorter or better. This depends on the background and tastes of the beholder. Geometers and analysts (as opposed to topologists) are likely to find the heat kernel method appealing, because K-theory, Bott periodicity and cobordism theory are avoided, not only for geometric operators which are expressible in terms of twisted Dirac operators, but also largely for more general elliptic pseudo-differential operators, as Melrose has done in [**Me93**]. Moreover, the heat method gives the index of a "geometric" elliptic differential operator naturally as the integral of a characteristic form (a polynomial of curvature forms) which is expressed solely in terms of the geometry of the operator itself (e.g., curvatures of metric tensors and connections). One does not destroy the geometry of the operator by taking advantage of the fact that it can be suitably deformed without altering the index. Rather, in the heat kernel approach, the invariance of the index under changes in the geometry of the operator is a consequence of the index formula itself rather than a means of proof. However, considerable analysis and effort are needed to obtain the heat kernel for $e^{-t\mathcal{D}^2}$ and to establish its asymptotic expansion as $t \to 0^+$. Also, it can be argued that in some respects the K-theoretical embedding/cobordism methods are more forceful and direct. Moreover, in [**LaMi**], we are cautioned that the index theorem for families (in its strong form) generally involves torsion elements in K-theory that are not detectable by cohomological means, and hence are not computable in terms of local densities produced by heat asymptotics. Nevertheless, when this difficulty does not arise, the K-theoretical expression for the topological index may be less appealing than the integral of a characteristic form, particularly for those who already understand and appreciate the geometrical formulation of characteristic classes. All disputes aside, the student who learns *both* approaches and formulations will be more accomplished (and probably older).

The classical geometric operators such as the Hirzebruch signature operator, the de Rham operator, the Dolbeaut operator and even the Yang-Mills operator can all be locally expressed in terms of chiral halves of twisted Dirac operators. Thus, we will focus on index theory for such operators. The index of any of these classical operators (and *their* twists) can then be obtained from the Local Index Theorem for twisted Dirac operators. This theorem supplies a well-defined n-form on M, whose integral is the index of the twisted Dirac operator. This n-form (or "index density") is expressed in terms of forms for characteristic classes which are polynomials in curvature forms. The Index Theorem thus obtained then becomes a formula that relates a global invariant quantity, namely the index of an operator, to the integral of a local quantity involving curvature. This is in the spirit of the Gauss-Bonnet Theorem which can be considered a special case.

DEFINITION 2.9. Let M be an oriented Riemannian n-manifold ($n = 2m$ even) with metric h, and oriented orthonormal frame bundle FM. Assume that M has a spin structure $P \to FM$, where P is a principal Spin(n)-bundle and the projection $P \to FM$ is a two-fold cover, equivariant with respect to $\mathrm{Spin}(n) \to \mathrm{SO}(n)$. Furthermore, let $E \to M$ be a Hermitian vector bundle with a unitary connection ε. The *twisted Dirac operator* $\mathcal{D}$ associated with the above data is

$$\mathcal{D} := (1 \otimes \mathbf{c}) \circ \nabla : C^\infty (E \otimes \Sigma (M)) \to C^\infty (E \otimes \Sigma (M)). \tag{2.5}$$

Here, $\Sigma(M)$ is the spin bundle over M associated to $P \to FM \to M$ via the spinor representation $\mathrm{Spin}(n) \to \mathrm{End}(\Sigma_n)$,

$$\mathbf{c} : C^\infty(\Sigma(M) \otimes TM^*) \to C^\infty(\Sigma(M))$$

is Clifford multiplication, and

$$\nabla : C^\infty(E \otimes \Sigma(M)) \to C^\infty(E \otimes \Sigma(M) \otimes TM^*)$$

is the covariant derivative determined by the connection ε and the spinorial lift to P of the Levi-Civita connection form, say θ, on FM.

Note that $\mathcal{D}$ here is a special case of an operator of Dirac type introduced in Section 1.3 with $\mathfrak{Cl}(M)$ acting on the second factor of $E \otimes \Sigma(M)$ which is playing the role of S. Let $\Sigma^\pm(M)$ denote the ± 1 eigenbundles of the complex Clifford volume element in $C^\infty(\mathfrak{Cl}(M))$, given at a point $x \in M$ by $i^m e_1 \cdots e_n$, where $e_1, \ldots, e_n$ is an oriented, orthonormal basis of T_xM. The $\Sigma^\pm(M)$ are the so-called chiral halves of $\Sigma(M) = \Sigma^+(M) \oplus \Sigma^+(M)$. Since

$$\begin{aligned}
&\nabla\left(C^\infty\left(E \otimes \Sigma^\pm(M)\right)\right) \subseteq C^\infty\left(E \otimes \Sigma^\pm(M) \otimes TM^*\right), \text{ and} \\
&(1 \otimes \mathbf{c})\left(C^\infty\left(E \otimes \Sigma^\pm(M) \otimes TM^*\right)\right) \subseteq C^\infty\left(E \otimes \Sigma^\mp(M)\right),
\end{aligned}$$

we have

$$\mathcal{D} = \mathcal{D}^+ \oplus \mathcal{D}^-, \quad \text{where } \mathcal{D}^\pm : C^\infty\left(E \otimes \Sigma^\pm(M)\right) \to C^\infty\left(E \otimes \Sigma^\mp(M)\right).$$

The symbol of the first-order differential operator $\mathcal{D}$ is computed as follows. For $\phi \in C^\infty(M)$ with $\phi(x) = 0$ and $\psi \in C^\infty(E \otimes \Sigma(M))$, we have at x

$$\begin{aligned}
(1 \otimes \mathbf{c}) \circ \nabla(\phi\psi) &= (1 \otimes \mathbf{c}) \circ ((d\phi)\psi + \phi\nabla\psi) = (1 \otimes \mathbf{c}) \circ (d\phi)\psi \\
&= (1 \otimes \mathbf{c}(d\phi))\psi.
\end{aligned}$$

Thus, the symbol $\sigma(\mathcal{D}) : T_xM^* \to \mathrm{End}(\Sigma(M))$ at the covector $\xi_x \in T_xM^*$ is given by

$$\sigma(\mathcal{D})(\xi_x) = 1 \otimes \mathbf{c}(\xi_x) \in \mathrm{End}(E_x \otimes \Sigma_x).$$

For $\xi_x \neq 0$, $\sigma(\mathcal{D})(\xi_x)$ is an isomorphism, since

$$\sigma(\mathcal{D})(\xi_x) \circ \sigma(\mathcal{D})(\xi_x) = 1 \otimes \mathbf{c}(\xi_x)^2 = -|\xi_x|^2 I.$$

Thus, $\mathcal{D}$ is an elliptic operator. Moreover, since $\sigma(\mathcal{D}^+)$ and $\sigma(\mathcal{D}^-)$ are restrictions of $\sigma(\mathcal{D})$, it follows that $\mathcal{D}^+$ and $\mathcal{D}^-$ are elliptic. It can be shown that $\mathcal{D}$ is formally self-adjoint, and $\mathcal{D}^+$ and $\mathcal{D}^-$ are formal adjoints of each other (see [**LaMi**]). We also have a pair of self-adjoint elliptic operators

$$\begin{aligned}
\mathcal{D}_+^2 &:= \mathcal{D}^2 | C^\infty\left(E \otimes \Sigma^+(M)\right) = \mathcal{D}^- \circ \mathcal{D}^+ \\
\mathcal{D}_-^2 &:= \mathcal{D}^2 | C^\infty\left(E \otimes \Sigma^-(M)\right) = \mathcal{D}^+ \circ \mathcal{D}^-.
\end{aligned}$$

For $\lambda \in \mathbb{C}$, let

$$V_\lambda\left(\mathcal{D}_\pm^2\right) := \left\{\psi \in C^\infty\left(E \otimes \Sigma^\pm(M)\right) \mid \mathcal{D}_\pm^2 \psi = \lambda\psi\right\}.$$

From the general theory of formally self-adjoint elliptic operators on compact manifolds, we know that

$$\operatorname{Spec}\left(\mathcal{D}_{\pm}^{2}\right)=\left\{\lambda\in\mathbb{C}\mid V_{\lambda}\left(\mathcal{D}_{\pm}^{2}\right)\neq\{0\}\right\}$$

consists of the eigenvalues of $\mathcal{D}_{\pm}^{2}$ and is a discrete subset of $[0,\infty)$, the eigenspaces $V_{\lambda}\left(\mathcal{D}_{\pm}^{2}\right)$ are finite-dimensional, and an $L^{2}\left(E\otimes\Sigma^{\pm}(M)\right)$-complete orthonormal set of vectors can be selected from the $V_{\lambda}\left(\mathcal{D}_{\pm}^{2}\right)$. Note that $\mathcal{D}^{+}\left(V_{\lambda}\left(\mathcal{D}_{+}^{2}\right)\right)\subseteq V_{\lambda}\left(\mathcal{D}_{-}^{2}\right)$, since for $\psi\in V_{\lambda}\left(\mathcal{D}_{+}^{2}\right)$ we have

$$\begin{aligned}\mathcal{D}_{-}^{2}\left(\mathcal{D}^{+}\psi\right)&=\left(\mathcal{D}^{+}\circ\mathcal{D}^{-}\right)\left(\mathcal{D}^{+}\psi\right)=\mathcal{D}^{+}\left(\left(\mathcal{D}^{-}\circ\mathcal{D}^{+}\right)(\psi)\right)\\&=\mathcal{D}^{+}\left(\mathcal{D}_{+}^{2}(\psi)\right)=\mathcal{D}^{+}(\lambda\psi)=\lambda\mathcal{D}^{+}(\psi),\end{aligned}$$

and similarly $\mathcal{D}^{-}\left(V_{\lambda}\left(\mathcal{D}_{-}^{2}\right)\right)\subseteq V_{\lambda}\left(\mathcal{D}_{+}^{2}\right)$. For $\lambda\neq 0$,

$$\mathcal{D}^{\pm}|V_{\lambda}\left(\mathcal{D}_{\pm}^{2}\right):V_{\lambda}\left(\mathcal{D}_{\pm}^{2}\right)\to V_{\lambda}\left(\mathcal{D}_{\mp}^{2}\right)$$

is an isomorphism, since it has the inverse $\frac{1}{\lambda}\mathcal{D}^{\mp}$. Thus the set of nonzero eigenvalues (and their multiplicities) of $\mathcal{D}_{+}^{2}$ coincides with that of $\mathcal{D}_{-}^{2}$. However, in general,

$$\dim V_{0}(\mathcal{D}_{+}^{2})-\dim V_{0}(\mathcal{D}_{-}^{2})=\dim\operatorname{Ker}(\mathcal{D}_{+}^{2})-\dim\operatorname{Ker}(\mathcal{D}_{-}^{2})=\operatorname{index}(\mathcal{D}^{+})\neq 0.$$

Since $\dim V_{\lambda}(\mathcal{D}_{+}^{2})-\dim V_{\lambda}(\mathcal{D}_{-}^{2})=0$ for $\lambda\neq 0$, obviously

$$\begin{aligned}\operatorname{index}(\mathcal{D}^{+})&=\dim V_{0}(\mathcal{D}_{+}^{2})-\dim V_{0}(\mathcal{D}_{-}^{2})\\&=\sum_{\lambda\in\operatorname{Spec}(\mathcal{D}_{+}^{2})}e^{-t\lambda}\left(\dim V_{\lambda}(\mathcal{D}_{+}^{2})-\dim V_{\lambda}(\mathcal{D}_{-}^{2})\right).\end{aligned}$$

This may seem like a very inefficient way to write index$(\mathcal{D}^{+})$, but the point is that the sum can be expressed as the integral of the supertrace of the heat kernel for the spinorial heat equation $\frac{\partial\psi}{\partial t}=-\mathcal{D}^{2}\psi$, from which the Local Index Theorem (Theorem 2.13 below) for $\mathcal{D}^{+}$ will eventually follow. However, first the existence of the heat kernel needs to be established.

Let the *positive* eigenvalues of $\mathcal{D}_{\pm}^{2}$ be placed in a sequence $0<\lambda_{1}\leq\lambda_{2}\leq\cdots$, where each eigenvalue is repeated according to its multiplicity. Let $u_{1}^{\pm},u_{2}^{\pm},\ldots$ be an L^{2}-orthonormal sequence in $C^{\infty}\left(E\otimes\Sigma^{+}(M)\right)$ with $\mathcal{D}_{\pm}^{2}\left(u_{j}^{\pm}\right)=\lambda_{j}u_{j}^{\pm}$ (i.e., $u_{j}^{\pm}\in V_{\lambda_{j}}\left(\mathcal{D}_{\pm}^{2}\right)$). We let $u_{0_{1}}^{+},\ldots,u_{0_{n^{+}}}^{+}$ be an L^{2}-orthonormal basis of $\operatorname{Ker}\mathcal{D}_{+}^{2}=\operatorname{Ker}\mathcal{D}_{+}$, and let $u_{0_{1}}^{-},\ldots,u_{0_{n^{-}}}^{-}$ be an L^{2}-orthonormal basis of $\operatorname{Ker}\mathcal{D}_{-}^{2}=\operatorname{Ker}\mathcal{D}_{-}$. We can pull back the bundle $E\otimes\Sigma^{\pm}(M)$ via either of the projections $M\times M\times(0,\infty)\to M$ given by $\pi_{1}(x,y,t):=x$ and $\pi_{2}(x,y,t):=y$, and take the tensor product of the results to form a bundle

$$\mathcal{K}^{\pm}:=\pi_{1}^{*}\left(E\otimes\Sigma^{\pm}(M)\right)\otimes\pi_{2}^{*}\left(E\otimes\Sigma^{\pm}(M)\right)\to M\times M\times(0,\infty).$$

Note that for $x\in M$, the Hermitian inner product $\langle\ ,\ \rangle_{x}$ on $\left(E\otimes\Sigma^{\pm}(M)\right)_{x}$ gives us a conjugate-linear map $\psi\mapsto\psi^{*}(\cdot):=\langle\cdot,\psi\rangle_{x}$ from $\left(E\otimes\Sigma^{\pm}(M)\right)_{x}$ to its dual space $(E\otimes\Sigma^{\pm}(M))^*_x$. Thus, we can (and do) make the identifications

$$\begin{aligned}\pi_1^*\left(E\otimes\Sigma^{\pm}(M)\right)&\otimes\pi_2^*\left(E\otimes\Sigma^{\pm}(M)\right)\\ &\cong\left(\pi_1^*\left(E\otimes\Sigma^{\pm}(M)\right)\right)^*\otimes\pi_2^*\left(E\otimes\Sigma^{\pm}(M)\right)\\ &\cong\operatorname{Hom}\left(\pi_1^*\left(E\otimes\Sigma^{\pm}(M)\right),\pi_2^*\left(E\otimes\Sigma^{\pm}(M)\right)\right).\end{aligned}$$

The full proof of the following Proposition 2.10 will be found in [**BlBo04**], but it is already contained in [**Gi95**] for readers of sufficient background.

PROPOSITION 2.10. *For $t>t_0>0$, the series $k'^{\pm}$ defined by*

$$k'^{\pm}(x,y,t):=\sum_{j=1}^{\infty}e^{-\lambda_j t}u_j^{\pm}(x)\otimes u_j^{\pm}(y)$$

converges uniformly in $C^q(\mathcal{K}^{\pm}|M\times M\times(t_0,\infty))$ for all $q\geq 0$. Hence $k'^{\pm}\in C^{\infty}(\mathcal{K}^{\pm})$, and (for $t>0$)

$$\frac{\partial}{\partial t}k'^{\pm}(x,y,t)=-\sum_{j=1}^{\infty}\lambda_j e^{-\lambda_j t}u_j^{\pm}(x)\otimes u_j^{\pm}(y)=-\mathcal{D}_{\pm}^2k'^{\pm}(x,y,t).\tag{2.6}$$

DEFINITION 2.11. The *positive and negative twisted spinorial heat kernels* $k^{\pm}\in C^{\infty}(\mathcal{K}^{\pm})$ (or the *heat kernels for $\mathcal{D}_{\pm}^2$*) are given, for $t>0$, by

$$\begin{aligned}k^{\pm}(x,y,t)&:=\sum_{i=1}^{n^{\pm}}u_{0_i}^{\pm}(x)\otimes u_{0_i}^{\pm}(y)+k'^{\pm}(x,y,t)\\ &=\sum_{i=1}^{n^{\pm}}u_{0_i}^{\pm}(x)\otimes u_{0_i}^{\pm}(y)+\sum_{j=1}^{\infty}e^{-\lambda_j t}u_j^{\pm}(x)\otimes u_j^{\pm}(y).\end{aligned}$$

The *total twisted spinorial heat kernel* (or the *heat kernel for $\mathcal{D}^2$*) is

$$k=\left(k^+,k^-\right)\in C^{\infty}\left(\mathcal{K}^+\right)\oplus C^{\infty}\left(\mathcal{K}^-\right)\cong C^{\infty}\left(\mathcal{K}^+\oplus\mathcal{K}^-\right)\subseteq C^{\infty}(\mathcal{K}),\tag{2.7}$$

where $\mathcal{K}:=\mathcal{K}^+\oplus\mathcal{K}^-=\operatorname{Hom}\left(\pi_1^*\left(E\otimes\Sigma(M)\right),\pi_2^*\left(E\otimes\Sigma(M)\right)\right)$.

The terminology is justified in view of the following result, whose proof is to be found in [**BlBo04**].

PROPOSITION 2.12. *Let $\psi_0^{\pm}\in C^{\infty}(E\otimes\Sigma^{\pm}(M))$, and let*

$$\psi^{\pm}(x,t)=\int_M\left\langle k^{\pm}(x,y,t),\psi_0^{\pm}(y)\right\rangle_y\nu_y.$$

Then for $t>0$, $\psi^{\pm}$ solves the heat equation with initial spinor field $\psi_0^{\pm}$:

$$\frac{\partial\psi^{\pm}}{\partial t}=-\mathcal{D}_{\pm}^2\psi,\ \ \textit{and}\ \lim_{t\to 0^+}\psi^{\pm}(\cdot,t)=\psi_0^{\pm}\ \textit{in}\ C^q\ \textit{for all}\ q\geq 0.$$

Moreover, for $\psi_0\in C^{\infty}(E\otimes\Sigma(M))$ and

$$\psi(x,t):=\int_M\left\langle k(x,y,t),\psi_0\right\rangle_y\nu_y,$$

we have $\dfrac{\partial \psi}{\partial t} = -\mathcal{D}^2\psi$, *and* $\lim_{t\to 0^+} \psi(\cdot,t) = \psi_0(\cdot)$ *in* C^q *for all* $q \geq 0$.

For any finite dimensional Hermitian vector space $(V, \langle\cdot,\cdot\rangle)$ with orthonormal basis $e_1, \ldots, e_N$, we have (for $v \in V$)

$$\begin{aligned}\operatorname{Tr}(v^* \otimes v) &= \sum_{i=1}^{N} \langle (v^* \otimes v)(e_i), e_i\rangle = \sum_{i=1}^{N} \langle v^*(e_i)\, v, e_i\rangle \\ &= \sum_{i=1}^{N} \langle \langle e_i, v\rangle v, e_i\rangle = \sum_{i=1}^{N} \langle e_i, v\rangle \langle v, e_i\rangle = \sum_{i=1}^{N} |\langle e_i, v\rangle|^2 = |v|^2.\end{aligned}$$

In particular, $k^{\pm}(x,x,t) \in \operatorname{End}((E \otimes \Sigma^{\pm}(M))_x)$, and

$$\operatorname{Tr}\left(k^{\pm}(x,x,t)\right) = \sum_{i=1}^{n^{\pm}} \left|u_{0_i}^{\pm}(x)\right|^2 + \sum_{j=1}^{\infty} e^{-\lambda_j t} \left|u_j^{\pm}(x)\right|^2.$$

Since this series converges uniformly and $\left\|u_{0_i}^{\pm}\right\|_{2,0} = \left\|u_j^{\pm}\right\|_{2,0} = 1$, we have

$$\int_M \operatorname{Tr}\left(k^{\pm}(x,x,t)\right) \nu_x = n^{\pm} + \sum_{j=1}^{\infty} e^{-\lambda_j t} < \infty.$$

For $t > 0$, we define the bounded operator $e^{-t\mathcal{D}_{\pm}^2} \in \operatorname{End}\left(L^2\left(E \otimes \Sigma^{\pm}(M)\right)\right)$ by

$$e^{-t\mathcal{D}_{\pm}^2}\left(\psi^{\pm}\right) = \sum_{i=1}^{n^{\pm}} \left(u_{0_i}^{\pm}, \psi_0^{\pm}\right) u_{0_i}^{\pm} + \sum_{j=1}^{\infty} e^{-\lambda_j t} \left(u_j^{\pm}, \psi_0^{\pm}\right) u_j^{\pm}.$$

Note that $e^{-t\mathcal{D}_{\pm}^2}$ is of trace class, since

$$\operatorname{Tr}\left(e^{-t\mathcal{D}_{\pm}^2}\right) = n^{\pm} + \sum_{j=1}^{\infty} e^{-\lambda_j t} = \int_M \operatorname{Tr}\left(k^{\pm}(x,x,t)\right) \nu_x < \infty.$$

Now, we have

$$\begin{aligned}\operatorname{index}(\mathcal{D}^+) &= \dim V_0(\mathcal{D}_+^2) - \dim V_0(\mathcal{D}_-^2) \\ &= n^+ - n^- + \sum_{j=1}^{\infty} \left(e^{-\lambda_j t} - e^{-\lambda_j t}\right) \\ &= n^+ + \sum_{j=1}^{\infty} e^{-\lambda_j t} - \left(n^- + \sum_{j=1}^{\infty} e^{-\lambda_j t}\right) \\ &= \int_M \left(\operatorname{Tr}\left(k^+(x,x,t)\right) - \operatorname{Tr}\left(k^-(x,x,t)\right)\right) \nu_x. \qquad (2.8)\end{aligned}$$

Since $\mathcal{D}^2 = \mathcal{D}_+^2 \oplus \mathcal{D}_-^2$, we also have the operator $e^{-t\mathcal{D}^2} \in \operatorname{End}\left(L^2\left(E \otimes \Sigma(M)\right)\right)$, which is of trace class and whose trace is given by

$$\operatorname{Tr}\left(e^{-t\mathcal{D}^2}\right) = \int_M \operatorname{Tr}(k(x,x,t))\nu_x = \int_M \left(\operatorname{Tr}(k^+(x,x,t)) + \operatorname{Tr}(k^-(x,x,t))\right)\nu_x.$$

The *supertrace* of $k(x,x,t)$ is defined by

$$\operatorname{Str}\left(k\left(x,x,t\right)\right) := \operatorname{Tr}\left(k^+\left(x,x,t\right)\right) - \operatorname{Tr}\left(k^-\left(x,x,t\right)\right),$$

and in view of (2.8), we have

$$\operatorname{index}\left(\mathcal{D}^+\right) = \int_M \operatorname{Str}\left(k\left(x,x,t\right)\right)\nu_x. \tag{2.9}$$

The left side is independent of t and so the right side is also independent of t. The main task now is to determine the behavior of $\operatorname{Str}\left(k\left(x,x,t\right)\right)$ as $t \to 0^+$. We suspect that for each $x \in M$, as $t \to 0^+$, $k\left(x,x,t\right)$ and $\operatorname{Str}\left(k\left(x,x,t\right)\right)$ are influenced primarily by the geometry (e.g., curvature form Ω^θ of M with metric h and Levi-Civita connection θ, and the curvature Ω^ε of the unitary connection for E) near x, since the heat sources of points far from x are not felt very strongly at x for small t. Indeed, we will give an outline a proof of the following Local Index Formula, the full proof of which will appear in [**BlBo04**].

THEOREM 2.13 (The Local Index Theorem). *In the notation of Definitions 2.9 and 2.11, let $\mathcal{D} : C^\infty\left(E \otimes \Sigma\left(M\right)\right) \to C^\infty\left(E \otimes \Sigma\left(M\right)\right)$ be a twisted Dirac operator and let $k \in C^\infty\left(\mathcal{K}\right)$ be the heat kernel for $\mathcal{D}^2$. If Ω^ε is the curvature form of the unitary connection ε for E and Ω^θ is the curvature form of the Levi-Civita connection θ for (M,h) with volume element ν, then*

$$\lim_{t\to 0^+} \operatorname{Str}\left(k\left(x,x,t\right)\right) = \left\langle \operatorname{Tr}\left(e^{i\Omega^\varepsilon/2\pi}\right) \wedge \det\left(\frac{i\Omega^\theta/4\pi}{\sinh(i\Omega^\theta/4\pi)}\right)^{\frac{1}{2}}, \nu_x\right\rangle. \tag{2.10}$$

REMARK 2.14. As will be explained below, the right side is really the inner product (with volume form ν at x) of the canonical form $\mathbf{ch}(E,\varepsilon) \smile \widehat{\mathbf{A}}(M,\theta)$ (depending on the connections ε for E and the Levi-Civita connection θ for the metric h) which represents $\mathbf{ch}(E) \smile \widehat{\mathbf{A}}(M)$. As a consequence, we obtain the Index Theorem for twisted Dirac operators from the Local Index Formula in Corollary 2.15 below. Thus the Local Index Formula is stronger than the Index Theorem for twisted Dirac operators. Indeed, the Local Index Formula yields the Index Theorem for elliptic operators which are locally expressible as twisted Dirac operators or direct sums of such.

COROLLARY 2.15 (Index formula for twisted Dirac operators). *For an oriented Riemannian n-manifold M (n even) with spin structure, and a Hermitian vector bundle $E \to M$ with unitary connection, let $\mathcal{D} = \mathcal{D}^+ \oplus \mathcal{D}^-$ be the twisted Dirac operator, with $\mathcal{D}^+ : C^\infty\left(E \otimes \Sigma^+(M)\right) \to C^\infty\left(E \otimes \Sigma^-(M)\right)$. We have*

$$\operatorname{index}(\mathcal{D}^+) = \left(\mathbf{ch}(E) \smile \widehat{\mathbf{A}}(M)\right)[M], \tag{2.11}$$

where $\mathbf{ch}(E)$ *is the total Chern character class of* E *and* $\widehat{\mathbf{A}}(M)$ *is the total* $\widehat{\mathbf{A}}$ *class of* M*, both defined below. In particular, we obtain:*

For $n = 2$*:*

$$\operatorname{index}\left(\mathcal{D}^+\right) = ch_1(E)[M] = c_1(E)[M].$$

For $n = 4$*:*

$$\begin{aligned}\operatorname{index}\left(\mathcal{D}^+\right) &= \big(\mathbf{ch}(E) \smile \widehat{\mathbf{A}}(M)\big)[M] \\ &= \big(- \dim E \cdot \tfrac{1}{24} p_1(TM) + ch_2(E)\big)[M] \\ &= \big(- \tfrac{1}{24} \dim E \cdot p_1(TM) + \tfrac{1}{2} c_1(E)^2 - c_2(E)\big)[M].\end{aligned}$$

PROOF. By (2.9), (2.10) and the above remark, we have

$$\begin{aligned}\operatorname{index}\left(\mathcal{D}^+\right) &= \int_M \operatorname{Str}(k(x,x,t))\, \nu_x \\ &= \lim_{t\to 0^+} \int_M \operatorname{Str}(k(x,x,t))\, \nu_x = \int_M \lim_{t\to 0^+} \operatorname{Str}(k(x,x,t))\, \nu_x \\ &= \int_M \big\langle \mathbf{ch}(E,\varepsilon)_x \smile \widehat{\mathbf{A}}(M,\theta)_x,\, \nu_x \big\rangle\, \nu_x = \big(\mathbf{ch}(E) \smile \widehat{\mathbf{A}}(M)\big)[M].\end{aligned}$$

□

We now explain the meaning of the form

$$\operatorname{Tr}\left(e^{i\Omega^\varepsilon/2\pi}\right) \wedge \det\left(\frac{i\Omega^\theta/4\pi}{\sinh(i\Omega^\theta/4\pi)}\right)^{\frac{1}{2}}.$$

The first part $\operatorname{Tr}\left(e^{i\Omega^\varepsilon/2\pi}\right)$ is relatively easy. We have (recall $\dim M = 2m$)

$$e^{i\Omega^\varepsilon/2\pi} := \sum_{k=0}^{\infty} \frac{1}{k!}\left(\frac{i}{2\pi}\right)^k \Omega^\varepsilon \wedge \overset{k}{\cdots} \wedge \Omega^\varepsilon = \sum_{k=0}^{m} \frac{1}{k!}\left(\frac{i}{2\pi}\right)^k \Omega^\varepsilon \wedge \overset{k}{\cdots} \wedge \Omega^\varepsilon, \tag{2.12}$$

where $\Omega^\varepsilon \wedge \overset{k}{\cdots} \wedge \Omega^\varepsilon \in \Omega^{2k}(\operatorname{End}(E))$. Also, $\operatorname{Tr}\left(i^k \Omega^\varepsilon \wedge \overset{k}{\cdots} \wedge \Omega^\varepsilon\right) \in \Omega^{2k}(M)$ and

$$\operatorname{Tr}\left(e^{i\Omega^\varepsilon/2\pi}\right) \in \bigoplus_{k=1}^{m} \Omega^{2k}(M).$$

This (by one of many equivalent definitions) is a representative of the total Chern character $\mathbf{ch}(E) \in \bigoplus_{k=1}^m H^{2k}(M,\mathbb{Q})$. The curvature Ω^θ of the Levi-Civita connection θ for the metric h has values in the skew-symmetric endomorphisms of TM; i.e., $\Omega^\theta \in \Omega^2(\operatorname{End}(TM))$. A skew-symmetric endomorphism of $\mathbb{R}^{2m}$, say $B \in \mathfrak{so}(n)$, has pure imaginary eigenvalues $\pm i r_k$, where $r_k \in \mathbb{R}$ $(1 \le k \le m)$. Thus, iB has real eigenvalues $\pm r_k$. Now $\frac{z/2}{\sinh(z/2)}$ is a power series in z with radius of convergence 2π. Thus, $\frac{isB/2}{\sinh(isB/2)}$ is defined for s sufficiently small and has eigenvalues $\frac{r_k s/2}{\sinh(r_k s/2)}$, each repeated twice. Hence

$$\det\left(\frac{isB/2}{\sinh(isB/2)}\right) = \prod_{k=1}^{m}\left(\frac{r_k s/2}{\sinh(r_k s/2)}\right)^2, \text{ and so}$$

$$\det\left(\frac{isB/2}{\sinh(isB/2)}\right)^{\frac{1}{2}} = \prod_{k=1}^{m}\frac{r_k s/2}{\sinh(r_k s/2)}.$$

The last product is a power series in s of the form

$$\prod_{k=1}^{m}\frac{r_k s/2}{\sinh(r_k s/2)} = \sum_{k=0}^{\infty} a_k\left(r_1^2,\dots,r_m^2\right)s^{2k}, \tag{2.13}$$

where each coefficient $a_k(r_1^2,\dots,r_m^2)$ is a homogeneous, symmetric polynomial in $r_1^2,\dots,r_m^2$ of degree k. One can always express any such a symmetric polynomial as a polynomial in the elementary symmetric polynomials $\sigma_1,\dots,\sigma_m$ in $r_1^2,\dots,r_m^2$, where

$$\sigma_1 = \sum_{i=1}^{m} r_i^2, \quad \sigma_2 = \sum_{i<j}^{m} r_i^2 r_j^2, \quad \sigma_3 = \sum_{i<j<k}^{m} r_i^2 r_j^2 r_k^2, \quad \dots .$$

These in turn may be expressed in terms of $\mathrm{SO}(n)$-invariant polynomials in the entries of $B \in \mathfrak{so}(n)$ via

$$\begin{aligned}\det(\lambda I - B) &= \prod_{j=1}^{m}(\lambda + ir_j)(\lambda - ir_j) = \prod_{j=1}^{m}\left(\lambda^2 + r_j^2\right)\\ &= \sum_{k=1}^{m}\sigma_k\left(r_1^2,\dots,r_m^2\right)\lambda^{2(m-k)}.\end{aligned}$$

On the other hand,

$$\det(\lambda I - B) = \sum_{k=1}^{m}\left(\frac{1}{(2k)!}\sum_{(i),(j)}\delta_{i_1\cdots i_{2k}}^{j_1\cdots j_{2k}}\, B_{j_1}^{i_1}\cdots B_{j_{2k}}^{i_{2k}}\right)\lambda^{2(m-k)},$$

and so

$$\sigma_k(r_1^2,\dots,r_m^2) = \frac{1}{(2k)!}\sum_{(i),(j)}\delta_{i_1\cdots i_{2k}}^{j_1\cdots j_{2k}}\, B_{j_1}^{i_1}\cdots B_{j_{2k}}^{i_{2k}},$$

where $(i) = (i_1,\cdots,i_{2k})$ is an ordered $2k$-tuple of distinct elements of $\{1,\dots,2m\}$ and (j) is a permutation of (i) with sign $\delta_{i_1\cdots i_{2k}}^{j_1\cdots j_{2k}}$. If we replace B_j^i with the 2-form $\frac{1}{2\pi}\left(\Omega^\theta\right)_j^i$ relative to an orthonormal frame field, we obtain the Pontryagin forms

$$p_k(\Omega^\theta) := \frac{1}{(2\pi)^{2k}(2k)!}\sum_{(i),(j)}\delta_{i_1\cdots i_{2k}}^{j_1\cdots j_{2k}}\,\Omega_{i_1 j_1}^\theta\wedge\cdots\wedge\Omega_{i_{2k}j_{2k}}^\theta,$$

which represent the Pontryagin classes of the $\mathrm{SO}(n)$ bundle FM. Note that $p_k(\Omega^\theta)$ is independent of the choice of framing by the *ad*-invariance of the polynomials

σ_k. If we express the $a_k(r_1^2, \ldots, r_m^2)$ as polynomials, say $\mathcal{A}_k(\sigma_1, \ldots, \sigma_k)$, in the σ_j $(j \leq k)$, we can ultimately write

$$\det \left(\frac{isB/2}{\sinh(isB/2)} \right)^{\frac{1}{2}} = \sum_{k=0}^{\infty} \mathcal{A}_k(\sigma_1, \ldots, \sigma_k)\, s^{2k}.$$

Formally replacing B by $\frac{1}{2\pi}\Omega^\theta$, we finally have the reasonable definition

$$\det \left(\frac{i\Omega^\theta/4\pi}{\sinh(i\Omega^\theta/4\pi)} \right)^{\frac{1}{2}} := \sum_{k=0}^{\infty} \mathcal{A}_k\big(p_1(\Omega^\theta), \ldots, p_k(\Omega^\theta)\big),$$

where the $p_j(\Omega^\theta)$ are multiplied via wedge product when evaluating the terms in the sum; the order of multiplication does not matter since $p_j(\Omega^\theta)$ is of even degree $4j$. Also, since $\mathcal{A}_k\left(p_1(\Omega^\theta), \ldots, p_k(\Omega^\theta)\right)$ is a $4k$-form, there are only a finite number of nonzero terms in the infinite sum. Abbreviating $p_j(\Omega^\theta)$ simply by p_j, one finds

$$\begin{aligned} \det \left(\frac{i\Omega^\theta/4\pi}{\sinh(i\Omega^\theta/4\pi)} \right)^{\frac{1}{2}} &= 1 - \frac{1}{24}p_1 + \frac{1}{5760}(7p_1^2 - 4p_2) \\ &\quad - \frac{1}{967\,680}(31p_1^3 - 44p_1p_2 + 16p_3) + \cdots . \end{aligned} \tag{2.14}$$

This (by one definition) represents the total $\widehat{A}$-class of M, denoted by

$$\widehat{\mathbf{A}}(M) \in \bigoplus_{k=1}^{m} H^{2k}(M, \mathbb{Q}), \tag{2.15}$$

where actually $\widehat{\mathbf{A}}(M)$ has only nonzero components in $H^{2k}(M, \mathbb{Q})$ when k is even (or $2k \equiv 0 \mod 4$). In (2.10) the multi-degree forms (2.12) and (2.14) have been wedged, and the top ($2m$-degree) component (relative to the volume form) has been harvested.

We now turn to our outline of the proof of Theorem 2.13. The well-known heat kernel (or fundamental solution) for the ordinary heat equation $u_t = \Delta u$ in the Euclidean space $\mathbb{R}^n$, is given by

$$e\,(x, y, t) = (4\pi t)^{-n/2} \exp\left(- |x - y|^2 / 4t \right). \tag{2.16}$$

Since $H\,(x, y, t)$ only depends on $r = |x - y|$ and t, it is convenient to write

$$e\,(x, y, t) = \mathcal{E}\,(r, t) := (4\pi t)^{-n/2} \exp\big(- r^2/4t \big). \tag{2.17}$$

We do not expect such a simple expression for the heat kernel $k = (k^+, k^-)$ from Definition 2.11. However, it can be shown that for $x, y \in M$ (of even dimension $n = 2m$) with sufficiently small Riemannian distance $r = d(x, y)$ from x to y, we have an asymptotic expansion as $t \to 0^+$ for $k\,(x, y, t)$ of the form

$$k(x, y, t) \sim H_Q(x, y, t) := \mathcal{E}(d(x, y), t) \sum_{j=0}^{Q} h_j(x, y) t^j, \tag{2.18}$$

for any fixed integer $Q > m+4$, where

$$h_j(x,y) \in \operatorname{Hom}\Big((E\otimes\Sigma(M))_x, (E\otimes\Sigma(M))_y\Big) \text{ for } j\in\{0,1,\dots,Q\}.$$

The meaning of $k(x,y,t) \sim H_Q(x,y,t)$ is that for $d(x,y)$ and t sufficiently small,

$$|k(x,y,t) - H_Q(x,y,t)| \le C_Q\,\mathcal{E}(d(x,y),t)t^{Q+1} \le C_Q\,t^{Q-m+1},$$

where C_Q is a constant, independent of (x,y,t). We then have

$$k(x,x,t) \sim (4\pi t)^{-m}\sum_{j=0}^{Q} h_j(x,x)t^j = (4\pi)^{-m}\sum_{j=0}^{Q} h_j(x,x)t^{j-m}. \tag{2.19}$$

Using (2.9), i.e., $\int_M \operatorname{Str}(k(x,x,t))\,\nu_x = \operatorname{index}(\mathcal{D}^+)$, and (2.19), we deduce that

$$\int_M \operatorname{Str}(h_j(x,x))\,\nu_x = 0 \text{ for } j\in\{0,1,\dots,m-1\}, \text{ while}$$

$$(4\pi)^{-m}\int_M \operatorname{Str}(h_m(x,x))\,\nu_x = \int_M \operatorname{Str}(k(x,x,t))\,\nu_x = \operatorname{index}(\mathcal{D}^+).$$

Thus, to prove the Local Index Formula, it suffices to show that

$$(4\pi)^{-m}\operatorname{Str}(h_m(x,x)) = \left\langle \operatorname{Tr}\left(e^{i\Omega^\varepsilon/2\pi}\right)\wedge\det\left(\frac{i\Omega^\theta/4\pi}{\sinh(i\Omega^\theta/4\pi)}\right)^{\frac{1}{2}}, \nu_x\right\rangle.$$

While this may not be the intellectual equivalent of climbing Mount Everest, it is not for the faint of heart.

We choose a normal coordinate system $(y^1,\dots,y^n)$ in a coordinate ball $\mathcal{B}$ centered at the fixed point $x\in M$, so that $(y^1,\dots,y^n)=0$ at x. The coordinate fields $\partial_1 := \partial/\partial y^1,\dots,\partial_n := \partial/\partial y^n$ are orthonormal at x, and for any fixed $y_0\in\mathcal{B}$ with coordinates $(y_0^1,\dots,y_0^n)$, the curve $t\mapsto t(y_0^1,\dots,y_0^n)$ is a geodesic through x. By parallel translating the frame $(\partial_1,\dots,\partial_n)$ at x along these radial geodesics, we obtain an orthonormal frame field $(E_1,\dots,E_n)$ on $\mathcal{B}$ which generally does not coincide with $(\partial_1,\dots,\partial_n)$ at points $y\in\mathcal{B}$ other than at x. The framing $(E_1,\dots,E_n)$ defines a particularly nice section $\mathcal{B}\to FM|B$ and we may lift this to a section $\mathcal{B}\to P|B$ of the spin structure, which enables us to view the space $C^\infty(\Sigma(M)|\mathcal{B})$ of spinor fields on $\mathcal{B}$ as $C^\infty(\mathcal{B},\Sigma_n)$, i.e., functions on $\mathcal{B}$ with values in the fixed spinor representation vector space $\Sigma_n = \Sigma_n^+\oplus\Sigma_n^-$. By similar radial parallel translation (with respect to the connection ε) of an orthonormal basis of the twisting bundle fiber E_x, we can identify $C^\infty(E|\mathcal{B})$ with $C^\infty(\mathcal{B},\mathbb{C}^N)$, where $N=\dim_{\mathbb{C}} E$. The coordinate expressions for the curvatures Ω^θ, Ω^ε, and $\mathcal{D}^2$ are as simple as possible in this so-called radial gauge.

With the above identifications, we proceed as follows. For $0\le Q\in\mathbb{Z}$, let $\Psi_Q\in C^\infty(\mathcal{B}\times(0,\infty),\mathbb{C}^N\otimes\Sigma_{2m})$ be of the form

$$\Psi_Q(y,t) := \mathcal{E}(r,t)\sum_{k=0}^{Q} U_k(y)t^k,$$

where $U_k \in C^\infty(\mathcal{B}, \mathbb{C}^N \otimes \Sigma_{2m})$. If $U_0(0) \in \mathbb{C}^N \otimes \Sigma_{2m}$ is arbitrarily specified, we seek a formula for $U_k(y)$, $k = 0, \ldots, Q$, such that

$$(\mathcal{D}^2 + \partial_t)\Psi_Q(y,t) = \mathcal{E}(r,t)t^Q\mathcal{D}^2(U_Q)(y). \tag{2.20}$$

By virtue of the generalized Lichnerowicz formula (see [**LaMi**, p. 164]), the square $\mathcal{D}^2$ of the Dirac operator $\mathcal{D}$ can be written as

$$\mathcal{D}^2\psi = -\Delta\psi + \tfrac{1}{2}\sum\nolimits_{j,k} \Omega^{\varepsilon}_{jk} E_j \cdot E_k \cdot \psi + \tfrac{1}{4}S\psi,$$

where "$\cdot$" is Clifford multiplication. It is convenient to define the 0-th order operator $\mathcal{F}$ on $C^\infty(\mathcal{B}, \mathbb{C}^N \otimes \Sigma_{2m})$ via

$$\mathcal{F}[\psi] := \tfrac{1}{2}\sum\nolimits_{j,k} \Omega^{\varepsilon}_{jk} E_j \cdot E_k \cdot \psi, \text{ so that}$$
$$\mathcal{D}^2 = -\Delta\psi + (\mathcal{F} + \tfrac{1}{4}S)[\psi].$$

The desired formula for the $U_k(y)$ involves the operator A on $C^\infty(\mathcal{B}, \mathbb{C}^N \otimes \Sigma_{2m})$ given by

$$A[\psi] := -h^{1/4}\mathcal{D}^2[h^{-1/4}\psi] = h^{1/4}\Delta[h^{-1/4}\psi] - (\mathcal{F} + \tfrac{1}{4}S)[\psi],$$

where $h^{1/4} := (\sqrt{\det h})^{1/2}$. For $s \in [0,1]$, let

$$A_s[\psi](y) := A[\psi](sy).$$

As is proved in [**Ble92**] or in the forthcoming book [**BlBo04**], we have:

PROPOSITION 2.16. *Let $U_0(0) \in \mathbb{C}^N \otimes \Sigma_{2m}$, and let $V_0 \in C^\infty(\mathcal{B}, \mathbb{C}^N \otimes \Sigma_{2m})$ be the constant function $V_0(y) \equiv U_0(0)$. Then the $U_k(y)$ which satisfy* (2.20) *are given by*

$$U_k(y) = h(y)^{-1/4}V_k(y), \quad \textit{where}$$
$$V_k(y) = \int_{I^k} \prod_{i=0}^{k-1} (s_i)^i \big(A_{s_{k-1}} \circ \cdots \circ A_{s_0}[V_0]\big)(y)\, ds_0 \ldots ds_{k-1}, \tag{2.21}$$

and where $I^k = \big\{(s_0, \ldots s_k) \mid s_i \in [0,1] \text{ for } i \in \{0, \ldots, k-1\}\big\}$.

Note that $U_0(0) \in \mathbb{C}^N \otimes \Sigma_{2m}$ may be arbitrarily specified, and once $U_0(0)$ is chosen, the $U_m(y)$ are uniquely determined via (2.21). Let $h_k(y) \in \mathrm{End}(\mathbb{C}^N \otimes \Sigma_{2m})$ be given by

$$h_k(y)(U_0(0)) := U_k(y) \tag{2.22}$$

(in particular, $h_0(0) = I \in \mathrm{End}(\mathbb{C}^N \otimes \Sigma_{2m})$), and

$$H_Q(0,y,t) := \mathcal{E}(r,t)\sum_{k=0}^{Q} h_k(y)t^k \in C^\infty(\mathcal{B}, \mathrm{End}(\mathbb{C}^N \otimes \Sigma_{2m})).$$

We may regard $H_Q(0,y,t)$ as

$$H_Q(x,y,t) \in \mathrm{Hom}\big(E_x \otimes \Sigma(M)_x, E_y \otimes \Sigma(M)_y\big),$$

where we recall that $x \in M$ is the point about which we have chosen normal coordinates. For y sufficiently close to x, we set

$$H_Q(x,y,t) := \mathcal{E}(d(x,y),t)\sum_{k=0}^{Q} h_k(x,y)t^k. \tag{2.23}$$

Of course, one expects that $H_Q(x,y,t)$ provides the desired asymptotic expansion (2.18). Although this is very plausible, it is not at all easy to prove honestly. When proofs are attempted in the literature, often steps are skipped, hands are waved, and errors are made. An extremely careful (and hence nearly unbearable) proof will be provided in [**BlBo04**], but we must forgo this here. Thus, we will only state here without proof that

$$k(x,y,t) \sim H_Q(x,y,t) := \mathcal{E}(d(x,y),t)\sum_{j=0}^{Q} h_j(x,y)t^j$$

for $d(x,y)$ sufficiently small, where the h_j are given in (2.22).

Using normal coordinates $(y^1,\dots,y^{2m}) \in B(r_0,0)$ about $x \in M$ and the radial gauge, and selecting $V_0 \in \mathbb{C}^N \otimes \Sigma_{2m}$, by Proposition 2.16, we have

$$h_m(x,x)(V_0) = \int_{I^m} \prod_{i=0}^{m-1}(s_i)^i\Big(\big(A_{s_{m-1}}\circ\cdots\circ A_{s_0}\big)[\widetilde{V}_0]\Big)(0)\,ds_0\dots ds_{m-1}, \tag{2.24}$$

where $\widetilde{V}_0 \in C^\infty(B(r_0,0),\mathbb{C}^N\otimes\Sigma_{2m})$ is the constant extension of V_0. Recall that for $\psi \in C^\infty(B(r_0,0),\mathbb{C}^N\otimes\Sigma_{2m})$, we have

$$A_s[\psi](y) := A[\psi](sy), \text{ where } A[\psi] := h^{1/4}\Delta[h^{-1/4}\psi] - (\mathcal{F}+\tfrac14 S)[\psi].$$

While the right side of (2.24) may seem unwieldy, there is substantial simplification due to facts that $\left(A_{s_{m-1}}\circ\cdots\circ A_{s_0}\right)[\widetilde{V}_0](y)$ is evaluated at $y=0$ in (2.24). Also, if $\gamma^1,\dots,\gamma^n$ denote the so-called gamma matrices for Clifford multiplication by $\partial_1,\dots,\partial_n$, only those terms of $\left(A_{s_{m-1}}\circ\cdots\circ A_{s_0}\right)[\widetilde{V}_0](0)$ which involve the product $\gamma_{n+1} := \gamma^1\cdots\gamma^n$ will survive when the supertrace $\mathrm{Str}\,(h_m(x,x))$ is taken. As a consequence, we have the following simplification (essentially contained in [**Ble92**], or better yet, to appear in [**BlBo04**])

PROPOSITION 2.17. *Let $R_{klji}(0) = h(\Omega^\theta(\partial_i,\partial_j)\partial_l,\partial_k)$ denote the components of the Riemann curvature tensor of h at x, and let $\Omega^\varepsilon_{ij}(0) := \Omega^\varepsilon(\partial_i,\partial_j)$ at x. Set*

$$\widetilde{\theta}^1(\partial_j) := \tfrac18\sum_{k,l,i} R_{klji}(0)\gamma^k\gamma^l y^i,$$

$$\mathcal{F}^0 := \tfrac12\sum_{i,j} F_{ij}\otimes\gamma^i\gamma^j = \tfrac12\sum_{i,j}\Omega^\varepsilon_{ij}(0)\otimes\gamma^i\gamma^j, \text{ and}$$

$$A^0 := \sum_i\left(\partial_i^2 + \widetilde{\theta}^1(\partial_i)^2\right) - \mathcal{F}_0.$$

For $V_0 \in \mathbb{C}^N \otimes \Sigma_{2m}$, *define*

$$(2.25)\quad h^0_m(0,0)(V_0) := \int_{I^k} \prod_{i=0}^{m-1} (s_i)^i \Big((A^0_{s_{m-1}} \circ \cdots \circ A^0_{s_0})[\widetilde{V}_0] \Big)(0)\, ds_0 \ldots ds_{m-1},$$

where $\widetilde{V}_0 \in C^\infty(B(r_0,0), \mathbb{C}^N \otimes \Sigma_{2m})$ *is the constant extension of* $V_0 \in \mathbb{C}^N \otimes \Sigma_{2m}$. *Then*

$$(2.26)\qquad \mathrm{Str}(h_m(0,0)) = \mathrm{Str}(h^0_m(0,0)).$$

In other words, in the computation of $\mathrm{Str}(h_m(0,0))$ *given by* (2.24), *we may replace* A *by* A^0.

This is a substantial simplification, not only in that A^0 is a second-order differential operator with coefficients which are at most quadratic in y, but it also shows that $\mathrm{Str}(h_m(x,x))$ only depends on the curvatures Ω^θ and Ω^ε at the point x. One might regard the gist of the Index Formula for twisted Dirac operator as exhibiting the global quantity index$(\mathcal{D}^+)$ as the integral of a form which may be locally computed. From this perspective, Proposition 2.17 does the job. Also, knowing in advance that index$(\mathcal{D}^+)$ is insensitive to perturbations in h and ε, one suspects that $\mathrm{Str}(h_m(x,x))\,\nu_x$ can be expressed in terms of the standard forms which represent characteristic classes for TM and E. The Local Index Formula confirms this. Moreover, for low values of m, say $m = 1$ or 2 (i.e., for 2 and 4-manifolds), one can directly compute $\mathrm{Str}(h^0_m(0,0))$ using (2.25), and thereby verify Theorem 2.13 and hence obtain Corollary 2.15 rather easily. For readers who have no use for the Local Index Theorem beyond dimension 4, this is sufficient. It requires more effort to prove Theorem 2.13 for general m. For lack of space, we cannot go into the details of this here, but they can be found in [**Ble92**] or [**BlBo04**]. It is well worth mentioning that the appearance of the sinh function in the Local Index Formula has its roots in Mehler's formula for the heat kernel

$$(2.27)\quad e_a(x,y,t) = \frac{1}{\sqrt{4\pi \frac{\sinh(2at)}{2a}}} \exp\left(- \frac{1}{4\frac{\sinh(2at)}{2a}} (\cosh(2at)(x^2+y^2) - 2xy) \right).$$

of the generalized 1-dimensional heat problem

$$u_t = u_{xx} - a^2x^2u, \quad u(x,t) \in \mathbb{R},\ (y,t) \in \mathbb{R}\times(0,\infty),$$
$$u(x,0) = f(x),$$

where $0 \neq a \in \mathbb{R}$ is a given constant. A solution of this problem is given by

$$u(x,t) = \int_{-\infty}^{\infty} e_a(x,y,t) f(y)\, dy,$$

and this reduces to the usual formula as $a \to 0$. The nice idea of using Mehler's formula in a rigorous derivation of the Local Index Theorem appears to be due to Getzler in [**Get83**] and [**Get86**], although it was at least implicitly involved in earlier heuristic supersymmetric path integral arguments for the Index Theorem. In the same vein, further simplifications and details can be found in [**BeGeVe92**],

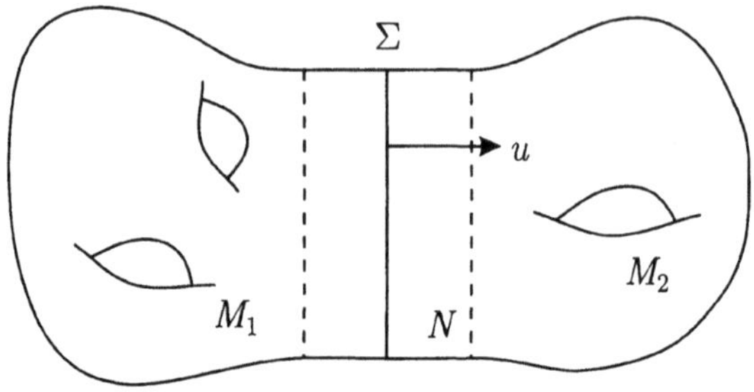

FIGURE 7. The partition of $M = M_1 \cup_\Sigma M_2$

[**Yu01**], and [**Ble92**], but the treatment to be found in [**BlBo04**] will be more self-contained and less demanding.

While the Local Index Theorem (Theorem 2.13) is stated for twisted Dirac operators, the same proof may be applied to obtain the index formulas for an elliptic operator, possibly on a *nonspin* manifold, which is only locally of the form of twisted Dirac operator $\mathcal{D}^+$. Indeed, if $\mathcal{A}$ is such an operator and k is the heat kernel for $\mathcal{A}^*\mathcal{A} \oplus \mathcal{A}\mathcal{A}^*$, then from the spectral resolution of $\mathcal{A}$, we can still deduce from the asymptotic expansion of k that

$$\operatorname{index}(\mathcal{A}) = (4\pi)^{-m} \int_M \operatorname{Str}(h_m(x,x))\, \nu_x,$$

where the supertrace Str is defined in the natural way. The crucial observation is that since $\mathcal{A}$ is locally in the form of a twisted Dirac operator, we can compute $\operatorname{Str}(h_m(x,x))$ in exactly the same way (i.e., locally) as we have done. Since it is not easy to find first-order elliptic operators of geometrical significance which are not expressible in terms of locally twisted Dirac operators (or 0-th order perturbations thereof), the Local Index Formula for twisted Dirac operators is much more comprehensive than it would appear at first glance.

2.3. Dirac type operators on manifolds with boundary. In this section we fix the notation and recall some basic properties of Dirac type operators on manifolds with boundary.

2.3.1. *The general setting for partitioned manifolds.* Let M be a compact smooth Riemannian *partitioned* manifold

$$M = M_1 \cup_\Sigma M_2 := M_1 \cup M_2,$$

where $M_1 \cap M_2 = \partial M_1 = \partial M_2 = \Sigma$, and Σ is a hypersurface (see Figure 7). We assume that $M \setminus \Sigma$ does not have a closed connected component (i.e. Σ intersects any connected component of M_1 and M_2). Let

$$\mathcal{D} : C^\infty(M;S) \to C^\infty(M;S)$$

be an *operator of Dirac type* acting on sections of a Hermitian bundle S of Clifford modules over M, i.e., $\mathcal{D} = \mathbf{c} \circ \nabla$ where $\mathbf{c}$ denotes the Clifford multiplication and ∇ is a connection for S. Unlike the more general case introduced in Section 1.3, we assume here that ∇ is compatible with $\mathbf{c}$ in the sense that $\nabla \mathbf{c} = 0$. From this compatibility assumption it follows that $\mathcal{D}$ is symmetric and essentially self-adjoint over M.

For even $n = \dim M$, the splitting $\mathfrak{Cl}(M) = \mathfrak{Cl}^+(M) \oplus \mathfrak{Cl}^-(M)$ of the Clifford bundles induces a corresponding splitting of $S = S^+ \oplus S^-$, and a *chiral decomposition*

$$\mathcal{D} = \begin{pmatrix} 0 & \mathcal{D}^- \\ \mathcal{D}^+ & 0 \end{pmatrix}, \quad \mathcal{D}^- = (\mathcal{D}^+)^*,$$

of the *total* Dirac operator. The *chiral Dirac operators* $\mathcal{D}^\pm$ are elliptic but not symmetric, and for that reason they may have nontrivial indices which provide us with important topological and geometric invariants, as it was found in the special case of twisted Dirac operators in Section 2.2.

Here we assume that all metric structures of M and S are product in a collar neighborhood $N = [-1,1] \times \Sigma$ of Σ. If u denotes the normal coordinate (running from M_1 to M_2), then

$$\mathcal{D}|_N = \sigma(\partial_u + \mathcal{B}), \text{ where } \sigma := \mathbf{c}\,(du)\,, \; \partial_u := \frac{\partial}{\partial u},$$

and $\mathcal{B}$ denotes the canonically associated Dirac operator over Σ, called the *tangential operator*. We have a similar product formula for the chiral Dirac operator. Here the point of the product structure is that then σ and $\mathcal{B}$ do not depend on the normal variable. Note that σ (Clifford multiplication by du) is a unitary mapping $L^2(\Sigma; S|_\Sigma) \to L^2(\Sigma; S|_\Sigma)$ with $\sigma^2 = -I$ and $\sigma\mathcal{B} = -\mathcal{B}\sigma$. In the non-product case, there are certain ambiguities in defining a 'tangential operator' which we shall not discuss here (but see also Equation (1.9)).

2.3.2. *Analysis tools: Green's formula.* In order to simplify the notation we set $X := M_2$. For greater generality, we consider the chiral Dirac operator

$$\mathcal{D}^+ : C^\infty(X; S^+) \to C^\infty(X; S^-)$$

and write $\mathcal{D}^-$ for its formally adjoint operator. The corresponding results follow at once for the total Dirac operator.

LEMMA 2.18. *Let $\langle \cdot, \cdot \rangle_\pm$ denote the scalar product in $L^2(X; S^\pm)$. Then we have*

$$\langle \mathcal{D}^+ f_+, f_- \rangle_- - \langle f_+, \mathcal{D}^- f_- \rangle_+ = -\int_\Sigma (\sigma \gamma_\infty f_+, \gamma_\infty f_-)\, \mathrm{dvol}_\Sigma$$

for any $f_\pm \in C^\infty(X; S^\pm)$.

Here,

$$\gamma_\infty : C^\infty(X; S^\pm) \to C^\infty(\Sigma; S^\pm|_\Sigma) \tag{2.28}$$

denotes the restriction of a section to the boundary Σ.

2.3.3. *Cauchy data spaces and the Calderón projection.* To explain the L^2 Cauchy data spaces we recall three additional, somewhat delicate and not widely known properties of operators of Dirac type on compact manifolds with boundary, cf. [**BoWo93**]:

1. The invertible extension to the double.
2. The Poisson type operator and the Calderón projection.
3. The twisted orthogonality of the Cauchy data spaces for chiral and total Dirac operators which gives the Lagrangian property in the symmetric case (i.e., for the total Dirac operator).

The idea and the properties of the Calderón projection were announced in Calderón [**Ca63**] and proved in Seeley [**Se66**] in great generality. In the following, we restrict ourselves to constructing the Calderón projection for operators of Dirac type (or, more generally, elliptic differential operators of first order) which simplifies the presentation substantially.

2.3.4. *Invertible extension.* First we construct the *invertible double.* Clifford multiplication by the inward normal vector gives a natural clutching of S^+ over one copy of X with S^- over a second copy of X to a smooth bundle $\widetilde{S^+}$ over the closed double $\widetilde{X}$. The product forms of $\mathcal{D}^+$ and $\mathcal{D}^- = (\mathcal{D}^+)^*$ fit together over the boundary and provide a new operator of Dirac type, namely

$$\widetilde{\mathcal{D}^+} := \mathcal{D}^+ \cup \mathcal{D}^- : C^\infty(\widetilde{X}, \widetilde{S^+}) \to C^\infty(\widetilde{X}, \widetilde{S^-}). \tag{2.29}$$

Clearly $(\mathcal{D}^+ \cup \mathcal{D}^-)^* = \mathcal{D}^- \cup \mathcal{D}^+$, and so $\operatorname{index} \widetilde{\mathcal{D}^+} = 0$. It turns out that $\widetilde{\mathcal{D}^+}$ is invertible with a pseudo-differential elliptic inverse $(\widetilde{\mathcal{D}^+})^{-1}$. Clearly, *local* solutions of the homogenous equation (here, solutions on *one* copy of X) do not extend to *global* solutions on $\widetilde{X}$ in general. As a matter of fact, the operator $\mathcal{D}^+$, even being the restriction of an invertible, locally defined differential operator, is not invertible in general, and we have $r^+(\widetilde{\mathcal{D}^+})^{-1}e^+\mathcal{D}^+ \neq I$, where $e^+ : L^2(X; S^+) \to L^2(\widetilde{X}; \widetilde{S^+})$ denotes the extension-by-zero operator and $r^+ : H^s(\widetilde{X}; \widetilde{S^+}) \to H^s(X; S^+)$ the natural restriction operator for Sobolev spaces for s real. The precise decomposition $L^2(\widetilde{X}; \widetilde{S^+}) = L^2(X; S^+) \times L^2(X; S^-)$ gives a different picture in case that a given operator T on one component can be extended to an invertible operator $\widetilde{T}$ on the whole space. This, indeed, would imply T invertible with $T^{-1} = r^+(\widetilde{T})^{-1}e^+$. The L^2 extension of $\widetilde{\mathcal{D}^+}$, however, is not a precise extension of the L^2 extension of $\mathcal{D}^+$. Therefore, in our case, the L^2-argument breaks down.

EXAMPLE 2.19. In the simplest possible two-dimensional case we consider the Cauchy-Riemann operator $\overline{\partial} : C^\infty(D^2) \to C^\infty(D^2)$ over the disc D^2, where $\overline{\partial} = \frac{1}{2}(\partial_x + i\partial_y)$. In polar coordinates, this operator has the form $\frac{1}{2}e^{i\varphi}(\partial_r + \frac{i}{r}\partial_\varphi)$. Therefore, after some small smooth perturbations (and modulo the factor $\frac{1}{2}$), we assume that $\overline{\partial}$ has the following form in a certain collar neighborhood of the boundary:

$$\overline{\partial} = e^{i\varphi}(\partial_r + i\partial_\varphi)$$

Now we construct the invertible double of the operator $\overline{\partial}$. By E^k, $k \in \mathbb{Z}$, we denote the bundle, which is obtained from two copies of $D^2 \times \mathbb{C}$ by the identification $(z, w) = (z, z^k w)$ near the equator. We obtain the bundle E^1 by gluing two halves of $D^2 \times \mathbb{C}$ by $\sigma(\varphi) = e^{i\varphi}$ and E^{-1} by gluing with the adjoint symbol. In such a way we obtain the operator

$$\widetilde{\overline{\partial}} := \overline{\partial} \cup (\overline{\partial})^* : C^\infty(S^2; E^1) \to C^\infty(S^2; E^{-1})$$

over the whole 2-sphere.

Let us analyze the situation more carefully. We fix $N := (-\varepsilon, +\varepsilon) \times S^1$, a bicollar neighborhood of the equator. The formal adjoint of $\overline{\partial}$ has the form

$$(\overline{\partial})^* = e^{-i\varphi}(-\partial_u + i\partial_\varphi + 1)$$

$(u = r - 1)$ in this cylinder. A section of E^1 is a couple (s_1, s_2) such that in N,

$$s_2(u, \varphi) = e^{i\varphi} s_1(u, \varphi)\,.$$

The couple $(\overline{\partial} s_1, (\overline{\partial})^* s_2)$ is a smooth section of E^{-1}. To show this, we check that $(\overline{\partial})^* s_2 = e^{-i\varphi}\overline{\partial} s_1$. In the neighborhood N, we have

$$\begin{aligned}(\overline{\partial})^* s_2 &= (\overline{\partial})^* \left(e^{i\varphi} s_1\right) = e^{-i\varphi}(-\partial_u + i\partial_\varphi + 1)\left(e^{i\varphi} s_1\right) \\ &= \partial_u s_1 + i e^{-i\varphi}\partial_\varphi\left(e^{i\varphi} s_1\right) + s_1 = (\partial_u + i\partial_\varphi) s_1 \\ &= e^{-i\varphi} e^{i\varphi}(\partial_u + i\partial_\varphi) s_1 = e^{-i\varphi}\left(\overline{\partial} s_1\right).\end{aligned}$$

Then the operator $\overline{\partial} \cup \overline{\partial}^*$ becomes injective and $\operatorname{index} \overline{\partial} \cup \overline{\partial}^* = 0$.

2.3.5. *The Poisson operator and the Calderón projection.* Next we investigate the solution spaces and their traces at the boundary. For a total or chiral operator of Dirac type over a smooth compact manifold with boundary Σ and for any real s we define the *null space*

$$\operatorname{Ker}(\mathcal{D}^+, s) := \{f \in H^s(X; S^+) \mid \mathcal{D}^+ f = 0 \text{ in } X \setminus \Sigma\}.$$

The null spaces consist of sections which are distributional for negative s; by elliptic regularity they are smooth in the interior; in particular they possess a smooth restriction on the hypersurface $\Sigma_\varepsilon = \{\varepsilon\} \times \Sigma$ parallel to the boundary Σ of X at a distance $\varepsilon > 0$. By a Riesz operator argument they can be shown to also possess a trace over the boundary. Of course, that trace is no longer smooth but belongs to $H^{s-\frac{1}{2}}(\Sigma; S^+|_\Sigma)$. More precisely, we have the following well-known *General Restriction Theorem* (for a proof see e.g. [**BoWo93**], Chap. 11 and 13):

THEOREM 2.20.

(a) *Let $s > \frac{1}{2}$. Then the restriction map γ_∞ of (2.28) extends to a bounded map*

$$\gamma_s : H^s(X; S^+) \to H^{s-\frac{1}{2}}(\Sigma; S^+|_\Sigma).$$

(b) *For $s \leq \frac{1}{2}$, the preceding reduction is no longer defined for arbitrary sections but only for solutions of the operator $\mathcal{D}^+$: let $f \in \operatorname{Ker}(\mathcal{D}^+, s)$ and let $\gamma_{(\varepsilon)} f$ denote the well-defined trace of f in $C^\infty(\Sigma_\varepsilon; S^+|_\Sigma)$. Then, as $\varepsilon \to 0_+$, the sections $\gamma_{(\varepsilon)} f$ converge to an element $\gamma_s f \in H^{s-\frac{1}{2}}(\Sigma; S^+|_\Sigma)$.*

(c) *Let $\widetilde{\mathcal{D}^+}$ denote the invertible double of $\mathcal{D}^+$, and let r_+ denote the restriction operator $r_+ : H^s(\widetilde{X};\widetilde{S}^+) \to H^s(X;S^+)$ and let γ_∞^* be the dual of γ_∞ in the distributional sense. For any $s \in \mathbb{R}$ the mapping* (Poisson type operator)

$$\mathcal{K} := r_+\left(\widetilde{\mathcal{D}^+}\right)^{-1}\gamma_\infty^*\sigma : C^\infty(\Sigma;S^+|_\Sigma) \to C^\infty(X;S^+)$$

extends to a continuous map $\mathcal{K}^{(s)} : H^{s-1/2}(\Sigma;S^+|_\Sigma) \to H^s(X;S^+)$ with

$$\operatorname{range}\mathcal{K}^{(s)} = \operatorname{Ker}(\mathcal{D}^+,s).$$

For $s = 0$, Theorem 2.20 can be reformulated in the following way:

COROLLARY 2.21. *For a constant C independent of f, we have*

$$\|\gamma(f)\|_{-\frac{1}{2}} \le C\left(\left\|\mathcal{D}^+ f\right\|_0 + \|f\|_0\right) \text{ for all } f \in D_{\max}\left(\mathcal{D}^+\right),$$

where $D_{\max}\left(\mathcal{D}^+\right) := \left\{f \in L^2(X;S) \mid \mathcal{D}^+ f \in L^2(X;S)\right\}$. So, the restriction

$$\gamma : D_{\max}\left(\mathcal{D}^+\right) \longrightarrow H^{-1/2}(\Sigma;S^+|_\Sigma)$$

is well defined and bounded.

Proofs of the previous theorem and corollary can be found, e.g., in [**BoWo93**], Theorems 13.1 and 13.8 for our situation ($\mathcal{D}^+$ is of order 1); and in Hörmander [**Ho66**] in greater generality (Theorem 2.2.1 and the Estimate (2.2.8), p. 194).

The composition

$$\mathcal{P}(\mathcal{D}^+) := \gamma_\infty \circ \mathcal{K} : C^\infty(\Sigma;S^+|_\Sigma) \to C^\infty(\Sigma;S^+|_\Sigma) \tag{2.30}$$

is called the *(Szegö-)Calderón projection.* It is a pseudo-differential projection (idempotent, but in general not orthogonal). We denote by $\mathcal{P}(\mathcal{D}^+)^{(s)}$ its extension to the s-th Sobolev space over Σ. It has the following geometric meaning.

We now have three options for defining the corresponding *Cauchy data* (or *Hardy*) *spaces*:

DEFINITION 2.22. For all real s we define

$$\Lambda(\mathcal{D}^+,s) := \gamma_s(\operatorname{Ker}(\mathcal{D}^+,s)),$$

$$\Lambda^{\text{clos}}(\mathcal{D}^+,s) := \overline{\gamma_\infty\left\{f \in C^\infty(X;S^+) \mid \mathcal{D}^+ f = 0 \text{ in } X\setminus\Sigma\right\}}^{H^{s-\frac{1}{2}}(\Sigma;S^+|_\Sigma)},$$

$$\Lambda^{\text{Cald}}(\mathcal{D}^+,s) := \operatorname{range}\mathcal{P}(\mathcal{D}^+)^{(s-\frac{1}{2})}.$$

The range of a projection is closed; the inclusions of the Sobolev spaces are dense; and $\operatorname{range}\mathcal{P}(\mathcal{D}^+) = \gamma_\infty\{f \in C^\infty(X;S^+) \mid \mathcal{D}^+ f = 0 \text{ in } X\setminus\Sigma\}$, as shown in [**BoWo93**]. So, the second and the third definition of the Cauchy data space coincide. Moreover, for $s > \frac{1}{2}$ one has $\Lambda(\mathcal{D}^+,s) = \Lambda^{\text{Cald}}(\mathcal{D}^+,s)$. This equality can be extended to the L^2 case ($s = \frac{1}{2}$, see also Theorem 2.31 below), and remains valid for any real s, as proved in Seeley, [**Se66**], Theorem 6. For $s \le \frac{1}{2}$, the result is somewhat counter-intuitive (see also Example 2.24 b in the following Section).

In conclusion, we have:

PROPOSITION 2.23. *For all $s \in \mathbb{R}$*

$$\Lambda(\mathcal{D}^+, s) = \Lambda^{\mathrm{clos}}(\mathcal{D}^+, s) = \Lambda^{\mathrm{Cald}}(\mathcal{D}^+, s).$$

2.3.6. *Calderón and Atiyah-Patodi-Singer projection.* The Calderón projection is closely related to another projection determined by the 'tangential' part of $\mathcal{D}^+$, described as follows. Let $\mathcal{B}$ denote the *tangential* symmetric elliptic differential operator over Σ in the product form

$$\mathcal{D}^+ = \sigma(\partial_u + \mathcal{B}) : C^\infty\left(N, S^+|N\right) \to C^\infty\left(N, S^-|N\right)$$

in a collar neighborhood N of Σ in X. It has discrete real eigenvalues and a complete system of L^2 orthonormal eigensections. Let $P_{\geq}(\mathcal{B})$ denote the spectral (Atiyah-Patodi-Singer) projection onto the subspace $L_+(\mathcal{B})$ of $L^2(\Sigma; S^+|_\Sigma)$ spanned by the eigensections corresponding to the nonnegative eigenvalues of $\mathcal{B}$. It is a pseudo-differential operator and its principal symbol p_+ is the projection onto the eigenspaces of the principal symbol $b(y,\zeta)$ of $\mathcal{B}$ corresponding to nonnegative eigenvalues. It turns out that p_+ coincides with the principal symbol of the Calderón projection.

We call the space of pseudo-differential projections with the same principal symbol p_+ the *Grassmannian* $\mathcal{G}r_{p_+}$ and equip it with the operator norm corresponding to $L^2(\Sigma; S^+|_\Sigma)$. It has countably many connected components; two projections P_1, P_2 belong to the same component, if and only if the *virtual codimension*

$$\mathbf{i}(P_2, P_1) := \operatorname{index}\{P_2 P_1 : \operatorname{range} P_1 \to \operatorname{range} P_2\} \tag{2.31}$$

of P_2 in P_1 vanishes; the higher homotopy groups of each connected component are given by Bott periodicity.

EXAMPLE 2.24 a. For the Cauchy-Riemann operator $\bar\partial$ on the unit disc $D^2 = \{|z| \leq 1\}$, the Cauchy data space is spanned by the eigenfunctions $e^{ik\theta}$ of the tangential operator ∂_θ over $S^1 = [0, 2\pi]/\{0, 2\pi\}$ for nonnegative k. So, the Calderón projection and the Atiyah-Patodi-Singer projection coincide in this case.

EXAMPLE 2.24 b. We now consider the cylinder $X^R = [0, R] \times \Sigma_0$ with $\mathcal{D}_R = \sigma(\partial_u + B)$. Here B denotes a symmetric elliptic differential operator of first order

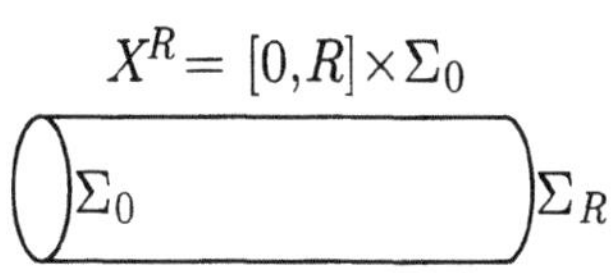

FIGURE 8. Cylinder of length R

acting on sections of a bundle E over Σ_0, and σ a unitary bundle endomorphism with $\sigma^2 = -I$ and $\sigma B = -B\sigma$. Let B be invertible (for the ease of presentation). Let $\{\varphi_k, \lambda_k\}$ denote the B's spectral resolution of $L^2(\Sigma_0; E)$ with

$$\cdots \leq \lambda_{-k} \leq \cdots \leq \lambda_{-1} < 0 < \lambda_1 \leq \cdots \leq \lambda_k \leq \cdots$$

Then

$$\begin{aligned} &B\varphi_k = \lambda_k \varphi_k \ \text{ for all } k \in \mathbb{Z} \setminus \{0\}, \\ &\lambda_{-k} = -\lambda_k\,,\ \sigma(\varphi_k) = \varphi_{-k}, \text{ and } \sigma(\varphi_{-k}) = -\varphi_k \ \text{ for } k > 0. \end{aligned} \tag{2.32}$$

We consider

$$\begin{aligned} f \in \operatorname{Ker}(\mathcal{D}_R, 0) &= \operatorname{span}\{e^{-\lambda_k u}\varphi_k\}_{k \in \mathbb{Z}\setminus\{0\}} \text{ in } L^2(X^R) \\ &= \operatorname{Ker} \mathcal{D}_{R\max} \quad \text{(kernel of maximal extension)} \end{aligned}$$

It can be written in the form

$$f(u,y) = f_>(u,y) + f_<(u,y),\ u \in [0,R]\,, \tag{2.33}$$

where

$$f_<(u,y) = \sum_{k<0} a_k e^{-\lambda_k u}\varphi_k(y) \quad \text{and} \quad f_>(u,y) = \sum_{k>0} a_k e^{-\lambda_k u}\varphi_k(y).$$

Because of

$$\langle f, f\rangle_{L^2(X^R)} < +\infty \iff \langle f_<, f_<\rangle < +\infty \ \text{ and } \langle f_>, f_>\rangle < +\infty\,,$$

the coefficients a_k satisfy the conditions

$$\sum_{k<0} |a_k|^2 \frac{e^{-2\lambda_k R} - 1}{2|\lambda_k|} < \infty \quad \text{or, equivalently,} \quad \sum_{k<0} |a_k|^2 \frac{e^{2|\lambda_k|R}}{|\lambda_k|} < \infty, \tag{2.34}$$

and

$$\sum_{k>0} |a_k|^2 \frac{1 - e^{-2\lambda_k R}}{2\lambda_k} < \infty \quad \text{or, equivalently,} \quad \sum_{k>0} \frac{|a_k|^2}{\lambda_k} < \infty. \tag{2.35}$$

We consider the space $\Lambda(\mathcal{D}_R, 0)$ consisting of all $\gamma(f)$ with $f \in \operatorname{Ker}(\mathcal{D}_R, 0)$. Here $\gamma(f)$ denotes the trace of f at the boundary

$$\Sigma = \partial X^R = -\Sigma_0 \sqcup \Sigma_R,$$

where Σ_R denotes a second copy of Σ_0. According to the spectral splitting (2.33), we have

$$\gamma(f) = (s^0_<, s^R_<) + (s^0_>, s^R_>),$$

where

$$s^0_> = f_>(0), \quad s^0_< = f_<(0), \quad s^R_> = f_>(R), \quad s^R_< = f_<(R).$$

Because of (2.34) and (2.35), we have

$$(s^0_<, s^R_>) \in C^\infty(\Sigma_0 \sqcup \Sigma_R) \quad \text{and} \quad (s^0_>, s^R_<) \in H^{-1/2}(\Sigma_0 \sqcup \Sigma_R).$$

Recall that

$$\sum a_k \varphi_k \in H^s(\Sigma_0) \iff \sum |a_k|^2 |k|^{2s/(m-1)} < \infty$$

and $|\lambda_k| \sim |k|^{\frac{1}{m-1}}$ for $k \to \pm\infty$, where $m-1$ denotes the dimension of Σ_0.

One notices that the estimate (2.34) for the coefficients of $s^0_<$ is stronger than the assertion that $\sum_{k<0} |a_k|^2 |\lambda_k|^N < \infty$ for all natural N. Thus our estimates confirm that not every smooth section can appear as initial value over Σ_0 of a solution of $\mathcal{D}_R f = 0$ over the cylinder.

To sum up the example, the Cauchy data space $\Lambda(\mathcal{D}_R, 0)$ can be written as the graph of an unbounded, densely defined, closed operator $T : \operatorname{Dom} T \to H^{-\frac{1}{2}}(\Sigma_R)$, mapping $s^0_< + s^0_> =: s^0 \mapsto s^R := s^R_< + s^R_>$ with $\operatorname{Dom} T \subset H^{-\frac{1}{2}}(\Sigma_0)$. To obtain a closed subspace of $L^2(\Sigma)$, one takes the range $\Lambda(\mathcal{D}_R, \frac{1}{2})$ of the L^2 extension $\mathcal{P}(\mathcal{D}_R)^{(0)}$ of the Calderón projection. It coincides with $\Lambda(\mathcal{D}_R, 0) \cap L^2(\Sigma)$ by Proposition 2.23. In Theorem 2.31 we show without use of the pseudo-differential calculus why the intersection $\Lambda(\mathcal{D}_R, 0) \cap L^2(\Sigma)$ must be closed in $L^2(\Sigma)$. See [**ScWo00**, p. 1214] for another description of the Cauchy data space $\Lambda(\mathcal{D}_R, \frac{1}{2})$, namely as the graph of a unitary elliptic pseudo-differential operator of order 0.

Since $\Sigma = -\Sigma_0 \sqcup \Sigma_R$, the tangential operator takes the form $\mathcal{B} = B \oplus (-B)$, and we obtain from (2.32)

$$\operatorname{range} P_>(\mathcal{B})^{(0)} = L_+(\mathcal{B}) = \operatorname{span}_{L^2(\Sigma)} \{(\varphi_k, \sigma(\varphi_k))\}_{k>0}\,.$$

For comparison, we have in this example

$$\operatorname{range} \mathcal{P}(\mathcal{D}_R)^{(0)} = \Lambda(\mathcal{D}_R, \tfrac{1}{2}) = \operatorname{span}_{L^2(\Sigma)} \{(\varphi_k, e^{-\lambda_k R} \varphi_k)\}_{k>0}\,,$$

hence $L_+(\mathcal{B})$ and $\Lambda(\mathcal{D}_R, \frac{1}{2})$ are transversal subspaces of $L^2(\Sigma)$. On the semi-infinite cylinder $[0,\infty) \times \Sigma$, however, we have only one boundary component Σ_0. Hence

$$\operatorname{range} P_>(B)^{(0)} = \operatorname{span}_{L^2(\Sigma_0)} \{\varphi_k\}_{k>0} = \lim_{R\to\infty} \operatorname{range} \mathcal{P}(\mathcal{D}_R)^{(0)}.$$

One can generalize the preceding example: For any smooth compact manifold X with boundary Σ and any real $R \geq 0$, let X^R denote the stretched manifold

$$X^R := ([-R, 0] \times \Sigma) \cup_\Sigma X.$$

Assuming product structures with $\mathcal{D} = \sigma(\partial_u + \mathcal{B})$ near Σ gives a well-defined extension $\mathcal{D}_R$ of $\mathcal{D}$. Nicolaescu proved in [**Ni95**] that the Calderón projection and the Atiyah-Patodi-Singer projection coincide up to a finite-dimensional component in the *adiabatic limit* ($R \to +\infty$ in a suitable setting). Even for finite R and, in particular for $R = 0$, one has the following interesting result. It was first proved by Scott [**Sco95**] (see also Grubb [**Gr99**] and the Appendix by Wojciechowski in [**DaKi99**], who both offered different proofs).

LEMMA 2.25. *For all $R \geq 0$, the difference $\mathcal{P}(\mathcal{D}_R) - P_\geq(\mathcal{B})$ is an operator with a smooth kernel.*

2.3.7. *Twisted orthogonality of Cauchy data spaces.* Green's formula (in particular the Clifford multiplication σ in the case of Dirac type operators) provides a symplectic structure for $L^2(\Sigma; S|_\Sigma)$ for linear symmetric elliptic differential operators of first order on a compact smooth manifold X with boundary Σ. For elliptic systems of second-order differential equations, various interesting results have been obtained in the 1970s by exploiting the symplectic structure of corresponding spaces (see e.g. [**LaSnTu75**]). Restricting oneself to first-order systems, the geometry becomes very clear and it turns out that the Cauchy data space $\Lambda(\mathcal{D}, \frac{1}{2})$ is a Lagrangian subspace of $L^2(\Sigma; S|_\Sigma)$.

More generally, in [**BoWo93**] we described the orthogonal complement of the Cauchy data space of the chiral Dirac operator $\mathcal{D}^+$ by

$$\sigma^{-1}(\Lambda(\mathcal{D}^-, \tfrac{1}{2})) = (\Lambda(\mathcal{D}^+, \tfrac{1}{2}))^\perp . \tag{2.36}$$

We obtained a short exact sequence

$$0 \to \sigma^{-1}(\Lambda(\mathcal{D}^-, s)) \hookrightarrow H^{s-\frac{1}{2}}(\Sigma; S^+|_\Sigma) \overset{\mathcal{K}(s)}{\to} \operatorname{Ker}(\mathcal{D}^+, s) \to 0.$$

For the total (symmetric) Dirac operator this means:

PROPOSITION 2.26. *The Cauchy data space $\Lambda(\mathcal{D}, \frac{1}{2})$ of the* total *Dirac operator is a Lagrangian subspace of the Hilbert space $L^2(\Sigma; S|_\Sigma)$ equipped with the symplectic form $\omega(\varphi, \psi) := (\sigma\varphi, \psi)$.*

2.3.8. *"Admissible" boundary value problems.* We refer to [**BoWo93**, Chap. 18], [**BrLe01**], and [**Sc01**], for a rigorous definition and treatment of large classes of admissible boundary value problems defined by pseudo-differential projections, Prominent examples belong to the Grassmannian $\mathcal{G}r_{p_+}$ (introduced above in Section 2.3.6). On even-dimensional manifolds, other prominent examples are local chiral projections and unitary modifications as explained in [**BoWo93**, p. 273].

For all admissible boundary conditions defined by a pseudo-differential projection R over Σ, the following features are common: (For simplicity, we suppress the distinction between total and chiral spinor bundle in this paragraph, and denote the bundle by E)

(i) For $f \in H^1(X; E)$ we have an estimate

$$\|f\|_1 \leq C\left(\left\|\mathcal{D}^+ f\right\|_0 + \|f\|_0 + \|R \circ i \circ \gamma(f)\|_{\frac{1}{2}} \right).$$

(ii) Defining a domain by

$$D = \operatorname{Dom}(\mathcal{D}^+_R) := \left\{ f \in H^1(X; E) \mid R(f|_\Sigma) = 0 \right\},$$

we obtain a closed Fredholm extension.

(iii) The restriction of $\operatorname{Dom}(\mathcal{D}^+_R)$ to the boundary makes a Fredholm pair with the Cauchy data space of $\mathcal{D}^+$.

(iv) The composition $R\mathcal{P}\mathcal{D}^+$ defines a Fredholm operator from the Cauchy data space to the range of R (see also Theorem 2.20).

(v) The space $\operatorname{Ker}\left(\mathcal{D}^+_R\right)$ consists only of smooth sections.

Warning: A new feature of operators of Dirac type on manifolds with boundary is that the index of admissible boundary value problems can jump under continuous or even smooth deformation of the coefficients. For instance, this is the case for the Atiyah-Patodi-Singer boundary problem, as follows from the next subsection.

A total compatible (and so symmetric) Dirac operator $\mathcal{D}$ and an orthogonal admissible projection R define a self-adjoint extension $\mathcal{D}_D^+ = \mathcal{D}^{+*}|_D$, if the projection R defines the domain D and satisfies the symmetry condition $I - R = \sigma^{-1}R\sigma$. We denote by $\mathcal{G}r^{\mathrm{sa}}_{p_+}$ or, shortly, $\mathcal{G}r^{\mathrm{sa}}$ the subspace of $\mathcal{G}r_{p_+}$ of orthogonal projections which satisfy the preceding condition and differ from the Atiyah-Patodi-Singer projection only by an infinitely smoothing operator.

Posing a suitable *well-posed* boundary value problem provides for a nicely spaced discrete spectrum near 0. Then, varying the coefficients of the differential operator and the imposed boundary condition suggests the use of the powerful topological concept of spectral flow. From (i) and a careful analysis of the corresponding parametrices, we see in [**BoLePh01**, Section 3] under which conditions the curves of the induced self-adjoint L^2 extensions become continuous curves in $\mathcal{CF}^{\mathrm{sa}}(L^2(X;E))$ in the gap topology so that their spectral flow is well-defined and truly homotopy invariant.

We summarize the main results. They depend strongly on the weak unique continuation property (either in the form of Section 1.4 or in the weaker form of (1.6) which is sufficient here) and the invertible extension (Section 2.3.4).

LEMMA 2.27. *For fixed $\mathcal{D}$ the mapping*

$$\mathcal{G}r^{\mathrm{sa}}(\mathcal{D}) \ni P \mapsto \mathcal{D}_P \in \mathcal{CF}^{\mathrm{sa}}(L^2(X;E))$$

is continuous from the operator norm to the gap metric.

THEOREM 2.28. *Let X be a compact Riemannian manifold with boundary. Let $\{\mathcal{D}_s\}_{s\in M}$, M a metric space, be a family of compatible operators of Dirac type. We assume that in each local chart, the coefficients of $\mathcal{D}_s$ depend continuously on s. Then we have:*

(a) *The Poisson operator $K_s : L^2(\Sigma; E|_\Sigma) \to H^{1/2}(X;E)$ of $\mathcal{D}_s$ depends continuously on s in the operator norm.*
(b) *The Calderón projector $P_+(s) : L^2(\Sigma; E_{|\Sigma}) \to L^2(\Sigma; E_{|\Sigma})$ of $\mathcal{D}_s$ depends continuously on s in the operator norm.*
(c) *The family*

$$M \ni s \mapsto (\mathcal{D}_s)_{P_+(s)} \in \mathcal{CF}^{\mathrm{sa}}(L^2(X;E))$$

is continuous.

(d) *Let $P_{t,\{t\in Y\}}$ be a norm-continuous path of orthogonal projections in $L^2(X;E)$. If*

$$P_t \in \bigcap_{s\in M} \mathcal{G}r^{\mathrm{sa}}(\mathcal{D}_s) \quad \textit{for every } t \in Y,$$

then

$$M \times Y \ni (s,t) \mapsto (\mathcal{D}_s)_{P_t} \in \mathcal{CF}^{\mathrm{sa}}(L^2(X;E))$$

is continuous.

Note that (b) is a pseudo-differential reformulation of the continuity of Cauchy data spaces which is valid in much greater generality (see our Section 1.2.2).

We close this section with a recent result of [**Gr02**]:

LEMMA 2.29 (G. Grubb 2002). *Let $\mathcal{D}$ be an operator of Dirac type over a compact manifold with boundary. Let P be an orthogonal projection which defines an 'admissible' self-adjoint boundary condition for $\mathcal{D}$. Then there exists an invertible operator of Dirac type $\mathcal{B}'$ over the boundary such that $P = P_{>}(\mathcal{B}')$.*

Grubb's Lemma shows that the Atiyah-Patodi-Singer boundary projection is the most general admissible self-adjoint boundary condition, in the specified sense.

2.4. The Atiyah-Patodi-Singer index theorem. Let X be a compact, oriented Riemannian manifold with boundary $Y = \partial X$ with $\dim X = n = 2m$ even. Let $\mathcal{D} : C^\infty(X,S) \to C^\infty(X,S)$ be a compatible operator of Dirac type where $S \to X$ is a bundle of Clifford modules. Relative to the splitting $S = S^+ \oplus S^-$ into chiral halves, we have the operators $\mathcal{D}^+ : C^\infty(X,S^+) \to C^\infty(X,S^-)$ and $\mathcal{D}^- : C^\infty(X,S^-) \to C^\infty(X,S^+)$ which are formal adjoints on sections with support in $X \setminus Y$. We assume that all structures (e.g., Riemannian metric, Clifford module, connection) are products on some collared neighborhood $N \cong [-1,1] \times Y$ of Y. Then $\mathcal{D}^+|_N := \mathcal{D}^+ : C^\infty(N, S^+|N) \to C^\infty(N, S^-|N)$ has the form

$$\mathcal{D}^+|_N = \sigma(\partial_u + \mathcal{B}).$$

Here $u \in [-1,1]$ is the normal coordinate (i.e., $N = \{(u,y) \mid y \in Y,\ u \in [-1,1]\}$) with $\partial_u = \frac{\partial}{\partial u}$ the inward normal), $\sigma = \mathbf{c}(du)$ is the (unitary) Clifford multiplication by du with $\sigma(S^+|N) = S^-|N$, and

$$\mathcal{B} : C^\infty(Y, S^+|_Y) \to C^\infty(Y, S^+|_Y)$$

denotes the canonically associated (elliptic, self-adjoint) Dirac operator over Y, called the *tangential operator*. Note that due to the product structure, σ and $\mathcal{B}$ do not depend on u. Let $P_{\geq}(\mathcal{B})$ denote the spectral (Atiyah-Patodi-Singer) projection onto the subspace $L_+(\mathcal{B})$ of $L^2(Y, S^+|_{\partial X})$ spanned by the eigensections corresponding to the nonnegative eigenvalues of $\mathcal{B}$. Let

$$\begin{aligned} C^\infty(X, S^+; P_\geq) &:= \left\{\psi \in C^\infty(X,S^+) \mid P_\geq(\mathcal{B})(\psi|_Y) = 0\right\}, \\ \mathcal{D}^+_{P_\geq} &:= \mathcal{D}^+|_{C^\infty(X,S^+;P_\geq)} : C^\infty(X,S^+;P_\geq) \to C^\infty(X,S^-). \end{aligned}$$

The eta function for $\mathcal{B}$ is defined by

$$\eta_{\mathcal{B}}(s) := \sum_{\lambda \in \operatorname{spec}\mathcal{B} - \{0\}} (\operatorname{sign}\lambda)\, m_\lambda\, |\lambda|^{-s},$$

for $\Re(s)$ sufficiently large, where m_λ is the multiplicity of λ. Implicit in the following result (originating in [**AtPaSi75**]) is that $\eta_{\mathcal{B}}$ extends to a meromorphic function on all $\mathbb{C}$, which is holomorphic at $s = 0$ so that $\eta_{\mathcal{B}}(0)$ is finite.

THEOREM 2.30 (Atiyah-Patodi-Singer index formula). *The operator* $\mathcal{D}^+_{P_\geq}$ *has a finite index given by*

$$\operatorname{index} \mathcal{D}^+_{P_\geq} = \int_X \left(\mathbf{ch}(S,\varepsilon) \wedge \widetilde{\mathbf{A}}(X,\theta)\right) - \frac{1}{2}\left(m_0 + \eta_{\mathcal{B}}(0)\right),$$

where $m_0 = \dim(\operatorname{Ker} \mathcal{B})$, $\mathbf{ch}(S,\varepsilon) \in \Omega^*(X,\mathbb{R})$ *is the total Chern character form of the complex vector bundle* S *with compatible, unitary connection* ε, *and* $\widetilde{\mathbf{A}}(X,\theta) \in \Omega^*(X,\mathbb{R})$ *is closely related to the total* $\widehat{\mathbf{A}}(X,\theta)$ *form relative to the Levi-Civita connection* θ, *namely* $\widetilde{\mathbf{A}}(X,\theta)_{4k} = 2^{2k-m}\widehat{\mathbf{A}}(X,\theta)_{4k}$.

PROOF. (outline) The proof found in [**AtPaSi75**] or [**BoWo93**] is based on the heat kernel method for computing the index, but the process is less straightforward than in the closed case because of the boundary condition. The appropriate heat kernel is constructed by means of Duhamel's method. Namely, an exact kernel is obtained from an approximate one by an iterative process initiated by writing the error as the integral of a derivative of the convolution of the true and approximate kernel; see (2.46) below. The initial approximate heat kernel is obtained by patching together two heat kernels, denoted by $\mathcal{E}_c$ and $\mathcal{E}_d$. Here, $\mathcal{E}_c$ is a heat kernel for a Dirac operator over an infinite extension $[0,\infty) \times Y$ of the collared neighborhood $N = [0,1] \times Y$ of Y in X, for which the boundary condition $P_\geq(\mathcal{B})(\psi|_Y) = 0$ is imposed. The heat kernel $\mathcal{E}_d$ is the usual one (without boundary conditions) for $e^{-t\widetilde{\mathcal{D}^-}\widetilde{\mathcal{D}^+}}$, where $\widetilde{\mathcal{D}^\pm}$ are chiral halves of the invertible Dirac operator (see (2.29)), namely

$$\widetilde{\mathcal{D}^\pm} := \mathcal{D}^\pm \cup \mathcal{D}^\mp : C^\infty(\widetilde{X}, \widetilde{S^\pm}) \to C^\infty(\widetilde{X}, \widetilde{S^\pm})$$

over the double $\widetilde{X}$, a closed manifold without boundary.

We begin with the construction of $\mathcal{E}_c$. We let

$$D := \partial_u + \mathcal{B} : C^\infty(N, S^+|N) \to C^\infty(N, S^+|N),$$

which has formal adjoint $D^* := -\partial_u + \mathcal{B}$. Define

$$\mathcal{D} := D|\operatorname{Dom}\mathcal{D} \quad \text{and} \quad \mathcal{D}^* = D^*|\operatorname{Dom}\mathcal{D}^*,$$

where

$$\begin{aligned}
\operatorname{Dom}\mathcal{D} &:= \{f \in H^1(\mathbb{R}_+ \times Y, \pi^*(S^+|Y)) \mid P_\geq(f|_{\{0\}\times Y}) = 0\},\\
\operatorname{Dom}\mathcal{D}^* &:= \{f \in H^1(\mathbb{R}_+ \times Y, \pi^*(S^-|Y)) \mid P_<(f|_{\{0\}\times Y}) = 0\}.
\end{aligned}$$

We also have the Laplacians given by

$$\Delta_c := \mathcal{D}^*\mathcal{D}|\operatorname{Dom}\mathcal{D}^*\mathcal{D} \quad \text{and} \quad \Delta_{c*} := \mathcal{D}\mathcal{D}^*|\operatorname{Dom}\mathcal{D}\mathcal{D}^*,$$

where

$$\begin{aligned}
\operatorname{Dom}\mathcal{D}^*\mathcal{D} &:= \{f \in \operatorname{Dom}\mathcal{D} \mid P_<(Df|_{\{0\}\times Y}) = 0\},\\
\operatorname{Dom}\mathcal{D}\mathcal{D}^* &:= \{f \in \operatorname{Dom}\mathcal{D}^* \mid P_\geq(D^*f|_{\{0\}\times Y}) = 0\}.
\end{aligned}$$

Let $\varphi_\lambda(y) \in C^\infty(Y, S^+|Y)$ be an eigensection of $\mathcal{B}$ with eigenvalue $\lambda \in \mathbb{R}$. Note that $g_\lambda(t; u, y) = f_\lambda(t, u)\varphi_\lambda(y)$ is a solution of the heat equation

$$\begin{aligned} 0 &= (\partial_t + \Delta_c)g_\lambda = \partial_t g_\lambda + (\partial_u + \mathcal{B})(-\partial_u + \mathcal{B})g_\lambda = (\partial_t - \partial_u^2 + \mathcal{B}^2)g_\lambda \\ &= (\partial_t f_\lambda - \partial_u^2 f_\lambda + \lambda^2 f_\lambda)\varphi_\lambda(y), \end{aligned}$$

with $g_\lambda(t; \cdot, \cdot) \in \operatorname{Dom} \mathcal{D}^*\mathcal{D}$, when $f_\lambda : (0, \infty) \times [0, \infty) \to \mathbb{R}$ solves the heat problem

$$\begin{aligned} &\partial_t f_\lambda = \partial_u^2 f_\lambda - \lambda^2 f_\lambda \text{ with boundary conditions} \\ &f_\lambda(t, 0) = 0 \text{ if } \lambda \geq 0, \\ &\partial_u f_\lambda(t, 0) + \lambda f_\lambda(t, 0) = D f_\lambda = 0 \text{ if } \lambda < 0. \end{aligned} \tag{2.37}$$

Recall that the *complementary error function* is

$$\operatorname{erfc}(x) := \frac{2}{\sqrt{\pi}} \int_x^\infty e^{-\xi^2} d\xi.$$

Of use to us are the facts

$$\begin{aligned} &\operatorname{erfc}'(x) = \frac{-2}{\sqrt{\pi}} e^{-x^2}, \text{ and} \\ &\operatorname{erfc}(x) \leq \frac{2}{\sqrt{\pi}} \frac{e^{-x^2}}{x + \sqrt{x^2 + \frac{4}{\pi}}} \leq e^{-x^2} \text{ for } x \geq 0. \end{aligned} \tag{2.38}$$

For $\lambda \geq 0$, the heat kernel for the problem (2.37) is (via the method of images)

$$e_\lambda(t; u, v) := \frac{e^{-\lambda^2 t}}{2\sqrt{\pi t}} \Big(e^{-\frac{(u-v)^2}{4t}} - e^{-\frac{(u+v)^2}{4t}} \Big) \text{ for } u, v \geq 0 \text{ and } t > 0.$$

For $\lambda < 0$, $u, v \geq 0$, and $t > 0$, the heat kernel is (using Laplace transforms)

$$e_\lambda(t; u, v) := \frac{e^{-\lambda^2 t}}{2\sqrt{\pi t}} \Big(e^{-\frac{(u-v)^2}{4t}} + e^{-\frac{(u+v)^2}{4t}} + \lambda e^{\lambda(u+v)} \operatorname{erfc}\big(\tfrac{u+v}{2\sqrt{t}} - \lambda\sqrt{t}\big) \Big).$$

The heat kernel for Δ_c (i.e., the kernel for $e^{-t\Delta_c}$) is then

$$\mathcal{E}_c(t; u, y; v, z) = \sum_\lambda e_\lambda(t; u, v)\varphi_\lambda(y) \otimes \varphi_\lambda(z)^* \tag{2.39}$$

Here λ ranges over $\operatorname{spec} \mathcal{B}$ and (by a convenient abuse of notation) $\varphi_\lambda(y) \otimes \varphi_\lambda(z)^*$ is really a sum $\sum_k \varphi_{\lambda,k}(y) \otimes \varphi_{\lambda,k}(z)^*$ where $\{\varphi_{\lambda,1}, \ldots, \varphi_{\lambda,k}\}$ is an orthonormal eigenbasis of the eigenspace for λ, and $\{\varphi^*_{\lambda,1}, \ldots, \varphi^*_{\lambda,k}\}$ is the dual basis. Similarly, the heat kernel for Δ_{c*} is

$$\mathcal{E}_{c*}(t; u, y; v, z) = \sum_\lambda e_{\lambda*}(t; u, v)\varphi_\lambda(y) \otimes \varphi_\lambda(z)^*, \tag{2.40}$$

where $e_{\lambda*}$ is the heat kernel for the problem

$$\begin{aligned} &\partial_t f_\lambda = \partial_u^2 f_\lambda - \lambda^2 f_\lambda \text{ with boundary conditions} \\ &f_\lambda(t, 0) = 0 \text{ if } \lambda < 0, \\ &\partial_u f_\lambda(t, 0) - \lambda f_\lambda(t, 0) = D^* f_\lambda = 0 \text{ if } \lambda \geq 0, \end{aligned} \tag{2.41}$$

where $f_\lambda : (0,\infty) \times [0,\infty) \to \mathbb{R}$. Thus, for $u, v \geq 0$ and $t > 0$, we have

$$e_{\lambda *}(t;u,v) := \frac{e^{-\lambda^2 t}}{2\sqrt{\pi t}}\Big(e^{-\frac{(u-v)^2}{4t}} - e^{-\frac{(u+v)^2}{4t}}\Big) \text{ for } \lambda < 0,$$

while for $\lambda \geq 0$ (note the change of sign in passing from (2.37) to (2.41)),

$$e_{\lambda *}(t;u,v) := \frac{e^{-\lambda^2 t}}{2\sqrt{\pi t}}\Big(e^{-\frac{(u-v)^2}{4t}} + e^{-\frac{(u+v)^2}{4t}} - \lambda e^{\lambda(u+v)} \operatorname{erfc}\big(\tfrac{u+v}{2\sqrt{t}} + \lambda\sqrt{t}\big)\Big).$$

Combining (2.39) and (2.40), we obtain the trace of the kernel $\mathcal{E}_c - \mathcal{E}_{c*}$ for the difference $e^{-t\Delta_c} - e^{-t\Delta_{c*}}$ evaluated at the point $(u,y;u,y)$ of the diagonal:

$$\begin{aligned}
\mathcal{K}(t;u,y) &:= \operatorname{Tr}\sum_\lambda (e_\lambda(t;u,u) - e_{*\lambda}(t;u,u))\varphi_\lambda(y)\otimes\varphi_\lambda^*(y) \\
&= \sum_{\lambda\geq 0}\Big(\frac{-e^{-\lambda^2 t}}{2\sqrt{\pi t}}e^{-u^2/t} - \frac{e^{-\lambda^2 t}}{2\sqrt{\pi t}}e^{-u^2/t} + \lambda e^{2\lambda u}\operatorname{erfc}\big(\tfrac{u}{\sqrt{t}} + \lambda\sqrt{t}\big)\Big)|\varphi_\lambda(y)|^2 \\
&\quad + \sum_{\lambda<0}\Big(\frac{e^{-\lambda^2 t}}{2\sqrt{\pi t}}e^{-u^2/t} + \lambda e^{-2\lambda u}\operatorname{erfc}\big(\tfrac{u}{\sqrt{t}} - \lambda\sqrt{t}\big) + \frac{e^{-\lambda^2 t}}{2\sqrt{\pi t}}e^{-u^2/t}\Big)|\varphi_\lambda(y)|^2 \\
&= \sum_\lambda \operatorname{sign}(\lambda)\Big(\frac{-e^{-\lambda^2 t}e^{-u^2/t}}{\sqrt{\pi t}} + |\lambda|e^{2|\lambda|u}\operatorname{erfc}\big(\tfrac{u}{\sqrt{t}} + |\lambda|\sqrt{t}\big)\Big)|\varphi_\lambda(y)|^2 \\
&= \sum_\lambda \operatorname{sign}(\lambda)\frac{\partial}{\partial u}\Big(\tfrac{1}{2}e^{2|\lambda|u}\operatorname{erfc}\big(\tfrac{u}{\sqrt{t}} + |\lambda|\sqrt{t}\big)\Big)|\varphi_\lambda(y)|^2.
\end{aligned}$$

We used $\operatorname{erfc}'(x) = \frac{-2}{\sqrt{\pi}}e^{-x^2}$ to obtain the last equality, and we have set $\operatorname{sign}(0) := 1$ for convenience. We compute (where the interchange of the sum and integral can be justified)

$$\begin{aligned}
\mathcal{K}(t) &:= \int_0^\infty\int_Y \mathcal{K}(t;u,y)\,dy\,du \\
&= \int_0^\infty\int_Y \sum_\lambda \operatorname{sign}(\lambda)\frac{\partial}{\partial u}\Big(\tfrac{1}{2}e^{2|\lambda|u}\operatorname{erfc}\big(\tfrac{u}{\sqrt{t}} + |\lambda|\sqrt{t}\big)\Big)|\varphi_\lambda(y)|^2\,dy\,du \\
&= \sum_\lambda m_\lambda \operatorname{sign}(\lambda)\int_0^\infty \frac{\partial}{\partial u}\Big(\tfrac{1}{2}e^{2|\lambda|u}\operatorname{erfc}\big(\tfrac{u}{\sqrt{t}} + |\lambda|\sqrt{t}\big)\Big)du \\
&= \sum_\lambda m_\lambda \operatorname{sign}(\lambda)\Big(\tfrac{1}{2}e^{2|\lambda|u}\operatorname{erfc}\big(\tfrac{u}{\sqrt{t}} + |\lambda|\sqrt{t}\big)\Big)\Big|_{u=0}^{\infty} \\
&= -\sum_\lambda \frac{1}{2}m_\lambda \operatorname{sign}(\lambda)\operatorname{erfc}\big(|\lambda|\sqrt{t}\big) \\
&= -\frac{1}{2}m_0 - \sum_{\lambda\neq 0}\frac{1}{2}m_\lambda \operatorname{sign}(\lambda)\operatorname{erfc}\big(|\lambda|\sqrt{t}\big).
\end{aligned}$$

Recall that m_λ is the multiplicity of λ. Thus,

$$\mathcal{K}(t) + \frac{1}{2}m_0 = -\sum_{\lambda\neq 0} \frac{1}{2} m_\lambda \operatorname{sign}(\lambda) \operatorname{erfc}(|\lambda|\sqrt{t}).$$

For $|\lambda| > 0$, one verifies using integration by parts and substitution that

$$\int_0^\infty \operatorname{erfc}(|\lambda|\sqrt{t})\, t^{s-1} dt = \frac{|\lambda|^{-2s}}{s\sqrt{\pi}} \Gamma(s+\tfrac{1}{2}).$$

Using this and the fact that $\mathcal{K}(t) + \frac{1}{2}h \leq Ce^{-\alpha t}$ for constants C and $\alpha > 0$,

$$\begin{aligned}
&\int_0^\infty \left(\mathcal{K}(t) + \tfrac{1}{2}m_0\right) t^{s-1} dt \\
&\qquad = -\int_0^\infty \Big(\sum_{\lambda\neq 0} \frac{1}{2} m_\lambda \operatorname{sign}(\lambda) \operatorname{erfc}(|\lambda|\sqrt{t})\Big) t^{s-1} dt \\
&\qquad = -\sum_{\lambda\neq 0} \frac{1}{2} m_\lambda \operatorname{sign}(\lambda) \int_0^\infty \operatorname{erfc}(|\lambda|\sqrt{t})\, t^{s-1} dt \\
&\qquad = -\sum_{\lambda\neq 0} \frac{1}{2} m_\lambda \operatorname{sign}(\lambda) \frac{|\lambda|^{-2s}}{s\sqrt{\pi}} \Gamma(s+\tfrac{1}{2}) \\
&\qquad = -\frac{\Gamma(s+\frac{1}{2})}{2s\sqrt{\pi}} \sum_{\lambda\neq 0} m_\lambda \operatorname{sign}(\lambda)|\lambda|^{-2s} \\
&\qquad = -\frac{\Gamma(s+\frac{1}{2})}{2s\sqrt{\pi}} \eta_B(2s).
\end{aligned}$$

Suppose that we have an asymptotic expansion

$$\mathcal{K}(t) \sim \sum_{k=-n+1}^{N} a_k t^{k/2} \quad \text{as } t \to 0^+; \text{ i.e.,}$$

$$\mathcal{K}(t) - \sum_{k=-n+1}^{N} a_k t^{k/2} = \mathrm{O}\left(t^{\frac{N+1}{2}}\right) \text{ as } t \to 0^+.$$

In (2.53) below, such an asymptotic expansion will eventually be produced (as was done in [**BoWo93**, p. 239f]) from the known asymptotic expansion (see [**Gi95**, p. 68]) of the heat kernel on a closed manifold, namely the double of M. Since $\Re(s) > -\frac{N+1}{2}$ implies $\Re(\frac{N+1}{2} + s - 1) > -1$, we then have that

$$f_1(s) := \int_0^1 \Big(\mathcal{K}(t) - \sum_{k=-n+1}^{N} a_k t^{k/2}\Big) t^{s-1} dt = \int_0^1 \mathrm{O}\left(t^{\frac{N+1}{2}+s-1}\right) ds$$

is holomorphic for $\Re(s) > -\frac{N+1}{2}$. We also have the entire function

$$f_\infty(s) := \int_1^\infty \left(\mathcal{K}(t) + \tfrac{1}{2}h\right) t^{s-1} dt$$

We claim that

$$-\frac{\Gamma(s+\frac{1}{2})}{2s\sqrt{\pi}}\eta_{\mathcal{B}}(2s) = \frac{m_0}{2s} + \sum_{k=-n+1}^{N}\left(\frac{a_k}{s+\frac{1}{2}k}\right) + f_1(s) + f_\infty(s).$$

Indeed, we have

$$\begin{aligned}
-&\frac{\Gamma(s+\frac{1}{2})}{2s\sqrt{\pi}}\eta_{\mathcal{B}}(2s)\\
&= \int_0^\infty \left(\mathcal{K}(t)+\tfrac{1}{2}h\right)t^{s-1}dt = \int_0^1 \left(\mathcal{K}(t)+\tfrac{1}{2}h\right)t^{s-1}dt + f_\infty(s)\\
&= \int_0^1 \left(\tfrac{1}{2}m_0 + \sum_{k=-n+1}^{N} a_k t^{k/2}\right)t^{s-1}dt + f_1(s) + f_\infty(s)\\
&= \int_0^1 \tfrac{1}{2}m_0 t^{s-1}dt + \sum_{k=-n+1}^{N} a_k \int_0^1 t^{(s+\frac{1}{2}k)-1}dt + f_1(s) + f_\infty(s)\\
&= \tfrac{1}{2}m_0 \frac{t^s}{s}\Big|_{t=0}^{t=1} + \sum_{k=-n+1}^{N} a_k \left(\frac{t^{s+\frac{1}{2}k}}{s+\frac{1}{2}k}\right)\Big|_{t=0}^{t=1} + f_1(s) + f_\infty(s)\\
&= \frac{m_0}{2s} + \sum_{k=-n+1}^{N}\left(\frac{a_k}{s+\frac{1}{2}k}\right) + f_1(s) + f_\infty(s).
\end{aligned}$$

Thus, where $\theta_N(s) = f_1(s) + f_\infty(s)$ is holomorphic for $\Re(s) > -\frac{N+1}{2}$,

$$\eta_{\mathcal{B}}(2s) = -\frac{2s\sqrt{\pi}}{\Gamma(s+\frac{1}{2})}\left(\frac{m_0}{2s} + \sum_{k=-n+1}^{N}\left(\frac{a_k}{\frac{1}{2}k+s}\right) + \theta_N(s)\right). \tag{2.42}$$

The heat kernels, say $\mathcal{E}_d$ for $\widetilde{\mathcal{D}^-}\widetilde{\mathcal{D}^+}$ and $\mathcal{E}_{d*}$ for $\widetilde{\mathcal{D}^+}\widetilde{\mathcal{D}^-}$, over the double $\widetilde{X}$ are more familiar. For $t > 0$,

$$\operatorname{Tr} e^{-t\widetilde{\mathcal{D}^-}\widetilde{\mathcal{D}^+}} - \operatorname{Tr} e^{-t\widetilde{\mathcal{D}^+}\widetilde{\mathcal{D}^-}} = \int_{\widetilde{X}} \operatorname{Tr}(\mathcal{E}_d(t;x,x) - \mathcal{E}_{d*}(t;x,x))\,dx,$$

and there is the asymptotic expansion

$$F(t;x) := \operatorname{Tr}(\mathcal{E}_d(t;x,x) - \mathcal{E}_{d*}(t;x,x)) \sim \sum_{k\geq -n} \alpha_k(x)t^{k/2}.$$

For $0 < a < b < 1$, let $\rho_{(a,b)} : [0,1] \to [0,1]$ be C^∞ and increasing with

$$\rho_{(a,b)}(u) = \begin{cases} 0 & \text{for } u \leq a, \\ 1 & \text{for } u \geq b. \end{cases}$$

Thinking of u as the normal coordinate function on $N = [0,1] \times Y \subset X$, and extending by the constant values 0 and 1, we can (and do) regard $\rho_{(a,b)}$ as a function on X. Let

$$
\begin{aligned}
Q(t;x,x') &:= \varphi_c(x)\,\mathcal{E}_c(t;x,x')\,\psi_c(x') + \varphi_d(x)\,\mathcal{E}_d(t;x,x')\,\psi_d(x') \\
&:= \big(1-\rho_{(5/7,6/7)}(x)\big)\,\mathcal{E}_c(t;x,x')\big(1-\rho_{(3/7,4/7)}(x')\big) \\
&\quad + \rho_{(1/7,2/7)}(x)\,\mathcal{E}_d(t;x,x')\,\rho_{(3/7,4/7)}(x').
\end{aligned}
$$

We have that $\psi_c + \psi_d = 1$ and $\{\psi_c, \psi_d\}$ is a partition of unity for the cover $\{u^{-1}(\frac{3}{7},\infty), u^{-1}[0,\frac{4}{7})\}$ of X. Moreover, for $j = c, d$,

$$
\text{(2.43)} \qquad \varphi_j|\operatorname{supp}\psi_j = 1 \quad \text{and} \quad \operatorname{dist}\big(\operatorname{supp}\partial_u^k\varphi_j, \operatorname{supp}\psi_j\big) \geq \frac{1}{7} \quad (\text{for } k \geq 1).
$$

Of course, we do not expect $Q(t;x,x')$ to equal the exact kernel for $\partial_t + \mathcal{D}^*\mathcal{D}$ throughout X. However, note that for $x, x' \in [0,1/7)\times Y$, $Q(t;x,x') = \mathcal{E}_c(t;x,x')$, so that $Q(t;x,x')$ meets the APS boundary condition.

Let $x' \in X$ be fixed. For $x \in X \setminus [0,6/7)\times Y$, we have

$$
Q(t;x,x') = \mathcal{E}_d(t;x,x')\psi_d(x'),
$$

while for $x \in [0,1/7)\times Y$,

$$
Q(t;x,x') = \mathcal{E}_c(t;x,x')\psi_c(x').
$$

Thus, for $x \in X \setminus ([\frac{1}{7},\frac{6}{7}]\times Y)$, $-(\partial_t + \mathcal{D}^*\mathcal{D})\,Q(\cdot;\cdot,x')|_{(t;x,x')} = 0$. Moreover, for $d(x,x') < 1/7$,

$$
\begin{aligned}
&\varphi_c(x)\psi_c(x') + \varphi_d(x)\psi_d(x') \\
&\quad = (1-\rho_{(5/7,6/7)}(x))(1-\rho_{(3/7,4/7)}(x')) + \rho_{(1/7,2/7)}(x)\rho_{(3/7,4/7)}(x') \\
&\quad = 1 - \rho_{(5/7,6/7)}(x) - \rho_{(3/7,4/7)}(x') \\
&\qquad\quad + \rho_{(5/7,6/7)}(x)\rho_{(3/7,4/7)}(x') + \rho_{(1/7,2/7)}(x)\rho_{(3/7,4/7)}(x') \\
&\quad = 1 - \rho_{(5/7,6/7)}(x) - \rho_{(3/7,4/7)}(x') + \rho_{(5/7,6/7)}(x) + \rho_{(3/7,4/7)}(x') \\
&\quad = 1.
\end{aligned}
$$

Note that $d(x,x') < 1/7$ implies $\rho_{(5/7,6/7)}(x)\rho_{(3/7,4/7)}(x') = \rho_{(5/7,6/7)}(x)$, and $\rho_{(1/7,2/7)}(x)\rho_{(3/7,4/7)}(x') = \rho_{(3/7,4/7)}(x)$. Because of this, we expect that

$$
\begin{aligned}
\lim_{t\to 0^+} Q(t;x,x') &= \lim_{t\to 0^+}\Big(\varphi_c(x)\mathcal{E}_c(t;x,x')\psi_c(x') + \varphi_d(x)\mathcal{E}_d(t;x,x')\psi_d(x')\Big) \\
&= \varphi_c(x)\delta(x,x')\psi_c(x') + \varphi_d(x)\delta(x,x')\psi_d(x') \\
&= \big(\varphi_c(x)\psi_c(x') + \varphi_d(x)\psi_d(x')\big)\delta(x,x') \\
&= \delta(x,x').
\end{aligned}
$$

Thus, on the operator level, we expect that $\lim_{t\to 0^+} Q_t = I$; i.e.,

$$
\lim_{t\to 0^+} Q_t(f)(x) = \lim_{t\to 0^+}\int_X Q(t;x,x')f(x')\,dx' = \int_X \delta(x,x')f(x')\,dx' = f(x).
$$

The extent to which $Q(\cdot;\cdot,x')$ fails to satisfy the heat equation in $[\frac{1}{7},\frac{6}{7}]\times Y$ is given by

$$
C(t;x,x') := -(\partial_t + \mathcal{D}^*\mathcal{D})\,Q(\cdot;\cdot,x')|_{(t;x,x')}.
$$

For $x' \in X$ fixed and $x = (u, y) \in N = [0,1] \times Y$, we have

$$
\begin{aligned}
-C(t;x,x') &= (\partial_t + \mathcal{D}^*\mathcal{D})\, Q(t;x,x') \\
&= (\partial_t + \mathcal{D}^*\mathcal{D})\Big(\sum_{j\in\{c,d\}} \varphi_j(x)\mathcal{E}_j(t;x,x')\psi_j(x') \Big) \\
&= \sum_{j\in\{c,d\}} \partial_u^2\big(\varphi_j(x)\big)\mathcal{E}_j(t;x,x')\psi_j(x') \\
&\qquad + 2\partial_u\big(\varphi_j(x)\big)\, \partial_u\mathcal{E}_j(t;x,x')\psi_j(x').
\end{aligned} \tag{2.44}
$$

For $d(x,x') < \frac{1}{7}$, we have $C(t;x,x') = 0$, since (2.43) implies that $\psi_c(x') = 0$ when $\partial_u^k(\varphi_j(x)) \neq 0$ and $d(x,x') < \frac{1}{7}$. Using this and the estimates

$$
\begin{aligned}
\mathcal{E}_j(t;x,x') &\le At^{-n/2}e^{-Bd(x,x')^2/t}, \\
\partial_u\mathcal{E}_j(t;x,x') &\le At^{-(n+1)/2}e^{-Bd(x,x')^2/t},
\end{aligned}
$$

for positive constants A and B, it follows from (2.44) that

$$
\begin{aligned}
|C(t;x,x')| &\le c_1 t^{-(n+1)/2}\, e^{-Bd(x,x')^2/t} \\
&\le c_1 t^{-(n+1)/2}\, e^{-B7^{-2}/t} = c_1\, e^{-c_2/t}
\end{aligned} \tag{2.45}
$$

for $0 < t < T_0 < \infty$, and positive constants c_1 and c_2.

Let $\mathcal{E}(t;x,x')$ be the exact heat kernel for $\mathcal{D}^*\mathcal{D}$ and let the corresponding operator be $\mathcal{E}(t) = \exp(-t\mathcal{D}^*\mathcal{D})$. On the operator level, we have

$$
\begin{aligned}
\mathcal{E}(t) - Q(t) &= \mathcal{E}(t)Q(0) - \mathcal{E}(0)Q(t) \\
&= \int_0^t \frac{d}{ds}(\mathcal{E}(s)Q(t-s))\, ds \\
&= \int_0^t \frac{d\mathcal{E}}{ds}Q(t-s) + \mathcal{E}(s)\frac{d}{ds}Q(t-s)\, ds \\
&= \int_0^t -\mathcal{D}^*\mathcal{D}\,\mathcal{E}(s)Q(t-s) + \mathcal{E}(s)\frac{d}{ds}Q(t-s)\, ds \\
&= \int_0^t \mathcal{E}(s)\Big(-\mathcal{D}^*\mathcal{D} - \frac{d}{d(t-s)}\Big)Q(t-s)\, ds \\
&= \int_0^t \mathcal{E}(s)C(t-s)\, ds.
\end{aligned} \tag{2.46}
$$

Hence on the kernel level,

$$
\begin{aligned}
\mathcal{E}(t;x,x') &= Q(t;x,x') + \int_0^t \int_X \mathcal{E}(s;x,z)C(t-s;z,x')\, dz ds \\
&= Q(t;x,x') + (\mathcal{E} * C)(t;x,x'),
\end{aligned} \tag{2.47}
$$

where the convolution operation is defined by

$$(\alpha * \beta)(t;x,x') := \int_0^t \int_X \alpha(s;x,z)\beta(t-s;z,x')dzds.$$

Note that (2.47) can be written as

$$\mathcal{E}(I - *C) = \mathcal{E} - \mathcal{E} * C = Q.$$

Thus, at least formally,

$$\begin{aligned} \mathcal{E} &= Q(I - *C(t))^{-1} = Q(I + \sum_{k=1}^{\infty}(*C)^k) \\ &= Q + Q * \sum_{k=1}^{\infty} C_k = Q + Q * \mathcal{C}, \end{aligned} \tag{2.48}$$

where

$$C_1 := C = -(\partial_t + \mathcal{D}^*\mathcal{D})Q, \quad C_{k+1} := C_1 * C_k, \quad \text{and } \mathcal{C} := \sum_{k=1}^{\infty} C_k.$$

The proof of the validity of the Levi sum (2.48) for $\mathcal{E}$ rests on the estimate (2.45) and we refer to [**BoWo93**, Chap. 22] for the details. The above constructions can be also be applied to obtain a kernel $\mathcal{E}_*(t;x,x')$ for $\exp(-t\mathcal{D}\mathcal{D}^*)$ starting from

$$Q_*(t;x,x') := \varphi_c(x)\mathcal{E}_{c*}(t;x,x')\psi_c(x') + \varphi_d(x)\mathcal{E}_{d*}(t;x,x')\psi_d(x').$$

Recall that for $(u,y) \in [0,\infty) \times Y$,

$$\begin{aligned} \mathcal{K}(t;u,y) &:= \mathrm{Tr}(\mathcal{E}_c(t;(u,y),(u,y)) - \mathcal{E}_{c*}(t;(u,y),(u,y))), \text{ and} \\ \mathcal{K}(t) &:= \int_0^{\infty} \int_Y \mathcal{K}(t;u,y)\, dydu. \end{aligned} \tag{2.49}$$

For $x \in \widetilde{X}$, let

$$F(t;x) := \mathrm{Tr}(\mathcal{E}_d(t;x,x) - \mathcal{E}_{d*}(t;x,x)).$$

Now,

$$\operatorname{index} \mathcal{D}^+_{P_\geq} = \mathrm{Tr}\left(e^{-t\mathcal{D}^*\mathcal{D}} - e^{-t\mathcal{D}\mathcal{D}^*}\right) = \mathrm{Tr}(\mathcal{E}(t) - \mathcal{E}_*(t)).$$

We use the following notation for functions being "exponentially close" as $t \to 0^+$:

$$f(t) \sim_{\exp} g(t) \iff |f(t) - g(t)| \leq ae^{-b/t} \text{ for } t \in (0,\varepsilon) \tag{2.50}$$

for some positive constants a, b, and ε. We claim (but must omit important details here – see [**BoWo93**, Chap. 22]) that

$$\mathrm{Tr}(\mathcal{E}(t) - \mathcal{E}_*(t)) \sim_{\exp} \mathrm{Tr}(Q(t) - Q_*(t)).$$

Since $\mathcal{E} = Q + Q * \mathcal{C}$ (resp., $\mathcal{E}_* = Q_* + Q_* * \mathcal{C}_*$), this result ultimately rests on (2.45) (resp., the corresponding result for C_*). Since $\varphi_c \psi_c = \psi_c$ and $\varphi_d \psi_d = \psi_d$,

$$\begin{aligned}\operatorname{Tr}(Q(t) - Q_*(t)) &= \int_X \operatorname{Tr}\left(Q(t;x,x) - Q_*(t;x,x)\right) dx \\ &= \int_X \operatorname{Tr}\Big(\varphi_c(x)(\mathcal{E}_c - \mathcal{E}_{c*})(t;x,x)\psi_c(x) \\ &\qquad + \varphi_d(x)(\mathcal{E}_d - \mathcal{E}_{d*})(t;x,x)\psi_d(x)\Big)\, dx \\ &= \int_0^1 \int_Y \mathcal{K}(t;u,y)\psi_c(u)\, dy du + \int_X F(t;x)\psi_d(x)\, dx.\end{aligned}$$

Let

$$\mathcal{K}_a(t) := \int_0^a \int_Y \mathcal{K}(t;u,y) dy du.$$

Note that

$$\begin{aligned}\mathcal{K}(t) - \mathcal{K}_a(t) &= \int_a^\infty \int_Y \mathcal{K}(t;u,y)\, dy du \\ &= \sum_\lambda m_\lambda \operatorname{sign}(\lambda)\Big(\tfrac{1}{2} e^{2|\lambda|u} \operatorname{erfc}\left(\tfrac{u}{\sqrt{t}} + |\lambda|\sqrt{t}\right)\Big)\Big|_{u=a}^\infty \\ &= -\frac{1}{2}\sum_\lambda m_\lambda \operatorname{sign}(\lambda)\, e^{2|\lambda|a} \operatorname{erfc}\left(\tfrac{a}{\sqrt{t}} + |\lambda|\sqrt{t}\right).\end{aligned}$$

Thus (using (2.38)) for some constant $c > 0$

$$\begin{aligned}|\mathcal{K}(t) - \mathcal{K}_a(t)| &\le \sum_\lambda m_\lambda e^{2|\lambda|a} \exp\Big(-\left(\tfrac{a}{\sqrt{t}} + |\lambda|\sqrt{t}\right)^2\Big) \\ &\le \frac{1}{2} e^{-a^2/t} \sum_\lambda m_\lambda\, e^{-|\lambda|^2 t} \sim_{t\to 0^+} c e^{-a^2/t} t^{-n/2},\end{aligned}$$

which says that $\mathcal{K}(t) \sim_{\exp} \mathcal{K}_a(t)$. Hence,

$$\begin{aligned}\operatorname{index} \mathcal{D}^+_{P_\ge} &= \operatorname{Tr}\left(\mathcal{E}(t) - \mathcal{E}_*(t)\right) \sim_{\exp} \operatorname{Tr}\left(Q(t) - Q_*(t)\right) \\ &= \int_0^1 \int_Y \mathcal{K}(t;u,y)\psi_c(u)\, dy du + \int_X F(t;x)\psi_d(x)\, dx \\ &\sim_{\exp} \Big(\int_0^\infty \int_Y \mathcal{K}(t;u,y)\, dy du + \int_X F(t;x)\psi_d(x)\, dx\Big) \\ &= \mathcal{K}(t) + \int_X F(t;x)\, dx.\end{aligned} \tag{2.51}$$

The last equality follows from (2.49) and the fact that

$$F(t;x) = \operatorname{Tr}\left(\mathcal{E}_d(t;x,x) - \mathcal{E}_{d*}(t;x,x)\right) = 0 \tag{2.52}$$

for $x \in N = [0,1] \times Y$, which is seen as follows. Since $\sigma^* = -\sigma$,

$$\widetilde{\mathcal{D}^+}|_N = \sigma(\partial_u + \mathcal{B}) : C^\infty(N, \widetilde{S^+}) \to C^\infty(N, \widetilde{S^-}) \text{ implies}$$
$$\widetilde{\mathcal{D}^-}|_N = (\sigma(\partial_u + \mathcal{B}))^* = (\partial_u - \mathcal{B})\sigma : C^\infty(N, \widetilde{S^-}) \to C^\infty(N, \widetilde{S^+}),$$

Thus, over N,

$$\begin{aligned}\widetilde{\mathcal{D}^-}\widetilde{\mathcal{D}^+} &= (\partial_u - \mathcal{B})\sigma\sigma(\partial_u + \mathcal{B}) = (-\partial_u^2 + \mathcal{B}^2)\big|C^\infty(N, \widetilde{S^+}),\\ \widetilde{\mathcal{D}^+}\widetilde{\mathcal{D}^-} &= \sigma(\partial_u + \mathcal{B})(\partial_u - \mathcal{B})\sigma = \sigma^2(\partial_u - \mathcal{B})(\partial_u + \mathcal{B})\\ &= (-\partial_u^2 + \mathcal{B}^2)\big|C^\infty(N, \widetilde{S^-}).\end{aligned}$$

Since $[(-\partial_u^2 + \mathcal{B}^2), \sigma] = 0$, σ maps the eigenspaces of $\widetilde{\mathcal{D}^-}\widetilde{\mathcal{D}^+}$ pointwise isometrically (over N) onto those of $\widetilde{\mathcal{D}^+}\widetilde{\mathcal{D}^-}$. Hence, we obtain (2.52) upon taking the trace of the eigensection expansion of $\mathcal{E}_d(t;x,x) - \mathcal{E}_{d*}(t;x,x)$ for $x \in N$. From (2.51), we obtain

$$\begin{aligned}\mathcal{K}(t) &= \operatorname{index} \mathcal{D}^+_{P_\geq} - \int_X F(t;x)\,dx\\ &\sim \operatorname{index} \mathcal{D}^+_{P_\geq} - \int_X \sum_{k\geq -n} \alpha_k(x) t^{k/2}\,dx\\ &= \operatorname{index} \mathcal{D}^+_{P_\geq} - \sum_{k\geq -n} A_k\, t^{k/2}\,dx\end{aligned} \tag{2.53}$$

for $A_k := \int_X \alpha_k(x)$, where $F(t;x) \sim \sum_{k\geq -n} \alpha_k(x)\, t^{k/2}$ is the asymptotic expansion of the trace $F(t;x)$ of the kernel for $e^{-t\widetilde{\mathcal{D}^-}\widetilde{\mathcal{D}^+}} - e^{-t\widetilde{\mathcal{D}^+}\widetilde{\mathcal{D}^-}}$, which is known for elliptic differential operators (e.g., $\widetilde{\mathcal{D}^-}\widetilde{\mathcal{D}^+}$ and $\widetilde{\mathcal{D}^+}\widetilde{\mathcal{D}^-}$) over closed manifolds such as $\widetilde{X}$ (see [**Gi95**, p. 68]). Thus, according to (2.42) with $a_k = -A_k$ for $k \neq 0$, and $a_0 = \operatorname{index} \mathcal{D}^+_{P_\geq} - A_0$, we get

$$\begin{aligned}\eta_{\mathcal{B}}(2s) &= -\frac{2s\sqrt{\pi}}{\Gamma(s+\frac{1}{2})}\Big(\frac{m_0}{2s} + \sum_{k=-n+1}^{N}\Big(\frac{a_k}{\frac{1}{2}k+s}\Big) + \theta_N(s)\Big)\\ &= -\frac{2s\sqrt{\pi}}{\Gamma(s+\frac{1}{2})}\Big(\frac{m_0}{2s} + \frac{1}{s}\operatorname{index} \mathcal{D}^+_{P_\geq} - \sum_{k=-n+1}^{N}\Big(\frac{A_k}{\frac{1}{2}k+s}\Big) + \theta_N(s)\Big).\end{aligned}$$

Setting $s = 0$ yields $\eta_{\mathcal{B}}(0) = -(m_0 + 2\operatorname{index} \mathcal{D}^+_{P_\geq} - 2A_0)$, or

$$\operatorname{index} \mathcal{D}^+_{P_\geq} = A_0 - \frac{1}{2}(m_0 + \eta_{\mathcal{B}}(0)).$$

□

2.5. Symplectic geometry of Cauchy data spaces. As we have seen in our Section 1.2.2 there exists a concept of Cauchy data spaces which solely is based on the concepts of minimal and maximal domain and which is more elementary and more general than the definitions provided in Section 2.3.5 which are based on pseudo-differential calculus.

In this section we stay in the real category and do not assume product structure near $\Sigma = \partial X$ unless otherwise stated. The operator $\mathcal{D}$ need not be of Dirac type. We only assume that it is a linear, elliptic, symmetric, differential operator of first order.

2.5.1. *The natural Cauchy data space.* Let $\mathcal{D}_0$ denote the restriction of $\mathcal{D}$ to the space $C_0^\infty(X;S)$ of smooth sections with support in the interior of X. As mentioned above, there is no natural choice of the order of the Sobolev spaces for the boundary reduction. Therefore, a systematic treatment of the boundary reduction may begin with the minimal closed extension $\mathcal{D}_{\min} := \overline{\mathcal{D}_0}$ and the adjoint $\mathcal{D}_{\max} := (\mathcal{D}_0)^*$ of $\mathcal{D}_0$. Clearly, $\mathcal{D}_{\max}$ is the maximal closed extension. This gives

$$D_{\min} := \operatorname{Dom}(\mathcal{D}_{\min}) = \overline{C_0^\infty(X;S)}^{\mathcal{G}} = \overline{C_0^\infty(X;S)}^{H^1(X;S)}$$

and

$$\begin{aligned} D_{\max} &:= \operatorname{Dom}(\mathcal{D}_{\max}) \\ &= \{u \in L^2(X;S) \mid \mathcal{D}u \in L^2(X;S) \text{ in the sense of distributions}\}. \end{aligned}$$

Here, the superscript $\mathcal{G}$ means the closure in the graph norm which coincides with the first Sobolev norm on $C_0^\infty(X;S)$. We form the space β of *natural boundary values* with the *natural trace map* γ as described in Section 1.2.

There we defined also the *natural Cauchy data space* $\Lambda(\mathcal{D}) := \gamma(\operatorname{Ker}\mathcal{D}_{\max})$ under the assumption that $\mathcal{D}$ has a self-adjoint L^2 extension with a compact resolvent. Such an extension always exists. Take for instance $\mathcal{D}_{\mathcal{P}(\mathcal{D})}$, the operator $\mathcal{D}$ with domain

$$\operatorname{Dom}_{\mathcal{P}(\mathcal{D})} := \{f \in H^1(X;S) \mid \mathcal{P}(\mathcal{D})^{(\frac{1}{2})}(f|_\Sigma) = 0\},$$

where $\mathcal{P}(\mathcal{D})$ denotes the Calderón projection defined in (2.30).

Clearly, $D_{\max}$ and $D_{\min}$ are $C^\infty(X)$ modules, and so the space β is a $C^\infty(\Sigma)$ module. This shows that β is *local* in the following sense: If Σ decomposes into r connected components $\Sigma = \Sigma_1 \sqcup \cdots \sqcup \Sigma_r$, then β decomposes into

$$\beta = \bigoplus_{j=1}^{r} \beta_j,$$

where

$$\beta_j := \gamma\big(\{f \in D_{\max} \mid \operatorname{supp} f \subset N_j\}\big)$$

with a suitable collared neighborhood N_j of Σ_j. Note that each β_j is a closed symplectic subspace of β and therefore a symplectic Hilbert space.

By Theorem 2.20a and, alternatively and in greater generality by Hörmander [**Ho66**] (Theorem 2.2.1 and the Estimate (2.2.8), p. 194), the space β is naturally embedded in the distribution space $H^{-\frac{1}{2}}(\Sigma; S|_\Sigma)$. Under this embedding we have $\Lambda(\mathcal{D}) = \Lambda(\mathcal{D}, 0)$, where the last space was defined in Definition 2.22.

If the metrics are product close to Σ, we can give a more precise description of the embedding of β, namely as a *graded* space of distributions. Let $\{\varphi_k, \Lambda_k\}$ be a spectral resolution of $L^2(\Sigma)$ by eigensections of $\mathcal{B}$. (Here and in the following we do not mention the bundle S). Once again, for simplicity, we assume $\operatorname{Ker}\mathcal{B} = \{0\}$ and have $\mathcal{B}\varphi_k = \lambda_k\varphi_k$ for all $k \in \mathbb{Z} \setminus \{0\}$, and $\lambda_{-k} = -\lambda_k$, $\sigma(\varphi_k) = \varphi_{-k}$, and $\sigma(\varphi_{-k}) = -\varphi_k$ for $k > 0$. In [**BoFu99**, Proposition 7.15] (see also [**BrLe01**] for a more general setting), it was shown that

$$\beta = \beta_- \oplus \beta_+ \quad \text{with}$$

$$\beta_- := \overline{[\{\varphi_k\}_{k<0}]}^{H^{\frac{1}{2}}(\Sigma)} \quad \text{and} \quad \beta_+ := \overline{[\{\varphi_k\}_{k>0}]}^{H^{-\frac{1}{2}}(\Sigma)}. \tag{2.54}$$

Then, β_- and β_+ are Lagrangian and transversal subspaces of β.

2.5.2. *Criss-cross reduction.* Let us define two Lagrangian and transversal subspaces $L_\pm$ of $L^2(\Sigma)$ in a similar way, namely by the closure in $L^2(\Sigma)$ of the linear span of the eigensections with negative, resp. with positive eigenvalue. Then L_+ is dense in β_+, and β_- is dense in L_-. This anti-symmetric relation may explain some of the well-observed delicacies of dealing with spectral invariants of continuous families of Dirac operators.

Moreover, $\gamma(D_{\text{aps}}) = \beta_-$, where

$$D_{\text{aps}} := \{f \in H^1(X) \mid P_>(f|_\Sigma) = 0\} \tag{2.55}$$

denotes the domain corresponding to the Atiyah-Patodi-Singer boundary condition. Note that a series $\sum_{k<0} c_k\varphi_k$ may converge to an element $\varphi \in L^2(\Sigma)$ without converging in $H^{\frac{1}{2}}(\Sigma)$. Therefore such $\varphi \in L_-$ can not appear as trace at the boundary of any $f \in D_{\max}$.

Recall Proposition 1.27 and note that $(\beta_-, \Lambda(\mathcal{D}))$ is a Fredholm pair.

This can all be achieved without the symbolic calculus of pseudo-differential operators. Therefore one may ask how the preceding approach to Cauchy data spaces and boundary value problems via the maximal domain and our symplectic space β is related to the approach via the Calderón projection, which we reviewed in the preceding section. How can results from the distributional theory be translated into L^2 results?

To relate the two approaches we recall a fairly general symplectic 'Criss-Cross' Reduction Theorem from [**BoFuOt01**, Theorem 1.2]. Let β and L be symplectic Hilbert spaces with symplectic forms ω_β and ω_L, respectively. Let

$$\beta = \beta_- \dotplus \beta_+ \quad \text{and} \quad L = L_- \dotplus L_+$$

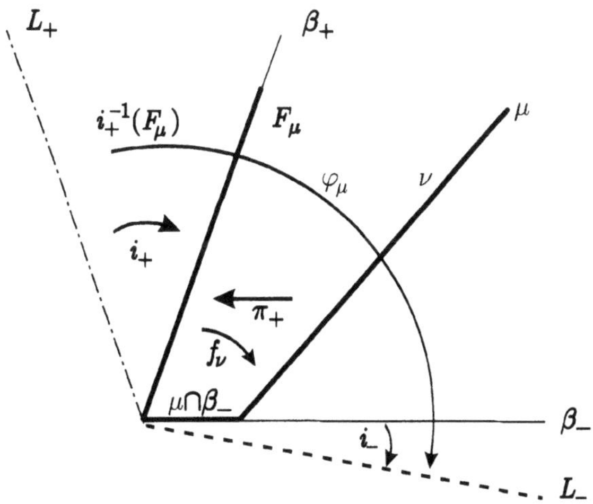

FIGURE 9. The mapping $\tau : \mathcal{FL}_{\beta_-}(\beta) \to \mathcal{FL}_{L_-}(L)$

be direct sum decompositions by transversal (not necessarily orthogonal) pairs of Lagrangian subspaces. We assume that there exist continuous, injective mappings

$$i_- : \beta_- \to L_- \quad \text{and} \quad i_+ : L_+ \to \beta_+$$

with dense images and which are compatible with the symplectic structures, i.e.,

$$\omega_L(i_-(x), a) = \omega_\beta(x, i_+(a)) \quad \text{for all } a \in L_+ \text{ and } x \in \beta_- .$$

Let $\mu \in \mathcal{FL}_{\beta_-}(\beta)$, for instance, $\mu = (\mu \cap \beta_-) \dotplus \nu$ with a suitable closed ν. Let us define (see also Figure 9)

$$\tau(\mu) := i_-(\mu \cap \beta_-) + \operatorname{graph}(\varphi_\mu), \tag{2.56}$$

where

$$\varphi_\mu : i_+^{-1}(F_\mu) \longrightarrow L_- ,$$
$$x \mapsto i_- \circ f_\nu \circ i_+(x).$$

Here F_μ denotes the image of μ under the projection π_+ from μ to β_+ along β_- and $f_\nu : F_\mu \to \beta_-$ denotes the uniquely determined bounded operator which yields ν as its graph. Then:

THEOREM 2.31. *The mapping* (2.56) *defines a continuous mapping*

$$\tau : \mathcal{FL}_{\beta_-}(\beta) \longrightarrow \mathcal{FL}_{\mathbf{L}_-}(L)$$

which maps the Maslov cycle $\mathcal{M}_{\beta_-}(\beta)$ *of* β_- *into the Maslov cycle* $\mathcal{M}_{L_-}(L)$ *of* L_- *and preserves the Maslov index*

$$\mathbf{mas}(\{\mu_s\}_{s\in[0,1]}, \beta_-) = \mathbf{mas}(\{\tau(\mu_s)\}_{s\in[0,1]}, L_-)$$

for any continuous curve $[0,1] \ni s \mapsto \mu_s \in \mathcal{FL}_{\beta_-}(\beta)$.

In the product case, the 'Criss-Cross' Reduction Theorem implies for our two types of Cauchy data that all results proved in the theory of natural boundary values (β theory) remain valid in the L^2 theory. In particular, we have:

COROLLARY 2.32. *The $L^2(\Sigma)$ part $\Lambda(\mathcal{D}) \cap L^2(\Sigma)$ of the natural Cauchy data space $\Lambda(\mathcal{D})$ is closed in $L^2(\Sigma)$. Actually, it is a Lagrangian subspace of $L^2(\Sigma)$ and it forms a Fredholm pair with the component L_-, defined at the beginning of this section.*

2.6. Non-additivity of the index.

2.6.1. *The Bojarski conjecture.* The *Bojarski conjecture* gives a quite different description of the index of an elliptic operator over a closed partitioned manifold $M = M_1 \cup_\Sigma M_2$. It relates the 'quantum' quantity index with a 'classical' quantity, the Fredholm intersection index of the Cauchy data spaces from both sides of the hypersurface Σ. It was suggested in [**Bo79**] and proved in [**BoWo93**] for operators of Dirac type.

PROPOSITION 2.33. *Let M be a partitioned manifold as before and let $\Lambda(\mathcal{D}_j^+, \frac{1}{2})$ denote the L^2 Cauchy data spaces, $j = 1, 2$ (see Definition 2.22). Then*

$$\operatorname{index} \mathcal{D}^+ = \operatorname{index}\left(\Lambda(\mathcal{D}_1^+, \tfrac{1}{2}), \Lambda(\mathcal{D}_2^+, \tfrac{1}{2})\right).$$

Recall that

$$\begin{aligned}\operatorname{index}\left(\Lambda(\mathcal{D}_1^+, \tfrac{1}{2}), \Lambda(\mathcal{D}_2^+, \tfrac{1}{2})\right) &:= \dim\left(\Lambda(\mathcal{D}_1^+, \tfrac{1}{2}) \cap \Lambda(\mathcal{D}_2^+, \tfrac{1}{2})\right) \\ &\quad - \dim\left(L^2(\Sigma; S|_\Sigma)/\left(\Lambda(\mathcal{D}_1^+, \tfrac{1}{2}) + \Lambda(\mathcal{D}_2^+, \tfrac{1}{2})\right)\right).\end{aligned}$$

It is equal to $\mathbf{i}(I - \mathcal{P}(\mathcal{D}_2^+), \mathcal{P}(\mathcal{D}_1^+))$, where $\mathcal{P}(\mathcal{D}_j^+)$ denotes the corresponding Calderón projections, and $\mathbf{i}(\cdot, \cdot)$ was defined in (2.31).

The proof of the above proposition depends on the unique continuation property for Dirac operators and the Lagrangian property of the Cauchy data spaces, more precisely the chiral twisting property (2.36).

2.6.2. *Generalizations for global boundary conditions.* On a smooth compact manifold X with boundary Σ, the solution spaces $\operatorname{Ker}(\mathcal{D}, s)$ depend on the order s of differentiability and they are infinite-dimensional. To obtain a finite index one must apply suitable boundary conditions (see [**BoWo93**] for local and global boundary conditions for operators of Dirac type). In this report, we restrict ourselves to boundary conditions of Atiyah-Patodi-Singer type (i.e., $P \in \mathcal{G}r(\mathcal{D})$), and consider the extension

$$\mathcal{D}_P : \operatorname{Dom}(\mathcal{D}_P) \to L^2(X; S) \tag{2.57}$$

of $\mathcal{D}$ defined by the domain

$$\operatorname{Dom}(\mathcal{D}_P) := \{f \in H^1(X; S) \mid P^{(0)}(f|_\Sigma) = 0\}. \tag{2.58}$$

It is a closed operator in $L^2(X; S)$ with finite-dimensional kernel and cokernel. We have an explicit formula for the adjoint operator

$$(\mathcal{D}_P)^* = \mathcal{D}_{\sigma(I-P)\sigma^*}. \tag{2.59}$$

In agreement with Proposition 1.27b, the preceding equation shows that an extension $\mathcal{D}_P$ is self-adjoint, if and only if $\operatorname{Ker} P^{(0)}$ is a Lagrangian subspace of the symplectic Hilbert space $L^2(\Sigma; S|_\Sigma)$.

Let us recall the *Boundary Reduction Formula for the Index* of (global) elliptic boundary value problems discussed in [**BoWo85**] (inspired by [**Se69**], see [**BoWo93**] for a detailed proof for Dirac operators). Like the Bojarski conjecture, the point of the formula is that it gives an expression for the index in terms of the geometry of the Cauchy data in the symplectic space of all (here L^2) boundary data.

PROPOSITION 2.34.

$$\operatorname{index} \mathcal{D}_P = \operatorname{index} \left\{ P\mathcal{P}(\mathcal{D}) : \Lambda(\mathcal{D}, \tfrac{1}{2}) \to \operatorname{range}(P^{(0)}) \right\}.$$

2.6.3. *Pasting formulas.* We shall close this section by mentioning a slight modification of the Bojarski conjecture/theorem, namely a non-additivity formula for the splitting of the index over partitioned manifolds.

From the Atiyah-Singer Index Theorem (here the expression of the index on the closed manifold M by an integral of the *index density*) and the Atiyah-Patodi-Singer Index Theorem (Theorem 2.30), we obtain at once

$$\operatorname{index} \mathcal{D} = \operatorname{index} (\mathcal{D}_1)_{P_\leq} + \operatorname{index} (\mathcal{D}_2)_{P_\geq} - \dim \operatorname{Ker}(\mathcal{B}).$$

Then an Agranovic-Dynin type correction formula (based on Proposition 2.34) yields:

THEOREM 2.35. *Let P_j be projections belonging to $\mathcal{G}r(\mathcal{D}_j)$, $j = 1, 2$. Then*

$$\operatorname{index} \mathcal{D} = \operatorname{index} (\mathcal{D}_1)_{P_1} + \operatorname{index} (\mathcal{D}_2)_{P_2} - \mathbf{i}\,(P_2, I - P_1)$$

It turns out that the boundary correction term $\mathbf{i}(P_2, I - P_1)$ equals the index of the operator $\sigma(\partial_t + \mathcal{B})$ on the cylinder $[0,1] \times \Sigma$ with the boundary conditions P_1 at $t = 0$ and P_2 at $t = 1$. A direct proof of Theorem 2.35 can be derived from Proposition 2.34 by elementary operations with the virtual indices $\mathbf{i}\,(P_1, \mathcal{P}\,(\mathcal{D}_1))$ and $\mathbf{i}\,(P_2, \mathcal{P}\,(\mathcal{D}_2))$; see, e.g., [**DaZh96**] in a more general setting.

REMARK 2.36.

(a) In this section we have not always distinguished between the *total* and the *chiral* Dirac operator because all the discussed index formulas are valid in both cases.

(b) Important index formulas for (global) elliptic boundary value problems for operators of Dirac type can also be obtained without analyzing the concept and the geometry of the Cauchy data spaces (see e.g. the celebrated Atiyah-Patodi-Singer Index Theorem 2.30 or [**Sc01**] for a recent survey of index formulas where there is no mention of the Calderón projection). The basic reason is that the index is an invariant represented by a local density inside the manifold plus a correction term which lives on the boundary and may be local or non-local as well. However, these formulas do not explain the simple origin of the index or the spectral flow, namely that *all* index information is naturally coded by the geometry of the Cauchy data spaces. To us it seems necessary to use the Calderón projection in order to understand (not calculate) the index of an elliptic boundary problem and the reason for the locality or non-locality.

2.7. Pasting of spectral flow.

2.7.1. *Spectral flow and the Maslov index.* Let $\{\mathcal{D}_t\}_{t\in[0,1]}$ be a continuous family of (from now on always *total*) Dirac operators with the same principal symbol and the same domain D. To begin with, we do not distinguish between the case of a closed manifold (when D is just the first Sobolev space and all operators are essentially self-adjoint) and the case of a manifold with boundary (when D is specified by the choice of a suitable boundary value condition).

We consider the *spectral flow* $\operatorname{SF}\{\mathcal{D}_{t,D}\}$ (see Section 1.1.4). We want a pasting formula for the spectral flow. To achieve that, one replaces the spectral flow of a continuous one-parameter family of self-adjoint Fredholm operators, which is a 'quantum' quantity, by the *Maslov index* of a corresponding path of Lagrangian Fredholm pairs, which is a 'quasi-classical' quantity. The idea is due to Floer and was worked out subsequently by Yoshida in dimension 3, by Nicolaescu in all odd dimensions, and pushed further by Cappell, Lee and Miller, Daniel and Kirk, and many other authors. For a survey, see [**BoFu98**], [**BoFu99**], [**DaKi99**], [**KiLe00**].

In this review we give two spectral flow formulas of that type. To begin with, we consider the case of a manifold with boundary. Then weak UCP for Dirac type operators (established in Section 1.4) implies weak inner UCP in the sense of (1.6), if the manifold is connected. Actually, it would suffice that there is no connected component without boundary. Let us fix the space β for the family. By Proposition 1.27c the corresponding family $\{\Lambda(\mathcal{D}_t)\}$ of natural Cauchy data spaces is continuous. Applying the *General Boundary Reduction Formula* for the spectral flow (Theorem 1.28) gives a family version of the Bojarski conjecture (our Proposition 2.33):

THEOREM 2.37. *The spectral flow of the family* $\{\mathcal{D}_{t,D}\}$ *and the Maslov index* $\mathbf{mas}\left(\{\Lambda(\mathcal{D}_t)\},\gamma(D)\right)$ *are well-defined, and we have*

$$\operatorname{SF}\{\mathcal{D}_{t,D}\} = \mathbf{mas}\left(\{\Lambda(\mathcal{D}_t)\},\gamma(D)\right). \tag{2.60}$$

We have various corollaries for the spectral flow on closed manifolds with fixed hypersurface (see [**BoFu99**]). Note that a partitioned manifold $M = M_1 \cup_\Sigma M_2$ can be considered as a manifold $M^\# = M_1 \sqcup M_2$ with boundary $\partial M^\# = -\Sigma \sqcup \Sigma$. Here $\sqcup$ denotes taking the disjoint union. Then any elliptic operator $\mathcal{D}$ over M defines an operator $\mathcal{D}^\#$ over $M^\#$ with natural Fredholm extension $\mathcal{D}^\#_D$ by fixing the domain

$$D := \{(f_1,f_2)\in H^1(M_1)\times H^1(M_2) \mid (f_1|_\Sigma, f_2|_\Sigma)\in\Delta\},$$

where Δ denotes the diagonal of $L^2\left(-\Sigma; S|_\Sigma\right)\times L^2\left(\Sigma; S|_\Sigma\right)$. By elliptic regularity we have

$$\operatorname{Ker}\mathcal{D}^\#_D = \operatorname{Ker}\mathcal{D} \quad\text{and}\quad \operatorname{SF}\left\{\mathcal{D}^\#_{t,D}\right\} = \operatorname{SF}\{\mathcal{D}_t\}.$$

For product structures near Σ, one can apply Theorem 2.31 and obtain an L^2 version of the preceding theorem which gives a new proof and a slight generalization of the Yoshida-Nicolaescu Formula (for details see [**BoFuOt01**, Section 3]). For safety reasons, we assume that $\mathcal{D}_t$ is a zero-order perturbation of $\mathcal{D}_0$, induced by a continuous change of the defining connection, or alternatively, $\mathcal{D}_t := \mathcal{D}_0 + C_t$ where C_t is a self-adjoint bundle morphism.

THEOREM 2.38.

$$\begin{aligned}\mathrm{SF}\{\mathcal{D}_t\} &= \mathbf{mas}\left(\left\{\Lambda_t^1 \cap L^2(-\Sigma) + \Lambda_t^2 \cap L^2(\Sigma)\right\}, \Delta\right)\\ &=: \mathbf{mas}\left(\{\Lambda_t^1 \cap L^2(-\Sigma)\}, \{\Lambda_t^2 \cap L^2(\Sigma)\}\right),\end{aligned}$$

where the Cauchy data spaces Λ_t^j are taken on each side of the hypersurface Σ.

REMARK 2.39. Here it is not compelling to use the symplectic geometry of the Cauchy data spaces (see Remark 2.36b). Actually, deep gluing formulas can and have been obtained for the spectral flow by coding relevant information not in the full infinite-dimensional Cauchy data spaces but in families of Lagrangian subspaces of suitable finite-dimensional symplectic spaces, like the kernel of the tangential operator (see [**CaLeMi96**] and [**CaLeMi00**]).

2.7.2. *Correction formula for the spectral flow.* Let D and D' be two domains between $D_{\min}$ and $D_{\max}$ such that both $\{\mathcal{D}_{t,D}\}$ and $\{\mathcal{D}_{t,D'}\}$ become families of self-adjoint Fredholm operators. We assume that D and D' differ only by a finite dimension, more precisely, we assume

$$\dim \gamma(D)/(\gamma(D) \cap \gamma(D')) = \dim \gamma(D')/(\gamma(D) \cap \gamma(D')) < \infty\,. \tag{2.61}$$

Then we find from Theorem 2.37 (for details see [**BoFu99**, Theorem 6.5]:

$$\begin{aligned}&\mathrm{SF}\{\mathcal{D}_{t,D}\} - \mathrm{SF}\{\mathcal{D}_{t,D'}\}\\ &\qquad= \mathbf{mas}\left(\{\Lambda(\mathcal{D}_t)\}, \gamma(D')\right) - \mathbf{mas}\left(\{\Lambda(\mathcal{D}_t)\}, \gamma(D)\right)\\ &\qquad= \sigma_{\text{Hör}}\left(\Lambda(\mathcal{D}_0), \Lambda(\mathcal{D}_1); \gamma(D'), \gamma(D)\right)\end{aligned} \tag{2.62}$$

(see Remark 1.26b). The assumption (2.61) is rather restrictive. The pair of domains, for instance, defined by the Atiyah-Patodi-Singer projection and the Calderón projection, may not always satisfy this condition. For the present proof, however, it seems indispensable.

3. The eta invariant

As mentioned in the introduction, a systematic functional analytical frame is missing for the eta invariant in contrast to the index and the spectral flow (see, however, [**CoMo95**] for an ambitious approach to establish analogues of Sobolev spaces, pseudo-differential operators, and zeta and eta functions in the context of noncommutative spectral geometry). Basically, however, the concept of the eta invariant of a (total and compatible, hence self-adjoint) Dirac operator is rather an immediate generalization of the index. Instead of measuring the chiral asymmetry of the zero eigenvalues we now measure the asymmetry of the *whole* spectrum.

Let $\mathcal{D}$ be an operator of Dirac type; i.e., roughly speaking, an operator with a real discrete spectrum which is nicely spaced without finite accumulation points and with an infinite number of eigenvalues on both sides of the real line. In close analogy with the definition of the zeta-function for essentially positive elliptic operators like the Laplacian, we set

$$\eta_{\mathcal{D}}(s) := \sum_{\lambda \in \operatorname{spec}(\mathcal{D}) \setminus \{0\}} \operatorname{sign}(\lambda)\, \lambda^{-s}\,.$$

Clearly, the formal sum $\eta_{\mathcal{D}}(s)$ is well defined for any complex number s with sufficiently large real part $\Re(s)$, and it vanishes for a symmetric spectrum (i.e., if for each $\lambda \in \operatorname{spec}(\mathcal{D})$ also $-\lambda \in \operatorname{spec}(\mathcal{D})$).

In Sections 2.2 and 2.4, we expressed the index by the difference of the traces of two related heat operators. Similarly, we also have a heat kernel expression for the eta function. Let $\mathcal{D}$ be any self-adjoint operator with compact resolvent and let $\{\lambda_k\}$ denote its eigenvalues ordered so that $\cdots \leq \lambda_{k-1} \leq \lambda_k \leq \lambda_{k+1} \leq \cdots$, each repeated according to its multiplicity. Formally, we have

$$\begin{aligned}
(3.1) \qquad \eta_{\mathcal{D}}(s) \cdot \Gamma\left(\frac{s+1}{2}\right) &= \sum_{\lambda_k \neq 0} \operatorname{sign}(\lambda_k) |\lambda_k|^{-s} \int_0^\infty r^{\frac{s-1}{2}} e^{-r}\, dr \\
&= \sum_{\lambda_k \neq 0} \lambda_k (\lambda_k^2)^{-\frac{s+1}{2}} \int_0^\infty (t\lambda_k^2)^{\frac{s-1}{2}} e^{-t\lambda_k^2}\, d(t\lambda_k^2) \\
&= \sum_{\lambda_k \neq 0} \int_0^\infty t^{\frac{s-1}{2}} \lambda_k e^{-t\lambda_k^2}\, dt = \int_0^\infty t^{\frac{s-1}{2}} \operatorname{Tr} e^{-t\mathcal{D}^2}\, dt.
\end{aligned}$$

For comparison we give the corresponding formula for the zeta function of the (positive) Dirac Laplacian $\mathcal{D}^2$:

$$(3.2) \qquad \zeta_{\mathcal{D}^2}(s) := \operatorname{Tr}(\mathcal{D}^2)^{-s} = \frac{1}{\Gamma(s)} \int_0^\infty t^{s-1} \operatorname{Tr} e^{-t\mathcal{D}^2}\, dt.$$

For the zeta function, we must assume that $\mathcal{D}$ has no vanishing eigenvalues (i.e. $\mathcal{D}^2$ is positive). Otherwise the integral on the right side is divergent. (The situation, however, can be cured by subtracting the orthogonal projection onto the kernel of $\mathcal{D}^2$ from the heat operator before taking the trace.) For the eta function, on the contrary, it clearly does not matter whether there are 0-eigenvalues and whether the summation is over all or only over the nonvanishing eigenvalues.

The derivation of (3.1) and (3.2) is completely elementary for $\Re(s) > \frac{1+\dim X}{2}$ and $\Re(s) > \frac{\dim X}{2}$, respectively, where X denotes the underlying manifold. It follows at once that $\eta(s)$ and $\zeta(s)$ admit meromorphic extensions to the whole complex plane. However, it is not clear at all how to characterize the operators for which the eta and the zeta functions have a finite value at $s = 0$; the *eta invariant* and the *zeta invariant*, respectively.

Historically, the eta invariant appeared for the first time in the 1970s as an error term showing up in the index formula for the APS spectral boundary value problem of a Dirac operator $\mathcal{D}$ on a compact manifold X with smooth boundary Σ (see our Section 2.4). More precisely, what arose was the eta invariant of the tangential operator (i.e., the induced Dirac operator over the closed manifold Σ). Even in that case it was hard to establish the existence and finiteness of the eta invariant.

Strictly speaking, one can define the eta invariant as the constant term in the Laurent expansion of the eta function around the point $s = 0$. For various applications this suffices. Many practical calculations, however, are much facilitated when we know the regularity *a priori*.

Basically, there are three different approaches to establish it: The original proof by Atiyah, Patodi, and Singer, worked out in [**BoWo93**, Corollary 22.9] and summarized in our Section 2.4; it is based on an assumption about the existence of a suitable asymptotic expansion for the corresponding heat kernel on the infinite cylinder $\mathbb{R}_+ \times \Sigma$. An intrinsic proof can be found in Gilkey [**Gi95**, Section 3.8]. It does not exploit that Σ bounds X, but requires strong topological means.

For a compatible (!) Dirac operator over a closed manifold Bismut and Freed [**BiFr86**] have shown that the eta function is actually a holomorphic function of s for $\Re(s) > -2$. They used the heat kernel representation (3.1) which implies that the eta invariant, when it exists, can be expressed as

$$\eta_{\mathcal{D}}(0) = \frac{1}{\sqrt{\pi}} \int_0^\infty \frac{1}{\sqrt{t}} \operatorname{Tr}\left(\mathcal{D}\, e^{-t\mathcal{D}^2}\right) dt.$$

It follows from (3.1) that the estimate

$$\left| \operatorname{Tr} \mathcal{D} e^{-t\mathcal{D}^2} \right| < c\sqrt{t}$$

implies the regularity of the eta function at $s = 0$. In fact, Bismut and Freed proved a sharper result, using nontrivial results from stochastic analysis. Inspired by calculations presented in Bismut and Cheeger [**BiCh89**, Section 3], Wojciechowski gave a purely analytic reformulation of the details of their proof. This provides a third and completely elementary way of proving the regularity of the eta function at $s = 0$. The essential steps are:

THEOREM 3.1. *Let $\mathcal{D} : C^\infty(\Sigma; E) \to C^\infty(\Sigma; E)$ denote a compatible Dirac operator over a closed manifold Σ of odd dimension m. Let $\mathrm{e}(t; x, x')$ denote the integral kernel of the heat operator $e^{-t\mathcal{D}^2}$. Then there exists a positive constant C such that*

$$\left| \operatorname{Tr}\left(\mathcal{D}_x e(t; x, x')\right)\big|_{x=x'} \right| < C\sqrt{t}$$

for all $x \in \Sigma$ and $0 < t < 1$.

We recall the definition of the 'local' η-function.

DEFINITION 3.2. Let $\{f_k; \lambda_k\}_{k\in\mathbb{Z}}$ be a discrete spectral resolution of $\mathcal{D}$. Then we define

$$\begin{aligned}\eta_{\mathcal{D}}(s;x) &:= \sum_{\lambda_k\neq 0} \operatorname{sign}(\lambda_k)|\lambda_k|^{-s} \langle f_k(x), f_k(x)\rangle_{E_x} \\ &= \frac{1}{\Gamma(\frac{s+1}{2})}\int_0^\infty t^{\frac{s-1}{2}}\Big(\sum_{\lambda_k\neq 0} \lambda_k e^{-t\lambda_k^2}\langle f_k(x), f_k(x)\rangle\Big)dt \\ &= \frac{1}{\Gamma(\frac{s+1}{2})}\int_0^\infty t^{\frac{s-1}{2}} \operatorname{Tr}\mathcal{D}\, \mathrm{e}(t;x,x)\,dt.\end{aligned}$$

COROLLARY 3.3. *Under the assumptions of the preceding theorem the 'local' eta function $\eta_{\mathcal{D}}(s;x)$ is holomorphic in the half plane $\Re(s) \geq -2$ for any $x \in \Sigma$.*

In the decade or so following 1975, it was generally believed that the existence of a finite eta invariant was a very special feature of operators of Dirac type on closed manifolds which are boundaries, and then, more generally, of Dirac type operators on all closed manifolds. Only after the seminal paper by Douglas and Wojciechowski [**DoWo91**] was it gradually realized that globally elliptic self-adjoint boundary value problems for operators of Dirac type also have a finite eta invariant. Once again, there are quite different approaches to obtain that result.

The work by Wojciechowski and collaborators is based on the Duhamel Principle and provides complete asymptotic expansions of the heat kernels for the self-adjoint Fredholm extension $\mathcal{D}_P$ where P belongs to the smooth self-adjoint Grassmannian. We recall from our review of the Atiyah-Patodi-Singer Index Theorem that the Duhamel Principle allows one to study the interior contribution and the boundary contribution separately and identify the singularities caused by the boundary contribution. It seems that Wojciechowski's method is only applicable if the metric structures close to the boundary are product.

Based on joint work with Seeley, Grubb [**Gr99**] also obtained nice asymptotic expansions of the trace of the heat kernels in (3.1) and (3.2). Contrary to the qualitative arguments of the Duhamel-Wojciechowski approach, Grubb's approach requires the explicit computation of certain coefficients in the expansion. Some of them have to vanish to guarantee the desired regularity.

For a slightly larger class of self-adjoint Fredholm extensions, Brüning and Lesch [**BrLe99a**] also studied the eta invariant. However, the authors had to deal with the residue of the eta function at $s = 0$ which is not present in the Duhamel-Wojciechowski approach.

In spite of the differences between the various approaches and types of results it seems that one consequence can be drawn immediately, namely there must be a deeper meaning of the eta invariant beyond its role as error term in the APS index formula. Happily, such a meaning was found by Singer in [**Si85**] for the eta invariant on closed manifolds and successively generalized by Wojciechowski and collaborators for manifolds with boundary, namely the identification of the eta invariant as the phase of the zeta function regularized determinant. We shall explain this now, and postpone our main topic, the pasting formulas for the eta invariant on partitioned manifolds.

3.1. Functional integrals and spectral asymmetry. Several important quantities in quantum mechanics and quantum field theory are expressed in terms of *quadratic functionals* and *functional integrals.* The concept of the determinant for Dirac operators arises naturally when one wants to evaluate the corresponding path integrals. As Itzykson and Zuber report in the chapter on *Functional Methods* of their monograph [**ItZu80**]: "The path integral formalism of Feynman and Kac provides a unified view of quantum mechanics, field theory, and statistical models. The original suggestion of an alternative presentation of quantum mechanical amplitudes in terms of path integrals stems from the work of Dirac (1933) and was brilliantly elaborated by Feynman in the 1940s. This work was first regarded with some suspicion due to the difficult mathematics required to give it a decent status. In the 1970s it has, however, proved to be the most flexible tool in suggesting new developments in field theory and therefore deserves a thorough presentation."

We shall restrict our discussion to the easiest variant of that complex matter by focusing on the partition function of a quadratic functional given by the Euclidean action of a *Dirac operator* which is assumed to be elliptic with imaginary time due to Wick rotation and coupled to continuously varying vector potentials (sources, fields, connections), for the ease of presentation in vacuum. We refer to Bertlmann, [**Be96**] and Schwarz, [**Sch93**] for an introduction to the quantum theoretic language for mathematicians and for a more extensive treatment of general aspects of quadratic functionals and functional integrals involving the relations to the Lagrangian and Hamiltonian formalism.

There are various alternative notions of "path integral" around, some more sophisticated than others. A mathematically rigorous formulation of the concept of "path integrals", as physicists typically "understand" it in quantum field theory is flimsy at best. For fields on Minkowski space, they are not mathematical integrals at all, because no measure is defined. For fields on Euclidean space (with imaginary time), one can construct genuine measures in limited settings which are not entirely realistic. Even then there is the issue of continuing the integrals back to real time which is generally ill-defined on a curved space-time. Many physicists do not care about such matters. Indeed, such physicists use path integrals primarily as compact generators of recipes to produce Feynman integrals which when regularized and renormalized yield coefficients in a formal power series in coupling constants for physical quantities of interest. However, no one has ever proved that these series converge, even for quantum electrodynamics (QED). Indeed the consensus of those who care is that these are only asymptotic series. Adding first few terms of these renormalized perturbation series yields remarkable 11-decimal point agreement with experiment, and QED is thereby hailed as a huge success. For many mathematicians and a few physicists, the great tragedy is that these recipes work so well without a genuine mathematical foundation. Although path integrals may not make precise sense per se, not only do they generate successful recipes in physics, but they can motivate fruitful ideas and precise concepts. In mathematics, path integrals have motivated the ζ-regularized determinant for the (Euclidean) Dirac operator, as a mathematically genuine canonical object, independent of particular choices made for regularization, which can be precisely calculated in principle.

A special feature of Dirac operators is that their determinants involve a *phase*, the imaginary part of the determinant's logarithm. As we will see now, this is a consequence of the fact that, unlike second-order semi-bounded Laplacians, first-order Dirac operators have an infinite number of both positive and negative eigenvalues. Then the phase of the determinant reflects the spectral asymmetry of the corresponding Dirac operator.

The simplest path integral we meet in quantum field theory takes the form of the *partition function* and can be written formally as the integral

$$Z(\beta) := \int_{\Gamma} e^{-\beta S(\omega)}\, d\omega\,, \tag{3.3}$$

where $d\omega$ denotes functional integration over the space $\Gamma := \Gamma(M;E)$ of sections of a Euclidean vector bundle E over a Riemannian manifold M.

In quantum theoretic language, M is space or space-time; a $\omega \in \Gamma$ is a position function of a particle or a spinor field. The scaling parameter β is a real or complex parameter, most often $\beta = 1$. The functional S is a quadratic real-valued functional on Γ defined by $S(\omega) := \langle \omega, T\omega \rangle$ with a fixed linear symmetric operator $T : \Gamma \to \Gamma$. Typically $T := \mathcal{D}$ is a Dirac operator and $S(\omega)$ is the action $S(\omega) = \int_M \langle \omega, \mathcal{D}\omega \rangle$.

Mathematically speaking, the integral (3.3) is an oscillating integral like the *Gaussian integral*. It is ill-defined in general because:

(I) As it stands, it is meaningless when $\dim \Gamma(M;E) = +\infty$ (i.e., when $\dim M \geq 1$).
(II) Even when $\dim \Gamma(M;E) < \infty$ (i.e. when $\dim M = 0$ and M consists of a finite set of points), the integral $Z(\beta)$ diverges unless $\beta S(\omega)$ is positive and nondegenerate.

Nevertheless, these expressions have been used and construed in quantum field theory. As a matter of fact, reconsidering the physicists' use and interpretation of these mathematically ill-defined quantities, one can describe certain formal manipulations which lead to normalizing and evaluating $Z(\beta)$ in a mathematically precise way.

We begin with a few calculations in *Case II*, inspired by Adams and Sen, [**AdSe96**], to show how *spectral asymmetry* is naturally entering into the calculations even in the finite-dimensional case and how this suggests a definition of the determinant in the infinite-dimensional case for the Dirac operator.

Then, let $\dim \Gamma = d < \infty$ and, for a symmetric endomorphism T, let

$$S(\omega) := \langle \omega, T\omega \rangle \text{ for all } \omega \in \Gamma.$$

Case 1. We assume that S positive and nondegenerate, i.e. T is strictly positive, say $\operatorname{spec} T = \{\lambda_1, \dots \lambda_d\}$ with all $\lambda_j > 0$. This is the classical case. We choose an orthonormal system of eigenvectors $(e_1, \dots, e_d)$ of T as basis for Γ. We have $S(\omega) = \sum \lambda_j x_j^2$ for $\omega = \sum x_j e_j$ and, for real $\beta > 0$, we get

$$\begin{aligned} Z(\beta) &= \int_G e^{-\beta S(\omega)}\, d\omega = \int_{\mathbb{R}^d} dx_1 \cdots dx_d\, e^{-\beta \sum \lambda_j x_j^2} \\ &= \int_{-\infty}^{\infty} e^{-\beta\lambda_1 x_1^2}\, dx_1 \int_{-\infty}^{\infty} e^{-\beta\lambda_2 x_2^2}\, dx_2 \cdots \int_{-\infty}^{\infty} e^{-\beta\lambda_d x_d^2}\, dx_d \\ &= \sqrt{\frac{\pi}{\beta\lambda_1}} \sqrt{\frac{\pi}{\beta\lambda_2}} \cdots \sqrt{\frac{\pi}{\beta\lambda_d}} = \pi^{d/2} \beta^{-d/2} (\det T)^{-1/2}. \end{aligned}$$

In that way the determinant appears when evaluating the simplest quadratic integral.

Case 2. If the functional S is positive and degenerate, $T \geq 0$, the partition function is given by

$$Z(\beta) = \pi^{\zeta/2} \beta^{-\zeta/2} (\det \omega T)^{-1/2} \operatorname{Vol}(\operatorname{Ker} T),$$

where $\zeta := \dim \Gamma - \dim \operatorname{Ker} T$ and $\widetilde{T} := T|_{(\operatorname{Ker} T)^\perp}$, but, of course $\operatorname{Vol}(\operatorname{Ker} T) = \infty$. For approaches to "renormalizing" this quantity in quantum chromodynamics, we refer to [**AdSe96**], [**BMSW97**], [**Sch93**]. One approach customary in physics is to take $\pi^{\zeta/2}\beta^{-\zeta/2}(\det \widetilde{T})^{-1/2}$ as the definition of the integral by setting the factor $\operatorname{Vol}(\operatorname{Ker} T)$ equal to 1.

Case 3. Now we assume that the functional S is nondegenerate, i.e. T invertible, but S is neither positive nor negative. We decompose $\Gamma = \Gamma_+ \times \Gamma_-$ and $T = T_+ \oplus T_-$ with $T_+, -T_-$ strictly positive on $\Gamma_\pm$. Formally, we obtain

$$\begin{aligned} Z(\beta) &= \left(\int_{\Gamma_+} d\omega_+\, e^{-\beta \langle \omega_+, T_+ \omega \rangle} \right) \left(\int_{\Gamma_-} d\omega_-\, e^{-(-\beta)\langle \omega_-, -T_- \omega \rangle} \right) \\ &= \pi^{d_+/2} \beta^{-d_+/2} (\det T_+)^{-1/2}\, \pi^{d_-/2} (-\beta)^{-d_-/2} (\det -T_-)^{-1/2} \\ &= \pi^{\zeta/2} \beta^{-d_+/2} (-\beta)^{-d_-/2} (\det |T|)^{-1/2}, \end{aligned}$$

where $d_\pm := \dim \Gamma_\pm$, hence $\zeta = d_+ + d_-$ and $|T| := \sqrt{T^2} = T_+ \oplus -T_-$.

Case 4. In the preceding formula, the term $(\beta)^{-d_+/2}(-\beta)^{-d_-/2}$ is undefined for $\beta \in \mathbb{R}_\pm$. We shall replace it by a more intelligible term for $\beta = 1$ by first expanding $Z(\beta)$ in the upper complex half plane and then formally setting $\beta = 1$. More precisely, let $\beta \in \mathbb{C}_+ = \{z \in \mathbb{C} \mid \Im z > 0\}$ and write $\beta = |\beta| e^{i\theta}$ with $\theta \in [0, \pi]$, hence $-\beta = |\beta| e^{i(\theta - \pi)}$ with $\theta - \pi \in [-\pi, 0]$. We set $\beta^a := |\beta|^a e^{i\theta a}$ and get

$$\begin{aligned} \beta^{-d_+/2}(-\beta)^{-d_-/2} &= (|\beta| e^{i\theta})^{-d_+/2} (|\beta| e^{i(\theta-\pi)})^{-d_-/2} \\ &= |\beta|^{-\zeta/2}\, e^{-i\frac{d_+}{2}\theta}\, e^{-i\frac{d_-}{2}\theta}\, e^{i\pi\frac{d_-}{2}}. \end{aligned}$$

Moreover,

$$-\frac{d_+}{2}\theta - \frac{d_-}{2}\theta + \pi\frac{d_-}{2} = -\frac{\theta}{2}(d_+ + d_-) + \frac{\pi}{2}\left(\frac{d_-}{2} + \frac{d_+}{2} + \frac{d_-}{2} - \frac{d_+}{2}\right)$$
$$= -\frac{\theta}{2}\zeta + \frac{\pi}{4}(\zeta - \eta) = -\frac{\pi}{4}\left(\frac{2\theta\zeta}{\pi} + (\eta - \zeta)\right),$$

where $\zeta := d_+ + d_-$ is the finite-dimensional equivalent of the ζ-invariant, counting the eigenvalues, and $\eta := d_+ - d_-$ is the finite-dimensional equivalent of the η-invariant, measuring the spectral asymmetry of T. We obtain

$$Z(\beta) = \pi^{\zeta/2} |\beta|^{-\zeta/2} e^{-i\frac{\pi}{4}(\frac{2\zeta\theta}{\pi} + (\eta - \zeta))} (\det |T|)^{-1/2}$$

and, formally, for $\beta = 1$, i.e., $\theta = 0$,

$$Z(1) = \pi^{\zeta/2} \underbrace{e^{-i\frac{\pi}{4}(\eta - \zeta)} (\det |T|)^{-1/2}}_{=: \det T} . \tag{3.4}$$

Equation (3.4) suggests a nonstandard definition of the determinant for the infinite-dimensional case.

REMARK 3.4.
(a) The methods and results of this section also apply to real-valued quadratic functionals on complex vector spaces. Since the integration in (3.3) in this case is over the real vector space underlying Γ, the expressions for the partition functions in this case become the square of those above.
(b) In the preceding calculations we worked with ordinary commuting numbers and functions. The resulting Gaussian integrals are also called *bosonic* integrals. If we consider *fermionic* integrals, we work with Grassmannian variables and obtain the determinant not in the denominator but in the nominator (see e.g. [**BeGeVe92**] or [**Be96**]).

3.2. The ζ-determinant for operators of infinite rank. Once again, our point of departure is finite-dimensional linear algebra. Let $T : \mathbb{C}^d \to \mathbb{C}^d$ be an invertible, positive operator with eigenvalues $0 < \lambda_1 \leq \lambda_2 \leq \cdots \leq \lambda_d$. We have the equality

$$\det T = \prod \lambda_j = \exp\left\{\sum \ln \lambda_j \, e^{-s\ln\lambda_j}\big|_{s=0}\right\}$$
$$= \exp\left(-\frac{d}{ds}\left(\sum \lambda_j^{-s}\right)\big|_{s=0}\right) = e^{-\frac{d}{ds}\zeta_T(s)|_{s=0}} ,$$

where $\zeta_T(s) := \sum_{j=1}^d \lambda_j^{-s}$.

We show that the preceding formula generalizes naturally, when T is replaced by a *positive definite self-adjoint elliptic* operator L (for the ease of presentation, of second order, like the Laplacian) acting on sections of a Hermitian vector bundle over a closed manifold M of dimension m. Then L has a discrete spectrum $\{\lambda_j\}_{j\in\mathbb{N}}$ with $0 < \lambda_1 \leq \lambda_2 \leq \cdots$, satisfying the asymptotic formula $\lambda_n \sim Cn^{m/2}$ for a constant $C > 0$ depending on L (see e.g. [**Gi95**], Lemma 1.6.3). We extend $\zeta_L(s) := \sum_{j=1}^{\infty} \lambda_j^{-s}$ in the complex plane by

$$\zeta_L(s) := \frac{1}{\Gamma(s)} \int_0^{\infty} t^{s-1} \operatorname{Tr} e^{-tL}\, dt$$

with $\Gamma(s) := \int_0^{\infty} t^{s-1} e^{-t} dt$. Note that e^{-tL} is the heat operator transforming any initial section f_0 into a section f_t satisfying the heat equation $\frac{\partial}{\partial t} f + Lf = 0$. Clearly $\operatorname{Tr} e^{-tL} = \sum e^{-t\lambda_j}$.

One shows that the original definition of $\zeta_L(s)$ yields a holomorphic function for $\Re(s)$ large and that its preceding extension is meromorphic in the entire complex plane with simple poles only. The point $s = 0$ is a regular point and $\zeta_L(s)$ is a holomorphic function at $s = 0$. From the asymptotic expansion of $\Gamma(s) \sim \frac{1}{s} + \gamma + sh(s)$ close to $s = 0$ with the Euler number γ and a suitable holomorphic function h we obtain an explicit formula

$$\zeta_L'(0) \sim \int_0^{\infty} \frac{1}{t} \operatorname{Tr} e^{-tL}\, dt - \gamma \zeta_L(0) .$$

This is explained in great detail, e.g., in [**Wo99**]. Therefore, Ray and Singer in [**RaSi71**] could introduce $\det_z(L)$ by defining:

$$\det{}_z L := e^{-\frac{d}{ds}\zeta_L(s)|_{s=0}} = e^{-\zeta_L'(0)} .$$

The preceding definition does not apply immediately to the main hero here, the Dirac operator $\mathcal{D}$, which has infinitely many positive λ_j and negative eigenvalues $-\mu_j$. Clearly by the preceding argument

$$\det{}_z \mathcal{D}^2 = e^{-\zeta_{\mathcal{D}^2}'} \quad \text{and} \quad \det{}_z |\mathcal{D}| = e^{-\zeta_{|\mathcal{D}|}'} = e^{-\frac{1}{2}\zeta_{\mathcal{D}^2}'} .$$

For the Dirac operator we set

$$\ln \det \mathcal{D} := -\frac{d}{ds} \zeta_{\mathcal{D}}(s)|_{s=0}$$

with, choosing[2] the branch $(-1)^{-s} = e^{i\pi s}$,

$$\begin{aligned} \zeta_{\mathcal{D}}(s) &= \sum \lambda_j^{-s} + \sum (-1)^{-s} \mu_j^{-s} = \sum \lambda_j^{-s} + e^{i\pi s} \sum \mu_j^{-s} \\ &= \frac{1}{2}\Big(\sum \lambda_j^{-s} + \sum \mu_j^{-s}\Big) + \frac{1}{2}\Big(\sum \lambda_j^{-s} - \sum \mu_j^{-s}\Big) \\ &\quad + e^{i\pi s} \left\{ \frac{1}{2}\Big(\sum \lambda_j^{-s} + \sum \mu_j^{-s}\Big) - \frac{1}{2}\Big(\sum \lambda_j^{-s} - \sum \mu_j^{-s}\Big) \right\} \\ &= \frac{1}{2} \left\{ \zeta_{\mathcal{D}^2}(\tfrac{s}{2}) + \eta_{\mathcal{D}}(s) \right\} + \frac{1}{2} e^{i\pi s} \left\{ \zeta_{\mathcal{D}^2}(\tfrac{s}{2}) - \eta_{\mathcal{D}}(s) \right\} , \end{aligned}$$

---

[2]Choosing the alternative representation, namely $(-1)^{-s} = e^{-i\pi s}$ yields the opposite sign of the phase of the determinant which may appear to be more natural for some quantum field models and also in view of (3.4). However, we follow the more common convention introduced by Singer in [Si85, p. 331] when defining the determinant of operators of Dirac type.

where $\eta_{\mathcal{D}}(s) := \sum \lambda_j^{-s} - \sum \mu_j^{-s}$. Later we will show that $\eta_{\mathcal{D}}(s)$ is a holomorphic function of s for $\Re(s)$ large with a meromorphic extension to the whole complex plane which is holomorphic in the neighborhood of $s = 0$. We obtain

$$\zeta'_{\mathcal{D}}(s) = \tfrac{1}{4}\zeta'_{\mathcal{D}^2}(\tfrac{s}{2}) + \tfrac{1}{2}\eta'_{\mathcal{D}}(s) + \tfrac{1}{2}i\pi e^{i\pi s}\{\zeta_{\mathcal{D}^2}(\tfrac{s}{2}) - \eta_{\mathcal{D}}(s)\} + \tfrac{1}{2}e^{i\pi s}\{\tfrac{1}{2}\zeta'_{\mathcal{D}^2}(\tfrac{s}{2}) - \eta'_{\mathcal{D}}(s)\}.$$

It follows that

$$\zeta'_{\mathcal{D}}(0) = \tfrac{1}{2}\zeta'_{\mathcal{D}^2}(0) + \tfrac{i\pi}{2}\left\{\zeta_{\mathcal{D}^2}(0) - \eta_{\mathcal{D}}(0)\right\},$$

and

$$\begin{aligned}\det{}_z\mathcal{D} &= e^{-\frac{1}{2}\zeta'_{\mathcal{D}^2}(0)}\, e^{-\frac{i\pi}{2}\{\zeta_{\mathcal{D}^2}(0)-\eta_{\mathcal{D}}(0)\}}\\ &= e^{-\frac{i\pi}{2}\{\zeta_{|\mathcal{D}|}(0)-\eta_{\mathcal{D}}(0)\}}\, e^{-\zeta'_{|\mathcal{D}|}(0)}\\ &= e^{-\frac{i\pi}{2}\{\zeta_{|\mathcal{D}|}(0)-\eta_{\mathcal{D}}(0)\}}\det{}_z|\mathcal{D}|\,.\end{aligned}$$

So, the Dirac operator's 'partition function' in the sense of (3.3) becomes

$$Z(1) = \pi^{\zeta_{|\mathcal{D}|}(0)}(\det{}_\zeta\mathcal{D})^{-\frac{1}{2}}\,.$$

3.3. Spectral invariants of different 'sensitivity'. We discuss the four spectral invariants of the Dirac operator $\mathcal{D}$ that enter in the preceding formulas.

3.3.1. *The index.* First recall that the index of arbitrary elliptic operators on closed manifolds and the spectral flow of 1-parameter families of self-adjoint elliptic operators are topological invariants and so stable under small variation of the coefficients and, by definition, solely depending on the multiplicity of the eigenvalue 0. In the theory of bounded or closed (not necessarily bounded) Fredholm operators and bounded or not necessarily bounded self-adjoint Fredholm operators and the related K and K^{-1} theory, we have a powerful functional analytical and topological frame for discussing these invariants. Moreover, index and spectral flow are local invariants, i.e., can be expressed by an integral where the integrand is locally expressed by the coefficients of the operator(s). Consequently, we have simple, precise pasting formulas for index and spectral flow on partitioned manifolds where the error term is localized along the separating hypersurface.

On manifolds with boundary the Calderón projection and its range, the Cauchy data space, change continuously when we vary the Dirac operator as shown in Section 2.3.8, exploiting the unique continuation property of operators of Dirac type. The same is not true for the Atiyah-Patodi-Singer projection: It can jump from one connected component of the Grassmannian to another component under small changes of the Dirac operator. Correspondingly, the index of the APS boundary problem can jump under small variation of the coefficients (i.e., of the defining connection or the underlying Riemannian or Clifford structure).

Regarding parity of the manifold, the index density (of all elliptic differential operators) vanishes on odd-dimensional manifolds for symmetry reason. Then, in the closed case the index vanishes, and on manifolds with boundary the APS Index Theorem takes the simple form $\operatorname{index}\mathcal{D}_{P_{\geq}} = -\dim\operatorname{Ker}B^+$ where we have the total Dirac operator on the left and a chiral component of the induced tangential operator on the right. Once again, the formula shows the instability of the index under small changes of the Dirac operator. This is no contradiction to the stability of the index on the spaces $\mathcal{F}$, respectively $\mathcal{CF}$, discussed in Section 1.1.3 because the graph norm distance between two APS realizations $\mathcal{D}_{P_{\geq}}$ and $\mathcal{D}'_{P_{\geq}}$ can remain bounded away from zero when $\mathcal{D}$ runs to $\mathcal{D}'$. This is the case if and only if the dimension of the kernel of the tangential operator changes under the deformation.

3.3.2. *The ζ-invariant.* Similarly, $\zeta_{\mathcal{D}^2}(0)$ is also local; i.e., it is given by the integral $\int_M \alpha(x)dx$, where $\alpha(x)$ denotes a certain coefficient in the heat kernel expansion and is locally expressed by the coefficients of $\mathcal{D}$. In particular, $\zeta_{\mathcal{D}^2}(0)$ remains unchanged for small changes of the spectrum. Actually, $\zeta_L(0)$ vanishes when L is the square of a self-adjoint elliptic operator on a closed manifold of odd dimension. It can be defined (and it vanishes, see [**PaWo02a**, Appendix]) for a large class of squares of operators of Dirac type with globally elliptic boundary conditions on compact, smooth manifolds (of odd dimension) with boundary. So, there are no nontrivial pasting formulas at all in such cases.

3.3.3. *The η-invariant.* Unlike the index and spectral flow on closed manifolds, we have neither an established functional analytical nor a topological frame for discussing $\eta_{\mathcal{D}}(0)$, nor is it given by an integral of a locally defined expression. On the circle, e.g., consider the operator

$$\mathcal{D}_a := -i\frac{d}{dx} + a = e^{-ixa}\mathcal{D}_0 e^{ixa}, \tag{3.5}$$

so that $\mathcal{D}_a$ and $\mathcal{D}_0$ have the same total symbol (coincide locally), but $\eta_{\mathcal{D}_a}(0) = -2a$ depends on a.

The η-invariant depends, however, only on finitely many terms of the symbol of the resolvent $(\mathcal{D}-\lambda)^{-1}$ and the real part (in $\mathbb{R}/\mathbb{Z}$) will not change when one changes or removes a finite number of eigenvalues. The integer part changes according to the net sign change occurring under removing or modifying eigenvalues. Moreover, the first derivative of the eta invariant of a smooth 1-parameter family of Dirac type operators is local, namely the spectral flow, as noted in our introduction. This leads again to precise (though not so simple) pasting formulas for the eta invariant on partitioned manifolds.

In even dimensions, the eta invariant vanishes on any closed manifold Σ for any Dirac type operator which is the tangential operator of a Dirac type operator on a suitable manifold which has Σ as its boundary because of the induced precise symmetry of the spectrum due to the anti-commutativity of the tangential operator with Clifford multiplication. For the study of eta of boundary value problems on even-dimensional manifolds see [**KlWo96**].

3.3.4. *The modulus of the determinant.* The number $\zeta'_{\mathcal{D}^2}(0)$ is the most delicate of the invariants involved: It is neither a local invariant, nor does it depend only on the total symbol of the Dirac operator. Even small changes of the eigenvalues will change the ζ' invariant and hence the determinant. Moreover, no precise pasting formulas are obtained but only adiabatic ones (i.e., by inserting a long cylinder around the separating hypersurface (see [**PaWo02a**], [**PaWo02b**], and [**PaWo02c**]).

Without proof we present the main results by Wojciechowski and collaborators, based on [**Wo99**] where the ζ-function regularized determinant was established for pseudo-differential boundary value conditions belonging to the smooth, self-adjoint Grassmannian. The first is a boundary correction formula, proved in [**ScWo00**] (see also the recent [**Sco02**]):

THEOREM 3.5 (Scott, Wojciechowski). *Let $\mathcal{D}$ be a Dirac operator over an odd-dimensional compact manifold M with boundary Σ and let $P \in \mathcal{G}r^{\mathrm{sa}}(\mathcal{D})$. Then the range of the Calderón projection $\mathcal{P}(\mathcal{D})$ (the Cauchy data space $\Lambda(\mathcal{D}, \frac{1}{2})$) and the range of P can be written as the graphs of unitary, elliptic, zero order operators K and T, respectively, which differ from the operator*

$$(B^+B^-)^{-1/2}B^+ : C^\infty(\Sigma; S^+|_\Sigma) \to C^\infty(\Sigma; S^-|_\Sigma)$$

by a smoothing operator. Moreover,

$$\det{}_\zeta \mathcal{D}_P = \det{}_\zeta \mathcal{D}_{\mathcal{P}(\mathcal{D})} \cdot \det{}_{\mathrm{Fr}} \tfrac{1}{2}(I + KT^{-1})\,. \tag{3.6}$$

The second result, in most simple form, is found in [**PaWo00**]:

THEOREM 3.6 (Park, Wojciechowski). *Let $R \in \mathbb{R}$ be positive, let M^R denote the stretched partitioned manifold $M^R = M_1 \cup_\Sigma [-R, 0] \times \Sigma \cup_\Sigma [0, R] \times \Sigma \cup_\Sigma M_2$, and let $\mathcal{D}_R$, $\mathcal{D}_{1,R}$, and $\mathcal{D}_{2,R}$ denote the corresponding Dirac operators. We assume that the tangential operator $\mathcal{B}$ is invertible. Then*

$$\lim_{R\to\infty} \frac{\det{}_\zeta \mathcal{D}_R^2}{\left(\det{}_\zeta (\mathcal{D}_{1,R})^2_{I-P_>}\right)\left(\det{}_\zeta (\mathcal{D}_{2,R})^2_{P_>}\right)} = 2^{-\zeta_{\mathcal{B}^2}(0)}.$$

Although Felix Klein in [**Kl27**] rated the determinant as *simplest example of an invariant*, today we must give an inverse rating. For the present authors, it is not the invariants that are stable under the largest transformation groups which deserve the highest interest, but rather (according to Dirac's approach to elementary particle physics) the *finest* invariants which exhibit *anomalies* under small perturbations. Correspondingly, the determinant and its amplitude are the most subtle and the most fascinating objects of our study. They are much more difficult to comprehend than the η-invariant 3.3.2; and the η-invariant is much more difficult to comprehend than the index.

3.4. Pasting formulas for the eta invariant – outlines. In the rest of this review, i.e. over the next 33 pages, we will prove a strikingly simple (to state) additivity property of the η-invariant. We fix the assumptions and the notation.

3.4.1. *Assumptions and notation.* **(a)** Let M be an odd-dimensional closed *partitioned* Riemannian manifold $M = M_1 \cup_\Sigma M_2$ with M_1, M_2 compact manifolds with common boundary Σ. Let S be a bundle of Clifford modules over M.

(b) To begin with we assume that $\mathcal{D}$ is a *compatible* Dirac operator over M. Thus, in particular, $\mathcal{D}$ is symmetric and has a unique self-adjoint extension in $L^2(M; S)$.

(c) We assume that there exists a bicollared cylindrical neighborhood (*a neck*) $N \simeq (-1,1) \times \Sigma$ of the separating hypersurface Σ, such that the Riemannian structure on M and the Hermitian structure on S are product in N; i.e., they do not depend on the normal coordinate u, when restricted to $\Sigma_u = \{u\} \times \Sigma$. Our convention for the orientation of the coordinate u is that it runs from M_1 to M_2; i.e., $M_1 \cap N = (-1, 0] \times \Sigma$ and $N \cap M_2 = [0,1) \times \Sigma$. Then the operator $\mathcal{D}$ takes the following form on N:

$$\mathcal{D}|_N = \sigma(\partial_u + B), \tag{3.7}$$

where the principal symbol in u-direction $\sigma : S|_\Sigma \to S|_\Sigma$ is a unitary bundle isomorphism (Clifford multiplication by the normal vector du) and the tangential operator $B : C^\infty(\Sigma; S|_\Sigma) \to C^\infty(\Sigma; S|_\Sigma)$ is the corresponding Dirac operator on Σ. Note that σ and B do not depend on the normal coordinate u in N and they satisfy the following identities

$$\sigma^2 = -I, \quad \sigma^* = -\sigma, \quad \sigma \cdot B = -B \cdot \sigma, \quad B^* = B. \tag{3.8}$$

Hence, σ is a skew-adjoint involution and S, the bundle of spinors, decomposes in N into $\pm i$-eigenspaces of σ, $S|_N = S^+ \oplus S^-$. It follows that (3.7) leads to the following representation of the operator $\mathcal{D}$ in N

$$\mathcal{D}|_N = \begin{pmatrix} i & 0 \\ 0 & -i \end{pmatrix} \cdot \left(\partial_u + \begin{pmatrix} 0 & B_- \\ B_+ & 0 \end{pmatrix} \right), \quad B_- = B_+^*,$$

where $B_+ : C^\infty(\Sigma; S^+) \to C^\infty(\Sigma; S^-)$ maps the spinors of positive chirality into the spinors of negative chirality.

(d) To begin with we consider only the case of $\operatorname{Ker} B = \{0\}$. That implies that B is an *invertible* operator. More precisely, there exists a pseudo-differential elliptic operator L of order -1 such that $BL = I_S = LB$ (see, e.g., [**BoWo93**, Prop. 9.5]).

(e) For real $R > 0$ we study the closed *stretched* manifold M^R which we obtain from M by inserting a cylinder of length $2R$, i.e., replacing the collar N by the cylinder $(-2R-1, +1) \times \Sigma$

$$M^R = M_1 \cup ([-2R, 0] \times \Sigma) \cup M_2 .$$

We extend the bundle S to the stretched manifold M^R in a natural way. The extended bundle will be also denoted by S. The Riemannian structure on M and the Hermitian structure on S are product on N. Hence we can extend them to smooth metrics on M^R in a natural way and, at the end, we can extend the operator $\mathcal{D}$ to an operator $\mathcal{D}^R$ on M^R by using formula (3.7). Then M^R splits into two manifolds with boundary: $M^R = M_1^R \cup M_2^R$ with

$$M_1^R = M_1 \cup ((-2R, -R] \times \Sigma),$$
$$M_2^R = ([-R, 0) \times \Sigma) \cup M_2,$$

and $\partial M_1^R = \partial M_2^R = \{-R\} \times \Sigma$. Consequently, the operator $\mathcal{D}^R$ splits into

$$\mathcal{D}^R = \mathcal{D}_1^R \cup \mathcal{D}_2^R.$$

We will impose spectral boundary conditions to obtain self-adjoint operators $\mathcal{D}_{1,P_<}$, $\mathcal{D}_{1,P_<}^R$, $\mathcal{D}_{2,P_>}$, and $\mathcal{D}_{2,P_>}^R$ in the L^2 spaces on the corresponding parts (see (3.9)).

(f) We also introduce the complete, noncompact Riemannian manifold *with cylindrical end*

$$M_2^\infty := ((-\infty, 0] \times \Sigma) \cup M_2$$

by gluing the half-cylinder $(-\infty, 0] \times \Sigma$ to the boundary Σ of M_2. Clearly, the Dirac operator $\mathcal{D}$ extends also to $C^\infty(M_2^\infty, S)$.

REMARK 3.7. Our presentation is somewhat simplified by our assumption (b) that $\mathcal{D}$ is compatible, and the assumption (d) that the tangential operator B is invertible. Both assumptions can be lifted, see [**Wo95**] and [**Wo99**].

We recall the following ideas in the big scheme from Section 2.5 of this review. Let $P_>$ (respectively $P_<$) denote the spectral projection of B onto the subspace of $L^2(\Sigma; S|_\Sigma)$ spanned by the eigensections corresponding to the positive (respectively negative) eigenvalues. Then $P_>$ is a self-adjoint elliptic boundary condition for the operator $\mathcal{D}_2 = \mathcal{D}|_{M_2}$ (see [**BoWo93**, Proposition 20.3]). This means that the operator $\mathcal{D}_{2,P_>}$ defined by

$$\begin{aligned} \mathcal{D}_{2,P_>} &= \mathcal{D}|_{M_2}, \\ \operatorname{Dom}(\mathcal{D}_{2,P_>}) &= \{s \in H^1(M_2; S|_{M_2}) \mid P_>(s|_\Sigma) = 0\}, \end{aligned} \tag{3.9}$$

is an unbounded self-adjoint operator in $L^2(M_2; S|_{M_2})$ with compact resolvent. In particular,

$$\mathcal{D}_{2,P_>} : \operatorname{Dom}(\mathcal{D}_{2,P_>}) \to L^2(M_2; S|_{M_2})$$

is a Fredholm operator with discrete real spectrum and the kernel of $\mathcal{D}_{2,P_>}$ consists of smooth sections of $S|_{M_2}$.

As mentioned before, the eta function of $\mathcal{D}_{2,P_>}$ is well defined and enjoys all properties of the eta function of the Dirac operator defined on a closed manifold. In particular, $\eta_{\mathcal{D}_{2,P_>}}(0)$, the eta invariant of $\mathcal{D}_{2,P_>}$, is well defined. Similarly, $P_<$ is a self-adjoint boundary condition for the operator $\mathcal{D}|_{M_1}$, and we define the operator $\mathcal{D}_{1,P_<}$ using a formula corresponding to (3.9).

To keep track of the various manifolds, operators, and integral kernels we refer to the following table where we have collected the major notations.

Manifolds	Operators	Integral kernels
$M = M_1 \cup_\Sigma M_2$	$e^{-t\mathcal{D}^2}$, $\mathcal{D}e^{-t\mathcal{D}^2}$	$\mathcal{E}(t;x,x')$
$M^R = M_1^R \cup_\Sigma M_2^R$	$e^{-t(\mathcal{D}^R)^2}$, $\mathcal{D}^R e^{-t(\mathcal{D}^R)^2}$	$\mathcal{E}^R(t;x,x')$
M_2	$e^{-t\mathcal{D}_{2,P_>}^2}$, $\mathcal{D}_2 e^{-t\mathcal{D}_{2,P_>}^2}$	$\mathcal{E}_2(t;x,x')$
$M_2^R = ([-R,0]\times\Sigma)\cup M_2$	$e^{-t(\mathcal{D}_{2,P_>}^R)^2}$, $\mathcal{D}_2^R e^{-t(\mathcal{D}_{2,P_>}^R)^2}$	$\mathcal{E}_2^R(t;x,x')$
$M_2^\infty = ((-\infty,0]\times\Sigma)\cup M_2$	$e^{-t(\mathcal{D}_2^\infty)^2}$, $\mathcal{D}_2^\infty e^{-t(\mathcal{D}_2^\infty)^2}$	$\mathcal{E}_2^\infty(t;x,x')$
$\Sigma_{\mathrm{cyl}}^\infty = (-\infty,+\infty)\times\Sigma$	$e^{-tD_{\mathrm{cyl}}^2}$, $D_{\mathrm{cyl}}e^{-tD_{\mathrm{cyl}}^2}$	$\mathcal{E}_{\mathrm{cyl}}(t;x,x')$
$\Sigma_{\mathrm{cyl}/2}^\infty = [0,+\infty)\times\Sigma$	$e^{-tD_{\mathrm{aps}}^2}$, $D_{\mathrm{aps}}e^{-tD_{\mathrm{aps}}^2}$	$\mathcal{E}_{\mathrm{aps}}(t;x,x')$

In addition, on M_2^R we have the operator $Q_2^R(t)$ with kernel $Q_2^R(t;x,x')$, and

$$C^R(t) = \left((\mathcal{D}_{2,P_>}^R)^2 + \frac{d}{dt}\right) Q_2^R(t) \text{ with kernel } C^R(t;x,x').$$

3.4.2. *The gluing formulas.* The most basic results for pasting η are the following theorem on the adiabatic limits of the η invariants and its additivity corollary:

THEOREM 3.8. *Attaching a cylinder of length $R > 0$ at the boundary of the manifold M_2, we can approximate the eta invariant of the spectral boundary condition on the prolonged manifold M_2^R by the corresponding integral of the 'local' eta function of the closed stretched manifold M^R:*

$$\lim_{R\to\infty}\left\{\eta_{\mathcal{D}_{2,P_>}^R}(0) - \int_{M_2^R}\eta_{\mathcal{D}^R}(0;x)\,dx\right\} \equiv 0 \mod \mathbb{Z}.$$

COROLLARY 3.9. $\eta_{\mathcal{D}}(0) \equiv \eta_{\mathcal{D}_{1,P_<}}(0) + \eta_{\mathcal{D}_{2,P_>}}(0) \mod \mathbb{Z}$.

REMARK 3.10. (a) With hindsight, it is not surprising that modulo the integers the preceding additivity formula for the η-invariant on a partitioned manifold is precise. An intuitive argument runs as follows. "Almost all" eigensections and eigenvalues of the operator $\mathcal{D}$ on the closed partitioned manifold $M = M_1 \cup M_2$ can be traced back, either to eigensections $\psi_{1,k}$ and eigenvalues $\mu_{1,k}$ of the spectral boundary problem $\mathcal{D}_{1,P_<}$ on the part M_1, or to eigensections $\psi_{2,\ell}$ and eigenvalues $\mu_{2,\ell}$ of the spectral boundary problem $\mathcal{D}_{2,P_>}$ on the part M_2. While we have no explicit exact correspondence, due to the product form of the Dirac operator in a neighborhood of the separating hypersurface, eigensections on one part M_1 or M_2 of the manifold M can be extended to smooth sections on the whole of M. These are not true eigensections of $\mathcal{D}$, but they have a relative error which is rapidly decreasing as $R \to \infty$ when we attach cylinders of length R to the part manifolds or, equivalently, insert a cylinder of length $2R$ in M. There is also a residual set $\{\mu_{0,j}\}$ of eigenvalues of $\mathcal{D}$ which can neither be traced back to eigenvalues of

$\mathcal{D}_{1,P_<}$ nor to those of $\mathcal{D}_{2,P_>}$. These eigenvalues can, however, be traced back to the kernel of the Dirac operators $\mathcal{D}_1^\infty$ and $\mathcal{D}_2^\infty$ on the part manifolds M_1^∞ and M_2^∞ with cylindrical ends. Because of Fredholm properties the residual set is finite and, hence (as noticed in Section 3.3) can be discarded for calculating the eta invariant modulo $\mathbb{Z}$.

Therefore, no R (i.e., no prolongation of the bicollar neighborhood N) enters the *formula*. Nevertheless, our *arguments* rely on an adiabatic argument to separate the spectrum of $\mathcal{D}$ into its three parts

$$\operatorname{spec}\mathcal{D} \sim \{\mu_{0,j}\} \cup \{\mu_{1,k}\} \cup \{\mu_{2,\ell}\}. \tag{3.10}$$

For the most part, however, we need not make all arguments explicit on the level of the single eigenvalue. It suffices to work on the level of the eta invariant for the following reason. Unlike the index, the eta invariant cannot be described by a local formula, as explained in Section 3.3. Nevertheless, it can be described by an integral over the manifold. The integrand, however, is not defined in local terms solely. In particular, when writing the eta function in integral form and decomposing the η integral

$$\eta_{\mathcal{D}}(s) = \int_M \eta_{\mathcal{D}}(s;x,x)\,dx = \int_{M_1} \eta_{\mathcal{D}}(s;x_1,x_1)\,dx_1 + \int_{M_2} \eta_{\mathcal{D}}(s;x_2,x_2)\,dx_2$$

there is no geometrical interpretation of the integrals on the right over the two parts of the manifold. This is very unfortunate. But for sufficiently large R, the integrals become intelligible and can be read as the η invariants of $\mathcal{D}^R_{1,P_<}$ and $\mathcal{D}^R_{2,P_>}$. That is the meaning of the adiabatic limit.

(b) Theorem 3.41 can be generalized to larger classes of boundary conditions by variational argument (see [**LeWo96**]) yielding the general gluing formula (in $\mathbb{R}/\mathbb{Z}$)

$$\eta_{\mathcal{D}}(0) = \eta_{\mathcal{D}_{1,P_1}}(0) + \eta_{\mathcal{D}_{2,P_2}}(0) + \eta^N_{P_1,I-P_2}(0),$$

where $\eta^N_{P_1,I-P_2}(0)$ denotes the η-invariant on the cylinder N, see also [**BrLe99**], [**DaFr94**], [**MaMe95**], and the recent review [**PaWo02e**]. G. Grubb's new result of [**Gr02**], mentioned in our introduction, may open an alternative route for proving the general gluing formula. The integer jump was calculated in [**KiLe00**] yielding (among other formulas)

$$\eta_{\mathcal{D}}(0) = \eta_{\mathcal{D}_{1,I-P}}(0) + \eta_{\mathcal{D}_{2,P}}(0) + 2\,\mathrm{SF}\{\mathcal{D}_{1,I-P_t}\} + 2\,\mathrm{SF}\{\mathcal{D}_{2,P_t}\},$$

where $\{P_t\}$ is a smooth curve in the Grassmannian from P to the Calderón projection $\mathcal{P}(\mathcal{D}_2)$.

(c) An interesting feature of [**KiLe00**] is that the pasting formula is derived from the Scott-Wojciechowski Comparison Formula (Theorem 3.5). This is quite analogous to the existence of two completely different proofs of the pasting formula for the index (Theorem 2.35), where also the derivation from the boundary reduction formula is much, much shorter than arguing via the Atiyah-Singer Index Theorem and the Atiyah-Patodi-Singer Index Theorem plus the Agranovic-Dynin Formula.

However, for this review we prefer the long way because of the many interesting insights about the gluing on the eigenvalue level which can be gained.

3.4.3. *Plan of the proof.* Let $\mathcal{E}_2^R(t)$ denote the integral kernel of the operator $\mathcal{D}_2^R\, e^{-t(\mathcal{D}_{2,P_>}^R)^2}$ defined on the manifold $M_2^R = ([-R,0]\times\Sigma)\cup M_2$. Without proof, we mentioned before that the eta invariant of the self-adjoint operator $\mathcal{D}_{2,P_>}^R$ is well defined and we have

$$\eta_{\mathcal{D}_{2,P_>}^R}(0) = \frac{1}{\sqrt{\pi}}\int_0^\infty \frac{dt}{\sqrt{t}}\int_{M_2^R} \operatorname{Tr}\mathcal{E}_2^R(t;x,x)\,dx$$

$$= \frac{1}{\sqrt{\pi}}\int_0^{\sqrt{R}} \frac{dt}{\sqrt{t}}\int_{M_2^R} \operatorname{Tr}\mathcal{E}_2^R(t;x,x)\,dx \tag{3.11}$$

$$+ \frac{1}{\sqrt{\pi}}\int_{\sqrt{R}}^\infty \frac{dt}{\sqrt{t}}\int_{M_2^R} \operatorname{Tr}\mathcal{E}_2^R(t;x,x)\,dx. \tag{3.12}$$

We first deal with the integral of (3.11) and show that it splits into an interior contribution and a cylinder contribution as $R\to\infty$. This will be done in Section 3.5 by first in 3.5.1 applying the Duhamel method which we introduced before in the proof of Theorem 2.30 (pp. 63ff). By Lemma 3.11, Lemma 3.12, Proposition 3.13, and Corollary 3.14, we can replace the heat kernel $\mathcal{E}_2^R$ of the Atiyah-Patodi-Singer boundary problem on the prolonged manifold M_2^R with boundary Σ by an artificially glued integral kernel Q_2^R which, near the boundary, is equal to the heat kernel of the APS problem on the half-infinite cylinder and in the interior equal to the heat kernel on the stretched closed manifold. We can do it in such a way that the original small-t integral (i.e., the integral from 0 to $\sqrt{R}$ in (3.11)) can be approximated by the new integral up to an error of order $O(e^{-cR})$. Then we will show in Lemma 3.15 and Lemma 3.16 (Paragraph 3.5.2) that the cylinder contribution is traceless. More precisely, we obtain that the trace $\operatorname{Tr} Q_2^R(T;x,x)$ can be replaced pointwise (for $x\in M_2^R$) by the trace $\operatorname{Tr}\mathcal{E}^R(t;x,x)$ of the integral kernel of the operator $\mathcal{D}^R e^{-t(\mathcal{D}^R)^2}$ which is defined on the stretched manifold M^R. Consequently, the small-t *chopped* η-invariant of the closed stretched manifold M^R coincides with the sum of the small-t chopped η-invariants of the APS problems on the two prolonged manifolds M_1^R and M_2^R up to $O\left(e^{-cR}\right)$.

Then we will show that the integral of (3.12) vanishes as $R\to\infty$. This is in Lemma 3.29 (p. 114f) a direct consequence of Theorem 3.17 which states that the eigenvalues of $\mathcal{D}_{2,P_>}^R$ are uniformly bounded away from 0. Theorem 3.17 is of independent interest. The proof is a long story stretching over Sections 3.6 and 3.7. First, in Paragraph 3.6.1 we will consider the operator D_{cyl} on the infinite cylinder $\Sigma_{\text{cyl}}^\infty$ in Definition 3.19 and Lemma 3.20. We obtain that D_{cyl} has no eigenvalues in the interval $(-\lambda_1,\lambda_1)$ where λ_1 denotes the smallest positive eigenvalue of the tangential operator B on the manifold Σ. Next, in Paragraph 3.6.2 we investigate the operator $\mathcal{D}_2^\infty$ on the manifold M_2^∞ with infinite cylindrical end in Lemma 3.21, Lemma 3.22, Lemma 3.23, and Proposition 3.24. Proposition 3.24 states that $\mathcal{D}_2^\infty$ has only finitely many eigenvalues in the aforementioned interval $(-\lambda_1,\lambda_1)$, each of finite multiplicity. Its proof is somewhat delicate and involves Lemma 3.26 and Corollary 3.27. Then, in Section 3.7 the proof of Theorem 3.17 follows with Lemma 3.28.

By then we will have that the sum of the η-invariants of the APS boundary problems on the prolonged manifolds M_1^R and M_2^R can be replaced by the small-t chopped η-invariant on the stretched closed manifold M^R with an error exponentially vanishing as $R \to \infty$. We then would like to repeat the preceding chain of arguments around Theorem 3.17 to show that the large-t chopped η-invariant of the operator $\mathcal{D}^R$ on M^R also vanishes. However, this can be done only in $\mathbb{R}/\mathbb{Z}$ because in general the eigenvalues of $\mathcal{D}^R$ are not bounded away from 0. So we shall present a different chain of arguments in Section 3.8. The main technical result is Theorem 3.30, once again of independent interest. It describes a partition of the eigenvalues into two subsets, the exponentially small ones and the eigenvalues bounded away from 0. The key for our arguments is a gluing construction (Definition 3.31). Then we first show by Lemma 3.32, Lemma 3.33 and Proposition 3.34 that $\mathcal{D}^R$ has at least $q := \dim \operatorname{Ker} \mathcal{D}_1^\infty + \dim \operatorname{Ker} \mathcal{D}_2^\infty$ exponentially small eigenvalues belonging to eigensections which we can approximate by pasting together L^2 solutions. Then we shall show in Lemma 3.35, Theorem 3.36, Lemma 3.37 and Proposition 3.38 that this makes the list of eigenvalues approaching 0 as $R \to \infty$ complete.

It follows in Lemma 3.39 (at the beginning of the closing Section 3.7) that the large-t chopped η-invariant on M^R vanishes asymptotically up to an integer error. This establishes Theorem 3.8. To arrive at Corollary 3.9, we must get rid of the adiabatic limit. This is a simple consequence of the locality of the derivative in R-direction of the η-invariants on the stretched part manifolds with APS boundary condition and on the stretched closed manifold (Proposition 3.40).

3.5. The adiabatic additivity of the small-t chopped η-invariant.

3.5.1. *Applying Duhamel's method to the small-t chopped η-invariant.* The simplest construction of a parametrix for $\mathcal{E}_2^R(t)$ (i.e., of an approximate heat kernel) is the following: we glue the kernel $\mathcal{E}$ of the operator $\mathcal{D}e^{-t\mathcal{D}^2}$ (given on the whole, closed manifold M) and the kernel $\mathcal{E}_{\text{aps}}^\infty$ of the L^2 extension of the operator $\sigma(\partial_u + B)e^{-t(\sigma(\partial_u+B))^2}$, given on the semi-infinite cylinder $[-R,\infty)\times\Sigma$ and subject to the Atiyah-Patodi-Singer boundary condition at the end $u=-R$. In that construction the gluing happens on the neck $N = [0,1)\times\Sigma$ with suitable cutoff functions.

Locally, the heat kernel is always of the form $(4\pi t)^{-m/2} e^{c_1 t} e^{-|x-x'|^2/4t}$. By Duhamel's Principle we get after gluing a similar global result for the kernel $\mathrm{e}_2^R(t;x,x')$ of the operator $e^{-t(\mathcal{D}_{2,P_>}^R)^2}$ and, putting a factor $t^{-1/2}$ in front, for the kernel of the combined operator $\mathcal{D}e^{-t(\mathcal{D}_{2,P_>}^R)^2}$ (e.g., see [**Gi95**, Lemma 1.9.1]). That yields two crucial estimates:

LEMMA 3.11. *There exist positive reals c_1, c_2, and c_3, which do not depend on R, such that for all $x, x' \in M_2^R$, and any $t > 0$ and $R > 0$,*

$$|\mathrm{e}_2^R(t;x,x')| \leq c_1\, t^{-\frac{m}{2}} e^{c_2 t}\, e^{-c_3 \frac{d^2(x,x')}{t}}, \tag{3.13}$$

$$|\mathcal{E}_2^R(t;x,x')| \leq c_1\, t^{-\frac{1+m}{2}} e^{c_2 t}\, e^{-c_3 \frac{d^2(x,x')}{t}}. \tag{3.14}$$

Here $d(x,x')$ denotes the geodesic distance.

Notice that exactly the same type of estimate is also valid for the kernel $\mathcal{E}^R(t;x,x')$ on the stretched closed manifold M^R and for the kernel $\mathcal{E}_{\mathrm{aps}}^\infty(t;x,x')$ on the infinite cylinder. For details see also [**BoWo93**, Theorem 22.14]. There, however, the term $e^{c_2 t}$ was suppressed in the final formula because the emphasis was on small time asymptotics.

As mentioned before, as $R \to \infty$, we want to separate the contribution to the kernel $\mathcal{E}_2^R$ which comes from the cylinder and the contribution from the interior by a gluing process. Unfortunately, the inequality 3.14 does not suffice to show that the contribution to the eta invariant, more precisely to the integral (3.11), which comes from the 'error' term vanishes with $R \to \infty$. Therefore, we introduce a different parametrix for the kernel $\mathcal{E}_2^R$.

Instead of gluing over the fixed neck $N = [0,1) \times \Sigma$, we glue over a segment N^R of growing length of the attached cylinder, say $N^R := (-\frac{4}{7}R, -\frac{3}{7}R) \times \Sigma$ (the reason for choosing these ratios will be clear soon). Thus, we choose a smooth partition of unity $\{\chi_{\mathrm{aps}}, \chi_{\mathrm{int}}\}$ on M_2^R suitable for the covering $\{U_{\mathrm{aps}}, U_{\mathrm{int}}\}$ with $U_{\mathrm{aps}} := [-R, -\frac{3}{7}R) \times \Sigma$ and $U_{\mathrm{int}} := \big((-\frac{4}{7}R, 0] \times \Sigma\big) \cup M_2$, hence $U_{\mathrm{aps}} \cap U_{\mathrm{int}} = N^R$. Moreover, we choose nonnegative smooth cutoff functions $\{\psi_{\mathrm{aps}}, \psi_{\mathrm{int}}\}$ such that

$$\psi_j \equiv 1 \ \text{ on } \left\{x \in M_2^R \mid \operatorname{dist}(x, \operatorname{supp}\chi_j) < \tfrac{1}{7}R\right\} \text{ and}$$
$$\psi_j \equiv 0 \ \text{ on } \left\{x \in M_2^R \mid \operatorname{dist}(x, \operatorname{supp}\chi_j) \geq \tfrac{2}{7}R\right\}$$

for $j \in \{\mathrm{aps}, \mathrm{int}\}$. We notice

$$\operatorname{dist}(\operatorname{supp}\psi_j', \operatorname{supp}\chi_j) = \operatorname{dist}(\operatorname{supp}\psi_j'', \operatorname{supp}\chi_j) \geq \frac{1}{7}R. \tag{3.15}$$

Moreover, we may assume that

$$\left|\frac{\partial^k \psi_j}{\partial u^k}\right| \leq \frac{c_0}{R}$$

for all k, where c_0 is a certain positive constant.

For any parameter $t > 0$, we define an operator $Q_2^R(t)$ on $C^\infty(M_2^R; S)$ with a smooth kernel, given by

$$Q_2^R(t;x,x') := \psi_{\mathrm{aps}}(x)\mathcal{E}_{\mathrm{aps}}^\infty(t;x,x')\chi_{\mathrm{aps}}(x') + \psi_{\mathrm{int}}(x)\mathcal{E}^R(t;x,x')\chi_{\mathrm{int}}(x'). \tag{3.16}$$

Recall that $\mathcal{E}^R$ denotes the kernel of the operator $\mathcal{D}^R e^{-t(\mathcal{D}^R)^2}$, given on the stretched closed manifold M^R. Notice that, by construction, $Q_2^R(t)$ maps the space $L^2(M_2^R; S)$ into the domain of the operator $\mathcal{D}_{2,P_>}^R$.

Then, for $x' \in U_{\text{aps}}$ with $\chi_{\text{aps}}(x') = 1$, we have by definition:

$$Q_2^R(t;x,x') = \begin{cases} \mathcal{E}_{\text{aps}}^{\infty}(t;x,x') & \text{if } d(x, \operatorname{supp}\chi_{\text{aps}}) < \frac{1}{7}R, \\ 0 & \text{if } d(x, \operatorname{supp}\chi_{\text{aps}}) \geq \frac{2}{7}R. \end{cases} \tag{3.17}$$

Correspondingly, we have for $x' \in U_{\text{int}}$ with $\chi_{\text{int}}(x') = 1$,

$$Q_2^R(t;x,x') = \begin{cases} \mathcal{E}^R(t;x,x') & \text{if } d(x, \operatorname{supp}\chi_{\text{int}}) < \frac{1}{7}R, \\ 0 & \text{if } d(x, \operatorname{supp}\chi_{\text{int}}) \geq \frac{2}{7}R. \end{cases} \tag{3.18}$$

For fixed $t > 0$, we determine the difference between the precise kernel $\mathcal{E}_2^R(t;x,x')$ and the approximate one $Q_2^R(t;x,x')$. Let $C^R(t)$ denote the operator

$$C^R(t) = \left((\mathcal{D}_{2,P_>}^R)^2 + \frac{d}{dt} \right) Q_2^R(t) \text{ with kernel } C^R(t;x,x').$$

By definition, we have $\left((\mathcal{D}_{2,P_>}^R)^2 + \frac{d}{dt} \right) \mathcal{E}_2^R(t) = 0$. Thus, $C^R(t)$ 'measures' the error we make when replacing the precise kernel $\mathcal{E}_2^R(t;x,x')$ by the glued, approximate one.

More precisely, we have by Duhamel's Formula

$$\mathcal{E}_2^R(t;x,x') - Q_2^R(t;x,x') = -\int_0^t ds \int_{M_2^R} dz\, \mathcal{E}_2^R(s;x,z) C^R(t-s;z,x')$$

with

$$\begin{aligned} C^R(t-s;z,x') &= \left((\mathcal{D}_{2,(z)}^R)^2 + \mathcal{D} \right) Q_2^R(t-s;z,x') \\ &= \left((\mathcal{D}_{2\,(z)}^R)^2 - \tfrac{d}{ds} \right) Q_2^R(t-s;z,x') \\ &= \psi_{\text{aps}}''(z) \mathcal{E}_{\text{aps}}^R(t-s;z,x') \chi_{\text{aps}}(x') + 2\psi_{\text{aps}}'(z) \frac{\partial}{\partial u} (\mathcal{E}_{\text{aps}}^R(t-s;z,x')) \chi_{\text{aps}}(x') \\ &\qquad + \psi_{\text{aps}}(z) \underbrace{\left(\mathcal{D}_{(z)}^2 - \tfrac{d}{ds} \right) \mathcal{E}_{\text{aps}}^R(t-s;z,x')}_{=0} \chi_{\text{aps}}(x') \\ &\quad + \psi_{\text{int}}''(z) \mathcal{E}^R(t-s;z,x') \chi_{\text{int}}(x') + 2\psi_{\text{int}}'(z) \frac{\partial}{\partial u} (\mathcal{E}^R(t-s;z,x')) \chi_{\text{int}}(x') \\ &\qquad + \psi_{\text{int}}(z) \underbrace{\left((\mathcal{D}_{(z)}^R)^2 - \tfrac{d}{ds} \right) \mathcal{E}^R(t-s;z,x')}_{=0} \chi_{\text{int}}(x'). \end{aligned}$$

Here, $\mathcal{D}_{(z)}$ denotes the operator $\mathcal{D}$ acting on the z variable; and in the partial derivative $\frac{\partial}{\partial u}$ the letter u denotes the normal coordinate of the variable z.

As stated in (3.15), the supports of χ_j and ψ_j' (and, equally, ψ_j'') are disjoint and separated from each other by a distance $R/7$ in the normal variable for $j \in \{\text{aps}, \text{int}\}$. Then the error term $C^R(t-s;z,x')$ vanishes both for the distance in the normal variable $d(z,x') < R/7$ and, actually, whenever z or x' are outside the segment $[-\frac{6}{7}R, \frac{1}{7}R] \times \Sigma$.

Let z and x' be on the cylinder and $|u-v| > R/7$ where u and v denote their normal coordinates. We investigate the error term $C^R(t-s;z,x')$ which consists of six summands. Two of them vanish as we have pointed out above. The remaining four summands involve the kernels $\mathcal{E}^\infty_{\rm aps}(t-s;z,x')$ on the infinite cylinder $[-R,\infty)\times\Sigma$ and $\mathcal{E}^R(t-s;z,x')$ on the stretched closed manifold M^R. We shall use that both kernels can be estimated according to inequality (3.14).

We estimate the first summand

$$|\psi''_{\rm aps}(z)\mathcal{E}^\infty_{\rm aps}(t-s;z,x')\chi_{\rm aps}(x')| \le \frac{c_0}{R}c_1(t-s)^{-\frac{1+m}{2}}e^{c_2t}e^{-c_3\frac{d^2(z,x')}{t-s}}$$
$$\le c_1'e^{c_2't}e^{-c_3'R^2/t}\,.$$

Here we have used $t\ge s\ge 0$, and

$$(t-s)^{-(1+m)/2}e^{-c_2\frac{d^2(z,x')}{(t-s)}} \le ct^{-(1+m)/2}e^{-c_2\frac{d^2(z,x')}{t}} \le \tilde{c}e^{-c_2\frac{d^2(z,x')}{2t}}.$$

Similarly we estimate the second summand

$$2\left|\psi'_{\rm aps}(z)\frac{\partial}{\partial u}\mathcal{E}^\infty_{\rm aps}(t-s;z,x')\chi_{\rm aps}(x')\right|$$
$$\le \frac{c_0c_1}{R}\frac{(t-s)^{-\frac{1+m}{2}}}{\sqrt{t}}\,e^{c_2t}\,e^{-c_3\frac{d^2(z,x')}{t-s}} \le c_1'\,e^{c_2't}\,e^{-c_3'R^2/t},$$

where the factor $1/\sqrt{t}$ comes from the differentiation of the kernel as explained before. The third and fourth summands, involving the kernel $\mathcal{E}^R$ of the closed stretched manifold M^R, are treated in exactly the same way. Altogether we have proved the following:

Lemma 3.12. *The error kernel $C^R(t;u,v)$ vanishes for $u\notin[-\frac{6}{7}R,-\frac{1}{7}R]$. Moreover, $C^R(t;u,v)$ vanishes whenever $|u-v|\le R/7$. For arbitrary $x,x'\in M_2^R$ we have the estimate*

$$|C^R(t;x,x')| \le c_1\,e^{c_2t}\,e^{-c_3R^2/t}$$

with constants c_1, c_2, c_3 independent of x,x',t,R.

We consider the pointwise error

$$\mathcal{E}_2^R(t;x,x)-Q_2^R(t;x,x) = \int_0^t ds\int_{M_2^R}dz\;\mathcal{E}_2^R(s;x,z)C^R(t-s;z,x).$$

We obtain the following proposition as a consequence of the preceding lemma.

Proposition 3.13. *For all $x\in M_2^R$ and all $t>0$ we have*

$$\operatorname{Tr}\mathcal{E}_2^R(t;x,x)-\operatorname{Tr}Q_2^R(t;x,x) = \operatorname{Tr}\left(\mathcal{E}_2^R(t;x,x)-Q_2^R(t;x,x)\right).$$

Moreover, there exist positive constants c_1, c_2, c_3, independent of R, such that the 'error' term satisfies the inequality

$$\left|\mathcal{E}_2^R(t;x,x)-Q_2^R(t;x,x)\right| \le c_1\,e^{c_2t}\,e^{-c_3(R^2/t)}.$$

PROOF. We estimate the error term

$$
\begin{aligned}
|\mathcal{E}_2^R(t;x,x) &- Q_2^R(t;x,x)| \\
&\leq \int_0^t ds \int_{M_2^R} dz\, |\mathcal{E}_2^R(s;x,z) C^R(t-s;z,x)| \\
&\leq c_1^2 e^{c_2 t} \int_0^t ds \int_{M_2^R} dz \left\{ s^{-\frac{d+1}{2}} e^{-c_3 \frac{d^2(x,z)}{s}} \right\} e^{-c_3 \frac{d^2(x,z)}{t-s}} \\
&\leq c_1^2 e^{c_2 t} \int_0^t ds \int_{\operatorname{supp}_z C^R(t-s;z,x)} dz\, e^{-c_4 \frac{d^2(x,z)}{s}} e^{-c_3 \frac{d^2(x,z)}{t-s}} \\
&\leq c_1^2 e^{c_2 t} \int_0^t ds \int_{\operatorname{supp}_z C^R(t-s;z,x)} dz\, e^{-c_5 \frac{tR^2}{s(t-s)}} \\
&\leq c_1^2 e^{c_2 t} cR \int_0^t ds\, e^{-c_5 \frac{tR^2}{s(t-s)}} \leq c_1^2 e^{c_2 t} 2cR \int_0^{t/2} ds\, e^{-c_5 \frac{tR^2}{s(t/2)}} \\
&= c_1^2 e^{c_2 t} 2cR \int_0^{t/2} ds\, e^{-2c_5 \frac{R^2}{s}} .
\end{aligned}
$$

Here we have used that $\operatorname{Vol}(\operatorname{supp}_z C^R(t-s;z,x)) \sim \operatorname{Vol}(\Sigma) \cdot R$ according to Lemma 3.12. We investigate the last integral.

$$
\begin{aligned}
\int_0^t e^{-\frac{c}{s}}\, ds &= -\int_0^t \frac{s^2}{c} e^{-\frac{c}{s}} (-\frac{c}{s^2})\, ds \\
&< -\int_0^t \frac{t^2}{c} e^{-\frac{c}{s}} (-\frac{c}{s^2})\, ds = -\frac{t^2}{c} \int_\infty^{\frac{c}{t}} e^{-r}\, dr = \frac{t^2}{c} e^{-\frac{c}{t}} .
\end{aligned}
$$

Thus we have

$$
|\mathcal{E}_2^R(t;x,x) - Q_2^R(t;x,x)| \leq c_1^2 e^{c_2 t}\, 2cR\, \frac{t^2}{c_6 R^2}\, e^{-\frac{c_6 R^2}{t}} \leq c_7\, e^{c_2 t} e^{-c_8 (R^2/t)} .
$$

□

The preceding result shows that, for t smaller than $\sqrt{R}$, the trace $\operatorname{Tr} \mathcal{E}_2^R(t;x,x)$ of the kernel of the operator $\mathcal{D}_2^R e^{-t(\mathcal{D}_{2,P_>}^R)^2}$ approaches the trace $\operatorname{Tr} Q_2^R(t;x,x)$ of the approximative kernel pointwise as $R \to \infty$. In particular, we have:

COROLLARY 3.14. *The following equality holds, as $R \to \infty$,*

$$
\begin{aligned}
\frac{1}{\sqrt{\pi}} \int_0^{\sqrt{R}} \frac{dt}{\sqrt{t}} &\int_{M_2^R} \operatorname{Tr} \mathcal{E}_2^R(t;x,x)\, dx \\
&= \frac{1}{\sqrt{\pi}} \int_0^{\sqrt{R}} \frac{dt}{\sqrt{t}} \int_{M_2^R} \operatorname{Tr} Q_2^R(t;x,x)\, dx + \mathrm{O}(e^{-cR}).
\end{aligned}
$$

PROOF. We have

$$\begin{aligned}
&\frac{1}{\sqrt{\pi}}\int_0^{\sqrt{R}}\frac{dt}{\sqrt{t}}\int_{M_2^R}\operatorname{Tr}\mathcal{E}_2^R(t;x,x)\,dx\\
&\quad=\frac{1}{\sqrt{\pi}}\int_0^{\sqrt{R}}\frac{dt}{\sqrt{t}}\int_{M_2^R}\operatorname{Tr}Q_2^R(t;x,x)\,dx\\
&\quad\quad+\frac{1}{\sqrt{\pi}}\int_0^{\sqrt{R}}\frac{dt}{\sqrt{t}}\int_{M_2^R}\operatorname{Tr}\left(\mathcal{E}_2^R(t;x,x)-Q_2^R(t;x,x)\right)\,dx,
\end{aligned}$$

and we have to show that the second summand on the right side is $\mathrm{O}(e^{-cR})$ as $R\to\infty$. We estimate

$$\begin{aligned}
&\left|\frac{1}{\sqrt{\pi}}\int_0^{\sqrt{R}}\frac{dt}{\sqrt{t}}\int_{M_2^R}\operatorname{Tr}\left(\mathcal{E}_2^R(t;x,x)-Q_2^R(t;x,x)\right)\,dx\right|\\
&\quad\le\frac{1}{\sqrt{\pi}}\int_0^{\sqrt{R}}\frac{dt}{\sqrt{t}}\int_{M_2^R}\left|\mathcal{E}_2^R(t;x,x)-Q_2^R(t;x,x)\right|dx\\
&\quad\le\frac{1}{\sqrt{\pi}}\int_0^{\sqrt{R}}\frac{dt}{\sqrt{t}}\int_{M_2^R}c_1e^{c_2t}e^{-c_3(R^2/t)}\,dx\\
&\quad\le\frac{c_1\operatorname{Vol}(M_2^R)}{\sqrt{\pi}}\int_0^{\sqrt{R}}\frac{e^{c_2t}e^{-c_3(R^2/t)}}{\sqrt{t}}\,dt\\
&\quad\le c_4\,R\int_0^{\sqrt{R}}e^{c_2\sqrt{R}}e^{-c_5R^{3/2}}\,dt\\
&\quad\le c_4\,R^{3/2}e^{-c_6R}\le c_7\,e^{-c_8R}.
\end{aligned}$$

□

3.5.2. *Discarding the cylinder contributions.* Corollary 3.14 shows that the essential part of the local eta function of the spectral boundary condition on the half manifold with attached cylinder of length R, i.e., the 'small-time' integral from 0 to $\sqrt{R}$ can be replaced, as $R\to\infty$, by the corresponding integral over the trace $\operatorname{Tr}Q_2^R(t;x,x)$ of the approximate kernel, constructed in (3.16). Now we show that $\operatorname{Tr}Q_2^R(t;x,x)$ can be replaced pointwise (for $x\in M_2^R$) by the trace $\operatorname{Tr}\mathcal{E}^R(t;x,x)$ of the kernel of the operator $\mathcal{D}^Re^{-t(\mathcal{D}^R)^2}$ which is defined on the stretched closed manifold M^R.

Consider the Dirac operator

$$\sigma(\partial_u+B):C^\infty([0,\infty)\times\Sigma;S)\to C^\infty([0,\infty)\times\Sigma;S),$$

on the semi-infinite cylinder with the domain

$$\left\{s\in C_0^\infty([0,\infty)\times\Sigma;S)\mid P_>(s|_{\{0\}\times\Sigma})=0\right\}.$$

It has a unique self-adjoint extension which we denote by D_{aps}. Recall that the integral kernel $\mathcal{E}_{\mathrm{aps}}^\infty$ of the operator $D_{\mathrm{aps}}\,e^{-t(D_{\mathrm{aps}})^2}$ enters in the definition of the approximative kernel Q_2^R as given in (3.16). We show that $\mathcal{E}_{\mathrm{aps}}^{\infty}(t;x,x)$ is traceless for all $x \in [0,\infty) \times \Sigma$. Then, it follows

$$\operatorname{Tr} Q_2^R(t;x,x) = \operatorname{Tr} \mathcal{E}^R(t;x,x) \quad \text{for all } x \in M_2^R. \tag{3.19}$$

To prove that a product TV is traceless, the following easy result can be used.

LEMMA 3.15. *Let σ be unitary with $\sigma^2 = -I$. We consider an operator V of trace class which is 'even', i.e., it commutes with σ. Moreover, T is odd, i.e., it anticommutes with σ. Then*

$$\operatorname{Tr}(TV) = 0.$$

PROOF. We have, by unitary equivalence,

$$\operatorname{Tr}(TV) = \operatorname{Tr}(-\sigma(TV)\sigma) = \operatorname{Tr}(-\sigma T\sigma V) = \operatorname{Tr}(\sigma^2 TV) = \operatorname{Tr}(-TV).$$

□

LEMMA 3.16. *Let $\chi : [0,\infty) \to \mathbb{R}$ be a smooth function with compact support and $t > 0$. Then the trace of the operator $\chi\, D_{\mathrm{aps}}\, e^{-t(D_{\mathrm{aps}})^2}$ vanishes. In particular,*

$$\int_{\Sigma} \operatorname{Tr} \mathcal{E}_{\mathrm{aps}}^{\infty}(t;u,y;u,y)\, dy = 0$$

for all $u \in [0,\infty)$.

PROOF. Clearly, $D_{\mathrm{aps}}^2 = (\sigma(\partial_u + B))^2 = -\partial_u^2 + B^2$ is even, hence also the power series $e^{-t(D_{\mathrm{aps}})^2}$ is even. On the other hand, as with B, σB is also odd. So,

$$\operatorname{Tr}\left(\chi\,\sigma B e^{-t(D_{\mathrm{aps}})^2}\right) = 0.$$

To show that

$$\operatorname{Tr}\left(\chi\,\sigma \partial_u e^{-t(D_{\mathrm{aps}})^2}\right) = 0,$$

we need a slightly more specific argument: Let $\mathrm{e}_{\mathrm{aps}}^{\infty}$ denote the heat kernel of the operator D_{aps}. For $u, v \in [0,\infty)$ and $y, z \in \Sigma$ it has the following form (see e.g. [**BoWo93**, Formulae 22.33 and 22.35]):

$$\mathrm{e}_{\mathrm{aps}}(t;u,y;v,z) = \sum_{k\in\mathbb{Z}} e_k(t;u,v)\varphi_k(y) \otimes \varphi_k^*(z)$$

for an orthonormal system $\{\varphi_k\}$ of eigensections of B. Hence,

$$\sigma\partial_u\, \mathrm{e}_{\mathrm{aps}}(t;u,y;v,z) = \sum_{k\in\mathbb{Z}} e_k'(t;u,v)\sigma\varphi_k(y) \otimes \varphi_k^*(z).$$

But $\langle \sigma\varphi_k; \varphi_k\rangle = 0$ on Σ since σ is skew-adjoint. □

3.6. Asymptotic vanishing of the large-t chopped η-invariant on stretched part manifold.

So far we have found

$$\eta_{\mathcal{D}^R_{2,P_>}}(0) = \frac{1}{\sqrt{\pi}} \int_0^{\sqrt{R}} \frac{dt}{\sqrt{t}} \int_{M_2^R} \operatorname{Tr} \mathcal{E}^R(t;x,x)\,dx + \mathrm{O}(e^{-cR}) + \frac{1}{\sqrt{\pi}} \int_{\sqrt{R}}^{\infty} \frac{dt}{\sqrt{t}} \int_{M_2^R} \operatorname{Tr} \mathcal{E}_2^R(t;x,x)\,dx$$

as $R \to \infty$. To prove Theorem 3.8, we still have to show

$$\frac{1}{\sqrt{\pi}} \int_{\sqrt{R}}^{\infty} \frac{dt}{\sqrt{t}} \int_{M_2^R} \operatorname{Tr} \mathcal{E}_2^R(t;x,x)\,dx = \mathrm{O}(e^{-cR}) \quad \text{and} \tag{3.20}$$

$$\frac{1}{\sqrt{\pi}} \int_{\sqrt{R}}^{\infty} \frac{dt}{\sqrt{t}} \int_{M_2^R} \operatorname{Tr} \mathcal{E}^R(t;x,x)\,dx = \mathrm{O}(e^{-cR}), \tag{3.21}$$

as $R \to \infty$. Recall that $\mathcal{E}_2^R(t;x,x')$ denotes the kernel of the operator $\mathcal{D}_2^R e^{-t(\mathcal{D}^R_{2,P_>})^2}$ on the compact manifold M_2^R with boundary $\{-R\} \times \Sigma$, and $\mathcal{E}^R(t;x,x')$ the kernel of the operator $\mathcal{D}^R e^{-t(\mathcal{D}^R)^2}$ on the closed stretched manifold M^R.

In the following we show (3.20), i.e., that we can neglect the contribution to the eta invariant of $\mathcal{D}^R_{2,P_>}$ which comes from the large t asymptotic of $\mathcal{E}_2^R(t;x,x')$. The key to that is that the eigenvalue of $\mathcal{D}^R_{2,P_>}$ with the smallest absolute value is uniformly bounded away from zero.

THEOREM 3.17. *Let $\mu_0(R)$ denote the smallest (in absolute value) nonvanishing eigenvalue of the operator $\mathcal{D}^R_{2,P_>}$ on the manifold M_2^R. Let us assume, as always in this section, that* $\operatorname{Ker} B = \{0\}$. *Then there exists a positive constant c_0, which does not depend on R, such that $\mu_0(R) > c_0$ for R sufficiently large.*

REMARK 3.18. As we will discover, this result indicates that the behavior of the small eigenvalues on M_2^R differs from that on the stretched, closed manifold M^R. On the manifold with boundary M_2^R with the attached cylinder of length R, the eigenvalues are bounded away from 0 when $R \to \infty$ due to the spectral boundary condition. That is the statement of Theorem 3.17 which we are going to prove in the next two sections. However, on M^R, the set of eigenvalues splits into one set of eigenvalues becoming exponentially small and another one of eigenvalues being uniformly bounded away from 0 as $R \to \infty$. This we are going to show further below. Roughly speaking, the reason for the different behavior is that on M_2^R the *eigensections* must satisfy the spectral boundary condition. Therefore they are exponentially decreasing on the cylinder, and the *eigenvalues* are bounded away from 0. But on M^R we have to cope with *eigensections* on a closed manifold which need not decrease, but require part of the *eigenvalues* to decrease exponentially (for details see Theorem 3.30 below).

3.6.1. *The cylindrical Dirac operator.* To prove Theorem 3.17 we first recall a few properties of the cylindrical Dirac operator $D_{\rm cyl} := \sigma(\partial_u + B)$ on the infinite cylinder $\Sigma^\infty_{\rm cyl} := (-\infty, +\infty) \times \Sigma$. A special feature of the cylindrical manifold $\Sigma^\infty_{\rm cyl}$ is that we may apply the theory of *Sobolev spaces* exactly as in the case of $\mathbb{R}^m$. The point is that we can choose a covering of the open manifold $\Sigma^\infty_{\rm cyl}$ by a *finite* number of coordinate charts. We can also choose a finite trivialization of the bundle $S|_{\Sigma^\infty_{\rm cyl}}$. Let $\{U_\iota, \kappa_\iota\}_{\iota=1}^K$ be such a trivialization, where $\kappa_\iota : S|_{U_\iota} \to V_\iota \times \mathbb{C}^N$ is a bundle isomorphism and V_ι an open (possibly non-compact) subset of $\mathbb{R}^m$. Let $\{f_\iota\}$ be a corresponding partition of unity. We assume that for any ι the derivatives of the function f_ι are bounded.

DEFINITION 3.19. We say that a section (or distribution) s of the bundle S over $\Sigma^\infty_{\rm cyl}$ belongs to the p-th *Sobolev space* $\mathcal{H}^p(\Sigma^\infty_{\rm cyl}; S)$, $p \in \mathbb{R}$, if and only if $f_\iota \cdot s$ belongs to the Sobolev space $\mathcal{H}^p(\mathbb{R}^m; \mathbb{C}^N)$ for any ι. We define the p-th *Sobolev norm*

$$\|s\|_p := \sum_{\iota=1}^K \left\| (I + \Delta_\iota)^{p/2} (f_\iota \cdot s) \right\|_{L^2(\mathbb{R}^m)},$$

where Δ_ι denotes the Laplacian on the trivial bundle $V_\iota \times \mathbb{C}^N \subset \mathbb{R}^m \times \mathbb{C}^N$.

LEMMA 3.20.

(a) *For the unique self-adjoint L^2 extension of $D_{\rm cyl}$ (denoted by the same symbol) we have*

$$\mathrm{Dom}(D_{\rm cyl}) = \mathcal{H}^1(\Sigma^\infty_{\rm cyl}; S).$$

(b) *Let λ_1 denote the smallest positive eigenvalue of the operator B on the manifold Σ. Then we have*

$$\langle (D_{\rm cyl})^2 s; s \rangle \geq \lambda_1^2 \|s\|^2$$

for all $s \in \mathrm{Dom}(D_{\rm cyl})$, and for any $\mu \in (-\lambda_1, +\lambda_1)$ the operator

$$D_{\rm cyl} - \mu : \mathcal{H}^1(\Sigma^\infty_{\rm cyl}; S) \to L^2(\Sigma^\infty_{\rm cyl}; S)$$

is an isomorphism of Hilbert spaces.

(c) *Let $\mathcal{R}_{\rm cyl}(\mu)$ denote the inverse of the operator $D_{\rm cyl} - \mu$. Then the family $\{\mathcal{R}_{\rm cyl}(\mu)\}_{\mu \in (-\lambda_1, \lambda_1)}$ is a smooth family of elliptic pseudo-differential operators of order -1.*

PROOF. (a) follows immediately from the corresponding result on the model manifold $\mathbb{R}^m$. To prove (b) we consider a spectral resolution $\{\varphi_k, \lambda_k\}_{k \in \mathbb{Z}\setminus 0}$ of $L^2(\Sigma; S)$ generated by the tangential operator B. Because of (3.8), we have $\lambda_{-k} = -\lambda_k$. We can assume $\varphi_{-k} = \sigma\varphi_k$ for $k \in \mathbb{N}$. We consider a section s belonging to the dense subspace $C_0^\infty(\Sigma^\infty_{\rm cyl}; S)$ of $\mathrm{Dom}(D_{\rm cyl})$, and expand it in terms of the preceding spectral resolution

$$s(u, y) = \sum_{k \in \mathbb{Z}\setminus\{0\}} f_k(u)\varphi_k(y).$$

Since $(D_{\mathrm{cyl}})^2 = -\partial_u^2 + B^2$, we obtain

$$(D_{\mathrm{cyl}})^2 s = \sum_k (\lambda_k^2 f_k - f_k'')\varphi_k,$$

hence

$$\begin{aligned}\langle (D_{\mathrm{cyl}})^2 s; s\rangle &= \sum_k \int_{-\infty}^{\infty} (\lambda_k^2 f_k(u) - f_k''(u))\bar{f}_k(u)\, du \\ &\geq \lambda_1^2 \|s\|^2 - \sum_k \int_{-\infty}^{\infty} f_k''(u)\bar{f}_k(u)\, du \\ &= \lambda_1^2 \|s\|^2 + \sum_k \int_{-\infty}^{\infty} f_k'(u)\bar{f}_k'(u)\, du \geq \lambda_1^2 \|s\|^2 .\end{aligned}$$

It follows that $(D_{\mathrm{cyl}})^2$ (and therefore D_{cyl}) has bounded inverse in $L^2(\Sigma_{\mathrm{cyl}}^\infty; S)$ and, more generally, that $(D_{\mathrm{cyl}})^2 - \mu$ is invertible for $\mu \in (-\lambda_1, \lambda_1)$.

To prove (c) we apply the symbolic calculus and construct a parametrix S for the operator D_{cyl}; i.e., an elliptic pseudo-differential operator of order -1 such that $SD_{\mathrm{cyl}} = I + T$, where T is a smoothing operator. Thus $D_{\mathrm{cyl}}^{-1} = S - TD_{\mathrm{cyl}}^{-1}$. The operator TD_{cyl}^{-1} is a smoothing operator, hence D_{cyl}^{-1} is an elliptic pseudo-differential operator of order -1. The same argument can be applied to the resolvent $\mathcal{R}_{\mathrm{cyl}}(\mu) = (D_{\mathrm{cyl}} - \mu)^{-1}$ for arbitrary $\mu \in (-\lambda_1, \lambda_1)$. The smoothness of the family follows by standard calculation. □

3.6.2. *The part manifolds with half-infinite cylindrical ends.* In order to prove Theorem 3.17 we need to refine the preceding results on the infinite cylinder $\Sigma_{\mathrm{cyl}}^\infty$ to the Dirac operator, naturally extended to the manifold $M_2^\infty = ((-\infty, 0] \times \Sigma) \cup M_2$ with cylindrical end. Let $C_0^\infty(M_2^\infty; S)$ denote the space of compactly supported smooth sections of S over M_2^∞. Then

$$\mathcal{D}_2^\infty|_{C_0^\infty(M_2^\infty;S)} : C_0^\infty(M_2^\infty; S) \to L^2(M_2^\infty; S) \tag{3.22}$$

is symmetric. Moreover, we have

LEMMA 3.21. *Let $s \in C^\infty(M_2^\infty; S)$ be an eigensection of $\mathcal{D}_2^\infty$. Then there exist $C, c > 0$ such that, on $(-\infty, 0] \times \Sigma$, one has $|s(u, y)| \leq Ce^{cu}$.*

PROOF. Let $\{\varphi_k, \lambda_k\}_{k \in \mathbb{Z}\setminus 0}$ be a spectral resolution of the tangential operator B. Because of (3.8) we have $\lambda_{-k} = -\lambda_k$ and we can assume that $\varphi_{-k} = \sigma\varphi_k$ for $k \in \mathbb{N}$. Then

$$\left\{\varphi_k^\pm = \tfrac{1}{\sqrt{2}}\left(\varphi_k \pm \sigma\varphi_k\right), \pm\lambda_k\right\}_{k\in\mathbb{N}} \tag{3.23}$$

is a spectral resolution of the composed operator σB on Σ. Notice that we have

$$\sigma\varphi_k^+ = -\varphi_k^- \quad \text{and} \quad \sigma\varphi_k^- = \varphi_k^+ . \tag{3.24}$$

Let $s \in C^\infty(M_2^\infty; S)$ and

$$\mathcal{D}_{L^2}^\infty \psi = \mu\psi \tag{3.25}$$

with $\mu \in \mathbb{R}$. We expand $s|_{(-\infty,0]\times\Sigma}$ in terms of the spectral resolution of σB just constructed:

$$s(u,y) = \sum_{k=1}^{\infty} f_k(u)\varphi_k^+(y) + g_k(u)\varphi_k^-(y)\,.$$

Because of (3.23), (3.24), and (3.25) the coefficients f_k, g_k must satisfy the system of ordinary differential equations

$$\begin{pmatrix} \lambda_k & \partial/\partial u \\ -\partial/\partial u & -\lambda_k \end{pmatrix} \begin{pmatrix} f_k \\ g_k \end{pmatrix} = \mu \begin{pmatrix} f_k \\ g_k \end{pmatrix}$$

or, equivalently,

$$\begin{pmatrix} f_k' \\ g_k' \end{pmatrix} = \mathbf{A} \begin{pmatrix} f_k \\ g_k \end{pmatrix} \quad \text{with } \mathbf{A} := \begin{pmatrix} 0 & -(\mu+\lambda_k) \\ \mu - \lambda_k & 0 \end{pmatrix}.$$

Since $s \in L^2$, of the eigenvalues $\pm\sqrt{\lambda_k^2 - \mu^2}$ of $\mathbf{A}$, only those which are on the positive real line enter in the construction of s by solving the preceding differential equation. In particular, all coefficients f_k, g_k must vanish identically for $\lambda_k \leq \mu$. Thus,

(3.26)

$$s(u,y) = \sum_{\lambda_k > \mu} a_k \left(\exp\left(\sqrt{\lambda_k^2 - \mu^2}\, u\right) \varphi_k^+ - \frac{\lambda_k - \mu}{\sqrt{\lambda_k^2 - \mu^2}} \exp\left(\sqrt{\lambda_k^2 - \mu^2}\, u\right) \varphi_k^- \right)$$

and, in particular,

$$|s(u,y)| \leq C \exp\left(\sqrt{\lambda_{k_0}^2 - \mu^2}\, \frac{u}{2}\right), \quad u < 0,$$

for some constant C, where λ_{k_0} denotes the smallest positive eigenvalue of B with $\lambda_{k_0} > |\mu|$. □

In spectral theory we are looking for self-adjoint L^2 extensions of a symmetric operator. We recall: On a *closed* manifold, the Dirac operator is *essentially self-adjoint*; i.e. its minimal closed extension is self-adjoint (and therefore there do not exist other self-adjoint extensions) and it is a Fredholm operator. On a compact manifold with boundary, the situation is much more complicated. There is a huge variety of dense domains to which the Dirac operator can be extended such that it becomes self-adjoint; and there is a smaller, but still large variety where the extension of the Dirac operator becomes self-adjoint *and* Fredholm (see Section 1.2.2 above and Booß-Bavnbek and Furutani [**BoFu98**]); a special type of self-adjoint and Fredholm domains are the domains specified by the boundary conditions belonging to the Grassmannian of all self-adjoint generalized Atiyah-Patodi-Singer projections.

Now we shall show that the situation on manifolds with (infinite) cylindrical ends resembles the situation on closed manifolds.

We recall the following simple lemma (see also Reed and Simon [**ReSi72**, Theorem VIII.3, Corollary, p. 257].

LEMMA 3.22. *Let A be a densely defined symmetric operator in a separable complex Hilbert space $\mathcal{H}$. We assume that* range$(A+iI)$ *is dense in $\mathcal{H}$. Then A is essentially self-adjoint.*

PROOF. Since A is symmetric, the operator $A+iI$ is injective and the operator $(A+iI)^{-1}$ is well defined and bounded on the dense subspace range$(A+iI)$ of $\mathcal{H}$. Then the closure R_i of $(A+iI)^{-1}$ has the whole space $\mathcal{H}$ as domain and R_i is bounded and injective. Now a standard argument of functional analysis (see e.g. Pedersen [**Pe89**], Proposition 5.1.7) says that the inverse R_i^{-1} of a densely defined, closed, and injective operator R_i has the same properties. Thus our R_i^{-1} is closed; and by construction it is the minimal closed extension of $A+iI$. Therefore, $R_i^{-1}-iI$ is symmetric and the minimal closed extension of A, hence self-adjoint and equal A^*. □

We apply the lemma for $\mathcal{H} = L^2(M_2^\infty; S)$ and take for A the operator of (3.22). To prove that the range $(\mathcal{D}_2^\infty + iI)(C_0^\infty(M_2^\infty; S))$ is dense in $L^2(M_2^\infty; S)$ we consider a section $s \in L^2(M_2^\infty; S)$ which is orthogonal to $(\mathcal{D}_2^\infty + iI)(C_0^\infty(M_2^\infty; S))$; i.e. the distribution $(\mathcal{D}_2^\infty - iI)s$ vanishes when applied to any test function, hence

$$(\mathcal{D}_2^\infty - iI)s = 0. \tag{3.27}$$

Since $\mathcal{D}_2^\infty - i$ is elliptic, by elliptic regularity s is smooth at all interior points, that is for our complete manifold in all points. On the cylinder $(-\infty, 0] \times \Sigma$ we expand s in terms of the eigensections of the composed operator σB on Σ. It follows that s satisfies an estimate of the form

$$|s(u,y)| \le Ce^{cu}, \quad (u,y) \in (-\infty, 0] \times \Sigma, \tag{3.28}$$

for some constants $C, c > 0$ (according to Lemma 3.21). On the manifold M_2^R with cylindrical end of finite length R we apply Green's formula and get

$$\langle \mathcal{D}_2^R s^R; s^R \rangle - \langle s^R; \mathcal{D}_2^R s^R \rangle = -\int_{\{-R\}\times\Sigma} (\sigma s|_{\{-R\}\times\Sigma}\, dy, s|_{\{-R\}\times\Sigma}), \tag{3.29}$$

where s^R denotes the restriction of s to the manifold M_2^R with boundary $\{-R\}\times\Sigma$. For $R \to \infty$, the right side of (3.29) vanishes; and the left side becomes $2i\,|s|^2$ by (3.27). Hence $s = 0$. Thus we have proved:

LEMMA 3.23. *The operator* (3.22) *is essentially self-adjoint.*

We denote the (unique) self-adjoint L^2 extension by the same symbol $\mathcal{D}_2^\infty$, and we define the Sobolev spaces on the manifold M_2^∞ as in Definition 3.19. Once again, the point is that manifolds with cylindrical ends, even if they are not compact but only complete, are like the infinite cylinder sufficiently simple to be covered by a finite system of local charts. Clearly

$$\mathrm{Dom}(\mathcal{D}_2^\infty) = \mathcal{H}^1(M_2^\infty; S) \text{ and } \mathcal{D}_2^\infty : \mathcal{H}^1(M_2^\infty; S) \to L^2(M_2^\infty; S)$$

is bounded. There are, however, substantial differences between the properties of the simple Dirac operator D_{cyl} on the infinite cylinder and the Dirac operator $\mathcal{D}_2^\infty$ on the manifold with cylindrical end. For instance, from B the discreteness of the spectrum and the regularity at 0 (i.e., 0 is not an eigenvalue) are passed on to D_{cyl}, but not to $\mathcal{D}_2^\infty$. Yet we can prove the following result:

PROPOSITION 3.24. *The operator*

$$\mathcal{D}_2^\infty : \mathrm{Dom}(\mathcal{D}_2^\infty) = \mathcal{H}^1(M_2^\infty; S) \to L^2(M_2^\infty; S)$$

is a Fredholm operator and its spectrum in the interval $(-\lambda_1, \lambda_1)$ *consists of finitely many eigenvalues of finite multiplicity. Here* λ_1 *denotes the smallest positive eigenvalue of* B.

REMARK 3.25. Actually, using more advanced methods one can show that the essential spectrum of $\mathcal{D}_2^\infty$ is equal to $(-\infty, -\lambda_1] \cup [\lambda_1, \infty)$ (see for instance Müller [**Mu94**, Section 4]).

Before proving the proposition we shall collect various criteria for the compactness of a bounded operator between Sobolev spaces on an open manifold. Let X be a complete (not necessarily compact) Riemannian manifold with a fixed Hermitian bundle. Recall the three cornerstones of the Sobolev analysis of Dirac operators for X *closed.*

Rellich Lemma: The inclusion $\mathcal{H}^1(X) \subset L^2(X)$ is compact.

Compact Resolvent: To each Dirac operator $\mathcal{D}$ we have a parametrix $\mathcal{R}$ which is an elliptic pseudo-differential operator of order -1 with principal symbol equal to the inverse of the principal symbol of $\mathcal{D}$. So $\mathcal{R}$ is a bounded operator from $L^2(X)$ to $\mathcal{H}^1(X)$, and hence compact in $L^2(X)$. In particular, for μ in the resolvent set the resolvent $(\mathcal{D}-\mu I)^{-1}$ is compact as operator in $L^2(X)$.

Smoothing Operator: Any integral operator over X with smooth kernel is a smoothing operator, i.e. it maps distributional sections of arbitrary low order into smooth sections. Moreover, it is of trace class and thus compact.

In the *general* case, i.e. for not necessarily compact X, the Rellich Lemma remains valid for sections with compact support. A compact resolvent is not attainable, hence the essential spectrum appears. Operators with smooth kernel remain smoothing operators, but in general they are no longer of trace class nor compact. We recall:

LEMMA 3.26. *Let X be a complete (not necessarily compact) Riemannian manifold with fixed Hermitian bundle. Let K be a compact subset of X.*

(a) *The injection $\mathcal{H}^1(X) \subset L^2(X)$ defines a compact operator when restricted to sections with support in K. In particular, for any cutoff function χ with support in K and any bounded operator $\mathcal{R} : L^2(X) \to \mathcal{H}^1(X)$ the operator $\chi\mathcal{R}$ is compact in $L^2(X)$.*

(b) *Let $T : L^2(X) \to L^2(X)$ be an integral operator with a kernel $k(x,y) \in L^2(X^2)$. Then the operator T is a bounded, compact operator (in fact it is of Hilbert-Schmidt class).*

(c) *Let* $T : L^2(X) \to L^2(X)$ *be a bounded compact operator and* $\mathcal{H}'$ *a closed subspace of* $L^2(X)$*, e.g.* $\mathcal{H}' := L^2(X')$ *where* X' *is a submanifold of* X *of codimension 0. Assume that* $T(\mathcal{H}') \subset \mathcal{H}'$*. Then* $T|_{\mathcal{H}'}$ *is compact as operator from* $\mathcal{H}'$ *to* $\mathcal{H}'$.

PROOF. (a) follows immediately from the local Rellich Lemma. (b) is the famous Hilbert–Schmidt Lemma. Also (c) is well known, see e.g. Hörmander [**Ho85**, Proposition 19.1.13] where (c) is proved within the category of trace class operators. □

In general an integral operator T with smooth kernel is not compact even if either $\operatorname{supp}_x k(x,x')$ or $\operatorname{supp}_{x'} k(x,x')$ are contained in a compact subset $K \subset X$. Consider for instance on $\Sigma^\infty_{\text{cyl}} = (-\infty, +\infty) \times \Sigma$ an integral operator T with a smooth kernel of the form

$$k(x,x') = \chi(x) d(x,x'),$$

where $d(x,x')$ denotes the distance and χ is a function with support in a ball of radius 1 (and equal 1 in a smaller ball). Then T is not a compact operator on $L^2(\Sigma^\infty_{\text{cyl}})$: choose a sequence $\{s_n\}$ of L^2 functions of norm 1 and with $\operatorname{supp} s_n$ contained in a ball of radius 1 such that $d(\operatorname{supp}\chi, \operatorname{supp} s_n) = n$. Then for any n we have $|Ts_n| > Cn$. Thus T is not compact, in fact not even bounded.

For the bounded resolvent (see Lemma 3.20)

$$\mathcal{R}_{\text{cyl}} : L^2(\Sigma^\infty_{\text{cyl}}; S) \to \mathcal{H}^1(\Sigma^\infty_{\text{cyl}}; S)$$

we have, however, the following corollary to the preceding lemma. It provides an example of a compact integral operator on an open manifold with a smooth kernel which is compactly supported only in one variable.

COROLLARY 3.27. *Let* χ *and* ψ *be smooth cutoff functions on* $\Sigma^\infty_{\text{cyl}}$ *with support contained in the half-cylinder* $(-\infty, 0) \times \Sigma$*. Let* $\operatorname{supp}\chi$ *be compact. Then the operators* $\chi\mathcal{R}_{\text{cyl}}\psi$ *and* $\psi\mathcal{R}_{\text{cyl}}\chi$ *are compact in* $L^2(\Sigma^\infty_{\text{cyl}}; S)$.

PROOF. The operator $\chi\mathcal{R}_{\text{cyl}}\psi$ is compact according to the preceding lemma, claim (a). Its adjoint operator is $\psi\mathcal{R}_{\text{cyl}}\chi$, since $\mathcal{R}_{\text{cyl}}$ is self-adjoint. Thus it is also compact (even if its range is *not* compactly supported). □

3.7. The estimate of the lowest nontrivial eigenvalue. In this section we prove Theorem 3.17. Recall that the tangential operator B is assumed to be nonsingular and that λ_1 denotes the smallest positive eigenvalue of B. So far, we have established that:

(I) The operator D_{cyl} on the infinite cylinder $\Sigma^\infty_{\text{cyl}}$ has no eigenvalues in the interval $(-\lambda_1, +\lambda_1)$.
(II) The operator $\mathcal{D}^\infty_2$ on the manifold M^∞_2 with infinite cylindrical end has only finitely many eigenvalues in the interval $(-\lambda_1, +\lambda_1)$, each of finite multiplicity.

We have to show that:

(III) The nonvanishing eigenvalues of $(\mathcal{D}_2^R)_{P_>}$ are bounded away from 0 by a bound independent of R.

PROOF OF THEOREM 3.17. The idea of the proof is the following. We define a positive constant μ_1 independent of R. Then let R be a positive real (more precisely $R > R_0$ for a suitable positive R_0), and $s \in L^2(M_2^R; S)$ any eigensection with eigenvalue $\mu \in (-\lambda_1/\sqrt{2}, +\lambda_1/\sqrt{2})$, i.e., $s \in \mathrm{Dom}(\mathcal{D}_2^R)_{P_>}$, i.e.,

$$P_>(s|_{\{-R\}\times\Sigma}) = 0 \quad \text{and} \quad \mathcal{D}_2^R s = \mu s.$$

Then we show that $\mu^2 > \mu_1/2$ for a certain real $\mu_1 > 0$ which is independent of R and s. A natural choice of μ_1 is

$$\mu_1 = \min\left\{ \tfrac{|\mathcal{D}_2^\infty \Psi|^2}{|\Psi|^2} \,\middle|\, \Psi \in \mathcal{H}^1(M_2^\infty; S) \text{ and } \Psi \perp \operatorname{Ker} \mathcal{D}_2^\infty \right\}.$$

Note that by (II) above (Proposition 3.24), $\operatorname{Ker} \mathcal{D}_2^\infty$ is of finite dimension. We shall define a certain extension $s^\infty \in \mathcal{H}^1(M_2^\infty; S)$ of s.

The reasoning would be easy, if we could extend s to an eigensection of $\mathcal{D}_2^\infty$ on all of M_2^∞. Then it would follow at once that the discrete part of the spectrum of $\mathcal{D}_2^\infty$ is not empty, μ belongs to it, $\sqrt{\mu_1}$ is the smallest eigenvalue > 0, and hence we would have $\mu^2 > \mu_1/2$ as desired. In general, such a convenient extension of the given eigensection s cannot be achieved. But due to the spectral boundary condition satisfied by s in the hypersurface $\{-R\} \times \Sigma$, the eigensection s over M_2^R can be continuously extended by a section over $(-\infty, -R] \times \Sigma$ on which the Dirac operator vanishes. By construction, both the enlargement α of the L^2 norm of s by the chosen extension and the cosine, say β, of the angle between s^∞ and $\operatorname{Ker} \mathcal{D}_2^\infty$ can be estimated independently of the specific choice of s and μ. It turns out that they both decrease exponentially with growing R.

Let $\{s_1, \ldots, s_q\}$ be an orthonormal basis of $\operatorname{Ker} \mathcal{D}_2^\infty$ and set

$$\widetilde{s} := s^\infty - \sum_{j=1}^{q} \langle s^\infty; s_j \rangle\, s_j\,.$$

Clearly, the section $\widetilde{s}$ belongs to $\mathcal{H}^1(M_2^\infty; S)$ and is orthogonal to $\operatorname{Ker} \mathcal{D}_2^\infty$.

Hence, on the one hand, $\frac{|\mathcal{D}_2^\infty \widetilde{s}|^2}{|\widetilde{s}|^2} \geq \mu_1$. On the other hand, we have by construction $|\mathcal{D}_2^\infty \widetilde{s}|^2 = |\mathcal{D}_2^\infty s^\infty|^2 = \left|\mathcal{D}_2^R s\right|^2_{M_2^R} = \mu^2$. Finally, we shall prove that

$$|\widetilde{s}| \to 1 \text{ as } R \to \infty. \tag{3.30}$$

Then the estimate

$$\mu^2 > \frac{\mu_1}{2}$$

follows for sufficiently large R. Since we have assumed that $\mu^2 < \frac{\lambda_1^2}{2}$, we also have $\mu_1 < \lambda_1^2$, hence μ_1 belongs to the discrete part of the spectrum of $(\mathcal{D}_2^\infty)^2$ and, by the *Min-Max Principle* (see e.g. [**ReSi78**]), it must be its smallest eigenvalue > 0.

Thus, to prove the Theorem 3.17 we are left with the task of first constructing a suitable extension $\widetilde{s}$ of s and then proving (3.30).

We expand $s|_{[-R,0]\times\Sigma}$ in terms of a spectral resolution

$$\{\varphi_k, \lambda_k; \sigma\varphi_k, -\lambda_k\}_{k\in\mathbb{N}}$$

of $L^2(\Sigma; S)$ generated by B:

$$s(u,y) = \sum_{k=1}^{\infty} f_k(u)\varphi_k(y) + g_k(u)\sigma\varphi_k(y)\,.$$

Since $\sigma(\partial_u + B)s|_{[-R,0]\times\Sigma} = 0$, the coefficients must satisfy the system of ordinary differential equations

$$\begin{pmatrix} f_k' \\ g_k' \end{pmatrix} = \mathbf{A}_k \begin{pmatrix} f_k \\ g_k \end{pmatrix} \quad \text{with } \mathbf{A}_k := \begin{pmatrix} -\lambda_k & \mu \\ -\mu & \lambda_k \end{pmatrix}. \tag{3.31}$$

Moreover, since $P_> s|_{\{-R\}\times\Sigma} = 0$, we have

$$f_k(-R) = 0 \ \text{ for any } k \geq 1.$$

Thus, for each k the pair (f_k, g_k) is uniquely determined up to a constant a_k. More explicitly, since the eigenvalues of $\mathbf{A}_k$ are $\pm(\lambda_k^2 - \mu^2)^{1/2}$, a suitable choice of the eigenvectors of $\mathbf{A}_k$ gives

$$f_k(u) = a_k \frac{\mu}{\sqrt{\lambda_k^2 - \mu^2}} \sinh\sqrt{\lambda_k^2 - \mu^2}(R+u),$$

$$g_k(u) = a_k\Big(\cosh\sqrt{\lambda_k^2 - \mu^2}(R+u) + \frac{\lambda_k}{\sqrt{\lambda_k^2-\mu^2}} \sinh(\lambda_k^2 - \mu^2)^{1/2}(R+u)\Big).$$

We assume $|s|_{L^2} = 1$. Then we have, with $v := (\lambda_k^2 - \mu^2)^{1/2}(R+u)$:

$$\begin{aligned}
1 \geq \int_{[-R,0]\times\Sigma} |s(u,y)|^2\, du dy &= \sum_{k=1}^{\infty} \int_{-R}^{0} \left(|f_k(u)|^2 + |g_k(u)|^2\right) du \\
&= \sum_{k=1}^{\infty} |a_k|^2 \frac{1}{(\lambda_k^2 - \mu^2)^{1/2}} \int_0^{(\lambda_k^2-\mu^2)^{1/2}R} \Bigg(\frac{\mu^2}{\lambda_k^2 - \mu^2}\sinh^2 v \\
&\qquad + \cosh^2 v + 2\frac{\lambda_k}{(\lambda_k^2-\mu^2)^{1/2}}\cosh v \sinh v + \frac{\lambda_k^2}{\lambda_k^2-\mu^2}\sinh^2 v\Bigg)\, dv \\
&= \sum_{k=1}^{\infty} |a_k|^2 \Bigg\{ -\frac{\lambda_k^2}{\lambda_k^2-\mu^2} R + (1/4)\frac{\mu^2}{(\lambda_k^2-\mu)^{3/2}}\sinh(2(\lambda_k^2-\mu^2)^{1/2}R) \\
&\qquad + (1/4)\left(1 + \frac{\lambda_k^2}{\lambda_k^2-\mu^2}\right)(\lambda_k^2-\mu^2)^{1/2}\sinh(2(\lambda_k^2-\mu^2)^{1/2}R) \\
&\qquad + \frac{\lambda_k^2}{\lambda_k^2-\mu^2}\cosh^2((\lambda_k^2-\mu^2)^{1/2}R)\Bigg\}.
\end{aligned}$$

Since $\lambda_k^2 \geq \lambda_1^2 > 2\mu^2$ we have $2(\lambda_k^2 - \mu^2)^{1/2} > \sqrt{2}\lambda_k > \lambda_k$. Moreover, we have for all $k \geq 1$

$$\begin{aligned}
-\Big(\frac{\lambda_k^2}{\lambda_k^2 - \mu^2}\Big)R + \Big(\frac{\lambda_k^2}{\lambda_k^2 - \mu^2}\Big)&\frac{(\lambda_k^2 - \mu^2)^{1/2}}{4}\sinh(2(\lambda_k^2 - \mu^2)^{1/2}R)\\
&> -R + \frac{(\lambda_1^2 - \mu^2)^{1/2}}{4}\sinh(2(\lambda_1^2 - \mu^2)^{1/2}R)\\
&> -R + \frac{\sqrt{2}}{8}\lambda_1 \sinh(\sqrt{2}\lambda_1 R) > 0,
\end{aligned}$$

if $R \geq R_0$ for some positive R_0 which depends only on λ_1 and not on μ, s and k.

Thus, for any k the sum in the braces can be estimated in the following way:

$$\{\dots\} > \Big(\frac{\lambda_k^2}{\lambda_k^2 - \mu^2}\Big)\cosh^2((\lambda_k^2 - \mu^2)^{1/2}R) > \frac{1}{4}e^{2(\lambda_k^2 - \mu^2)^{1/2}R} > \frac{1}{4}e^{\lambda_k R}.$$

Hence, we have

$$\sum_{k=1}^{\infty} |a_k|^2 e^{\lambda_k R} \leq 4\,. \tag{3.32}$$

Note that the preceding estimate does not depend on R (provided that $R > R_0$), k or the specific choice of s, and that R_0 only depends on λ_1.

According to (3.32) the absolute value of the coefficients a_k is rapidly decreasing in such a way that, in particular, we can extend the eigensection s of $\mathcal{D}_{2,P_>}^R$, given on M_2^R to a continuous section on M_2^∞ by the formula

$$s^\infty(x) := \begin{cases} s(x) & \text{for } x \in M_2^R,\\ \sum_{k=1}^\infty a_k e^{\lambda_k(R+u)}\sigma\varphi_k(y) & \text{for } x = (u,y) \in (-\infty, -r] \times \Sigma. \end{cases}$$

By construction, s^∞ is smooth on $M_2^\infty \setminus (\{-R\} \times \Sigma)$ and belongs to the Sobolev space $\mathcal{H}^1(M_2^\infty; S)$. It follows from (3.32) that

$$\begin{aligned}
|s^\infty|_{L^2}^2 &= |s|_{L^2}^2 + \sum_{k=1}^\infty |a_k|^2 \int_{-\infty}^{-R} e^{2\lambda_k(R+u)}\,du = 1 + \sum_{k=1}^\infty |a_k|^2 \frac{1}{2\lambda_k}\\
&\leq 1 + \frac{1}{2\lambda_1}\sum_{k=1}^\infty |a_k|^2 \leq 1 + \frac{1}{2\lambda_1}\Big(\sum_{k=1}^\infty |a_k|^2 e^{\lambda_k R}\Big)e^{-\lambda_1 R}\\
&\leq 1 + \frac{2}{\lambda_1}e^{-\lambda_1 R}.
\end{aligned} \tag{3.33}$$

Next, let $\Psi \in \operatorname{Ker}\mathcal{D}_2^\infty$ and assume that $|\Psi| = 1$. By (3.26), on $(-\infty, 0] \times \Sigma$ the section Ψ has the form

$$\Psi(u,y) = \sum_{k=1}^\infty b_k e^{\lambda_k u} G(y)\varphi_k(y) \tag{3.34}$$

with

$$\sum_{k=1}^{\infty}\int_{-\infty}^{0}|b_k|^2 e^{2\lambda_k u}\,du=\sum_{k=1}^{\infty}\frac{1}{2\lambda_k}|b_k|^2<\infty.$$

Set $l := \Psi|_{M_2^R}$. Then l satisfies the equations

$$\mathcal{D}_2^R l=0 \quad\text{and}\quad P_>(l|_{\{-R\}\times\Sigma})=0.$$

Hence, l belongs to $\operatorname{Ker}\mathcal{D}^R_{2,P_>}$. This implies the following equality:

$$\begin{aligned}\int_{M_2^R}\langle s^\infty(x);\Psi(x)\rangle\,dx&=\frac{1}{\mu}\int_{M_2^R}\langle\mathcal{D}_2^R s^\infty(x);l(x)\rangle\,dx\\&=\frac{1}{\mu}\int_{M_2^R}\langle s^\infty(x);\mathcal{D}_2^R l(x)\rangle\,dx-\frac{1}{\mu}\int_{\Sigma}\langle\sigma s^\infty(-R,y);l(-R,y)\rangle\,dx.\end{aligned}$$

On the other hand,

$$\int_{(-\infty,-r]\times\Sigma}\langle s^\infty(x);\Psi(x)\rangle\,dx=\sum_{k>0}\frac{a_k\overline{b_k}}{2\mu}e^{-\lambda_k R}\le C_1e^{-\lambda_1 R}.$$

Therefore,

$$|\langle s^\infty;\Psi\rangle|\le C_1e^{-\lambda_1 R}. \tag{3.35}$$

Hence, we have proved:

LEMMA 3.28. *Any eigensection $s\in\mathcal{H}^1(M_2^R;S)$ of $\mathcal{D}^R_{2,P_>}$ with eigenvalue $\mu\in(-\lambda_1/\sqrt{2},\lambda_1/\sqrt{2})$ can be extended to a continuous section s^∞ on M_2^∞ which is smooth on $M_2^\infty\setminus(\{-R\}\times\Sigma)$ and belongs to the first Sobolev space $\mathcal{H}^1(M_2^\infty;S)$. Moreover, the enlargement of the norm of s by the extension and the cosine of the angle between s^∞ and $\operatorname{Ker}\mathcal{D}_2^\infty$ are exponentially decreasing by* (3.33) *and* (3.35).

The final step in proving the Theorem 3.17 follows at once from the preceding lemma. We recall: By definition of μ_1 we have $|\mathcal{D}_2^\infty\widetilde{s}|^2/|\widetilde{s}|^2\ge\mu_1$ and by construction of $\widetilde{s}$ we have $|\mathcal{D}_2^\infty\widetilde{s}|^2=\mu^2$. Thus, we have $\mu^2\ge|\widetilde{s}|^2\mu_1$. To get the desired bound $\mu^2>\mu_1/2$, it remains to show that $|\widetilde{s}|^2>1/2$ for sufficiently large R.

Since $\widetilde{s}$ is the orthogonal projection of s^∞ onto $(\operatorname{Ker}\mathcal{D}_2^\infty)^\perp$ and the basis $\{s_1,\dots,s_q\}$ of $\operatorname{Ker}\mathcal{D}_2^\infty$ is orthonormal, we have

$$|\widetilde{s}|^2=|s^\infty|^2-\sum_{j=1}^{q}|\langle s^\infty;s_j\rangle|^2\le 1+\frac{2}{\lambda_1}e^{-\lambda_1 R}-\sum_{j=1}^{q}|\langle s^\infty;s_j\rangle|^2.$$

Thus, Theorem 3.17 follows from

$$\big||\widetilde{s}|^2-1\big|\le\frac{2}{\lambda_1}e^{-\lambda_1 R}+qC_1e^{-2\lambda_1 R}\le C_2e^{-C_3 R}.$$

□

We finish this section by proving the asymptotic estimate (3.20). Recall that $\mathcal{E}_2^R(t;x,x')$ denotes the kernel of the operator $\mathcal{D}_{2,P_>}^R\, e^{-t(\mathcal{D}_{2,P_>}^R)^2}$, where $\mathcal{D}_{2,P_>}^R$ denotes the Dirac operator over the manifold M_2^R with the spectral boundary condition at the boundary $\{-R\}\times\Sigma$. Then we have:

LEMMA 3.29.

$$\frac{1}{\sqrt{\pi}}\int_{\sqrt{R}}^{\infty}\frac{dt}{\sqrt{t}}\int_{M_2^R}\operatorname{Tr}\mathcal{E}_2^R(t;x,x)\,dx = \mathrm{O}(e^{-cR}).$$

PROOF. For any eigenvalue $\mu\neq 0$ of $\mathcal{D}_{2,P_>}^R$ and $R>0$ sufficiently large ($R\cdot c_0^2\geq 1$, where c_0 denotes the lower uniform bound for μ^2 of Theorem 3.17) we have

$$\begin{aligned}\left|\int_{\sqrt{R}}^{\infty}\frac{1}{\sqrt{t}}\mu e^{-t\mu^2}\,dt\right| &\leq \int_{\sqrt{R}}^{\infty}\frac{1}{\sqrt{t}}|\mu|e^{-t\mu^2}\,dt = \int_{|\mu|R^{1/4}}^{\infty}e^{-\tau^2}\,d\tau\\ &\leq \int_{|\mu|R^{1/4}}^{\infty}\tau e^{-\tau^2}\,d\tau = \Big[-\frac{1}{2}e^{-\tau^2}\Big]_{|\mu|R^{1/4}}^{\infty}\\ &= \frac{1}{2}e^{-\mu^2\sqrt{R}},\end{aligned}\tag{3.36}$$

which gives

$$\begin{aligned}&\left|\int_{\sqrt{R}}^{\infty}\frac{dt}{\sqrt{t}}\int_{M_2^R}\operatorname{Tr}\left(\mathcal{D}_{2,P_>}^R\, e^{-t(\mathcal{D}_{2,P_>}^R)^2}\right)\right| \leq \int_{\sqrt{R}}^{\infty}\frac{dt}{\sqrt{t}}\int_{M_2^R}\sum_{\mu\neq 0}|\mu|e^{-t\mu^2}\,dt\\ &\leq \frac{1}{2}\sum_{\mu\neq 0}e^{-\mu^2\sqrt{R}} = \frac{1}{2}\sum_{\mu\neq 0}e^{-(\sqrt{R}-1)\mu^2}e^{-\mu^2}\\ &\leq C_1\, e^{-\sqrt{R}\mu_0^2}\operatorname{Tr}\left(e^{-\left(\mathcal{D}_{2,P_>}^R\right)^2}\right)\leq C_2\, e^{-\sqrt{R}\mu_0^2}\operatorname{Vol}(M_2^R)\\ &\leq C_3\, e^{-\sqrt{R}\mu_0^2}\leq C_3\, e^{-C_4\sqrt{R}}.\end{aligned}$$

Here we have exploited that the heat kernel $\mathrm{e}_2^R(t;x,x')$ of the operator $\mathcal{D}_{2,P_>}^R$ can be estimated by

$$|\mathrm{e}_2^R(t;x,x')|\leq c_1 t^{-\frac{m}{2}}e^{c_2 t}e^{-c_3\frac{d^2(x,x')}{t}}$$

according to (3.13). Thus,

$$\left|\operatorname{Tr}\left(e^{-(\mathcal{D}_{2,P_>}^R)^2}\right)\right|\leq\int_{M_2^R}|\operatorname{Tr}\mathrm{e}_2^R(1;x,x)|\,dx\leq c_1 e^{c_2}\int_{M_2^R}dx.\tag{3.37}$$

□

3.8. The spectrum on the closed stretched manifold. Thus far, we have proved the asymptotic equation

$$\frac{1}{\sqrt{\pi}}\int_0^{\sqrt{R}}\frac{dt}{\sqrt{t}}\int_{M_2^R}\operatorname{Tr}\mathcal{E}^R(t;x,x)\,dx+\mathrm{O}(e^{-cR})=\eta_{\mathcal{D}^R_{2,P_>}}(0)$$

as $R\to\infty$. It follows that

$$\lim_{R\to\infty}\eta_R=\lim_{R\to\infty}\left(\eta_{\mathcal{D}^R_{1,P_<}}(0)+\eta_{\mathcal{D}^R_{2,P_>}}(0)\right),$$

where

$$\eta_R:=\frac{1}{\sqrt{\pi}}\int_0^{\sqrt{R}}\frac{dt}{\sqrt{t}}\int_{M^R}\operatorname{Tr}\mathcal{E}^R(t;x,x)\,dx.$$

To prove Theorem 3.8, we still have to show (3.21), i.e., that we can extend the integration from $\sqrt{R}$ to infinity:

$$\frac{1}{\sqrt{\pi}}\int_{\sqrt{R}}^{\infty}\frac{dt}{\sqrt{t}}\int_{M_2^R}\operatorname{Tr}\mathcal{E}^R(t;x,x)\,dx=\mathrm{O}(e^{-cR})\text{ as }R\to\infty.$$

Recall that $\mathcal{E}^R(t;x,x')$ denotes the kernel of the operator $\mathcal{D}^Re^{-t(\mathcal{D}^R)^2}$ on the closed stretched manifold M^R.

Formally, our *task* of proving the preceding estimate is reminiscent of our previous task of proving the corresponding estimate for the kernel $\mathcal{E}_2^R(t;x,x')$ of the operator $\mathcal{D}_2^Re^{-t(\mathcal{D}^R_{2,P_>})^2}$ (see Lemma 3.29). Both integrals are over the same prolonged compact manifold M_2^R with boundary $\{-R\}\times\Sigma$. However, the methods we can apply are different: In the previous case, we had a uniform positive bound for the absolute value of the smallest nonvanishing eigenvalue of the boundary value problem $\mathcal{D}^R_{2,P_>}$ for sufficiently large R.

As mentioned above in Remark 3.18, such a bound does not exist for the Dirac operator $\mathcal{D}^R$ on the closed stretched manifold M^R. Moreover, for the spectral boundary condition we shall show

$$\dim\operatorname{Ker}\mathcal{D}^R_{2,P_>}=\dim\operatorname{Ker}\mathcal{D}_{2,P_>}\quad\text{and}\quad\eta_{\mathcal{D}^R_{2,P_>}}(0)=\eta_{\mathcal{D}_{2,P_>}}(0)$$

for any R (see Proposition 3.40 below). For $\mathcal{D}^R$, on the contrary, the dimension of the kernel can change and, thus, $\eta_{\mathcal{D}^R}$ can admit an integer jump in value as $R\to\infty$. This is due to the presence of 'small' eigenvalues created by L^2 solutions of the operators $\mathcal{D}_1^\infty$ and $\mathcal{D}_2^\infty$ on the half-manifolds with cylindrical ends. We use a straightforward analysis of small eigenvalues inspired by the proof of Theorem 3.17 to prove the following result.

THEOREM 3.30. *There exist $R_0>0$ and positive constants a_1, a_2, and a_3, such that for any $R>R_0$, the eigenvalue μ of the operator $\mathcal{D}^R$ is either bounded away from 0 with $a_1<|\mu|$, or it is exponentially small $|\mu|<a_2e^{-a_3R}$. Let $\mathcal{W}^R$ denote the subspace of $L^2(M^R;S)$ spanned by the eigensections of $\mathcal{D}^R$ corresponding to the exponentially small eigenvalues. Then* $\dim\mathcal{W}^R=q$, *where* $q=\dim(\operatorname{Ker}\mathcal{D}_1^\infty)+\dim(\operatorname{Ker}\mathcal{D}_2^\infty)$.

Recall from Proposition 3.24 that the operator $\mathcal{D}_j^\infty$, acting on the first Sobolev space $\mathcal{H}^1(M_j^\infty; S)$, is an (unbounded) self-adjoint Fredholm operator in $L^2(M_j^\infty; S)$ which has a discrete spectrum in the interval $(-\lambda_1, +\lambda_1)$ where λ_1 denotes the smallest positive eigenvalue of the tangential operator B. Thus, the space $\operatorname{Ker} \mathcal{D}_j^\infty$ of L^2 solutions is of finite dimension.

To prove the theorem we first investigate the small eigenvalues of the operator $\mathcal{D}^R$ and the pasting of L^2 solutions. Let $R > 0$. We reparametrize the normal coordinate u such that $M_1^R = M_1 \cup ((-R, 0] \times \Sigma)$ and $M_2^R = ([0, R) \times \Sigma) \cup M_2$, and introduce the subspace $\mathcal{V}^R \subset L^2(M^R; S)$ spanned by L^2 solutions of the operators $\mathcal{D}_j^\infty$. We choose an auxiliary smooth real function $f^R = f_1^R \cup f_2^R$ on M^R with $f^R = 1$ outside the cylinder $[-R, R] \times \Sigma$, and where f^R is a function of the normal variable u on the cylinder. Moreover, we assume $f^R(-u) = f^R(u)$ (i.e., $f_1^R(-u) = f_2^R(u)$), and that f_2^R is an increasing function of u with

$$f_2^R(u) = \begin{cases} 0 & \text{for } 0 \leq u \leq \frac{R}{4}, \\ 1 & \text{for } \frac{R}{2} \leq u \leq R. \end{cases}$$

We also assume that there exists a constant $\gamma > 0$ such that $|\frac{\partial^p f_2^R}{\partial u^p}(u)| < \gamma R^{-p}$. If $s_j \in C^\infty(M_j^\infty; S)$, we define $s_1 \cup_{f^R} s_2$ by the formula

$$\left(s_1 \cup_{f^R} s_2\right)(x) := \begin{cases} f_1^R(x) s_1(x) & \text{for } x \in M_1^R, \\ f_2^R(x) s_2(x) & \text{for } x \in M_2^R. \end{cases}$$

Clearly, we have

$$\begin{aligned} s_1 \cup_{f^R} s_2 &= s_1 \cup_{f^R} 0 + 0 \cup_{f^R} s_2, \\ \mathcal{D}^R(s_1 \cup_{f^R} s_2) &= (\mathcal{D}_1^\infty s_1) \cup_{f^R} (\mathcal{D}_2^\infty s_2) + s_1 \cup_{g^R} s_2, \quad \text{and} \\ \left|s_1 \cup_{f^R} s_2\right|^2 &= \left|s_1 \cup_{f^R} 0\right|^2 + \left|0 \cup_{f^R} s_2\right|^2, \end{aligned} \tag{3.38}$$

where $g^R := g_1^R \cup g_2^R$ with $g_j^R(u, y) = \sigma(y) \frac{\partial f_j^R}{\partial u}(u, y)$ and $|\cdot|$ denotes the L^2 norm on the manifold M^R.

DEFINITION 3.31. The subspace $\mathcal{V}^R \subset C^\infty(M^R; S)$ is defined by

$$\mathcal{V}^R := \operatorname{span}\{s_1 \cup_{f^R} s_2 \mid s_j \in \operatorname{Ker} \mathcal{D}_j^\infty\}.$$

Let $\{s_{1,1}, \ldots, s_{1,q_1}\}$ be a basis of $\operatorname{Ker} \mathcal{D}_1^\infty$ and let $\{s_{2,1}, \ldots, s_{2,q_2}\}$ be a basis of $\operatorname{Ker} \mathcal{D}_2^\infty$. Then the $q = q_1 + q_2$ sections $\{s_{1,\nu_1} \cup_{f^R} 0\} \cup \{0 \cup_{f^R} s_{2,\nu_2}\}$ form a basis of $\mathcal{V}^R$. We want to show that $\mathcal{V}^R$ approximates the space $\mathcal{W}^R$ of eigensections of $\mathcal{D}^R$ corresponding to the 'small' eigenvalues, for R sufficiently large. We begin with an elementary result:

LEMMA 3.32. *There exists R_0, such that for any $R > R_0$ and any $s \in \mathcal{V}^R$, the following estimate holds*

$$\left|\mathcal{D}^R s\right| \leq e^{-\lambda_1 R} |s| .$$

PROOF. It suffices to prove the estimate for basis sections of $\mathcal{V}^R$. Thus, let $s = s_1 \cup_{f^R} 0$ with $s_1 \in \operatorname{Ker} \mathcal{D}_1^\infty$. By (3.38) we have

$$\mathcal{D}^R s(x) = \begin{cases} 0 & \text{for } x \in M_1 \cup M_2, \\ \sigma(y)\frac{\partial f_1^R}{\partial u}(u,y)s_1(u,y) & \text{for } x = (u,y) \in [-R,R] \times \Sigma. \end{cases} \tag{3.39}$$

Here f_1^R is continued in a trivial way on the whole cylinder $[-R,R] \times \Sigma$. Now, s_1 is a L^2 solution of $\mathcal{D}_1^\infty$, hence $s_1(u,y) = \sum_k c_k e^{-(R+u)\lambda_k}\varphi_k(y)$ on this cylinder where $\{\varphi_k, \lambda_k; \sigma\varphi_k, -\lambda_k\}_{k\in\mathbb{N}}$ is, as above, a spectral resolution of $L^2(\Sigma; S)$ for B. We estimate the norm of $\mathcal{D}^R s$:

$$\begin{aligned}
\left|\mathcal{D}^R s\right|^2 &= \left|\frac{\partial f_1^R}{\partial u} s_1\right|^2 \\
&= \sum_k \int_{-\frac{R}{2}}^{-\frac{R}{4}} \int_\Sigma \left(\frac{\partial f_1^R}{\partial u}\right)^2 |c_k|^2 e^{-2(R+u)\lambda_k} (\varphi_k(y); \varphi_k(y))\, dy\, du \\
&= \int_{-\frac{R}{2}}^{-\frac{R}{4}} \left(\frac{\partial f_1^R}{\partial u}\right)^2 \sum_k |c_k|^2 e^{-2(R+u)\lambda_k} 1\, du \\
&\le \frac{\gamma^2}{R^2} \sum_k \left(|c_k|^2 \int_{-\frac{R}{2}}^{-\frac{R}{4}} e^{-2(R+u)\lambda_k}\, du\right) \\
&= \frac{\gamma^2}{R^2} \sum_k \left(|c_k|^2 \int_{R\lambda_k}^{\frac{3}{2}R\lambda_k} e^{-v}\, \frac{dv}{2\lambda_k}\right) \\
&\le \frac{\gamma^2}{R^2} \sum_k |c_k|^2 \left(\frac{e^{-R\lambda_k} - e^{-\frac{3}{2}R\lambda_k}}{2\lambda_k}\right) \\
&\le \frac{\gamma^2}{R^2} \sum_k \frac{e^{-R\lambda_k}}{2\lambda_k} |c_k|^2 \le \frac{\gamma^2}{R^2} e^{-R\lambda_1} \sum_k \frac{|c_k|^2}{2\lambda_k}.
\end{aligned}$$

On the other hand, we have the elementary inequality

$$\begin{aligned}
|s|^2 = \left|s_1 \cup_{f^R} 0\right|^2 &\ge \int_{-R}^{-R+1} \int_\Sigma |s_1(u,y)|^2\, dydu \\
&= \sum_k |c_k|^2 \left(\frac{1 - e^{-2\lambda_k}}{2\lambda_k}\right) \ge d \sum_k \frac{|c_k|^2}{2\lambda_k},
\end{aligned}$$

with $0 < d \le 1 - e^{-2\lambda_1}$. Thus, we have the following estimate for any $s \in \mathcal{V}^R$ of the form $s_1 \cup_{f^R} 0$ and for sufficiently large R

$$\left|\mathcal{D}^R s\right|^2 \le \frac{\gamma^2}{R^2 d} e^{-R\lambda_1} d \sum_k \frac{|c_k|^2}{2\lambda_k} \le \frac{\gamma^2}{R^2 d} e^{-R\lambda_1} |s|^2 \le e^{-R\lambda_1} |s|^2.$$

For $s = 0 \cup_{f^R} s_2$, we estimate the norm of $\mathcal{D}_R s$ in the same way, in view of the fact that s_2 has the form $s_2(u,y) = \sum_k d_k e^{(u+R)\lambda_k}\sigma(y)\varphi_k(y)$ on the cylinder. □

Let $\{\lambda_k;\psi_k\}$ denote a spectral decomposition of the space $L^2(M^R;S)$ generated by the operator $\mathcal{D}^R$. For $a>0$, let P_a denote the orthogonal projection onto the space $\mathcal{H}_a := \operatorname{span}\{\psi_k \mid |\lambda_k|>a\}$.

LEMMA 3.33. *For sufficiently large R, we have the estimate*

$$|(I-P_{e^{-R\lambda_1/4}})\,s| \le e^{-R\lambda_1/2}\,|s| \quad \textit{for all } s\in\mathcal{V}^R.$$

PROOF. We represent s as the series $s=\sum_k a_k\psi_k$. We have

$$\begin{aligned}|(I-P_{e^{-R\lambda_1/4}})\,s| &= \sum_{\lambda_k^2>e^{-R\lambda_1/2}} a_k^2 \le \sum_{\lambda_k^2>e^{-R\lambda_1/2}} e^{\frac{R\lambda_1}{2}}\lambda_k^2a_k^2\\ &\le \sum_k e^{R\lambda_1/2}\lambda_k^2a_k^2 = e^{R\lambda_1/2}\left|\mathcal{D}^Rs\right|^2\\ &\le e^{R\lambda_1/2}e^{-R\lambda_1}\,|s|^2 = e^{-R\lambda_1/2}\,|s|^2\,.\end{aligned}$$

□

PROPOSITION 3.34. *The spectral projection $P_{e^{-R\lambda_1/4}}$ restricted to the subspace $\mathcal{V}^R$ is an injection. In particular, $\mathcal{D}^R$ has at least q eigenvalues ρ such that $|\rho|\le e^{-R\lambda_1/4}$, where q is the sum of the dimensions of the spaces* $\operatorname{Ker}\mathcal{D}_j^\infty$ *of L^2 solutions of the operators $\mathcal{D}_1^\infty$ and $\mathcal{D}_2^\infty$.*

PROOF. Let $s\in\mathcal{V}^R$, and assume that $P_{e^{-R\lambda_1/4}}(s)=0$. We have

$$|s| = |(I-P_{e^{-R\lambda_1/4}})\,s| \le e^{-\frac{R\lambda_1}{2}}\,|s| \le \frac{1}{2}\,|s|\,,$$

for R sufficiently large. □

The proposition shows that the operator $\mathcal{D}^R$ has at least q exponentially small eigenvalues with corresponding eigensections, which we can approximate by pasting together L^2 solutions. Now we will show that this makes the list of eigenvalues approaching 0 as $R\to+\infty$ complete.

Let ψ be an eigensection of $\mathcal{D}^R$ corresponding to an eigenvalue μ, where $|\mu|<\lambda_1$. As in the proof of Theorem 3.17, we expand $\psi|_{[-R,R]\times\Sigma}$ in terms of a spectral resolution

$$\{\varphi_k,\lambda_k;\sigma\varphi_k,-\lambda_k\}_{k\in\mathbb{N}}$$

of $L^2(\Sigma;S)$ generated by B:

$$\psi(u,y) = \sum_{k=1}^{\infty} f_k(u)\varphi_k(y) + g_k(u)\sigma\varphi_k\,,$$

where the coefficients satisfy the system of ordinary differential equations of (3.31)

$$\begin{pmatrix} f_k' \\ g_k'\end{pmatrix} = \mathbf{A}_k\begin{pmatrix} f_k \\ g_k\end{pmatrix} \quad \text{with } \mathbf{A}_k := \begin{pmatrix} -\lambda_k & \mu \\ -\mu & \lambda_k\end{pmatrix}.$$

For the eigenvalues $\pm\sqrt{\lambda_k^2-\mu^2}$ of $\mathbf{A}_k$ and the eigenvectors

$$\begin{pmatrix}\lambda_k+\sqrt{\lambda_k^2-\mu^2}\\ \mu\end{pmatrix} \quad\text{and}\quad \begin{pmatrix}\mu\\ \lambda_k+\sqrt{\lambda_k^2-\mu^2}\end{pmatrix},$$

we get a natural splitting of $\psi(u,y)$ in the form $\psi(u,y)=\psi_+(u,y)+\psi_-(u,y)$ with

$$\psi_+(u,y)=\sum_k a_k e^{-\sqrt{\lambda_k^2-\mu^2}u}\left\{\left(\lambda_k+\sqrt{\lambda_k^2-\mu^2}\right)\varphi_k(y)+\mu\sigma(y)\varphi_k(y)\right\},$$

$$\psi_-(u,y)=\sum_k b_k e^{\sqrt{\lambda_k^2-\mu^2}u}\left\{\mu\varphi_k(y)+\left(\lambda_k+\sqrt{\lambda_k^2-\mu^2}\right)\sigma(y)\varphi_k(y)\right\}.$$

Then we have the following estimate of the L^2 norm of ψ in the y direction on the cylinder:

LEMMA 3.35. *Assume that $|\psi|=1$. There exist positive constants c_1, c_2 such that $\left|\psi_{|\{u\}\times\Sigma}\right|\le c_1e^{-c_2R}$ for $-\frac{3}{4}R\le u\le\frac{3}{4}R$.*

PROOF. We have

$$\begin{aligned}
&\left|\psi_{|\{-R+r\}\times\Sigma}\right|^2\\
&\quad\le e^{-2r\sqrt{\lambda_k^2-\mu^2}}\left|\sum_k a_k e^{-R\sqrt{\lambda_k^2-\mu^2}}\left\{\left(\lambda_k+\sqrt{\lambda_k^2-\mu^2}\right)f_k+\mu\sigma f_k\right\}\right|^2\\
&\quad= e^{-2r\sqrt{\lambda_k^2-\mu^2}}\left|\psi_{|\{-R\}\times\Sigma}\right|^2.
\end{aligned}$$

In the same way we get $\left|\psi_{|\{R-r\}\times\Sigma}\right|^2\le e^{-2r\sqrt{\lambda_k^2-\mu^2}}\left|\psi_{|\{R\}\times\Sigma}\right|^2$.

Let us observe that, in fact, the argument used here proves that

$$\left|\psi_{+|\{r\}\times\Sigma}\right|\le e^{-(r-s)\sqrt{\lambda_k^2-\mu^2}}\left|\psi_{+|\{s\}\times\Sigma}\right| \quad\text{and}$$
$$\left|\psi_{-|\{s\}\times\Sigma}\right|\le e^{-(r-s)\sqrt{\lambda_k^2-\mu^2}}\left|\psi_{-|\{r\}\times\Sigma}\right|$$

for any $-R<s<r<R$. We also have another elementary inequality

$$\left|\psi_{|\{r\}\times\Sigma}\right|^2\ge\left|\psi_{+|\{r\}\times\Sigma}\right|^2-2\left|\psi_{+|\{r\}\times\Sigma}\right|\left|\psi_{-|\{r\}\times\Sigma}\right|.$$

This helps estimate the L^2 norm of $\psi_\pm$ in the y direction. We have

$$\begin{aligned}
|\psi|^2&\ge\int_{-R}^{-R+1}\left|\psi_{|\{u\}\times\Sigma}\right|^2du\\
&\ge\int_{-R}^{-R+1}\left(\left|\psi_{+|\{u\}\times\Sigma}\right|^2-2\left|\psi_{+|\{u\}\times\Sigma}\right|\left|\psi_{-|\{u\}\times\Sigma}\right|\right)du\\
&\ge\left|\psi_{+|\{-R\}\times\Sigma}\right|^2\\
&\qquad-2\int_{-R}^{-R+1}\left|\psi_{+|\{-R\}\times\Sigma}\right|e^{-2R\sqrt{\lambda_k^2-\mu^2}}\left|\psi_{-|\{R\}\times\Sigma}\right|du\\
&\ge\left|\psi_{+|\{-R\}\times\Sigma}\right|^2-2e^{-2R\sqrt{\lambda_k^2-\mu^2}}\left|\psi_{+|\{-R\}\times\Sigma}\right|\left|\psi_{-|\{R\}\times\Sigma}\right|.
\end{aligned}$$

In the same way we obtain

$$|\psi|^2 \geq \left|\psi_{-|\{R\}\times\Sigma}\right|^2 - 2e^{-2R\sqrt{\lambda_k^2-\mu^2}} \left|\psi_{+|\{-R\}\times\Sigma}\right| \left|\psi_{-|\{R\}\times\Sigma}\right|.$$

We add the last two inequalities and use

$$2\left|\psi_{+|\{-R\}\times\Sigma}\right| \; \left|\psi_{-|\{R\}\times\Sigma}\right| \leq \left|\psi_{+|\{-R\}\times\Sigma}\right|^2 + \left|\psi_{-|\{R\}\times\Sigma}\right|^2$$

to obtain

$$2|\psi|^2 \geq \left(1 - e^{-2R\sqrt{\lambda_k^2-\mu^2}}\right)\left(\left|\psi_{+|\{-R\}\times\Sigma}\right|^2 + \left|\psi_{-|\{R\}\times\Sigma}\right|^2\right).$$

This gives us the inequality we need, namely

$$\left|\psi_{\pm|\{\mp R\}\times\Sigma}\right|^2 \leq 4|\psi|^2.$$

Now we finish the proof of the lemma.

$$\begin{aligned}\left|\psi_{|\{u\}\times\Sigma}\right| &= \left|\psi_{+|\{u\}\times\Sigma} + \psi_{-|\{u\}\times\Sigma}\right| \\ &\leq e^{-(u+R)\sqrt{\lambda_k^2-\mu^2}}\left|\psi_{+|\{-R\}\times\Sigma}\right| + e^{-(R-u)\sqrt{\lambda_k^2-\mu^2}}\left|\psi_{-|\{R\}\times\Sigma}\right| \\ &\leq 2\left(e^{-(u+R)\sqrt{\lambda_k^2-\mu^2}} + e^{-(R-u)\sqrt{\lambda_k^2-\mu^2}}\right)|\psi| \leq c_1 e^{-c_2 R}\end{aligned}$$

for certain positive constants c_1, c_2 when $-\frac{3}{4}R \leq u \leq \frac{3}{4}R$. □

We are ready to state the technical main result of this section.

THEOREM 3.36. *Let ψ denote an eigensection of the operator $\mathcal{D}^R$ corresponding to an eigenvalue μ, where $|\mu| < \lambda_1$. Assume that ψ is orthogonal to the subspace $P_{e^{-R\lambda_1/4}}\mathcal{V}^R \subset L^2(M^R;S)$. Then there exists a positive constant c, such that $|\mu| > c$.*

To prove the theorem we may assume that $|\psi| = 1$. We begin with an elementary consequence of Lemma 3.33.

LEMMA 3.37. *For any $s \in \mathcal{V}^R$ we have*

$$|\langle\psi\,;s\rangle| \leq e^{-R\lambda_1/2}|s|.$$

PROOF. We have

$$\begin{aligned}|\langle\psi;s\rangle| &= |\langle\psi;P_{e^{-R\lambda_1/4}}(s) + (s - P_{e^{-R\lambda_1/4}}(s))\rangle| = |\langle\psi;P_{e^{-R\lambda_1/4}}(s)\rangle| \\ &\leq |\psi|\;|P_{e^{-R\lambda_1/4}}(s)| \leq e^{-R\lambda_1/2}|s|.\end{aligned}$$

□

We want to compare ψ with the eigensections on the corresponding manifolds with cylindrical ends. We use ψ to construct a suitable section on the manifold $M_2^\infty = ((-\infty,R]\times\Sigma)\cup M_2$. (Note the reparametrization compared with the convention chosen in the beginning of this chapter.) Let $h : M_2^\infty \to \mathbb{R}$ be a smooth increasing function such that h is equal to 1 on M_2 and h is a function of the normal variable on the cylinder, equal to 0 for $u \leq \frac{1}{2}R$, and equal to 1 for $\frac{3}{4}R \leq u$. We also assume, as usual, that $|\frac{\partial^p h}{\partial u^p}| \leq \gamma R^{-p}$ for some constant $\gamma > 0$. We define

$$\psi_2^\infty(x) := \begin{cases} h(x)\psi(x) & \text{for } x \in M_2^R, \\ 0 & \text{for } x \in (-\infty, 0] \times \Sigma. \end{cases}$$

PROPOSITION 3.38. *There exist positive constants c_1 and c_2, such that*

$$|\langle \psi_2^\infty; s\rangle| \leq c_1 e^{-c_2 R}\, |s|$$

for any $s \in \operatorname{Ker} \mathcal{D}_2^\infty$.

PROOF. For a suitable cutoff function f_2^R we have

$$\begin{aligned} |\langle \psi_2^\infty; s\rangle| &= \left| \int_{M_2^\infty} (\psi_2^\infty(x); s(x))\, dx \right| = \left| \int_{M_2^R} \big(h(x)\psi(x); f_2^R(x)s(x)\big)\, dx \right| \\ &\leq \left| \int_{M_2^R} \big(\psi(x); f_2^R(x)s(x)\big)\, dx \right| + \left| \int_{M_2^R} \big((1-h(x))\psi(x); f_2^R(x)s(x)\big)\, dx \right|. \end{aligned}$$

We use Lemma 3.37 to estimate the first summand:

$$\begin{aligned} \left| \int_{M_2^R} \big(\psi(x); f_2^R(x)s(x)\big)\, dx \right| &= \left| \int_{M_2^R} \big(\psi(x); \big(0 \cup_{f^R} s\big)(x)\big)\, dx \right| \\ &= \big| \big\langle \psi; 0 \cup_{f^R} s \big\rangle \big| \leq e^{-R\lambda_1/2}\, |s|. \end{aligned}$$

We use Lemma 3.35 to estimate the second summand:

$$\begin{aligned} &\left| \int_{M_2^R} \big((1-h(x))\psi(x); f_2^R(x)s(x)\big)\, dx \right| \\ &\qquad \leq \int_{M_2^R} \big|\big((1-h(x))\psi(x); f_2^R(x)s(x)\big)\big|\, dx \\ &\qquad \leq \int_{M_2^R} |((1-h(x))\psi(x))|\, \big|f_2^R(x)s(x)\big|\, dx \\ &\qquad \leq \left(\int_{M_2^R} |((1-h(x))\psi(x))|^2\, dx \right)^{\frac{1}{2}} |s| \\ &\qquad \leq \left(\int_0^{\frac{3}{4}R} \left|\psi_{|\{u\}\times\Sigma}\right|^2 \right)^{\frac{1}{2}} |s|\, du \\ &\qquad \leq \big(c_1^2\, e^{-2c_2 R}\, \tfrac{3}{4}R\big)^{\frac{1}{2}}\, |s| \leq c_3\, e^{-c_4 R} |s|. \end{aligned}$$

□

PROOF OF THEOREM 3.36. Now we estimate μ^2 from below by following the proof of Theorem 3.17. We choose $\{s_k\}_{k=1}^{q_2}$ an orthonormal basis of the kernel of the operator $\mathcal{D}_2^\infty$. Let us define

$$\widetilde{\psi} := \psi_2^\infty - \sum_{k=1}^{q_2} \langle \psi_2^\infty, s_k \rangle\, s_k.$$

Then $\widetilde{\psi}$ is orthogonal to $\operatorname{Ker}\mathcal{D}_2^\infty$, and it follows from Proposition 3.38 that

$$\left|\widetilde{\psi}\right| \geq \frac{1}{3}\left|\psi_2^\infty\right| > \kappa > 0,$$

for R large enough, where κ is independent of R, of the specific choice of the eigensection ψ, and of the cutoff function h. Let μ_1^2 denote the smallest nonzero eigenvalue of the operator $(\mathcal{D}_2^\infty)^2$. Once again, it follows from the Min-Max Principle, that $\left\langle (\mathcal{D}_2^\infty)^2\,\widetilde{\psi};\widetilde{\psi}\right\rangle \geq \mu^2\kappa^2$. We have

$$\begin{aligned}
\mu^2 = \langle(\mathcal{D}^R)^2\psi;\psi\rangle &\geq \int_{M^R} |\mathcal{D}^R\psi(x)|^2\,dx \\
&= \int_{M^R} |\mathcal{D}^R\big(h(x)\psi(x) + (1-h(x))\psi(x)\big)|^2\,dx \\
&\geq \int_{M^R} |\mathcal{D}^R h(x)\psi(x)|^2 dx - \int_{M^R} |\mathcal{D}^R\big((1-h(x))\psi(x)\big)|^2\,dx.
\end{aligned}$$

It is not difficult to estimate the first term from below. We have

$$\int_{M_2^\infty} \left|(\mathcal{D}_2^\infty\psi_2^\infty)(x)\right|^2 dx = \langle(\mathcal{D}_2^\infty)^2\psi_2^\infty;\psi_2^\infty\rangle = \langle(\mathcal{D}_2^\infty)^2\widetilde{\psi};\widetilde{\psi}\rangle \geq \mu_1^2\kappa^2.$$

We estimate the second term as follows:

$$\begin{aligned}
\int_{M^R} \left|\mathcal{D}^R(1-h(x)\psi(x))\right|^2 dx &= \int_{M_2^R} \left|(1-h(x))(\mathcal{D}^R\psi)(x) - \sigma(x)\tfrac{\partial h}{\partial u}(x)\psi(x)\right|^2 dx \\
&\leq \int_{M_2^R} \left(\left|\mu(1-h(x))\psi(x)\right|^2 + 2\left|\mu(1-h(x))\psi(x)\right| + \left|\sigma(x)\tfrac{\partial h}{\partial u}(x)\psi(x)\right|^2\right) dx
\end{aligned}$$

Now we use Lemma 3.35 successively to estimate each summand on the right side by $c_1e^{-c_2R}$. This gives us

$$\int_{M_2^R} \left|\mathcal{D}^R\big((1-h)\psi\big)(x)\right|^2 dx \leq c_3e^{-c_4R},$$

and finally we have $\mu^2 \geq \mu_1^2\kappa^2 - c_3e^{-c_4R} \geq \frac{1}{2}\mu_1^2\kappa^2$ for R large enough. □

Theorem 3.30 is an easy consequence of Theorem 3.36.

3.9. The additivity for spectral boundary conditions. In this section we finish the proof of Theorem 3.8. We still have to show equation (3.21):

LEMMA 3.39. *We have* $\eta^R = \mathrm{O}(e^{-cR}) \mod \mathbb{Z}$ *where*

$$\eta^R := \frac{1}{\sqrt{\pi}}\int_{\sqrt{R}}^\infty \frac{1}{\sqrt{t}}\operatorname{Tr}\left(\mathcal{D}^R e^{-t\mathcal{D}_R^2}\right) dt.$$

PROOF. It follows from Theorem 3.30 that we have 'exponentially small' eigenvalues corresponding to the eigensections from the subspace $\mathcal{W}^R$ and the eigenvalues μ bounded away from 0, with $|\mu| \geq a_1$, corresponding to the eigensections from the orthogonal complement of $\mathcal{W}^R$. First we show that we can neglect the contribution due to the eigenvalues that are bounded away from 0. We are precisely in the same situation as with the large t asymptotic of the corresponding integral for the Atiyah-Patodi-Singer boundary problem on the half manifold with the cylinder attached. Literally, we can repeat the proof of Lemma 3.29 by replacing $\mathcal{D}^R_{2,P_>}$ by $\mathcal{D}^R$ and the uniform bound for the smallest positive eigenvalue of $\mathcal{D}^R_{2,P_>}$ by our present bound a_1. Thus, we have

$$\begin{aligned}
&\left|\int_{\sqrt{R}}^{\infty} \frac{1}{\sqrt{t}} \operatorname{Tr}\left(\mathcal{D}^R e^{-t\mathcal{D}_R^2}|_{(\mathcal{W}^R)^\perp}\right) dt\right| \\
&\qquad \le \int_{\sqrt{R}}^{\infty} \frac{1}{\sqrt{t}} \left\{ \sum_{|\mu|\ge a_1} |\mu| e^{-t\mu^2} \right\} dt \\
&\qquad \le \int_{\sqrt{R}}^{\infty} \frac{1}{\sqrt{t}} \left\{ \sum_{|\mu|\ge a_1} e^{-(t-1)\mu^2} \right\} dt \\
&\qquad \le \int_{\sqrt{R}}^{\infty} \frac{1}{\sqrt{t}} \left\{ \sum_{|\mu|\ge a_1} e^{-\mu^2} \right\} e^{-(t-2)a_1^2}\, dt \\
&\qquad \le e^{2a_1^2} \operatorname{Tr}\left(e^{-t\mathcal{D}_R^2}\right) \int_{\sqrt{R}}^{\infty} \frac{1}{\sqrt{t}} e^{-ta_1^2}\, dt \\
&\qquad = e^{2a_1^2} \operatorname{Tr}\left(e^{-t\mathcal{D}_R^2}\right) \frac{1}{a_1} \int_{\sqrt{R}}^{\infty} \frac{1}{\sqrt{t}} e^{-ta_1^2}\, a_1\, dt \\
&\qquad \le \frac{e^{2a_1^2}}{2a_1} \operatorname{Tr}\left(e^{-t\mathcal{D}_R^2}\right) e^{-a_1^2\sqrt{R}}.
\end{aligned}$$

For the last inequality see (3.36). A standard estimate on the heat kernel of the operator $\mathcal{D}^R$ gives (as in (3.37)) the inequality $\operatorname{Tr}\left(e^{-t\mathcal{D}_R^2}\right) \le b_3 \cdot \operatorname{Vol}(M^R) \le b_4 R$, which implies that

$$\left|\int_{\sqrt{R}}^{\infty} \frac{1}{\sqrt{t}} \operatorname{Tr}\left(\mathcal{D}^R e^{-t\mathcal{D}_R^2}|_{(\mathcal{W}^R)^\perp}\right) dt\right| \le b_5\, e^{-b_6\sqrt{R}}. \tag{3.40}$$

This proves that the contribution from the large eigenvalues disappears as $R \to \infty$. The essential part of η^R comes from the subspace $\mathcal{W}^R$:

$$\begin{aligned}
&\frac{1}{\sqrt{\pi}} \int_{\sqrt{R}}^{\infty} \frac{1}{\sqrt{t}} \operatorname{Tr}\left(\mathcal{D}^R e^{-t\mathcal{D}_R^2}|_{\mathcal{W}^R}\right) dt \\
&\qquad = \sum_{|\mu|<a_1} \frac{1}{\sqrt{\pi}} \int_{\sqrt{R}}^{\infty} \frac{1}{\sqrt{t}} \mu e^{-t\mu^2}\, dt \\
&\qquad = \sum_{|\mu|<a_1} \operatorname{sign}(\mu) \frac{2}{\sqrt{\pi}} \int_{|\mu|R^{1/4}}^{\infty} e^{-v^2}\, dv.
\end{aligned} \tag{3.41}$$

It follows from Theorem 3.30 that $\lim_{R\to\infty} |\mu|\, R^{1/4} = 0$. Thus, the last term in (3.41) is equal to

$$\operatorname{sign}_R(\mathcal{D}) := \sum_{|\mu|<a_1} \operatorname{sign}(\mu) \tag{3.42}$$

plus the smooth error term which is rapidly decreasing as $R \to \infty$. □

Thus we have proved Theorem 3.8. In particular, we have proved

$$\lim_{R\to\infty} \eta_{\mathcal{D}^R}(0) \equiv \lim_{R\to\infty} \left\{\eta_{\mathcal{D}^R_{1,P_<}}(0) + \eta_{\mathcal{D}^R_{2,P_>}}(0)\right\} \mod \mathbb{Z}.$$

To establish the true additivity assertion of Corollary 3.9, we show that the preceding eta invariants do not depend on R modulo integers.

PROPOSITION 3.40 (W. Müller). *The eta invariant* $\eta_{\mathcal{D}^R_{2,P_>}}(0) \in \mathbb{R}/\mathbb{Z}$ *is independent of the cylinder length R.*

PROOF. Near to the boundary of M_2^R we parametrize the normal coordinate $u \in [-R, 1)$ with the boundary at $u = -R$. First we show that $\dim \operatorname{Ker} \mathcal{D}^R_{2,P_>}$ is independent of R. Let $s \in \operatorname{Ker} \mathcal{D}^R_{2,P_>}$, namely

$$s \in C^\infty(M_2^R; S), \quad \mathcal{D}_2^R s = 0, \quad \text{and} \quad P_>(s|_{\{-R\}\times\Sigma}) = 0. \tag{3.43}$$

As in equation (3.26) (and in equation (3.34) of the proof of Theorem 3.17) we may expand $s|_{[-R,0]\times\Sigma}$ in terms of the eigensections of the tangential operator B:

$$s(u,y) = \sum_{k=1}^{\infty} e^{\lambda_k u} \sigma(y)\varphi_k(y).$$

Let $R' > R$. Then s can be extended in the obvious way to $\widetilde{s} \in \operatorname{Ker} \mathcal{D}^{R'}_{2,P_>}$, and the map $s \mapsto \widetilde{s}$ defines an isomorphism of $\operatorname{Ker} \mathcal{D}^R_{2,P_>}$ onto $\operatorname{Ker} \mathcal{D}^{R'}_{2,P_>}$. Next, observe that there exists a smooth family of diffeomorphisms $f_R : [0,1) \to [-R, 1)$ which have the following cutoff properties

$$f_R(u) = \begin{cases} u & \text{for } \frac{2}{3} < u < 1, \\ u + R & \text{for } 0 \le u < \frac{1}{3}. \end{cases}$$

Let $\psi_R : [0,1) \times \Sigma \to [-R, 1) \times \Sigma$ be defined by $\psi_R(u,y) := (f_R(u), y)$, and extend ψ_R to a diffeomorphism $\psi_R : M_2 \to M_2^R$ in the canonical way, i.e., ψ_R becomes the identity on $M_2 \setminus ((0,1) \times \Sigma)$. There is also a bundle isomorphism which covers ψ_R. This induces an isomorphism $\psi_R^* : C^\infty(M_2^R; S) \to C^\infty(M_2; S)$. Let

$$\widetilde{\mathcal{D}_2^R} := \psi_R^* \circ \mathcal{D}_2^R \circ (\psi_R^*)^{-1}.$$

Then $\{\widetilde{\mathcal{D}_2^R}\}_R$ is a family of Dirac operators on M_2, and $\widetilde{\mathcal{D}_2^R} = \sigma(\partial_u + B)$ near Σ. We pick the self-adjoint L^2 extension defined by $\operatorname{Dom} \widetilde{\mathcal{D}^R_{2,P_>}} := \psi_R^* \left(\operatorname{Dom} \mathcal{D}^R_{2,P_>}\right)$.

Hence,

$$\eta_{\widetilde{\mathcal{D}^R_{2,P_>}}}(s) = \eta_{\mathcal{D}^R_{2,P_>}}(s) \quad \text{and} \quad \operatorname{Ker} \widetilde{\mathcal{D}^R_{2,P_>}} = \psi_R^* \operatorname{Ker} \mathcal{D}^R_{2,P_>}.$$

In particular, $\dim \widetilde{\mathcal{D}^R_{2,P_>}}$ is constant, and we apply variational calculus to get

$$\frac{d}{dR}\left(\eta_{\mathcal{D}^R_{2,P_>}}(0)\right) = -\frac{2}{\sqrt{\pi}} c_m(R),$$

where $c_m(R)$ is the coefficient of $t^{-1/2}$ in the asymptotic expansion of

$$\operatorname{Tr} \dot{A}_R e^{-tA_R^2} \sim \sum_{j=0}^{\infty} c_j(R) t^{(j-m-1)/2}$$

with $A_R := \widetilde{\mathcal{D}^R_{2,P_>}}$ and $m := \dim M$. Now let S^1_R denote the circle of radius $2R$. We lift the Clifford bundle from Σ to the torus $\mathbf{T}_R := S^1_R \times \Sigma$. We define the action of $\widehat{D^R} : C^\infty(\mathbf{T}_R, S) \to C^\infty(\mathbf{T}_R, S)$ by $\widehat{D^R} = \sigma(\partial_u + B)$. Since $c_m(R)$ is locally computable, it follows in the same way as above that

$$\frac{d}{dR}\left(\eta_{\widehat{D^R}}(0)\right) = -\frac{2}{\sqrt{\pi}} c_m(R).$$

But a direct computation shows that the spectrum of $\widehat{D^R}$ is symmetric. Hence $\eta_{\widehat{D^R}}(s) = 0$ and, therefore, $c_m(R) = 0$. □

In the same way we show that $\eta_{\mathcal{D}^R}$ is independent of R. This proves the additivity assertion of Corollary 3.9. In fact, we have proved a little bit more:

THEOREM 3.41. *The following formula holds for R large enough*

$$\eta_{\mathcal{D}}(0) = \eta_{\mathcal{D}_{1,P_<}}(0) + \eta_{\mathcal{D}_{2,P_>}}(0) + \operatorname{sign}_R(\mathcal{D}),$$

where $\operatorname{sign}_R(\mathcal{D}) := \sum_{|\mu|<a_1} \operatorname{sign}(\mu)$*; see* (3.42).

Theorem 3.41 has an immediate corollary which describes the case in which our additivity formula holds in $\mathbb{R}$, not just in $\mathbb{R}/\mathbb{Z}$.

COROLLARY 3.42. *If* $\operatorname{Ker} \mathcal{D}_1^\infty = \{0\} = \operatorname{Ker} \mathcal{D}_2^\infty$, *then*

$$\eta_{\mathcal{D}}(0) = \eta_{\mathcal{D}_{1,P_<}}(0) + \eta_{\mathcal{D}_{2,P_>}}(0).$$

References

[AdSe96] D.H. Adams and S. Sen, *Partition Function of a quadratic functional and semiclassical approximation for Witten's 3–manifold invariant*, Preprint, Trinity College, Dublin, 1996.

[Ah53] L. Ahlfors, *Complex Analysis*, McGraw-Hill, New York, 1953.

[AlSi80] A. Alonso and B. Simon, *The Birman-Krein-Vishik theory of self-adjoint extensions of semibounded operators*, J. Operator Theory **4** (1980), 251–270.

[AlSi81] ———, *Addenda to "The Birman-Krein-Vishik theory of self-adjoint extensions of semibounded operators"*, J. Operator Theory **6** (1981), 407.

[Ar67] V.I. Arnold, *Characteristic class entering in quantization conditions*, Funkcional Anal. i Priložen. **1** (1967), 1–14. (Russian; English translation in: Functional Anal. Appl. **1** (1967), 1–13; French translation Complément 1 to V.P. Maslov, "Theorie des perturbations et méthodes asymptotiques", Dunod, Gauthier-Villars, Paris, 1972, 341–361).

[Ar57] N. Aronszajn, *A unique continuation theorem for solutions of elliptic partial differential equations or inequalities of second order*, J. Math. Pures Appl. **36** (1957), 235–249.

[ABP73] M.F. Atiyah, R. Bott, and V.K. Patodi, *On the heat equation and the index theorem* (Russ. transl., Moscow, 1973), Invent. Math. **19** (1973), 279–330 (and Errata, **28** (1975), 277–280).

[AtPaSi75] M.F. Atiyah, V.K. Patodi, and I.M. Singer, *Spectral asymmetry and Riemannian geometry. I*, Math. Proc. Cambridge Phil. Soc. **77** (1975), 43–69.

[AtSi68] M.F. Atiyah and I.M. Singer, *The index of elliptic operators I* (Russ. transl. in: Yonexn Matem. Hayk **23** (1968), Nr. 5 (143), 99–142), Ann. of Math. **87** (1968), 484–530.

[AtSi69] ———, *Index theory for skew-adjoint Fredholm operators*, Publ. Math. Inst. Hautes Etudes Sci. **37** (1969), 305–325.

[BeGeVe92] N. Berline, E. Getzler, and M. Vergne, *Heat Kernels and Dirac Operators*, Springer-Verlag, Berlin–Heidelberg–New York, 1992.

[Be96] R.A. Bertlmann, *Anomalies in Quantum Field Theory*, Clarendon Press, Oxford, 1996.

[BiCh89] J.-M. Bismut and J. Cheeger, *η–invariants and their adiabatic limits*, J. Amer. Math. Soc. **2** (1989), 33–70.

[BiFr86] J.-M. Bismut and D.S. Freed, *The analysis of elliptic families. II. Dirac operators, eta invariants, and the holonomy theorem*, Comm. Math. Phys. **107** (1986), 103–163.

[Ble92] D.D. Bleecker, *The supertrace of the steady asymptotic of the spinorial heat kernel*, J. Math. Phys. **33** (1992), 2053–2070.

[BlBo04] D.D. Bleecker and B. Booss–Bavnbek, Index Theory with Applications to Mathematics and Physics, International Publishers, Boston (in preparation).

[Bo79] B. Bojarski, *The abstract linear conjugation problem and Fredholm pairs of subspaces*, in: "In Memoriam I.N. Vekua", Tbilisi Univ, Tbilisi, 1979, pp. 45–60 (Russian).

[Boo72] B. Booss, *Elliptische Topologie von Transmissionsproblemen*, Bonner Mathematische Schriften Nr. 58, Bonn, 1972.

[Boo00] B. Booss–Bavnbek, *The unique continuation property for Dirac operators – revisited*, Geometry and Topology: Aarhus (1998), Contemp. Math. **258** (2000), 21–32.

[BoFu98] B. Booss–Bavnbek and K. Furutani, *The Maslov index – a functional analytical definition and the spectral flow formula*, Tokyo J. Math. **21** (1998), 1–34.

[BoFu99] ———, *Symplectic functional analysis and spectral invariants*, Geometric Aspects of Partial Differential Equations: Roskilde (1998), Contemp. Math. **242** (1999), 53–83.

[BoFuOt01] B. Booss–Bavnbek, K. Furutani, and N. Otsuki, *Criss-cross reduction of the Maslov index and a proof of the Yoshida–Nicolaescu theorem*, Tokyo J. Math. **24** (2001), 113–128.

[BoLePh01] B. Booss–Bavnbek, M. Lesch, and J. Phillips, *Unbounded Fredholm operators and spectral flow*, Preprint math.FA/0108014, to appear in Canad. J. Math.

[BoMaWa02] B. Booss–Bavnbek, M. Marcolli, and B.L. Wang, *Weak ucp and perturbed monopole equations*, Internat. J. Math. **13** (2002), 987–1008.

[BMSW97] B. Booss–Bavnbek, G. Morchio, F. Strocchi, and K.P. Wojciechowski, *Grassmannian and chiral anomaly*, J. Geom. Phys. **22** (1997), 219–244.

[BoScWo98] B. Booss–Bavnbek, S.G. Scott, and K.P. Wojciechowski, *The ς-determinant and C-determinant on the Grassmannian in dimension one*, Letters in Math. Phys. **45** (1998), 353–362.

[BoWo85] B. Booss–Bavnbek and K.P. Wojciechowski, *Desuspension of splitting elliptic symbols, Part I*, Ann. Glob. Analysis and Geometry **3** (1985), 337–383. Part II, Ann. Glob. Analysis and Geometry **4** (1986), 349–400. Addendum: Ann. Glob. Analysis and Geometry **5** (1987), 87.

[BoWo93] ———, *Elliptic Boundary Problems for Dirac Operators*, Birkhäuser, Boston, 1993.

[BrLe99a] J. Brüning and M. Lesch, *On the η-invariant of certain nonlocal boundary value problems*, Duke Math. J. **96** (1999), 425–468.

[BrLe99] ———, *Spectral theory of boundary value problems for Dirac type operators*, Geometric Aspects of Partial Differential Equations: Roskilde (1998), Contemp. Math. **242** (1999), 203–215.

[BrLe01] ———, *On boundary value problems for Dirac type operators. I. Regularity and self-adjointness*, J. Funct. Anal. **185** (2001), 1–62.

[Bu95] U. Bunke, *On the gluing problem for the η-ivariant*, J. Diff. Geom. **41** (1995), 397–448.

[Ca63] A.P. Calderón, *Boundary value problems for elliptic equations*, in: Outlines of the Joint Soviet–American Symposium on Partial Differential Equations, Novosibirsk, 1963, pp. 303–304.

[CaLeMi94] S.E. Cappell, R. Lee, and E.Y. Miller, *On the Maslov index*, Comm. Pure Appl. Math. **47** (1994), 121–186.

[CaLeMi96] ———, *Selfadjoint elliptic operators and manifold decompositions Part I: Low eigenmodes and stretching*, Comm. Pure Appl. Math. **49** (1996), 825–866. *Part II: Spectral flow and Maslov index*, Comm. Pure Appl. Math. **49** (1996), 869–909. *Part III: Determinant line bundles and Lagrangian intersection*, Comm. Pure Appl. Math. **52** (1999), 543–611.

[CaLeMi00] ———, *Surgery formulae for analytical invariants of manifolds*, Preprint, 2000.

[Ca33] T. Carleman, *Sur les systèmes linéaires aux dérivées partielles du premier ordre à deux variables*, Comptes Rendus **197** (1933), 471–474.

[Ca39] ———, *Sur un problème d'unicité pour les systèmes d'équations aux dérivées partielles à deux variables indépendentes*, Ark. Mat. Astr. Fys. **26 B**, No. 17 (1939), 1–9.

[CodLev] E.A. Coddington and N. Levinson, *Theory of Ordinary Differential Equations*, McGraw–Hill Book Company, New York, 1955.

[CoMo95] A. Connes and H. Moscovici, *The local index formula in noncommutative geometry*, GAFA – Geom. Funct. Anal. **5** (1995), 174–243.

[Co56] H.O. Cordes, *Über die eindeutige Bestimmtheit der Lösungen elliptischer Differentialgleichungen durch Anfangsvorgaben*, Nachr. Akad. Wiss. Göttingen Math.-Phys. Kl. IIa, Nr. **11** (1956), 239–258.

[CorLab] H.O. Cordes and J.P. Labrousse, *The invariance of the index in the metric space of closed operators*, J. Math. Mech. **12** (1963) 693–719.

[CoHi] R. Courant and D. Hilbert, *Methods of Mathematical Physics I and II* (trans. from German), Interscience Publishers, New York, 1953/62.

[DaFr94] X. Dai and D. Freed, *η-invariants and determinant lines*, J. Math. Phys. **35** (1994), 5155–5195.

[DaZh96] X. Dai and W. Zhang, *Splitting of the family index*, Comm. Math. Phys. **182** (1996), 303–318.

[DaKi99] M. Daniel and P. Kirk, with an appendix by K.P. Wojciechowski, *A general splitting formula for the spectral flow*, Michigan Math. Journal **46** (1999), 589–617.

[DoWo91] R.G. Douglas and K.P. Wojciechowski, *Adiabatic limits of the η-invariants. The odd-dimensional Atiyah-Patodi-Singer problem*, Comm. Math. Phys. **142** (1991), 139–168.

[Get83] E. Getzler, *Pseudodifferential operators on supermanifolds and the index theorem*, Comm. Math. Phys. **92** (1983), 163–178.

[Get86] ———, *A short proof of the Atiyah-Singer index theorem*, Topology **25** (1986), 111–117.

[Gi73] P.B. Gilkey, *Curvature and eigenvalues of the Laplace operator for elliptic complexes*, Advances in Math. **10** (1973), 344–382.

[Gi95] ———, *Invariance Theory, the Index Theorem, and the Heat Equation*, CRC Press, Boca Raton–Ann Arbor–London–Tokyo, 1995.

[GoKr57] I.Z. Gohberg and M.G. Krein, *The basic propositions on defect numbers, root numbers, and indices of linear operators* (translation from Upehi Mat. Nauk **12**, Nr. 2, (1957) 44–118), Amer. Math. Soc. Transl. (2) **13** (1960) 185–264.

[Go97] M. de Gosson, *Maslov Classes, Metaplectic Representation and Lagrangian Quantization*, Mathematical Research vol. **95**, Akademie Verlag, Berlin 1997.

[Gr99] G. Grubb, *Trace expansions for pseudodifferential boundary problems for Dirac-type operators and more general systems*, Ark. Mat. **37** (1999), 45–86.

[Gr02] ———, *Spectral boundary conditions for generalizations of Laplace and Dirac operators*, Comm. Math. Phys. **240** (2003), 243–280. (math.AP/0302286)

[Gu93] E. Guentner, *K-homology and the index theorem*, Index Theory and Operator Algebras (Boulder 1991), Contemp. Math. **148**, (1993), 47–66.

[Ha52] W. Haack, *Probleme höherer Charakteristik für ein System von zwei elliptischen Differentialgleichungen*, Math. Nachr. **8** (1952), 123–132.

[Har64] P. Hartman, *Ordinary differential equations*, Wiley, New York, 1964.

[He52] G. Hellwig, *Das Randwertproblem eines linearen elliptischen Systems*, Math. Zeitschrift **56** (1952), 388–408.

[Ho63] L. Hörmander, *Linear Partial Differential Equations*, Springer-Verlag, Berlin–Heidelberg–New York, 1963/1969.

[Ho66] L. Hörmander, *Pseudo-differential operators and non-elliptic boundary problems*, Ann. of Math. **83** (1966), 129–209.

[Ho85] L. Hörmander, *The Analysis of Linear Partial Differential Operators III*, Grundlehren d. Mathem. **274**, Springer-Verlag, Berlin–Heidelberg–New York, 1985.

[ItZu80] C. Itzykson and J.-B. Zuber, *Quantum Field Theory*, McGraw–Hill, New York, 1980.

[Kat] T. Kato, *Perturbation Theory for Linear Operators*, Springer-Verlag, Berlin, 1966, 2d ed., 1976.

[KiLe00] P. Kirk and M. Lesch, *The η-invariant, Maslov index, and spectral flow for Dirac-type operators on manifolds with boundary*, Preprint math.DG/0012123, to appear in Forum Math.

[Kl27] F. Klein, *Vorlesungen über die Entwicklung der Mathematik im 19. Jahrhundert. Teil II. Die Grundbegriffe der Invariantentheorie und ihr Eindringen in die Mathematische Physik*, Springer, Berlin, 1927.

[KlWo96] S. Klimek and K.P. Wojciechowski, *Adiabatic cobordism theorems for analytic torsion and η-invariant*, J. Funct. Anal. **136** (1996), 269–293.

[LaSnTu75] B. Lawruk, J. Śniatycki, and W.M. Tulczyjew, *Special symplectic spaces*, J. Differential Equations **17** (1975), 477–497.

[LaMi] H.B. Lawson and M. Michelsohn, *Spin Geometry*, Princeton Univ. Press, Princeton, 1989.

[Le78] J. Leray, *Analyse Lagrangiénne et mécanique quantique: Une structure mathématique apparentée aux développements asymptotiques et à l'indice de Maslov*, Série Math. Pure et Appl., I.R.M.P., Strasbourg, 1978 (English translation 1981, MIT Press).

[LeWo96] M. Lesch and K.P. Wojciechowski, *On the η-invariant of generalized Atiyah-Patodi-Singer problems*, Illinois J. Math. **40** (1996), 30–46.

[MaMe95] R.R. Mazzeo and R.B. Melrose, *Analytic surgery and the eta invariant*, GAFA – Geom. Funct. Anal. **5** (1995), 14–75.

[McK-Si] H.P. McKean, Jr. and I.M. Singer, *Curvature and the eigenvalues of the Laplacian*, J. Differential Geometry **1** (1967), 43–69.

[Me93] R.B. Melrose, *The Atiyah-Patodi-Singer Index Theorem*, A. K. Peters, Boston, 1993.

[Mu94] W. Müller, *Eta invariants and manifolds with boundary*, J. Diff. Geom. **40** (1994), 311–377.

[Ni95] L. Nicolaescu, *The Maslov index, the spectral flow, and decomposition of manifolds*, Duke Math. J. **80** (1995), 485–533.

[Ni00] ———, *On the Space of Fredholm Operators*, Preprint math.DG/0005089.

[Ni73] L. Nirenberg, *Lectures on Linear Partial Differential Equations*, Conference Board of the Mathematical Sciences (CBMS), Regional Conference Series in Mathematics no. 17, Amer. Math. Soc., Providence R.I., 1973.

[No01] S.P. Novikov, *Interview*, European Mathematical Society – Newsletter, Issue **42** (December 2001), 17–20.

[PaWo00] J. Park and K.P. Wojciechowski, *Relative ζ-determinant and adiabatic decomposition for the ζ-determinant of the Dirac Laplacian*, Letters in Math. Phys. **52** (2000), 329–337.

[PaWo02a] ———, with Appendix by Y. Lee, *Adiabatic decomposition of the ζ-determinant of the Dirac Laplacian I. The case of invertible tangential operator*, Comm. in Partial Differential Equations **27** (2002), 1407–1435.

[PaWo02b] ———, *Analytic surgery of the ζ-determinant of the Dirac operator*, Nuclear Physics B. Proc. Supp. **104** (2002), 89–115.

[PaWo02c] ———, *Scattering theory and adiabatic decomposition of the ζ-determinant of the Dirac Laplacian*, Math. Res. Lett. **9** (2002), 17–25.

[PaWo02d] ———, *Adiabatic decomposition of the ζ-determinant and scattering theory*, MPI Preprint, Bonn, 2002.

[PaWo02e] ———, *Scattering theory, the adiabatic decomposition of the ζ-determinant and the Dirac to Neumann operator*, IUPUI Preprint, Indianapolis, 2002.

[Pa71] V.K. Patodi, *Curvature and the eigenforms of the Laplace operator*, J. Diff. Geom. **5** (1971), 233–249.

[Pe89] G.K. Pedersen, *Analysis Now*, Springer-Verlag, New York, 1989.

[Ph96] J. Phillips, *Self-adjoint Fredholm operators and spectral flow*, Canad. Math. Bull. **39** (1996), 460–467.

[Pl61] A. Pliś, *A smooth linear elliptic differential equation without any solution in a sphere*, Comm. Pure Appl. Math. **14** (1961), 599–617.

[RaSi71] D. Ray and I.M. Singer, *R-torsion and the Laplacian on Riemannian manifolds*, Adv. Math. **7** (1971), 145–210.

[ReSi72] M. Reed and B. Simon, *Methods of Modern Mathematical Physics, I. Functional Analysis*, Academic Press, NewYork, 1972.

[ReSi78] ———, *Methods of Modern Mathematical Physics, IV. Analysis of Operators*, Academic Press, NewYork, 1978.

[Sc01] B.W. Schulze, *An algebra of boundary value problems not requiring Shapiro-Lopatinskij conditions*, J. Funct. Anal. **197** (2001), 374–408.

[Sc56] L. Schwartz, *Ecuaciones differenciales elipticas*, Bogota, Universidad Nacional de Colombia, 1956.

[Sch93] A.S. Schwarz, *Quantum Field Theory and Topology*, Springer, Berlin–Heidelberg–New York, 1993. (Russian original: Kvantovaya teoriya polya i topologiya, Nauka, Moscow 1989).

[Sco95] S.G. Scott, *Determinants of Dirac boundary value problems over odd-dimensional manifolds*, Comm. Math. Phys. **173** (1995), 43–76.

[Sco02] ———, *Zeta determinants on manifolds with boundary*, J. Funct. Anal. **192** (2002), 112–185.

[ScWo00] S.G. Scott and K.P.Wojciechowski, *The ς-determinant and Quillen determinant for a Dirac operator on a manifold with boundary*, GAFA – Geom. Funct. Anal. **10** (2000), 1202–1236.

[Se66] R.T. Seeley, *Singular integrals and boundary value problems*, Amer. J. Math. **88** (1966), 781–809.

[Se67] ———, *Complex powers of an elliptic operator*, Proc. Symp. Pure Math. X. AMS Providence, 1967, 288–307.

[Se69] ———, *Topics in pseudo-differential operators*, in: CIME Conference on Pseudo-Differential Operators (Stresa 1968). Ed. Cremonese, Rome, 1969, pp. 167–305.

[Si85] I.M. Singer, *Families of Dirac operators with applications to physics*, Asterisque, hors série (1985), 323–340.

[Sw78] R.C. Swanson, *Fredholm intersection theory and elliptic boundary deformation problems, I*, J. Differential Equations **28** (1978), 189–201.

[Ta96] M.E. Taylor, *Partial Differential Equations I – Basic Theory*, Springer, New York, 1996.

[Tr80] F. Treves, *Pseudodifferential and Fourier Integral Operators I*, Plenum Press, New York, 1980.

[Ve56] I.N. Vekua, *Systeme von Differentialgleichungen erster Ordnung vom elliptischen Typus und Randwertaufgaben* (transl. from Russian), VEB Deutscher Verlag der Wissenschaften, Berlin, 1956.

[We82] N. Weck, *Unique continuation for some systems of partial differential equations*, Applicable Analysis **13** (1982), 53–63.

[Wo94] K.P. Wojciechowski, *The additivity of the η-invariant: The case of an invertible tangential operator*, Houston J. Math. **20** (1994), 603–621.

[Wo95] ———, *The additivity of the η-invariant. The case of a singular tangential operator*, Comm. Math. Phys. **169** (1995), 315–327.

[Wo99] ———, *The ζ-determinant and the additivity of the η-invariant on the smooth, self-adjoint Grassmannian*, Comm. Math. Phys. **201** (1999), 423–444.

[Yu01] Y.L. Yu, *The Index Theorem and the Heat Equation Method*, Nankai Tracts in Mathematics, vol. 2, World Scientific Publishing Co., River Edge, NJ, 2001.

Department of Mathematics, University of Hawaii, Honolulu, HI 96822, USA
E-mail address: bleecker@math.hawaii.edu

Institut for matematik og fysik, Roskilde Universitetscenter, Postboks 260, DK-4000 Roskilde, Denmark
E-mail address: booss@ruc.dk

Operator Theory:
Advances and Applications, Vol. 151, 131–169

Index theory of Dirac operators on manifolds with corners up to codimension two

Paul Loya

ABSTRACT. In this expository article, we survey index theory of Dirac operators using the Gauss-Bonnet formula as the catalyst to discuss index formulas on manifolds with and without boundary. Considered in detail are the Atiyah-Singer and Atiyah-Patodi-Singer index theorems, their heat kernel proofs, and their generalizations to manifolds with corners of codimension two via the method of 'attaching cylindrical ends'.

1. Introduction: The Gauss-Bonnet formula and index theory

The purpose of this paper is to serve as an overview of index theory for Dirac operators on manifolds with corners with emphasis on the b-geometry approach of Melrose [**59**] to such a theory. The underlying theme of this paper is that index formulas are basically generalizations of the classical Gauss-Bonnet formula.

This paper is organized as follows. First, to understand what index theory is and why it is important, we recall the Gauss-Bonnet formula. In particular, we interpret the Gauss-Bonnet formula as an index formula. This interpretation leads us naturally to the Atiyah-Singer index formula for Dirac operators on manifolds without boundary published in 1963 [**6**], which we discuss in Section 2. In 1973, Atiyah, Patodi, and Singer in the seminal paper [**4**] extended the Atiyah-Singer formula for Dirac operators to manifolds with smooth boundary. We present this formula from the 'cylindrical end' point of view in Section 3. We also reformulate the Atiyah-Patodi-Singer (henceforth APS) problem using the language and notation of the b-geometry. In Section 4, we present a b-geometric proof of the APS index formula. Currently, there is no direct analog of the 'APS index formula' for manifolds with corners of codimension two, except under certain restrictive nondegeneracy conditions [**51**], [**67**], which we discuss in Section 5. However, in joint

2000 *Mathematics Subject Classification.* Primary 58J20; Secondary 58J28, 47A53.
Key words and phrases. Index theory; b-pseudodifferential operators; Dirac operators.
Supported by a Ford Foundation Fellowship administered by the National Research Council.

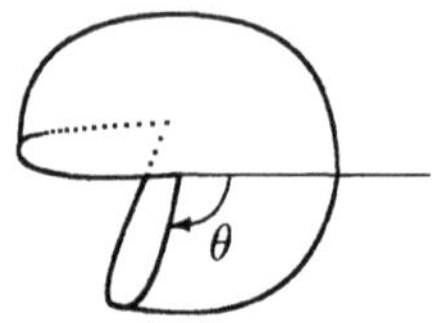

FIGURE 1. The manifold P_θ, where $-\pi < \theta < \pi$, is the unit sphere with a wedge removed.

work with Melrose [**53**], these restrictions are removed by perturbing Dirac operators using *b*-smoothing operators. We discuss such perturbations and the resulting index formulas for the perturbed Dirac operators in Section 6.

Finally, I thank Gerd Grubb, Rafe Mazzeo, and Richard Melrose for helping to make this research possible. I also thank the referees for valuable suggestions.

1.1. The classical Gauss-Bonnet formula. Let M be a compact, oriented, two-dimensional Riemannian manifold without boundary. Then the Gauss-Bonnet theorem states that

$$\chi(M) = \frac{1}{2\pi}\int_M K, \tag{1.1}$$

where $\chi(M)$ is the Euler characteristic of M and K is the Gaussian curvature of M. The interesting aspect of the Gauss-Bonnet formula is that the left-hand side is a topological/combinatorial object while the right-hand side is a geometric object.[1] This formula was proved by Bonnet in 1848, but is attributed also to Gauss because he proved a special case of it earlier. See [**68**, Ch. 8] for a proof of the Gauss-Bonnet formula.

A natural question to ask is: Does the Gauss-Bonnet formula continue to hold if M has a smooth boundary, or more generally, if M has corners; that is, has a crooked boundary? To answer this question, we consider a concrete example. Cut out a wedge from the unit sphere producing the manifold P_θ, where $-\pi < \theta < \pi$, as shown in Figure 1. This manifold is an example of a manifold with corners of codimension two. Let us check if (1.1) holds verbatim for P_θ. Note that P_θ is topologically equivalent to a disk and hence to a triangle and so has Euler characteristic equal to one (since the number of vertices $-$ edges $+$ faces $= 1$ for a triangle). Since the Gaussian curvature of the unit sphere is one, $\int_{P_\theta} K = \text{Area}(P_\theta)$. Thus $\int_{P_\theta} K$ changes with θ: It is approximately 4π when θ is close to $-\pi$ and it decreases to 0 as θ approaches $+\pi$. Hence,

$$\chi(P_\theta) \neq \frac{1}{2\pi}\int_{P_\theta} K \quad \text{for general } \theta.$$

Thus the Gauss-Bonnet formula (1.1) does not hold verbatim when M has corners. Intuitively, one might guess that the formula does not hold because of the presence

[1]The right-hand side of (1.1) turns out to be topological as well since $-K/2\pi$ defines the Chern class of M.

of the boundary and corners. In fact, one can verify directly that the following formula does hold:

$$\chi(P_\theta) = \frac{1}{2\pi}\int_{P_\theta} K + \frac{1}{2\pi}(2\theta). \tag{1.2}$$

The angle 2θ is called the total exterior angle of the corners. This formula is a special case of the general Gauss-Bonnet formula:

THEOREM 1.1 (Gauss-Bonnet, 1878). *Given a compact, oriented, two-dimensional Riemannian manifold M with corners, we have*

$$\begin{aligned}\chi(M) = &\frac{1}{2\pi}\int_M K \\ &+ \frac{1}{2\pi}(\textit{total geodesic curvature of } \partial M) \\ &+ \frac{1}{2\pi}(\textit{sum of the exterior angles at the corners}).\end{aligned} \tag{1.3}$$

Here, the geodesic curvature of ∂M measures the deviation of the smooth components of ∂M from being geodesics. There is no middle term in the formula (1.2) since the smooth components of ∂P_θ are great circles, which are geodesics on the sphere. Hence, the total geodesic curvature of ∂P_θ is zero.

The Gauss-Bonnet formula in (1.3) is very beautiful as it bridges topology, geometry, and now linear algebra: The left-hand side belongs to combinatorial topology while the first two terms on the right are geometrical and the last term is linear algebraic since it has to do with angles between vectors at the corners. Functional analysis also comes into the picture when we interpret the Gauss-Bonnet formula as an index formula. We remark that comparing the Gauss-Bonnet formula (1.1) for a manifold without boundary to the general formula (1.3), we see that the second and third terms on the right in (1.3) can be thought of as correction terms coming from the smooth boundary components and corners respectively.

1.2. The Gauss-Bonnet formula as an index formula. We now explain how the Gauss-Bonnet formula can be interpreted as an index formula. We first need to introduce the Gauss-Bonnet operator. Let M be a compact, oriented, two-dimensional Riemannian manifold without boundary. Let

$$d : C^\infty(M, \Lambda^k) \to C^\infty(M, \Lambda^{k+1})$$

be the exterior derivative, where $C^\infty(M, \Lambda^k)$ denotes the space of smooth k-forms on M, and let

$$d^* : C^\infty(M, \Lambda^{k+1}) \to C^\infty(M, \Lambda^k)$$

be the adjoint of d with respect to the natural L^2 inner product on k-forms given by integration with respect to the Riemannian volume form. Let $\Lambda^{ev} = \Lambda^0 \oplus \Lambda^2$ be the even form bundle and $\Lambda^{odd} = \Lambda^1$ be the odd form bundle. Then both d and d^* map $C^\infty(M, \Lambda^{ev})$ into $C^\infty(M, \Lambda^{odd})$. The operator

$$D_{GB} = d + d^* : C^\infty(M, \Lambda^{ev}) \to C^\infty(M, \Lambda^{odd})$$

is called the *Gauss-Bonnet operator.* By definition of the (nonnegative) Laplacian,

$$D_{GB}^{*} D_{GB} = \Delta,$$

where Δ is the Laplacian on the even forms. Thus, D_{GB} represents in some respects a square root of the Laplacian. For this reason D_{GB} is called a *Dirac operator* after the physicist Paul Dirac who in the 1920's was searching for, and found, a square root of the Laplacian in his efforts to quantize the electron. However, in his case, he was working with a Lorentz metric rather than a Riemannian metric.

Before presenting the index formula interpretation of the Gauss-Bonnet formula, we recall two results from Hodge theory. We denote the Sobolev space of order k by H^k. So, $H^k(M, \Lambda^{ev})$ consists of those even-degree forms u on M such that for each $0 \leq j \leq k$, $(d + d^*)^j u$ is square integrable. Then $H^1(M, \Lambda^{ev})$ is the natural domain of D_{GB}.

THEOREM 1.2. *The operator* $D_{GB} : H^1(M, \Lambda^{ev}) \to L^2(M, \Lambda^{odd})$ *is Fredholm, which means that it is 'almost invertible' in the sense that*

(1) D_{GB} *has a finite dimensional kernel;* $\dim \ker D_{GB} < \infty$.
(2) D_{GB} *has a finite dimensional cokernel;*

$$\dim \operatorname{coker} D_{GB} = \dim \left(L^2(M, \Lambda^{odd}) / \operatorname{Im}(D_{GB}) \right) < \infty.$$

The first condition means that D_{GB} is 'almost injective' in the sense that it is injective up to a finite dimensional space, and the second condition means that D_{GB} is 'almost surjective' in the sense that it is surjective up to a finite dimensional space. The *index* is the difference between the dimensions of the kernel and cokernel:

$$\operatorname{ind} D_{GB} = \dim \ker D_{GB} - \dim \operatorname{coker} D_{GB} \in \mathbb{Z}.$$

Theorem 1.2 follows from the fact that D_{GB} is *elliptic*, and can be proved in a variety of ways, for instance, using pseudodifferential operators [**34**], by embedding properties of Sobolev spaces [**74**], or my favorite proof via the heat operator [**9**]. The second result we need is:

THEOREM 1.3. *The index of* D_{GB} *is the Euler characteristic of* M,

$$\operatorname{ind} D_{GB} = \chi(M).$$

This result can be proved using the Hodge theorem, which is described as follows. Define the k-th deRham cohomology of M by

$$H^k_{\mathrm{dR}}(M) = \{\alpha \in C^\infty(M, \Lambda^k)\,;\, d\alpha = 0\} / \{d\beta\,;\, \beta \in C^\infty(M, \Lambda^{k-1})\}.$$

The Hodge theorem states that given a deRham cohomology class $[\alpha] \in H^k_{\mathrm{dR}}(M)$ there exists a unique representative of this class $\beta \in [\alpha]$ such that $(d + d^*)\beta = 0$. It is worthwhile mentioning that although the exterior derivative d is canonical, the operator d^* depends on the Riemannian metric chosen on the manifold. The work of Connes with Gromov and Moscovici [**25**] treats a branch of index theory which deals with the analytic difficulties involved in not having a suitable invariant inner product. In any case, the Hodge theorem implies the important fact

$$\ker(d+d^*) \text{ on } k \text{ forms} \cong H^k_{\mathrm{dR}}(M).$$

Now using the description of the Euler characteristic in terms of the cohomology:

$$\begin{aligned}\chi(M) &= \sum_{k \text{ even}} \dim H^k_{\mathrm{dR}}(M) - \sum_{k \text{ odd}} \dim H^k_{\mathrm{dR}}(M) \\ &= \dim H^0_{\mathrm{dR}}(M) + \dim H^2_{\mathrm{dR}}(M) - \dim H^1_{\mathrm{dR}}(M),\end{aligned}$$

and the fact that $\operatorname{coker} D_{GB} \cong \ker D^*_{GB}$, one gets Theorem 1.3. In view of the Gauss-Bonnet formula (1.1), we finally have

THEOREM 1.4 (Index version of Gauss-Bonnet). *For the Gauss-Bonnet operator on a compact, oriented, two-dimensional Riemannian manifold M without boundary, we have*

$$\operatorname{ind} D_{GB} = \frac{1}{2\pi}\int_M K.$$

The profound aspect about this version of the Gauss-Bonnet formula is that the left-hand side is a functional analytic object related to the existence and uniqueness of solutions to the equation $D_{GB}u = f$, while the right-hand side is a topological/geometric object. Hence, this formula implies the important fact that the topology/geometry of the manifold can be investigated using functional analysis. We now discuss a far-reaching generalization of this formula to higher dimensional manifolds without boundary. We discuss generalizations of the Gauss-Bonnet formula (1.3) for manifolds with corners in Sections 3, 5, and 6.

2. The Atiyah-Singer index formula

According to Hirzebruch (cf. [**14**, p. vii]) the Atiyah-Singer formula is "one of the deepest and hardest results in mathematics", "probably has wider ramifications in topology and analysis than any other single result". Although the proof of the Atiyah-Singer formula is difficult, understanding it is not if one keeps in mind that it is basically a higher dimensional analog of the Gauss-Bonnet formula.

In this section, we describe the Atiyah-Singer index formula and we outline its proof using the 'heat kernel method'.

2.1. Statement of the Atiyah-Singer index theorem. There are two ingredients to the Atiyah-Singer index formula. The first is topological/geometric data: Let M be an even-dimensional, compact, oriented, Riemannian manifold without boundary, and let E and F be Hermitian vector bundles over M. The second ingredient is functional analytic/geometric data: Let

$$D : C^\infty(M,E) \to C^\infty(M,F)$$

be a *Dirac type operator*. This means that D is an elliptic first-order differential operator such that "$D^*D = \Delta$" in the sense that the principal symbol of D^*D is just the metric $\sigma(D^*D)(\xi) = |\xi|^2$ for all cotangent vectors ξ.[2]

The simplest example of a Dirac type operator is the Cauchy-Riemann operator. Let $M = \mathbb{R}^2$ with its usual Euclidean metric and let $E = F = \mathbb{C}$. Then

$$D_{CR} = \partial_x + i\partial_y$$

is the *Cauchy-Riemann operator*. In this case,

$$D_{CR}^* D_{CR} = \big(- \partial_x + i\partial_y\big)\big(\partial_x + i\partial_y\big) = -\big(\partial_x^2 + \partial_y^2\big)$$

is exactly the Laplacian.

Another example is the higher-dimensional Gauss-Bonnet operator. Let M be an even-dimensional, compact, oriented, Riemannian manifold without boundary, and let E and F be the even and odd degree form bundles, respectively:

$$E = \Lambda^{ev} = \bigoplus_{k \text{ even}} \Lambda^k \quad \text{and} \quad F = \Lambda^{odd} = \bigoplus_{k \text{ odd}} \Lambda^k.$$

Then,

$$D_{GB} = d + d^* : C^\infty(M, \Lambda^{ev}) \to C^\infty(M, \Lambda^{odd})$$

is called the *Gauss-Bonnet operator*. By definition of the Laplacian on forms, we have $D_{GB}^* D_{GB} = \Delta$, and so D_{GB} is a Dirac operator. Since the even and odd form bundles are real bundles, they of course are technically not Hermitian, but we can always make them so by complexifying them. Regardless, the index formula below still applies to the Gauss-Bonnet operator.

The celebrated Atiyah-Singer index theorem is the following.

THEOREM 2.1 (Atiyah-Singer, 1963). *Let $D : C^\infty(M, E) \to C^\infty(M, F)$ be a Dirac type operator on an even-dimensional, compact, oriented, Riemannian manifold without boundary. Then, $D : H^1(M, E) \to L^2(M, F)$ is Fredholm and*

$$\operatorname{ind} D = \int_M K_{AS}, \tag{2.1}$$

where the 'Atiyah-Singer integrand' K_{AS} is an explicitly defined polynomial in the curvature forms of the manifold M and the vector bundles E and F.

The integral of the polynomial K_{AS} is by definition the integral of the volume form component of K_{AS}. For those readers familiar with characteristic classes, the polynomial K_{AS} is the product of the $\hat{A}$ polynomial of M and the (relative) Chern polynomials of E and F, see [**9**, Ch. 4]. Hence, K_{AS} is both a topological and a geometric object. The reason why we assume that M is even-dimensional is that for odd-dimensional manifolds it turns out that both sides of (2.1) are zero.

Note the similarity between the index version of the Gauss-Bonnet formula given in Theorem 1.4 and the Atiyah-Singer formula for the index of a Dirac type

[2]In some occasions, we will need D to be a *compatible* Dirac operator, which means that it is associated to a 'unitary Clifford connection'. For simplicity, we leave this notion undefined, and at the few places where we actually need this extra hypothesis, we will state so in a footnote.

operator: The Gaussian curvature in the Gauss-Bonnet formula is replaced by a polynomial in the curvature of the manifold and vector bundles. As with the Gauss-Bonnet formula, the profound feature of the Atiyah-Singer formula is that the left-hand side of (2.1) is a functional analytic object related to the existence and uniqueness of solutions to the equation $Du = f$, while the right-hand side is a topological/geometric object. In particular, the Atiyah-Singer formula has the following deep consequence: It implies that the topology/geometry of a manifold can be investigated using functional analytic tools, cf. [**72**], [**55**], and Section 2.3.

For an application of the Atiyah-Singer index theorem, consider the Gauss-Bonnet operator defined above. As in the two-dimensional case explained before Theorem 1.4, for a general even-dimensional, compact, oriented, Riemannian manifold without boundary, Hodge theory implies that

$$\operatorname{ind} D_{GB} = \chi(M).$$

On the other hand, see [**9**, Ch. 4] for the details, working out the explicitly defined polynomial K_{AS} for $E = \Lambda^{ev}$ and $F = \Lambda^{odd}$ gives, after a little bit of algebra,

$$K_{AS} = e(M),$$

where $e(M)$ is the Euler, or Pfaffian, polynomial defined by taking the n-th power of the Riemannian curvature tensor of M and multiplying it by $1/n! \times (-1/2\pi)^n$, where $2n$ is the dimension of M. Thus, the Atiyah-Singer index formula implies

$$\chi(M) = \int_M e(M).$$

This generalization of the Gauss-Bonnet formula is due to Chern [**23**].

Another important corollary of the Atiyah-Singer formula is Hirzebruch's formula for the signature of a manifold. Assume now that M is $4k$ dimensional. Then the map

$$H^{2k}_{\mathrm{dR}}(M) \times H^{2k}_{\mathrm{dR}}(M) \ni ([\alpha],[\beta]) \longmapsto \int_M \alpha \wedge \beta \in \mathbb{R}$$

is a well-defined symmetric bilinear map (here, $[\cdot]$ denotes the corresponding cohomology class). We can represent this map by a matrix by choosing any basis of the finite dimensional real vector space $H^{2k}_{\mathrm{dR}}(M)$. The signature of this matrix, the number of positive eigenvalues minus the number of negative ones, is defined independent of the basis chosen; the signature of the manifold, $\operatorname{sign}(M)$, is by definition the signature of the matrix with respect to any such basis. Hirzebruch [**42**] gives a formula for the signature:

$$\operatorname{sign}(M) = \int \mathcal{L}(M),$$

where $\mathcal{L}(M)$ is the $\mathcal{L}$-class polynomial in the curvature tensor of M. This formula is a simple corollary of the Atiyah-Singer index theorem. In this case, D is the 'signature operator', which is equal to $d + d^*$ like the Gauss-Bonnet operator, but with the vector bundles E and F being essentially certain eigenspaces of the Hodge star operator. The details can be found in [**9**]. For those interested readers, the $\mathcal{L}$-class polynomial is given by

$$\mathcal{L}(M) = \sqrt{\det\Big(\frac{R/2\pi i}{\tanh(R/2\pi i)}\Big)},$$

where R is the Riemannian curvature tensor.

Yet another classical formula that is a simple corollary of the Atiyah-Singer index formula is the Riemann-Roch formula and its generalization to complex manifolds, see [**9**], [**34**], [**74**].

2.2. Outline of the proof of the Atiyah-Singer formula. We outline a proof of the Atiyah-Singer index theorem based on the heat kernel approach of Mckean and Singer [**57**] as exploited by Atiyah, Bott, and Patodi in [**2**]. For accessible versions of the proof, see [**9**], [**34**], [**82**]. I especially like the exposition by Roe [**74**].

To start off, we need the heat operators e^{-tD^*D} and e^{-tDD^*}. Consider for instance e^{-tD^*D}. Then, for each $t > 0$,

$$e^{-tD^*D} : C^\infty(M, E) \to C^\infty(M, E),$$

and it is the solution operator to the heat equation for D^*D in the sense that for each $u \in C^\infty(M, E)$, $u_t = e^{-tD^*D}u$ is the unique solution to the heat equation

$$(\partial_t + D^*D)u_t = 0, \quad t > 0; \qquad u_0 = u.$$

The heat operator e^{-tD^*D} can be defined by means of the resolvent and the functional calculus (cf. [**34**], [**51**]), it can be constructed asymptotically via Hadamard's method (cf. [**9**], [**59**]), or it can be defined using the spectrum as follows. Let $\{\lambda_j\}$ be the eigenvalues of the self-adjoint operator D^*D. Then,

$$e^{-tD^*D} = \sum_j e^{-t\lambda_j}\pi_j,$$

where π_j is the orthogonal projection onto the eigenspace associated to the eigenvalue λ_j. This sum converges uniformly and absolutely and in fact, it can be used to show that the heat operator is, for each $t > 0$, a smoothing operator; that is, for each $t > 0$ the heat operator is an integral operator with a smooth Schwartz kernel [**74**]. In particular, for each $t > 0$, the heat operator is trace class and

$$e^{-tD^*D} = \pi_{\ker D^*D} + F(t), \tag{2.2}$$

where the remainder $F(t) \to 0$ exponentially in the space of smoothing operators as $t \to \infty$. A similar formula holds for e^{-tDD^*}.

The key steps of the Mckean-Singer proof are to consider the function

$$h(t) = \mathrm{Tr}(e^{-tD^*D}) - \mathrm{Tr}(e^{-tDD^*})$$

and to prove the following amazing properties:

(1) $\lim_{t\to\infty} h(t) = \operatorname{ind} D$,

(2) $\lim_{t\to 0} h(t) = \int_M K_{AS}$,

(3) $\frac{d}{dt}h(t) = 0$ so that $h(t)$ is constant.

Equating the values of the constant function $h(t)$ at $t = 0$ and $t = \infty$ proves the index formula. Consider property (1). The formula (2.2) implies that

$$\begin{aligned}\lim_{t\to\infty} h(t) &= \operatorname{Tr}(\pi_{\ker D^*D}) - \operatorname{Tr}(\pi_{\ker DD^*})\\ &= \dim\ker D^*D - \dim\ker DD^*.\end{aligned}$$

Integration by parts shows that $\ker D^*D = \ker D$ and $\ker DD^* = \ker D^*$. Indeed, clearly $\ker D \subset \ker D^*D$ and if $(\cdot,\cdot)$ denotes the L^2 inner product, then

$$D^*Du = 0 \;\Rightarrow\; (D^*Du, u) = 0 \;\Rightarrow\; (Du, Du) = 0 \;\Rightarrow\; Du = 0.$$

Thus $\ker D^*D \subset \ker D$ and so $\ker D^*D = \ker D$. Similarly, $\ker DD^* = \ker D^*$. Thus, as $\operatorname{coker} D \cong \ker D^*$, we obtain

$$\lim_{t\to\infty} h(t) = \operatorname{ind} D.$$

To determine the limit as $t \to 0$ of $h(t)$, we use the trace formulas:

$$\operatorname{Tr}(e^{-tD^*D}) = \int_M \operatorname{tr} e^{-tD^*D}(p,p)\,dg, \quad \operatorname{Tr}(e^{-tDD^*}) = \int_M \operatorname{tr} e^{-tDD^*}(p,p)\,dg,$$

obtained by integrating the pointwise trace of the heat kernels restricted to the diagonal. Now the *local index theorem* states that[3]

$$\lim_{t\to 0}\left\{\operatorname{tr} e^{-tD^*D}(p,p) - \operatorname{tr} e^{-tDD^*}(p,p)\right\} = K_{AS}(p)$$

uniformly in t, where the right-hand side really represents the coefficient of the volume form component of the differential form $K_{AS}(p)$. This result was proved originally by Mckean and Singer [**57**] for dimension two, generalized to higher dimensions by Gilkey [**32**] using invariance theory, and by Patodi [**70**] using a super-symmetry trick which was further developed by Alvarez-Gaumé [**1**] in the setting of path integrals and by Getzler [**30**] in a pseudodifferential setting. Thus,

$$\lim_{t\to 0} h(t) = \int_M K_{AS}. \tag{2.3}$$

Hence, by the fundamental theorem of calculus, we have

$$\operatorname{ind} D - \int_M K_{AS} = \int_0^\infty \frac{d}{dt}h(t)\,dt. \tag{2.4}$$

---

[3]This formula technically only applies to compatible Dirac operators, and not to arbitrary Dirac type operators. In general, the left-hand side of (2.3) has an asymptotic expansion as $t \to 0$ starting with negative powers of t and the right-hand side is the constant term in the expansion.

We now show that $\frac{d}{dt}h(t) = 0$. We first claim that $D^*De^{-tD^*D} = D^*e^{-tDD^*}D$. To see this, let $u \in C^\infty(M,E)$. Then, $v_t = D^*De^{-tD^*D}u$ and $w_t = D^*e^{-tDD^*}Du$ agree at $t = 0$ and they both satisfy the equation $(\partial_t + D^*D)u_t = 0$. By uniqueness of solutions to the heat equation [**59**, p. 271], we must have $v_t = w_t$; hence, $D^*De^{-tD^*D} = D^*e^{-tDD^*}D$. Thus

$$\begin{aligned} \frac{d}{dt}h(t) &= \operatorname{Tr}\big(- D^*De^{-tD^*D} + DD^*e^{-tDD^*}\big) \\ &= \operatorname{Tr}\big(- D^*e^{-tDD^*}D + DD^*e^{-tDD^*}\big) \\ &= \operatorname{Tr}\big([D, D^*e^{-tDD^*}]\big), \end{aligned} \tag{2.5}$$

where $[D, D^*e^{-tDD^*}]$ is the commutator of D and $D^*e^{-tDD^*}$. Using the well-known fact that the trace vanishes on commutators of pseudodifferential operators when at least one factor is smoothing implies that $\frac{d}{dt}h(t) = 0$. Hence, according to (2.4) we have

$$\operatorname{ind} D = \int_M K_{AS},$$

which is the Atiyah-Singer formula!

2.3. Some remarks on the Atiyah-Singer index theorem. The Atiyah-Singer index formula can be generalized to elliptic pseudodifferential operators using K-theory. However, in this generality, the form K_{AS} occurring on the right-hand side of the index formula is not explicitly defined in terms of the curvature forms. The fact that K_{AS} is explicitly defined in terms of the curvature forms for Dirac type operators is a very special property of Dirac operators and is one of the reasons why Dirac operators are important in applications. The original proof of the Atiyah-Singer index theorem as sketched in [**6**] used cobordism theory, cf. [**69**]. A few years later, the proof was reworked in a series of papers [**7, 8**]. The 'heat kernel proof' appeared in [**2**]. See [**14**] for a comparison of the various proofs. The Atiyah-Singer formula has been generalized to many different contexts, for example, to families of Dirac operators by Bismut [**11**], see [**74**], [**9**], and especially [**34**, Ch. 5] for other generalizations.

The Atiyah-Singer index theorem has far-reaching applications (see [**45**, Ch. 4]) that include group actions on manifolds, immersions into Euclidean space, integrality and divisibility of certain characteristic numbers, existence of metrics with positive scalar curvature [**37**], twisted signature and Riemann-Roch-Hirzebruch formulas, and formal dimensions of certain moduli spaces [**27, 36**].

3. The Atiyah-Patodi-Singer index formula

Now we ask a similar question concerning the Atiyah-Singer index formula as we did for the Gauss-Bonnet formula in the introduction: Does the Atiyah-Singer formula, $\operatorname{ind} D = \int_M K_{AS}$, continue to hold verbatim if M has a smooth boundary? From our experience with the Gauss-Bonnet formula, we expect that the answer is "no"; there should be a correction term added to the right-hand side

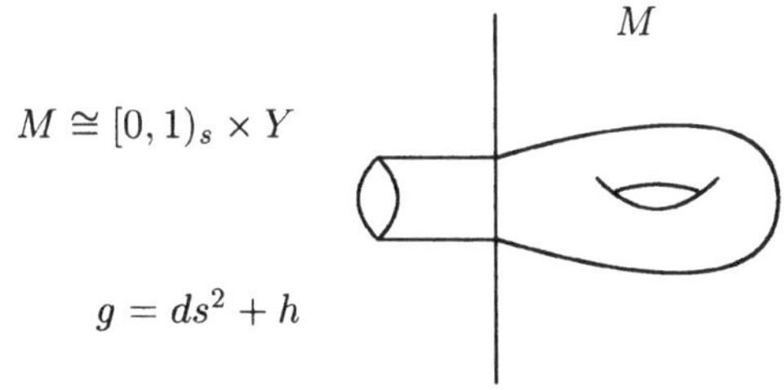

FIGURE 2. The manifold M with a collar neighborhood near its boundary over which all geometric structures are of product type.

due to the presence of the boundary. This is in fact the case. It turns out that the correction term is a *spectral invariant* of the boundary.

In this section, we describe the Atiyah-Patodi-Singer (or APS) index formula [**5**], which extends the Atiyah-Singer formula to manifolds with boundary. For manifolds with boundary, there are various ways to develop an index theory, for instance, introducing boundary conditions or 'attaching a cylindrical end' to the boundary. We focus on the latter method. For the BVP point of view, see [**15**]. Finally, we reformulate the index problem in terms of Melrose's b-geometric objects.

3.1. Attaching a cylindrical end. The ingredients of the APS index formula include topological/geometric data: Let M be an even-dimensional, compact, oriented, Riemannian manifold with boundary and let E and F be Hermitian vector bundles over M. For simplicity, we assume that M has a collar neighborhood $M \cong [0,1)_s \times Y$ where the metric is a product $g = ds^2 + h$ with h a metric on $Y = \partial M$, and where E and F are isomorphic to their restrictions E_0 and F_0 respectively to Y over this collar. See Figure 2.

We are also given functional analytic/geometric data: Let

$$D : C^\infty(M, E) \to C^\infty(M, F)$$

be a *Dirac type operator*, a first-order elliptic differential operator such that the principal symbol of D^*D is the metric $\sigma(D^*D)(\xi) = |\xi|^2$ for all cotangent vectors ξ. We assume that D is of product type on the collar of the following sort:

$$D = \Gamma(\partial_s + D_Y),$$

where

$$D_Y : C^\infty(Y, E_0) \to C^\infty(Y, E_0)$$

is a self-adjoint Dirac type operator on the odd-dimensional manifold Y, and where Γ is a unitary isomorphism from E_0 onto F_0. With these hypotheses, one might think that $D : H^1(M, E) \to L^2(M, F)$ is Fredholm. This however is not the case.

THEOREM 3.1. *The Dirac type operator*

$$D : H^1(M, E) \to L^2(M, F)$$

is never Fredholm. In fact, its kernel is infinite dimensional!

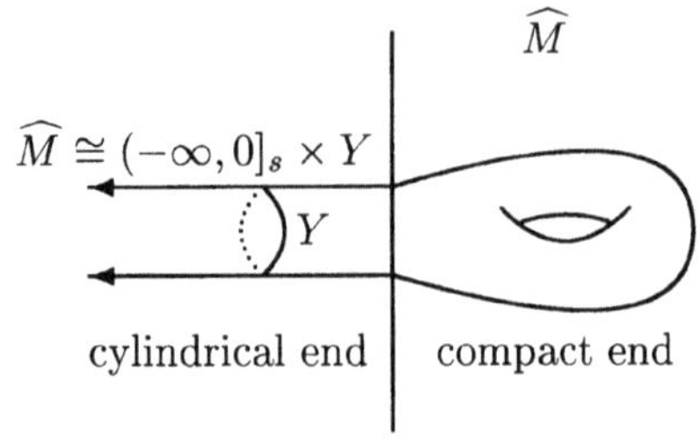

FIGURE 3. Attaching an infinite cylinder to M produces the manifold with cylindrical end $\widehat{M}$.

A proof of Theorem 3.1 can be found in [**15**]. To see why this theorem holds, consider the following simple example. Let $M_0 = [0,1] \times \mathbb{S}^1$ with metric $g = ds^2 + d\theta^2$, let $E = F = \mathbb{C}$, and let

$$D_0 = \partial_s + i\partial_\theta.$$

Certainly, the manifold and operator are of product type. Moreover, $\ker D_0$ consists of all functions $f(s,\theta)$ that are holomorphic in $z = s + i\theta$ for $0 \le s \le 1$ and periodic in θ with period 2π. Of course, there are infinitely many such functions, for example e^{kz} where $k \in \mathbb{Z}$. Thus, $\dim\ker D_0 = \infty$.

Since D is not Fredholm, it might look like our hopes for an index formula are crushed. By the way, it turns out that in general, $D : H^1(M,E) \to L^2(M,F)$ is surjective [**15**]. Thus, the problem with D is its kernel on $H^1(M,E)$. There are various ways that have been developed to 'tame' the infinite dimensional kernel. One successful method is the theory of boundary value problems pioneered by Calderón [**19**] and Seeley [**77**] as explained in [**15**]. However, we will focus on the method of attaching a cylindrical end, which is described as follows.

Consider $\widehat{D}_0 = \partial_s + i\partial_\theta$ on the enlarged manifold $\widehat{M_0} = (-\infty,\infty)_s \times \mathbb{S}^1$ rather than on $M_0 = [0,1]_s \times \mathbb{S}^1$. Here, $\widehat{M_0}$ has the naturally extended metric $g = ds^2 + d\theta^2$. We claim that on $\widehat{M_0}$, we have $\ker \widehat{D}_0 = 0$ on $H^1(\widehat{M_0})$. Indeed, $\ker \widehat{D}_0$ consists of all functions $f(s,\theta) \in H^1(\widehat{M_0})$ that are holomorphic in $z = s + i\theta$ for $s \in \mathbb{R}$ and periodic in θ with period 2π. By Sobolev embedding, f is bounded in s and hence is a bounded holomorphic function on $\mathbb{C}$, so is constant by Liouville's theorem. Since $f \in H^1(\widehat{M_0})$, the constant must be zero.

With this example as motivation, in the general case we enlarge the compact manifold with boundary M to a noncompact manifold $\widehat{M}$ as follows: Let $\widehat{M}$ be the manifold formed by taking the infinite cylinder $(-\infty,0]_s \times Y$ and gluing it onto the end of the collar $[0,1)_s \times Y$ of M as shown in Figure 3:

$$\widehat{M} = (-\infty, 0]_s \times Y \sqcup_{\partial M} M.$$

Since all the geometric structures and the Dirac operator were of product type on the collar of M, they all have natural extensions to the manifold $\widehat{M}$. We denote these extended structures on $\widehat{M}$ using the same notations for the original objects on

M; however, since the extended Dirac operator on $\widehat{M}$ has a completely different domain than the Dirac operator on M, we denote the extension of the Dirac operator by $\widehat{D}$. Note that the natural domain of $\widehat{D}$ is $H^1(\widehat{M}, E)$, which consists of those sections u on $\widehat{M}$ such that $\widehat{D}u$ is square integrable with respect to the measure dg on $\widehat{M}$. Now we ask: Does this idea work? Is the operator $\widehat{D}$ Fredholm on its natural domain? The answer is: sometimes. It turns out that the boundary operator D_Y, which can be considered the model operator for $\widehat{D}$ 'at infinity' on the cylindrical end, determines the Fredholm condition.

THEOREM 3.2. *The Dirac type operator*

$$\widehat{D} : H^1(\widehat{M}, E) \to L^2(\widehat{M}, F)$$

is Fredholm if and only if the boundary operator $D_Y : H^1(Y, E_0) \to L^2(Y, E_0)$ *is invertible; that is, if it has zero kernel.*

It turns out that the kernel of $\widehat{D}$ is always finite dimensional, so the enlargement of M to $\widehat{M}$ did tame the infinite dimensional kernel of D as expected, but the cokernel of $\widehat{D}$ is infinite dimensional unless D_Y is invertible. For a proof of Theorem 3.2, see [**59**, Th. 5.60]. There is a general principle underlying the Fredholm properties of Dirac operators on noncompact manifolds:

(3.1) **General Principle:** A Dirac operator on a noncompact manifold is Fredholm if and only if it is invertible 'at infinity'.

Recall that a Fredholm operator is an operator that is 'almost invertible'. Roughly speaking, a Dirac operator is always 'almost invertible' on the 'compact end' of a noncompact manifold simply because a Dirac operator is elliptic so we can always construct a parametrix for it on the compact end; however, to construct a global parametrix for the Dirac operator, we need to invert the Dirac operator 'at infinity'.

We now show that Dirac operators can *always* be made Fredholm on weighted Sobolev spaces. To see this, extend the coordinate function s on the cylindrical end of $\widehat{M}$ into the compact end of $\widehat{M}$ to be a positive function there. Let $\alpha \in \mathbb{R}$. Then observe that on the cylindrical end we have

$$e^{-\alpha s} \widehat{D}\, e^{\alpha s} = \Gamma(\partial_s + D_Y + \alpha),$$

and $D_Y + \alpha$ is invertible for $|\alpha| > 0$ less than the smallest absolute value of a nonzero eigenvalue of D_Y. Hence, the 'General Principle' implies that

$$e^{-\alpha s} \widehat{D} e^{\alpha s} : H^1(\widehat{M}, E) \to L^2(\widehat{M}, F)$$

is Fredholm for all $|\alpha| > 0$ sufficiently small, which is equivalent to

$$\widehat{D} : e^{\alpha s} H^1(\widehat{M}, E) \to e^{\alpha s} L^2(\widehat{M}, F)$$

is Fredholm on weighted Sobolev spaces. Thus we have the following:

THEOREM 3.3. *There exists a $\delta > 0$ such that for all $0 < |\alpha| < \delta$, the Dirac type operator*

$$\widehat{D} : e^{\alpha s} H^1(\widehat{M}, E) \to e^{\alpha s} L^2(\widehat{M}, F)$$

is Fredholm.

For the proof of this theorem, see [**59**, Th. 5.60]. We next state the APS formula for the index of the operator $\widehat{D}$ on weighted Sobolev spaces.

3.2. Statement of the Atiyah-Patodi-Singer index theorem. Before stating the APS index theorem for the operator $\widehat{D}$, we first need to define the eta invariant. Since D_Y is a self-adjoint elliptic operator on the closed compact manifold Y, it has discrete spectrum $\{\lambda_j\} \subset \mathbb{R}$. The eta function, $\eta(z)$, is the holomorphic function

$$\eta(z) = \sum_{\lambda_j \neq 0} \frac{\operatorname{sign} \lambda_j}{|\lambda_j|^z}.$$

One of the main results of [**5**] was that $\eta(z)$ defines a meromorphic function on $\mathbb{C}$ that is regular at $z = 0$. The *eta invariant* of D_Y is the value of the eta function at zero, $\eta(D_Y) = \eta(0)$, which represents a formal signature of the operator D_Y:

$$\text{“}\eta(D_Y) = \sum_{\lambda_j \neq 0} \operatorname{sign} \lambda_j = \#\{\lambda_j > 0\} - \#\{\lambda_j < 0\}\text{”}.$$

Thus, $\eta(D_Y)$ is a measurement of the spectral asymmetry of D_Y. Another way to express the eta function is through the heat operator:

$$\eta(z) = \frac{1}{\Gamma(\frac{z+1}{2})} \int_0^\infty t^{\frac{z-1}{2}} \operatorname{Tr}(D_Y \, e^{-tD_Y^2}) \, dt, \tag{3.2}$$

where $\Gamma(z)$ is the Gamma function. This formula follows from the fact that

$$\operatorname{Tr}(D_Y \, e^{-tD_Y^2}) = \sum_{\lambda_j \neq 0} \lambda_j \, e^{-t\lambda_j^2}$$

and that

$$\frac{1}{\Gamma(\frac{z+1}{2})} \int_0^\infty t^{\frac{z-1}{2}} \lambda_j \, e^{-t\lambda_j^2} \, dt = \frac{\lambda_j}{|\lambda_j|^{z+1}} \frac{1}{\Gamma(\frac{z+1}{2})} \int_0^\infty t^{\frac{z-1}{2}} e^{-t} \, dt = \frac{\operatorname{sign} \lambda_j}{|\lambda_j|^z},$$

where we made the change of variables $t \mapsto t/|\lambda_j|^2$. Moreover, the *local index theorem for odd-dimensional manifolds* proved by Bismut and Freed [**10**] states that $\operatorname{Tr}(D_Y e^{-tD_Y^2})$ is a smooth function of $t^{1/2}$ vanishing at $t = 0$, and so the formula (3.2) can be used to prove that[4]

$$\eta(D_Y) = \frac{1}{\sqrt{\pi}} \int_0^\infty t^{-1/2} \operatorname{Tr}(D_Y\, e^{-tD_Y^2})\, dt. \tag{3.3}$$

We are now ready to state the Atiyah-Patodi-Singer index theorem.

THEOREM 3.4 (Atiyah-Patodi-Singer, 1973). *Let D be a Dirac type operator on an even-dimensional, compact, oriented, Riemannian manifold with boundary with product type structures specified. Then there exists a $\delta > 0$ such that for all $0 < |\alpha| < \delta$, the Dirac type operator*

$$\widehat{D} : e^{\alpha s} H^1(\widehat{M}, E) \to e^{\alpha s} L^2(\widehat{M}, F)$$

is Fredholm and if its index is denoted by $\operatorname{ind}_\alpha \widehat{D}$, then

$$\operatorname{ind}_\alpha \widehat{D} = \int_M K_{AS} - \frac{1}{2}\Big\{\eta(D_Y) + \operatorname{sign}\alpha \cdot \dim\ker D_Y\Big\},$$

where K_{AS} is the Atiyah-Singer integrand and $\eta(D_Y)$ is the eta invariant.

The APS theorem in [**5**] technically only applies to the case of $\alpha > 0$; as presented above, the theorem is due to Melrose [**59**]. We prove this theorem using Melrose's *b*-geometry approach in Section 4. An important corollary is the notable generalization of Hirzebruch's signature formula to manifolds with boundary: If D is the signature operator, then using the fact that $\operatorname{ind}_\alpha \widehat{D} + \operatorname{ind}_{-\alpha} \widehat{D} = 2\operatorname{sign}(M)$ for $\alpha > 0$ sufficiently small (see [**59**, Sec. 9.3]) gives

$$2\operatorname{sign}(M) = \int_M \mathcal{L}(M) - \frac{1}{2}\Big\{\eta(D_Y) + \dim\ker D_Y\Big\} + \int_M \mathcal{L}(M) - \frac{1}{2}\Big\{\eta(D_Y) - \dim\ker D_Y\Big\},$$

or

$$\operatorname{sign}(M) = \int_M \mathcal{L}(M) - \frac{1}{2}\eta(D_Y).$$

Hirzebruch's conjecture for the signature-defect, the difference between the signature of M and the integral of the $\mathcal{L}$-class polynomial, was the original motivation of Atiyah, Patodi, and Singer in the discovery of the eta invariant [**4, 5**].

[4]Actually, the local index theorem for odd-dimensional manifolds is not true for arbitrary D_Y but only for those associated to a 'unitary Clifford connection', cf. the discussion in footnote (3) concerning the local index theorem for even-dimensional manifolds. In general, $\operatorname{Tr}(D_Y e^{-tD_Y^2})$ only has an asymptotic expansion as $t \to 0$ starting from negative powers of t. In this case, the lower limit 0 in the integral (3.3) must be replaced by $\varepsilon > 0$ and the resulting integral has an asymptotic expansion as $\varepsilon \to 0$. The right-hand side of the above equation represents the constant term in the expansion.

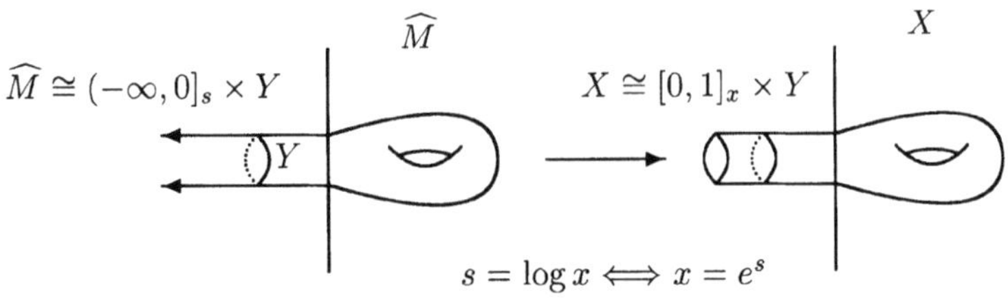

FIGURE 4. The compact manifold with boundary X is the compactification of the manifold with cylindrical end $\widehat{M}$.

3.3. Interpretation as b-objects. One of the primary tools used to prove Fredholm properties of Dirac operators, or elliptic differential operators in general, on compact manifolds without boundary is the algebra of pseudodifferential operators. Such an algebra would be useful on a manifold with cylindrical end in order to prove Fredholm properties of Dirac operators on such manifolds. For various classes of operators defined on noncompact manifolds, see Lockhart and McOwen [**49**], Rabinovič [**73**], or Schrohe [**76**]. One usually requires the noncompact manifold to have a finite atlas with control at infinity of the coordinate changes and special estimates at infinity on the symbols of the operators considered on the manifold. In particular, a direct definition of pseudodifferential operators on manifolds with a cylindrical end might be considered unbalanced as the analysis is treated in distinctly differently ways on the cylindrical end and the compact end.

Melrose's novel idea was to unify the analysis on these two ends by making the whole manifold compact; that is, compactifying the cylindrical end forming a *compact* manifold with boundary. On this new compact manifold with boundary, he defines a space of pseudodifferential operators, imitating as close as possible, the *global* geometric definition of pseudodifferential operators on compact manifolds without boundary in terms of their Schwartz kernels as discussed in, for example, Hörmander [**43**]. The resulting operators are called *b*-pseudodifferential operators. For excellent introductions to this subject, see Grieser [**35**], Mazzeo [**56**], or Melrose [**59**].

We now explain the compactification. On the cylindrical end $(-\infty, 0]_s \times Y$ of $\widehat{M}$ we make the change of variables $x = e^s$. As $s \to -\infty$, $x \to 0$. Thus, under this change of variables, $\widehat{M}$ becomes the interior of the compact manifold with boundary X, where X has the same compact end as $\widehat{M}$ but with the cylindrical end $(-\infty, 0]_s \times Y$ replaced with the compact manifold $[0,1]_x \times Y$, see Figure 4. Since $x = e^s$, we have $ds = dx/x$ and $\partial_s = x\partial_x$. Thus the geometric objects on the manifold with cylindrical end transform into corresponding singular geometric '*b*-objects' on the compact manifold with boundary:

$$g = ds^2 + h \rightsquigarrow g = \left(\frac{dx}{x}\right)^2 + h \quad (b\text{-metric}),$$

$$dg = ds\,dh \rightsquigarrow dg = \frac{dx}{x}\,dh \quad (b\text{-measure}),$$

and consequently,

$$H^k(\widehat{M}) \rightsquigarrow H_b^k(X) \quad (b\text{-Sobolev space}),$$
$$\widehat{D} = \Gamma(\partial_s + D_Y) \rightsquigarrow \widehat{D} = \Gamma(x\partial_x + D_Y) \quad (b\text{-differential operator}).$$

Although the manifold X is topologically compact, its interior is geometrically a manifold with cylindrical end since X inherited all its geometric structures from $\widehat{M}$. In particular, the boundary of X is geometrically at infinity. The fact that X is compact is key to the definition of b-pseudodifferential operators since these operators are defined using only the usual classes of smooth functions and distributions on compact manifolds with boundary. Of course, there is a trade off: The distributions defining the Schwartz kernels of b-pseudodifferential operators are required to have a special structure, which takes some time getting used to [**59**, Ch. 4].

We repeat the statement of the APS index theorem in the current context.

THEOREM 3.5. *With the same hypotheses as in Theorem 3.4, but now in the b-geometry context, there exists a* $\delta > 0$ *such that for all* $0 < |\alpha| < \delta$*, the Dirac type operator*

$$\widehat{D} : x^\alpha H_b^1(X, E) \to x^\alpha L_b^2(X, F)$$

is Fredholm and if its index is denoted by $\operatorname{ind}_\alpha \widehat{D}$*, then*

$$\operatorname{ind}_\alpha \widehat{D} = \int_M K_{AS} - \frac{1}{2}\Big\{\eta(D_Y) + \operatorname{sign}\alpha \cdot \dim\ker D_Y\Big\},$$

where K_{AS} *is the Atiyah-Singer integrand and* $\eta(D_Y)$ *is the eta invariant*

$$\eta(D_Y) = \frac{1}{\sqrt{\pi}} \int_0^\infty t^{-1/2}\operatorname{Tr}(D_Y\, e^{-tD_Y^2})\, dt.$$

In Section 4, we prove this theorem.

3.4. Some remarks on the Atiyah-Patodi-Singer index theorem. The original Atiyah-Patodi-Singer index theorem was proved in the context of (pseudo-differential) boundary value problems. A nice introduction to these methods can be found in the book by Booß-Bavnbek and Wojciechowski [**15**]. The 'direct approach' to the APS boundary value problem based on asymptotic expansions of the heat operator was initiated by Grubb and Seeley [**39**], see also Grubb's book [**38**]. The approach of 'attaching cylinders' was mentioned in [**5**], but was not developed. Besides attaching a cylinder to the boundary, another way to develop an index theory on manifolds with boundary is by attaching a cone, see Cheeger [**22**]; for other generalizations, see Atiyah, Donnelly, and Singer [**3**], Müller [**64**], Stern [**81**], Brüning [**16**], Fedosov and Schulze [**29**], Schulze, Sternin, and Shatalov [**79**], and Carron [**21**]. Melrose introduced the b-geometry in the seminal paper [**58**], and these ideas were developed by Melrose and Mendoza in [**60**]. The APS index formula was generalized to Fredholm b-pseudodifferential operators by Piazza [**71**].

The eta invariant itself has become a topic of much interest. In particular, the extension of the eta function and its regularity properties for pseudodifferential operators has been examined by Gilkey [**33**] and Wodzicki [**85**] among others. For a survey of various topics on the eta invariant, including its decomposition under gluing of manifolds, see Mazzeo and Piazza [**55**].

The index theorem for families of Dirac operators was extended to the case when the fibers are manifolds with boundary by Bismut and Cheeger [**12, 13**] under the assumption that the Dirac operators on the fibers of the boundary fibration are invertible. Later, this result was generalized by Melrose and Piazza [**62, 63**] with no assumptions on the boundary Dirac operators. For further extensions of the index formula, see Getzler [**31**], Wu [**86**], Melrose and Nistor [**61**], and Leichtnam, Lott, and Piazza [**46**].

Other expository articles incorporating recent developments on the APS index theorem include Müller [**66**], Piazza [**72**], and Seeley [**78**]. Because of space constraints, we have only touched the surface on some of the many extensions and generalizations of the Atiyah-Patodi-Singer problem, a list of references can be found in the book by Gilkey [**34**, Ch. 5].

4. Melrose's b-geometry proof of the Atiyah-Patodi-Singer theorem

Let D be a Dirac type operator on an even-dimensional, compact, oriented, Riemannian manifold with boundary M with product type structures near the boundary as described in Section 3.1. For simplicity, we assume that the boundary Dirac operator D_Y is invertible. Now form the manifold with cylindrical end $\widehat{M}$ and then compactify it under the change of variables $x = e^s$, where s is the variable on the cylinder, to form the manifold X as described in Section 3.3. Then $\widehat{D}$ defines an operator on X such that

$$\widehat{D} : H^1_b(X, E) \to L^2_b(X, F)$$

is Fredholm. We now give the b-geometry proof of the APS index formula in Theorem 3.5 modeling, as close as possible, the proof the Atiyah-Singer formula given in Section 2.2. We shall see that there are certain variations to the proof that need to be fleshed out in order to make the proof work.

4.1. The proof of APS with details left out. As with the Atiyah-Singer proof given in Section 2.2, we would like to define the Mckean-Singer function

$$\text{“}\operatorname{Tr}(e^{-t\widehat{D}^*\widehat{D}}) - \operatorname{Tr}(e^{-t\widehat{D}\widehat{D}^*})\text{”}.$$

Here, we meet our first variation to the Atiyah-Singer proof – the reason for the quotation marks is that the heat operators are not trace class, and so the traces are not even defined! Basically, the heat operators are not trace class because X has infinite volume (is geometrically not compact) which implies that the heat kernels restricted to the diagonal are not integrable. Thus, we cannot prove the APS formula by imitating the proof of the Atiyah-Singer formula verbatim. However, in Section 4.3, we define a natural extension of the trace called the b-trace, denoted by ${}^b\mathrm{Tr}$, such that the heat operators are b-trace class. We can now define a modified Mckean-Singer function

$$h(t) = {}^b\mathrm{Tr}(e^{-t\widehat{D}^*\widehat{D}}) - {}^b\mathrm{Tr}(e^{-t\widehat{D}\widehat{D}^*}),$$

and continue as in Section 2.2. As in the manifold without boundary case, the following limits hold:

$$\lim_{t\to\infty} h(t) = \operatorname{ind}\widehat{D},$$

and[5]

$$\lim_{t\to 0} h(t) = \int_M K_{AS}.$$

In fact, using b-pseudodifferential operators, the proofs of these two results are not much different from the corresponding proofs in the manifold without boundary case, see Chapters 7 and 8 of [**59**] for the proofs. Continuing as in Section 2.2, we find that

$$\operatorname{ind}\widehat{D} = \int_M K_{AS} + \int_0^\infty \frac{d}{dt}h(t)\,dt,$$

where repeating the same algebraic calculation as before, we have

$$\frac{d}{dt}h(t) = {}^b\mathrm{Tr}\left([\widehat{D}, \widehat{D}^* e^{-t\widehat{D}\widehat{D}^*}]\right).$$

Here, we meet our second variation – in the proof of the Atiyah-Singer index formula, this expression is zero, in this present case it is not. Figuratively speaking, the b-trace is a trace on the interior of X and only fails to be a trace on the boundary of X. Thus intuitively, ${}^b\mathrm{Tr}\left([\widehat{D}, \widehat{D}^* e^{-t\widehat{D}\widehat{D}^*}]\right)$ should be a boundary integral of some sort. This is in fact the case; in Section 4.4 we compute that

$${}^b\mathrm{Tr}\left([\widehat{D}, \widehat{D}^* e^{-t\widehat{D}\widehat{D}^*}]\right) = -\frac{1}{\sqrt{4\pi t}}\,\mathrm{Tr}(D_Y\, e^{-tD_Y^2}) = \frac{1}{2\sqrt{\pi}}\, t^{-1/2}\,\mathrm{Tr}(D_Y\, e^{-tD_Y^2}).$$

Hence,

$$\operatorname{ind}\widehat{D} = \int_X K_{AS} - \frac{1}{2}\eta(D_Y),$$

where

$$\eta(D_Y) = \frac{1}{\sqrt{\pi}}\int_0^\infty t^{-1/2}\,\mathrm{Tr}(D_Y\, e^{-tD_Y^2})\,dt,$$

and the Atiyah-Patodi-Singer index formula is proved!

---

[5]The same discussion as in footnote (3) concerning the local index theorem on manifolds without boundary applies in this situation too. The integral of K_{AS} is over M because the product type assumption implies that the volume form component of K_{AS} is supported on the manifold M regarded as a subset of X.

4.2. Some facts about the heat kernels. To implement the proof in Section 2.2, we need the heat operators

$$e^{-t\widehat{D}^*\widehat{D}} : L_b^2(X,E) \to H_b^2(X,E) \quad \text{and} \quad e^{-t\widehat{D}\widehat{D}^*} : L_b^2(X,F) \to H_b^2(X,F).$$

It turns out that these heat operators are b-smoothing operators; that is, they are b-pseudodifferential operators of order $-\infty$ [**59**, Ch. 7], which implies a couple of useful results. These results can be proved using other methods, but the theory of b-pseudodifferential operators gives these results more or less 'for free'. First, the Schwartz kernels of these heat operators are smooth on the interior of X^2 vanishing to infinite order at ∂X^2 except at $\partial X \times \partial X$. The second result is that these heat operators have a simple structure on the collar of X described as follows. For concreteness, we focus on $e^{-t\widehat{D}\widehat{D}^*}$. On the collar $[0,1]_x \times Y$ of X we have

$$\widehat{D} = \Gamma(x\partial_x + D_Y),$$

where Γ is a unitary isomorphism of E_0 onto F_0. Thus, on the collar,

$$\widehat{D}\widehat{D}^* = \Gamma(x\partial_x + D_Y)(-x\partial_x + D_Y)\Gamma^* = \Gamma((xD_x)^2 + D_Y^2)\Gamma^*,$$

where $D_x = i^{-1}\partial_x$. This suggests that near ∂X we have

$$e^{-t\widehat{D}\widehat{D}^*} = \Gamma e^{-t(xD_x)^2} e^{-tD_Y^2}\Gamma^* + \mathcal{O}(x),$$

where $e^{-t(xD_x)^2}$ is the heat operator for xD_x on $[0,\infty)_x$, and where $\mathcal{O}(x)$ is an operator smooth in x and vanishing at $x = 0$. In fact, even a stronger result is true. Under the change of variables $s = \log x$, which takes the interior of $[0,\infty)_x$ onto $(-\infty,\infty)_s$, we have $xD_x = D_s$. Since the Schwartz kernel of $e^{-tD_s^2}$ on $(-\infty,\infty)_s$ is given by the well-known formula

$$\widetilde{K}_1(s,s',t) = \frac{1}{\sqrt{4\pi t}} e^{-(s-s')^2/4t},$$

the Schwartz kernel of $e^{-t(xD_x)^2}$ is obtained by setting $s = \log x$:

$$K_1(x,x',t) = \frac{1}{\sqrt{4\pi t}} e^{-(\log x - \log x')^2/4t}.$$

The second result is that the Schwartz kernel of $e^{-t\widehat{D}\widehat{D}^*}$ near $\partial X \times \partial X$ is given by

$$e^{-t\widehat{D}\widehat{D}^*}(x,y,x',y') = \Gamma(y)K_1(x,x',t)\, e^{-tD_Y^2}(y,y')\,\Gamma(y')^* + \mathcal{O}(x), \tag{4.1}$$

where $\mathcal{O}(x)$ is a smooth function of the variables x, $\log x - \log x'$, y, and y' that vanishes at $x = 0$.

4.3. Filling in the details for the b-trace. The simple fact that dx/x is not integrable over $[0,1]_x$ implies that the heat operators are not trace class. Indeed, consider the heat operator $e^{-t\widehat{D}\widehat{D}^*}$. By (4.1), on the collar $[0,1]_x \times Y$ we have

$$\operatorname{tr} e^{-t\widehat{D}\widehat{D}^*}(x,y,x,y) = \frac{1}{\sqrt{4\pi t}} \operatorname{tr} e^{-tD_Y^2}(y,y) + \mathcal{O}(x), \tag{4.2}$$

where we used that $\Gamma\Gamma^* = \mathrm{Id}$ since Γ is a unitary isomorphism, and where $\mathcal{O}(x)$ is smooth in x and vanishes at $x = 0$. Since $dg = (dx/x)\, dh$ on this collar and

$$\int_{[0,1]\times Y} \operatorname{tr} e^{-tD_Y^2}(y,y)\, \frac{dx}{x} dh = \left(\int_0^1 \frac{dx}{x}\right) \cdot \int_Y \operatorname{tr} e^{-tD_Y^2}(y,y)\, dh,$$

the following trace formula does *not* make sense:

$$\operatorname{Tr}(e^{-t\widehat{D}\widehat{D}^*}) = \int_X \operatorname{tr} e^{-t\widehat{D}\widehat{D}^*}(p,p)\, dg.$$

Similarly, the corresponding integral for $e^{-t\widehat{D}^*\widehat{D}}$ does not exist. Although the trace formula above does not make sense, we can 'force' it to make sense by the considering another notion of trace and integral as we now describe. Note that for $\operatorname{Re} z > 0$, x^z is integrable with respect to dx/x over $[0,1]_x$. Extend the coordinate function x on the collar of X to be a smooth function on X which is positive off the collar. Then it follows that $x^z e^{-t\widehat{D}\widehat{D}^*}$ is trace class for $\operatorname{Re} z > 0$ with trace given by

$$\operatorname{Tr}(x^z e^{-t\widehat{D}\widehat{D}^*}) = \int_X x^z \operatorname{tr} e^{-t\widehat{D}\widehat{D}^*}(p,p)\, dg, \quad \operatorname{Re} z > 0.$$

This argument is the basis for defining a new functional called the b-trace, which we introduce after the following lemma.

LEMMA 4.1. *Let $f \in C^\infty(X)$. Then for all complex numbers z with* $\operatorname{Re} z > 0$, *the integral*

$$F(z) = \int_X x^z f\, dg$$

exists, and it extends from $\operatorname{Re} z > 0$ *to define a meromorphic function on all of* $\mathbb{C}$. *The b-integral of f is by definition the regular value of $F(z)$ at $z = 0$:*

$$ {}^b\!\!\int_X f\, dg = \operatorname{Reg}_{z=0} F(z). \tag{4.3}$$

Finally, the residue of $F(z)$ at $z = 0$ given by

$$\operatorname{Res}_{z=0} F(z) = \int_Y f(0,y)\, dh.$$

To understand why this lemma is true, note that $x^z = e^{z\log x}$ is an entire function of z for $x > 0$. Thus, we may assume that f is supported on the collar $[0,1]_x \times Y$ of X. Then $F(z)$ is well defined for $\operatorname{Re} z > 0$ since $x^z f(x,y)$ is integrable with respect to the measure $(dx/x)\, dh$ as long as $\operatorname{Re} z > 0$. Now expand $f(x,y)$ in Taylor series at $x = 0$: $f(x,y) \sim \sum_{k=0}^\infty x^k f_k(y)$. Since

$$\int_{[0,1]\times Y} x^{z+k} f_k(y) \frac{dx}{x} dh = \frac{1}{z+k} \int_Y f_k(y)\, dh$$

it follows that $F(z)$ extends from $\operatorname{Re} z > 0$ to be a meromorphic function on $\mathbb{C}$ with only simple poles at $z = \{0, -1, -2, \ldots\}$ with residue at $z = 0$ given by

$$\int_Y f_0(y)\, dh = \int_Y f(0, y)\, dh.$$

To see why the notion of the b-integral is natural, note that if $f(0, y) = 0$, then $F(z)$ is regular at $z = 0$, and

$${}^b\!\!\int_X f\, dg = \operatorname{Reg}_{z=0} F(z) = F(0) = \int_X f\, dg,$$

which is the usual integral of f.

The *b-trace* of the heat operator $e^{-t\widehat{D}\widehat{D}^*}$ is by definition

$${}^b\mathrm{Tr}(e^{-t\widehat{D}\widehat{D}^*}) = {}^b\!\!\int_X \operatorname{tr} e^{-t\widehat{D}\widehat{D}^*}(p, p)\, dg,$$

where the b-integral of the function $\operatorname{tr} e^{-t\widehat{D}\widehat{D}^*}(p, p)$ is defined by (4.3):

$${}^b\!\!\int_X \operatorname{tr} e^{-t\widehat{D}\widehat{D}^*}(p, p)\, dg = \operatorname{Reg}_{z=0} \mathrm{Tr}(x^z e^{-t\widehat{D}\widehat{D}^*}).$$

The b-trace of $e^{-t\widehat{D}^*\widehat{D}}$ is defined similarly.

4.4. Filling in the details for the eta invariant. In this section, we show that

$${}^b\mathrm{Tr}\left([\widehat{D}, \widehat{D}^* e^{-t\widehat{D}\widehat{D}^*}]\right) = -\frac{1}{\sqrt{4\pi t}}\, \mathrm{Tr}(D_Y\, e^{-tD_Y^2}).$$

The proof is just a computation using the definition of the b-trace,

$${}^b\mathrm{Tr}\left([\widehat{D}, \widehat{D}^* e^{-t\widehat{D}\widehat{D}^*}]\right) = \operatorname{Reg}_{z=0} \mathrm{Tr}\, x^z [\widehat{D}, \widehat{D}^* e^{-t\widehat{D}\widehat{D}^*}],$$

where $\mathrm{Tr}\, x^z[\widehat{D}, \widehat{D}^* e^{-t\widehat{D}\widehat{D}^*}]$ is meromorphically extended from $\operatorname{Re} z > 0$. Observe that

$$x^z[\widehat{D}, \widehat{D}^* e^{-t\widehat{D}\widehat{D}^*}] = [x^z, \widehat{D}]\widehat{D}^* e^{-t\widehat{D}\widehat{D}^*} + [\widehat{D}, x^z \widehat{D}^* e^{-t\widehat{D}\widehat{D}^*}].$$

Since the trace vanishes on commutators, we have $\mathrm{Tr}[\widehat{D}, x^z \widehat{D}^* e^{-t\widehat{D}\widehat{D}^*}] = 0$ for $\operatorname{Re} z > 0$, and thus its meromorphic extension to all of $\mathbb{C}$ is also zero. Hence,

$${}^b\mathrm{Tr}\left([\widehat{D}, \widehat{D}^* e^{-t\widehat{D}\widehat{D}^*}]\right) = \operatorname{Reg}_{z=0} \mathrm{Tr}[x^z, \widehat{D}]\widehat{D}^* e^{-tDD^*} = \operatorname{Reg}_{z=0} \mathrm{Tr}\, x^z a(z),$$

where

$$a(z) = x^{-z}[x^z, \widehat{D}]\widehat{D}^* e^{-t\widehat{D}\widehat{D}^*} = \widehat{D}\widehat{D}^* e^{-t\widehat{D}\widehat{D}^*} - x^{-z}\widehat{D} x^z \widehat{D}^* e^{-t\widehat{D}\widehat{D}^*}.$$

Note that $a(0) = 0$. We claim that $a(z)$ is an entire function of z. Indeed, since x^z is an entire function of z for $x > 0$, we may consider $a(z)$ on the collar $[0, 1]_x \times Y$ over which $\widehat{D} = \Gamma(x\partial_x + D_Y)$. In this case, $x^{-z}\widehat{D}x^z = \Gamma(x\partial_x + D_Y + z)$ which is entire, so $a(z)$ is entire. Since $a(z)$ is entire and vanishes at $z = 0$, we can write

$$a(z) = zA + \mathcal{O}(z^2),$$

where A is independent of z. It follows that

$$^b\mathrm{Tr}\left([\widehat{D}, \widehat{D}^* e^{-t\widehat{D}\widehat{D}^*}]\right) = \mathrm{Reg}_{z=0} \mathrm{Tr}\, x^z a(z) = \mathrm{Res}_{z=0} \mathrm{Tr}\, x^z A,$$

and so by Lemma 4.1,

$$^b\mathrm{Tr}\left([\widehat{D}, \widehat{D}^* e^{-t\widehat{D}\widehat{D}^*}]\right) = \int_Y \mathrm{tr}\, A(x,y,x,y)|_{x=0}\, dh. \tag{4.4}$$

To calculate this integral, we work over the collar $[0,1]_x \times Y$. Here,

$$x^{-z}\widehat{D}x^z = \Gamma(x\partial_x + D_Y + z) = \widehat{D} + z\,\Gamma,$$

and thus

$$a(z) = \widehat{D}\widehat{D}^* e^{-t\widehat{D}\widehat{D}^*} - x^{-z}\widehat{D}x^z\widehat{D}^* e^{-t\widehat{D}\widehat{D}^*} = -z\Gamma\widehat{D}^* e^{-t\widehat{D}\widehat{D}^*},$$

which implies that over the collar,

$$A = -\Gamma\widehat{D}^* e^{-t\widehat{D}\widehat{D}^*}.$$

As $\widehat{D}^* = (-x\partial_x + D_Y)\Gamma^*$ and $\Gamma^*\Gamma = \mathrm{Id}$, by (4.1) we have

$$A = -\Gamma(y)(-x\partial_x + D_Y)\frac{1}{\sqrt{4\pi t}} e^{-(\log x - \log x')^2/4t} e^{-tD_Y^2}(y,y')\,\Gamma(y')^*$$

modulo a term that vanishes on the boundary. It follows that

$$A(x,y,x,y)|_{x=0} = -\frac{1}{\sqrt{4\pi t}}\,\Gamma(y) D_Y e^{-tD_Y^2}(y,y)\,\Gamma(y)^*,$$

which in view of (4.4) gives

$$\begin{aligned} ^b\mathrm{Tr}\left([\widehat{D}, \widehat{D}^* e^{-t\widehat{D}\widehat{D}^*}]\right) &= -\frac{1}{\sqrt{4\pi t}} \int_Y \mathrm{tr}\, D_Y e^{-tD_Y^2}(y,y)\, dh \\ &= -\frac{1}{\sqrt{4\pi t}}\, \mathrm{Tr}(D_Y\, e^{-tD_Y^2}). \end{aligned}$$

5. Index theory on manifolds with corners of codimension two

In this section, we describe an extension of the APS index formula to manifolds with corners of codimension two. As we discussed for manifolds with boundary, there were various ways to develop an index theory for Dirac operators, e.g., introducing boundary conditions or attaching a 'cylindrical end' to the boundary. For manifolds with corners, it turns out that there is *no* well-developed theory of boundary value problems for Dirac operators. However, we can still formulate an index problem by attaching 'multi-cylindrical ends' and considering an L^2-index problem.

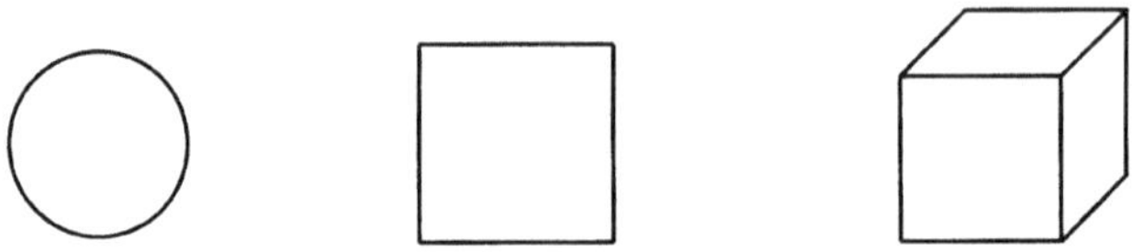

FIGURE 5. Examples of manifolds with corners of codimension one, two, and three respectively.

5.1. Dirac operators on manifolds with corners. We begin by defining manifolds with corners. An n-dimensional manifold with corners X is a paracompact Hausdorff topological space with local charts of the form $[0,1)^k \times (-1,1)^{n-k}$, where k can run anywhere between 0 and n depending on where the chart is located on the manifold, such that the transition maps between any two charts is smooth. A codimension k face Z is a connected closed subset of X such that given any interior point of Z there is a coordinate patch on X centered at the point of the form $[0,1)^k \times (-1,1)^{n-k}$. The largest codimension face that occurs is called the codimension of X. A boundary hypersurface is the same as a codimension one face. For technical reasons we assume that each boundary hypersurface has a boundary defining function; that is, for each hypersurface H of X, there is a nonnegative smooth function $\rho_H \in C^\infty(X)$ which vanishes only on H where it has a nonzero differential. Note that a manifold with corners of codimension one is just a manifold with boundary. Examples of manifolds with corners are found in Figure 5. The disk is a manifold with corners of codimension one. The square is a manifold with corners of codimension two; its edges are boundary hypersurfaces and its corners are codimension two faces. Lastly, the solid cube is a manifold with corners of codimension three; its sides are boundary hypersurfaces, its edges are codimension two faces, and its corners are codimension three faces.

To build a geometric index theory, we first need topological/geometric data. We focus on manifolds with corners of codimension two. Thus let M be an even-dimensional, compact, oriented, Riemannian manifold with corners of codimension two, and let E and F be Hermitian vector bundles over M. For simplicity, we assume that M has exactly two boundary hypersurfaces that intersect in exactly one codimension two face Y. We fix a labeling M_1 and M_2 of the hypersurfaces. Near each hypersurface M_i, we assume that M has a collar neighborhood $M \cong [0,1)_{s_i} \times M_i$ where the metric is a product $g = ds_i^2 + h_i$ with h_i a metric on M_i, and where E and F are isomorphic to their restrictions E_i and F_i respectively to M_i. For compatibility we assume that the product decompositions near each M_i give a common decomposition $M \cong [0,1)_{s_1} \times [0,1)_{s_2} \times Y$ near the corner where the metric is a product $g = ds_1^2 + ds_2^2 + h$ with h a metric on Y, and where E and F are isomorphic to their restrictions E_0 and F_0, respectively, to Y. See Figure 6.

Next, we need analytic/geometric data: Let

$$D : C^\infty(M,E) \to C^\infty(M,F)$$

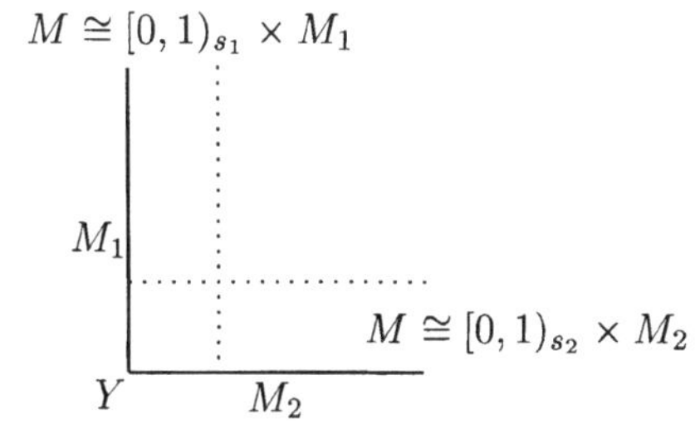

FIGURE 6. The manifold M near the corner Y. On the common intersection of the collars, M is isomorphic to $[0,1)_{s_1} \times [0,1)_{s_2} \times Y$.

be a *Dirac type operator*, a first-order elliptic differential operator such that the principal symbol of D^*D is the metric $\sigma(D^*D)(\xi) = |\xi|^2$ for all cotangent vectors ξ, which is of product type near each hypersurface:

$$D = \Gamma_i(\partial_{s_i} + D_i) \tag{5.1}$$

on the collar $M \cong [0,1)_{s_i} \times M_i$, where Γ_i is a unitary isomorphism from E_i onto F_i, and where

$$D_i : C^\infty(M_i, E_i) \to C^\infty(M_i, E_i)$$

is a (formally) self-adjoint Dirac type operator on the odd-dimensional manifold with boundary M_i. We assume that on the product decomposition near the corner, $M \cong [0,1)_{s_1} \times [0,1)_{s_2} \times Y$, the Dirac operator takes the form

$$D = \Gamma_1 \partial_{s_1} + \Gamma_2 \partial_{s_2} + B, \tag{5.2}$$

where

$$B : C^\infty(Y, E_0) \to C^\infty(Y, F_0)$$

is a Dirac type operator on the even-dimensional manifold without boundary Y.

Hidden in these assumptions are some interesting algebraic consequences at the corner as we now describe. Comparing (5.1) and (5.2), we find that on the collar $M \cong [0,1)_{s_1} \times [0,1)_{s_2} \times Y$, we have

$$\Gamma_i(\partial_{s_i} + D_i) = \Gamma_1 \partial_{s_1} + \Gamma_2 \partial_{s_2} + B, \quad i = 1, 2.$$

Multiplying each side by $\Gamma_i^* = \Gamma_i^{-1}$ and solving for D_i gives

$$D_1 = \Gamma_1^* \Gamma_2 \partial_{s_2} + \Gamma_1^* B \quad \text{and} \quad D_2 = \Gamma_2^* \Gamma_1 \partial_{s_1} + \Gamma_2^* B. \tag{5.3}$$

Since each D_i is assumed (formally) self-adjoint: $D_i^* = D_i$, we must have $\Gamma_1^* \Gamma_2 = -\Gamma_2^* \Gamma_1$ and $\Gamma_i^* B = B^* \Gamma_i$; that is,

$$\Gamma_i^* \Gamma_j + \Gamma_j^* \Gamma_i = 2\,\delta_{ij}, \quad B^* \Gamma_i = \Gamma_i^* B, \tag{5.4}$$

where δ_{ij} is the Kronecker delta. The reader familiar with Clifford multiplication might recognize the left equality as representing a 'Clifford two structure' at the corner. Factoring out the $\Gamma_1^* \Gamma_2$ from the right-hand side of D_1 in (5.3) gives

$$D_1 = \Gamma(\partial_{s_2} + D_Y), \quad \text{where } \Gamma = \Gamma_1^* \Gamma_2, \;\; D_Y = \Gamma_2^* B.$$

We call D_Y the *induced Dirac operator* on Y. We really should call D_Y the induced operator from D_1. However, the induced operator from D_2 is related to D_Y in a simple way. Indeed, one can verify that

$$D_2 = -\Gamma(\partial_{s_1} + \widetilde{D}_Y), \quad \widetilde{D}_Y = \Gamma D_Y.$$

The induced Dirac operator on Y refers only to D_Y and not $\widetilde{D}_Y$.

As part of the 'Clifford two package', the induced operator D_Y has a nice splitting property as we now describe. First, the left-hand identity in (5.4) implies that

$$\Gamma^2 = -\mathrm{Id}.$$

Hence, $\Gamma : E_0 \to E_0$ has eigenvalues $\pm i$. Let $E_0^\pm$ denote the eigenspaces corresponding to the eigenvalues $\pm i$; these are subbundles of E_0 and

$$E_0 = E_0^+ \oplus E_0^-$$

is an orthogonal decomposition since Γ is unitary. Also, a short computation utilizing (5.4) gives

$$D_Y \, \Gamma = -\Gamma \, D_Y.$$

Thus D_Y is odd with respect to Γ; hence odd with respect to the $\mathbb{Z}_2$-grading $E_0 = E_0^+ \oplus E_0^-$. We summarize this property in the following lemma.

LEMMA 5.1. *With respect to the orthogonal decomposition $E_0 = E_0^+ \oplus E_0^-$, where $E_0^\pm$ are the $\pm i$ eigenspaces of $\Gamma = \Gamma_1^* \Gamma_2$, the induced Dirac operator $D_Y = \Gamma_2^* B$ takes the following form*

$$\begin{bmatrix} 0 & D_Y^- \\ D_Y^+ & 0 \end{bmatrix} : C^\infty(Y, E_0^+ \oplus E_0^-) \to C^\infty(Y, E_0^+ \oplus E_0^-),$$

where $D_Y^\pm$ are the restrictions of D_Y to $C^\infty(Y, E_0^\pm)$.

Note that since D_Y is self-adjoint, we have $(D_Y^+)^* = D_Y^-$. The following theorem follows from the cobordism theorem of Atiyah-Singer, which is published in Palais' book [**69**]. The cobordism theorem was one of the key steps in the original proof of the Atiyah-Singer index theorem [**6**].

THEOREM 5.2. *The index of the Dirac type operator*

$$D_Y^+ : H^1(Y, E_0^+) \to L^2(Y, E_0^-)$$

on the even-dimensional manifold without boundary Y is zero: $\operatorname{ind} D_Y^+ = 0$. *Since $(D_Y^+)^* = D_Y^-$, it follows that* $\dim\ker D_Y^+ = \dim\ker D_Y^-$.

5.2. Attaching multi-cylindrical ends. As in the manifold with boundary case, we cannot build an index theory of D with its natural domain on the manifold with corners M:

THEOREM 5.3. *The Dirac type operator*

$$D : H^1(M, E) \to L^2(M, F)$$

is never Fredholm. In fact, $\dim\ker D = \infty$.

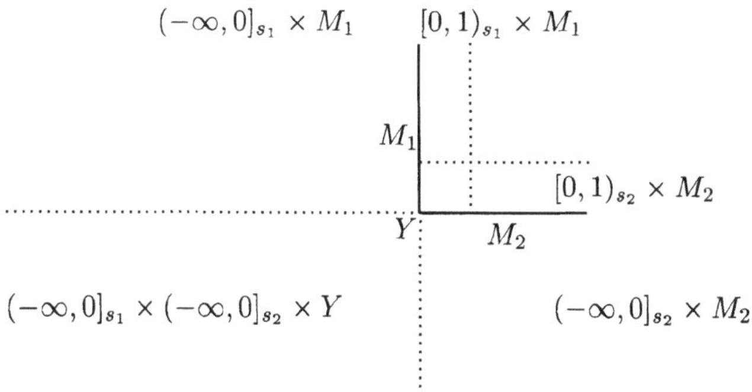

FIGURE 7. A view of $\widehat{M}$ near Y. The manifold with 'multi-cylindrical ends' $\widehat{M}$ is obtained by gluing multiple cylinders onto M.

The corresponding theorem for manifolds with boundary (Theorem 3.1) can be used to prove this result. For a concrete example consider the Cauchy-Riemann operator $D_{CR} = \partial_x + i\partial_y$ on the square $[-1,1]_x \times [-1,1]_y$. Certainly, the manifold and operator are both of product type. Then, $\ker D_{CR}$ is infinite dimensional since the kernel of the D_{CR} consists of all holomorphic functions on the square.[6]

Currently there is no suitable theory of elliptic boundary value problems for manifolds with corners of codimension two because the Calderón projector [**19**] in this context is not understood. However, in analogy with the case of a manifold with boundary, we can build an index theory through attaching multi-cylinders and then compactifying again forming the corresponding b-theory.

We attach multi-cylinders as follows: Let $\widehat{M}$ be the manifold formed by taking the infinite cylinder $(-\infty, 0]_{s_1} \times M_1$ and attaching it onto the collar $[0,1)_{s_1} \times M_1$ of M, then taking $(-\infty, 0]_{s_2} \times M_2$ and attaching it onto the collar $[0,1)_{s_2} \times M_2$, and finally taking $(-\infty, 0]_{s_1} \times (-\infty, 0]_{s_2} \times Y$ and attaching it onto the remaining open quadrant; see Figure 7. Since all the geometric structures and the Dirac operator are of product type near the boundary of M, they all have natural extensions to the manifold $\widehat{M}$. We denote these extended structures on $\widehat{M}$ using the same notations as were used for the original objects on M, except for the Dirac operator which we denote by $\widehat{D}$.

The 'General Principle' (3.1) gives the following theorem (see [**53**] for a proof).

THEOREM 5.4. *The Dirac type operator*

$$\widehat{D} : H^1(\widehat{M}, E) \to L^2(\widehat{M}, F)$$

is Fredholm if and only if $\widehat{D}_i : H^1(\widehat{M}_i, E_i) \to L^2(\widehat{M}_i, E_i)$ *for* $i = 1, 2$, *and the corner operator* $D_Y : H^1(Y, E_0) \to L^2(Y, E_0)$ *are each invertible.*

---

[6] Although the square has four corners instead of one, this example illustrates the point of the theorem.

Here, $\widehat{M_i}$ is the manifold with cylindrical end formed by attaching an infinite cylinder to the odd-dimensional compact manifold with boundary M_i, and $\widehat{D}_i$ is the natural extension of the Dirac operator D_i to $\widehat{M_i}$.

From Theorem 3.3, we know that the Dirac operator on a manifold with a cylindrical end formed from a manifold with boundary can *always* be made Fredholm by considering it on weighted Sobolev spaces. For a manifold with corners of codimension two, this is not the case.

THEOREM 5.5. *There exists a $\delta > 0$ such that for all $0 < |\alpha| < \delta$, the Dirac type operator*

$$\widehat{D} : e^{\alpha s} H^1(\widehat{M}, E) \to e^{\alpha s} L^2(\widehat{M}, F)$$

is Fredholm if and only if the corner operator $D_Y : H^1(Y, E_0) \to L^2(Y, E_0)$ is invertible (has zero kernel).

See [**53**] or [**52**] for a proof. Here, each coordinate function s_i is extended into the rest of $\widehat{M}$ to be a positive bounded function there, $\alpha = (\alpha_1, \alpha_2)$ is a pair of real numbers, $0 < |\alpha| < \delta$ means that $0 < |\alpha_i| < \delta$ for $i = 1, 2$, and finally, $e^{\alpha s} = e^{\alpha_1 s_1} e^{\alpha_2 s_2}$. We remark that in many cases the kernels of Dirac operators represent topological quantities. In these cases, the invertibility of the corner Dirac operator would require certain cohomology groups of the corner Y to vanish. Thus the Fredholm condition in Theorem 5.5 is actually very restrictive.

5.3. Müller's generalization of the APS index formula. We now explain Müller's generalization [**67**] of the APS formula in Theorem 3.4 to manifolds with corners of codimension two under the assumption that the corner Dirac operator D_Y is invertible. We remove this assumption in Section 6.

We first need to introduce the *b*-eta invariants of the operators $\widehat{D}_1$ and $\widehat{D}_2$, cf. [**59**, Sec. 9.7]. Consider the operator $\widehat{D}_1$ on $\widehat{M_1}$. Here, $\widehat{M_1}$ is the manifold with cylindrical end formed by attaching an infinite cylinder to the odd-dimensional compact manifold with boundary M_1. The operator $\widehat{D}_1$ turns out to have continuous spectrum, and not discrete spectrum, due to the fact that $\widehat{M_1}$ has infinite volume. Thus its eta invariant cannot be defined as a regularized signature in the same way as in the case of a manifold with boundary considered in Section 3.2. However, since the heat operator of $\widehat{D}_1^2$ does exist, we can still try to define the eta invariant via the integral (3.3):

$$\text{“}\eta(\widehat{D}_1) = \frac{1}{\sqrt{\pi}} \int_0^\infty t^{-1/2} \operatorname{Tr}(\widehat{D}_1 \, e^{-t\widehat{D}_1^2}) \, dt\text{”}.$$

Unfortunately, the operator $\widehat{D}_1 \, e^{-t\widehat{D}_1^2}$ is not trace class, cf. Section 4.3, so the right-hand side is not defined, which is the reason for the quotes. However, the *b*-trace of $\widehat{D}_1 \, e^{-t\widehat{D}_1^2}$ is defined.[7] Replacing Tr with ${}^b\mathrm{Tr}$ in the above formula defines

[7]We first compactify $\widehat{M_1}$, and then define the *b*-trace as in Section 4.3.

the *b-eta invariant*[8],

$$ {}^b\eta(\widehat{D}_1) = \frac{1}{\sqrt{\pi}} \int_0^\infty t^{-1/2}\, {}^b\mathrm{Tr}(\widehat{D}_1\, e^{-t\widehat{D}_1^2})\, dt. $$

The *b*-eta invariant of $\widehat{D}_2$ is defined in the same way. The APS formula in Theorem 3.4 generalizes as follows.

THEOREM 5.6 (Müller, 1996). *Let D be a Dirac type operator on an even-dimensional, compact, oriented, Riemannian manifold with corners of codimension two with exactly two boundary hypersurfaces intersecting in exactly one corner and with product type structures specified. Then there exists a $\delta > 0$ such that for all $0 < |\alpha| < \delta$, the Dirac type operator*

$$ \widehat{D} : e^{\alpha s} H^1(\widehat{M}, E) \to e^{\alpha s} L^2(\widehat{M}, F) $$

is Fredholm if and only if the corner operator $D_Y : H^1(Y, E_0) \to L^2(Y, E_0)$ is invertible (has zero kernel); in which case,

$$ \mathrm{ind}_\alpha \widehat{D} = \int_M K_{AS} - \frac{1}{2} \sum_{i=1,2} \Big\{ {}^b\eta(\widehat{D}_i) + \mathrm{sign}\,\alpha \cdot \dim\ker \widehat{D}_i \Big\}. $$

Müller's theorem in [**67**] technically only applies to the case of $\alpha > 0$; in the generality presented above, the theorem is due to Melrose, cf. [**53**]. The formula for $\mathrm{ind}_\alpha \widehat{D}$ is almost exactly the same as the APS formula in Theorem 3.4. In fact, using the *b*-calculus, the proof of Theorem 5.6 proceeds in almost identical fashion as the proof of Theorem 3.4. The only 'hard' part is defining the appropriate generalization of *b*-pseudodifferential operators and the *b*-trace to manifolds with corners of codimension two. Once this machinery is set up, the proof of the APS index formula can be used to prove Theorem 5.6.

5.4. b-version of Müller's theorem. In analogy with the case of a manifold with boundary, we now compactify the manifold $\widehat{M}$ by introducing the change of variables $x_1 = e^{s_1}$ and $x_2 = e^{s_2}$. As $s_i \to -\infty$, $x_i \to 0$, and so this change of variables compactifies $\widehat{M}$ to be the interior of a compact manifold with corners of codimension two, which we denote by X. Moreover, since $ds_i = dx_i/x_i$ and $\partial_{s_i} = x_i \partial_{x_i}$, the geometric objects on $\widehat{M}$ transform into corresponding singular geometric '*b*-objects' on the compact manifold with corners:

$$ g = ds_1^2 + ds_2^2 + h \ \rightsquigarrow\ g = \Big(\frac{dx_1}{x_1}\Big)^2 + \Big(\frac{dx_2}{x_2}\Big)^2 + h \quad (b\text{-metric}), $$

$$ dg = ds_1\, ds_2\, dh \ \rightsquigarrow\ dg = \frac{dx_1}{x_1} \frac{dx_2}{x_2}\, dh \quad (b\text{-measure}), $$

$$ H^k(\widehat{M}) \ \rightsquigarrow\ H_b^k(X) \quad (b\text{-Sobolev space}), $$

---

[8]The same discussion as in footnote (4) concerning the local index theorem on odd-dimensional manifolds without boundary applies in this situation too.

and finally,

$$\widehat{D} = \Gamma_1 \partial_{s_1} + \Gamma_2 \partial_{s_2} + B \rightsquigarrow \widehat{D} = \Gamma_1 x_1 \partial_{x_1} + \Gamma_2 x_2 \partial_{x_2} + B,$$

a *b*-differential operator. We repeat the statement of Müller's theorem in the present context.

THEOREM 5.7. *Let D be a Dirac type operator on an even-dimensional, compact, oriented, Riemannian manifold with corners of codimension two with exactly two boundary hypersurfaces intersecting in exactly one corner and with product type structures specified. Then there exists a $\delta > 0$ such that for all $0 < |\alpha| < \delta$, the Dirac type operator*

$$\widehat{D} : x^\alpha H_b^1(X, E) \to x^\alpha L_b^2(X, F)$$

is Fredholm if and only if the corner operator $D_Y : H^1(Y, E_0) \to L^2(Y, E_0)$ is invertible (has zero kernel); in which case,

$$\operatorname{ind}_\alpha \widehat{D} = \int_M K_{AS} - \frac{1}{2} \sum_{i=1,2} \left\{ {}^b\eta(\widehat{D}_i) + \operatorname{sign}\alpha \cdot \dim\ker \widehat{D}_i \right\}.$$

As already mentioned, with a proper generalization of *b*-pseudodifferential operators and the *b*-trace to manifolds with corners of codimension two, the proof of Theorem 5.7 proceeds in almost identical fashion as the proof of Theorem 3.4. In fact, the above theorem and its proof generalize to not only Dirac type operators but also *b*-pseudodifferential on manifolds with corners of *arbitrary* codimension (see [**50, 51, 52, 53**]). These generalizations are due to Melrose (for Dirac operators) and the author (for *b*-pseudodifferential operators), cf. Lauter and Moroianu [**44**] for the cusp case. The ability to handle arbitrary codimensions is a nice feature of using *b*-pseudodifferential operators to attack index problems on manifolds with corners.

5.5. Some remarks on index theory on manifolds with corners. In [**75**] Salomenson builds an index theory for Dirac operators on manifolds with corners of codimension two by attaching cylinders in a very different way than considered here. Instead of attaching separate cylinders to each hypersurface M_i and then filling in the lower quadrant with a product cylinder as shown in Figure 7, he notes that ∂M has a natural smooth structure and attaches the cylinder $(-\infty, 0] \times \partial M$ onto M. This creates a manifold with cylindrical end like in the case of a manifold with boundary, except that it has a 'wedge singularity' at the original corner Y. Results of Cheeger [**22**] or Chou [**24**] can be used to handle the wedge singularity.

In a different direction, Hassel, Mazzeo, and Melrose [**41**] prove a signature formula for manifolds with corners of codimension two. Unlike the signature formulas for manifolds with and without boundary, which are direct corollaries of index formulas on such manifolds, the HMM formula is *not* a consequence of an index formula on manifolds with corners of codimension two. Instead, they round off the corner and consider X as a limit as $\varepsilon \to 0$ of manifolds with smooth boundary X_ε.

The resulting signature formula is obtained by a careful analysis of the limit of the APS signature formulas of each X_ε. They rely on 'analytic surgery' techniques in [**40**] to identify the limiting formula. Wall [**84**] considers a manifold with boundary divided into two parts, each a manifold with corners of codimension two (e.g., a disk divided into two half wedges). Although not an index formula *per se*, Wall gives a formula for the signature of the manifold with boundary in terms of the signatures of the two manifolds with corners of codimension two and a correction term given by the Maslov index of certain Lagrangian subspaces, cf. Section 6.

6. Perturbations of Dirac operators on manifolds with corners

We now consider the APS index formula for manifolds with corners of codimension two dropping the invertibility assumption on the corner Dirac operator. From our experience with the Gauss-Bonnet formula in the introduction, we expect there to be a correction term added to the right-hand side of the APS formula due to the presence of the corners. Theorem 5.6 did not have a corner contribution, essentially because the invertibility of the corner Dirac operator D_Y makes the Dirac operator $\widehat{D}$ not 'notice' the presence of the corners. For the Gauss-Bonnet formula, the correction term was given by the exterior angles of the corners. For the APS formula without the invertibility assumption on the corner operator, there is a correction term in the index formula and it represents an 'exterior angle' of sorts between certain Lagrangian vector spaces. In this section, we use the same notation as in Section 5.

6.1. Fredholm perturbation of Dirac operators. By Theorem 5.5, there exists a $\delta > 0$ such that for all $0 < |\alpha| < \delta$, the Dirac type operator

$$\widehat{D} : x^\alpha H_b^1(X, E) \to x^\alpha L_b^2(X, F)$$

is Fredholm if and only if the corner operator $D_Y : H^1(Y, E_0) \to L^2(Y, E_0)$ is invertible (has zero kernel). This nondegeneracy condition is actually very restrictive since in many cases the kernels of Dirac operators represent cohomology. However, we now show that it is *always* possible to make $\widehat{D}$ Fredholm on weighted Sobolev spaces by perturbation with b-smoothing operators.

To define these perturbations we recall some notation from Section 5.1. The manifold with corners of codimension two M is assumed to have exactly two boundary hypersurfaces M_1 and M_2 that intersect in exactly one codimension two face Y. Near the corner Y the Dirac type operator D takes the form

$$D = \Gamma_1 \partial_{s_1} + \Gamma_2 \partial_{s_2} + B,$$

where

$$B : C^\infty(Y, E_0) \to C^\infty(Y, F_0)$$

is a Dirac type operator on the even-dimensional manifold without boundary Y. The induced operator D_1 on the hypersurface M_1 takes the form

$$D_1 = \Gamma(\partial_{s_2} + D_Y), \quad \Gamma = \Gamma_1^* \Gamma_2, \quad D_Y = \Gamma_2^* B, \tag{6.1}$$

and the operator D_2 on M_2 takes the form

$$D_2 = -\Gamma(\partial_{s_1} + \widetilde{D}_Y), \quad \widetilde{D}_Y = \Gamma D_Y. \tag{6.2}$$

The minus sign in front of Γ and the fact that $\widetilde{D}_Y = \Gamma D_Y$ will come into play later. Also, see Lemma 5.1, $E_0 = E_0^+ \oplus E_0^-$ where $E_0^\pm$ are the $\pm i$ eigenspaces of $\Gamma = \Gamma_1^* \Gamma_2$, and the induced Dirac operator $D_Y = \Gamma_2^* B$ is odd with respect to Γ and so decomposes as

$$\begin{bmatrix} 0 & D_Y^- \\ D_Y^+ & 0 \end{bmatrix} : C^\infty(Y, E_0^+ \oplus E_0^-) \to C^\infty(Y, E_0^+ \oplus E_0^-),$$

where $D_Y^\pm$ are the restrictions of D_Y to $C^\infty(Y, E_0^\pm)$. Moreover, see Theorem 5.2,

$$D_Y^+ : H^1(Y, E_0^+) \to L^2(Y, E_0^-)$$

has index zero; that is, $\dim \ker D_Y^+ = \dim \ker D_Y^-$.

We now define the perturbations. Since the kernel of the Dirac type operator D_Y is exactly the obstruction to $\widehat{D}$ being Fredholm on weighted Sobolev spaces, the perturbations are chosen to be isomorphisms on the kernel. Since D_Y is odd with respect to Γ, we only consider isomorphisms on $\ker D_Y$ having the same property. Thus, let $T : \ker D_Y \to \ker D_Y$ be a self-adjoint unitary isomorphism that is odd with respect to Γ. Hence T decomposes as an odd matrix

$$\begin{bmatrix} 0 & T^- \\ T^+ & 0 \end{bmatrix} : \ker D_Y^+ \oplus \ker D_Y^- \to \ker D_Y^+ \oplus \ker D_Y^-,$$

where $T^\pm : \ker D_Y^\pm \to \ker D_Y^\mp$ are unitary isomorphisms with respect to the L^2 inner product on $\ker D_Y \subset L^2(Y, E_0)$. Such an operator T exists because $\dim \ker D_Y^+ = \dim \ker D_Y^-$. We can define T explicitly as follows. Let $\{u_j\}_{j=1}^N$ and $\{v_j\}_{j=1}^N$ be orthonormal bases of $\ker D_Y^+$ and $\ker D_Y^-$, respectively. By elliptic regularity, u_j, $v_j \in C^\infty(Y, E_0)$ for every j. Then,

$$T = \sum_{j=1}^N u_j \otimes \overline{v}_j + \sum_{j=1}^N v_j \otimes \overline{u}_j$$

defines a self-adjoint unitary isomorphism on $\ker D_Y$ that is odd with respect to Γ and any such T can be written in this way for some choice of bases. Moreover, this formula shows that T is a smoothing operator on Y. Obviously,

$$D_Y - T : H^1(Y, E_0) \to L^2(Y, E_0)$$

is invertible. This suggests that if we can extend T to an operator $\widehat{T}$ on X, then

$$\widehat{D} - \widehat{T} : x^\alpha H_b^1(X, E) \to x^\alpha L_b^2(X, F)$$

is Fredholm for all $0 < |\alpha| < \delta$ for some $\delta > 0$. To extend T, we first define $\widehat{T}$ on the manifold with multi-cylindrical ends $\widehat{M}$. Let $\chi \in C_c^\infty([0,1)^2)$ be such that

$\chi(x) = 1$ for x near zero. Then $\chi(e^s) = \chi(e^{s_1}, e^{s_2})$ can be regarded as a smooth function on $\widehat{M}$ supported on the cylindrical end (cf. Figure 7)

$$(-\infty, 0]_{s_1} \times (-\infty, 0]_{s_2} \times Y.$$

Let $\varphi(\xi_1, \xi_2) = e^{-\xi_1^2 - \xi_2^2}$. Then, given any $u \in C_c^\infty(\widehat{M}, E)$, we define

$$\widehat{T}u(s,y) = \chi(e^s)\,\Gamma_2\,\frac{1}{(2\pi)^2}\int_{\mathbb{R}^2} e^{is\cdot\xi}\,\varphi(\xi)\,T\,\widehat{\chi u}(\xi, y)\,d\xi, \tag{6.3}$$

where $\widehat{\chi u}$ is the Fourier transform of $\chi(e^s)\,u(s,y)$ with respect to s:

$$\widehat{\chi u}(\xi, y) = \frac{1}{(2\pi)^2}\int_{\mathbb{R}^2} e^{-is\cdot\xi}\,\chi(e^s)\,u(s,y)\,ds.$$

The reason for the factor of Γ_2 on the right-hand side of $\widehat{T}u$ is that $\widehat{T}$ is required to map sections of E to sections of F and the Γ_2 factor provides this property. Note that the operator T in the definition of $\widehat{T}$ only acts on the y variable of $\widehat{\chi u}(\xi, y)$. Regarded as an operator on the compactified manifold X under the change of variables $x_i = e^{s_i}$, the operator $\widehat{T}$ is an example of a b-smoothing operator; that is, a b-pseudodifferential operator of order $-\infty$. The mapping properties of such operators (cf. [**59**, Ch. 5]) imply that

$$\widehat{T} : x^\alpha H_b^k(X, E) \to x^\alpha H_b^\ell(X, F) \quad \text{for all } \alpha, k, \ell.$$

The next result follows from the properties of b-pseudodifferential operators and the fact that $D_Y - T : H^1(Y, E_0) \to L^2(Y, E_0)$ is invertible, cf. [**52, 53**].

LEMMA 6.1. *There exists a $\delta > 0$ such that for all $0 < |\alpha| < \delta$,*

$$\widehat{D} - \widehat{T} : x^\alpha H_b^1(X, E) \to x^\alpha L_b^2(X, F)$$

is Fredholm.

6.2. An index formula for perturbed Dirac operators. We now give a formula for the index of $\widehat{D} - \widehat{T}$. Before doing so, we need to review the 'scattering Lagrangian' of each operator $\widehat{D}_i$. Consider the operator $\widehat{D}_1$ on $\widehat{M}_1$. Recall that $\widehat{M}_1$ is formed by attaching an infinite cylinder $(-\infty, 0]_{s_2} \times Y$ to the odd-dimensional compact manifold with boundary M_1. The set

$$\Lambda_{C_1} = \left\{ \lim_{s_2 \to -\infty} u(s_2, y)\,;\, u \in C^\infty(\widehat{M}_1, E) \text{ is bounded, and } \widehat{D}_1 u = 0 \right\}$$

is called the *scattering Lagrangian* of $\widehat{D}_1$. It turns out that $\Lambda_{C_1} \subset \ker D_Y$ and the dimension of Λ_{C_1} is exactly one-half the dimension of $\ker D_Y$, cf. [**59**, Sec. 6.5]. The *scattering matrix* of $\widehat{D}_1$ is the operator $C_1 : \ker D_Y \to \ker D_Y$ defined by $C_1 = +1$ on Λ_{C_1} and $C_1 = -1$ on $\Lambda_{C_1}^\perp$, where '$\perp$' means the orthogonal complement with respect to the L^2 inner product. Then C_1 is odd with respect to Γ, cf. [**65**, Sec. 4]. The scattering Lagrangian Λ_{C_2} and the matrix C_2 of $\widehat{D}_2$ are defined in the same way. In [**53**] we give the following formula for the index of $\widehat{D} - \widehat{T}$:

THEOREM 6.2 (Loya-Melrose, 2002). *Let D be a Dirac type operator on an even-dimensional, compact, oriented, Riemannian manifold with corners of codimension two with exactly two boundary hypersurfaces intersecting in exactly one corner and with product type structures specified. Let $T : \ker D_Y \to \ker D_Y$ be a self-adjoint unitary isomorphism that is odd with respect to Γ and let $\widehat{T}$ be the perturbation defined by* (6.3). *Then there exists a $\delta > 0$ such that for all $0 < |\alpha| < \delta$, the perturbed Dirac operator*

$$\widehat{D} - \widehat{T} : x^\alpha H_b^1(X, E) \to x^\alpha L_b^2(X, F)$$

is Fredholm. Moreover, if its index is denoted by $\operatorname{ind}_\alpha(\widehat{D} - \widehat{T})$, *then*

$$\begin{aligned} \operatorname{ind}_\alpha(\widehat{D} - \widehat{T}) = \int_M K_{AS} - \frac{1}{2} \sum_{i=1,2} \left\{ {}^b\eta(\widehat{D}_i) + \operatorname{sign} \alpha \cdot \dim \ker \widehat{D}_i \right\} \\ - \frac{1}{2} c_\alpha(\Lambda_T, \Lambda_{C_1}, \Lambda_{C_2}). \end{aligned} \tag{6.4}$$

The first line on the right-hand side is the same as in Theorem 5.7; the third 'corner correction term' is described as follows. First, $\Lambda_T \subset \ker D_Y$ is the $+1$ eigenspace of the matrix T (since T is a self-adjoint unitary isomorphism, $T^2 = \mathrm{Id}$, so T has eigenvalues ± 1). Then,

$$c_\alpha(\Lambda_T, \Lambda_{C_1}, \Lambda_{C_2}) = \operatorname{sign} \alpha \cdot \left\{ \dim(\Lambda_T \cap \Lambda_{C_1}) + \dim(\Lambda_{\Gamma T} \cap \Lambda_{C_2}) \right\} + \eta(D_{\mathcal{G}}), \tag{6.5}$$

where $\Lambda_{\Gamma T} \subset \ker D_Y$ is the $+1$ eigenspace of the self-adjoint unitary isomorphism ΓT and $\eta(D_{\mathcal{G}})$ is the eta invariant of a Dirac operator on a directed graph $\mathcal{G}$ defined as follows, cf. [**41**], [**55**], [**20**], [**47**]. This graph has two vertices v_1 and v_2 representing the hypersurfaces M_1 and M_2, respectively, and two edges e_{12} and e_{21} connecting the vertices representing the single corner Y. To put a manifold structure on this graph, we identify e_{jk} with the interval $[-1, 1]_t$, where the vertex v_j corresponds to $t = -1$ and the vertex v_k to $t = +1$. We consider $V = \ker D_Y \oplus \ker D_Y$ as a 'vector bundle' over $\mathcal{G}$ where the first and second factors of $\ker D_Y$ are 'fibers' over the edges e_{12} and e_{21}, respectively. Thus a section of this vector bundle is a sum $s_{12} \oplus s_{21}$, where $s_{jk} : e_{jk} = [-1, 1] \to \ker D_Y$. We define a Dirac operator $D_{\mathcal{G}}$ acting on sections of $\mathcal{G}$ by

$$D_{\mathcal{G}} = \Gamma \frac{d}{dt} \oplus \left(- \Gamma \frac{d}{dt} \right).$$

The minus sign in the second term stems from the minus sign in (6.2). The domain of $D_{\mathcal{G}}$ consists of those sections $s_{12} \oplus s_{21}$ such that $s_{12}(v_1) \in \Lambda_T$, $s_{12}(v_2) \in \Lambda_{C_1}$, and $s_{21}(v_2) \in \Lambda_{\Gamma T}$, $s_{21}(v_1) \in \Lambda_{C_2}$. The Lagrangian $\Lambda_{\Gamma T}$ paired with the scattering Lagrangian Λ_{C_2} stems from the fact that $\widetilde{D}_Y = \Gamma D_Y$ in (6.2). The term $\eta(D_{\mathcal{G}})$ appearing in (6.5) is then the eta invariant of $D_{\mathcal{G}}$.

Lesch and Wojciechowski [**47**] give the following linear-algebraic form for the eta term:

$$\eta(D_{\mathcal{G}}) = m(\Lambda_T, \Lambda_{C_1}) - m(\Lambda_{\Gamma T}, \Lambda_{C_2}),$$

where

$$m(\Lambda_{L_1}, \Lambda_{L_2}) = -\frac{1}{i\pi} \sum_{\substack{e^{i\theta} \in \operatorname{spec}(-L_1^- L_2^+) \\ \theta \in (-\pi, \pi)}} i\theta$$

for any given self-adjoint unitary isomorphisms L_1, L_2 on $\ker D_Y$ that are odd with respect to Γ, and with $\Lambda_{L_i} \subset \ker D_Y$ denoting the $+1$ eigenspace of L_i. The number $m(\Lambda_{L_1}, \Lambda_{L_2})$ can be interpreted as an 'exterior angle' of sorts between Λ_{L_1} and Λ_{L_2}, cf. [**48**], [**20**], [**18**]. Hassel, Mazzeo, and Melrose [**40, 41**] give a somewhat more sophisticated linear-algebraic description of the eta term.

Theorem 6.2 is proved as follows: First, following the proof of Theorem 5.7, which uses similar arguments found in Section 4, produces the formula

$$\operatorname{ind}_\alpha(\widehat{D} - \widehat{T}) = \int_M K_{AS} - \frac{1}{2} \sum_{i=1,2} \left\{ {}^b\eta(\widehat{D}_i - \widehat{T}_i) + \operatorname{sign}\alpha \cdot \dim\ker(\widehat{D}_i - \widehat{T}_i) \right\},$$

where $\widehat{T}_i$ is an operator naturally induced by $\widehat{T}$ on $\widehat{M}_i$. The second and most difficult part of the proof is to show that the terms involving $\widehat{D}_i - \widehat{T}_i$ in this formula decompose as in (6.4). To do this, we show that

$$\dim\ker(\widehat{D}_i - \widehat{T}_i) = \dim\ker \widehat{D}_i + \dim(\Lambda_{T_i} \cap \Lambda_{C_i}),$$

where $T_1 = T$ and $T_2 = \Gamma T$, and that

$${}^b\eta(\widehat{D}_i - \widehat{T}_i) = {}^b\eta(\widehat{D}_i) \pm m(\Lambda_{T_i}, \Lambda_{C_i}),$$

where the sign is positive or negative if $i = 1$ or $i = 2$, respectively. The decomposition of the b-eta invariants uses techniques that have been developed by many authors concerning gluing/splitting formulas for eta invariants, e.g. Brüning and Lesch [**17**] (cf. Vishik [**83**]), Douglas and Wojciechowski [**28**], Lesch and Wojciechowski [**47**], and Müller [**65**]. For related works on the eta invariant, see Singer [**80**], Bunke [**18**], Dai and Freed [**26**], and Hassel, Mazzeo, and Melrose [**40**].

6.3. Some concluding remarks. If M has more than one corner, the result from Theorem 6.2 still holds (with minor changes in the index formula accounting for the various faces and corners) as long as we *assume* that each corner Dirac operator has index zero [**53**]. This assumption allows us to construct separate perturbations for each corner, then sum these perturbations producing a b-smoothing operator giving a Fredholm perturbation of the Dirac operator. Melrose and Nistor show that it is in fact necessary that each corner operator have index zero for the existence of a b-smoothing Fredholm perturbation. However, using a slightly larger class of perturbations called 'overblown' b-smoothing operators, it is possible to make Fredholm perturbations without any assumptions at the corners [**54**].

Current plans include relating the index of the perturbed Dirac operator to an index of the Dirac operator on a domain depending on the choice of perturbation. The index should be a type of Carron index [**21**]. In future work, we expect to generalize the program of 'overblown' b-smoothing Fredholm perturbations of Dirac operators to manifolds with corners of arbitrary codimension. Finally, one of the ultimate goals of this project is to derive via an index computation an analytic formula for the topological signature of any compact manifold with corners of arbitrary codimension in terms of geometric and other types of invariants of the manifold and its boundary faces.

References

[1] L. Alvarez-Gaumé, *Supersymmetry and the Atiyah-Singer index theorem*, Comm. Math. Phys. **90** (1983), no. 2, 161–173.

[2] M. F. Atiyah, R. Bott, and V. K. Patodi, *On the heat equation and the index theorem*, Invent. Math. **19** (1973), 279–330.

[3] M. F. Atiyah, H. Donnelly, and I. M. Singer, *Eta invariants, signature defects of cusps, and values of L-functions*, Ann. of Math. (2) **118** (1983), no. 1, 131–177.

[4] M. F. Atiyah, V. K. Patodi, and I. M. Singer, *Spectral asymmetry and Riemannian geometry*, Bull. London Math. Soc. **5** (1973), 229–234.

[5] ———, *Spectral asymmetry and Riemannian geometry. I*, Math. Proc. Cambridge Philos. Soc. **77** (1975), 43–69.

[6] M.F. Atiyah and I.M. Singer, *The index of elliptic operators on compact manifolds*, Bull. Amer. Math. Soc. **69** (1963), 422–433.

[7] ———, *The index of elliptic operators I*, Ann. Math. **87** (1968), 484–530.

[8] ———, *The index of elliptic operators II*, Ann. Math. **87** (1968), 531–545.

[9] N. Berline, E. Getzler, and M. Vergne, *Heat kernels and Dirac operators*, Springer-Verlag, New York, 1992.

[10] J.-M. Bismut and D.S. Freed, *The analysis of elliptic families. II. Dirac operators, eta invariants, and the holonomy theorem*, Comm. Math. Phys. **107** (1986), no. 1, 103–163.

[11] J.-M. Bismut, *The Atiyah-Singer index theorem for families of Dirac operators: two heat equation proofs*, Invent. Math. **83** (1985), no. 1, 91–151.

[12] J.-M. Bismut and J. Cheeger, *Families index for manifolds with boundary, superconnections, and cones. I. Families of manifolds with boundary and Dirac operators*, J. Funct. Anal. **89** (1990), no. 2, 313–363.

[13] ———, *Families index for manifolds with boundary, superconnections and cones. II. The Chern character*, J. Funct. Anal. **90** (1990), no. 2, 306–354.

[14] B. Booss and D. D. Bleecker, *Topology and analysis*, Springer-Verlag, New York, 1985, The Atiyah-Singer index formula and gauge-theoretic physics, Translated from the German by Bleecker and A. Mader.

[15] B. Booß-Bavnbek and K. P. Wojciechowski, *Elliptic boundary problems for Dirac operators*, Birkhäuser Boston Inc., Boston, MA, 1993.

[16] J. Brüning, *L^2-index theorems on certain complete manifolds*, J. Differential Geom. **32** (1990), no. 2, 491–532.

[17] J. Brüning and M. Lesch, *On the η-invariant of certain nonlocal boundary value problems*, Duke Math. J. **96** (1999), no. 2, 425–468.

[18] U. Bunke, *On the gluing problem for the η-invariant*, J. Differential Geom. **41** (1995), no. 2, 397–448.

[19] A.-P. Calderón, *Boundary value problems for elliptic equations*, Outlines Joint Sympos. Partial Differential Equations (Novosibirsk, 1963), Acad. Sci. USSR Siberian Branch, Moscow, 1963, pp. 303–304.

[20] S. E. Cappell, R. Lee, and E. Y. Miller, *On the Maslov index*, Comm. Pure Appl. Math. **47** (1994), no. 2, 121–186.

[21] G. Carron, *Théorèmes de l'indice sur les variétés non-compactes*, J. Reine Angew. Math. **541** (2001), 81–115.

[22] J. Cheeger, *Spectral geometry of singular Riemannian spaces*, J. Differential Geom. **18** (1983), no. 4, 575–657 (1984).

[23] S.S. Chern, *A simple intrinsic proof of the Gauss-Bonnet formula for closed Riemannian manifolds*, Ann. of Math. (2) **45** (1944), 747–752.

[24] A. W. Chou, *The Dirac operator on spaces with conical singularities and positive scalar curvatures*, Trans. Amer. Math. Soc. **289** (1985), no. 1, 1–40.

[25] A. Connes, M. Gromov, and H. Moscovici, *Conjecture de Novikov et fibrés presque plats*, C. R. Acad. Sci. Paris Sér. I Math. **310** (1990), no. 5, 273–277.

[26] X. Dai and D. S. Freed, *η-invariants and determinant lines*, J. Math. Phys. **35** (1994), no. 10, 5155–5194, Topology and physics.

[27] S. K. Donaldson and P. B. Kronheimer, *The geometry of four-manifolds*, The Clarendon Press Oxford University Press, New York, 1990, Oxford Science Publications.

[28] R. G. Douglas and K. P. Wojciechowski, *Adiabatic limits of the η-invariants. The odd-dimensional Atiyah-Patodi-Singer problem*, Comm. Math. Phys. **142** (1991), no. 1, 139–168.

[29] B. V. Fedosov and B.-W. Schulze, *On the index of elliptic operators on a cone*, Schrödinger operators, Markov semigroups, wavelet analysis, operator algebras, Akademie Verlag, Berlin, 1996, pp. 348–372.

[30] E. Getzler, *A short proof of the local Atiyah-Singer index theorem*, Topology **25** (1986), no. 1, 111–117.

[31] ———, *Cyclic homology and the Atiyah-Patodi-Singer index theorem*, Index theory and operator algebras (Boulder, CO, 1991), Amer. Math. Soc., Providence, RI, 1993, pp. 19–45.

[32] P. B. Gilkey, *Curvature and the eigenvalues of the Laplacian for elliptic complexes*, Advances in Math. **10** (1973), 344–382.

[33] ———, *The residue of the global η function at the origin*, Adv. in Math. **40** (1981), no. 3, 290–307.

[34] ———, *Invariance theory, the heat equation, and the Atiyah-Singer index theorem*, second ed., CRC Press, Boca Raton, FL, 1995.

[35] D. Grieser, *Basics of the b-calculus*, Approaches to singular analysis (Berlin, 1999), Birkhäuser, Basel, 2001, pp. 30–84.

[36] M. Gromov, *Pseudoholomorphic curves in symplectic manifolds*, Invent. Math. **82** (1985), no. 2, 307–347.

[37] M. Gromov and H. B. Lawson, Jr., *Positive scalar curvature and the Dirac operator on complete Riemannian manifolds*, Inst. Hautes Études Sci. Publ. Math. (1983), no. 58, 83–196 (1984).

[38] G. Grubb, *Functional calculus of pseudodifferential boundary problems*, second ed., Progress in Math., Birkhäuser, Boston, 1996.

[39] G. Grubb and R.T. Seeley, *Weakly parametric pseudodifferential operators and Atiyah-Patoti-Singer operators*, Invent. Math. **121** (1995), 481–529.

[40] A. Hassell, R. Mazzeo, and R. B. Melrose, *Analytic surgery and the accumulation of eigenvalues*, Comm. Anal. Geom. **3** (1995), no. 1-2, 115–222.

[41] ———, *A signature formula for manifolds with corners of codimension two*, Topology **36** (1997), no. 5, 1055–1075.

[42] F. Hirzebruch, *Topological methods in algebraic geometry*, Springer-Verlag, Berlin, 1995, Translated from the German and Appendix One by R. L. E. Schwarzenberger, With a preface to the third English edition by the author and Schwarzenberger, Appendix Two by A. Borel, Reprint of the 1978 edition.

[43] L. Hörmander, *The analysis of linear partial differential operators. III*, second ed., Springer-Verlag, Berlin, 1985, Pseudo-Differential Operators.

[44] R. Lauter and S. Moroianu, *The index of cusp operators on manifolds with corners*, Ann. Global Anal. Geom. **21** (2002), no. 1, 31–49.

[45] H. B. Lawson, Jr. and M.-L. Michelsohn, *Spin geometry*, Princeton University Press, Princeton, NJ, 1989.

[46] E. Leichtnam, J. Lott, and P. Piazza, *On the homotopy invariance of higher signatures for manifolds with boundary*, J. Differential Geom. **54** (2000), no. 3, 561–633.

[47] M. Lesch and K. P. Wojciechowski, *On the η-invariant of generalized Atiyah-Patodi-Singer boundary value problems*, Illinois J. Math. **40** (1996), no. 1, 30–46.
[48] G. Lion and M. Vergne, *The Weil representation, Maslov index and theta series*, Birkhäuser Boston, Mass., 1980.
[49] R. B. Lockhart and R. C. McOwen, *Elliptic differential operators on noncompact manifolds*, Ann. Scuola Norm. Sup. Pisa Cl. Sci. (4) **12** (1985), no. 3, 409–447.
[50] P. Loya, *On the b-pseudodifferential calculus on manifolds with corners*, Ph.D. thesis, MIT, 1998.
[51] ______, *Tempered operators and the heat kernel and complex powers of elliptic pseudodifferential operators*, Comm. Partial Differential Equations **26** (2001), no. 7 & 8, 1253–1321.
[52] ______, *The index of b-pseudodifferential operators on manifolds with corners*, preprint, 2002.
[53] P. Loya and R.B. Melrose, *Fredholm perturbations of Dirac operators on manifolds with corners*, preprint, 2002.
[54] ______, *General Fredholm perturbations of Dirac operators on manifolds with corners*, in preparation.
[55] R. Mazzeo and P. Piazza, *Dirac operators, heat kernels and microlocal analysis. II. Analytic surgery*, Rend. Mat. Appl. (7) **18** (1998), no. 2, 221–288.
[56] R. Mazzeo, *Elliptic theory of differential edge operators. I*, Comm. Partial Differential Equations **16** (1991), no. 10, 1615–1664.
[57] H.P. McKean and I.M. Singer, *Curvature and the eigenvalues of the Laplacian*, J. Diff. Geometry **1** (1967), 43–69.
[58] R. B. Melrose, *Transformation of boundary problems*, Acta Math. **147** (1981), no. 3-4, 149–236.
[59] ______, *The Atiyah-Patodi-Singer Index Theorem*, A.K. Peters, Wellesley, 1993.
[60] R.B. Melrose and G.A. Mendoza, *Elliptic pseudodifferential operators of totally characteristic type*, MSRI preprint, 1983.
[61] R.B. Melrose and V. Nistor, *Homology of pseudodifferential operators I. Manifolds with boundary*, preprint, 1996.
[62] R. B. Melrose and P. Piazza, *Families of Dirac operators, boundaries and the b-calculus*, J. Differential Geom. **46** (1997), no. 1, 99–180.
[63] ______, *An index theorem for families of Dirac operators on odd-dimensional manifolds with boundary*, J. Differential Geom. **46** (1997), no. 2, 287–334.
[64] W. Müller, *Signature defects of cusps of Hilbert modular varieties and values of L-series at* $s = 1$, J. Differential Geom. **20** (1984), no. 1, 55–119.
[65] ______, *Eta invariants and manifolds with boundary*, J. Differential Geom. **40** (1994), no. 2, 311–377.
[66] ______, *The eta invariant (some recent developments)*, Astérisque (1995), no. 227, Exp. No. 787, 5, 335–364, Séminaire Bourbaki, Vol. 1993/94.
[67] ______, *On the L^2-index of Dirac operators on manifolds with corners of codimension two. I*, J. Differential Geom. **44** (1996), 97–177.
[68] B. O'Neill, *Elementary differential geometry*, Academic Press, New York, 1966.
[69] R.S. Palais, *Seminar on the Atiyah-Singer Index Theorem*, Annals of Math. Studies No. 57, Princeton, NJ, 1965.
[70] V. K. Patodi, *Curvature and the eigenforms of the Laplace operator*, J. Differential Geometry **5** (1971), 233–249.
[71] P. Piazza, *On the index of elliptic operators on manifolds with boundary*, J. of Func. Anal. **117** (1993), 308–359.
[72] ______, *Dirac operators, heat kernels and microlocal analysis. I. Family index theory*, Proceedings of the Workshop on Differential Geometry and Topology (Palermo, 1996), no. 49, 1997, pp. 187–201.

[73] V. S. Rabinovič, *Pseudodifferential operators on a certain class of noncompact manifolds*, Mat. Sb. (N.S.) **89(131)** (1972), 46–60, 165.
[74] J. Roe, *Elliptic operators, topology and asymptotic methods*, second ed., Longman, Harlow, 1998.
[75] G. Salomonsen, *Atiyah-Patodi-Singer type index theorems for manifolds with splitting of η-invariants*, Geom. Funct. Anal. **11** (2001), no. 5, 1031–1095.
[76] E. Schrohe, *Complex powers on noncompact manifolds and manifolds with singularities*, Math. Ann. **281** (1988), no. 3, 393–409.
[77] R. T. Seeley, *Singular integrals and boundary value problems*, Amer. J. Math. **88** (1966), 781–809.
[78] ———, *Asymptotic expansions for Atiyah-Patodi-Singer problems*, Harmonic analysis and partial differential equations (Chicago, IL, 1996), Univ. Chicago Press, Chicago, IL, 1999, pp. 275–283.
[79] Bert-Wolfgang Schulze, Boris Sternin, and Victor Shatalov, *On the index of differential operators on manifolds with conical singularities*, Ann. Global Anal. Geom. **16** (1998), no. 2, 141–172.
[80] I. M. Singer, *The eta invariant and the index*, Mathematical aspects of string theory, World Scientific, Singapore, 1988, pp. 239–258.
[81] M. Stern, *L^2-index theorems on locally symmetric spaces*, Invent. Math. **96** (1989), no. 2, 231–282.
[82] M. E. Taylor, *Partial differential equations. II*, Springer-Verlag, New York, 1996, Qualitative studies of linear equations.
[83] S. M. Vishik, *Generalized Ray-Singer conjecture. I. A manifold with a smooth boundary*, Comm. Math. Phys. **167** (1995), no. 1, 1–102.
[84] C. T. C. Wall, *Non-additivity of the signature*, Invent. Math. **7** (1969), 269–274.
[85] M. Wodzicki, *Spectral asymmetry and zeta functions*, Invent. Math. **66** (1982), no. 1, 115–135.
[86] F. Wu, *The Chern-Connes character for the Dirac operator on manifolds with boundary*, *K*-Theory **7** (1993), no. 2, 145–174.

Department of Mathematics, Binghamton University, Binghamton, NY 13902-6000
E-mail address: paul@math.binghamton.edu

Operator Theory:
Advances and Applications, Vol. 151, 170–238

Index defects in the theory of spectral boundary value problems

Anton Savin and Boris Sternin

ABSTRACT. We study index defects in spectral boundary value problems for elliptic operators. Explicit analytic expressions for index defects in various situations are given. The corresponding topological indices are computed as homotopy invariants of the principal symbol.

Introduction

The classical Hirzebruch formula

$$\operatorname{sign} M = \int_M \widehat{L}(p_1, p_2, \dots, p_k) \tag{0.1}$$

expresses the signature of a closed oriented $4k$-dimensional manifold in terms of its Pontryagin characteristic classes. From the viewpoint of elliptic operator theory, formula (0.1) expresses the index of a specific elliptic operator (later called the Hirzebruch operator) via stable homotopy invariants of its principal symbol. (For the Hirzebruch operator, these invariants coincide with the Pontryagin classes of the manifold.)

Unfortunately, formula (0.1) has no immediate analog for manifolds with boundary: there are examples showing that the signature of such a manifold cannot be expressed in terms of Pontryagin classes.

In 1973, Hirzebruch [**Hir73**] considered a class of manifolds with boundary (arising from algebraic-geometric considerations on Hilbert modular varieties) that have naturally defined relative Pontryagin classes. Although the right-hand side of (0.1) makes sense in this case, the equality in (0.1) fails. The difference between the right- and left-hand sides was called the *signature defect*, and the problem was

2000 *Mathematics Subject Classification.* Primary 58J20, 58J32; Secondary 58J28, 46L85.

The work was partially supported by the Russian Foundation for Basic Research under grants Nos. 02-01-00118, 02-01-00928, and 03-02-16336, by Arbeitsgruppe Partielle Differentialgleichungen und Komplexe Analysis, Institut für Mathematik, Universität Potsdam and Deutsche Forschungsgemeinschaft.

to compute it, i.e., find a function f of the boundary of the manifold such that the difference

$$\operatorname{sign} M - f(\partial M)$$

can be expressed via Pontryagin classes of the manifold, or, in the language of elliptic theory, via the principal symbol of the Hirzebruch operator.

Hirzebruch conjectured a formula for $f(\partial M)$ and proved it in a number of examples. A complete solution of the signature defect problem was given later by Atiyah–Donnelly–Singer [**ADS83**] and Müller [**Mül84**].

The aim of the present survey is to describe index defects for some natural classes of general elliptic operators on manifolds with boundary. We consider only boundary value problems. Note, however, that index defects also occur in completely different situations, e.g., for elliptic operators in pseudodifferential subspaces of Sobolev spaces (rather than in Sobolev spaces themselves) on compact closed manifolds (see [**SS99, SS00**]). Here we deal with two classes of operators important in applications, namely, operators satisfying Gilkey's parity condition [**Gil89a**] and operators on $\mathbb{Z}_n$-manifolds in the sense of Freed and Melrose [**FM92**], and give explicit index defect formulas in both cases.

The classical theory: The Atiyah–Singer and Atiyah–Bott index formulas.

The Atiyah–Singer formula on closed manifolds. Let D be an elliptic operator, say, in Sobolev spaces on a closed manifold M. It is well known that D is Fredholm. The celebrated Atiyah–Singer theorem [**AS68**] gives a topological formula for the index $\operatorname{ind} D$ in terms of the principal symbol $\sigma(D)$. By applying the difference construction to the principal symbol, one obtains an element

$$[\sigma(D)] \in K_c(T^*M)$$

in the K-group with compact supports of the cotangent bundle T^*M. The Atiyah–Singer formula reads

$$\operatorname{ind} D = \operatorname{ind}_t[\sigma(D)], \tag{0.2}$$

where $\operatorname{ind}_t[\sigma(D)]$ is a functional of the principal symbol of the operator which can be written out in closed form. In other words, the Atiyah–Singer formula expresses an analytic invariant of the operator (the index) in terms of topological invariants of the principal symbol.

The Atiyah–Bott index formula for boundary value problems. If the boundary ∂M is not empty, then the operator D is no longer Fredholm (one can show that it always has an infinite-dimensional kernel), and one should equip it with boundary conditions to obtain a well-posed problem. The classical boundary conditions are most natural.

A *classical boundary value problem* is a system of equations of the form[1]

$$\begin{cases} Du = f, \\ B\left(u|_{\partial M}\right) = g, \end{cases} \tag{0.3}$$

where u and f are functions on M and g is a function on ∂M. The operator B in the boundary condition is a differential operator; it is applied to the restriction of the unknown function to the boundary.

The ellipticity condition for problem (0.3) (see [**Hör85**]) can be stated in terms of the principal symbols $\sigma(D)$ of the operator and $\sigma(B)$ of the boundary condition. Atiyah and Bott [**AB64**] showed that the index theory of classical boundary value problems is similar to that of elliptic operators on closed manifolds. Namely, under the ellipticity condition, problem (0.3) defines a difference element

$$[\sigma(D,B)] \in K_c(T^*(M \setminus \partial M)),$$

where $T^*(M \setminus \partial M)$ is the cotangent bundle over the interior of M, and the index of the corresponding Fredholm operator is given by the formula

$$\operatorname{ind}(D,B) = \operatorname{ind}_t[\sigma(D,B)], \tag{0.4}$$

similar to (0.2). An index formula of the same structure can be obtained for boundary value problems in the pseudodifferential setting introduced by Boutet de Monvel [**BdM71**]. A detailed exposition can be found in the book [**RS82**]. For simplicity we consider only classical boundary value problems.

However, the theory of classical boundary value problems has an essential drawback. *For some operators, there are no well-posed[2] classical boundary conditions at all!*

Atiyah and Bott showed that the obstruction to the existence of well-posed boundary conditions is of topological nature and computed it. The obstruction proves to be nonzero for most geometric operators: the Dirac operator, the Hirzebruch operator, and the Cauchy–Riemann operator. In other words, there are no Fredholm classical boundary value problems for these operators.

It is still possible to sidestep the obstruction and, in particular, equip the above-mentioned operators with well-posed boundary conditions. To this end, one has to consider a more general class of boundary value problems, namely, so-called *problems in subspaces*, which are described in the next subsection.

Boundary value problems in subspaces. Spectral problems. A *boundary value problem in subspaces* is a boundary value problem of the form

$$\begin{cases} Du = f, \\ B\left(u|_{\partial M}\right) = g, \quad g \in \operatorname{Im} P, \end{cases} \tag{0.5}$$

[1]To simplify the presentation, we consider first-order operators and occasionally speak of functions instead of sections of vector bundles.

[2]That is, defining a Fredholm problem for the original differential expression D in suitable spaces.

where the right-hand side g of the boundary condition lies in the range

$$\operatorname{Im} P \subset C^\infty(\partial M, G)$$

of a pseudodifferential projection operator

$$P : C^\infty(\partial M, G) \to C^\infty(\partial M, G)$$

in the function space on the boundary and the boundary operator B is also pseudodifferential. This class of boundary value problems was introduced in [**SSS98**] and further studied in [**NSSS98, SSS01, SSS99b**]. In particular, it was shown that the ellipticity condition can be stated in terms of the principal symbols of D, B, and P, just as in the classical case. However, from the topological point of view these problems are opposite to classical boundary value problems.

The two most important differences are as follows.

I. *There exists a Fredholm boundary value problem in subspaces for an arbitrary elliptic operator.* An example is given by the *spectral Atiyah–Patodi–Singer* boundary value problem [**APS75**][3]

$$\begin{cases} Du = f, \\ \Pi_+(A)u|_{\partial M} = g, \quad g \in \operatorname{Im}\Pi_+(A). \end{cases} \tag{0.6}$$

Here D is a first-order operator assumed to have the form

$$D\big|_{U_{\partial M}} \simeq \frac{\partial}{\partial t} + A \tag{0.7}$$

in a collar neighborhood $U_{\partial M}$ of the boundary, where A is an elliptic self-adjoint operator called the *tangential operator* of D, and $\Pi_+(A)$ is the spectral projection of A on $\overline{\mathbb{R}}_+$, i.e., the orthogonal projection on the subspace spanned by eigenvectors of A with nonnegative eigenvalues.

An arbitrary boundary value problem in subspaces can be reduced to a spectral problem by a stable homotopy (see [**SS99, SSS99b**]). Therefore, topologically, problem (0.6) can be viewed as the general case of a problem in subspaces. Thus, for simplicity we consider only spectral problems (0.6) for operators D satisfying (0.7). By $\operatorname{ind}(D, \Pi_+(A))$ we denote the index of problem (0.6).

II. *The index of a boundary value problem in subspaces is not determined by the principal symbol of the operator D.* To illustrate this, consider a deformation of lower-order terms of D such that some eigenvalue of the tangential operator changes its sign. At this point, the spectral projection $\Pi_+(A)$ experiences a jump, so that the index of the problem may change. On the other hand, the index remains constant as long as the deformation produces continuously varying spectral projections. Let us give a simple example.

Consider the zero-order deformation

$$D_\tau = D + \tau\chi(t)$$

[3]Atiyah, Patodi, and Singer used only homogeneous boundary conditions. However, problem (0.6) is equivalent to the corresponding homogeneous problem as far as the solvability and the index problem are concerned.

of the Cauchy–Riemann operator

$$D = \frac{\partial}{\partial t} + i\frac{\partial}{\partial \varphi}$$

on the cylinder $\mathbb{S}^1 \times [0,1]$, where $\chi(t)$ is a smooth function such that $\chi(0) = 1$ and $\chi(1) = 0$. The tangential operator of the family D_τ depends on τ only on one of the bases of the cylinder, namely, on $\mathbb{S}^1 \times \{0\}$, where it has the form

$$A_\tau = i\frac{\partial}{\partial \varphi} + \tau.$$

The eigenvalues of A_τ are given by the formula $\tau + 2\pi n, n \in \mathbb{Z}$. As τ passes through zero, one of the eigenvalues changes its sign, so that the spectral projection undergoes a jump. The index is also discontinuous:

$$\operatorname{ind}(D_\tau, \Pi_+(A_\tau)) = \begin{cases} -2, & -2\pi < \tau < 0, \\ -1, & \tau = 0, \\ 0, & 0 < \tau < 2\pi. \end{cases}$$

This example makes it clear that one cannot obtain an index formula similar to (0.4) for spectral boundary value problems; in other words, a topological computation of the index in this case is impossible in principle.

REMARK 0.1. The spectral boundary value problems are intimately related with the theory of elliptic operators on manifolds with singularities. Indeed, Atiyah–Patodi–Singer showed that if the tangential operator is invertible then the index of the spectral problem (0.6) coincides with the index of the extension $\widetilde{D}$ of D to M with the infinite cylinder attached to the boundary:

$$\widetilde{M} = M \cup [0,\infty) \times \partial M$$

and the operator is translation invariant on the cylinder. If we now make the change of variables $r = e^t$ then $\widetilde{D}$ becomes a Fuchs degenerate operator at $r = 0$:

$$r\frac{\partial}{\partial r} + A.$$

Operators of this form are studied in the elliptic theory on manifolds with isolated conical singularities (see e.g. [**Les97, Mel93, Sch91**]).

The index and the index defect for spectral problems. The aim of this survey is to show that in many cases of interest one obtains a homotopy invariant of the principal symbol of the operator by adding some analytic invariant to the index of a spectral boundary value problem.

The correction term is naturally called an *index defect* of the problem, since it is this term that restores the homotopy invariance of the index. It is natural to require that the correction term be determined solely by the structure of the operator in a neighborhood of the boundary, for on closed manifolds the analytic index itself is homotopy invariant and zero can be taken for the correction term.

Therefore, we introduce the following statement of the index defect problem.

The index defect problem for spectral boundary value problems. Construct a functional def-ind(D) of elliptic operators D on a manifold with boundary such that

(1) the sum $\operatorname{ind}(D, \Pi_+(A)) + \text{def-ind}\, D$ is a homotopy invariant of D;[4]
(2) def-ind(D) is determined solely by the tangential operator A.

A functional with these properties will be called an *index defect*. Conditions (1) and (2) imply that the *index defect is determined by the spectral projection and is its homotopy invariant.*

Needless to say, to obtain an actual defect formula (which is our main problem), we should compute the homotopy invariant in (1) topologically, i.e., express it in the form

$$\operatorname{ind}(D, \Pi_+(A)) + \text{def-ind}\, D = \operatorname{ind}_t(\sigma(D)), \tag{0.8}$$

where $\operatorname{ind}_t(\sigma(D))$ is a functional on the set of homotopy classes of elliptic principal symbols.

In the remaining part of the introduction, we explain the main methods that can be used to define index defects and describe approaches to the proof of the corresponding index defect formulas (0.8). However, prior to proceeding to these topics, we consider the following phenomenon of utmost importance.

The obstruction to index defect formulas. The desired index defect formula (0.8) can be viewed as a decomposition of the index of a spectral boundary value problem into a finite-dimensional contribution of the principal symbol and an (infinite-dimensional) contribution of the tangential operator:

$$\operatorname{ind}(D, \Pi_+(A)) = f_1(\sigma(D)) + f_2(A), \tag{0.9}$$

where the functional f_1 is a homotopy invariant of the principal symbol.

It turns out that *there is no index decomposition of the form* (0.9) *on the set of all elliptic operators* (see [**SSS99a**]). Therefore, index defect formulas and decompositions of the form (0.9) can be sought only in some *subsets* of the space of elliptic operators on a manifold with boundary.

The obstruction to the existence of decompositions (0.9) was computed in [**SSS99a**]. It is the one-dimensional cohomology class of the space of elliptic operators whose value on a cycle is equal to the Atiyah–Patodi–Singer spectral flow of the corresponding family of tangential operators. There exists a decomposition (0.9) on a subspace Σ of the space of elliptic operators if and only if the restriction of this cohomology class to Σ is trivial.

The cited result, unfortunately, proves only the *existence* of a decomposition (0.9) and does not give a satisfactory formula for an index defect. To study index defects, one has to use other methods.

[4]It readily follows that the sum is a homotopy invariant of the principal symbol of D.

Approaches to the definition of index defects. Let us briefly describe two methods useful in defining index defects.

A. The geometric index formula of Atiyah–Patodi–Singer. In 1975, Atiyah–Patodi–Singer [**APS76a**] obtained the formula

$$\operatorname{ind}(D, \Pi_+(A)) = \int_M a(D) - \eta(A) \tag{0.10}$$

for the index of spectral boundary value problems, where the density $a(D)$ is determined by the coefficients of D, just as in the case of closed manifolds. The new contribution to the index is given by the spectral η-invariant $\eta(A)$ of the tangential operator A.

This formula is often called a *geometric index formula*, since for geometric operators (the Hirzebruch, Dirac, Todd and Euler operators) the integrand on the right-hand side is determined by the metric and coincides with the local Atiyah–Singer density in the case of closed manifolds (i.e., with the L-form for the Hirzebruch operator, the A-form for the Dirac operator, etc.). An extension of the Atiyah–Patodi–Singer formula has been obtained by Melrose and Nistor [**MN96**] (see also [**LM02**]) for elliptic operators on manifolds with isolated singularities. In particular, an extension of the η-invariant to pseudodifferential operators is known (see [**Mel95**]).

Unfortunately, the Atiyah–Patodi–Singer formula does not define an index defect, since neither of the terms on the right-hand side is a homotopy invariant of D. However, formula (0.10) can be used to define index defects as follows.

We have already pointed out that the index of the spectral boundary value problem experiences jumps under homotopies of D. However, the sum

$$\operatorname{ind}(D, \Pi_+(A)) + \eta(A) \tag{0.11}$$

varies smoothly by the Atiyah–Patodi–Singer theorem. Moreover, the sum (0.11) is homotopy invariant if and only if for an arbitrary homotopy D_τ (with parameter τ) in our class of operators the derivative

$$\frac{d}{d\tau} \int_M a(D_\tau) \tag{0.12}$$

is zero. Since the density $a(D_\tau)$ is given by a closed-form expression involving the coefficients of D_τ, one can use a detailed analysis of $a(D_\tau)$ to construct classes of operators for which the derivative is zero and hence the η-invariant of the tangential operator is the desired index defect.

B. Operator algebras. Another method for defining index defects relies on K-theory of operator algebras.

As was mentioned already, any index defect is determined by the spectral projection $\Pi_+(A)$ and is its homotopy invariant. For simplicity, consider the class of matrix projections

$$\Pi_+(A) : C^\infty(\partial M, \mathbb{C}^N) \longrightarrow C^\infty(\partial M, \mathbb{C}^N) \tag{0.13}$$

and assume that the matrix entries lie in some algebra $\mathcal{A}$ of operators on the boundary.

Then index defects admit an alternative description as homotopy invariants of projections.

Recall that the *K-group $K_0(\mathcal{A})$ of the algebra $\mathcal{A}$* (e.g., see [**Bla98**]) is just the Grothendieck group of the Abelian semigroup of homotopy classes of projections (0.13) with entries in $\mathcal{A}$ (for arbitrary N). Thus, an index defect functional defines a homomorphism

$$d : K_0(\mathcal{A}) \longrightarrow \mathbb{R}$$

of the K-group into real numbers.

Moreover, a simple computation shows that the homotopy invariance of the sum $\operatorname{ind}(D, \Pi_+(A)) + d(\Pi_+(A))$ is equivalent to either of the following two conditions for the functional d (provided that $\mathcal{A}$ contains the ideal of finite rank operators).

(1) The functional d is a dimension type invariant of projections. More precisely, for two arbitrary projections

$$P_{1,2}, \quad \operatorname{Im} P_1 \subset \operatorname{Im} P_2, \quad \dim \operatorname{Im}(P_2 - P_1) < \infty,$$

differing by a finite rank projection $P_2 - P_1$, one has

$$d(P_2) - d(P_1) = \dim \operatorname{Im}(P_2 - P_1).$$

(2) The functional d defines a commutative diagram

(0.14)
$$\begin{array}{ccc} K_0(\mathcal{K}) = \mathbb{Z} & & \\ \downarrow & \searrow & \\ K_0(\mathcal{A}) & \xrightarrow{\ d\ } & \mathbb{R}, \end{array}$$

where the diagonal arrow is the natural embedding $\mathbb{Z} \subset \mathbb{R}$.

Examples. Now we consider some explicit implementations of the methods. Each example starts with a description of the class of operators for which the index defect is to be considered.

EXAMPLE 0.2. **Problems in even subspaces** [**SS99**]. Consider operators D such that the symbol of the tangential operator is even in the momentum variables ξ:

$$\sigma(A)(x, \xi) = \sigma(A)(x, -\xi).$$

This condition singles out the subalgebra of operators with even principal symbol in the algebra of pseudodifferential operators on the boundary. In this subalgebra, consider the subalgebra $\mathcal{A}$ of zero-order operators. If ∂M is *even-dimensional*, then the vertical arrow in diagram (0.14) is a monomorphism, which implies the existence of the desired functional d. One can also prove that d is unique under some natural conditions.

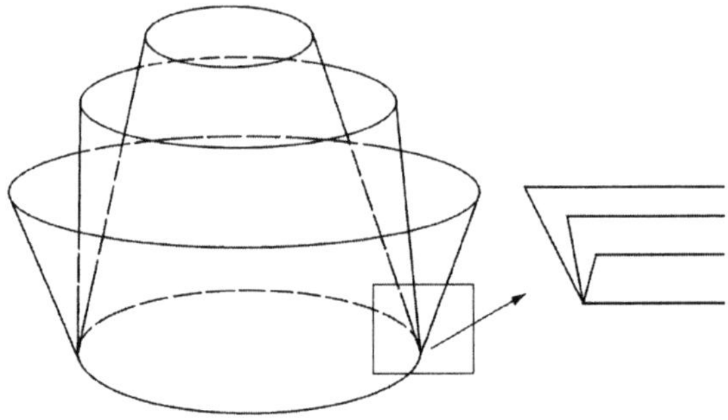

FIGURE 1. A singular space

Thus, there is a well-defined homotopy invariant $d(P) \in \mathbb{R}$ on the set of pseudodifferential projections P with even principal symbol. In contrast with the index of elliptic operators, this functional is not integer; it can take arbitrary rational values whose denominators are powers of 2 (dyadic values) see [**SS03**].

Thus, the value of the dimension functional d on the spectral projection can be taken as the index defect for operators D whose tangential operator A has an even principal symbol. In other words, the sum

$$\operatorname{ind}(D, \Pi_+(A)) + d(\Pi_+(A))$$

is a homotopy invariant of D. This poses a problem of computing this invariant in terms of the principal symbol of D. In the next subsection, we explain the main ideas underlying the computation of this invariant.

EXAMPLE 0.3. **Spectral problems on $\mathbb{Z}_n$-manifolds** [**SS01**]. Consider a manifold M whose boundary is represented as the total space of a covering

$$\pi : \partial M \longrightarrow X$$

over a smooth base X. Geometrically, such a manifold can be viewed as a smooth model of the singular space $\overline{M}^{\pi}$ (known as a $\mathbb{Z}_n$-*manifold*, where n is the number of sheets of the covering) obtained by identifying the points in each fiber of π. A neighborhood of a singular point looks like an open book with n sheets (see Fig. 1), and $X \subset \overline{M}^{\pi}$ is the edge, where the sheets meet.

On M we consider elliptic operators whose tangential operator A is the lift by π of some operator A_0 on the base of the covering. (The lift is well defined, since π is a local diffeomorphism.)

For a trivial covering, this condition guarantees that the index of the corresponding spectral boundary value problem viewed as the $\bmod n$-residue

$$\bmod n\text{-}\operatorname{ind} D \in \mathbb{Z}_n$$

is a homotopy invariant of D. Freed and Melrose [**FM92**] obtained the $\bmod n$*-index formula*

$$\bmod n\text{-}\operatorname{ind} D = \operatorname{ind}_t [\sigma(D)],$$

where

$$[\sigma(D)] \in K_c\left(\overline{T^*M}^\pi\right)$$

is the difference element defined by the principal symbol of D in the K-group of the singular space $\overline{T^*M}^\pi$ obtained from the cotangent bundle T^*M by identifying the points in each fiber of the covering $\partial T^*M \longrightarrow T^*X \times \mathbb{R}$, by analogy with the definition of $\overline{M}^\pi$.

However, the $\bmod n$-index *is not* an invariant of the principal symbol if the covering is nontrivial. The index defect in this case turns out to be given by the difference

$$\text{def-ind}(D) \stackrel{\text{def}}{=} \eta(A) - n\eta(A_0) \in \mathbb{R}/n\mathbb{Z}.$$

An expression of this type is known as the *relative Atiyah–Patodi–Singer* η*-invariant* [**APS76b**] of an operator A_0 with coefficients in a flat bundle. The index defect problem in this case is to compute the homotopy invariant

$$\operatorname{inv} D = \bmod n\text{-}\operatorname{ind} D + \text{def-ind}\, D \in \mathbb{R}/n\mathbb{Z}$$

as a residue modulo n.

Approaches to index defect formulas. To state and prove an index defect formula, one can apply all methods that are useful in the proof of ordinary index formulas. We content ourselves with describing only two approaches.

Method 1. Homotopy classification.

Roughly speaking, the method consists of two steps.

(1) First, one carries out the homotopy classification of elliptic operators D to be considered, or, more technically, computes the group of stable homotopy classes of these operators.

(2) Second, one finds a generating set of this group, so that the proof of an index defect formula is reduced to its verification for the generators.

This scheme goes back to the first proof of the Atiyah–Singer formula [**AS63**], where elliptic operators are classified (modulo stable homotopies) by elements of the K-group $K_c(T^*M)$ of the cotangent bundle at step (1), the Hirzebruch operator with coefficients in various vector bundles gives a rational generating set of the K-group (on an orientable even-dimensional manifold) at step (2), and finally cobordism theory is used to compare the analytic and topological index of these geometric operators.

Method 2. Poincaré duality.

Another method for proving index formulas is based on Poincaré duality in K-theory. Let us illustrate this method using the classical Atiyah–Singer theorem as an example.

On a closed manifold, there is an index homomorphism

$$\operatorname{ind}_a : K_c(T^*M) \longrightarrow \mathbb{Z}, \tag{0.15}$$

which takes each element of K-theory to the (analytic) index of the corresponding elliptic operator. On the other hand, Poincaré duality in K-theory gives the pairing

$$\begin{aligned} \langle\, , \rangle : K_c(T^*M) \times K(M) &\longrightarrow \mathbb{Z}, \\ (x, y) &\longmapsto p_!(xy), \end{aligned} \tag{0.16}$$

which is nonsingular on the free parts of the groups. Here

$$p_! : K_c(T^*M) \longrightarrow \mathbb{Z}$$

is the direct image mapping induced by the projection $p : M \to pt$ to a one-point space. It follows that the homomorphism (0.15) can be represented as the pairing with some element $y \in K(M)$; i.e.,

$$\operatorname{ind} D = \langle [\sigma(D)], y \rangle$$

for all elliptic operators D, where y is uniquely determined by M modulo torsion. Therefore, to obtain an index formula, it suffices to compute the element y. In these terms, the Atiyah–Singer formula states that one can take $y = 1 \in K(M)$, the element corresponding to the trivial line bundle.

Let us show how one can apply these methods to find and prove index defect formulas.

Continuation of Example 0.2 (an index defect formula in even subspaces).

We consider operators D with even principal symbol of the tangential operator on an even-dimensional manifold M. The homotopy classification of the corresponding spectral boundary value problems turns out to be isomorphic (modulo 2-torsion) to that of classical boundary value problems, i.e., to the group $K_c(T^*(M \setminus \partial M))$. Therefore, to obtain a topological formula for the homotopy invariant

$$\operatorname{inv} D = \operatorname{ind}(D, \Pi_+(A)) + d(\Pi_+(A)),$$

it suffices to generalize the Atiyah–Bott topological index (0.4) to boundary value problems in even subspaces. Such a generalization was obtained in [**SS99**]. We point out that the topological index in this formula proves to be a half-integer, and a topological consequence of this formula is the half-integrality of the index defect. The index defect formula has a number of applications. For example, it enabled the authors [**SS02**] to solve Gilkey's problem on the nontriviality of η-invariants of even-order operators on odd-dimensional manifolds.

Continuation of Example 0.3 (an index defect formula on $\mathbb{Z}_n$-manifolds). For a manifold whose boundary is an n-sheeted covering, the sum

$$\operatorname{inv} D \stackrel{\text{def}}{=} \operatorname{mod} n\text{-}\operatorname{ind} D + \text{def-ind}\, D$$

can be viewed as a homomorphism

$$\operatorname{inv} : K_c(\overline{T^*M}^{\pi}) \longrightarrow \mathbb{R}/n\mathbb{Z}, \tag{0.17}$$

The two main differences between (0.17) and (0.15) are as follows:

(1) now we use the group $\mathbb{R}/n\mathbb{Z}$ rather than $\mathbb{R}$;
(2) the space $\overline{T^*M}^{\pi}$ has singularities.

To take account of (1), one should replace the classical Poincaré duality (0.16) by "Poincaré duality with coefficients," i.e., *Pontryagin duality*

$$\langle , \rangle : K_c(T^*M) \times K(M, \mathbb{R}/\mathbb{Z}) \longrightarrow \mathbb{R}/\mathbb{Z}. \tag{0.18}$$

(See [**SS02**]; here $K(M, \mathbb{R}/\mathbb{Z})$ is the K-group with coefficients in the compact group $\mathbb{R}/\mathbb{Z}$.)

To tackle (2), one needs an extension of Poincaré duality to singular manifolds like $\overline{M}^{\pi}$. We point out that duality in the classical sense may not be valid on a manifold with singularities. However, the desired duality can be obtained in the framework of Connes' noncommutative geometry [**Con94**]. A detailed exposition is given in Appendix C, and now we describe only the main ideas involved.

To a singular $\mathbb{Z}_n$-manifold $\overline{M}^{\pi}$, one assigns a noncommutative C^*-algebra $\mathcal{A}_{M,\pi}$, which should be regarded as the "function algebra" on $\overline{M}^{\pi}$. Now Poincaré duality gives a pairing

$$\langle \, , \, \rangle : K_c(\overline{T^*M}^{\pi}) \times K_0(\mathcal{A}_{M,\pi}) \longrightarrow \mathbb{Z}, \tag{0.19}$$

where the second group is the K-group of the C^*-algebra $\mathcal{A}_{M,\pi}$. As a special case of this construction for $\partial M = \varnothing$, one obtains the pairing (0.16), since in this case the algebra $\mathcal{A}_{M,\pi}$ coincides with the algebra of continuous functions on M and the K-group of the algebra of continuous functions on a space coincides with the K-group of the space.

Applying the constructions (0.18) and (0.19) to the mapping (0.17), one finds that

$$\operatorname{inv} D = \langle [\sigma(D)], y \rangle \in \mathbb{Q}/n\mathbb{Z},$$

just as in the case of a closed manifold, for some element

$$y \in K_0(\mathcal{A}_{M,\pi}, \mathbb{Q}/n\mathbb{Z})$$

depending only on the manifold. The index defect theorem for $\mathbb{Z}_n$-manifolds in the authors's paper [**SS01**] gives an explicit formula for this element.

Acknowledgements. The authors wish to express their keen gratitude to Professor B.-W. Schulze of Potsdam University, where the paper was written, for his kind hospitality. We would like to thank V. Nazaikinskii, who read a preliminary version of the paper and made a number of important remarks. Finally, we thank the referee for the remarks and valuable suggestions.

The contents of the paper are as follows. In the first section, we define spectral boundary value problems and prove a theorem on index decompositions. Section 2 deals with index defects in even subspaces. In Section 3, we study index defects on $\mathbb{Z}_n$-manifolds. The paper concludes with three appendices, the first dealing with the Atiyah–Patodi–Singer η-invariant and the remaining two with Poincaré duality in K-theory on smooth manifolds and $\mathbb{Z}_n$-manifolds.

1. Spectral boundary value problems and their index

1.1. Atiyah–Patodi–Singer spectral boundary value problems. We start by introducing some notation. Let M be a smooth compact manifold with boundary and

$$D : C^\infty(M,E) \longrightarrow C^\infty(M,F)$$

an elliptic first-order differential operator on M acting in the spaces of sections of vector bundles $E, F \in \operatorname{Vect}(M)$.

We choose some collar neighborhood

$$U_{\partial M} \simeq \partial M \times [0,1)$$

of the boundary ∂M. The normal coordinate on the half-open interval $[0,1)$ is denoted by t. Then D can be represented in this neighborhood in the form

$$D\big|_{U_{\partial M}} \simeq \frac{\partial}{\partial t} + A(t) \tag{1.1}$$

(here $\simeq$ stands for equality up to a vector bundle isomorphism) for a smooth family $A(t)$ of elliptic first-order differential operators on ∂M. This representation can be obtained as follows. In the collar neighborhood of the boundary, we take some isomorphisms $\pi^*(E|_{\partial M}) \simeq E|_{\partial M\times[0,1)}$ and $\pi^*(F|_{\partial M}) \simeq F|_{\partial M\times[0,1)}$, where $\pi : \partial M \times [0,1) \to \partial M$ is the natural projection. Then D is isomorphic in $U_{\partial M}$ to an operator in the spaces

$$C^\infty\Big(\partial M \times [0,1), \pi^*(E|_{\partial M})\Big) \longrightarrow C^\infty\Big(\partial M \times [0,1), \pi^*(F|_{\partial M})\Big),$$

where the operator $\partial/\partial t$ is well defined, and we obtain a decomposition

$$D|_{U_{\partial M}} \simeq \Gamma(t)\frac{\partial}{\partial t} + A'(t),$$

where $\Gamma(t)$ is a vector bundle homomorphism. By ellipticity, $\Gamma(t)$ is an isomorphism, and we arrive at (1.1).

For simplicity, we also assume that for small t the family $A(t)$ is independent of t and coincides with a given self-adjoint operator

$$A : C^\infty(\partial M, E|_{\partial M}) \longrightarrow C^\infty(\partial M, E|_{\partial M})$$

on the boundary. The operator A is called the *tangential operator* of D.

DEFINITION 1.1. The *Atiyah–Patodi–Singer spectral boundary value problem* for an elliptic operator D is the boundary value problem

$$\begin{cases} Du = f, \\ \Pi_+(A)u|_{\partial M} = g, \quad g \in \operatorname{Im} \Pi_+(A), \end{cases} \tag{1.2}$$

where $\Pi_+(A)$ is the nonnegative spectral projection of A:

$$\Pi_+(A)e_\lambda = \begin{cases} e_\lambda, & \lambda \geq 0, \\ 0, & \lambda < 0, \end{cases}$$

for any eigenvector e_λ of A with eigenvalue λ.

For an arbitrary elliptic operator D, the spectral boundary value problem defines a bounded Fredholm operator in the spaces

$$(D, \Pi_+(A)) : H^s(M, E) \longrightarrow H^{s-1}(M, F) \oplus \overline{\operatorname{Im} \Pi_+(A)}, \quad s > 1/2, \tag{1.3}$$

where $\overline{\operatorname{Im} \Pi_+(A)}$ is the closure of the range of $\Pi_+(A)$ in $H^{s-1/2}(\partial M, E|_{\partial M})$ (see [**APS75**]).

As usual, the index $\operatorname{ind}(D, \Pi_+(A))$ of the spectral boundary value problem is independent of the Sobolev smoothness exponent s and can be computed in spaces of smooth functions.

Note that, in contrast with the index of elliptic operators on closed manifolds, the index of the Atiyah–Patodi–Singer problem is not invariant under homotopies of D. Indeed, consider a smooth homotopy

$$D_\tau : C^\infty(M, E) \longrightarrow C^\infty(M, F), \quad \tau \in [0, 1],$$

of elliptic operators, i.e. an elliptic operator family with coefficients smoothly depending on τ. Suppose that an eigenvalue of the tangential operator A_τ changes its sign at some point $\tau = \tau'$. Then the corresponding spectral projection experiences a jump, and consequently, the spaces in which the operator acts change discontinuously. This intuitive argument is stated in Theorem 1.4 below in terms of the spectral flow.

1.2. The spectral flow. Consider a smooth family $\{A_\tau\}_{\tau\in[0,1]}$ of elliptic self-adjoint operators on a closed manifold and assume that the operators at the endpoints $t = 0$ and $t = 1$ are invertible.

DEFINITION 1.2. *The spectral flow* $\operatorname{sf}\{A_\tau\}_{\tau\in[0,1]}$ of the family $\{A_\tau\}_{\tau\in[0,1]}$ is the net number of eigenvalues of A_τ that change their sign from minus to plus as the parameter τ increases from 0 to 1 (see Fig. 2).

Unfortunately, this definition does not makes sense for general families. The spectral flow of an arbitrary family $\{A_\tau\}$ can be defined by different methods (see [**Phi96**], [**Mel93**], [**Sal95**], [**BBW93**], [**DZ98**], [**NSS99**], and other papers). For example, one can slightly deform the straight line $\lambda = 0$ in the (λ, τ)-plane in a way such that the spectral curves of the operators A_τ intersect the perturbed line transversally. Then the spectral flow can be defined as the intersection number

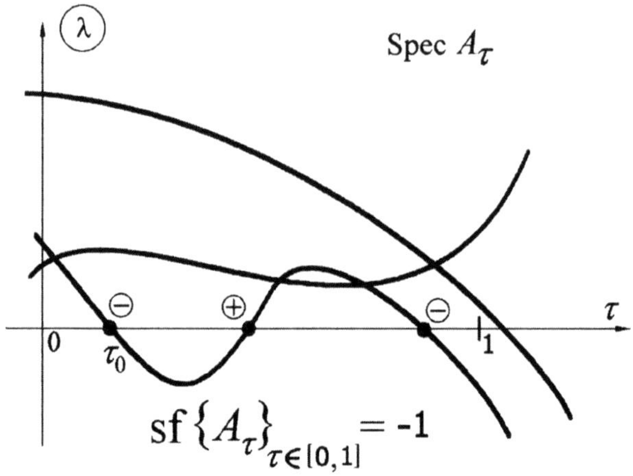

FIGURE 2. Spectral flow

of the perturbed line with the graph of the spectrum of the family. The desired perturbation can be constructed explicitly as follows (e.g., see [**Mel93**]).

A continuous family of elliptic self-adjoint operators has the following property. For an arbitrary $\tau' \in [0,1]$, there exists a number $\lambda_{\tau'}$ that is not an eigenvalue of A_τ in an ε-neighborhood of τ'. This follows from the fact that the spectrum of A_τ is discrete.

Now we choose a finite partition

$$0 = \tau_0 < \tau_1 < \ldots < \tau_N = 1 \tag{1.4}$$

of the interval $[0,1]$ and numbers $\{\lambda_i\}_{i=0,N-1}$, referred to as weights, such that λ_i is not an eigenvalue of the family A_τ on the interval $[\tau_i, \tau_{i+1}]$. We also assume that $\lambda_0 = \lambda_N = 0$.

DEFINITION 1.3. The *spectral flow* of the family $\{A_\tau\}_{\tau\in[0,1]}$ is the number

$$\operatorname{sf}\{A_\tau\}_{\tau\in[0,1]} = \sum_{i=1}^{N-1} \operatorname{ind}(\Pi_{\lambda_i}(A_{\tau_i}), \Pi_{\lambda_{i-1}}(A_{\tau_i})), \tag{1.5}$$

where $\Pi_\lambda(A)$ is the spectral projection of a self-adjoint operator A corresponding to eigenvalues greater than or equal to λ and

$$\operatorname{ind}(\Pi_\lambda(A), \Pi_\mu(A)) = \operatorname{sgn}(\mu - \lambda) \dim \Lambda_{\lambda,\mu},$$

is the relative index of two projections. Here $\Lambda_{\lambda,\mu}$ is the spectral subspace of A corresponding to eigenvalues in the interval $[\min\{\lambda,\mu\}, \max\{\lambda,\mu\})$.

One can show that the spectral flow (1.5) is well defined, i.e., is independent of the choice of the partition $\{\tau_i\}$ and the weights $\{\lambda_i\}$.

Let us now return to our original problem and consider a smooth homotopy D_τ, $\tau \in [0,1]$, of elliptic operators on a manifold with boundary. The corresponding family of tangential operators will be denoted by $\{A_\tau\}$. We shall now give a formula for the difference of indices at the endpoints of the homotopy. It turns out that the difference is equal to the spectral flow of the family of tangential operators.

THEOREM 1.4. (The spectral flow theorem.) *One has*

$$\operatorname{ind}(D_0, \Pi_+(A_0)) - \operatorname{ind}(D_1, \Pi_+(A_1)) = \operatorname{sf}\{A_t\}_{t\in[0,1]}. \tag{1.6}$$

REMARK 1.5. This result does not contradict the homotopy invariance of the index, since we change not only the operator D, but also the spaces where the index is computed.

PROOF. 1. One can readily obtain (1.6) if the family A_τ is invertible for all τ. Indeed, in this case the right-hand side of (1.6) is zero by the definition of the spectral flow. Let us verify that the indices on the left-hand side are equal. To this end, we note that the family of nonnegative spectral projections is smooth. Consider the Cauchy problem

$$\dot{U}_\tau = \left[\dot{\Pi}_+(A_\tau), \Pi_+(A_\tau)\right] U_\tau, \quad U_0 = \mathrm{Id}.$$

One can readily verify that the solution U_τ is a unitary elliptic operator specifying an isomorphism

$$U_\tau : \operatorname{Im}\Pi_+(A_0) \longrightarrow \operatorname{Im}\Pi_+(A_\tau)$$

of the subspaces defined by the pseudodifferential projections $\Pi_+(A_0)$ and $\Pi_+(A_\tau)$. The composition

$$\begin{pmatrix} 1 & 0 \\ 0 & U_\tau^{-1} \end{pmatrix} (D_\tau, \Pi_+(A_\tau))^t : C^\infty(M,E) \longrightarrow C^\infty(M,F) \oplus \operatorname{Im}\Pi_+(A_0)$$

has the same index as the original problem $(D_\tau, \Pi_+(A_\tau))$. On the other hand, the composition acts in spaces independent of t. Thus, by the standard index invariance, its index does not change.

2. Now consider the general case in which the operators of the tangential family A_τ may be noninvertible. To this end, we choose some partition (1.4). Using the argument in the previous part of the proof, on the first interval $[\tau_0, \tau_1]$ we obtain

$$\operatorname{ind}(D_0, \Pi_{\lambda_0}(A_{\tau_0})) = \operatorname{ind}(D_\tau, \Pi_{\lambda_0}(A_\tau)), \quad \tau \in [\tau_0, \tau_1].$$

Considering this equation for $\tau = \tau_1$ and replacing the projection $\Pi_{\lambda_0}(A_{\tau_1})$ by $\Pi_{\lambda_1}(A_{\tau_1})$ (they differ by a finite-dimensional projection), we obtain[5]

$$\operatorname{ind}(D_0, \Pi_{\lambda_0}(A_{\tau_0})) = \operatorname{ind}(D_{\tau_1}, \Pi_{\lambda_1}(A_{\tau_1})) + \operatorname{ind}(\Pi_{\lambda_1}(A_{\tau_1}), \Pi_{\lambda_0}(A_{\tau_1})).$$

A similar modification of $\operatorname{ind}(D_{\tau_1}, \Pi_{\lambda_1}(A_{\tau_1}))$ at τ_2 gives

$$\begin{aligned}\operatorname{ind}(D_0, \Pi_{\lambda_0}(A_{\tau_0})) &= \operatorname{ind}(D_{\tau_2}, \Pi_{\lambda_2}(A_{\tau_2})) \\ &\quad + \operatorname{ind}(\Pi_{\lambda_1}(A_{\tau_1}), \Pi_{\lambda_0}(A_{\tau_1})) + \operatorname{ind}(\Pi_{\lambda_2}(A_{\tau_2}), \Pi_{\lambda_1}(A_{\tau_2})).\end{aligned}$$

[5]To justify this and subsequent index computations, one uses the fact that the relative index $\operatorname{ind}(P,Q)$ of projections coincides with the index of the Fredholm operator $Q : \operatorname{Im} P \to \operatorname{Im} Q$.

Proceeding similarly at the subsequent points $\tau_3, \tau_4, \ldots$, we obtain the desired equation

$$\operatorname{ind}(D_0, \Pi_+(A_0)) = \operatorname{ind}(D_1, \Pi_+(A_1)) + \operatorname{sf}\{A_\tau\}_{\tau \in [0,1]}.$$

□

1.3. A theorem on index decompositions. It follows from the spectral flow theorem in the previous subsection that the index of the spectral boundary value problem $(D, \Pi_+(A))$ is uniquely determined by the principal symbol $\sigma(D)$ and the tangential operator A. There arises a natural question: Is it possible to decompose the index of the spectral boundary value problem into the sum

$$\operatorname{ind}(D, \Pi_+(A)) = f_1(\sigma(D)) + f_2(A) \tag{1.7}$$

of a homotopy invariant $f_1(\sigma(D))$ of the principal symbol of the operator and a functional $f_2(A)$ of the tangential operator? If this representation is possible, how to find it?

We shall sometimes refer to Eq. (1.7) as an "index decomposition."

REMARK 1.6. The functional f_2 is not a homotopy invariant of the tangential operator in general. Indeed, the set of operators with a given principal symbol $\sigma(D)$ contains operators with arbitrary index. Thus, f_2 can take infinitely many values. A more precise analysis shows that f_2 is a homotopy invariant of the corresponding spectral projection.

It turns out that there is an *obstruction* to the index decomposition.

Indeed, suppose that (1.7) is valid. Consider a homotopy D_τ of elliptic operators such that the homotopy of their tangential operators A_τ is periodic: $A_0 = A_1$. We claim that in this case the indices of the spectral boundary value problems for D_0 and D_1 are equal. Indeed,

$$f_1(\sigma(D_0)) = f_1(\sigma(D_1)),$$

since the symbols are homotopic, and

$$f_2(A_0) = f_2(A_1),$$

since the tangential operators A_0 and A_1 coincide by assumption.

On the other hand, by virtue of the spectral flow theorem, the difference of indices at the endpoints of the homotopy is equal to the spectral flow of the periodic family of tangential operators. Thus, we obtain the following result.

PROPOSITION 1.7. *If the index decomposition is valid, then for an arbitrary homotopy of tangential operators A_τ, $\tau \in \mathbb{S}^1$, one has*

$$\operatorname{sf}\{A_\tau\}_{\tau \in \mathbb{S}^1} = 0. \tag{1.8}$$

It is well known that there exist periodic families of elliptic self-adjoint operators with nontrivial spectral flow (1.8). (Simple examples can be found in [**Sav99**].) Therefore, *the index decomposition* (1.7) *does not exist on the space of all elliptic operators.*

REMARK 1.8. A reader familiar with the theorem on the cobordism invariance of the index [**Pal65, MP97**] might expect that the spectral flow is always zero. Indeed, according to Atiyah–Patodi–Singer [**APS76a**] the spectral flow of a periodic family is equal to the index of the operator $\partial/\partial t + A_t$ on the torus $\mathbb{S}^1 \times X$. Therefore, if A_t's are tangential operators of a periodic family D_t of elliptic operators, then by the cobordism invariance the spectral flow has to vanish. However, the periodicity of D_t is not assumed in the proposition.

In other words, to achieve (1.7), one has to consider subspaces rather than the entire space of elliptic operators. Using this idea, one can prove a result similar to Proposition 1.7, where one considers only homotopies of tangential operators within some given class of operators.

Namely, let Σ be a subspace of the space of all elliptic Hermitian symbols acting in the restriction of the bundle E to the boundary. In the space $\mathrm{Ell}(M, E, F)$ of elliptic operators on M acting between the spaces of sections of vector bundles E and F, we consider the subspace $\mathrm{Op}(\Sigma)$ of operators such that the symbols of the corresponding tangential operators belong to Σ:

$$\mathrm{Op}(\Sigma) = \Big\{ D \in \mathrm{Ell}(M, E, F) \;\Big|\; \sigma(A) \in \Sigma \Big\}.$$

We assume that Σ is nondegenerate in the following sense: the natural mapping $\mathrm{Op}(\Sigma) \to \Sigma$ taking an elliptic operator on the manifold with boundary to the principal symbol of its tangential operator is surjective. In other words, every element of Σ can be realized as the symbol of the tangential operator for some elliptic operator on M.

DEFINITION 1.9. The class $\mathrm{Op}(\Sigma)$ *admits an index decomposition* if there exist two functionals

$$f_{1,2} : \mathrm{Op}(\Sigma) \longrightarrow \mathbb{R}$$

such that

(1) the first functional is a homotopy invariant of the principal symbol of the operator, i.e., $f_1(D) = f_1(\sigma(D))$;
(2) the second functional is determined by the tangential operator, i.e., $f_2(D) = f_2(A)$;
(3) for $D \in \mathrm{Op}(\Sigma)$, one has

$$\mathrm{ind}(D, \Pi_+(A)) = f_1(\sigma(D)) + f_2(A).$$

We shall state a *necessary and sufficient condition* for the existence of a decomposition (1.7) in terms of the following condition on the class Σ of symbols.

DEFINITION 1.10. The class $\mathrm{Op}(\Sigma)$ is said to be *admissible* if for an arbitrary periodic family $\{A_\tau\}_{\tau \in \mathbb{S}^1}$ of elliptic self-adjoint operators on ∂M one has

$$\mathrm{sf}\{A_\tau\}_{\tau \in \mathbb{S}^1} = 0$$

provided that $\sigma(A_\tau) \in \Sigma$ for all τ.

THEOREM 1.11. (The index decomposition theorem.) *The class* Op(Σ) *admits an index decomposition if and only if it is admissible.*

PROOF. Necessity can be proved by analogy with Proposition 1.7. The proof of sufficiency can be found in [**SSS99a**]. □

The admissibility condition can be verified effectively. Indeed, the principal symbol of an elliptic self-adjoint operator A on a manifold X (in our case, $X = \partial M$) defines an element

$$[\sigma(A)] = [\operatorname{Im}\Pi_+\sigma(A)] \in K^0(S^*X)$$

in the K-group, where $\operatorname{Im}\Pi_+\sigma(A) \in \operatorname{Vect}(S^*X)$ is the subbundle generated by the positive spectral subspaces of the principal symbol $\sigma(A)$ on the cosphere bundle S^*X (with respect to some Riemannian metric). Then the spectral flow of a periodic family $A = \{A_t\}_{t\in\mathbb{S}^1}$ of elliptic self-adjoint operators can be computed by the Atiyah–Patodi–Singer formula [**APS76a**]

$$\operatorname{sf}\{A_t\}_{t\in\mathbb{S}^1} = \left\langle \operatorname{ch}[\sigma(A)] \cup \operatorname{Td}(T^*X\otimes\mathbb{C}), [S^*X\times\mathbb{S}^1]\right\rangle. \tag{1.9}$$

Here $\operatorname{ch}[\sigma(A)] \in H^{ev}\left(S^*X\times\mathbb{S}^1\right)$ is the Chern character of the element

$$[\sigma(A)] = [\operatorname{Im}\Pi_+\sigma(A)] \in K^0\left(S^*X\times\mathbb{S}^1\right)$$

defined by the principal symbol of the family, Td is the Todd class, and $\langle,\rangle$ is the pairing between homology and cohomology.

REMARK 1.12. The obstruction to the index decomposition given in Theorem 1.11 has the following cohomological interpretation. Note that the spectral flow of a periodic family of tangential operators with symbols in Σ defines a homomorphism

$$\begin{aligned} \operatorname{sf}: \pi_1(\Sigma) &\longrightarrow \mathbb{Z}, \\ \{\sigma(A_t)\}_{t\in\mathbb{S}^1} &\longmapsto \left\langle \operatorname{ch}[\sigma(A)] \cup \operatorname{Td}(T^*X\otimes\mathbb{C}), [S^*X\times\mathbb{S}^1]\right\rangle, \end{aligned}$$

of the fundamental group into integers. It vanishes on commutators. Therefore, by the well-known isomorphism $H_1(X) = \pi_1(X)/[\pi_1(X),\pi_1(X)]$, the spectral flow defines a cohomology class

$$[\operatorname{sf}] \in H^1(\Sigma,\mathbb{R}).$$

Now the admissibility condition is equivalent to the vanishing of this cohomology class.

Theorem 1.11 shows that if the integral (1.9) is zero for an arbitrary periodic homotopy in some class Σ, then the corresponding class of spectral boundary value problems is admissible and the index admits a decomposition.

Now consider examples in which this condition is satisfied.

1.4. Examples.

1. Even subspaces. Consider the antipodal involution

$$\alpha : S^*\partial M \longrightarrow S^*\partial M, \quad \alpha(x,\xi) = (x,-\xi)$$

of the cosphere bundle of the boundary ∂M. For a vector bundle $E \in \mathrm{Vect}(M)$, we consider elliptic Hermitian symbols

$$a : \pi^* E|_{\partial M} \longrightarrow \pi^* E|_{\partial M}, \quad \pi : S^*\partial M \to \partial M,$$

invariant under the involution

$$a(x,-\xi) = a(x,\xi). \tag{1.10}$$

In this case, the contributions to (1.9) from antipodal points $\pm\xi$ in the integral over the cosphere $S^*_x\partial M$ corresponding to any $x \in \partial M$ cancel provided that ∂M is odd-dimensional. Here we have used the fact that α preserves (or reverses) the orientation of $S^*\partial M$ depending on the parity of the dimension of ∂M. Thus, Eq. (1.8) is satisfied in the case of an *even-dimensional* manifold M for operators whose tangential operators have even principal symbols (see Eq. (1.10)), and the index decomposition for the corresponding spectral boundary value problems is possible. In the following section, we obtain the index defect formula for this case.

2. Odd subspaces. For an *odd-dimensional* manifold M, the antipodal involution α preserves the orientation of $S^*\partial M$. In this case, one should consider odd symbols a antiinvariant under α:

$$a(x,-\xi) = -a(x,\xi).$$

A computation shows that the contributions of antipodal points to (1.9) cancel modulo a form lifted from the base of the cosphere bundle. Therefore, the integral is zero, and this class of boundary value problems has an index decomposition. This "odd" case is largely analogous to the even case. Some new phenomena appear in this case. We touch on this theory only briefly at the end of Subsection 2.2. For a detailed exposition, we refer the reader to [**SS00**].

3. Manifolds whose boundary is a covering. Suppose that the boundary is a smooth n-sheeted covering

$$\pi : \partial M \longrightarrow X.$$

We consider the class of operators adapted to this structure in the sense that their tangential operators are lifted from the base of the covering. The lift is defined by the local diffeomorphism π.

Let us compute the spectral flow of a periodic family of tangential operators $\{A_\tau\}_{\tau\in\mathbb{S}^1}$. By assumption, this family is the lift of some family $\{\widetilde{A}_\tau\}_{\tau\in\mathbb{S}^1}$ of elliptic self-adjoint operators from the base X. Since formula (1.9) is local, we obtain

$$\mathrm{sf}\{A_\tau\}_{\tau\in\mathbb{S}^1} = n\,\mathrm{sf}\{\widetilde{A}_\tau\}_{\tau\in\mathbb{S}^1} \in n\mathbb{Z}.$$

This is zero as a residue modulo n. Therefore, the assumption of Theorem 1.11 is satisfied modulo n, and the index of the corresponding boundary value problems as a residue modulo n admits the desired decomposition.[6] The index defect in this situation will be studied in Section 3.

2. Index defects for problems with parity conditions

In this section, we describe index defects for spectral boundary value problems in even subspaces (see Example 0.2 in the introduction). The methods of K-theory of operator algebras are very effective in this case. In the framework of this approach, the index defect appears naturally as a dimension type functional of subspaces determined by spectral projections.

In this section, we first define a dimension-type functional of subspaces and then prove the defect formula.

2.1. The dimension functional for even subspaces.

DEFINITION 2.1. A pseudodifferential operator

$$P : C^{\infty}(X, E) \to C^{\infty}(X, E)$$

is said to be *even* if its principal symbol $\sigma(P)$ is invariant under the antipodal involution:

$$\sigma(P) = \alpha^*\sigma(P), \qquad \alpha : S^*X \to S^*X, \quad \alpha(x,\xi) = (x, -\xi). \tag{2.1}$$

PROPOSITION 2.2. *Let A be an even-order elliptic self-adjoint differential operator. Then the spectral projection $\Pi_+(A)$ is even.*

PROOF. The principal symbol of a differential operator of order n has the property

$$\alpha^*\sigma(A) = (-1)^n \sigma(A).$$

Since

$$\Pi_+(A) = \frac{A + |A|}{2|A|}$$

(for an invertible A), we see that the principal symbol of $\Pi_+(A)$ is given by

$$\sigma(\Pi_+(A)) = \Pi_+(\sigma(A)),$$

where $\Pi_+(\sigma(A))$ is the orthogonal projection on the nonnegative spectral bundle of the symbol $\sigma(A)$. (Here we use the following result due to Seeley [**See67**]: the principal symbol of a function of a self-adjoint operator is equal to the same function of the symbol.)

The last formula gives the desired equality

$$\sigma(\Pi_+(A)) = \alpha^*\sigma(\Pi_+(A))$$

for even-order operators. □

---

[6] We use a theorem similar to Theorem 1.11, where one considers the index modulo n and the spectral flow modulo n; this result can be proved by the same method.

We say that two even projections P_1 and P_2 are *isomorphic*, if there is an isomorphism of the corresponding ambient vector bundles (possibly after adding zero projections in some bundles to P_1 and P_2) interchanging the projections. Two projections in the sections of the same bundle are *homotopic*, if there is a homotopy joining them. The Abelian semigroup of even pseudodifferential projections on X modulo homotopy and isomorphism is denoted by $\widehat{\mathrm{Even}}\,(X)$. The corresponding Grothendieck group [7] will be denoted by $K\big(\widehat{\mathrm{Even}}\,(X)\big)$. Let us give an alternative description of this group.

Consider the algebra $\Psi_{ev}(X)$ of scalar even pseudodifferential operators of order zero. The Grothendieck group of the semigroup of homotopy classes of projections in matrix algebras over $\Psi_{ev}(X)$ is denoted by $K_0(\Psi_{ev}(X))$ and called the *even K-group of the algebra* $\Psi_{ev}(X)$ (e.g., see [**Bla98**]).

LEMMA 2.3. *One has*

$$K_0(\Psi_{ev}(X)) \simeq K\big(\widehat{\mathrm{Even}}\,(X)\big).$$

PROOF. The proof is immediate from the definitions of the Grothendieck group $K\big(\widehat{\mathrm{Even}}\,(X)\big)$ and the K-group of an algebra. Both are Grothendieck groups of Abelian semigroups generated by homotopy classes of even pseudodifferential projections. □

Let us compute the K-group of $\Psi_{ev}(X)$.

THEOREM 2.4. *If X is odd-dimensional, then there is an isomorphism*

$$(\mathbb{Z} \oplus K(X)) \otimes \mathbb{Z}\left[\frac{1}{2}\right] \longrightarrow K_0(\Psi_{ev}(X)) \otimes \mathbb{Z}\left[\frac{1}{2}\right]. \tag{2.2}$$

Here the mapping takes an integer $k \in \mathbb{Z}$ to a projection of rank k and a vector bundle $E \in \mathrm{Vect}(X)$ to a projection defining E as a subbundle of some trivial bundle. By $\mathbb{Z}[1/2]$ we denote the ring of dyadic rationals.

PROOF. 1. Let $\overline{\Psi}_{ev}(X)$ be the C^*-algebra obtained as closure of $\Psi_{ev}(X)$ in the L^2-norm. The closure does not change the K-group:

$$K_0(\Psi_{ev}(X)) \simeq K_0(\overline{\Psi}_{ev}(X)).$$

2. The algebra $\overline{\Psi}_{ev}(X)$ contains the ideal $\mathcal{K}$ of compact operators, and one has the exact sequence of algebras

$$0 \longrightarrow \mathcal{K} \longrightarrow \overline{\Psi}_{ev}(X) \longrightarrow C(P^*X) \longrightarrow 0. \tag{2.3}$$

Here the projection takes each operator to its principal symbol. We also treat even symbols as continuous functions on the projectivization $P^*X = S^*X/\mathbb{Z}_2$ of the cosphere bundle. The sequence is well defined by virtue of the well-known norm estimates

$$\inf_{K\in\mathcal{K}} \|D+K\|_{L^2(X)\to L^2(X)} = \sup_{(x,\xi)\in S^*X} |\sigma(D)(x,\xi)|.$$

[7] Recall that the *Grothendieck group* S of an Abelian semigroup $\widetilde{S}$ is generated by formal differences $[a]-[b], a, b \in \widetilde{S}$ modulo the relation $[a+c]-[b+c]=[a]-[b]$.

Furthermore, the sequence (2.3) induces the long exact sequence

$$\to K_1(C(P^*X)) \xrightarrow{\text{ind}} K_0(\mathcal{K}) \longrightarrow K_0(\overline{\Psi}_{ev}(X)) \xrightarrow{\text{smbl}} K_0(C(P^*X)) \to K_1(\mathcal{K}) \to$$

of K-groups, which is in our case reduced to

$$K^1(P^*X) \xrightarrow{\text{ind}} \mathbb{Z} \longrightarrow K_0(\overline{\Psi}_{ev}(X)) \xrightarrow{\text{smbl}} K(P^*X) \to 0. \tag{2.4}$$

Here we have substituted the well-known equalities $K_0(\mathcal{K}) = \mathbb{Z}$ and $K_1(\mathcal{K}) = 0$ and replaced the K-groups of the function algebra $C(P^*X)$ by the topological K-groups of P^*X.

Let us describe the mappings in (2.4). The first mapping takes an elliptic even symbol to the index of the corresponding operator. The second mapping takes a positive integer k to a projection of rank k. (Such a projection is even, since its symbol is zero.) Finally, the mapping smbl takes a pseudodifferential projection P to the range $\operatorname{Im}\sigma(P) \in \operatorname{Vect}(P^*X)$ of its principal symbol treated as a vector bundle over P^*X.

For the existence of a functional d making the diagram

$$\begin{array}{ccc} K_0(\mathcal{K}) = \mathbb{Z} & & \\ \downarrow & \searrow & \\ K_0(\overline{\Psi}_{ev}(X)) & \xrightarrow{d} & \mathbb{R} \end{array} \tag{2.5}$$

commute, it is necessary that the vertical arrow be a monomorphism or, equivalently, the index mapping in (2.4) be zero. This condition is satisfied if X is odd-dimensional. Indeed, it is well known (e.g., see [**Pal65**]) that the index of operators with even principal symbol is zero on such manifolds. Therefore, the sequence can finally be rewritten as

$$0 \longrightarrow \mathbb{Z} \longrightarrow K_0(\overline{\Psi}_{ev}(X)) \longrightarrow K(P^*X) \longrightarrow 0. \tag{2.6}$$

3. Let us slightly simplify this sequence further. To this end, we note that the natural projection $p : P^*X \to X$ for an odd-dimensional X induces an isomorphism in K-theory modulo 2-torsion. More precisely, the following is valid.

PROPOSITION 2.5. [**Gil89b**] *The projection $P^*X \to X$ induces an isomorphism*

$$p^* : K^*(X) \otimes \mathbb{Z}\left[\frac{1}{2}\right] \longrightarrow K^*(P^*X) \otimes \mathbb{Z}\left[\frac{1}{2}\right].$$

SKETCH OF PROOF. One first verifies the statement over a point $x \in X$: here $K^1(\mathbb{RP}^{2n}) = 0$ and $\widetilde{K}^0(\mathbb{RP}^{2n}) = \mathbb{Z}_{2^n}$, which shows that

$$p^* : K^*(\{x\}) \otimes \mathbb{Z}\left[\frac{1}{2}\right] \longrightarrow K^*(P^*_xX) \otimes \mathbb{Z}\left[\frac{1}{2}\right]$$

is an isomorphism. Now the isomorphism for the entire space can be proved by the Mayer–Vietoris principle ([**BT82**]). □

Taking a tensor product of the sequence (2.6) by the ring of dyadic rationals (which does not violate the exactness), we obtain the exact sequence

$$0 \longrightarrow \mathbb{Z}\left[\frac{1}{2}\right] \longrightarrow K_0(\overline{\Psi}_{ev}(X)) \otimes \mathbb{Z}\left[\frac{1}{2}\right] \longrightarrow K(X) \otimes \mathbb{Z}\left[\frac{1}{2}\right] \longrightarrow 0.$$

4. This exact sequence has a splitting mapping

$$K^0(X) \otimes \mathbb{Z}\left[\frac{1}{2}\right] \longrightarrow K\big(\widehat{\text{Even}}\,(X)\big) \otimes \mathbb{Z}\left[\frac{1}{2}\right]$$

taking a vector bundle to a projection on the space of its sections in $C^\infty(X, \mathbb{C}^N)$ for sufficiently large N.

5. The splitting obviously gives the desired isomorphism (2.2). □

The first component of the isomorphism (2.2) will be called the *dimension functional* for even pseudodifferential projections. It is useful to restate Theorem 2.4 in the following way.

THEOREM 2.6. [**SS99**] (A theorem on the dimension functional.) *On an odd-dimensional manifold X, there exists a unique group homomorphism*

$$d : K\big(\widehat{\text{Even}}\,(X)\big) \longrightarrow \mathbb{Z}\left[\frac{1}{2}\right]$$

with the property

$$d(P) = \text{rank} P \tag{2.7}$$

for a finite rank projection P under the normalization

$$d(P_F) = 0,$$

for all projections $P_F : C^\infty(X, E) \longrightarrow C^\infty(X, E)$ on the space of sections of a subbundle $F \subset E$. Moreover, for an arbitrary even projection P and for a sufficiently large N the projection $2^N P$ is homotopic to the direct sum of a projection on the space of sections of a subbundle and a finite rank projection.

2.2. The index defect formula. Let us apply the dimension functional d of even projections to the theory of spectral boundary value problems.

1. The dimension d as an index defect. We shall consider elliptic operators D with even tangential operator A. In this case, the spectral projection $\Pi_+(A)$ is even as well (see Proposition 2.2).

REMARK 2.7. Clearly, this condition cannot be satisfied for first-order differential operators. Thus, in this subsection we consider the more general class of operators that are pseudodifferential far from the boundary and have the form (1.1) with a first-order pseudodifferential operator A on ∂M near the boundary. For this class of operators, the spectral boundary value problems are also well defined and have the Fredholm property. Such differential operators with respect to the normal variable were considered in detail in [**Hör85**], Chapter 20.

LEMMA 2.8. *The sum*

$$\operatorname{ind}(D,\Pi_+(A)) + d(\Pi_+(A))$$

is a homotopy invariant of the operator D.

PROOF. Consider a homotopy D_τ, $\tau \in [0,1]$, and the corresponding homotopy A_τ of tangential operators. By the homotopy invariance of the dimension functional and property (2.7), it follows that the variation of the dimension is equal to the spectral flow

$$d(\Pi_+(A_1)) - d(\Pi_+(A_0)) = \operatorname{sf}\{A_\tau\}_{\tau\in[0,1]}.$$

(The equality can be obtained by analogy with the proof of the theorem on the spectral flow with the use of a partition of the interval $[0,1]$ and some choice of weights on the intervals of the partition.) A similar formula holds for the variation of the index:

$$\operatorname{ind}(D_1,\Pi_+(A_1)) - \operatorname{ind}(D_0,\Pi_+(A_0)) = -\operatorname{sf}\{A_\tau\}_{\tau\in[0,1]}.$$

Combining the two expressions, we obtain the desired homotopy invariance of the sum indicated in the lemma. □

This homotopy invariant will be denoted by

$$\operatorname{inv} D \stackrel{\text{def}}{=} \operatorname{ind}(D,\Pi_+(A)) + d(\Pi_+(A)).$$

Let us give a topological formula for this analytic invariant. This can be done by generalizing the Atiyah–Singer topological index to the case of spectral boundary value problems in even subspaces.

2. The topological index. It turns out that the principal symbol $\sigma(D)$ has a natural extension to the double

$$2M = M \bigcup\nolimits_{\partial M} M$$

of the manifold M. The double is obtained by gluing two copies of M along the common boundary.

To construct the desired extension, we take the symbol $\sigma(D)$ on the first copy of M and the symbol $\alpha^*\sigma\,(D)$ on the second copy. Here

$$\alpha : S^*M \longrightarrow S^*M$$

is the antipodal involution. Then in a neighborhood of the boundary the symbols $\sigma(D)$ and $\alpha^*\sigma(D)$ have the form

$$i\tau + a(x,\xi) \quad \text{and} \quad -i\tau + a(x,\xi),$$

respectively. They are taken into one another by the coordinate transformation

$$x \to x, \quad t \to -t$$

in a neighborhood of the boundary. Therefore, taken together, they define an elliptic symbol $\sigma(D) \cup \alpha^*\sigma(D)$ on the double of M. The difference element of this symbol will be denoted by

$$[\sigma(D) \cup \alpha^*\sigma(D)] \in K_c(T^*2M).$$

Then the *topological index* of D is defined as half the Atiyah–Singer topological index of this element on the double:

$$\operatorname{ind}_t[\sigma(D)] \stackrel{\text{def}}{=} \frac{1}{2} \operatorname{ind}_t[\sigma(D) \cup \alpha^* \sigma(D)].$$

3. The index defect formula.

THEOREM 2.9. [**SS99**] *Let D be an elliptic operator on an even-dimensional manifold with even tangential operator A. Then*

$$\operatorname{inv} D = \operatorname{ind}_t[\sigma(D)]. \tag{2.8}$$

PROOF. We know that both sides of the formula are determined by the principal symbol of the operator. Thus, to prove the equality it suffices to simplify the principal symbol by a homotopy and then check the equality for the simplified symbol.

For sufficiently large N, Theorem 2.6 in the previous subsection gives a homotopy π_τ, $\tau \in [0,1]$, of the direct sum of 2^N copies of the symbol $\sigma(\Pi_+(A))$ to a projection on some subbundle $E_0 \subset 2^N E|_{\partial M}$. One can lift this homotopy of projections to a homotopy $\sigma(D_\tau)$ of elliptic symbols with the properties

$$\sigma(D_0) = 2^N \sigma(D), \quad \sigma(\Pi_+(A_\tau)) = \pi_\tau.$$

By the homotopy invariance of both sides of Eq. (2.8), it suffices to prove the equality only for the symbol $\sigma(D_1)$ obtained at the end of the homotopy. This symbol in a neighborhood of the boundary depends only on the absolute value of the covector ξ. Thus, one can consider an elliptic operator D_1 with this symbol and the corresponding spectral boundary value problem (D_1, Π_+) such that the spectral projection Π_+ is actually a projection on the space of sections of the subbundle E_0. The spectral boundary value problem in this case is a classical boundary value problem; the index defect (i.e. the dimension functional) is zero, and Eq. (2.8) follows from the Atiyah–Bott formula. □

REMARK 2.10. (On the dimension functional of odd subspaces.) A similar functional was constructed in [**SS00**] on the space of "odd projections" with principal symbols satisfying

$$\alpha^* \sigma(P) = 1 - \sigma(P).$$

Such projections arise as spectral projections of odd-order differential operators (cf. Proposition 2.2). However, the methods of K-theory cannot be applied directly to odd projections, since odd symbols do not form an algebra. Moreover, the geometry of odd projections differs from the geometry of even ones: for example, an odd projection on a manifold of dimension $2k$ can act only in the space of sections of a vector bundle whose dimension is a multiple of 2^{k-1} (see [**SS00**]). In the cited paper, we constructed a dimension functional and proved the defect formula. Let us only mention that the topological index in the odd case can be obtained if on the double $2M$ one considers the symbol $\sigma(D) \cup \alpha^* \sigma(D)^{-1}$.

REMARK 2.11. One can show **[SSS99b]** that the value $d(\Pi_+(A))$ of the dimension functional and the element $[\sigma(D) \cup \alpha^*\sigma(D)]$ form a complete system of stable homotopy invariants of the spectral boundary value problem $(D, \Pi_+(A))$, i.e., classify these problems modulo stable homotopy.

2.3. The dimension functional and the η-invariant. The dimension functional for even pseudodifferential projections was constructed in Theorem 2.6. More precisely, the theorem claims the existence of the functional. In this subsection, we address the question of an explicit analytic expression for this functional. It turns out that such a description can be given in terms of the spectral Atiyah–Patodi–Singer η-invariant. The reader can find some details about the η-invariant in Appendix A.

Gilkey **[Gil89a]** noted that the η-invariant is rigid within the class of differential operators satisfying the *parity condition*

$$\operatorname{ord} A + \dim X \equiv 1 (\operatorname{mod} 2), \tag{2.9}$$

which relates the order of the operator to the dimension of the manifold. Rigidity is understood in the sense that the fractional part of the η-invariant is not only spectrally invariant but also homotopy invariant.

Condition (2.9) coincides with the conditions under which there exists a functional d of the corresponding spectral projections. (Recall that in the previous subsections this functional was considered for even projections on odd-dimensional manifolds and odd projections on even-dimensional manifolds.) This is not a mere coincidence. In fact, the η-invariant of an arbitrary elliptic self-adjoint operator A coincides with the value of the dimension functional on the spectral projection $\Pi_+(A)$.

THEOREM 2.12. **[SS99, SS00]** *If an elliptic self-adjoint differential operator satisfies the parity condition* (2.9), *then*

$$\eta(A) = d(\Pi_+(A)).$$

PROOF. For simplicity, we consider only opperators of even order $2l$.

1. It was shown in Theorem 2.6 that the dimension d of even projections can be defined as the unique homomorphism

$$d : K(\widehat{\text{Even}}(X)) \longrightarrow \mathbb{R}$$

with the following two properties: (a) $d(P) = \operatorname{rank} P$ for a finite rank projection P, (b) $d(P_F) = 0$ for projections on the spaces of sections of vector bundles $F \in \operatorname{Vect}(X)$. Therefore, to prove the theorem it suffices to show that the η-invariant defines a similar homomorphism and enjoys the same properties.

2. Thus, we should consider the η-invariant for general even pseudodifferential operators. Unfortunately, the rigidity of the η-invariant is lost in this class, since lower-order terms of the operator contribute to the η-invariant. However, the invariance still holds (see Appendix A) in the class of $\mathbb{R}_*$-*invariant* operators, i.e.,

operators with complete symbol having an asymptotic expansion

$$\mathrm{smbl} A \sim a_{2l}(x,\xi) + a_{2l-1}(x,\xi) + a_{2l-2}(x,\xi) + \dots,$$

where each homogeneous term $a_k(x,\xi)$ of order k has the following parity with respect to the momentum variables ξ:

$$a_k(x,-\xi) = (-1)^k a_k(x,\xi).$$

For $\mathbb{R}_*$-invariant elliptic self-adjoint operators, the spectral projection $\Pi_+(A)$ is either even or odd according to the parity of the order of the operator. It is easy to see that the dimension functional is also uniquely characterized by (a) and (b) also on the space of $\mathbb{R}_*$-invariant projections (technically, this is a consequence of the isomorphism $K_0(\Psi_{ev}(X)) = K_0(\Psi_{\mathbb{R}_*})$, where $\Psi_{\mathbb{R}_*}$ is the algebra of $\mathbb{R}_*$-invariant operators).

3. We claim that the η-invariant defines a homomorphism:

$$\begin{array}{rcl} \eta : K(\widehat{\mathrm{Even}}(X)) & \longrightarrow & \mathbb{R} \\ P & \mapsto & \eta(A), \end{array}$$

which takes a class represented by an $\mathbb{R}_*$-invariant projection to the η invariant of an $\mathbb{R}_*$-invariant elliptic self-adjoint operator A of positive order such that $\Pi_+(A) = P$. To prove that this homomorphism is well defined we show (1) $\eta(A)$ is determined by the projection P and (2) $\eta(A)$ is invariant under the equivalence relations defining $K(\widehat{\mathrm{Even}}(X))$.

To prove η is a well-defined invariant of the projection, we first note that two such operators A of the same order are linearly homotopic. Thus, their η-invariants are equal by virtue of the rigidity property of the η-invariant proved in Appendix A. Finally, the order of A is not essential, since $\eta(A) = \eta(A^{2k+1})$. To prove that η extends to the homomorphism of the K-group, it suffices to show that it is homotopy invariant (this also follows from rigidity) and it takes a zero projection to zero. This last fact is indeed true, since for an invertible Laplacian Δ the η-invariant is equal to the ζ-invariant

$$\eta\left(-\Delta^l\right) = -\zeta(\Delta^l), \quad \mathrm{Im}\,\Pi_+\left(-\Delta^l\right) = \{0\} \subset C^\infty(X,E).$$

But the ζ-invariant of the differential operator Δ is zero in odd dimensions (see the remark following Theorem 4.7 in Appendix A).

4. Let us now verify property (a) for the η-invariant. For a nonnegative Laplacian Δ with k-dimensional kernel, we obtain

$$\dim \mathrm{Im}\,\Pi_+(-\Delta) = k, \quad \eta(-\Delta) = k - \zeta(\Delta).$$

This proves property (a):

$$\eta(-\Delta) = \dim \mathrm{Im}\,\Pi_+(-\Delta) = d(\Pi_+(-\Delta)).$$

Similarly, one obtains (b):

$$\eta(\Delta) = 0, \quad \mathrm{Im}\,\Pi_+(\Delta) = C^\infty(X,E).$$

By the characterization of the dimension functional, this proves that the η-invariant of an operator A coincides with the dimension d of the corresponding spectral projection. □

3. Index defects on twisted $\mathbb{Z}_n$-manifolds

In this section, we consider another geometric situation in which index defects naturally arise. This situation comes from elliptic theory on manifolds with singularities. More precisely, we consider so-called twisted $\mathbb{Z}_n$-manifolds. Such manifolds look like a book with n sheets near the singularity; the singular set is the edge where the sheets meet.

The index theorem was stated and proved by Freed and Melrose [**FM92**] for the special case in which a neighborhood of the singular set consists of n sheets globally. Note that in this case it is natural to view the index as a residue modulo n, and the index theorem computes this index-residue.

We consider general twisted $\mathbb{Z}_n$-manifolds. In this case, the index-residue is no longer a homotopy invariant and an index defect arises. Intuitively, the proof of the index defect formula follows the scheme explained in Subsection , so here we give only the main steps of the proof. We first define a special element in the K-theory of a $\mathbb{Z}_n$-manifold in Subsection 3.4. Then the topological index is defined as the pairing with this element. The last subsection contains an application of the defect formula.

3.1. Twisted $\mathbb{Z}_n$-manifolds and elliptic operators.
1. Twisted $\mathbb{Z}_n$-manifolds

DEFINITION 3.1. A *twisted $\mathbb{Z}_n$-manifold,* where n is a positive integer, is a smooth compact manifold M with boundary ∂M equipped with the structure of the total space of an n-sheeted covering

$$\pi : \partial M \longrightarrow X \tag{3.1}$$

over a smooth base X (see Fig. 3).

Geometrically, a twisted $\mathbb{Z}_n$-manifold (M,π) naturally defines the singular space

$$\overline{M}^{\pi} = M/\sim,$$

obtained by the identification of points in each fiber of the covering (see Fig. 1); the corresponding equivalence relation $\sim$ is defined as

$$x \sim y \Longleftrightarrow x = y \text{ or } \{x, y \in \partial M \text{ and } \pi(x) = \pi(y)\}. \tag{3.2}$$

Sullivan [**Sul70**] introduced the notion of a $\mathbb{Z}_n$-manifold. These manifolds correspond to the structure of a *trivial* covering π. One of the motivations showing the interest in such manifolds is that (in the orientable case) the manifold $\overline{M}^{\pi}$ has a fundamental class in homology with coefficients in $\mathbb{Z}_n$:

$$\left[\overline{M}^{\pi}\right] \in H_m\left(\overline{M}^{\pi}, \mathbb{Z}_n\right), \qquad m = \dim M.$$

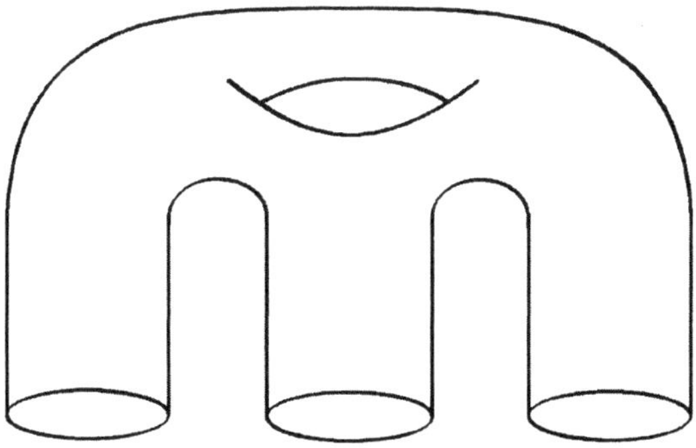

FIGURE 3. $\mathbb{Z}_3$-manifold

These manifolds with singularities were also used to give a geometric realization of bordism theory with coefficients in the group $\mathbb{Z}_n$. Further developments in this direction can be found, e.g., in [**Bot92**].

For brevity, we frequently omit the word "twisted" in what follows.

We point out that while most technical constructions of the theory deal with the smooth model (M, π), the most important results, e.g., the defect formula in Subsection 3.5, are stated in terms of the singular space $\overline{M}^{\pi}$ itself.

2. Natural mappings associated with coverings. Let us recall some natural mappings induced by the covering (3.1). First, one has the *direct image* mapping

$$\pi_! : \operatorname{Vect}(\partial M) \longrightarrow \operatorname{Vect}(X)$$

taking a vector bundle $E \in \operatorname{Vect}(\partial M)$ to the bundle $\pi_! E \in \operatorname{Vect}(X)$ with fiber

$$(\pi_! E)_x = C^\infty\left(\pi^{-1}(x), E\right), \quad x \in X$$

— the space of functions on the fiber $\pi^{-1}(x)$. This is indeed a vector bundle as can be seen either noting that $C^\infty(X, E)$ is a finitely generated projective module over $C^\infty(X)$, or computing explicitly the transition functions of $\pi_! E$. The transition functions on the intersection of two small trivializing sets in X are block-diagonal matrices with entries given by the transition functions of the restrictions of E to the leaves of the covering.

This definition leads to a natural isomorphism

$$\beta_E : C^\infty(\partial M, E) \xrightarrow{\simeq} C^\infty(X, \pi_! E)$$

of section spaces on the total space and the base of the covering. For example, scalar functions on the total space correspond to sections of the bundle $\pi_! 1 \in \operatorname{Vect}(X)$ on the base. (Here 1 stands for the trivial line bundle.)

This isomorphism enables one to transfer operators acting on ∂M to X and vice versa. More precisely, for an operator

$$P : C^\infty(\partial M, E) \longrightarrow C^\infty(\partial M, E)$$

on ∂M, by $\pi_! P$ we denote its *direct image* given by the formula

$$\pi_! P = \beta_E P \beta_E^{-1} : C^\infty(X, \pi_! E) \longrightarrow C^\infty(X, \pi_! E).$$

One also has the *inverse image*

$$\pi^! P' = \beta_E^{-1} P' \beta_E : C^\infty(\partial M, E) \longrightarrow C^\infty(\partial M, E) \tag{3.3}$$

of operators P' on X.

On twisted $\mathbb{Z}_n$-manifolds, we consider the following class of operators adapted to the structure of the singularity.

3. The class of operators. For a pair (M, π), we consider elliptic differential operators

$$D : C^\infty(M, E) \longrightarrow C^\infty(M, F),$$

which in a neighborhood of the boundary are lifted from the base of the covering. More precisely, we suppose that they satisfy the following assumption.

Assumption 1. The restrictions of E and F to the boundary are lifted from the base:

$$E|_{\partial M} \simeq \pi^* E_0, \quad F|_{\partial M} \simeq \pi^* F_0, \qquad E_0, F_0 \in \operatorname{Vect}(X);$$

moreover, the corresponding isomorphisms are given. For some operator

$$D_0 : C^\infty(X \times [0,1), E_0) \longrightarrow C^\infty(X \times [0,1), F_0)$$

on the cylinder with base X, the direct image $(\pi \times 1)_! D$ in a neighborhood of the boundary can be inserted in the commutative diagram

$$\begin{array}{ccc} C^\infty(X \times [0,1), (\pi \times 1)_! E) & \xrightarrow{(\pi \times 1)_! D} & C^\infty(X \times [0,1), (\pi \times 1)_! F) \\ \simeq \downarrow & & \downarrow \simeq \\ C^\infty(X \times [0,1), E_0 \otimes \pi_! 1) & \xrightarrow{D_0 \otimes 1_{\pi_! 1}} & C^\infty(X \times [0,1), F_0 \otimes \pi_! 1). \end{array} \tag{3.4}$$

Here we suppose that we have chosen a diffeomorphism

$$U_{\partial M} \simeq [0,1) \times \partial M \tag{3.5}$$

of a collar neighborhood of the boundary and extended the projection π to it as $\pi \times 1$. Then $D_0 \otimes 1_{\pi_! 1}$ is the operator D_0 with coefficients in the flat bundle $\pi_! 1$ (e.g., see [**APS76a**]).

Recall that a bundle is called *flat* if it is defined in terms of locally constant transition functions (for equivalent definitions and constructions of flat bundles see the above cited paper or [**Gil95**]).

We also assume for simplicity that the operator D_0 in a neighborhood of the base of the cylinder $X \times [0,1)$ has the form (1.1), i.e.,

$$D_0|_{X\times[0,\varepsilon)} = \Gamma\left(\frac{\partial}{\partial t} + A_0\right),$$

where Γ is some vector bundle isomorphism and the tangential operator A_0 is a first-order elliptic self-adjoint operator on X.

It follows from this assumption that in a neighborhood of the boundary the operator D has the form

$$\frac{\partial}{\partial t} + \pi^!\left(A_0 \otimes 1_{\pi_! 1}\right)$$

(up to a vector bundle isomorphism); i.e., the tangential operator A is equal to $\pi^!\left(A_0 \otimes 1_{\pi_! 1}\right)$.

We note that the classical operators always satisfy Assumption 1 in an appropriate geometric setting. For example, the Hirzebruch operator on an oriented even-dimensional Riemannian manifold M satisfies the assumption if (1) in a collar neighborhood of the boundary we take a metric pulled back from the cylinder $X \times [0,1]$; (2) the covering $\pi : \partial M \to X$ is oriented. Similar statements hold also for the Dirac and Todd operators; we leave them to the reader.

4. The difference element. Assumption 1 is closely related to the above-discussed manifolds with singularities. Indeed, note that the total space T^*M of the cotangent bundle is a (noncompact) $\mathbb{Z}_n$-manifold. It follows that the principal symbol of an elliptic operator D defines a difference element

$$[\sigma(D)] \in K_c\left(\overline{T^*M}^{\pi}\right)$$

in the K-group, since the commutative diagram (3.4) shows that the restriction of $\sigma(D)$ to the boundary is isomorphic to a symbol lifted from the base of the covering $T^*M|_{\partial M} \to T^*X \times \mathbb{R}$.

3.2. The Freed–Melrose index theorem modulo n. This subsection deals with the index theory on $\mathbb{Z}_n$-manifolds corresponding to trivial coverings π; i.e., the boundary of the corresponding smooth manifold M is a disjoint union of n copies of the base of the covering.

1. The index modulo n. It turns out that in this case operators satisfying Assumption 1 have a nontrivial homotopy invariant. Namely, let

$$\operatorname{mod} n\text{-}\operatorname{ind} D \in \mathbb{Z}_n$$

be the index of the spectral boundary value problem $(D, \Pi_+(A))$ treated as a residue modulo n.

PROPOSITION 3.2. *The index-residue* $\operatorname{mod} n\text{-}\operatorname{ind} D$ *is a homotopy invariant of the operator* D.

PROOF. Consider a homotopy $\{D_t\}_{t\in[0,1]}$. By the spectral flow theorem,

$$\operatorname{ind}(D_1, \Pi_+(A_1)) - \operatorname{ind}(D_0, \Pi_+(A_0)) = -\operatorname{sf}\{A_t\}_{t\in[0,1]}.$$

On the other hand, by assumption, the family A_t of tangential operators is the direct sum of n copies of the family $\{A_{0,t}\}$ of tangential operators on X. Therefore,

$$\mathrm{sf}\{A_t\}_{t\in[0,1]} = n\mathrm{sf}\{A_{0,t}\}_{t\in[0,1]} \equiv 0(\mathrm{mod} n).$$

This shows that the spectral flow vanishes as a residue modulo n and proves the homotopy invariance of the index-residue. □

This homotopy invariant index residue was computed in terms of the principal symbol by Freed and Melrose. In Subsection 3.5, we obtain a more general formula, and now we briefly recall the Freed–Melrose formula, just to make the exposition complete.

2. The Freed–Melrose theorem. Consider the category of $\mathbb{Z}_n$-manifolds with morphisms given by embeddings that take the boundary to the boundary and the fibers of the coverings to the fibers. The direct image mapping in K-theory extends to this category. More precisely, for an embedding f of a pair (M,π) in (N,π_N) there is a direct image mapping

$$f_! : K_c\big(\overline{T^*M}^{\pi}\big) \longrightarrow K_c\big(\overline{T^*N}^{\pi_N}\big).$$

On the other hand, there is a universal space in which one can embed an arbitrary $\mathbb{Z}_n$-manifold corresponding to a *trivial* covering. The universal space can be defined as follows. From $\mathbb{R}^L$ we cut away the union of n disjoint discs. We denote the resulting manifold with boundary by M_n. It can be viewed as a $\mathbb{Z}_n$-manifold, since its boundary consists of n diffeomorphic spheres. (The diffeomorphisms are given by translations.) One can readily compute the K-group of the cotangent bundle of this space:

$$K_c\big(\overline{T^*M_n}^{\pi'}\big) \simeq \mathbb{Z}_n.$$

Freed and Melrose proved the following index theorem.

THEOREM 3.3. [**FM92**] *One has*

$$\mathrm{mod}\, n\text{-ind} D = f_!\left[\sigma\left(D\right)\right],$$

where the direct image

$$f_! : K_c\big(\overline{T^*M}^{\pi}\big) \longrightarrow K_c\big(\overline{T^*M_n}^{\pi'}\big) \simeq \mathbb{Z}_n$$

is induced by an embedding

$$f : M \longrightarrow M_n.$$

The proof models the K-theoretic proof of the Atiyah–Singer index theorem based on embeddings. The cornerstone of the proof is the statement that the analytic index is preserved under the direct image mapping, that is, the diagram

$$\begin{array}{ccc} K_c(\overline{T^*M}^{\pi}) & \xrightarrow{\quad f_! \quad} & K_c(\overline{T^*N}^{\pi_N}) \\ & \searrow^{\mathrm{mod}\, n\text{-ind}} \qquad {}^{\mathrm{mod}\, n\text{-ind}}\swarrow & \\ & \mathbb{Z}_n & \end{array}$$

commutes for an embedding f of M in N.

3.3. The index defect problem on twisted $\mathbb{Z}_n$-manifolds. In contrast to the case of $\mathbb{Z}_n$-manifolds corresponding to trivial coverings, considered in the previous subsection, the index modulo n is no longer a homotopy invariant if the covering is *nontrivial.* By way of illustration, consider the following simple example.

On the cylinder $\mathbb{S}^1 \times [T_1, T_2]$ with coordinates (φ, t), consider the scalar elliptic operator

$$D = \frac{\partial}{\partial t} + \left(-i\frac{\partial}{\partial \varphi} + t\right).$$

Clearly, the tangential operator is lifted from the base of any of the coverings

$$\pi_n : \mathbb{S}^1 \to \mathbb{S}^1, \quad \varphi \mapsto n\varphi.$$

On the other hand, if T_1 or T_2 passes through zero, then the index of the corresponding spectral boundary value problem jumps by 1 and is not constant as a residue modulo n for any $n \neq 1$.

1. The index defect. The example shows that the index-residue is not homotopy invariant and cannot be computed topologically. The following theorem gives the index defect in this situation.

THEOREM 3.4. [**SS01**] *On a twisted $\mathbb{Z}_n$-manifold, the difference*

$$\eta(A) - n\eta(A_0) \in \mathbb{R}/n\mathbb{Z}$$

is an index defect. Here n is the number of sheets of the covering, and $\eta(A)$ and $\eta(A_0)$ are the spectral Atiyah–Patodi–Singer η-invariants of the tangential operators A and A_0, respectively. In other words, the sum

$$\operatorname{mod} n\text{-ind}\,(D, \Pi_+(A)) + \eta(A) - n\eta(A_0) \in \mathbb{R}/n\mathbb{Z}, \tag{3.6}$$

is a homotopy invariant of D.

PROOF. Consider a homotopy D_t of operators. The corresponding homotopy of tangential operators will be denoted by A_t and the homotopy of operators on the base X by $A_{0,t}$.

As t varies, the index $\operatorname{ind}(D_t, \Pi_+(A_t))$ can jump. The geometric Atiyah–Patodi–Singer index formula (see Appendix A) shows that the sum

$$\operatorname{ind}(D_t, \Pi_+(A_t)) + \eta(A_t)$$

is a smooth function of the parameter. Furthermore, the derivative of the sum with respect to t is a local invariant; namely, it is equal to the integral over the manifold of an expression determined by the complete symbol of the family A_t of tangential operators. On the other hand, the complete symbols of A_t and $A_{0,t}$ locally coincide by Assumption 1. Therefore, the jumps in the sum

$$\operatorname{ind}(D_t, \Pi_+(A_t)) + \eta(A_t) - n\eta(A_{0,t}) \tag{3.7}$$

of three terms come only from the third term. These jumps, however, are multiples of n. Hence, reducing (3.7) modulo n, we obtain a quantity independent of t. This proves the theorem. □

2. A relation to known invariants. We denote the homotopy invariant constructed in the theorem by

$$\mathrm{inv} D \stackrel{\mathrm{def}}{=} \bmod n\text{-ind}\,(D, \Pi_+(A)) + \eta\,(A) - n\eta\,(A_0) \in \mathbb{R}/n\mathbb{Z}.$$

In some special cases, it can be reduced to the following invariants.

(1) For a trivial covering, we have an isomorphism

$$A \simeq \underbrace{A_0 \oplus A_0 \oplus \ldots \oplus A_0}_{n \text{ copies}}.$$

Hence $\eta(A) = n\eta(A_0)$, and $\mathrm{inv} D$ is thereby reduced to the Freed–Melrose $\bmod n$-index [**FM92**]

$$\bmod n\text{-ind}(D, \Pi_+) \in \mathbb{Z}_n \subset \mathbb{R}/n\mathbb{Z}.$$

(2) On the other hand, if we consider only the fractional part $\{\}$ of (3.6), then the index vanishes and we obtain the Atiyah–Patodi–Singer *relative η-invariant* [**APS76b, APS76a**]

$$\{\eta\,(A_0 \otimes 1_{\pi_! 1}) - n\eta\,(A_0)\} \in \mathbb{R}/\mathbb{Z}$$

of the operator A_0 with coefficients in the flat bundle $\pi_! 1 \in \mathrm{Vect}\,(X)$.

3. The invariant $\mathrm{inv} D$ **as an obstruction**. Let us give an interpretation of the invariant inv as an obstruction.

Suppose that the manifold M itself is the total space of a covering $\widetilde{\pi}$ over some base Y and the induced covering at the boundary coincides with π:

$$\begin{array}{ccc} \partial M & \subset & M \\ \pi \downarrow & & \downarrow \widetilde{\pi} \\ X & \subset & Y. \end{array}$$

PROPOSITION 3.5. *If an operator*

$$D : C^\infty(M, E) \to C^\infty(M, F) \tag{3.8}$$

satisfying Assumption 1 *is the lift of some operator D_0 from Y, then*

$$\mathrm{inv} D = 0.$$

PROOF. By the Atiyah–Patodi–Singer formula, the sum

$$\mathrm{ind}\,(D, \Pi_+(A)) + \eta\,(A)$$

can be expressed as the integral over the manifold of an expression depending on the complete symbol of D. On the other hand, the operators D and D_0 on the total space and on the base locally coincide. Thus, we obtain

$$\mathrm{ind}\,(D, \Pi_+(A)) + \eta\,(A) = n\,(\mathrm{ind}\,(D_0, \Pi_+(A_0)) + \eta\,(A_0)).$$

Transposing the term $n\eta\,(A_0)$ to the left-hand side, we obtain $\mathrm{inv} D = 0$. □

REMARK 3.6. We point out that an arbitrary operator D is not induced by some elliptic operator D_0 on the base in general. Proposition 3.5 gives a necessary condition for the existence of such operators: if D_0 exists, then inv$D$ is zero.

The principal symbol of D defines an element in the K-group of the space $\overline{T^*M}^{\pi}$. To associate some topological index with this element, in the following subsection we use Poincaré duality on manifolds with singularities of the above-described type. For the reader's convenience, the survey contains Appendices B and C, where we discuss Poincaré duality on smooth manifolds and on singular $\mathbb{Z}_n$-manifolds.

3.4. The element of K-theory defined by a manifold whose boundary is a covering. In this subsection, we show that a pair (M, π) consisting of a compact manifold M and a covering $\pi : \partial M \to X$ defines a special element in the K-group

$$K_0(\mathcal{A}_{M,\pi}, \mathbb{Q}/n\mathbb{Z})$$

with coefficients in the group $\mathbb{Q}/n\mathbb{Z}$. Here $\mathcal{A}_{M,\pi}$ is the C^*-algebra of the twisted $\mathbb{Z}_n$-manifold. This algebra is defined in Appendix C. To define our element, we first need a geometric description of elements of the K-group. The coefficient group $\mathbb{Q}/n\mathbb{Z}$ looks slightly complicated, and hence we first consider a geometric realization of the K-groups with finite coefficient group $\mathbb{Z}_n$. After this, we return to the complicated coefficients.

1. K-theory with coefficients $\mathbb{Z}_n$. There are several equivalent approaches to the definition of these K-groups (e.g., see [**Bla98**]). We use the definition coming from topological K-theory. Namely, the group $\mathbb{Z}_n$ has the corresponding *Moore space.* It is the 2-dimensional complex M_n obtained from the unit disk by the identification of points on its boundary under the natural action of the group $\mathbb{Z}_n$:

$$M_n = \{ z \in \mathbb{C} \mid |z| \leq 1 \} / \left\{ e^{i\varphi} \sim e^{i\left(\varphi + \frac{2\pi k}{n}\right)} \right\}.$$

In particular, for $n = 2$ this space is the real projective plane $M_2 = \mathbb{RP}^2$.

The K-group with coefficients $\mathbb{Z}_n$ for an algebra $\mathcal{A}$ can be defined as

$$K_0\left(\mathcal{A}, \mathbb{Z}_n\right) = K_0\left(\widetilde{C}_0\left(M_n, \mathcal{A}\right)\right), \tag{3.9}$$

where $\widetilde{C}_0\left(M_n, \mathcal{A}\right)$ is the algebra of $\mathcal{A}$-valued continuous functions on the Moore space vanishing at a marked point $pt \in M_n$. This definition is similar to the definition of the topological K-groups with coefficients:

$$K^0\left(Y, \mathbb{Z}_n\right) = K^0\left(Y \times M_n, Y \times pt\right)$$

for a space Y.

PROPOSITION 3.7. *The group $K_0\left(\widetilde{C}_0\left(M_n, \mathcal{A}_{M,\pi}\right)\right)$ is isomorphic to the group of stable homotopy classes of triples (E', F', σ'), where $E', F' \in \mathrm{Vect}\left(M \times M_n\right)$ and*

$$\sigma' : \pi_! \, E'|_{\partial M} \longrightarrow \pi_! \, F'|_{\partial M}$$

is an isomorphism over the product $X \times M_n$. Here trivial triples are understood as triples induced by a vector bundle isomorphism defined on $M \times M_n$.

This result is a generalization of Lemma 6.10 to families. The proof is similar to that of the lemma.

The triples introduced in this proposition can sometimes be written out starting from explicit geometric data on the $\mathbb{Z}_n$-manifold with the use of the following proposition.

PROPOSITION 3.8. *A triple (E, F, σ), where*

$$E \in \operatorname{Vect}(M), \quad F \in \operatorname{Vect}(X), \qquad \pi_!\left(E|_{\partial M}\right) \overset{\sigma}{\simeq} kF, \tag{3.10}$$

and σ is an isomorphism, defines an element of $K_0(\mathcal{A}_{M,\pi}, \mathbb{Z}_k)$.

PROOF. On the Moore space M_k, we take a line bundle ε such that $[\varepsilon] - 1 \in \widetilde{K}(M_k) \simeq \mathbb{Z}_k$ is the generator of the reduced group. We also choose a trivialization $\rho : k\varepsilon \to \mathbb{C}^k$.

Now consider the triple (E, F, σ). With it we associate the following element (in the sense of Proposition 3.7):

$$[E \otimes \varepsilon, E, \sigma'] \in K_0\big(\widetilde{C}_0(M_k, \mathcal{A}_{M,\pi})\big),$$

where the isomorphism σ' is defined as the composition

$$\pi_!\, E|_{\partial M} \otimes \varepsilon \overset{\sigma\otimes 1}{\longrightarrow} kF \otimes \varepsilon \simeq F \otimes k\varepsilon \overset{1\otimes\rho}{\longrightarrow} F \otimes \mathbb{C}^k \simeq kF \overset{\sigma^{-1}\otimes 1}{\longrightarrow} \pi_!\, E|_{\partial M}. \tag{3.11}$$

□

2. K-theory with coefficients $\mathbb{Q}/n\mathbb{Z}$. Now we treat $\mathbb{Q}/n\mathbb{Z}$ as the direct limit of finite groups $\mathbb{Z}_{nN}$ corresponding to the embeddings

$$\mathbb{Z}_{nN} \subset \mathbb{Q}/n\mathbb{Z}, \quad x \mapsto x/N.$$

Then the K-groups with coefficients $\mathbb{Q}/n\mathbb{Z}$ can also be defined as the direct limit

$$K_0(\mathcal{A}_{M,\pi}, \mathbb{Q}/n\mathbb{Z}) = \varinjlim K_0(\mathcal{A}_{M,\pi}, \mathbb{Z}_{nN}).$$

3. The K-theory element of a $\mathbb{Z}_n$-manifold. The bundle $\pi_! 1 \in \operatorname{Vect}(X)$ is flat, and therefore, for large N there exists a trivialization

$$N\pi_! 1 \overset{\alpha}{\simeq} \mathbb{C}^{nN}.$$

Then the triple $[\mathbb{C}^N, \mathbb{C}, \alpha]$ defines an element of $K_0(\mathcal{A}_{M,\pi}, \mathbb{Z}_{Nn})$ by virtue of Proposition 3.8. By letting $N \to \infty$, we obtain an element of the K-group with coefficients $\mathbb{Q}/n\mathbb{Z}$.

Unfortunately, this construction is ambiguous in the choice of the trivialization α, and different trivializations give different elements.

It turns out that there is a canonical choice of α. Namely, for the covering $\pi : \partial M \to X$ consider the mapping

$$f : X \longrightarrow BS_n$$

to the classifying space BS_n of the permutation group on n elements. (Here we treat the bundle π as an associated bundle of the principal S_n-bundle; then f is the classifying mapping for this principal bundle.) Moreover, $\pi_! 1 \in \mathrm{Vect}(X)$ is the pullback of the universal bundle $\gamma_n \in \mathrm{Vect}_n(BS_n)$ over the classifying space. Assume that the range of f is contained in a finite skeleton $(BS_n)_{N'}$ of the classifying space. The space BS_n enjoys the Mittag-Leffler condition (see [**Ati61**]): for given N', there exists an $l \geq 0$ such that

(3.12)
$$\mathrm{Im}\left[K^1\left((BS_n)_{N'+l}\right) \to K^1((BS_n)_{N'})\right] = \mathrm{Im}\left[K^1\left((BS_n)_{N'+m+l}\right) \to K^1((BS_n)_{N'})\right]$$

for all nonnegative m.

Now let us choose an N such that the sum of N copies of the restriction of the universal bundle γ_n to $(BS_n)_{N'+l}$ is trivial. We take some trivialization

$$N\gamma_n \overset{\alpha'}{\simeq} \mathbb{C}^{nN}.$$

Then on M we choose the induced trivialization

$$\alpha = f^*\alpha'.$$

One can show that for this special trivialization the K-theory element defined by the triple $(\mathbb{C}^N, \mathbb{C}, \alpha)$ is independent of the choice of α' and f. We denote this element by

$$[\widetilde{\pi_! 1}] \in K_0\left(\mathcal{A}_{M,\pi}, \mathbb{Q}/n\mathbb{Z}\right).$$

3.5. The index defect formula.

1. The general formula.

THEOREM 3.9. [**SS01**] *For an elliptic operator D on a twisted $\mathbb{Z}_n$-manifold (M,π), one has*

$$\mathrm{inv} D = \left\langle [\sigma(D)], [\widetilde{\pi_! 1}] \right\rangle,$$

where $\langle , \rangle$ is Poincaré duality with coefficients:

$$\langle , \rangle : K^0_c\left(\overline{T^*M}^{\pi}\right) \times K_0\left(\mathcal{A}_{M,\pi}, \mathbb{Q}/n\mathbb{Z}\right) \longrightarrow \mathbb{Q}/n\mathbb{Z}. \tag{3.13}$$

SKETCH OF PROOF. It is well known in index theory how to compute fractional homotopy invariants (see [**APS76a**]). The idea is to express the fractional invariant as an index of some family of elliptic operators. Then it suffices to apply the Atiyah–Singer formula for families. Let us use this idea in the present situation.

1. Consider the family

$$D^* \oplus (D \otimes 1_\varepsilon) : C^\infty(M, F \oplus E \otimes \varepsilon) \longrightarrow C^\infty(M, E \oplus F \otimes \varepsilon)$$

of first-order elliptic operators on M parametrized by the Moore space M_{nN}. (The number N will be chosen later on.) Here D^* is the adjoint operator, and $D \otimes 1_\varepsilon$ stands for operator D with coefficients in the bundle $\varepsilon \in \mathrm{Vect}(M_{nN})$. On the other hand, consider the direct sum of N copies of this family. It turns out that the sum admits well-posed *classical boundary conditions.* Indeed, for sufficiently large N

the direct sum of N copies of the flat bundle $\pi_! 1$ is trivial. Let us choose some trivialization

$$N\pi_! 1 \overset{\alpha}{\simeq} \mathbb{C}^{nN}. \tag{3.14}$$

Then we obtain an isomorphism

$$\pi_! \left(N \ E|_{\partial M}\right) \simeq \pi_! N \otimes E_0 \xrightarrow{\alpha \otimes 1} \mathbb{C}^{nN} \otimes E_0$$

on the base of the covering and a similar isomorphism

$$\pi_! \left(N \ E|_{\partial M}\right) \otimes \varepsilon \simeq \pi_! N \otimes E_0 \otimes \varepsilon \xrightarrow{\alpha \otimes 1} \mathbb{C}^{nN} \otimes E_0 \otimes \varepsilon \simeq nN\varepsilon \otimes E_0 \xrightarrow{\rho \otimes 1} \mathbb{C}^{nN} \otimes E_0.$$

The corresponding isomorphisms of the spaces of sections are denoted by

$$B_1 = \alpha \otimes 1 : C^\infty \left(X, \pi_! \left(N \ E|_{\partial M}\right)\right) \longrightarrow C^\infty \left(X, \mathbb{C}^{nN} \otimes E_0\right),$$

$$B_2 = (\rho \otimes 1)(\alpha \otimes 1) : C^\infty \left(X, \pi_! \left(N \ E|_{\partial M}\right) \otimes \varepsilon\right) \longrightarrow C^\infty \left(X, \mathbb{C}^{nN} \otimes E_0\right).$$

In this notation, we define the following family of boundary value problems:

$$\begin{cases} ND^* u = f_1, \qquad N\left(D \otimes 1_\varepsilon\right) v = f_2, \\ B_1 \beta_E \ u|_{\partial M} + B_2 \beta_E \ v|_{\partial M} = g, \qquad g \in C^\infty \left(X, \mathbb{C}^{nN} \otimes E_0\right). \end{cases} \tag{3.15}$$

Note that this is *not* a family of classical boundary value problems, since the boundary value g is not a function on the boundary. The boundary condition is a nonlocal one — it relates the values of u and v over the whole fiber of π. However, in a neighborhood of the boundary (3.15) is equivalent to a classical boundary value problem on the cylinder $[0, \infty) \times X$ if one applies isomorphisms of the form

$$C^\infty(\partial M \times [0,1), E) \simeq C^\infty(X \times [0,1), (\pi \times 1)_! E)$$

to u, v, f_1, f_2. The latter classical boundary value problem is elliptic. Thus, (3.15) is a family of Fredholm boundary value problems (more on this class of nonlocal boundary value problems can be found in [**SS01**]). By $\Phi_\alpha(D)$ we denote the family (3.15) for some given trivialization α.

2. By virtue of the embedding $\mathbb{Z}_{nN} \subset \mathbb{Q}/n\mathbb{Z}$, we can treat the index-residue $\operatorname{ind} \Phi_\alpha(D) \in \widetilde{K}(M_{nN}) \simeq \mathbb{Z}_{nN}$ of the family as a fractional rational number. The main step in the proof of the defect formula is the following result.

LEMMA 3.10. *One has*

$$\operatorname{inv} D = \operatorname{ind} \Phi_\alpha(D) \in \mathbb{R}/n\mathbb{Z}$$

(*provided the trivialization α is chosen as in Subsection* 3.4).

PROOF OF THE LEMMA. It is well known that the index of a boundary value problem, say (D, B), can be decomposed as the sum of the index of the spectral boundary problem for the operator D and the index of the boundary operator B

restricted to the spectral subspace of the tangential operator. For the family (3.15) this gives

$$(3.16)\quad \operatorname{ind}\Phi_\alpha(D) = N\operatorname{ind}(D,\Pi_+(A))([\varepsilon]-1) \\ +\operatorname{ind}\bigl((B_1+B_2)\beta_E : N\operatorname{Im}\Pi_-(A)\oplus N\operatorname{Im}\Pi_+(A)\otimes\varepsilon \longrightarrow C^\infty\left(X,\mathbb{C}^{nN}\otimes E_0\right)\bigr)$$

Let us now compute the index of the boundary family. To this end, we decompose the target space as the direct sum

$$C^\infty\left(X,\mathbb{C}^{nN}\otimes E_0\right)\simeq\left(nN\operatorname{Im}\Pi_-(A_0)\right)\oplus\left(nN\varepsilon\otimes\operatorname{Im}\Pi_+(A_0)\right).$$

Here the isomorphism is defined by the formula

$$\begin{array}{ccc}\left(nN\operatorname{Im}\Pi_-(A_0)\right)\oplus\left(nN\varepsilon\otimes\operatorname{Im}\Pi_+(A_0)\right) & \longrightarrow & C^\infty\left(X,\mathbb{C}^{nN}\otimes E_0\right)\\ u,v & \mapsto & u+(\rho\otimes 1)v.\end{array}$$

This enables us to decompose the operator family $(B_1+B_2)\beta_E$ in Eq. (3.16) as the direct sum (modulo compact operators). Hence, the index of this family is also a sum

$$(3.17)\quad \operatorname{ind}(B_1+B_2)\beta_E = \operatorname{ind}\beta_E + \operatorname{ind}\bigl(N\operatorname{Im}\Pi_-(\pi_!A)\overset{\Pi_-(A_0)B_1}{\longrightarrow} nN\operatorname{Im}\Pi_-(A_0)\bigr) \\ +\operatorname{ind}\bigl(N\operatorname{Im}\Pi_+(\pi_!A)\overset{\Pi_+(A_0)B_1}{\longrightarrow} nN\operatorname{Im}\Pi_+(A_0)\bigr)[\varepsilon] \\ = 0+\operatorname{ind}\bigl[N\operatorname{Im}\Pi_+(\pi_!A)\overset{\Pi_+(A_0)B_1}{\longrightarrow} nN\operatorname{Im}\Pi_+(A_0)\bigr](-1+[\varepsilon])$$

The index of an elliptic operator (not a family!) in the square brackets can be expressed in terms of η-invariants by the Atiyah–Patodi–Singer "index formula for flat bundles" [**APS76a**]

$$(3.18)\quad \operatorname{ind}\bigl(N\operatorname{Im}\Pi_+(\pi_!A)\overset{\Pi_+(A_0)B_1}{\longrightarrow} nN\operatorname{Im}\Pi_+(A_0)\bigr) \\ = N\eta(A)-nN\eta(A_0)+\langle[\sigma(A_0)],[\pi_!1]\rangle,$$

where the brackets stand for the pairing

$$(3.19)\qquad \langle,\rangle : K_c^1(T^*X)\times K^1(X,\mathbb{Q})\longrightarrow\mathbb{Q}$$

of the difference element of the elliptic self-adjoint operator A_0

$$[\sigma(A_0)]\in K^1(T^*X)$$

and the element $[\pi_!1]\in K^1(X,\mathbb{Q})$ corresponding to the trivialized flat bundle $N\pi_!1$. (More details on this formula can be found in [**Gil95**].)

Substituting (3.18) into Eq. (3.16) and transposing the η-invariants to the left-hand side, we obtain

$$\operatorname{inv}D = \operatorname{ind}\Phi_\alpha(D)+\langle[\sigma(A_0)],[\pi_!1]\rangle.$$

2) It remains to show that for a special choice of the trivialization (3.14) the last term in (3.18) is zero.

Indeed, consider the classifying mapping

$$f : X\to(BS)_{N'}.$$

The computation of the pairing (3.19) can be moved to the classifying space:

$$\langle [\sigma(A_0)], [\pi_! 1] \rangle = \langle f_! [\sigma(A_0)], [\gamma_n] \rangle, \quad [\pi_! 1] = f^* [\gamma_n] \in K^1(X, \mathbb{Q}), \tag{3.20}$$

where $[\gamma_n] \in K^1((BS_n)_{N'}) \otimes \mathbb{Q}$ is the element defined by the trivialized flat bundle $N\gamma_n$. The embedding $(BS_n)_{N'} \subset (BS_n)_{N'+L'}$ induces the commutative diagram

$$\begin{array}{ccc} K_c^1(T^*(BS_n)_{N'}) \times K^1((BS_n)_{N'+L'}, \mathbb{Q}) & \to K_c^1(T^*(BS_n)_{N'}) \times K^1((BS_n)_{N'}, \mathbb{Q}) \\ \downarrow & \downarrow \\ K_c^1(T^*(BS_n)_{N'+L'}) \times K^1((BS_n)_{N'+L'}, \mathbb{Q}) \to & \mathbb{Q}. \end{array}$$

It follows from this diagram that the pairing (3.20) gives zero, since the range of the mapping

$$K_c^1(T^*(BS_n)_{N'}) \to K_c^1(T^*(BS_n)_{N'+L'})$$

is contained in the torsion subgroup.

Therefore, the expression for $\mathrm{ind}\Phi_\alpha(D)$ is reduced to the desired relation

$$\mathrm{ind}\Phi_\alpha(D) = \mathrm{inv} D.$$

This proves Lemma 3.10. □

3. To complete the proof of the theorem, it suffices to relate the families index of $\Phi_\alpha(D)$ to the Poincaré duality pairing. In the index theory of classical boundary value problems, there is a well-known operation of "order reduction," which reduces elliptic boundary value problems to zero-order operators (see [**Hör85**] or [**SS99**]). This operation preserves the index. Applying this operation to the family $\Phi_\alpha(D)$, one can show that the result is a family of admissible operators in the sense of Subsection 6.2 of Appendix C. A computation shows that this family of admissible operators coincides with the family corresponding to the product of the elements

$$[\sigma(D)] \in K_c(\overline{T^*M}^\pi), \quad [\widetilde{\pi_! 1}] \in K_0(\mathcal{A}_{M,\pi}, \mathbb{Q}/n\mathbb{Z}).$$

Since the Poincaré duality pairing of two elements is defined as the index, we obtain the desired formula for the index of the problem as the Poincaré pairing with coefficients:

$$\mathrm{ind}\,\Phi_\alpha(D) = \left\langle [\sigma(D)], [\widetilde{\pi_! 1}] \right\rangle.$$

Together with the equality in Lemma 3.10, this completes the proof of the theorem. □

REMARK 3.11. Using the results of Appendix C, one can compute Poincaré duality topologically for a regular covering π. In this case, we obtain an index defect formula in topological terms.

2. The index defect in the G-equivariant case. In a number of cases, the invariant inv can be computed effectively with the use of Lefschetz theory. Suppose we are given an action of a finite group G on M such that the action is free on the boundary ∂M. As a covering π, we take the natural projection to the quotient space:

$$\pi : \partial M \to \partial M/G.$$

Note that we do not require that the action be free in the interior of M.

Consider a G-equivariant elliptic operator D on M. By $L(D,g) \in \mathbb{C}$, $g \in G$, we denote the usual contribution to the Lefschetz formula (see [**Don78**]) of the fixed point set of an element g.

PROPOSITION 3.12. *One has*

$$\operatorname{inv} D \equiv -\sum_{g \neq e} L(D,g) \pmod{n}. \tag{3.21}$$

PROOF. Consider the equivariant index $\operatorname{ind}_g(D,\Pi_+)$ of the spectral boundary value problem and the equivariant η-function (see [**Don78**]) of the tangential operator A on the boundary.

By $(D,\Pi_+)^G$ and A^G we denote the restrictions of the corresponding operators to the spaces of G-invariant sections. Clearly, A^G is equivalent to A_0 on the base of the covering. On the other hand, one can express the usual invariants in terms of their equivariant counterparts:

$$\operatorname{ind}(D,\Pi_+)^G = \frac{1}{|G|}\sum_{g \in G} \operatorname{ind}_g(D,\Pi_+), \qquad \eta\left(A^G\right) = \frac{1}{|G|}\sum_{g\in G}\eta(A,g).$$

This is easy to check with the use of character theory. By virtue of these expressions, we can rewrite $\operatorname{inv} D$ as

$$\operatorname{inv} D = \operatorname{ind}_e(D,\Pi_+) - \sum_{g\neq e}\eta(A,g).$$

The η-invariants here can be expressed by the equivariant Atiyah–Patodi–Singer formula (see [**Don78**]):

$$-\eta(A,g) = \operatorname{ind}_g(D,\Pi_+) - L(D,g).$$

Thus, we obtain

$$\operatorname{inv} D = |G| \operatorname{ind}(D,\Pi_+)^G - \sum_{g\neq e} L(D,g).$$

This gives the desired equation (3.21). □

3.6. An application to η-invariants. The index defect formula enables one to express the fractional part of the η-invariant in the following situation.

Let M be an even-dimensional spin manifold such that its boundary is a covering with spin structure coinciding with that induced from the base. Let us also choose a vector bundle $E \in \operatorname{Vect}(M)$ that at the boundary is the lift of some bundle $E_0 \in \operatorname{Vect}(X)$. In a collar neighborhood of the boundary, we take a product metric on M that is the lift of some metric from the base. Finally, we choose a similar connection in E.

PROPOSITION 3.13. *The Dirac operator D_M on M with coefficients in E satisfies the assumptions of Theorem* 3.9, *and one has the formula*

$$\{\eta(D_X \otimes 1_{E_0})\} = \frac{1}{n}\left(\int_M \widehat{A}(M)\operatorname{ch}E - \left\langle [\sigma(D_M)], \left[\widetilde{\pi_! 1}\right]\right\rangle\right) \in \mathbb{R}/\mathbb{Z},$$

for the fractional part of the η-invariant of the self-adjoint Dirac operator D_X with coefficients in the bundle E_0 on the base of the covering.

PROOF. This formula follows from the defect formula, where the index of the spectral boundary value problem is expanded by the Atiyah–Patodi–Singer formula

$$\operatorname{ind}(D_M, \Pi_+) = \int_M \widehat{A}(M)\,\mathrm{ch}E - \eta(D_{\partial M}).$$

□

4. Appendix A. The Atiyah–Patodi–Singer η-invariant

In this appendix, we give a brief overview of the spectral η-invariant. Most of the results were proved in the original paper [**APS76a**], and there is also a very stimulating exposition in [**BBW93**]. Therefore, here we either omit the proofs or only indicate the main idea.

4.1. The geometric index formula and the η-invariant. Atiyah–Patodi–Singer [**APS75**] gave a formula for the index of spectral boundary value problems for geometric first-order operators. Namely, using the heat equation method [**ABP73**], they obtained the formula

$$\operatorname{ind}(D, \Pi_+(A)) = \int_X a(D) - \eta(A)$$

for the index of the spectral boundary value problem on a manifold X for an operator having the form (1.1) near the boundary. The first contribution is defined by the constant term $a(D)$ in the local asymptotic expansion of the heat kernel

$$\operatorname{tr}(e^{-tD^*D}(x,x)) - \operatorname{tr}(e^{-tDD^*}(x,x))$$

as $t \to 0$. This term is determined just as in the case of operators on closed manifolds as some algebraic expression in the coefficients of the operator. The new feature of the spectral boundary value problem is the so-called *η-invariant* of the tangential operator A. Let us recall its definition.

Let A be an elliptic self-adjoint operator of some positive order on a closed manifold M. The *η-function* of A is defined by the formula

$$\eta(A,s) = \sum_{\lambda_j \in \mathrm{Spec}A, \lambda_j \neq 0} \mathrm{sgn}\lambda_i \,|\lambda_i|^{-s} \equiv \operatorname{Tr}\left(A\left(A^2\right)^{-s/2-1/2}\right).$$

It is analytic in the half-space $\operatorname{Re} s > \dim M/\mathrm{ord}D$ (where the series is absolutely convergent). This spectral function is a generalization of the ζ-function

$$\zeta(A,s) = \sum_{\lambda_j \in \mathrm{Spec}A} |\lambda_i|^{-s}$$

of positive definite elliptic operators. By analogy with the ζ-invariant

$$\zeta(A) = \frac{1}{2}\zeta(A,0),$$

it is natural to introduce the following definition.

DEFINITION 4.1. The *η-invariant* of A is the number

$$\eta(A) = \frac{1}{2}(\eta(A,0) + \dim\ker A) \in \mathbb{R}. \tag{4.1}$$

Of course, for this definition to make sense, it is necessary to have an analytic extension of the η-function to the point $s = 0$. Analytic methods show that the η-function extends meromorphically to the entire complex plane, possibly with a simple pole at the origin. Atiyah–Patodi–Singer [**APS76a**] for odd-dimensional manifolds and Gilkey [**Gil81**] for even-dimensional manifolds proved, using global topological methods, that the residue at the point $s = 0$ is nevertheless zero. Thus, the η-function is holomorphic at $s = 0$ and the η-invariant is well defined.

The η-invariant is by definition only a spectral invariant, and it can vary for a deformation of the operator. Consider an example.

EXAMPLE 4.2. On the circle of length 2π with coordinate φ, consider the operator

$$A_t = -i\frac{d}{d\varphi} + t.$$

Here t is some real constant. Let us compute the η-invariant. Since the spectrum is given by the lattice $t + \mathbb{Z}$ (with period one), it follows that the η-invariant is a periodic function of the parameter t. Thus, we can suppose that $0 < t < 1$. Collecting the eigenvalues in pairs, we can rewrite the η-function as

$$\eta(A_t, s) = \sum_{n\geq 1}\left[(n+t)^{-s} - (n-t)^{-s}\right] + t^{-s}.$$

Let us show that this series absolutely converges on the real line for $s > 0$ and compute the limit as $s \to +0$. By the Taylor formula, we have

$$\left[(n+t)^{-s} - (n-t)^{-s}\right] = -2tsn^{-s} + O\left(\frac{s}{n^{2+s}}\right).$$

Thus, as $s \to +0$ we obtain

$$\sum_{n>1}\left[(n+t)^{-s} - (n-t)^{-s}\right] \sim -2ts\sum_{n\geq 1} n^{-s} \sim -2ts\int_1^\infty x^{-s-1}dx = -2t,$$

and for the η-invariant we have

$$\eta(A_t) = \frac{\eta(A_t, 0) + \dim\ker A_t}{2} = \frac{1}{2} - \{t\},$$

where $\{\,\} \in [0,1)$ is the fractional part of a number. Hence, for a smooth family A_t of elliptic operators the corresponding family of η-invariants is only piecewise smooth. In addition, the jumps (which are integral!) occur for parameter values such that some eigenvalue changes its sign. Let us also mention the half-integer parameter values $t \in \mathbb{Z} + 1/2$, where the spectrum of A_t is symmetric with respect to the origin. For these values, the η-function vanishes identically. In this case, one says that the operator A_t has a spectral symmetry, and the η-invariant is regarded as a *measure of spectral asymmetry* of the operator.

4.2. The derivative of the η-invariant. It turns out that the η-invariant is a piecewise smooth function of the parameter in the general case as well. More precisely, the following result is valid.

PROPOSITION 4.3. [**APS76a**] *For a smooth family $\{A_t\}_{t\in[0,1]}$ of elliptic self-adjoint operators, the following assertions hold.*

(1) *If the operator A_{t_0} is invertible, then the function $\eta(A_t)$ is smooth in a neighborhood of t_0 and its derivative can be expressed in terms of the derivative of the ζ-function of the auxiliary family*

$$B_{t,t_0} = |A_{t_0}| + (t-t_0)\left(\frac{d}{d\tau}A_\tau\right)\Big|_{\tau=t_0}.$$

Namely,

$$\frac{d}{dt}\eta(A_t)\Big|_{t=t_0} = \frac{d}{dt}\zeta(B_{t,t_0})\Big|_{t=t_0}. \tag{4.2}$$

(2) *In the general case, $\eta(A_t)$ is piecewise smooth and can be decomposed in the form*

$$\begin{gathered}\eta(A_{t'})-\eta(A_0)=\int_0^{t'}\omega(t_0)\,dt_0+\operatorname{sf}(A_t)_{t\in[0,t']},\\ \omega(t_0)=\tfrac{d}{dt}\zeta(B_{t,t_0})\big|_{t=t_0}\in C^\infty[0,1],\end{gathered} \tag{4.3}$$

as the sum of a smooth function of the parameter and a piecewise constant function sf *called the spectral flow (see Subsection* 1.2*). Here*

$$B_{t,t_0} = |A_{t_0}| + P_{\ker A_{t_0}} + (t-t_0)\left(\frac{d}{d\tau}A_\tau\right)\Big|_{\tau=t_0},$$

and $P_{\ker A_{t_0}}$ is the orthogonal projection on the kernel of A_{t_0}.

COROLLARY 4.4. *The fractional part*

$$\{\eta(A_t)\}\in\mathbb{R}/\mathbb{Z}\simeq\mathbb{S}^1$$

of the η-invariant is a smooth function of the parameter t for a smooth family A_t.

In the general case, the η-invariant is not homotopy invariant and can take arbitrary real values.

4.3. The homotopy invariance of the η-invariant. It turns out, however, that in some special operator classes the η-invariant possesses homotopy invariance. To this end, it is necessary that the two components in the decomposition (4.3) vanish. The easiest way to eliminate the second component, i.e. the spectral flow, is to consider only the fractional part $\{\eta(A)\}$ of the η-invariant. To obtain the vanishing of the second component, it is convenient to use the formula for the derivative on the left-hand side in Eq. (4.2). Seeley [**See67**] proved (see also [**Agr94**] and [**GS95**]) that the value of the ζ-function at the origin can be expressed via the complete symbol of the operator. Explicitly, for a positive self-adjoint operator A with complete symbol having the asymptotic expansion

$$\sigma(A) \sim a_m + a_{m-1} + a_{m-2} + \ldots,$$

the ζ-invariant is computed by the following procedure. Let us define the symbols b_{-m-j}, $j \geq 0$, by the recursion relations

$$(4.4)\quad b_{-m-j}(x,\xi,\lambda)(a_m(x,\xi)-\lambda) + \sum_{\substack{k+l+|\alpha|=j,\\ l>0}} \frac{1}{\alpha!}(-i\partial_\xi)^\alpha b_{-m-k}(x,\xi,\lambda)(-i\partial_x)^\alpha a_{m-l}(x,\xi) = 0.$$

The symbols in (4.4) depend on the coordinates x, momenta ξ, and additionally on the parameter λ. In this notation, the ζ-invariant is given by

$$(4.5)\qquad 2\zeta(A) = \frac{1}{(2\pi)^{\dim M}\operatorname{ord}A}\int_{S^*M} dx d\xi \int_0^\infty b_{-\dim M-\operatorname{ord}A}(x,\xi,-\lambda)\,d\lambda.$$

Note the following properties of this formula.

(1) (Locality.) For two locally isomorphic operators A and A', their ζ-invariants coincide:

$$\zeta(A) = \zeta(A').$$

(Operators are said to be *locally isomorphic* if their complete symbols coincide in a neighborhood of every point of the manifold in some coordinate system for some trivializations of vector bundles.)

(2) (Homogeneity.) The terms b_j are positively homogeneous functions:

$$b_j(x,t\xi,t^m\lambda) = (-1)^j b_j(x,\xi,\lambda), \qquad t>0.$$

These properties enable one to find classes of operators where the derivative of the η-invariant in Eq. (4.2) is zero. Let us introduce some of the known classes.

First, we define a class of pseudodifferential operators that generalize differential operators.

DEFINITION 4.5. A classical pseudodifferential operator A with complete symbol

$$\sigma(A) \sim a_m + a_{m-1} + a_{m-2} + \ldots$$

is said to be $\mathbb{R}_*$-*invariant* if the components of its complete symbol are homogeneous functions

$$a_j(x,t\xi) = t^j a_j(x,\xi), \qquad t\in\mathbb{R}_*,$$

with respect to the group of nonzero real numbers $\mathbb{R}_*$.

To define the second class of operators, recall that a *flat bundle* $\gamma \in \operatorname{Vect}(M)$ is a vector bundle with locally constant (i.e. constant on connected subsets) transition functions. For an operator

$$A : C^\infty(M,E) \longrightarrow C^\infty(M,F),$$

one can define an *operator with coefficients in the flat bundle*, denoted by

$$A \otimes 1_\gamma : C^\infty(M, E\otimes\gamma) \longrightarrow C^\infty(M, F\otimes\gamma).$$

This is locally isomorphic to the direct sum of $\dim\gamma$ copies of A. One can globally define this operator by gluing the local complete symbols of A with the use of a partition of unity. We also require that the transition functions of a flat bundle be unitary.

EXAMPLE 4.6. On the circle $\mathbb{S}^1$ with coordinate φ, consider the vector bundle γ with the transition function $e^{2\pi i t}$. Then the operator

$$-i\frac{d}{d\varphi} + t$$

from Example 4.2 is locally isomorphic to $-i\frac{d}{d\varphi}\otimes 1_\gamma$. The isomorphism

$$e^{-ti\varphi}\left(-i\frac{d}{d\varphi}\right)e^{it\varphi} = -i\frac{d}{d\varphi} + t$$

is given by the trivialization $e^{it\varphi}$ of γ.

THEOREM 4.7. [**APS76a**], [**Gil89a**] *The fractional part of the η-invariant is homotopy invariant in the following two classes of elliptic self-adjoint operators:*

(1) *the class of direct sums*

$$A\otimes 1_\gamma \oplus (-\dim\gamma A)$$

with a given flat bundle $\gamma \in \operatorname{Vect}(X)$;

(2) *the class of $\mathbb{R}_*$-invariant operators if the following parity condition is satisfied :*

$$\dim A + \operatorname{ord} M \equiv 1 \pmod 2. \tag{4.6}$$

PROOF. 1) The main idea of the proof is to use the locality of the ζ-invariant. More precisely, for the fractional part of η one has

$$\{\eta(A\otimes 1_\gamma \oplus (-nA))\} = \{\eta(A\otimes 1_\gamma)\} - \{n\eta(A)\}, \qquad n = \dim\gamma.$$

Then for a smooth homotopy A_t this gives (see Eq. (4.2))

$$\frac{d}{dt}\{\eta(A_t)\} = \frac{d}{dt}\zeta(B_t), \quad B_t = |A_{t_0}| + (t-t_0)\left(\frac{d}{d\tau}A_\tau\right)\Big|_{\tau=t_0}.$$

A similar formula

$$\frac{d}{dt}\{\eta(A_t\otimes 1_\gamma)\} = \frac{d}{dt}\zeta(B_t')$$

is valid for the derivative of the operators with coefficients in the flat bundle. Recall that $A\otimes 1_\gamma$ and nA are locally isomorphic. Therefore, the positive definite operators B_t and B_t' are locally isomorphic as well. Therefore, the locality of the ζ-invariant gives the desired relation

$$\frac{d}{dt}\left\{\eta\left(A_t\otimes 1_\gamma\right)\right\}=\frac{d}{dt}\left\{n\eta\left(A_t\right)\right\}.$$

This proves the homotopy invariance.

2) Consider first the case of even-order operators. It is easy to see that for a homotopy A_t the corresponding positive definite operators B_t (see Eq. (4.2)) are also $\mathbb{R}_*$-invariant. By induction, this gives the $\mathbb{R}_*$-homogeneity

$$b_j\left(x,-\xi,\lambda,t\right)=(-1)^j\,b_j\left(x,\xi,\lambda,t\right) \tag{4.7}$$

of the coefficients corresponding to B_t (see the recursion relations (4.4)). Hence,

$$\frac{d}{dt}\left\{\eta\left(A_t\right)\right\}=\frac{d}{dt}\zeta\left(B_t\right)=\mathrm{Const}\frac{d}{dt}\Bigg(\int\limits_{S^*M}dxd\xi'\int\limits_0^\infty b_{-\dim M-\mathrm{ord}A}\left(x,\xi,-\lambda,t\right)d\lambda\Bigg).$$

Using the homogeneity (4.7) and the assumption that the manifold is odd-dimensional, we see that the integrand $b_{-\dim M-\mathrm{ord}A}\left(x,\xi,-\lambda,t\right)$ is an odd function on the sphere S_x^*M. Therefore, the integral is zero, and we obtain the desired relation

$$\frac{d}{dt}\left\{\eta\left(A_t\right)\right\}=0.$$

For odd-order operators, one obtains the different homogeneity

$$b_j\left(x,-\xi,\lambda,-t\right)=(-1)^{j+1}\,b_j\left(x,\xi,\lambda,t\right).$$

Substituting these homogeneous functions into the expression for the ζ-invariant, we obtain

$$\frac{d}{dt}\zeta\left(B_t\right)=\frac{d}{dt}\zeta\left(B_{-t}\right).$$

This gives the desired homotopy invariance of the fractional part of the η-invariant:

$$\frac{d}{dt}\left\{\eta\left(A_t\right)\right\}=\frac{d}{dt}\zeta\left(B_t\right)=0.$$

□

REMARK 4.8. Note that in the proof we also obtained the vanishing of both the derivative and the ζ-invariant itself for $\mathbb{R}_*$-invariant even-order operators on odd-dimensional manifolds.

5. Appendix B. Elliptic operators and Poincaré duality. Smooth theory

In this appendix, we show that Poincaré duality in K-theory on smooth manifolds can naturally be described in terms of elliptic operators. We consider both closed manifolds and manifolds with boundary. Using the Poincaré isomorphism, we construct Poincaré duality as a nonsingular pairing. In this context, the Atiyah–Singer index theorem can be used to make the pairing effectively computable. The topics covered in this appendix can also be found in the recent book [**HR00**]. Our approach is closer to differential equations.

5.1. The Poincaré isomorphism on a closed manifold.

1. Atiyah's generalized elliptic operators. It is well known that for a sufficiently nice topological space X (e.g., a finite CW-complex) there is a pairing

$$H_i(X,\mathbb{Z}) \times H^i(X,\mathbb{Z}) \longrightarrow \mathbb{Z} \tag{5.1}$$

of homology and cohomology groups. (The pairing is nondegenerate on the free parts of the groups.) A similar pairing can be constructed in K-theory on a smooth closed manifold M. Namely, we consider the pairing

$$\langle , \rangle : K_c^0(T^*M) \times K^0(M) \longrightarrow \mathbb{Z} \tag{5.2}$$

taking the difference element $[\sigma(D)] \in K_c^0(T^*M)$ of an elliptic operator

$$D : C^\infty(M,E) \longrightarrow C^\infty(M,F)$$

and a vector bundle $G \in \mathrm{Vect}(M)$ to the index of D with coefficients in G.[8] The pairing (5.2) is nondegenerate on the free parts of the groups. (This can be proved by using the Atiyah–Singer index formula and by passing with the use of the Chern character to cohomology.) Comparing (5.1) and (5.2), we can make a guess that the *K-homology* groups of M can be defined in terms of topological K-theory:

$$K_0(M) \stackrel{\text{def}}{=} K_c^0(T^*M). \tag{5.3}$$

Unfortunately, this definition has a significant drawback, since the right-hand side of the formula does not make sense for a singular space M. Nevertheless, the right-hand side of (5.3) can be defined for an arbitrary compact space provided that we interpret $K_c^0(T^*M)$ as the group of stable homotopy classes of elliptic operators on M.

Atiyah [**Ati69**] suggested the following abstract notion of an elliptic operator.

DEFINITION 5.1. A *generalized elliptic operator* over a compact space X is a triple (D, H_1, H_2), where

$$D : H_1 \longrightarrow H_2$$

is a Fredholm operator acting in the Hilbert spaces H_1, H_2, these spaces are modules over the C^*-algebra $C(X)$ of continuous complex-valued functions on X, and D almost commutes with the module structure; i.e.,

$$[D,f] \in \mathcal{K}$$

for all $f \in C(X)$, where $\mathcal{K}$ is the space of compact operators from H_1 to H_2.

[8] Recall that an operator with coefficients in a vector bundle has the principal symbol

$$\sigma(D) \otimes 1_G : \pi^*(E \otimes G) \longrightarrow \pi^*(F \otimes G).$$

Any operator with this symbol is denoted by $D \otimes 1_G$.

Note that the compactness of the commutator in this definition originates from the fundamental property of differential operators on manifolds: for a smooth function f, the order of the commutator $[D, f]$ is at most the order of D minus one.

Atiyah showed that generalized elliptic operators on a compact space X define elements in the K-homology group $K_0(X)$ of the space, where K_0 is the generalized homology theory dual to topological K-theory. Atiyah also conjectured that the K-homology groups can be defined solely in terms of generalized elliptic operators. This conjecture was proved by Kasparov [**Kas73**] and independently by Brown–Douglas–Fillmore [**BDF77**]. It turned out more convenient to work with a certain modification of Atiyah's original definition.

We start from the description of the even group $K_0(X)$. (The odd groups $K_1(X)$ will be introduced later.)

2. Even cycles and the group $K_0(X)$. This group is generated by so-called *even cycles.*

DEFINITION 5.2. An *even cycle* in the K-homology of a space X is a pair (F, H) given by a $\mathbb{Z}_2$-graded Hilbert space

$$H = H_0 \oplus H_1,$$

where the components $H_{0,1}$ are $*$-modules over the C^*-algebra $C(X)$, and an odd (with respect to the grading) bounded operator

$$F : H \to H.$$

The operator and the module structure have the properties

$$f(F - F^*) \sim 0, \quad f(F^2 - 1) \sim 0, \quad [F, f] \sim 0 \quad \text{for all } f \in C(X), \tag{5.4}$$

where $\sim$ means equality modulo compact operators.

The even K-homology group $K_0(X)$ can be obtained by introducing some equivalence relation on the cycles. The simplest one is the *stable operator homotopy.* (See [**Bla98**], chapter VIII, where one can find a number of other equivalence relations on cycles. These relations are pairwise equivalent.) Let us define this relation.

Two cycles are said to be *isomorphic* if the corresponding $C(X)$-modules H are isomorphic and the operators F coincide under this isomorphism. Two cycles are *homotopic* if they become isomorphic after some homotopy of the operators F. A *trivial cycle* is a cycle for which all relations in (5.4) are satisfied exactly. Finally, two cycles are *stably operator homotopic* (for short, stably homotopic) if they become homotopic after adding some trivial cycles to each of them.

One can show that the set of stable homotopy classes of even cycles is an abelian group with respect to the direct sum. This group is denoted by $K_0(X)$. It is called the *K-homology group* of the space X.

REMARK 5.3. The K-homology groups behave covariantly for continuous mappings. Consider a continuous mapping $f : X \to Y$. Then a cycle (F, H) over X

can be treated as a cycle over Y if the $C(Y)$-module structure on H is defined as the composition of the induced mapping $f^* : C(Y) \to C(X)$ with the original $C(X)$-module structure on H. This mapping of cycles induces a homomorphism

$$f_! : K_0(X) \longrightarrow K_0(Y).$$

After these definitions, it is almost a tautology to say that ordinary elliptic operators on closed manifolds define K-homology elements. However, because of its importance, we describe the corresponding construction in detail.

Namely, consider an elliptic pseudodifferential operator

$$D : C^\infty(M, E) \longrightarrow C^\infty(M, F)$$

acting on sections of some bundles E and F on a closed manifold M. Then its partial isomorphism part

$$D' = (P_{\ker D} + D^* D)^{-1/2} D$$

($P_{\ker D}$ is the projection on the kernel of D) in the polar decomposition defines a Fredholm operator

$$D' : L^2(M, E) \longrightarrow L^2(M, F),$$

where both spaces are $C(M)$-modules. (The module structure is the pointwise product of functions.) Moreover, D' commutes with the module structure up to compact operators. For smooth functions, this follows from the composition formulas for pseudodifferential operators. Then the general case follows by continuity. We define the matrix operator

$$F = \begin{pmatrix} 0 & D'^* \\ D' & 0 \end{pmatrix}.$$

Then F is a self-adjoint odd operator on the naturally $\mathbb{Z}_2$-graded $C(M)$-module $H = L^2(M, E) \oplus L^2(M, F)$. Thus, the pair (F, H) is an even cycle. The corresponding element in K-homology is denoted by

$$[D] \in K_0(M). \tag{5.5}$$

3. The quantization mapping. This construction can be interpreted in quite a different way. Namely, the principal symbol of D

$$\sigma(D) : \pi^* E \longrightarrow \pi^* F, \quad \pi : T^*M \to M,$$

is a vector bundle isomorphism over T^*M except for the zero section. The corresponding difference element in K-theory with compact supports is denoted by

$$[\sigma(D)] \in K_c^0(T^*M). \tag{5.6}$$

Considering two elements (5.5) and (5.6) together, one can readily show that there is a well-defined mapping

$$\begin{aligned} Q : K_c^0(T^*M) &\longrightarrow K_0(M), \\ [\sigma(D)] &\mapsto [D], \end{aligned} \tag{5.7}$$

which sends symbols to the corresponding operators. This mapping is naturally called the *quantization mapping*. It is well defined on homotopy classes, since operators with the same symbol are homotopic.

4. Odd cycles. We have defined the group $K_0(X)$ in terms of even cycles. The odd group $K_1(X)$ is generated by odd cycles.

DEFINITION 5.4. An *odd cycle* in the K-homology of a space X is given by a Hilbert space H that is a $*$-module over $C(X)$ and a bounded operator

$$F : H \to H$$

such that

$$f(F - F^*) \sim 0, \quad f(F^2 - 1) \sim 0, \quad [F, f] \sim 0 \tag{5.8}$$

for an arbitrary function f.

The only difference between odd and even cycles is in the $\mathbb{Z}_2$-grading structure.

The set of stable homotopy classes of odd cycles on X is denoted by $K_1(X)$. Let us show how odd cycles arise from elliptic *self-adjoint* operators on closed manifolds. Consider an elliptic self-adjoint operator

$$A : C^\infty(M, E) \longrightarrow C^\infty(M, F).$$

We can extend it to the L^2-spaces of sections (if the order of A is zero). This gives the odd cycle $(L^2(M, E), (1 + A^2)^{-1/2}A)$, if $\operatorname{ord} A > 0$. The corresponding K-homology class is denoted by

$$[A] \in K_1(M).$$

On the other hand, the principal symbol of A defines a difference element

$$[\sigma(A)] \in K_c^1(T^*M).$$

Let us recall its definition (see [**APS76a**]). The principal symbol of an elliptic self-adjoint operator is an invertible Hermitian endomorphism of a bundle over the cosphere bundle S^*M. Hence, over S^*M we can consider the vector bundle denoted by $\operatorname{Im} \sigma(\Pi_+(A))$ and generated at a point $(x, \xi) \in S^*M$ by the eigenvectors of the symbol $\sigma(A)(x, \xi)$ corresponding to positive eigenvalues. Then the difference element is defined by the formula

$$[\sigma(A)] = \partial[\operatorname{Im} \Pi_+(A)] \in K_c^1(T^*M),$$

where

$$\partial : K(S^*M) \to K_c^1(T^*M)$$

is the coboundary mapping in the K-theory of the pair $S^*M \subset B^*M$. (Here B^*M is the bundle of unit balls in T^*M with respect to some Riemannian metric, and the cosphere bundle S^*M is realized as its boundary.)

By analogy with the even case, there is a well-defined quantization mapping

$$\begin{array}{rcl} Q : K_c^1(T^*M) & \longrightarrow & K_1(M), \\ {[\sigma(A)]} & \mapsto & [A]. \end{array} \tag{5.9}$$

This is well defined, since the coboundary ∂ induces an isomorphism

$$K(S^*M)/K(M) \simeq K^1_c(T^*M).$$

A classical theorem (see [**Kas88**]) claims that the quantization mapping is an isomorphism. This isomorphism is called the *Poincaré isomorphism.*

THEOREM 5.5. (The Poincaré isomorphism.) *The quantization mappings given by* (5.7) *and* (5.9) *define isomorphisms*

$$Q : K^*_c(T^*M) \longrightarrow K_*(M),$$

where the index $*$ *is either* 0 *or* 1.

The proof can be found in [**Kas88**].

5.2. Duality and the topological index. As was already mentioned, K-homology is dual to topological K-theory. The most important manifestation of this duality is the pairing

$$\langle , \rangle : K_i(X) \times K^i(X) \longrightarrow \mathbb{Z}, \quad i = 0, 1, \tag{5.10}$$

which we define in this subsection. The pairing (5.10) will be defined as the composition of the product

$$K_i(X) \times K^i(X) \longrightarrow K_0(X) \tag{5.11}$$

with the mapping

$$p_! : K_0(X) \longrightarrow K_0(pt) = \mathbb{Z}$$

induced by the projection of X into a one-point space.

Let us start by saying that $p_!$ takes a Fredholm operator to its analytic index. Indeed, an even cycle over a point is just a Fredholm operator, and Fredholm operators have only one stable homotopy invariant, namely, the index. This shows that $K_0(pt) = \mathbb{Z}$.

It remains to define the product (5.11). Let us consider it for $i = j = 0$.

The formula for the product (e.g., see [**Ati69**]) for a general element of the group $K_0(X)$ mimics the construction of an operator with coefficients in a vector bundle. Namely, this mapping takes a cycle (F, H) and a vector bundle G to the even cycle

$$(F \otimes 1_{\mathbb{C}^N}, 1_H \otimes P(H \otimes \mathbb{C}^N)),$$

where $P : \mathbb{C}^N \to \mathbb{C}^N$ is some projection over X defining G and

$$1_H \otimes P(H \otimes \mathbb{C}^N) \subset H \otimes \mathbb{C}^N$$

is the range of the projection $1_H \otimes P$. Thus, the pairing $\langle [D], [G] \rangle$ of an elliptic operator D with a bundle G can be computed by applying the Atiyah–Singer formula to $D \otimes 1_G$:

$$\langle [D], [G] \rangle = \operatorname{ind}_t(D \otimes 1_G).$$

Let us also mention that the product of odd groups

$$K_1(X) \times K^1(X) \longrightarrow K_0(X)$$

is defined in the spirit of the theory of Toeplitz type operators (see [**BD82**]).

Comparing the constructions of the present section with the de Rham theory, one can make the following glossary of similar terms:

de Rham theory	Elliptic theory
cohomology $H^*(M)$	topological K-theory $K^*(M)$
homology $H_*(M)$	K-homology $K_*(M)$
cocycle ω, $d\omega = 0$	vector bundle E
cycle γ, $\partial\gamma = 0$	elliptic operator D
integral $\int_\gamma \omega$	index $\operatorname{ind}(D \otimes 1_E)$

5.3. Poincaré duality on manifolds with boundary. Absolute and relative cycles. Let us now consider duality for manifolds with boundary.

Let M be a compact smooth manifold with boundary ∂M. In this case, one also has Poincaré duality, frequently called Poincaré–Lefschetz duality. In (co)homology (for oriented M), the duality amounts to two group isomorphisms

$$H^i(M) \longrightarrow H_{n-i}(M, \partial M), \quad H^i(M, \partial M) \longrightarrow H_{n-i}(M), \quad n = \dim M.$$

Thus, duality relates the ordinary groups to the so-called relative groups.

Let us define similar isomorphisms in K-theory. For a manifold M with boundary, one has two natural K-homology groups: the ordinary group $K_*(M)$ and the so-called *relative K-homology group* $K_*(M, \partial M)$. The relative groups are defined as follows.

It is clear that in Definition 5.2 we have used only the algebra $C(M)$ of functions on M. At the same time, a manifold with boundary has another natural C^*-algebra, namely, the algebra $C_0(M \setminus \partial M)$ of functions vanishing at the boundary. We denote the group generated by cycles over $C_0(M \setminus \partial M)$ by $K_*(M, \partial M)$ and call it the *relative K-homology* of the manifold.

Let us show how elements of these groups arise from elliptic operators on manifolds with boundary. On the analogy with homology theory, we refer to cycles over the algebra $C(M)$ as *absolute cycles* and cycles over $C_0(M \setminus \partial M)$ as *relative cycles.*

We first describe the ordinary (absolute) cycles.

1. Elliptic operators and absolute cycles. In this case, we consider elliptic operators D induced near the boundary by vector bundle isomorphisms. Just as in the case of closed manifolds, such operators almost commute with functions $f \in C(M)$:

$$[D, f] \sim 0.$$

Therefore, the construction of the previous subsection can be carried out word for word in this case; i.e., D defines a K-homology class

$$[D] \in K_*(M).$$

On the other hand, the principal symbol of such an operator is an isomorphism on T^*M *everywhere near the boundary.* Hence, the operator D has a difference element

$$[\sigma(D)] \in K_c^*(T^*(M \setminus \partial M)).$$

REMARK 5.6. This class of zero-order operators induced in a neighborhood of the boundary by vector bundle isomorphisms naturally arises in index theory of classical boundary value problems. Namely, the classical procedure of order reduction (see [**Hör85**] or [**SSS99b**]) reduces a boundary value problem for a differential operator to a pseudodifferential operator of order zero of precisely the form considered here. Moreover, the reduction preserves the index.

2. Elliptic operators and relative cycles. Relative cycles arise as follows. An elliptic operator D of order one on M defines an element

$$[D] \in K_*(M, \partial M),$$

which can be constructed as follows. Consider an embedding $M \subset \widetilde{M}$ of M in some closed manifold $\widetilde{M}$ of the same dimension. Let $\widetilde{D}$ be an arbitrary extension of D to this closed manifold as an elliptic operator. On $\widetilde{M}$ we consider the zero-order operator

$$\widetilde{F} = \left(1 + \widetilde{D}^*\widetilde{D}\right)^{-1/2} \widetilde{D}.$$

The restriction of this operator to M is defined as

$$F = i^*\widetilde{F}i_* : L^2(M, E) \longrightarrow L^2(M, F), \tag{5.12}$$

where $i_* : L^2(M) \to L^2\left(\widetilde{M}\right)$ is the extension by zero and $i^* : L^2\left(\widetilde{M}\right) \to L^2(M)$ is the restriction operator.

If D is symmetric, then one can readily verify that F satisfies the properties

$$F - F^* \sim 0, \quad f\left(F^2 - 1\right) \sim 0, \quad [F, f] \sim 0$$

for a function $f \in C_0(M \backslash \partial M)$ vanishing on the boundary. These relations show that F defines an element in $K_1(M, \partial M)$.

If D is not symmetric, then instead of the operator F one considers the corresponding matrix operator as in Subsection 5.1. In this case, the pair (F, H) defines an element of the group $K_0(M, \partial M)$.

3. The quantization mapping and the Poincaré isomorphism. For elliptic operators defining absolute cycles on manifolds with boundary, we obtain two elements

$$[\sigma(D)] \in K_c^*(T^*(M \setminus \partial M)), \quad [D] \in K_*(M).$$

Similarly, operators corresponding to relative cycles also define a pair of elements

$$[\sigma(D)] \in K_c^*(T^*M), \quad [D] \in K_*(M, \partial M).$$

(Here $[\sigma(D)]$ is the Atiyah–Singer difference element of D.) One can show that the corresponding *quantization mappings*

$$
\begin{array}{rcc}
Q : K_c^*(T^*(M \setminus \partial M)) & \longrightarrow & K_*(M), \\
K_c^*(T^*M) & \longrightarrow & K_*(M, \partial M), \\
{[\sigma(D)]} & \mapsto & [D]
\end{array}
\tag{5.13}
$$

are well defined. It was proved in [**Kas88**] that both quantization mappings are isomorphisms. They are called the *Poincaré isomorphisms in K-theory on manifolds with boundary.*

4. The exact sequence of a pair in K-homology. It is well known that in K-homology for an embedding $\partial M \subset M$ there is a six-term exact sequence

$$
\begin{array}{ccccc}
 & K_0(\partial M) & \longrightarrow & K_0(M) & \\
\nearrow & & & & \searrow \\
K_1(M, \partial M) & & & & K_0(M, \partial M). \\
\nwarrow & & & & \swarrow \\
 & K_1(M) & \longleftarrow & K_1(\partial M) &
\end{array}
\tag{5.14}
$$

The boundary maps in this sequence can be explicitly computed (see [**BDT89**] or [**MP92**]). It turns out that this sequence is isomorphic to the exact sequence of the pair $T^*M|_{\partial M} \subset T^*M$ in topological K-theory:

$$
\begin{array}{ccccc}
 & K_c(T^*\partial M) & \longrightarrow & K_c(T^*(M \setminus \partial M)) & \\
\nearrow & & & & \searrow \\
K_c^1(T^*M) & & & & K_c(T^*M). \\
\nwarrow & & & & \swarrow \\
 & K_c^1(T^*(M \setminus \partial M)) & \longleftarrow & K_c^1(T^*\partial M) &
\end{array}
\tag{5.15}
$$

The isomorphism can be obtained by applying the quantization mappings term by term to the sequence (5.15).

6. Appendix C. Poincaré duality on $\mathbb{Z}_n$-manifolds

In this section, we construct Poincaré duality for $\mathbb{Z}_n$-manifolds (see [**SS01**]). Unfortunately, these manifolds with singularities have no duality in the framework of the usual topological K-groups. Quite remarkably, however, the duality can be restored if one applies the approach of noncommutative geometry [**Con94**] to this problem and states duality in terms of K-theory of some noncommutative algebras of functions on the corresponding spaces. We show that the approach of noncommutative geometry, at least for $\mathbb{Z}_n$-manifolds, can be completely described in terms of elliptic operator theory.

Before we proceed to the description of duality for $\mathbb{Z}_n$-manifolds, we recall the main features of the passage from topological K-theory to K-theory of algebras. This passage goes as follows.

(1) The topological group $K^0(X)$ of a compact space X can be identified with the Grothendieck group of homotopy classes of projections in matrix algebras over $C(X)$. Clearly, this definition makes sense for an arbitrary unital C^*-algebra $\mathcal{A}$. The corresponding group is denoted by $K_0(\mathcal{A})$. Similarly, the odd group $K^1(X)$ is identified with the group of stable homotopy classes of unitary operators in matrix algebras over $C(X)$. The similar group for a C^*-algebra $\mathcal{A}$ is denoted by $K_1(\mathcal{A})$. Summarizing, the new groups are reduced in the commutative case to the topological K-groups
$$K^*(X) \simeq K_*(C(X)).$$
(2) K-homology. It is clear from Definition 5.2 that the definition of the K-homology group of a space uses only the C^*-algebra of functions on it. Therefore, the same definition for an arbitrary C^*-algebra $\mathcal{A}$ gives two groups $K^0(\mathcal{A})$ and $K^1(\mathcal{A})$. These groups are generated, respectively, by even and odd cycles over $\mathcal{A}$.

REMARK 6.1. We note the change of variance of the functors (and the corresponding change in the position of indices). For example, an analog of the topological K-groups is the group $K_*(\mathcal{A})$, which is a covariant functor with respect to algebra homomorphisms. Similarly, the K-homology of spaces translates into the K-cohomology (a contravariant functor) of algebras.

For a nonunital algebra $\mathcal{A}$, the K_0-group is defined as the kernel of the natural surjective mapping
$$K_0(\mathcal{A}) \stackrel{\text{def}}{=} \ker\left(K_0(\mathcal{A}^+) \longrightarrow K_0(\mathbb{C})\right),$$
where $\mathcal{A}^+ = \mathcal{A} \oplus \mathbb{C}$ is the algebra with attached unit. The mapping is induced by the algebra homomorphism $\mathcal{A} \oplus \mathbb{C} \to \mathbb{C}$. The K_1-group is defined slightly simpler:
$$K_0(\mathcal{A}) \stackrel{\text{def}}{=} K_1(\mathcal{A}^+).$$
The construction is similar to the definition of topological K-theory for locally compact spaces (see [**Ati89**]).

6.1. Relative cycles. The C^*-algebra of a $\mathbb{Z}_n$-manifold. Since a $\mathbb{Z}_n$-manifold for $n = 1$ is just a smooth manifold with boundary, it is natural to construct duality separately for relative and absolute cycles.

Consider a twisted $\mathbb{Z}_n$-manifold (M, π) (see Definition 3.1), or a $\mathbb{Z}_n$-manifold for short. On M, we take some elliptic operator
$$D : C^\infty(M, E) \longrightarrow C^\infty(M, F)$$

satisfying Assumption 1. Its principal symbol defines a difference element

$$[\sigma(D)] \in K_c(\overline{T^*M}^{\pi}) \tag{6.1}$$

of the topological K-group of the $\mathbb{Z}_n$-manifold (T^*M,π). Note that (6.1) includes additional information (compared with the image of $[\sigma(D)]$ in $K_c(T^*M)$) related to the structure of the space $\overline{T^*M}^{\pi}$. Let us define a similar element in K-homology. Note that by assumption, the operator D almost commutes both with functions $f \in C_0(M \setminus \partial M)$ and with transpositions of the sheets of π in a neighborhood of the boundary.

By $\mathcal{A}_{M,\pi}$ we denote the C^*-subalgebra of operators in $L^2(M)$ generated by multiplications by functions $f \in C_0(M \setminus \partial M)$ and transposition operators in a neighborhood of the boundary. If we take a neighborhood of the boundary where transpositions of the sheets are allowed and denote the complement of this neighborhood by M', then the algebra can be explicitly described in the form

$$\mathcal{A}_{M,\pi} = \left\{ (u,v) \;\middle|\; \begin{array}{c} u \in C_0(M'), v \in C_0(X \times (0,1], \mathrm{End}\pi_!1) \\ \beta(u|_{\partial M'})\beta^{-1} = v|_{t=1} \end{array} \right\}$$

as a subalgebra in the direct sum $C_0(M') \oplus C_0(X \times (0,1], \mathrm{End}\pi_!1)$. Here $\mathrm{End}\pi_!1$ is the vector bundle of endomorphisms of $\pi_!1 \in \mathrm{Vect}(X)$.

REMARK 6.2. The algebra $\mathcal{A}_{M,\pi}$ can also be defined as the C^*-algebra of the etale-groupoid [**BN94**] corresponding to the extension of the equivalence relation (3.2) to the neighborhood of the boundary.

Under Assumption 1, one can prove the following result.

LEMMA 6.3. *The spaces $L^2(M,E)$ and $L^2(M,F)$ have a natural module structure over the algebra $\mathcal{A}_{M,\pi}$ such that D almost commutes with this module structure. Thus, D defines a class $[D] \in K^*(\mathcal{A}_{M,\pi})$.*

PROOF. By Assumption 1, in a neighborhood of the boundary one has vector bundle isomorphisms

$$E \simeq \pi^* E_0, \quad E_0 \in \mathrm{Vect}(X \times [0,1]).$$

In a similar way, for the spaces of sections we have

$$C(\partial M \times [0,1], E) \simeq C(X \times [0,1], \pi_!1 \otimes E_0).$$

The latter space has a natural $C_0(X \times (0,1], \mathrm{End}\pi_!1)$-module structure. This module structure can be glued with the $C(M)$-module structure on the entire manifold, thus defining the desired module structure over the algebra $\mathcal{A}_{M,\pi}$. A similar construction applies to F. The desired almost commutativity of D with the elements of $\mathcal{A}_{M,\pi}$ follows from the equivariance of D with respect to transpositions of sheets of the covering. □

6.2. Absolute cycles. Nonlocal operators. Clearly, the relative cycles on a $\mathbb{Z}_n$-manifold were obtained in the previous subsection by restricting the class of operators to those almost commuting with the algebra $\mathcal{A}_{M,\pi}$. In this subsection, we enlarge the class of absolute cycles on M requiring only that the operators in question almost commute with elements of the algebra of continuous functions on the singular space $\overline{M}^{\pi}$ but not of the larger algebra

$$C(M) \supset C(\overline{M}^{\pi}).$$

1. Definition of the class of operators. Needless to say, ordinary absolute cycles on M (represented by elliptic operators of order zero induced by vector bundle isomorphisms near the boundary) are cycles for the algebra $C(\overline{M}^{\pi})$ as well. However, there are also more general cycles on the space $\overline{M}^{\pi}$. To find them, let us replace $C(\overline{M}^{\pi})$ by a homotopic algebra of functions on M that are lifted from the base of the covering in a given neighborhood of the boundary. We denote this algebra by the same symbol. Here is an "heuristic derivation" of relative cycles.

Consider some operator D on M. In a neighborhood of the boundary, its direct image $(\pi \times 1)_! D$ acts on the base of the covering $\pi \times 1 : \partial M \times [0,1) \to X \times [0,1)$. The spaces in which the operator acts have the usual $C(X)$-module structure (determined by the pointwise product of functions). Therefore, D almost commutes with functions in $C(\overline{M}^{\pi})$ if (a) it is a pseudodifferential operator far from the boundary and (b) its direct image in a neighborhood of the boundary is a pseudodifferential operator on the base. Thus, we arrive at the following definition.

DEFINITION 6.4. An *admissible operator* on M is an operator of the form

$$D = \psi' D' \varphi' + \psi'' \left(\pi^! D''\right) \varphi'', \tag{6.2}$$

where

$$D' : C^\infty(M, E) \to C^\infty(M, F)$$

is a pseudodifferential operator of order zero on M,

$$D'' : C^\infty(X \times [0,1], (\pi \times 1)_! E) \longrightarrow C^\infty(X \times [0,1], (\pi \times 1)_! F)$$

is a pseudodifferential operator on $X \times [0,1]$, the cutoff functions φ' and ψ' vanish in a neighborhood of the boundary, and φ'' and ψ'' vanish in a neighborhood of the subset $X \times \{1\} \subset X \times [0,1]$. Finally, the operator D'' is induced by a vector bundle isomorphism in the neighborhood $X \times [0,1/2) \subset X \times [0,1]$ on the cylinder.

REMARK 6.5. Inverse images like $\pi^! D''$ are in general *nonlocal* operators. By way of illustration, consider the trivial covering

$$\partial M = \underbrace{X \sqcup X \sqcup \ldots \sqcup X}_{n \text{ copies}} \longrightarrow X$$

of n copies of X. On ∂M, we choose a trivial bundle $E = \mathbb{C}$. Then $\pi_! E = \mathbb{C}^n$; the direct image of a differential operator on the total space is always a diagonal operator

$$\pi_! P = \operatorname{diag}\left(P|_{X_1}, \ldots, P|_{X_n}\right) : C^\infty\left(X, \mathbb{C}^n\right) \longrightarrow C^\infty\left(X, \mathbb{C}^n\right),$$

whereas the inverse image of an operator on the base is a matrix operator of general form. Therefore, near the boundary we admit usual pseudodifferential operators as well as operators corresponding to transpositions of values of a function on the sheets of the covering. In a sense, admissible operators are obtained as an extension of the algebra of pseudodifferential operators by the nonlocal operators corresponding to elements of $\mathcal{A}_{M,\pi}$.

2. Symbols of admissible operators. One can readily define the symbol of an admissible operator and state the ellipticity condition. Namely, the symbol of an admissible operator D is a pair (σ_M, σ_X) of usual symbols, where

$$\sigma_M : p_M^* \, E|_{M'} \longrightarrow p_M^* \, F|_{M'}, \quad p_M : S^*M \longrightarrow M,$$

is a symbol on the part of the manifold where the operator is pseudodifferential and

$$\sigma_X : p_X^* \, (\pi_! E)|_{X\times[0,1]} \longrightarrow p_X^* \, (\pi_! F)|_{X\times[0,1]}, \quad p_X : S^*(X\times[0,1]) \longrightarrow X\times[0,1],$$

is a symbol on the cylinder $X \times [0,1]$. The symbols satisfy the compatibility condition

$$\pi_! \left(\sigma_M|_{\partial M'}\right) = \sigma_X.$$

An operator is said to be *elliptic* if both components of its symbol are invertible. An elliptic operator is Fredholm in appropriate Sobolev spaces.

Summarizing, we see that admissible elliptic operators D on a $\mathbb{Z}_n$-manifold define absolute cycles

$$[D] \in K_*(\overline{M}^{\pi}).$$

Let us show that the symbol of an elliptic admissible operator defines an element in K-theory. In other words, let us define an analog of the Atiyah–Singer difference construction for admissible elliptic operators.

3. The difference construction for admissible operators. Let us cut M into two parts

$$M' = M \backslash \{\partial M \times [0,1)\} \qquad \text{and} \qquad \partial M \times [0,1].$$

Then the symbol $\sigma(D)$ of an admissible elliptic operator D is naturally represented as a pair (σ_M, σ_X). Each symbol in this pair has the corresponding difference element

$$[\sigma_M] \in K_c^0(T^*M'), \qquad [\sigma_X] \in K_c^0(T^*(X\times(0,1])).$$

In the latter case, we use the difference construction for an absolute cycle with symbol σ_X of order zero.

However, a single element of some topological K-group cannot be constructed from these data, since the manifolds T^*M' and $T^*(X\times(0,1])$ cannot be glued smoothly (their boundaries may be nondiffeomorphic). Nevertheless, the pasting can be done if we glue the algebras of these manifolds rather than the manifolds themselves.

Let $\mathcal{A}_{T^*M,\pi}$ be the C^*-algebra of the $\mathbb{Z}_n$-manifold (T^*M,π). It is the subalgebra

$$\mathcal{A}_{T^*M,\pi} \subset C_0\left(T^*M'\right) \oplus C_0\left(T^*\left(X\times(0,1]\right), \operatorname{End} p^*\pi_!1\right),$$

defined by the compatibility condition

$$\mathcal{A}_{T^*M,\pi} = \left\{u \oplus v \mid \beta\, u|_{\partial M'} \beta^{-1} = v|_{t=1}\right\}$$

on the boundaries.

The *difference construction* for admissible operators is a mapping

$$\chi : \operatorname{Ell}(M,\pi) \longrightarrow K_0\left(\mathcal{A}_{T^*M,\pi}\right) \tag{6.3}$$

of the group $\operatorname{Ell}(M,\pi)$ of stable homotopy classes of elliptic admissible operators on M into the K_0-group of the algebra $\mathcal{A}_{T^*M,\pi}$.

Let us explicitly describe the element $\chi[D]$ corresponding to an elliptic admissible operator

$$D : C^\infty(M,E) \longrightarrow C^\infty(M,F)$$

with symbol $\sigma(D) = (\sigma_M, \sigma_X)$. By stabilization, one can always assume that the bundle F is trivial: $F = \mathbb{C}^k$. Now let us choose some embeddings of E and $\mathbb{C}^k$ in trivial vector bundles and some projections

$$P_E, P_{\mathbb{C}^k} : \mathbb{C}^{N+L} \longrightarrow \mathbb{C}^{N+L}$$

defining E and $\mathbb{C}^k$ as

$$E \simeq \operatorname{Im} P_E \subset \mathbb{C}^N \oplus 0, \qquad \mathbb{C}^k \simeq \operatorname{Im} P_{\mathbb{C}^k} \subset 0 \oplus \mathbb{C}^L.$$

Let $P_{\pi_!E}, P_{\pi_!\mathbb{C}^k}$ be the direct images of these projections in a neighborhood of the boundary. Clearly, these projections define the direct images of the corresponding subbundles.

Then the difference construction of D is defined by the formula

$$\chi[D] = [P_1 \oplus P_2] - \left[P_{\mathbb{C}^k} \oplus P_{\pi_!\mathbb{C}^k}\right],$$

where the projection P_1 over T^*M' is defined by

$$\begin{cases} P_E \cos^2|\xi| + P_{\mathbb{C}^k}\sin^2|\xi| + \left(\sigma_M^{-1}P_{\mathbb{C}^k} + \sigma_M P_E\right)\sin|\xi|\cos|\xi|, & |\xi| \le \pi/2, \\ P_{\mathbb{C}^k}, & |\xi| > \pi/2 \end{cases} \tag{6.4}$$

(we assume that the symbol σ_M is homogeneous of order zero in the covariable ξ), and the projection P_2 over $T^*(X\times[0,1])$ is defined by one of the expressions

$$\begin{gathered} P_{\pi_!E}\cos^2|\xi| + P_{\pi_!\mathbb{C}^k}\sin^2|\xi| + 1/2\left(\sigma_X^{-1}P_{\pi_!\mathbb{C}^k} + \sigma_X P_{\pi_!E}\right)\sin 2|\xi|, \\ P_{\pi_!E}\cos^2\varphi + P_{\pi_!\mathbb{C}^k}\sin^2\varphi + 1/2\left(\sigma_X^{-1}P_{\pi_!\mathbb{C}^k} + \sigma_X P_{\pi_!E}\right)\sin 2\varphi, \\ P_{\pi_!\mathbb{C}^k}. \end{gathered}$$

Here the first case is taken for the parameter values $x' \in X\times[1/2,1], |\xi| \le \pi/2$, the second case for $x' \in X\times[0,1/2], |\xi| < \pi t$, and the third expression applies to the remaining points. Here for brevity we write

$$\varphi = |\xi| + \pi/2\,(1-2t).$$

REMARK 6.6. Geometrically, these projections define a subbundle that coincides over the zero section (i.e. for $|\xi| = 0$) with $E \subset \mathbb{C}^{N+L}$, coincides for $|\xi| \geq \pi/2$ with the orthogonal subbundle $\mathbb{C}^k \subset \mathbb{C}^{N+L}$, and is a rotation of one of the subbundles towards the other via the isomorphisms defined by the elliptic symbol $\sigma(D)$ at the intermediate points.

PROPOSITION 6.7. *The difference construction* (6.3) *is a well-defined group isomorphism.*

PROOF. The difference construction χ preserves the equivalence relations on the groups $\mathrm{Ell}^0(M,\pi)$ and $K(\mathcal{A}_{T^*M,\pi})$. This can be proved by observing that a homotopy of operators gives a homotopy of symbols and hence a homotopy of the corresponding projections $P_{1,2}$. Furthermore, $\chi[D]$ is independent of the choice of embeddings in a trivial bundle, since all such embeddings are pairwise homotopic. The proof of the fact that χ defines a one-to-one mapping can be obtained along the same lines. □

6.3. The Poincaré isomorphism. Using the difference construction of the previous subsection, we define quantization mappings on $\mathbb{Z}_n$-manifolds: one for absolute cycles

$$K_*(\mathcal{A}_{T^*M,\pi}) \longrightarrow K_*(\overline{M}^\pi),$$

corresponding to the theory of admissible nonlocal operators, and another for relative cycles

$$K_c^*(\overline{T^*M}^\pi) \longrightarrow K^*(\mathcal{A}_{M,\pi}),$$

corresponding to the operators discussed in Section 3 and Subsection 6.1.

THEOREM 6.8. (The Poincaré isomorphism.) *The quantization mappings on $\mathbb{Z}_n$-manifolds are isomorphisms.*

PROOF. 1) The algebra $\mathcal{A}_{T^*M,\pi}$ has the ideal

$$I = C_0(T^*(X \times (0,1)), \mathrm{End}p^*\pi_! 1)$$

with quotient $\mathcal{A}_{T^*M,\pi}/I \simeq C_0(T^*M)$. Consider the exact sequence

$$\to K_c(T^*X) \to K_0(\mathcal{A}_{T^*M,\pi}) \to K_c(T^*M) \to K_c^1(T^*X) \to \tag{6.5}$$

of this pair. (To obtain this sequence, we use the natural isomorphisms

$$K_*(C_0(Y, \mathrm{End}G)) \simeq K_*(C_0(Y)) \simeq K^*(Y)$$

for a vector bundle $G \in \mathrm{Vect}(Y)$.) Now consider the commutative diagram

$$\begin{array}{ccccccccc}
\to & K_c^*(T^*X) & \to & K_*(\mathcal{A}_{T^*M,\pi}) & \to & K_c^*(T^*M) & \to & K_c^{*+1}(T^*X) & \to \\
 & \downarrow & & \downarrow & & \downarrow & & \downarrow & \\
\to & K_*(X) & \to & K_*(\overline{M}^\pi) & \to & K_*(M,\partial M) & \to & K_{*+1}(X) & \to
\end{array}$$

The second row is the exact sequence of the pair $X \subset \overline{M}^\pi$ in K-homology. The vertical arrows (except for the second) are Poincaré isomorphisms on closed manifolds and manifolds with boundary. Thus, by the 5-lemma, the mapping

$$K_*(\mathcal{A}_{T^*M,\pi}) \longrightarrow K_*(\overline{M}^\pi)$$

is an isomorphism as well.

2) The proof of the second isomorphism can be carried out in a similar way. In this case, the following diagram is relevant:

$$\begin{array}{ccccccccc}
\leftarrow & K^1(T^*X) & \leftarrow & K^0\left(\overline{T^*M}^{\pi}\right) & \leftarrow & K^0\left(T^*\left(M\backslash\partial M\right)\right) & \leftarrow & K^0(T^*X) & \leftarrow \\
 & \downarrow & & \downarrow & & \downarrow & & \downarrow & \\
\leftarrow & K_1(X) & \leftarrow & K^0\left(\mathcal{A}_{M,\pi}\right) & \leftarrow & K_0(M) & \leftarrow & K_0(X) & \leftarrow
\end{array}$$

(Here the first row is the exact sequence of the pair $\mathbb{R}\times T^*X\subset\overline{T^*M}^{\pi}$.) □

6.4. Poincaré duality.

1. Definition of the pairing. Let us now define Poincaré–Lefschetz duality on $\mathbb{Z}_n$-manifolds following the same scheme as in the case of smooth manifolds. Since we have relative and absolute cycles, two dualities are expected:

$$K_c^i\left(\overline{T^*M}^{\pi}\right)\times K_i\left(\mathcal{A}_{M,\pi}\right)\longrightarrow\mathbb{Z} \tag{6.6}$$

and

$$K_i\left(\mathcal{A}_{T^*M,\pi}\right)\times K^i\left(\overline{M}^{\pi}\right)\longrightarrow\mathbb{Z}. \tag{6.7}$$

To save space, we consider only the first duality. The second can be considered in a similar way.

Let us define the pairing (6.6) as follows: we act on the first argument by the Poincaré isomorphism

$$Q:K_c^*\left(\overline{T^*M}^{\pi}\right)\longrightarrow K^*\left(\mathcal{A}_{M,\pi}\right)$$

and then apply the index pairing

$$K^*\left(\mathcal{A}_{M,\pi}\right)\times K_*\left(\mathcal{A}_{M,\pi}\right)\longrightarrow\mathbb{Z}$$

of K-groups of opposite variance.

THEOREM 6.9. (Poincaré duality.)[**SS01**] *On a $\mathbb{Z}_n$-manifold, the pairings*

$$K_c^i\left(\overline{T^*M}^{\pi}\right)\times K_i\left(\mathcal{A}_{M,\pi}\right)\longrightarrow\mathbb{Z},\quad i=0,1, \tag{6.8}$$

are nonsingular on the free parts of the groups.

PROOF. Fixing the first argument of the pairing, we obtain a mapping

$$K_c^i\left(\overline{T^*M}^{\pi}\right)\otimes\mathbb{Q}\longrightarrow K_i'\left(\mathcal{A}_{M,\pi}\right),$$

where we write $G'=\operatorname{Hom}(G,\mathbb{Q})$. This mapping occurs in the commutative diagram

$$\begin{array}{ccccccc}
K_c^1(T^*X,\mathbb{Q}) & \longleftarrow & K_c^0\left(\overline{T^*M}^{\pi},\mathbb{Q}\right) & \longleftarrow & K_c^0\left(T^*(M\backslash\partial M),\mathbb{Q}\right) & \longleftarrow & K_c^0(T^*X,\mathbb{Q}) \\
\downarrow & & \downarrow & & \downarrow & & \downarrow \\
K^{1\prime}(X) & \longleftarrow & K_0'\left(\mathcal{A}_{M,\pi}\right) & \longleftarrow & K^{0\prime}(M) & \longleftarrow & K^{0\prime}(X).
\end{array}$$

All vertical arrows except for the second are isomorphisms (by Poincaré duality on smooth manifolds). Thus, by the 5-lemma the remaining mapping is also an isomorphism. Hence, (6.8) is nondegenerate in the first factor.

The nondegeneracy in the second factor can be proved in a similar way. □

2. An application to spin^c**-manifolds.** Consider a pair (M,π), where M has a *spin*^{c}-structure induced over the boundary by a *spin*^{c}-structure on the base X of the covering π. Then the group $K_c^*\left(\overline{T^*M}^\pi\right)$ is a free $K^{*+n}\left(\overline{M}^\pi\right)$-module with one generator (where $n=\dim M$). A generator is given by the difference element

$$[\sigma(D)] \in K_c^n\left(\overline{T^*M}^\pi\right)$$

of the Dirac operator on M. (This can be proved by analogy with the case of closed manifolds; e.g., see [**LM89**].) Therefore, one can define the Poincaré duality pairing

$$K^{*+n}\left(\overline{M}^\pi\right) \times K_*\left(\mathcal{A}_{M,\pi}\right) \longrightarrow \mathbb{Z}$$

using the composition $K^{*+n}\left(\overline{M}^\pi\right) \to K_c\left(\overline{T^*M}^\pi\right)$. It follows from Theorem 6.9 that this pairing is nonsingular on the free parts of the groups.

3. Computation of the pairing. Our aim is to find a computable formula for the Poincaré duality pairing. To be definite, we consider only the case $i=0$ in (6.6). To this end, we start from an explicit geometric realization of the groups. Let us first give a realization of the group $K_0\left(\mathcal{A}_{M,\pi}\right)$.

LEMMA 6.10. *The group $K_0\left(\mathcal{A}_{M,\pi}\right)$ is isomorphic to the group of stable homotopy classes of triples*

$$(E,F,\sigma), \qquad E,F\in \operatorname{Vect}(M),\ \sigma: \pi_!\, E|_{\partial M} \longrightarrow \pi_!\, F|_{\partial M},$$

where σ is a vector bundle isomorphism. Here trivial triples are those induced by a global vector bundle isomorphism over M.

PROOF. This lemma is similar to Proposition 6.7: both give a topological realization of the K_0-group of the C^*-algebra of the $\mathbb{Z}_n$-manifolds. Hence, a triple $\left(E,\mathbb{C}^k,\sigma\right)$ (note that an arbitrary triple can be reduced to this form) defines the element

$$[P_E\oplus P_2]-\left[P_{\mathbb{C}^k}\oplus P_{\pi_!\mathbb{C}^k}\right]\in K_0\left(\mathcal{A}_{M,\pi}\right)$$

of the K_0–group, where the projection P_2 over $X\times[0,1]$ is defined as

$$P_2 = P_{\pi_!E}\cos^2\varphi + P_{\pi_!\mathbb{C}^k}\sin^2\varphi + \left(\sigma^{-1}(x)\,P_{\pi_!\mathbb{C}^k} + \sigma(x)\,P_{\pi_!E}\right)\frac{\sin 2\varphi}{2},\quad \varphi=\frac{\pi}{2}(1-t),$$

and P_E and $P_{\mathbb{C}^k}\subset\mathbb{C}^N$ are some projections on subbundles isomorphic to E and $\mathbb{C}^k$, respectively. One must also assume that these subbundles are mutually orthogonal.

The proof of the fact that this construction gives a well-defined isomorphism with $K_0\left(\mathcal{A}_{M,\pi}\right)$ can be carried out by analogy with the proof of Proposition 6.7. □

This realization enables one to define the product

$$K_c^0\left(\overline{T^*M}^\pi\right)\times K_0\left(\mathcal{A}_{M,\pi}\right)\longrightarrow K_0\left(\mathcal{A}_{T^*M,\pi}\right)$$

geometrically in terms of operators with coefficients in vector bundles. More precisely, for two elements

$$[\sigma] \in K_c^0\left(\overline{T^*M}^\pi\right), \qquad \text{and} \quad [E, F, \sigma'] \in K_0\left(\mathcal{A}_{M,\pi}\right)$$

we consider the symbol

$$\sigma \otimes 1_E \oplus \sigma^{-1} \otimes 1_F \tag{6.9}$$

on M. The direct image of the restriction of this symbol to the boundary is

$$\begin{aligned}\pi_! \left(\sigma \otimes 1_E \oplus \sigma^{-1} \otimes 1_F\big|_{\partial M}\right) &= \pi_!\sigma \otimes 1_{\pi_!\, E|_{\partial M}} \oplus \pi_!\sigma^{-1} \otimes 1_{\pi_!\, F|_{\partial M}} \\ &\simeq \left(\pi_!\sigma \oplus \pi_!\sigma^{-1}\right) \otimes 1_{\pi_!\, E|_{\partial M}}.\end{aligned}$$

The last isomorphism is induced by the vector bundle isomorphism

$$\pi_!\, E|_{\partial M} \overset{\sigma'}{\simeq} \pi_!\, F|_{\partial M}.$$

There is a standard homotopy of the symbol $\left(\pi_!\sigma \oplus \pi_!\sigma^{-1}\right) \otimes 1_{\pi_!\, E|_{\partial M}}$ to the identity:

$$\begin{pmatrix} \pi_!\sigma & 0 \\ 0 & 1 \end{pmatrix} \begin{pmatrix} \cos\tau & \sin\tau \\ -\sin\tau & \cos\tau \end{pmatrix} \begin{pmatrix} 1 & 0 \\ 0 & \pi_!\sigma^{-1} \end{pmatrix} \begin{pmatrix} \cos\tau & -\sin\tau \\ \sin\tau & \cos\tau \end{pmatrix}, \qquad \tau \in [0, \pi/2].$$

Therefore, we have constructed an extension of the symbol (6.9) to an elliptic symbol of some admissible operator on M. Finally, the desired product

$$[\sigma] \times [E, F, \sigma'] \in K_0\left(\mathcal{A}_{T^*M,\pi}\right)$$

can be defined as the difference element of this symbol. One can prove that the pairing

$$\langle , \rangle : K_c^0\left(\overline{T^*M}^\pi\right) \times K_0\left(\mathcal{A}_{M,\pi}\right) \longrightarrow K_0\left(\mathcal{A}_{T^*M,\pi}\right) \xrightarrow{\text{ind}} \mathbb{Z} \tag{6.10}$$

defined as a composition of this product with the index mapping coincides with Poincaré duality defined earlier.

6.5. A topological index for $\mathbb{Z}_n$-manifolds. Clearly, definition (6.10) of duality in the previous subsection still contains one component defined analytically. This is the index mapping

$$\text{ind} : K_0(\mathcal{A}_{T^*M,\pi}) \longrightarrow \mathbb{Z}.$$

A topological formula for it was obtained in [**SS01**]. Let us briefly recall this result.

We suppose that the following geometric condition is satisfied: there is a free action of a finite group G on ∂M by diffeomorphisms such that π is the projection to the quotient space $\partial M/G$.

Consider two pairs (M, π) and (M', π').

DEFINITION 6.11. An *embedding* f of the pair (M, π) in (M', π') is an embedding of manifolds with boundary such that $f : M \to M'$, $f(\partial M) \subset \partial M'$ and the restriction of f to the boundary is a G-equivariant mapping.

An embedding induces the direct image mapping

$$f_! : K_*(\mathcal{A}_{T^*M,\pi}) \longrightarrow K_*(\mathcal{A}_{T^*M',\pi'})$$

of the K-groups.

For these $\mathbb{Z}_n$-manifolds, one can give a universal space in which an arbitrary $\mathbb{Z}_n$-manifold can be embedded, the embedding being unique up to homotopy. To this end, by $\pi_N : EG_N \longrightarrow BG_N$ we denote the N-universal bundle for the group G such that the spaces EG_N and BG_N are smooth compact manifolds without boundary. (For the existence of such models, e.g., see [**LM98**].)

PROPOSITION 6.12. *For a sufficiently large N, there exists an embedding f of (M,π) in $(EG_N \times [0,\infty), \pi_N)$. The embedding is unique up to homotopy.*

SKETCH OF PROOF. By the N-universality of the covering π_N, there exists an equivariant mapping $\partial M \to EG_N$. If the dimension of the space EG_N is sufficiently large, then a general position argument shows that slightly deforming this mapping one obtains a smooth embedding. Then, by virtue of the N-connectedness of EG_N, this extends to a mapping $M \to EG_N$. In turn, by a small deformation outside the boundary ∂M, this mapping can be made an embedding globally. □

The index theorem can be stated with the use of an embedding in the universal space. To this end, we introduce the K-group of the (infinite-dimensional) classifying space for $\mathbb{Z}_n$-manifolds as the direct limit

$$K_*(\mathcal{A}_{T^*EG\times[0,\infty),\pi_\infty}) = \varinjlim K_*(\mathcal{A}_{T^*EG_N\times[0,\infty),\pi_N}) \tag{6.11}$$

of the K-groups corresponding to the filtration of EG by skeletons.

THEOREM 6.13. [**SS01**] *The even K-group of the classifying space is isomorphic to $\mathbb{Z}$, and the index can be computed in terms of the direct image mapping:*

$$\operatorname{ind} D = f_![\sigma(D)],$$

$$f_![\sigma(D)] \in K_0(\mathcal{A}_{T^*EG\times[0,\infty),\pi_\infty}) \simeq \mathbb{Z}.$$

References

[Agr94] M. S. Agranovich, Elliptic operators on closed manifolds, In *Partial differential equations. VI*, volume 63 of *Encycl. Math. Sci.*, pp. 1–130, 1994.

[Ati61] M. F. Atiyah, *Characters and cohomology of finite groups*, Publ. Math. IHES **9** (1961), 23–64.

[Ati69] ———, Global theory of elliptic operators, In *Proc. of the Int. Symposium on Functional Analysis*, pp. 21–30, Tokyo, 1969. University of Tokyo Press.

[Ati89] ———, K-Theory, The Advanced Book Program. Addison–Wesley, Inc., second edition, 1989.

[AB64] M. F. Atiyah and R. Bott, The index problem for manifolds with boundary, In *Bombay Colloquium on Differential Analysis*, pp. 175–186, Oxford, 1964. Oxford University Press.

[ABP73] M. F. Atiyah, R. Bott, and V. K. Patodi, *On the heat equation and the index theorem*, Invent. Math. **19** (1973), 279–330.

[ADS83] M. F. Atiyah, H. Donnelly, and I. M. Singer, *Eta invariants, signature defects of cusps and values of L-functions*, Annals Math. **118** (1983), 131–177.

[APS75] M. Atiyah, V. Patodi, and I. Singer, *Spectral asymmetry and Riemannian geometry I*, Math. Proc. Cambridge Philos. Soc. **77** (1975), 43–69.

[APS76b] ———, *Spectral asymmetry and Riemannian geometry II*, Math. Proc. Cambridge Philos. Soc. **78** (1976), 405–432.

[APS76a] ———, *Spectral asymmetry and Riemannian geometry III*, Math. Proc. Cambridge Philos. Soc. **79** (1976), 71–99.

[AS63] M. F. Atiyah and I. M. Singer, *The index of elliptic operators on compact manifolds*, Bull. Amer. Math. Soc. **69** (1963), 422–433.

[AS68] ———, *The index of elliptic operators I*, Ann. of Math. **87** (1968), 484–530.

[BD82] P. Baum and R. G. Douglas, Toeplitz operators and Poincare duality, In *Toeplitz Centennial*, volume 4 of *Operator Theory, Adv. Appl.*, pp. 137–166, Toeplitz Mem. Conf. Tel Aviv 1981, 1982.

[BDT89] P. Baum, R. G. Douglas, and M. E. Taylor, *Cycles and relative cycles in analytic K-homology*, J. Differ. Geom. **30** (1989), no. 3, 761–804.

[Bla98] B. Blackadar, K-Theory for Operator Algebras, Number 5 in Mathematical Sciences Research Institute Publications, Cambridge University Press, 1998. Second edition.

[BBW93] B. Booß-Bavnbek and K. Wojciechowski, Elliptic Boundary Problems for Dirac Operators, Birkhäuser, Boston–Basel–Berlin, 1993.

[BT82] R. Bott and L. Tu, Differential Forms in Algebraic Topology, volume 82 of *Graduate Texts in Mathematics*, Springer–Verlag, Berlin–Heidelberg–New York, 1982.

[Bot92] B. Botvinnik, Manifolds with singularities and the Adams-Novikov spectral sequence, volume 170 of *London Mathematical Society Lecture Note Series*, Cambridge University Press, Cambridge, 1992.

[BdM71] L. Boutet de Monvel, *Boundary problems for pseudodifferential operators*, Acta Math. **126** (1971), 11–51.

[BDF77] L. Brown, R. Douglas, and P. Fillmore, *Extensions of C^*-algebras and K-homology*, Ann. Math. II **105** (1977), 265–324.

[BN94] J.-L. Brylinski and V. Nistor, *Cyclic cohomology of etale groupoids*, K-theory **8** (1994), 341–365.

[Con94] A. Connes, Noncommutative geometry, Academic Press Inc., San Diego, CA, 1994.

[DZ98] X. Dai and W. Zhang, *Higher spectral flow*, J. Funct. Anal. **157** (1998), no. 2, 432–469.

[Don78] H. Donnelly, *Eta-invariants for G-spaces*, Indiana Univ. Math. J. **27** (1978), 889–918.

[FM92] D. Freed and R. Melrose, *A mod k index theorem*, Invent. Math. **107** (1992), no. 2, 283–299.

[Gil81] P. B. Gilkey, *The residue of the global eta function at the origin*, Adv. in Math. **40** (1981), 290–307.

[Gil89a] ———, The geometry of spherical space form groups, In *Series in Pure Mathematics*, volume 7, Singapore, 1989, World Sientific Publ. Co. Pte. Ltd.

[Gil89b] ———, The eta invariant of even order operators, *Lecture Notes in Mathematics*, vol. 1410, pp. 202–211, 1989.

[Gil95] ———, Invariance theory, the heat equation, and the Atiyah-Singer index theorem, Studies in Advanced Mathematics. CRC Press, Boca Raton, FL, second edition, 1995.

[GS95] G. Grubb and R. Seeley, *Weakly parametric pseudodifferential operators and Atiyah–Patodi–Singer boundary problems*, Invent. Math. **121** (1995), 481–529.

[HR00] N. Higson and J. Roe, Analytic K-homology, Oxford University Press, Oxford, 2000.

[Hir73] F. Hirzebruch, *Hilbert modular surfaces*, Enseignement Math. **19** (1973), 183–281.

[Hör85] L. Hörmander, The Analysis of Linear Partial Differential Operators. III, Springer–Verlag, Berlin-Heidelberg-New York-Tokyo, 1985.

[Kas73] G. G. Kasparov, *The generalized index of elliptic operators*, Funct. Anal. Appl. **7** (1973), 238–240.

[Kas88] ———, *Equivariant KK-theory and the Novikov conjecture*, Invent. Math. **91** (1988), no. 1, 147–201.

[LM02] R. Lauter and S. Moroianu, *The index of cusp operators on manifolds with corners*, Ann. Global Anal. Geom. **21** (2002), no. 1, 31–49.

[LM89] H. B. Lawson and M. L. Michelsohn, Spin geometry, Princeton Univ. Press, Princeton, 1989.

[Les97] M. Lesch, Differential Operators of Fuchs Type, Conical Singularities, and Asymptotic Methods, volume 136 of *Teubner–Texte zur Mathematik*, B. G. Teubner Verlag, Stuttgart–Leipzig, 1997.

[LM98] G. Luke and A. S. Mishchenko, Vector bundles and their applications, volume 447 of *Mathematics and its Applications*, Kluwer Academic Publishers, Dordrecht, 1998.

[Mel93] R. Melrose, The Atiyah–Patodi–Singer Index Theorem, Research Notes in Mathematics, A. K. Peters, Boston, 1993.

[Mel95] ———, *The eta invariant and families of pseudodifferential operators*, Math. Research Letters **2** (1995), no. 5, 541–561.

[MN96] R. Melrose and V. Nistor, Homology of pseudodifferential operators I. Manifolds with boundary, Preprint, 1996.

[MP92] R. Melrose and P. Piazza, *Analytic K-theory on manifolds with corners*, Adv. in Math. **92** (1992), no. 1, 1–26.

[MP97] ———, *Families of Dirac operators, boundaries and the b-calculus*, J. of Diff. Geometry **46** (1997), no. 1, 99–180.

[Mül84] W. Müller, *Signature defects of cusps of Hilbert modular varieties and values of L-series at* $s = 1$, J. Diff. Geometry **20** (1984), 55–119.

[NSS99] V. Nazaikinskii, B.-W. Schulze, and B. Sternin, On the Homotopy Classification of Elliptic Operators on Manifolds with Singularities, Univ. Potsdam, Institut für Mathematik, Potsdam, Oktober 1999, Preprint no. 99/21.

[NSSS98] V. Nazaikinskii, B.-W. Schulze, B. Sternin, and V. Shatalov, *Spectral boundary value problems and elliptic equations on singular manifolds*, Differents. Uravnenija **34** (1998), no. 5, 695–708, English transl.: Differential Equations **34** (1998), no. 5, 696–710.

[Pal65] R. S. Palais, Seminar on the Atiyah–Singer index theorem, Princeton Univ. Press, Princeton, NJ, 1965.

[Phi96] J. Phillips, *Self-adjoint Fredholm operators and spectral flow*, Canad. Math. Bull. **39** (1996), no. 4, 460–467.

[RS82] S. Rempel and B.-W. Schulze, Index Theory of Elliptic Boundary Problems, Akademie–Verlag, Berlin, 1982.

[Sal95] D. Salamon, *The spectral flow and the Maslov index*, Bull. LMS **27** (1995), no. 1, 1–33.

[Sav99] A. Yu. Savin, *On operators that allow the decomposition of index formulas for spectral boundary value problems*, Doklady Mathematics **60** (1999), no. 2, 220–222.

[SSS99a] A. Savin, B.-W. Schulze, and B. Sternin, *On invariant index formulas for spectral boundary value problems*, Differentsial'nye uravnenija **35** (1999), no. 5, 709–718.

[SSS99b] ———, The Homotopy Classification and the Index of Boundary Value Problems for General Elliptic Operators, Univ. Potsdam, Institut für Mathematik, October 1999, Preprint no. 99/20, arXiv: math/9911055.

[SSS01] ———, On the homotopy classification of elliptic boundary value problems, In *Partial differential equations and spectral theory: PDE 2000 Conference in Clausthal, Germany*, volume 126 of *Operator theory: advances and applications*, pp. 299–306, Birkhäuser Basel-Boston-Berlin, 2001.

[SS99] A. Yu. Savin and B. Yu. Sternin, *Elliptic operators in even subspaces*, Matem. sbornik **190** (1999), no. 8, 125–160, English transl.: Sbornik: Mathematics **190** (1999), no. 8, 1195–1228, arXiv: math/9907027.

[SS00] ———, *Elliptic operators in odd subspaces*, Matem. sbornik **191** (2000), no. 8, 89–112, English transl.: Sbornik: Mathematics **191** (2000), no. 8, arXiv: math/9907039.

[SS01] ———, Index Defects in the Theory of Nonlocal Boundary Value Problems and the η-Invariant, Univ. Potsdam, Institut für Mathematik, Potsdam, November 2001, Preprint no. 01/31, arXiv: math/0108107.

[SS02] ———, *The eta-invariant and Pontryagin duality in K-theory*, Math. Notes **71** (2002), no. 2, 245–261, arXiv: math/0006046.

[SS03] ———, *The eta invariant and parity conditions*, Adv. in Math. (2003), (to appear), preliminary version at http://www.math.uni-potsdam.de/a_partdiff/prepr/2000_21.zip.

[Sch91] B.-W. Schulze, Pseudodifferential Operators on Manifolds with Singularities, North–Holland, Amsterdam, 1991.

[SSS98] B.-W. Schulze, B. Sternin, and V. Shatalov, *On general boundary value problems for elliptic equations*, Math. Sb. **189** (1998), no. 10, 145–160, English transl.: Sbornik: Mathematics **189** (1998), no. 10, 1573–1586.

[See67] R. T. Seeley, *Complex powers of an elliptic operator*, Proc. Sympos. Pure Math. **10** (1967), 288–307.

[Sul70] D. Sullivan, Geometric Topology. Localization, Periodicity and Galois Symmetry, MIT, Cambridge, Massachusets, 1970.

Independent University of Moscow, Moscow 121002, Bolshoi Vlasevsky Pereulok, Dom 11, Russia
E-mail address: antonsavin@mtu-net.ru

Independent University of Moscow, Moscow 121002, Bolshoi Vlasevsky Pereulok, Dom 11, Russia
E-mail address: sternin@mail.ru

Operator Theory:
Advances and Applications, Vol. 151, 239–264

Cyclic homology and pseudodifferential operators, a survey

Moulay-Tahar Benameur, Jacek Brodzki, and Victor Nistor

ABSTRACT. We present a brief introduction to Hochschild and cyclic homology designed for researchers interested in pseudodifferential operators and their applications to index theory, spectral invariants, and asymptotics.

Introduction

Singular cohomology is often used in Algebraic Topology to obtain invariants of topological spaces. In the same spirit, Hochschild and cyclic homology often provide interesting invariants of algebras. A possible important application of these algebra invariants is the study of spaces with additional structures; these include, for instance, spaces with singularities or spaces endowed with group actions.

This paper is a rapid survey of Hochschild and cyclic homology, designed for mathematicians and physicists interested in pseudodifferential operators and their applications to index theory, spectral invariants, and asymptotics. We assume only very little familiarity with the subject, including homological algebra. Moreover, we have also included some short proofs when we felt that they are particularly helpful to the reader.

Hochschild homology is usually thought of as a generalization of the notion of differential forms and (periodic) cyclic homology as a generalization of (de Rham) cohomology, from smooth compact manifolds to essentially arbitrary algebras. The main guiding principle of this development is the correspondence

$$\text{“}\mathit{Space}\text{”} \quad \leftrightarrow \quad \text{“}\mathit{Algebra\ of\ functions\ on\ that\ space}\text{”},$$

which is at the heart of non-commutative geometry [**5, 20, 21, 31, 34, 37, 38, 42, 43, 44**]. This principle had been used before in Algebraic Geometry where this becomes the correspondence (contravariant equivalence of categories) between affine algebraic varieties over a field $\mathfrak{k}$ and commutative, reduced, finitely generated

2000 *Mathematics Subject Classification.* 58J42.

Benameur was partially supported by CNRS grant. Nistor was partially supported by NSF Grant DMS-9971951 and a “collaborative research” grant.

algebras over $\mathfrak{k}$. In Functional Analysis this principle is illustrated by the Gelfand–Naimark equivalence between the categories of compact topological spaces and of commutative, unital C^*-algebras. One of the main ideas of non-commutative geometry is that this correspondence somehow extends also to (classes of) non-commutative algebras. The theory of Quantum groups [**39, 42**] is a truly remarkable example of how this principle works in practice. Another remarkable example is that the singular cohomology of spaces and the long exact cohomology sequences of pairs of spaces extend to algebras – this is the Excision theorem in periodic cyclic homology [**23**].

Our main example of how this principle works in practice will be centered upon algebras like $\mathcal{C}^\infty(M)$ and $\Psi^\infty(M)$, where M is a smooth compact manifold and $\Psi^\infty(M)$ denotes the algebra of classical pseudodifferential operators of integral order on M. We shall see that the Hochschild homology of $\mathcal{C}^\infty(M)$ can be identified with the space of differential forms on M, whereas the periodic cyclic homology of $\mathcal{C}^\infty(M)$ is the same as the de Rham cohomology of M. The results on $\mathcal{C}^\infty(M)$ are interesting not only on their own, but also because they lead to a calculation of the Hochschild, cyclic, and periodic cyclic homology of algebras of pseudodifferential operators.

The contents of the paper are as follows. In Section 1, we define Hochschild homology and give some elementary properties and concrete computations. In Section 2, we introduce the cyclic homology and the Connes-Karoubi Chern character. We also recall the Cuntz-Quillen Excision theorem together with an abstract cyclic cohomological index theorem. In Section 3, we extend our homologies and the complexes defining them to the setting of Fréchet algebras and discover, in particular, that when dealing with topological algebras, choice of a suitable topological tensor product becomes important. Section 4 deals with algebras of complete pseudodifferential symbols on groupoids with corners. There, we give the computation of periodic cyclic homology and Hochschild homology for such algebras. Finally, Section 5 applies these results to some geometric examples in non-commutative geometry. In particular, it relates Hochschild homology computations with local residues. No results in this paper are new.

We have not attempted to provide a comprehensive list of references, restricting instead to the sources most relevant to the results presented here. The main sources of general information on cyclic type homology theories are the books [**9, 21, 31, 37**], which contain detailed bibliographic information.

We thank the referee for carefully reading our manuscript and for making several useful suggestions.

1. Hochschild homology

We begin by recalling the definitions of Hochschild homology groups of an algebra A, which we shall assume to be defined over the field of complex numbers. Later, we shall treat the case of algebras equipped with topology.

Let A be an algebra, not necessarily with unit. A *trace* on A is a linear map $\tau : A \to \mathbb{C}$ such that

$$\tau(a_0a_1 - a_1a_0) = 0$$

for all $a_0, a_1 \in A$. This definition captures an important property of the usual matrix trace Tr, defined on the algebra $A = M_n(\mathbb{C})$ by $Tr([a_{ij}]) = \sum a_{ii}$. The space of all traces on A is denoted $\mathrm{HH}^0(A)$, and is also called the *0th Hochschild cohomology group*, for reasons that we explain now.

We shall need two differential complexes associated with the algebra A, denoted $\mathcal{H}(A)$ and $\mathcal{H}'(A)$. The space of n-chains for both will be the $n+1$-fold tensor product of A by itself. The two complexes are equipped with differentials b' and b defined as follows:

$$(1.1)\qquad \begin{aligned} b'(a_0 \otimes a_1 \otimes \ldots \otimes a_n) &= \sum_{i=0}^{n-1}(-1)^i a_0 \otimes \ldots \otimes a_i a_{i+1} \otimes \ldots \otimes a_n, \\ b(a_0 \otimes a_1 \otimes \ldots \otimes a_n) &= b'(a_0 \otimes a_1 \otimes \ldots \otimes a_n) + (-1)^n a_n a_0 \otimes \ldots \otimes a_{n-1}, \end{aligned}$$

where $a_0, a_1, \ldots, a_n \in A$. Thus b and b' define linear maps

$$b,\, b' \,:\, A^{\otimes n+1} \to A^{\otimes n}.$$

By definition, *Hochschild homology groups* of the algebra A, denoted $\mathrm{HH}_n(A)$, are the homology groups of the complex

$$(1.2)\qquad \mathcal{H}(A) := (A^{\otimes n+1}, b).$$

By contrast, the complex

$$(1.3)\qquad \mathcal{H}'(A) := (A^{\otimes n+1}, b')$$

is often acyclic. This happens, for example when A has a unit, because

$$s(a_0 \otimes a_1 \otimes \ldots \otimes a_n) = 1 \otimes a_0 \otimes a_1 \otimes \ldots \otimes a_n$$

is a homotopy between 0 and the identity:

$$b's + sb' = 1.$$

In this case the complex $\mathcal{H}'(A)$ is a resolution of A by free A-bimodules. It has been noted by Wodzicki [**61**] that nonunital algebras whose $\mathcal{H}'(A)$ complex is acyclic have useful properties. Algebras with this property are now called *H-unital* (homologically unital).

The cohomology groups of the dual complex

$$\mathcal{H}^*(A) := (\mathrm{Hom}(A^{\otimes n+1}, \mathbb{C}), b^*)$$

are called the *Hochschild cohomology groups* of A and denoted $\mathrm{HH}^n(A)$. Since τ is a trace if, and only if, $\tau \circ b = 0$, we obtain that $\mathrm{HH}^0(A)$ is indeed the space of traces on A.

Clearly the groups $\mathrm{HH}_n(A)$ are *covariant* functors in A, in the sense that any algebra morphism $\phi : A \to B$ induces a morphism

$$\phi_* : \mathrm{HH}_n(A) \to \mathrm{HH}_n(B)$$

for any integer $n \geq 0$. Similarly, we also obtain a morphism $\phi^* : \mathrm{HH}^n(B) \to \mathrm{HH}^n(A)$. In other words, Hochschild cohomology is a *contravariant* functor. It is interesting to note that if Z is the center of A, then $\mathrm{HH}_n(A)$ is also a Z-module, where, at the level of complexes the action is given by

$$z(a_0 \otimes a_1 \otimes \ldots \otimes a_n) = za_0 \otimes a_1 \otimes \ldots \otimes a_n. \tag{1.4}$$

for all $z \in Z$. As z is in the center of A, this action will commute with the Hochschild differential b.

When computing the Hochschild homology of an algebra A, it is often useful to replace the original complex $\mathcal{H}(A)$ with an equivalent complex. This can be achieved, for example, using the next Proposition, which is a standard result from homological algebra (see [**40, 41**], for instance). Before we can state that proposition, we need to recall the following terminology. An *A-bimodule* is an abelian group M which is at the same time a left- and right- A-module. A *free* A-bimodule is one of the form $A \otimes V \otimes A$, where V is a vector space and

$$a_0\big(a_0' \otimes v \otimes a_1'\big)a_1 = (a_0a_0') \otimes v \otimes (a_1'a_1).$$

Let $f : A \otimes V_1 \otimes A \to A \otimes V_0 \otimes A$ be a morphism of free bimodules, and assume that $f(1 \otimes v \otimes 1) = \sum a_k \otimes w_k \otimes a_k'$. We shall denote by $f \otimes 1$ the map $A \otimes V_1 \to A \otimes V_0$ defined by

$$(f \otimes 1)(a \otimes v) = \sum a_k' a a_k \otimes w_k$$

Let A^{opp} be A as a vector space but with multiplication $\star$ given by $a \star a' := a'a$. Let $A^e := A \otimes A^{opp}$. The reader familiar with balanced tensor products will note that the above construction amount to taking the tensor product $- \otimes_{A^e} A$ of f and the respective free modules. (Note that a free A-bimodule is the same thing as a free A^e-module.)

PROPOSITION 1.1. *Let A be a unital algebra and*

$$A \leftarrow A \otimes V_0 \otimes A \xleftarrow{d_1} A \otimes V_1 \otimes A \xleftarrow{d_2} \ldots$$

be an exact complex of A-bimodules. Then the homology groups of the complex

$$A \otimes V_0 \xleftarrow{d_1 \otimes 1} A \otimes V_1 \xleftarrow{d_2 \otimes 1} \ldots$$

are naturally isomorphic to the groups $\mathrm{HH}_n(A)$.

For example, when A is unital,[1] the complex $\mathcal{H}'(A)$ is exact, and when we apply to it the above proposition we obtain the Hochschild complex $\mathcal{H}(A)$. Another useful corollary of this proposition is derived when we use the exactness of the localization functor. The following result is due to Brylinski [**14**].

PROPOSITION 1.2. *Let S be a multiplicative subset of the center Z of the algebra A. Then* $\mathrm{HH}_*(S^{-1}A) \cong S^{-1}\,\mathrm{HH}_*(A)$.

[1]we assumed A unital in the definition of $f \otimes 1$

Let us use now Proposition 1.1 to determine the Hochschild homology of the algebra $\mathcal{O}[\mathbb{C}^n] := \mathbb{C}[X_1, \ldots, X_n]$ of polynomials in n-variables. We first need to recall the algebraic de Rham complex. Let W^* be the complex vector space generated by the 1-forms $dX_1, \ldots, dX_n$:

$$W^* = \mathbb{C}dX_1 \oplus \ldots \oplus \mathbb{C}dX_n$$

We define the space of *q-forms on $\mathbb{C}^n$* to be $\Omega^q(\mathbb{C}^n) = \mathcal{O}[\mathbb{C}^n] \otimes \Lambda^q W^*$. Here $\mathbb{C}^n$ is of course regarded as an affine algebraic variety. Then the de Rham differential

$$d : \Omega^q(\mathbb{C}^n) \to \Omega^{q+1}(\mathbb{C}^n)$$

is defined by

$$d\Big(PdX_{i_1} \ldots dX_{i_q}\Big) = \Big(\frac{\partial P}{\partial X_1}dX_1 + \ldots + \frac{\partial P}{\partial X_n}dX_n\Big)dX_{i_1} \ldots dX_{i_q}.$$

(We omitted the $\wedge$ in the products above.) We can now define the *Hochschild-Kostant-Rosenberg-Connes map (HKRC-map)*

$$\chi : A^{\otimes q+1} \to \Omega^q(\mathbb{C}^n),$$

$A = \mathcal{O}[\mathbb{C}^n] = \mathbb{C}[X_1, \ldots, X_n]$, by the formula

$$\chi(a_0 \otimes \ldots \otimes a_q) = \frac{1}{q!}a_0 da_1 \ldots da_q. \tag{1.5}$$

PROPOSITION 1.3. *We have $\chi \circ b = 0$ and hence χ defines an isomorphism $\chi_* : \mathrm{HH}_q(A) \to \Omega^q(\mathbb{C}^n)$.*

The proof of this proposition is a paradigm for the computation of the Hochschild homology of commutative algebras, so we present it in detail.

PROOF. Our main tool will be the Koszul complex, which provides a general recipe for building resolutions of $A = \mathcal{O}[\mathbb{C}^n]$. We construct it as follows. We fix the sequence $(v_1, v_2, \ldots, v_n)$, $v_j = X_j \otimes 1 - 1 \otimes X_j$ of elements of $A \otimes A$. (This is a *regular sequence*, see **[27]** for the definition). Let V be the vector space with basis the elements v_j of this regular sequence. Also, let $\mathcal{K}_q = A \otimes \Lambda^q V \otimes A$. The spaces $\mathcal{K}_q$ will form the Koszul complex relative to the sequence $(v_1, \ldots, v_n)$. The differentials of the Koszul complex are given by the following formula

$$\delta(a \otimes v_{i_1} \wedge \ldots \wedge v_{i_q} \otimes a') := \sum_{j=1}^{q} (-1)^{j-1} \Big[aX_{i_j} \otimes v_{i_1} \wedge \ldots \wedge \widehat{v_{i_j}} \wedge \ldots \wedge v_{i_q} \otimes a' - a \otimes v_{i_1} \wedge \ldots \wedge \widehat{v_{i_j}} \wedge \ldots \wedge v_{i_q} \otimes X_{i_j} a'\Big]$$

where $\widehat{v}$ means that the symbol v is to be omitted.

The Koszul complex $\mathcal{K}$ provides a resolution of A by free $A \otimes A$-modules (this follows from the fact that the sequence v_j is regular). In other words, the complex $\mathcal{K}$ has vanishing homology, except in dimension zero, where its homology is isomorphic to A via the multiplication map

$$\mathcal{K}_0 = A \otimes A \ni a \otimes a' \to aa' \in A\,.$$

Let $w = v_{i_1} \wedge \ldots \wedge v_{i_q} \in \Lambda^q V$. Define $f : \mathcal{K} \to \mathcal{H}'(A) := (A^{\otimes q+2}, b')$ by

$$f(a \otimes w \otimes a') := \sum_{\sigma \in S_q} \operatorname{sign}(\sigma) a \otimes v_{i_{\sigma(1)}} \otimes v_{i_{\sigma(2)}} \otimes \ldots \otimes v_{i_{\sigma(q)}} \otimes a',$$

where the sum is taken over the symmetric group on q letters and $\operatorname{sign}(\sigma)$ denotes the sign of the permutation σ. The map f is a chain map, which means that it commutes with the differentials:

$$f\delta = b'f,$$

and hence it induces a morphism of complexes (or, in this case, resolutions). The morphism f is the identity on $\mathcal{K}_0 = A \otimes A = \mathcal{H}'_0(A)$ and hence it induces the identity on the homology groups in dimension zero, that is, on the algebra A. (This is a standard homological fact that can easily be proved as an exercise.)

Using the remark just before Proposition 1.1 we see that the map

$$f \otimes 1 : \mathcal{K} \otimes_{A^e} A \to \mathcal{H}'(A) \otimes_{A^e} A \cong \mathcal{H}(A)$$

will also induce an isomorphism in homology.

The result then follows because $\delta \otimes 1 = 0$ and $\chi \circ f$ is the identity, if we identify V with W^* by $v_j \to dX_j$. □

The same argument can be used to prove the following result (see [**29, 38**]).

THEOREM 1.4. *Let Y be a smooth, complex algebraic variety and $\mathcal{O}[Y]$ be the ring of regular functions on Y and $\Omega^q(Y)$ be the space of algebraic q-forms on Y. Then the HKRC map χ induces isomorphisms*

$$\chi : \mathrm{HH}_q(\mathcal{O}[Y]) \to \Omega^q(Y)$$

for all $q \geq 0$.

Another important class of examples deals with algebras associated with groups. The simplest situation is probably the following. Let Γ be a group (we do not assume that Γ has topology) and let $\mathbb{C}[\Gamma]$ be the group algebra of Γ. By definition, $\mathbb{C}[\Gamma]$ is the space of finite linear combinations $\sum_\gamma a_\gamma \gamma$, $\gamma \in \Gamma$, $a_\gamma \in \mathbb{C}$ equipped with the product $(a\gamma)(a'\gamma') := aa'(\gamma\gamma')$. For any $\gamma \in \Gamma$ we denote by

$$\Gamma_\gamma := \{g \in \Gamma, g\gamma = \gamma g\},$$

the *centralizer* of γ. Let $\mathrm{H}_q(\Gamma_\gamma)$ be the qth homology group of Γ_γ with complex coefficients (it is the homology with complex coefficients of any CW-complex whose fundamental group is Γ_γ and all other homotopy groups vanish). Let $\langle\Gamma\rangle$ denote the set of conjugacy classes of Γ and $\langle\gamma\rangle$ denote the conjugacy class of $\gamma \in \Gamma$. Then [**17**]:

THEOREM 1.5. *Let Γ be a discrete group. Then*

$$\mathrm{HH}_q(\mathbb{C}[\Gamma]) \simeq \bigoplus_{\langle\gamma\rangle \in \langle\Gamma\rangle} \mathrm{H}_q(\Gamma_\gamma).$$

In the above sum, we take exactly one γ from each conjugacy class of Γ. Also, note that the groups $\mathrm{H}_q(\Gamma_x)$ and $\mathrm{H}_q(\Gamma_y)$ are *canonically* isomorphic if x and y are conjugate in Γ.

These results extend to crossed products. See for example [**16, 18**] for the calculation of the Hochschild, cyclic, and periodic cyclic homology of crossed product algebras of the form $\mathcal{C}^\infty(M) \rtimes \Gamma$, where Γ is a discrete group acting on M by diffeomorphisms. Also, see [**3**] for some applications of these results to orbifold cohomology as well as [**24, 51**] for results on general crossed products $A \rtimes \Gamma$, with A not necessarily commutative. (When [**51**] was being written, the paper [**24**] was not available in Romanian libraries, which explains the overlap in results between the two papers.)

Hochschild homology is also compatible with completions, in the following sense. Let A be a finitely generated module over its center $\mathfrak{k}$ and let $\hat{M}$ denote the completion of a $\mathfrak{k}$–module M with respect to the topology defined by an ideal $I \subset \mathfrak{k}$, then we have [**33**]

THEOREM 1.6. *Denote by* $\mathrm{HH}_q^{top}(\hat{A})$ *the homology of the completion of the Hochschild complex of* A *with respect to the natural filtration defined by* I^kA. *Then*

$$\mathrm{HH}_q^{top}(\hat{A}) \simeq \widehat{\mathrm{HH}}_q(A).$$

2. Cyclic homology

We now define cyclic homology, which, as was the case with Hochschild homology, will appear as the homology of an explicitly defined differential complex. The cyclic complex will be introduced as the total complex of a double complex constructed using the Hochschild complex $\mathcal{H}(A)$, together with a new differential B whose definition uses in a crucial way the action of a cyclic group of a suitable size.

Let us assume first that A is a unital algebra. We shall denote by t the (signed) generator of cyclic permutations:

$$t(a_0 \otimes a_1 \otimes \ldots \otimes a_n) = (-1)^n a_n \otimes a_0 \otimes \ldots \otimes a_{n-1} \tag{2.1}$$

Using this operator and the contracting homotopy s of the complex $\mathcal{H}'(A)$ we construct the operator B of degree $+1$ in two steps. First define

$$B_0(a_0 \otimes a_1 \otimes \ldots \otimes a_n) = s \sum_{k=0}^{n} t^k(a_0 \otimes a_1 \otimes \ldots \otimes a_n) \tag{2.2}$$

and then put $B = (1-t)B_0$. Then B is a differential, $B^2 = 0$ and it anticommutes with the Hochschild differential b: $[b, B]_+ := bB + Bb = 0$.

We have thus defined a double complex which fills up an octant in the plane, see Figure 1. Note that columns in this complex are copies of the Hochschild complex $\mathcal{H}(A)$. The cyclic complex $\mathcal{C}(A)$ is by definition the total complex of this

$$
\begin{array}{ccccc}
\downarrow b & & \downarrow b & & \downarrow b \\
A^{\otimes 3} & \xleftarrow{B} & A^{\otimes 2} & \xleftarrow{B} & A \\
\downarrow b & & \downarrow b & & \\
A^{\otimes 2} & \xleftarrow{B} & A & & \\
\downarrow b & & & & \\
A & & & &
\end{array}
$$

FIGURE 1. The cyclic bicomplex of the algebra A.

double complex. Thus the space of cyclic n-chains is defined by

$$\mathcal{C}(A)_n = \bigoplus_{k\geq 0} \mathcal{H}_{n-2k}(A), \tag{2.3}$$

we see that $(\mathcal{C}(A), b + B)$, is a complex, called *the cyclic complex* of A, whose homology is by definition the *cyclic homology* of A (see [**20, 31, 60**]).

There is a canonical operator $S : \mathcal{C}_n(A) \to \mathcal{C}_{n-2}(A)$, called the *Connes periodicity operator*, which shifts the cyclic complex down by two degrees, and which is defined by $S(\omega_n, \omega_{n-2}, \omega_{n-4}, \dots) \mapsto (\omega_{n-2}, \omega_{n-4}, \dots)$, where $\omega_i \in \mathcal{H}_i(A)$. This operator induces the short exact sequence of complexes

$$0 \to \mathcal{H}(A) \xrightarrow{I} \mathcal{C}(A) \xrightarrow{S} \mathcal{C}(A)[2] \to 0,$$

where the map I is the inclusion of the Hochschild complex as the first column of the cyclic complex and where on the right the cyclic complex has been shifted by two degrees. Applying the homology functor to this sequence we obtain the following exact sequence of Connes and Tsygan:

$$\dots \to \mathrm{HH}_n(A) \xrightarrow{I} \mathrm{HC}_n(A) \xrightarrow{S} \mathrm{HC}_{n-2}(A) \xrightarrow{B} \mathrm{HH}_{n-1}(A) \xrightarrow{I} \dots, \tag{2.4}$$

where B is the differential defined above, see [**20, 38**] for more details. Anticipating a little bit, this exact sequence exists whether or not A is endowed with a topology. The *periodic cyclic complex* of an algebra A is the complex

$$\mathcal{C}^{\mathrm{per}}(A) := \varprojlim \mathcal{C}(A),$$

the inverse limit being taken with respect to the periodicity morphism S. It is a $\mathbb{Z}/2\mathbb{Z}$-graded complex, whose chains are (possibly infinite) sequences of Hochschild chains with degrees of the same parity.

More generally, cyclic homology groups can be defined for "mixed complexes," [**32**]. A *mixed complex* $(\mathcal{X}, b, B)$ is a $\mathbb{Z}_+$-graded complex vector space $\mathcal{X} = (\mathcal{X}_n)_{n\geq 0}$

endowed with two differentials b and B, $b : \mathcal{X}_n \to \mathcal{X}_{n-1}$ and $B : \mathcal{X}_n \to \mathcal{X}_{n+1}$, satisfying

$$b^2 = B^2 = bB + Bb = 0.$$

The *cyclic complex*, $\mathcal{C}(\mathcal{X})$, associated to a mixed complex $(\mathcal{X}, b, B)$ is the complex

$$\mathcal{C}_n(\mathcal{X}) = \mathcal{X}_n \oplus \mathcal{X}_{n-2} \oplus \mathcal{X}_{n-4} \dots = \bigoplus_k \mathcal{X}_{n-2k}$$

with differential $b + B$. It is the total complex associated to the bicomplex shown in Figure 2. The cyclic homology groups of the mixed complex $\mathcal{X}$ are the homology groups of the cyclic complex of $\mathcal{X}$,

$$\mathrm{HC}_n(\mathcal{X}) = \mathrm{H}_n(\mathcal{C}(\mathcal{X}), b + B).$$

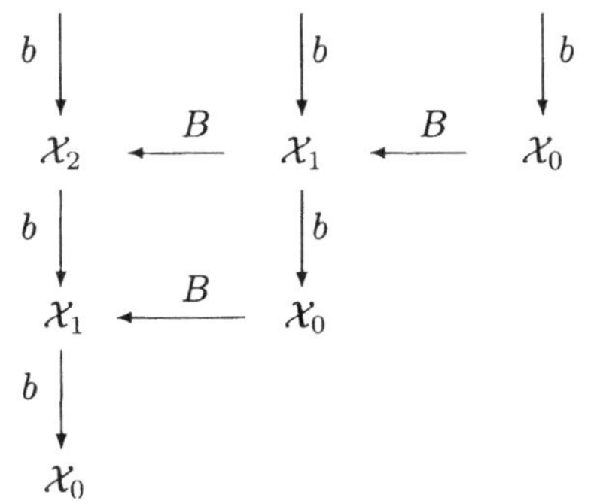

FIGURE 2. The bicomplex associated with a mixed complex.

We review below a few necessary results on mixed complexes. The interested reader can find more information in [**30**].

The homology of the complex $(\mathcal{X}_n, b)$ is called the Hochschild homology of the mixed complex $(\mathcal{X}, b, B)$ and is denoted $\mathrm{HH}_*(\mathcal{X})$. If A is an algebra with unit and if we let $\mathcal{X}_n = A^{\otimes n+1}$, with the differentials b and B defined above, then we recover the usual definitions of the Hochschild and cyclic homology groups of A. If A is an arbitrary algebra, denote by $A^+ = A + \mathbb{C}$ the algebra with an adjoined unit and let $\mathcal{X}_n(A) = \ker(\mathcal{X}_n(A^+) \to \mathcal{X}_n(\mathbb{C}))$. The cyclic and Hochschild homology groups of the resulting mixed complex $(\mathcal{X}_n(A), b, B)$ are, by definition, the cyclic and Hochschild homology groups of the not necessarily unital algebra A. When A is a unital algebra, the two ways of defining Hochschild and cyclic homology groups agree. Cyclic and Hochschild homology are covariant functors on the category of mixed complexes: a morphism $f : A \to B$ of algebras induces a morphism of the corresponding mixed complexes and hence it induces morphisms $f_* : \mathrm{HC}_*(A) \to \mathrm{HC}_*(B)$ and $f_* : \mathrm{HH}_*(A) \to \mathrm{HH}_*(B)$.

The cyclic complex of a mixed complex $(\mathcal{X}, b, B)$ has a periodicity endomorphism $S : \mathcal{C}_n(\mathcal{X}) \to \mathcal{C}_{n-2}(\mathcal{X})$ which maps $\mathcal{X}_n$ to 0 and is the identity on the other factors, as was seen in the case of the cyclic complex $\mathcal{C}(A)$. This defines a map $\mathrm{HC}_n(\mathcal{X}) \longrightarrow \mathrm{HC}_{n-2}(\mathcal{X})$, also denoted by S. The homology groups $\mathrm{HP}_i(\mathcal{X})$ of the projective limit complex

$$\mathcal{C}_n^{\mathrm{per}}(\mathcal{X}) = \lim_{\leftarrow} \mathcal{C}_{n+2k}(\mathcal{X}) \tag{2.5}$$

with respect to the periodicity operator S are called the *periodic cyclic homology* groups of $\mathcal{X}$:

$$\mathrm{HP}_i(\mathcal{X}) = \mathrm{H}_i(\mathcal{C}_n^{\mathrm{per}}(\mathcal{X}))\,. \tag{2.6}$$

Periodic cyclic homology is a $\mathbb{Z}/2\mathbb{Z}$–graded homology theory, $\mathrm{HP}_i = \mathrm{HP}_{i+2}$. Standard homological algebra shows that cyclic and periodic cyclic homology fit into the following $\lim^1$ exact sequence

$$0 \longrightarrow \lim_{\leftarrow}{}^1 \mathrm{HC}_{m+1}(\mathcal{X}) \longrightarrow \mathrm{HP}_m(\mathcal{X}) \longrightarrow \lim_{\leftarrow} \mathrm{HC}_m(\mathcal{X}) \longrightarrow 0. \tag{2.7}$$

We shall write $\mathrm{HP}_m(A)$ for the periodic cyclic homology of the (mixed complex associated to the) algebra A. Again, functorial properties of mixed complexes show that the periodic cyclic homology of an algebra is a covariant functor.

Cyclic cohomology and *periodic cyclic cohomology* are defined by duality and, in case there is no topology, they are the duals of the corresponding homology groups.

The following standard lemma is useful for many calculations.

LEMMA 2.1. *Let $f : X \to X'$ be a morphism of mixed complexes (*i.e. *commuting with b and B) that induces an isomorphism of the Hochschild homology groups. Then f induces an isomorphism of the cyclic homology, periodic cyclic homology, Hochschild cohomology, cyclic cohomology, and periodic cyclic cohomology groups.*

PROOF. Consider filtrations of the cyclic complexes associated to X and X' by the columns of the corresponding double complexes. These filtrations define convergent spectral sequences and f is an isomorphism of these spectral sequences. The standard comparison theorem for spectral sequences is now enough to prove the isomorphisms of the cyclic homology groups of X and X'. (Alternatively, one can use here the SBI-exact sequence.) For periodic cyclic homology, use also the $\lim^1$ exact sequence above relating cyclic and periodic cyclic homologies.

The cohomology groups are dual (in this topology free setting) of the corresponding homology groups. □

Here is an application of this lemma. Consider

$$Tr_* : \mathcal{H}_q(M_N(A)) \to \mathcal{H}_q(A), \quad q \in \mathbb{Z}_+,$$

the map defined by $Tr_*(b_0 \otimes \ldots \otimes b_q) = Tr(m_0 m_1 \ldots m_q) a_0 \otimes \ldots \otimes a_q$, if $b_k = m_k \otimes a_k \in M_N(\mathbb{C}) \otimes A = M_N(A)$. Also consider the (unital) inclusion $\iota : A \to M_N(A)$ and ι_* be the morphism induced on the Hochschild complexes.

PROPOSITION 2.2. *The map Tr_* is a morphism of mixed complexes. Both ι_* and Tr_* induce isomorphisms on Hochschild, cyclic, and periodic cyclic homologies and cohomologies such that $(\iota_*)^{-1} = N^{-1} Tr_*$.*

PROOF. An easy calculation shows that Tr_* is a morphism of mixed complexes. The map ι_* induces an isomorphism in Hochschild homology, by the Künneth formula for Hochschild homology [**37**, 1.0.16, 1.2.4]. Thus ι_* induces an isomorphism of cyclic and periodic cyclic homology as well. The equation $Tr_*\iota_* = N$ is immediate and gives the rest of the proposition. $\square$

For commutative algebras, cyclic homology can be calculated using the Hochschild homology and the equation $\chi \circ B = d_{DR} \circ \chi$, relating the de Rham differential d_{DR} and the differential B via the HKRC map χ. Let $A = \mathcal{O}[Y]$ be the algebra of regular functions on a smooth complex algebraic variety, as in the statement of Theorem 1.4. To compute $\mathrm{HC}_q(A)$ and the other groups associated to A, we first notice that χ defines a morphism

$$(A^{\otimes q+1}, b, B) \to (\Omega^q(Y), 0, d_{DR})$$

of mixed complexes, which is an isomorphism on Hochschild homology, by the same theorem of Loday and Quillen used above. This then gives the following.

THEOREM 2.3. *Let Y be a smooth complex algebraic variety and $\mathrm{H}_q(Y)$ be its singular homology groups with complex coefficients. If $A = \mathcal{O}[Y]$, then*

$$\mathrm{HC}_q(A) \simeq \Omega^q(Y)/d\Omega^{q-1}(Y) \oplus \bigoplus_{j=1}^{[q/2]} \mathrm{H}_{q-2j}(Y),$$

with the periodicity morphism S identifying with the natural projection. In particular, $\mathrm{HP}_q(A) \simeq \bigoplus_{j\in\mathbb{Z}} \mathrm{H}_{q-2j}(Y)$.

Let us mention that the result on *periodic* cyclic homology of the above theorem extends to the case when Y is not necessarily smooth by a result of [**25**], but the proof is more difficult. (See [**33**] for a proof of this result in the spirit of this paper, using Theorem 1.6.)

We shall use these calculations to construct Chern characters. By taking Y to be a point, we obtain that $\mathrm{HC}_{2q}(\mathbb{C}) \simeq \mathbb{C}$ and $\mathrm{HC}_{2q+1}(\mathbb{C}) \simeq 0$. We can take these isomorphisms to be compatible with the periodicity operator S and such that for $q = 0$ it reduces to

$$\mathrm{HC}_0(\mathbb{C}) = \mathrm{HH}_0(\mathbb{C}) = \mathbb{C}/[\mathbb{C}, \mathbb{C}] = \mathbb{C}.$$

We shall denote by $\eta_q \in \mathrm{HC}_{2q}(\mathbb{C})$ the unique element such that

$$S^q \eta_q = 1 \in \mathbb{C} = \mathrm{HC}_0(\mathbb{C}).$$

If $e \in M_N(A)$ is a projection, it will induce a (non-unital) morphism $\psi : \mathbb{C} \to M_N(A)$ by $\lambda \mapsto \lambda e$. Then the *Connes-Karoubi Chern character of e in cyclic homology* [**20, 31**] is defined by

$$Ch_q([e]) = Tr_*(\psi_*(\eta_q)) \in \mathrm{HC}_{2q}(A). \tag{2.8}$$

This map can be shown to depend only on the class of e in K-theory and to define a morphism

$$Ch_q : K_0(A) \to \mathrm{HC}_{2q}(A).$$

One can define similarly the Chern character in periodic cyclic homology and the Chern character on K_1 (algebraic K-theory). For the Connes-Karoubi Chern character on K_1, we use instead $Y = \mathbb{C}^*$, whose algebra of regular functions is $\mathcal{O}[Y] \simeq \mathbb{C}[z, z^{-1}]$, the algebra of Laurent polynomials in z and z^{-1} (this algebra, in turn, is isomorphic to the group algebra of $\mathbb{Z}$). Then $\mathrm{HC}_q(\mathcal{O}[Y]) \simeq \mathbb{C}$, for any $q \geq 1$. We are interested in the odd groups, which will be generated by elements $v_k \in \mathrm{HC}_{2k+1}(\mathcal{O}[Y])$, which can be chosen to satisfy $v_1 = z^{-1} \otimes z$ and $S^k v_{2k+1} = v_1$.

Then, if $u \in M_N(A)$ is an invertible element, it defines a morphism ψ : $\mathbb{C}[\mathbb{C}^*] \to M_N(A)$. The *Connes-Karoubi Chern character of u in cyclic homology* is thus defined by

$$Ch_q([u]) = Tr_*(\psi_*(v_q)) \in \mathrm{HC}_{2q+1}(A). \tag{2.9}$$

Again, this map can be shown to depend only on the class of u in K-theory and to define a morphism

$$Ch_q : K_1(A) \to \mathrm{HC}_{2q+1}(A).$$

Both the Chern character on K_0 and on K_1 are functorial, by construction.

Cyclic homology behaves to a large extent like Hochschild homology. Periodic cyclic homology however has a few additional properties that often make it easier to compute. The most important among them is the excision property, which we state here, although we shall not need it in this paper. The proof of excision in periodic cyclic homology relies in a crucial way on the following theorem of Goodwillie. (see [**26**]).

THEOREM 2.4. *If $I \subset A$ is a nilpotent two-sided ideal, then the quotient morphism $A \to A/I$ induces an isomorphism* $\mathrm{HP}_*(A) \to \mathrm{HP}_*(A/I)$.

A deep and far reaching consequence of Goodwillie's theorem, and also of Wodzicki's Excision theorem in Hochschild homology [**61**], is the Excision theorem in periodic cyclic homology established by Cuntz and Quillen [**23**].

THEOREM 2.5. *Any two-sided ideal J of an algebra A over a characteristic* 0 *field gives rise to a periodic six-term exact sequence*

$$\begin{array}{ccccc} \mathrm{HP}_0(J) & \longrightarrow & \mathrm{HP}_0(A) & \longrightarrow & \mathrm{HP}_0(A/J) \\ \partial\uparrow & & & & \downarrow\partial \\ \mathrm{HP}_1(A/J) & \longleftarrow & \mathrm{HP}_1(A) & \longleftarrow & \mathrm{HP}_1(J). \end{array} \tag{2.10}$$

It is interesting to note that this theorem holds in much greater generality in periodic cyclic homology than, for instance, in cyclic homology, where one needs to assume that the ideal J is an H-unital algebra.

Excision in periodic cyclic homology is compatible with excision in K-theory, which is seen from the following result [**53**].

THEOREM 2.6. *Let $I \subset A$ the a two-sided ideal of a complex algebra A. Then the diagram*

$$\begin{array}{ccccccccccc} K_1(I) & \longrightarrow & K_1(A) & \longrightarrow & K_1(A/I) & \xrightarrow{\partial} & K_0(I) & \longrightarrow & K_0(A) & \longrightarrow & K_0(A/I) \\ \downarrow & & \downarrow & & \downarrow & & \downarrow & & \downarrow & & \downarrow \\ \mathrm{HP}_1(I) & \longrightarrow & \mathrm{HP}_1(A) & \longrightarrow & \mathrm{HP}_1(A/I) & \xrightarrow{\partial} & \mathrm{HP}_0(I) & \longrightarrow & \mathrm{HP}_0(A) & \longrightarrow & \mathrm{HP}_0(A/I), \end{array}$$

in which the vertical arrows are induced by the Chern characters $Ch : K_i \to \mathrm{HP}_i$, for $i = 0, 1$, commutes.

It is useful to mention here that periodic cyclic homology can be also defined using a variant of the cyclic complex, the X-complex, which is obtained as follows. Let $TA \to A$ be the canonical surjection defined on the free algebra generated by A, regarded as a vector space. Denote by I the kernel of this morphism and complete the usual periodic cyclic complex of TA with respect to the powers of I. Since TA is free, a theorem of Loday and Quillen [**38**] allows us to replace its Hochschild complex with a canonical complex of length two. This descends to the completion. We obtain in this way a simpler complex for the periodic cyclic homology of A, which gives the same cohomology, by Goodwillie's result 2.4. See [**22, 23, 46, 56**].

This result was used in [**53, 54**] to establish a link with index theory, which we now illustrate with the following simple example. Let M be a smooth, compact manifold. We shall denote by C_p the space of bounded linear maps T such that $Tr(|T|^p) < \infty$, $p \geq 1$. Also, we shall denote by $\Psi^m(M)$ the space of classical pseudodifferential operators on M of order m. It is known that $\Psi^{-1}(M) \subset C_p$, if $p > \dim M$. Since C_p is an ideal of the algebra $\mathcal{L}(\mathcal{H})$ of bounded operators on $L^2(M)$ and $\Psi^0(M)$ consists of bounded operators, we can define the algebra $\mathcal{E} := \Psi^0(M) + C_p$, which will contain C_p as a two-sided ideal. Denote by $S^*M = (T^*M \setminus 0)/\mathbb{R}_+^*$, the cosphere bundle of M. Since $C_p \cap \Psi^0(M) = \Psi^{-1}(M)$ and the principal symbol $\sigma^{(0)} : \Psi^0(M)/\Psi^{-1}(M) \to \mathcal{C}^\infty(S^*M)$ is an isomorphism, we obtain the following exact sequence

$$0 \to C_p \longrightarrow \mathcal{E} \xrightarrow{\sigma^{(0)}} C(S^*M) \to 0.$$

The trace defines a cyclic cocycle $Tr_*(a_0, \ldots, a_{2k+1}) = c_k \times Tr(a_0 a_1 \ldots a_{2k+1})$ on C_p, if $2k+1 \geq p$, [**20**]. The connecting map ∂ in K-theory in the above diagram, when composed with Tr_* gives the analytic index

$$\mathrm{Index}_a = Tr_* \circ \partial : K_1(C(S^*(M)) \longrightarrow K_0(C_p) \simeq \mathbb{Z}.$$

The Atiyah-Singer index theorem [**1**] can then be rephrased using Theorem 2.6 to say that

$$\partial[Tr_*] \simeq \pi^* \operatorname{Td}(M) \cap [S^*M] \in \bigoplus_k \mathrm{H}_{2k+1}(S^*M) = \mathrm{HP}^1(\mathcal{C}^\infty(S^*M)),$$

where $\partial : \mathrm{HP}^0(C^\infty(M\times M)) \to \mathrm{HP}^1(\mathcal{A}(M))$ is the boundary map in periodic cyclic cohomology, $\mathrm{Td}(M)$ is the Todd class of M and $[Tr_*]$ is the class of Tr_* in periodic cyclic cohomology. See also Theorem 3.1. The reader should be warned that there exists no proof of the Atiyah-Singer index theorem based on Theorem 2.6, although this would certainly be very interesting. This example can be generalized to give a proof of the Connes-Moscovici index theorem for coverings [**54**].

The last two theorems underscore strong similarities between periodic cyclic homology and K-theory. It is worth mentioning however that periodic cyclic homology is much easier to compute, and hence it is known for many more algebras, than K-theory.

3. Introducing topology

In applications one is often interested in algebras with topology. Then in the definition of Hochschild and cyclic (co)homology one has to take the topology into account, and this means to complete $A^{\otimes q+1}$ in a suitable sense.

A locally convex algebra is a locally convex vector space A over $\mathbb{C}$ equipped with a separately continuous multiplication. In many examples the locally convex algebra A will be a Fréchet algebra, in which case the multiplication map will be jointly continuous. Given that cyclic type homology theories of an algebra are defined using complexes constructed from the tensor algebra of A it is clear that topology on A will force one to decide which topological tensor product to use.

The inductive tensor product $\otimes_i$ solves the universal problem for separately continuous bilinear maps in the sense that any such map $E \times F \to G$, where E, F, G are locally convex topological vector spaces, extends to a continuous linear map $E \otimes_i F \to G$. The projective tensor product $\otimes_\pi$ solves the same universal problem for jointly continuous bilinear maps. When E is a Fréchet space, then $E \otimes_i E \simeq E \otimes_\pi E$, as any separately continuous map is automatically jointly continuous, by [**8**, III.30, Corollary 1]. When E is also a *nuclear* space, a further simplification takes place as then there is a unique topological tensor product which is compatible with the algebraic tensor product.

A basic example of this situation is provided by the algebra A of smooth functions on a compact manifold. As a locally convex vector space, A is a complete, nuclear Fréchet space. In this case, a suitable topological tensor product to consider is the *completed projective tensor* product $\hat{\otimes}$ [**28**, Définition 2, p. 32]. In the remainder of this paper we shall restrict our attention to Fréchet algebras and so our topological tensor product of choice will be the completed projective tensor product. More details on the related issues can be found in [**11, 22**].

Let us assume that A is a Fréchet algebra. We can define then

$$\mathcal{H}_q(A) := A\hat{\otimes}A\hat{\otimes}\ldots\hat{\otimes}A, \qquad (q+1) \text{ factors },$$

where $\hat{\otimes}$ denotes as before the completed projective tensor product, which is the standard definition of the Hochschild complex for topological algebras. (The differentials b and b' then extend to the projective tensor product because of the continuity of the multiplication.) We shall denote the homology of the Hochschild complex $\mathcal{H}(A) := (\mathcal{H}_q(A), b)$ by $\mathrm{HH}_q(A)$, as before.

The following theorem from [**20**] generalizes Theorem 1.4 to the topological case. This theorem is one of the main reasons why Hochschild homology is regarded as a generalization of differential forms and why periodic cyclic homology is regarded as a generalization of de Rham cohomology.

Note that the HKRC-map extends to the projective tensor products.

THEOREM 3.1. *Let M be a smooth compact manifold and let $\Omega^q(M)$ be the space of smooth q-forms on M. Then the HKRC map induces isomorphisms*

$$\chi : \mathrm{HH}_q(\mathcal{C}^\infty(M)) \to \Omega^q(M).$$

Consequently, $\mathrm{HP}_q(\mathcal{C}^\infty(M)) \simeq \bigoplus_{j\in\mathbb{Z}} \mathrm{H}^{q+2j}(M)$. *The dual identifications are valid for the corresponding cohomologies.*

PROOF. Let $A = \mathcal{C}^\infty(M)$. We begin by noticing that [**28**, Ch. 2, Example 1, p. 80]

$$\mathcal{H}_q(A) := A^{\hat{\otimes} q+1} = \mathcal{C}^\infty(M^{q+1}). \tag{3.1}$$

Assume now that M is (diffeomorphic to) a compact ball in a Euclidean space. Then the proof of our theorem is exactly as that of Proposition 1.3, but only after replacing A^e by $\mathcal{C}^\infty(M \times M)$. (It is implicit here that the theory of resolutions and derived functors extends to topological algebras. Indeed, recall that this is achieved by requiring all maps to have closed, complemented images.)

Then we need a localization principle that will help us reduce the proof to the case of a closed ball in a Euclidean space. To this end, we shall adapt an argument from [**16**]. Let F_k be the set of functions $f : M^{q+1} \to \mathbb{C}$ that vanish on a neighborhood of the set $\{(x, x, \dots, x, x_1, \dots, x_{q-k})\} \subset M^{q+1}$, ($k+1$ repetitions of x), for any $x, x_1, \dots, x_{q-k} \in M$.

Then $F_0 \subset F_1 \subset \dots$ and $bF_k \subset F_k$. Let $F_{-1} = 0$. We claim that each complex F_k/F_{k-1} is acyclic. Indeed, fix a metric on M and choose for any $\epsilon > 0$ a smooth function $g_\epsilon : M^2 \to \mathbb{C}$ such that $g_\epsilon(x, y) = 0$ if the distance between x and y is less than $\epsilon/2$ and $g_\epsilon(x, y) = 1$ if the distance between x and y is greater than ϵ. Let

$$(L_\epsilon f)(x_0, x_1, \dots, x_{q+1}) := g_\epsilon(x_{k-1}, x_k) f(x_0, \dots, x_{k-1}, x_{k+1}, \dots, x_{q+1}).$$

Then $(bL_\epsilon + L_\epsilon b)f - (-1)^k g(x_{k-1}, x_k) f \in F_{k-1}$ and $g(x_{k-1}, x_k) f = f$ for $\epsilon > 0$ small enough. This verifies our claim that F_k/F_{k-1} is acyclic.

Let $F_\infty = \bigcup F_n$. The spectral sequence associated to the resulting filtration of F_∞ is a first quadrant spectral sequence, and hence F_∞ is acyclic.

Denote by $\widetilde{\mathcal{H}}(M) := \mathcal{H}(M)/F_\infty$. Then the natural projection

$$\mathcal{H}(M) \to \mathcal{H}(M)/F_\infty$$

induces an isomorphism in homology (*i.e.* it is a quasi-isomorphism). We can thus replace $\mathcal{H}(M)$ with $\widetilde{\mathcal{H}}(M)$ in our calculation. This has the advantage that

$$\phi\widetilde{\mathcal{H}}(M) \simeq \phi\widetilde{\mathcal{H}}(B)$$

if $B \subset M$ is a closed manifold (possibly with boundary) such that ϕ has support in the interior of B. The action of ϕ was defined in Equation (1.4). Since the cohomology of $\widetilde{\mathcal{H}}(B)$ is known, determined by the HKRC-map χ and $\chi(\phi x) = \phi\chi(x)$ for any x in a Hochschild complex, a partition of unity argument can be used to complete the proof.

The corresponding result for cohomology follows by observing that the range of b in the Hochschild complex is closed, by the result we have just proved for homology. This is enough to conclude that the fact that χ is an isomorphism in homology implies that the dual map χ' is an isomorphism in cohomology. □

The resulting isomorphism helps identify the Chern characters on K_0 and K_1 with the classical Chern characters (up to normalization factors involving $2\pi\imath$), see [**12, 52**].

Our final remark in this section is that many general results proved for cyclic type homology theories for algebras without topology carry over to the topological case. The most important of these is excision. The method of Cuntz and Quillen has been extended by Cuntz to the case of topological locally multiplicatively convex algebras [**22**], and more recently, to the category of bornological algebras [**46**].

The periodic cyclic homology of algebras of this type satisfies the excision property with respect to extensions

$$0 \to J \to A \to A/J \to 0$$

of algebras from one of the categories mentioned above where the surjection on the right has a continuous (or bounded, in the bornological case) linear section. This implies that, as a topological vector space, the image of the ideal J is complemented in A. This requirement can be relaxed at the cost of requiring that J be H-unital [**10**].

4. Algebras of pseudodifferential operators

The $K-$theory of many important algebras is difficult to compute [**4**]. Often a good substitute, especially when one is interested in index theorems, is provided by periodic cyclic homology.

If M is a smooth compact manifold, then an interesting algebra is the algebra $\mathcal{A}(M)$ of complete classical symbols on M. The computation of the homologies of this algebra was carried out in [**15**], and the result for the periodic cyclic homology is:

$$\mathrm{HP}_q(\mathcal{A}(M)) \simeq \bigoplus_{j\in\mathbb{Z}} \mathrm{H}^{q+2j}(S^*M \times S^1), \quad q = 0, 1,$$

where $\pi : S^*M \to M$ is the cosphere bundle of M.

Other interesting examples are provided by algebras of pseudodifferential operators on groupoids [**47, 55**]. Let $\mathcal{G}$ be a longitudinally smooth groupoid with corners with space of units a manifold with corners M. Denote by $d, r : \mathcal{G} \to M$ the domain and range map of $\mathcal{G}$. So $\mathcal{G}$ is itself a manifold with corners but we assume that the fibres of d are smooth manifolds (without boundary or corners). The algebra

$$\Psi^{\infty}(\mathcal{G}) = \bigcup_{m\in\mathbb{Z}} \Psi^m(\mathcal{G})$$

of (scalar) classical uniformly supported pseudodifferential operators on $\mathcal{G}$ is then defined, see [**36, 47, 55**]. For basic facts about pseudodifferential operators, see one of the many monographs available, for example [**50**], [**57**], or [**58**].

The algebra of smoothing operators is by definition

$$\Psi^{-\infty}(\mathcal{G}) := \bigcap_{m\in\mathbb{Z}} \Psi^m(\mathcal{G}).$$

The algebra

$$\mathcal{A}(\mathcal{G}) := \Psi^{\infty}(\mathcal{G})/\Psi^{-\infty}(\mathcal{G})$$

of complete symbols on $\mathcal{G}$ is endowed with an inductive limit topology. In order to provide a natural framework for a homological study of such algebras, one is led to a wider category of algebras with topology that we shall call *topologically filtered algebras* [**5, 6**].

The reader not familiar with groupoids can assume that our algebras of pseudodifferential operators are the algebras of classical pseudodifferential operators on a smooth, compact manifold. This corresponds to the case $\mathcal{G} = M \times M$.

We shall now study the homology of the algebras $\mathcal{A}(M)$, and, more generally, $\mathcal{A}(\mathcal{G})$. A convenient approach is provided by topologically filtered algebras, which are algebras $\mathcal{A}$ endowed with bifiltrations $F_n^m\mathcal{A} \subset \mathcal{A}$ satisfying suitable conditions. Before formally formulating the definition of a topologically filtered algebra, let us just say that in the case of $\mathcal{A} = \mathcal{A}(M)$, M compact, $F_n^m\mathcal{A}$ is independent of m. The second index, or filtration (with respect to m), is needed in order to treat algebras of compactly supported complete symbols on a non-compact manifold.

Recall that an algebra $\mathcal{A}$ with a given topology, is a *topologically filtered algebra* if there exists an increasing bifiltration $F_n^m\mathcal{A} \subset \mathcal{A}$,

$$F_n^m\mathcal{A} \subset F_{n'}^{m'}\mathcal{A}, \quad \text{if } n \le n' \text{ and } m \le m',$$

by closed, complemented subspaces, satisfying the following properties:

(1) $\mathcal{A} = \cup_{n,m} F_n^m\mathcal{A}$ and $F_n^{-1}\mathcal{A} = 0$;
(2) The union $\mathcal{A}_n := \cup_m F_n^m\mathcal{A}$ is a closed subspace such that

$$F_n^m\mathcal{A} = \mathcal{A}_n \cap \big(\cup_j F_j^m\mathcal{A}\big);$$

(3) Multiplication maps $F_n^m\mathcal{A} \otimes F_{n'}^{m'}\mathcal{A}$ to $F_{n+n'}^{m+m'}\mathcal{A}$;
(4) The maps

$$F_n^m\mathcal{A}/F_{n-j}^{m'}\mathcal{A} \otimes F_{n'}^{m'}\mathcal{A}/F_{n'-j}^{m'}\mathcal{A} \longrightarrow F_{n+n'}^{m+m'}\mathcal{A}/F_{n+n'-j}^{m+m'}\mathcal{A}$$

induced by multiplication are continuous;
(5) The quotient $F_n^m\mathcal{A}/F_{n-j}^m\mathcal{A}$ is a nuclear Fréchet space in the induced topology;

(6) The natural map

$$F_n^m \mathcal{A} \longrightarrow \varprojlim F_n^m / F_{n-j}^m \mathcal{A}, \quad j \to \infty$$

is a homeomorphism; and

(7) The topology on $\mathcal{A}$ is the strict inductive limit of the subspaces $F_n^n \mathcal{A}$, as $n \to \infty$ (recall that $F_n^n \mathcal{A}$ is assumed to be closed in $F_{n+1}^{n+1} \mathcal{A}$).

As the referee has pointed out, we can replace conditions (4), (5), and (7) with the condition that $\mathcal{A}$ be a nuclear LF-space with a separately continuous multiplication.

For topologically filtered algebras, the multiplication is not necessarily jointly continuous, and the definition of the Hochschild and cyclic homologies using the projective tensor product of the algebra $\mathcal{A}$ with itself, as in the previous section, is not very useful. For this reason, we change the definition of the spaces $\mathcal{H}_m(\mathcal{A})$ defining the Hochschild complexes as follows.

Consider

$$F_p' = \lim_{m \to \infty} \sum_{k_0 + \ldots + k_q = p} \widehat{\otimes}_{j=0}^q F_{k_j}^m \mathcal{A}, \tag{4.1}$$

(projective tensor products) which defines an increasing sequence of subspaces (*i.e.* filtration) of $\mathcal{A}^{\widehat{\otimes} q+1}$. We use this filtration to define $\mathcal{H}_q(\mathcal{A})$ as a completion. Namely,

$$F_p \mathcal{H}_q(\mathcal{A}) := \varprojlim F_p'/F_{p-j}' \quad \text{and} \quad \mathcal{H}_q(\mathcal{A}) := \cup_p F_p \mathcal{H}_q(\mathcal{A}) \tag{4.2}$$

where $j \to \infty$ in the projective limit. The operators b and B extend to well defined maps, still denoted b and B, defined on $\mathcal{H}_q(\mathcal{A})$, for any q, which allows us to define the cyclic complex and the cyclic homology of the algebra $\mathcal{A}$ as the homology of the complex $(\mathcal{C}_*(\mathcal{A}), b+B)$, with $\mathcal{C}_q(\mathcal{A}) := \oplus \mathcal{H}_{q-2k}(\mathcal{A})$, as for topological algebras.

For any topologically filtered algebra, we denote

$$Gr(\mathcal{A}) := \oplus_n \mathcal{A}_n / \mathcal{A}_{n-1}$$

the *graded algebra* associated to $\mathcal{A}$, where $\mathcal{A}_n$ is the union $\cup_m F_n^m \mathcal{A}$, as before. Its topology is that of an inductive limit of Fréchet spaces:

$$Gr(\mathcal{A}) \simeq \lim_{N, m \to \infty} \oplus_{n=-N}^N F_n^m \mathcal{A} / F_{n-1}^m \mathcal{A},$$

which makes sense by (2) in the definition of the topologically filtered algebra $\mathcal{A}$.

For the algebras like $Gr(\mathcal{A})$, we need yet a third way of topologizing its iterated tensor products. For our purposes, the correct definition is then

$$\mathcal{H}_q(Gr(\mathcal{A})) := \lim_{N, m \to \infty} \left(\oplus_{n=-N}^N F_n^m \mathcal{A} / F_{n-1}^m \mathcal{A} \right)^{\widehat{\otimes} q+1}.$$

Note that this is not intended to be a "topological tensor product," but just a vector space, which happens to suit our purposes. This corrects the definition in [**5**], which is algebraically not so convenient as this one, although it does give a topological tensor product.

The Hochschild homology of $Gr(\mathcal{A})$ is the homology of $(\mathcal{H}_*(Gr(\mathcal{A})),b)$. The operator B again extends to a map $B : \mathcal{H}_q(Gr(\mathcal{A})) \to \mathcal{H}_{q+1}(Gr(\mathcal{A}))$ and we can define the cyclic homology of $Gr(\mathcal{A})$ as above. The operators S, B and I associated to $\mathcal{H}_q(Gr(\mathcal{A}))$ are the graded operators associated with the corresponding operators (also denoted S, B and I) on $\mathcal{H}_q(\mathcal{A})$.

The Hochschild and cyclic complexes of the algebra $Gr(\mathcal{A})$ decompose naturally as direct sums of complexes indexed by $p \in \mathbb{Z}$. For example, $\mathcal{H}_q(Gr(\mathcal{A}))$ is the direct sum of the subspaces $\mathcal{H}_q(Gr(\mathcal{A}))_p$, where

$$\mathcal{H}_q(Gr(\mathcal{A}))_p = \lim_{m,N\to\infty} \bigoplus_{k_j} \left(\widehat{\otimes}_{j=0}^{q} F^m_{k_j}\mathcal{A}/F^m_{k_j-1}\mathcal{A}\right),$$
$$\text{where } k_0 + k_1 + \ldots + k_q = p \text{ and } -N \leq k_j \leq N.$$

The corresponding subcomplexes of the cyclic complex are defined similarly. We denote by $\mathrm{HH}_*(Gr(\mathcal{A}))_p$ and $\mathrm{HC}_*(Gr(\mathcal{A}))_p$ the homologies of the corresponding complexes (Hochschild and, respectively, cyclic).

The following two results are consequences of standard results in homological algebra (for topologically filtered algebras they were proved in [**5**]).

LEMMA 4.1. *Let $\mathcal{A}$ be a topologically filtered algebra. Then the natural filtrations on the Hochschild and cyclic complexes of $\mathcal{A}$ define spectral sequences $\mathrm{EH}^r_{k,h}$ and $\mathrm{EC}^r_{k,h}$ such that*

$$\mathrm{EH}^1_{k,h} \simeq \mathrm{HH}_{k+h}(Gr(\mathcal{A}))_k \quad \textit{and} \quad \mathrm{EC}^1_{k,h} \simeq \mathrm{HC}_{k+h}(Gr(\mathcal{A}))_k.$$

Moreover, the periodicity morphism S induces a morphism $S' : \mathrm{EC}^r_{k,h} \to \mathrm{EC}^r_{k,h-2}$ of spectral sequences. For $r = 1$, the morphism S' is the graded map associated to the periodicity operator $S : \mathrm{HC}_n(Gr(\mathcal{A})) \to \mathrm{HC}_{n-2}(Gr(\mathcal{A}))$ and the natural filtration of the groups $\mathrm{HC}_n(Gr(\mathcal{A}))$.

Let us go back now to the algebra of complete symbols on our groupoid $\mathcal{G}$. We shall denote by $\mathcal{O}(M)$ the space of smooth functions on the interior of M that have only rational singularities at the boundary faces. If every hyperface H of M has a defining function x_H, then $\mathcal{O}(M)$ is the ring generated by $\mathcal{C}^\infty(M)$ and x_H^{-1}. Let then

$$\mathcal{A}_{\mathcal{L}}(\mathcal{G}) := \mathcal{O}(M)\mathcal{A}(\mathcal{G}).$$

PROPOSITION 4.2. [**5**] *Assume that $\mathcal{G}$ and M are as above and that M is σ-compact. Then the quotients $\mathcal{A}(\mathcal{G})$ and $\mathcal{A}_{\mathcal{L}}(\mathcal{G})$ are topologically filtered algebras.*

The multifiltrations are given by the order of the symbols, an exhaustive sequence of compact subsets and the degree of the rational singularities. See [**5, 6**] for more precise constructions.

Let $A\mathcal{G}$ be the Lie algebroid of $\mathcal{G}$ and let $S^*\mathcal{G}$ be the sphere bundle of $A^*\mathcal{G}$, that is, the set of unit vectors in the dual of the Lie algebroid of $\mathcal{G}$. Denote $\mathrm{H}_c^{[q]} = \oplus_{k\in\mathbb{Z}} \mathrm{H}_c^{q+2k}$ the singular cohomology with compact support and coefficients in $\mathbb{C}$. The de Rham cohomology of compactly supported differential forms with only rational singularities at the corners will be denoted by $\mathrm{H}^{[q]}_{c,\mathcal{L}} = \oplus_{k\in\mathbb{Z}} \mathrm{H}^{q+2k}_{c,\mathcal{L}}$ and will be called *the Laurent-de Rham cohomology.*

The periodic cyclic homology of the algebras $\mathcal{A}(\mathcal{G})$ and $\mathcal{A}_{\mathcal{L}}(\mathcal{G})$ can be computed without any further assumption on the groupoid $\mathcal{G}$ [**5**].

THEOREM 4.3. *Assume that the base M is σ-compact, then we have:*

$$\mathrm{HP}_m(\mathcal{A}(\mathcal{G})) \simeq \mathrm{H}^{[m]}_c(S^*\mathcal{G} \times S^1) \quad \textit{and} \quad \mathrm{HP}_m(\mathcal{A}_{\mathcal{L}}(\mathcal{G})) \simeq \mathrm{H}^{[m]}_{c,\mathcal{L}}(S^*\mathcal{G} \times S^1).$$

This theorem generalizes several earlier calculations. For instance, let $F \subset TM$ be an integrable subbundle of the tangent bundle TM to M and assume for simplicity that M is smooth. Then the algebra of complete symbols along the foliation defined by F coincides with Connes' algebra [**19, 48**], except that we require transverse smoothness [**55**]. We denote this algebra by $\mathcal{A}(M, F)$.

THEOREM 4.4. *The periodic cyclic homology of the algebra $\mathcal{A}(M, F)$ is given by:*

$$\mathrm{HP}_q(\mathcal{A}(M, F)) \simeq \mathrm{H}^{[q]}_c(S^*F \times S^1).$$

Finer differential invariants can be captured by computing Hochschild homology. Recall that $A^*\mathcal{G} \to M$, the dual of the Lie algebroid of $\mathcal{G}$, is a Poisson manifold with corners. We shall then denote the Poisson differential associated with this structure by δ, see [**13**]. If $A^*\mathcal{G} \smallsetminus 0$ denotes the open submanifold of $A^*\mathcal{G}$ which is the complement of the zero section, then the radial action of $\mathbb{R}^*_+$ allows us to consider $l-$homogeneous $k-$forms $\Omega^k(A^*\mathcal{G} \smallsetminus 0)_l$ and Laurent type homogeneous forms $\Omega^k_{\mathcal{L}}(A^*\mathcal{G} \smallsetminus 0)_l$. It is then easy to check that the Poisson differential sends $\Omega^k(A^*\mathcal{G} \smallsetminus 0)_l$ (resp. $\Omega^k_{\mathcal{L}}(A^*\mathcal{G} \smallsetminus 0)_l$) to $\Omega^{k-1}(A^*\mathcal{G} \smallsetminus 0)_{l-1}$ (resp. $\Omega^{k-1}_{\mathcal{L}}(A^*\mathcal{G} \smallsetminus 0)_{l-1}$). We shall denote the resulting homology spaces by $\mathrm{H}^{\delta}_k(A^*\mathcal{G} \smallsetminus 0)_l$ and $\mathrm{H}^{\delta}_{k,\mathcal{L}}(A^*\mathcal{G} \smallsetminus 0)_l$.

PROPOSITION 4.5. *The algebra $\mathcal{A}_{\mathcal{L}}(\mathcal{G})$ is H-unital. Let*

$$\chi : \mathrm{HH}_l(Gr(\mathcal{A}))_d \to \Omega^l(A^*(\mathcal{G}) \smallsetminus 0)_d$$

be the HKRC-isomorphism, and let $d_1 : \mathrm{EH}^1_{k,h} \to \mathrm{EH}^1_{k-1,h}$ be the first differential of the spectral sequence associated to $\mathcal{A}$ by Lemma 4.1. Then $\chi \circ d_1 \circ \chi^{-1} = -\sqrt{-1}\delta$, and hence $\mathrm{EH}^2_{k,h} \simeq \mathrm{H}^{\delta}_{k+h}(A^*(\mathcal{G}) \smallsetminus 0)_k$.

PROOF. As we have already observed the term $\mathrm{EH}^1_{k,h}$ of the above spectral sequence coincides with the Hochschild homology space $\mathrm{HH}_{k+h}(Gr(\mathcal{A}_{\mathcal{L}}(\mathcal{G})))_k$. Since the graded algebra $Gr(\mathcal{A}_{\mathcal{L}}(\mathcal{G}))$ is commutative, we can use the HKRC-map to identify $\mathrm{EH}^1_{k,h}$ with the space of $(k+h)-$forms on $A^*\mathcal{G} \smallsetminus 0$ that are $k-$homogeneous with respect to the radial action of $\mathbb{R}^*_+$.

The rest of the proof consists in identifying the differential

$$d^1 : \mathrm{EH}^1_{k,h} \longrightarrow \mathrm{EH}^1_{k-1,h}$$

under the HKRC–map.

Let q be a quantization map $q : S^m(A^*\mathcal{G}) \to \Psi^m(\mathcal{G})$ (this map is a local inverse for the principal symbol). Choose an anti-symmetric tensor in the last m-variables

$$\eta = \sum \operatorname{sign}(\sigma) f_0 \otimes f_{\sigma(1)} \otimes \ldots \otimes f_{\sigma(m)},$$

with $f_j \in S^\infty(A^*\mathcal{G})$. We denote by

$$q(\eta) = \sum \operatorname{sign}(\sigma) q(f_0) \otimes q(f_{\sigma(1)}) \otimes \ldots \otimes q(f_{\sigma(m)})$$

the *quantization* of η. Let $k = \deg f_0 + \ldots + \deg f_m$ be the total degree. Because

$$[q(a), q(b)] = -\sqrt{-1} q(\{a, b\}) + \ldots,$$

where the dots represent terms of order at most $\deg a + \deg b - 2$, the quantity $b \circ q(\eta)$ is of total order at most $k - 1$ and hence, modulo terms of order $k - 2$, $\chi \circ b \circ q(\eta))$ is easily checked to be exactly $\delta(\eta)$. □

5. Applications and examples

5.1. Manifolds with corners. When the groupoid $\mathcal{G}$ is the groupoid describing the b-calculus (*i.e.* the "stretched product," M_b^2, in Melrose's terminology) on the manifold with corners M, the spectral sequence associated with Hochschild homology satisfies

$$E^2_{k,h} = 0 \quad h \neq n,$$

and hence it collapses at EH^2. The asymptotic completeness of the algebras of complete symbols shows that this spectral sequence also converges (this is part of a more general result on topologically filtered algebras, [**6**]). Therefore, for these algebras, the computation of Hochschild homology is complete.

It turns out that many algebras of complete symbols on manifolds with corners become isomorphic when introducing Laurent coefficients [**6, 45, 49, 35**]. Because of this, we shall denote the algebra of complete symbols on the groupoid M_b^2 simply by $\mathcal{A}_{\mathcal{L}}(M)$.

Denote by $H^q_{\mathcal{L}}(X)$ the homology of the complex $(\mathcal{O}(X)\Omega^*(X), d)$ of de Rham differential forms with Laurent singularities at the boundary.

THEOREM 5.1. *Let M be a manifold with corners. We have*

$$\mathrm{HH}_q(\mathcal{A}_{\mathcal{L}}(M)) \simeq \mathrm{H}^{2n-q}_{c,\mathcal{L}}(S^*M \times S^1).$$

As an easy consequence of this theorem, we obtain the dimension of the space of residue traces on manifolds with corners.

COROLLARY 5.2. *The dimension of the space of traces of $\mathcal{A}_{\mathcal{L}}(M)$ is the number of minimal faces of M.*

We now turn to the computation of cyclic homology. A direct inspection shows that the operator B is trivial in this case. This allows us to deduce also the cyclic homology of $\mathcal{A}_{\mathcal{L}}(M)$.

THEOREM 5.3. *Let M be a smooth, compact manifold with corners. We have*

$$\mathrm{HC}_q(\mathcal{A}_{\mathcal{L}}(M)) \simeq \bigoplus_{k\geq 0} \mathrm{HH}_{q-2k}(\mathcal{A}_{\mathcal{L}}(M)).$$

5.2. Fibrations by manifolds with corners. Assume now that the groupoid $\mathcal{G}$ describes the vertical pseudodifferential calculus on a connected fibration with corners $\pi : M \to B$ over a smooth manifold B. Then we shall denote $\mathcal{A}_{\mathcal{L}}(\mathcal{G})$ by $\mathcal{A}_{\mathcal{L}}(M|B)$, because these algebras turn out to be isomorphic under some pretty general conditions.

When the manifold M has no corners and $\mathcal{G} = M \times_B M$ is the space of pairs of points with the same projection on B, we recover the Atiyah-Singer algebra of families of smooth complete symbols along the fibers of the fibration $M \to B$, described for example in [**2, 5**]. Denote by n the dimension of M, by p the dimension of the fibers and by q the dimension of B.

Denote by $\mathcal{F}^j$ the local coefficient system over B defined by

$$\mathcal{F}^j(b) := \mathrm{HH}_{2p-j}(\mathcal{A}_{\mathcal{L}}(\pi^{-1}(b))) \simeq \mathrm{H}^j_{c,\mathcal{L}}(\pi_0^{-1}(b)),$$

where $\pi_0 : S^*_{vert}M \times S^1 \to B$ is the natural projection and $S^*_{vert}M$ is the vertical cosphere bundle.

THEOREM 5.4. *Let $M \to B$ be a fibration of manifolds with corners, with B smooth, then*

$$\mathrm{HH}_m(\mathcal{A}_{\mathcal{L}}(M|B)) \simeq \bigoplus_{k+h=m} \Omega^h_c(B, \mathcal{F}^{2p-k}).$$

As an easy consequence and in the case without corners for simplicity, we obtain that (when $p \geq 2$), residue traces are in one-one correspondence with distributions on the base manifold B.

5.3. Longitudinal symbols on foliations. Foliations provide several examples in noncommutative geometry. We shall look hence at the case of complete symbols on the holonomy groupoid of a foliation, that is the case of the algebra of complete symbols along the leaves of a foliation.

We first need some definitions and notations. Let (X, F) be a smooth manifold X of dimension n_0 equipped with a regular smooth foliation F. The transverse bundle to the foliation (X, F) is the quotient vector bundle $\nu = TX/F$. We denote by p_0 the dimension of F. The codimension will be denoted by q_0 so that $n_0 = p_0 + q_0$.

The space $\Omega^{k,h}(X, F)$ denotes the space of differential forms of bidegree (k, h), *i.e.* of smooth sections of the bundle $\Lambda^k F^* \otimes \Lambda^h \nu^*$. A choice of a supplementary subbundle H to F in TX induces the splittings

$$\Theta_H : T^*X \cong F^* \bigoplus \nu^* \quad \text{and} \quad \Omega^d(X) \cong \bigoplus_{k+h=d} \Omega^{k,h}(X, F). \tag{5.1}$$

The splitting (5.1) endows $\Omega^*(X)$ with a bigrading so that the de Rham differential decomposes as a sum of three bihomogeneous components

$$d = d_F + d_\dashv + \partial,$$

where d_F is the $(1,0)$ component called the longitudinal differential, $d_\dashv$ is the $(0,1)$ component and ∂ is an extra component which can be shown to have bidegree $(-1,2)$ [**59**].

For $s = 0, \cdots, q_0$ we get the longitudinal de Rham complexes $(\Omega^{*,s}(X,F), d_F)$

$$0 \to \Omega^{0,s}(X,F) \xrightarrow{d_F} \Omega^{1,s}(X,F) \xrightarrow{d_F} \dots \xrightarrow{d_F} \Omega^{p_0,s}(X,F) \to 0.$$

Let $\mathrm{H}^{r,s}(X,F)$ denote the rth homology space of the longitudinal de Rham complex. The de Rham cohomology spaces of the smooth manifold X will be denoted by $\mathrm{H}^k(X)$ as is customary.

Dual to (k,h)-differential forms, we define a (k,h)-current as a generalized section of the bundle $\Lambda^{p_0-k}(F^*) \otimes \Lambda^{q_0-h}(\nu^*) \otimes \mathbb{C}_\nu$, where $\mathbb{C}_\nu$ is the orientation line bundle of ν. We denote the space of (k,h)-currents by $A_{k,h}(X,F)$. By choosing a transverse distribution H, we can view any (k,h)-current as a continuous linear form on the space of compactly supported differential (k,h)-forms with respect to H. By duality, we define a longitudinal differential on (k,h)-currents, still denoted d_F, and get in this way longitudinal complexes $(A_{*,h}(X,F), d_F)_{0\leq h\leq q_0}$:

$$0 \to A_{p_0,h} \xrightarrow{d_F} A_{p_0-1,h} \xrightarrow{d_F} \dots \xrightarrow{d_F} A_{0,h} \to 0.$$

The homology of this complex is denoted $\mathrm{H}_{*,h}(X,F)$.

Let now $(M,\mathcal{F})$ be a new smooth connected foliated manifold with $\dim(M) = n$ and $\dim(\mathcal{F}) = p$. We assume that the bundle $\mathcal{F}$ is oriented and we denote by q the codimension of the foliation, so that $n = p + q$. The above manifold X will be a total space of some fibration over M with an induced foliation F, as we shall see.

If we denote by $\Psi^m(M,\mathcal{F})$ the set of (compactly supported) pseudodifferential operators of integer order $\leq m$ along the leaves of $\mathcal{F}$, then $\Psi^{-\infty}(M,\mathcal{F}) = \cap_{m\in\mathbb{Z}}\Psi^m(M,\mathcal{F})$ is isomorphic to the smooth convolution algebra $\mathcal{C}_c^\infty(\mathcal{G})$ of the holonomy groupoid $\mathcal{G}$ associated to $\mathcal{F}$, [**55**]. We shall denote as usual by $\Psi^\infty(M,\mathcal{F}) = \cup_{m\in\mathbb{Z}}\Psi^m(M,\mathcal{F})$ the set of all classical pseudodifferential operators of integer order along the leaves of $\mathcal{F}$. Then we obtain the usual exact sequence

$$0 \to \mathcal{C}_c^\infty(\mathcal{G}) \longrightarrow \Psi^\infty(M,\mathcal{F}) \longrightarrow \mathcal{A}(M,\mathcal{F}) \to 0,$$

with the quotient $\mathcal{A}(M,\mathcal{F})$ being the algebra of *complete symbols along the leaves of* $\mathcal{F}$.

We endow $X = S^*\mathcal{F}^* \times S^1$ with the foliation F whose leaves are the total spaces of the restriction of $X \to M$ to the leaves of $(M,\mathcal{F})$. In particular, (X,F) has the same codimension as $(M,\mathcal{F})$.

THEOREM 5.5. [**7**] *Let $(M,\mathcal{F})$ be a foliated manifold, and let $(\mathrm{EH}^r, d^r)_{r\geq 1}$ be the spectral sequence associated with the Hochschild homology of the algebra $\mathcal{A}(M,\mathcal{F})$ of complete symbols along the leaves of $\mathcal{F}$, as before.*

Then the spectral sequence $(\mathrm{EH}^r, d^r)_{r\geq 1}$ *converges to the Hochschild homology of* $\mathcal{A}(M,\mathcal{F})$, *and we have*

$$\mathrm{EH}^2_{k,h} \simeq \mathrm{H}^{p-k,h-p}(S^*F \times S^1, F).$$

The space of residue traces along the leaves of $(M,\mathcal{F})$ can then be deduced.

COROLLARY 5.6. $\mathrm{HH}_0(\mathcal{A}(M,\mathcal{F}))$ *is isomorphic to the space* $\mathrm{H}^{2p,0}(S^*\mathcal{F} \times S^1, F)$. *Moreover when the dimension of the foliation* $\mathcal{F}$ *is* ≥ 2,

$$\mathrm{HH}_0(\mathcal{A}(M,\mathcal{F})) \simeq \mathrm{H}^{p,0}(M,\mathcal{F}).$$

Thus, the space of residue traces on the foliation $(M,\mathcal{F})$ is isomorphic to the space of $(2p,0)$-invariant currents on $(S^*\mathcal{F} \times S^1, F)$. More precisely, for $p \geq 2$ we get

$$\mathrm{HH}^0(\mathcal{A}(M,\mathcal{F})) \simeq \mathrm{H}_{p,0}(M,\mathcal{F}).$$

References

[1] M.F. Atiyah and I. M Singer, *The index of elliptic operators, I*, Ann. Math. **87** (1968), 484-530.

[2] M.F. Atiyah and I. M Singer, *The index of elliptic operators, IV*, Ann. Math. **93** (1971), 119–138.

[3] V. Baranovsky, *Orbifold cohomology as periodic cyclic homology*, math.AG/0206256.

[4] P. Baum and A. Connes, *Geometric K-theory for Lie groups and foliations*, Preprint IHES, Paris (1982).

[5] M-T. Benameur and V. Nistor, Homology of complete symbols and noncommutative geometry, In *Quantization of Singular Symplectic Quotients*, N.P. Landsman, M. Pflaum, and M. Schlichenmaier, ed., *Progress in Mathematics* **198**, pages 21–46, Birkhäuser, Basel - Boston - Berlin, 2001.

[6] M.-T. Benameur and V. Nistor, *Homology of algebras of families of pseudodifferential operators*, To appear in J. Funct. Anal.

[7] M.-T. Benameur and V. Nistor, *Homology and residues for foliations*, submitted for publication.

[8] N. Bourbaki, *Topological Vector Spaces*, Springer Verlag, Berlin, 1981.

[9] J. Brodzki, An introduction to K-theory and cyclic cohomology, Advanced Topics in Mathematics. PWN—Polish Scientific Publishers, Warsaw, 1998.

[10] J. Brodzki, Z. Lykova, *Excision in cyclic type homology of Fréchet algebras*, Bull. London Math. Soc. 33 (2001), no. 3, 283–291.

[11] J. Brodzki, R. Plymen, *Periodic cyclic homology of certain nuclear algebras*, C. R. Acad. Sci., Paris, Sér. I, Math. 329, No.8, 671-676 (1999).

[12] J. Brodzki, R. Plymen, *Chern character for the Schwartz algebra of p-adic GL(n)*, Bulletin of the LMS, 34, (2002) 219-228.

[13] J.-L. Brylinski, *A differential complex for Poisson manifolds*, J.Diff. Geom. **28** (1988), 93–114.

[14] J.-L. Brylinski, *Central localization in Hochschild homology*, J. Pure Appl. Algebra **57** (1989), 1–4.

[15] J.-L. Brylinski and E. Getzler, *The homology of Algebras of Pseudo-differential Symbols and the Noncommutative Residue*, K-Theory **1** (1987), 385–403.

[16] J.-L. Brylinski and V. Nistor, *Cyclic cohomology of etale groupoids*, K-Theory **8** (1994), 341–365.

[17] D. Burghelea, *The cyclic homology of the group rings*, Comment. Math. Helv. 60 (1985), no. 3, 354–365.

[18] M. Crainic, *Cyclic homology of etale groupoids: the general case*, K-theory **17** (1999), 319–362.

[19] A. Connes, Sur la théorie non commutative de l'intégration, In *Algèbres d'opérateurs*, Lect. Notes Math. 725, Springer Verlag, 19–143, 1979.

[20] A. Connes, *Noncommutative differential geometry*, Publ. Math. IHES **62** (1985), 41–144.

[21] A. Connes, Noncommutative Geometry, Academic Press, New York - London, 1994.

[22] J. Cuntz, *Excision in periodic cyclic theory for topological algebras*, In: *Cyclic cohomology and noncommutative geometry (Waterloo, ON, 1995)*, 43–53, Amer. Math. Soc., Providence, RI, 1997.

[23] J. Cuntz and D. Quillen, *Excision in bivariant periodic cyclic cohomology*, Invent. Math. **127** (1997), 67–98.

[24] B. Feigin and B. L. Tsygan, Cyclic homology of algebras with quadratic relations, universal enveloping algebras and group algebras, in *K-theory, arithmetic and geometry (Moscow, 1984-1986)*, 210–239 Lect. Notes Math. **1289**, Springer, Berlin, 1987.

[25] B. Feigin and B. L. Tsygan, *Additive K-Theory and cristaline cohomology*, Funct. Anal. Appl. **19** (1985), 52–62.

[26] T. G. Goodwillie, *Cyclic homology, derivations, and the free loopspace*, Topology **24** (1985), 187–215.

[27] P. Griffiths and J. Harris, Principles of algebraic geometry, Reprint of the 1978 original, Wiley Classics Library, John Wiley and Sons, New York, 1994, xiv+813 pp.

[28] A. Grothendieck, *Produits tensoriels topologiques et espaces nucléaires*, Memoirs of the AMS, 16, 1955.

[29] G. Hochschild, B. Kostant, and A. Rosenberg, *Differential forms on regular affine algebras*, Trans. AMS **102** (1962), 383–408.

[30] John D. Jones and C. Kassel, *Bivariant cyclic theory*, K-Theory **3** (1989), 339–365.

[31] M. Karoubi, *Homologie cyclique et K-théorie* Asterisque **149** (1987), 1–147.

[32] C. Kassel, *Cyclic homology, comodules, and mixed complexes*, J. Algebra **107** (1987), 195–216.

[33] D. Kazhdan, V. Nistor, and P. Schneider, *Hochschild and cyclic homology of finite type algebras*, Selecta Math. (N.S.) **4** (1998), 321–359.

[34] M. Kontsevich and A. Rosenberg, *Noncommutative smooth spaces*, The Gelfand Mathematical Seminar, 1996–1999, 85–108, Birkhäuser Boston, Boston, Ma, 2000.

[35] R. Lauter and S. Moroianu, *Homology of pseudo-differential operators on manifolds with fibered boundaries*, Mainz University preprint 2000.

[36] R. Lauter and V. Nistor, Analysis of geometric operators on open manifolds: a groupoid approach, In *Quantization of Singular Symplectic Quotients*, N.P. Landsman, M. Pflaum, and M. Schlichenmaier, ed., *Progress in Mathematics* **198**, pages 181–229. Birkhäuser, Basel - Boston - Berlin, 2001.

[37] J.-L. Loday, Cyclic Homology, Springer-Verlag, Berlin-Heidelberg-New York, 1992.

[38] J.-L. Loday and D. Quillen, *Cyclic homology and the Lie homology of matrices*, Comment. Math. Helv. **59** (1984), 565–591.

[39] G. Lusztig, Introduction to quantum groups, Progress in Mathematics, 110, Birkhäuser Boston, Inc., Boston MA, 1993, xii+341.

[40] S. Mac Lane, *Homology*, Springer-Verlag, Berlin-Heidelberg-New York, 1995.

[41] S. Mac Lane and I. Moerdijk, *Sheaves in geometry and logic. A first introduction to topos theory*, Universitext, Springer-Verlag, Berlin-Heidelberg-New York, 1994.

[42] Yu. Manin, Topics in noncommutative geometry, M. B. Porter Lectures, Princeton University Press, Princeton, NJ, 1991, viii+164pp.

[43] Yu. Manin, *Real multiplication and non-commutative geometry*, Lectures at M.P.I, May 2001.

[44] Yu. Manin and M. Marcolli, *Continued fractions, modular symbols, and noncommutative geometry*, Selecta Math. New Ser., **8** (2002), 475–521.
[45] R. Melrose and V. Nistor, *Homology of pseudodifferential operators I.Manifolds with boundary*, preprint.
[46] R. Meyer, *Analytic Cyclic Homology*, Ph. D. Thesis, Münster University, 1999, preprint math.KT/9906205.
[47] B. Monthubert, *Pseudodifferential calculus on manifolds with corners and groupoids*, Proc. Amer. Math. Soc. **12** (1999), 2871–2881.
[48] C.C Moore and C. Schochet, *Global analysis on Foliated spaces*, Springer-Verlag 1988.
[49] S. Moroianu, *Residue functionals on the algebra of adiabatic pseudo-differential operators*, MIT thesis, 1999.
[50] V. E. Nazaikinskii, V. E. Shatalov, and B. Yu. Sternin, *Methods of Noncommutative Analysis*, de Gruyter Studies in Mathematics 22, 1996.
[51] V. Nistor, *Group cohomology and the cyclic cohomology of crossed products*, Invent. Math. **99** (1990), 411–424.
[52] V. Nistor, *A bivariant Chern–Connes character*, Ann. Math. **138** (1993), 555–590.
[53] V. Nistor, *Higher index theorems and the boundary map in cyclic homology*, Documenta **2** (1997), 263–295.
[54] V. Nistor, *An index theorem for foliated bundles*, J. Funct.Anal. **141** (1996), 421–434.
[55] V. Nistor and A. Weinstein, and Ping Xu, *Pseudodifferential operators on groupoids*, Pacific J. Math. **189** (1999), 117–152.
[56] M. Puschnigg, *Excision in cyclic homology theories*, Invent. Math. **143** (2001), no. 2, 249–323.
[57] M. A. Shubin, Pseudodifferential operators and spectral theory, Springer Verlag, Berlin-Heidelberg-New York, 1987.
[58] M.E. Taylor, Partial differential equations, Applied Mathematical Sciences, vol. I–III, Springer-Verlag, New York, (1995-1997).
[59] P. Tondeur, Geometry of foliations, Monographs in Mathematics, Birkhauser, (1997).
[60] B. L. Tsygan, *Homology of matrix Lie algebras over rings and Hochschild homology*, Uspekhi Math. Nauk., **38** (1983), 217–218.
[61] M. Wodzicki, *Excision in cyclic homology and in rational algebraic K-theory*, Annals of Mathematics **129** (1989), 591–640.

Pennsylvania State University and Inst. Desargues, Univ. Lyon 1, France
E-mail address: benameur@igd.univ-lyon1.fr

University of Southampton, U.K.
E-mail address: j.brodzki@maths.soton.ac.uk

Inst. Desargues and Pennsylvania State University, University Park, PA 16802, USA
E-mail address: nistor@math.psu.edu

Operator Theory:
Advances and Applications, Vol. 151, 265–341

Index and secondary index theory for flat bundles with duality

Ulrich Bunke and Xiaonan Ma

ABSTRACT. We discuss some aspects of index and secondary index theory for flat bundles with duality. This theory was first developed by Lott. Our main purpose in the present paper is provide a modification with better functorial properties.

1. Introduction to the paper

This article has its origin in the work of J. Lott [**21**], where he develops two versions of secondary index theory for flat vector bundles.

The basic relation in the first version is the transition from a complex of flat real vector bundles to its cohomology. The primary characteristic classes in this case are the Bott-Chern classes. The secondary analytic information is given by the analytic torsion forms. The main building block of the theory is the construction of a push-forward operation (primary and secondary index map) for fibre bundles with closed fibres. In a sense it is given by considering the fibre-wise de Rham complex twisted by a flat bundle as an infinite-dimensional object of the theory. Then one applies the equivalence relation (take the cohomology) to reduce to finite-dimensional vector bundles again. As shown in [**14**] (based on the analytic results of [**24**]) this first version of secondary index theory has the expected functorial properties with respect to iterated fibre bundles.

Lott's second version of secondary index theory involves flat real vector bundles with parallel non-degenerate quadratic or symplectic forms (flat duality bundles). The basic relation in Lott's approach was that hyperbolic forms were considered to be trivial. The primary characteristic classes in this case are Chern classes and the secondary information is given by eta-forms. Again, the main building block of the theory was the construction of a push-forward for fibre bundles with closed oriented even-dimensional fibres. The infinite-dimensional object in this case is

2000 *Mathematics Subject Classification.* Primary: 58J28; Secondary: 58J35,19L10, 19K56.
Part of this work was done while the second author was a member of SFB 288. The second author would like to thank Humboldt Universität zu Berlin for hospitality.

the signature operator twisted by the flat vector bundle. By applying the relation one again reduces to cohomology and therefore to the finite-dimensional world. It turns out that Lott's definition is not functorial with respect to iterated fibre bundles already on the primary level. In order to repair this defect one must enlarge the equivalence relation. We replace triviality of hyperbolic bundles by lagrangian reduction. Note that a hyperbolic bundle admits an lagrangian sub-bundle which has a complementary lagrangian sub-bundle. It is the existence of a complementary lagrangian sub-bundle that we must give up.

The set of flat duality bundles together with the secondary information and taken modulo equivalence is now organized in primary L- and secondary $\bar{L}$-groups. These are quotients of the corresponding groups introduced by Lott. For maps out of these groups we are going to use the same formulas involving generators as in Lott's work. One of the achievement of the present paper is the verification that these constructions are still well-defined, i.e., they factor over the enlarged equivalence relation. The other main result is that our version now also enjoys functoriality with respect to iterated fibres bundles.

In Section 2 we define the L-functor on spaces. It only depends on the fundamental group of the space. If the representation theory of the fundamental group is sufficiently well-known then we can explicitely compute the L-group. Then we study certain relations between flat duality bundles which hold in L-theory. This information is needed later in the proof of functoriality with respect to iterated fibre bundles. We also discuss a natural transformation from L- to K-theory.

In Section 3 we introduce the secondary counterpart $\bar{L}$. We relate it with secondary K-theory ($K^{-1}_{\mathbb{R}/\mathbb{Z}}$ in Lott's notation). Then we show how the relations which were already investigated in the primary case now extend to the secondary situation. In fact, the knowledge of many relations in $\bar{L}$ helps in arguments showing well-definedness and functoriality for maps with values in $\bar{L}$.

In Section 4 we study an η-homomorphism from $\bar{L}$ to $\mathbb{R}/\mathbb{Z}$ ("η" since its analytic definition involves η-invariants) and its lift to $\mathbb{R}$. The homomorphism to $\mathbb{R}/\mathbb{Z}$ comes from the natural transformation to $K^{-1}_{\mathbb{R}/\mathbb{Z}}$-theory and the usual pairing with K-homology. In Lott's version it has a lift to $\mathbb{R}$. Unfortunately this lift does not factor over our enlarged equivalence relation. In order to repair this defect we introduce an extended version of L and $\bar{L}$-theory. This extended version now admits a real-valued η-homomorphism. It is possible to define a push-forward for the extended L and $\bar{L}$-groups, but note that these are not contravariant functors of the underlying space. The properties of the extended L-groups deserve further study.

In Section 5 we define the secondary index map. The verification of well-definedness and functoriality is based on the behavior of η-forms under adiabatic limits. So we first state these results without proof and then turn to the details of the secondary index map. The results of this section were the main goal of the present paper.

The last Section 6 is devoted to the proof of the adiabatic limits result needed and stated in Section 5. In contrast to the preceeding sections, where we tried to give complete proofs, in this last section we will only sketch the main steps. It should be clear that keeping the level of depth also across the last section would expand the paper by a factor of three. The arguments in the last sections are in fact very similar to the corresponding proofs for analytic torsion forms. For a specialist it should not be too complicated (but by the experience of the first author it is also not easy) to take the stated theorems, find their counterparts for analytic torsion forms together with the proof in the indicated literature, and then correspondingly modify this proof to show the statements for the eta-forms.

2. The functor L_ϵ

2.1. Introduction and summary. The main object of this section is functor L_ϵ from the category *Top* of topological spaces and continuous maps to the category of $\mathbb{Z}_2$-graded rings and ring homomorphisms. For a space X the elements of the ring $L_\epsilon(X)$ are locally constant sheaves of (anti-)symmetric forms over $\mathbb{R}$ considered up to isotropic reduction. The ring operations are induced by the direct sum and the tensor product.

It turns out that the functor L_ϵ factors over the homotopy category *hTop*. For a path-connected space X the ring $L_\epsilon(X)$ only depends on the fundamental group of X (see Subsection 2.3 for more details).

On nice spaces a locally constant sheaf of finite-dimensional $\mathbb{R}$-modules gives rise to a real vector bundle. The form on the sheaf induces a form on the bundle. This observation leads to a natural transformation from L_ϵ to the complex K-theory functor K^0. In this way we consider L_ϵ as a refinement of K^0.

Given a K-oriented morphism $\pi\colon X \to B$ in *Top*, say a locally trivial fibre bundle with fibre a closed even-dimensional manifold which admits a vertical Spin$_c$-structure, there is a wrong-way homomorphism of groups $\pi_!^{\mathrm{Spin}_c}\colon K^0(X) \to K^0(B)$. Analytically, it is given by the index of the twisted fibrewise Spin$_c$-Dirac operator.

If the fibres are merely oriented, then we can use the twisted fibrewise signature operator to define the wrong-way homomorphism $\pi_!^{\mathrm{sign}}\colon K^0(X) \to K^0(B)$. The interesting point about the functor L is now that $\pi_!^{\mathrm{sign}}$ can be lifted to a group homomorphism $\pi_*^L\colon L_\epsilon(X) \to L_{\epsilon'}(B)$. It is essentially given by taking the fibrewise cohomology of the locally constant sheaf on X. This yields a locally constant sheaf on B. Using fibrewise Poincaré duality, we define the (anti-)symmetric form on the cohomology sheaf.

It turns out that this wrong-way map is functorial with respect to iterated fibre bundles and natural with respect to pull-back of fibre bundles. A similar functor $L_\epsilon^{\mathrm{Lott}}$ was previously defined by Lott [**21**] using a smaller equivalence relation so that $L_\epsilon(X)$ is a quotient of $L_\epsilon^{\mathrm{Lott}}(X)$. The corresponding wrong-way maps $\pi_*^{L,\mathrm{Lott}}$ are not functorial with respect to iterated fibre bundles.

In the first three sections we will denote by X a topological space and by M a manifold.

2.2. Definition and first properties.

2.2.1. *Definition of* L_ϵ. We now give details of the definition of the contravariant functor L_ϵ from the category *Top* of topological spaces and continuous maps to $\mathbb{Z}_2$-graded rings.

Let X be a topological space. If R is a ring and E is a R-module, then the constant sheaf of R-modules $\underline{E}_X$ with stalk E is the associated sheaf to the presheaf which is associates to any non-empty open subset $U \subset X$ the space of sections E such that the restriction to subsets is given by the identity. A sheaf $\mathcal{F}$ of R-modules over X is called locally constant, if there is an open covering $\{U_\lambda\}$ of X such that $\mathcal{F}\big|_{U_\lambda}$ is a constant sheaf for all λ.

If R is a field, then we say that $\mathcal{F}$ is a locally constant sheaf of finite-dimensional R-modules if there is a suitable open covering such that $\mathcal{F}\big|_{U_\lambda}$ is the constant sheaf with the stalk being a finite-dimensional vector space over R. If $\mathcal{F}$ is a locally constant sheaf of finite-dimensional R-modules over X, then let $\mathcal{F}^* := \mathrm{Hom}_R(\mathcal{F}, \underline{R}_X)$ be its dual. If $q\colon \mathcal{F} \to \mathcal{E}$ is a homomorphism between two such sheaves, then we have an adjoint $q^*\colon \mathcal{F}^* \to \mathcal{E}^*$.

From now on we consider the case $R := \mathbb{R}$. Let $\epsilon \in \mathbb{Z}_2 = \{-1, 1\}$. An ϵ–*symmetric duality structure* on $\mathcal{F}$ is an isomorphism of sheaves $q\colon \mathcal{F} \xrightarrow{\sim} \mathcal{F}^*$ satisfying $q^* = \epsilon q$.

To define the group $L_\epsilon(X)$ we first consider an abelian semigroup $\hat{L}_\epsilon(X)$ with zero element. Then we construct $L_\epsilon(X)$ by introducing a relation. An element of the semigroup $\hat{L}_\epsilon(X)$ is an isomorphism class of a pair $(\mathcal{F}, q)$ consisting of a locally constant sheaf of finite-dimensional $\mathbb{R}$-modules and an ϵ–symmetric duality structure q. The operation in $\hat{L}_\epsilon(X)$ is given by direct sum of representatives

$$(\mathcal{F}, q) + (\mathcal{F}', q') := (\mathcal{F} \oplus \mathcal{F}', q \oplus q') \; .$$

The relation on $\hat{L}_\epsilon(X)$ is generated by *lagrangian reduction.* If $i\colon \mathcal{L} \hookrightarrow \mathcal{F}$ is an inclusion of a locally constant subsheaf, then we can consider the sheaf $\mathcal{L}^\perp := \ker(i^* \circ q)$. This sheaf is again a locally constant subsheaf of $\mathcal{F}$. The sheaf $\mathcal{L}$ is called *lagrangian* if it is isotropic, i.e., $\mathcal{L} \subset \mathcal{L}^\perp$, and coisotropic, i.e., $\mathcal{L}^\perp \subset \mathcal{L}$.

We say that the element $(\mathcal{F}, q)$ is equivalent to zero by lagrangian reduction, $(\mathcal{F}, q) \sim 0$, if it admits a locally constant lagrangian subsheaf. The equivalence relation on $\hat{L}_\epsilon(X)$ is now the minimal equivalence relation which is compatible with the semigroup structure and contains lagrangian reductions.

Definition 2.1. We define $L_\epsilon(X) := \hat{L}_\epsilon(X)/\sim$.

The class of $(\mathcal{F}, q)$ in $L_\epsilon(X)$ will be denoted by $[\mathcal{F}, q]$.

Lemma 2.2. *$L_\epsilon(X)$ is a group.*

PROOF. We have $[\mathcal{F},q]+[\mathcal{F},-q]=0$. Indeed, consider the diagonal embedding $\mathcal{F}\to\mathcal{F}\oplus\mathcal{F}$. Its image is a locally constant lagrangian subsheaf of $(\mathcal{F}\oplus\mathcal{F},q\oplus -q)$. □

The $\mathbb{Z}_2$-graded ring structure $L_\epsilon(X)\otimes L_{\epsilon'}(X)\to L_{\epsilon\epsilon'}(X)$ is induced by the tensor product:

$$[\mathcal{F},q][\mathcal{F}',q']:=\frac{\sqrt{\epsilon}\sqrt{\epsilon'}}{\sqrt{\epsilon\epsilon'}}[\mathcal{F}\otimes\mathcal{F}',q\otimes q'].$$

The sign-convention is made such that later we have a natural transformation of rings from L to complex K-theory K^0.

If $f\colon Y\to X$ is a morphism in *Top*, then $f^*\colon L_\epsilon(X)\to L_\epsilon(Y)$ is defined by $f^*[\mathcal{F},q]=[f^*\mathcal{F},f^*q]$. It is easy to check that f^* is well-defined. Furthermore, it follows from the fact that we work with locally constant sheaves that the map f^* only depends on the homotopy class of f. Therefore, L factors over the homotopy category *hTop* in which maps are considered up to homotopy.

REMARK 2.3. A version $L^{\mathrm{Lott}}_\epsilon(X)$ of this ring was first introduced by Lott [**21**]. His definition differs from ours since our relation "lagrangian reduction" is replaced by "hyperbolic reduction" in the definition of Lott. Here a pair $(\mathcal{F},q)$ is called hyperbolic if there is an lagrangian subsheaf $\mathcal{L}\subset\mathcal{F}$ such that this embedding extends to an isomorphism $(\mathcal{L}\oplus\mathcal{L}^*,q_{\mathrm{can}})\cong(\mathcal{F},q)$, where

$$q_{\mathrm{can}}:=\begin{pmatrix}0 & \mathrm{id}_{\mathcal{L}}\\ \epsilon\,\mathrm{id}_{\mathcal{L}} & 0\end{pmatrix}.$$

In particular, $L_\epsilon(X)$ is a quotient of $L^{\mathrm{Lott}}_\epsilon(X)$.

2.2.2. *Some simple properties.* In the definition of the functor L_ϵ we tried to generate the equivalence relation in a certain minimal way. This simplifies the check of the well-definedness of a transformation out of $L_\epsilon(X)$ which is given on representatives. To check the well-definedness of a transformation with values in $L_\epsilon(X)$ it is useful to know some list of further relations which hold in $L_\epsilon(X)$.

For $(\mathcal{F},q)$ as above, we also denote $q(x,y):=q(x)(y)$. If $N\in\mathrm{End}(\mathcal{F})$, then we define its adjoint with respect to q by $N':=q^{-1}\circ N^*\circ q$.

DEFINITION 2.4. A $\mathbb{Z}$-grading of $(\mathcal{F},q)$ of length $n\in\mathbb{Z}$ is a semisimple element $N\in\mathrm{End}(\mathcal{F})$ such that N has integral eigenvalues in $\{0,\dots,n\}$ and $N'=n-N$. We set $\mathcal{F}^k:=\ker(N-k)$. An element $v\in\mathrm{End}(\mathcal{F})$ is called a compatible differential if it is of degree one with respect to the grading, $v^2=0$, and $v'=-v$, i.e., $q(vx,y)+q(x,vy)=0$ for $x,y\in\mathcal{F}$.

If $(\mathcal{F},q)$ has a $\mathbb{Z}$-grading of length n, then it is the sum of the subsheaves $\mathcal{F}^k$. The duality pairs $\mathcal{F}^k$ with $\mathcal{F}^{n-k}$. If it has in addition a compatible differential v, then we can consider the cohomology $\mathcal{H}:=\ker(v)/\mathrm{im}(v)$. It is again a locally constant sheaf of finite-dimensional $\mathbb{R}$-modules with an induced duality structure $q_{\mathcal{H}}$ and $\mathbb{Z}$-grading of length n.

LEMMA 2.5. (1) *If $(\mathcal{F},q)$ admits a $\mathbb{Z}$-grading of length n, then in $L_\epsilon(X)$ we have*

$$[\mathcal{F},q]=\begin{cases}0 & n \text{ odd},\\ \left[\mathcal{F}^{n/2},q|_{\mathcal{F}^{n/2}}\right] & n \text{ even}.\end{cases}$$

(2) *If $(\mathcal{F},q)$ admits in addition a compatible differential, then in $L_\epsilon(X)$ we have*

$$[\mathcal{F},q]=[\mathcal{H},q_{\mathcal{H}}].$$

PROOF. To show the first assertion note that for $k<n/2$ we can take off summands of the form $(\mathcal{F}^k\oplus\mathcal{F}^{n-k},q|_{\mathcal{F}^k\oplus\mathcal{F}^{n-k}})$. These summands represent trivial elements of $L_\epsilon(X)$, since they contain the lagrangian subsheaves $\mathcal{F}^k\subset\mathcal{F}^k\oplus\mathcal{F}^{n-k}$.

To show the second assertion first note that $-[\mathcal{H},q_{\mathcal{H}}]=[\mathcal{H},-q_{\mathcal{H}}]$. Hence $[\mathcal{F},q]-[\mathcal{H},q_{\mathcal{H}}]=[\mathcal{F}\oplus\mathcal{H},q\oplus(-q_{\mathcal{H}})]$. Let $i\colon \ker(v)\to\mathcal{F}\oplus\mathcal{H}$ be given by $i(x)=x\oplus[x]$. Then $\mathrm{im}(i)$ is a locally constant lagrangian subsheaf of $(\mathcal{F}\oplus\mathcal{H},q\oplus(-q_{\mathcal{H}}))$. Therefore, $[\mathcal{F}\oplus\mathcal{H},q\oplus(-q_{\mathcal{H}})]=0$. □

Let $\left(F^k\mathcal{F}\right)_{k=0,\dots,n+1}$ be a decreasing filtration of $\mathcal{F}$ with $F^0\mathcal{F}=\mathcal{F}$ and $\mathcal{F}^{n+1}F=0$. We obtain a dual filtration $\left(F^l\mathcal{F}^*\right)_{l=0,\dots n+1}$ by setting $F^l\mathcal{F}^*=\mathrm{Ann}(F^{n+1-l}\mathcal{F})=\left\{x\in\mathcal{F}^*\mid x(y)=0 \text{ for any } y\in F^{n+1-l}\mathcal{F}\right\}$. Let $\mathrm{Gr}^k(\mathcal{F})=F^k\mathcal{F}/F^{k+1}\mathcal{F}$. We have a natural isomorphism $\mathrm{Gr}(\mathcal{F}^*)\cong\mathrm{Gr}(\mathcal{F})^*$ which identifies $\mathrm{Gr}^k(\mathcal{F})^*$ with $\mathrm{Gr}^{n-k}(\mathcal{F}^*)$.

DEFINITION 2.6. A compatible decreasing filtration of $(\mathcal{F},q)$ of length n is a decreasing filtration $\left(F^k\mathcal{F}\right)_{k=0,\dots,n+1}$ by locally constant subsheaves such that $q\colon\mathcal{F}\to\mathcal{F}^*$ preserves the filtrations.

Given a compatible filtration of $(\mathcal{F},q)$ we obtain an induced ϵ-symmetric duality structure $\mathrm{Gr}(q)\colon\mathrm{Gr}(\mathcal{F})\to\mathrm{Gr}(\mathcal{F}^*)\xrightarrow{\cong}\mathrm{Gr}(\mathcal{F})^*$.

LEMMA 2.7. *In $L_\epsilon(X)$ we have $[\mathcal{F},q]\sim[\mathrm{Gr}(\mathcal{F}),\mathrm{Gr}(q)]$.*

PROOF. Note that $[\mathrm{Gr}(\mathcal{F}),\mathrm{Gr}(q)]$ has a $\mathbb{Z}$-grading of length n. In view of Lemma 2.5 (1), it suffices to show that

$$[\mathcal{F},q]=\begin{cases}0 & n\text{ odd},\\ [\mathrm{Gr}^{n/2}(\mathcal{F}),\mathrm{Gr}(q)|_{\mathrm{Gr}^{n/2}(\mathcal{F})}] & n\text{ even}.\end{cases}$$

By the following procedure we can decrease the length of the filtration by two. Note that

$$[\mathcal{F},q]=\left[\mathcal{F}\oplus(\mathrm{Gr}^0(\mathcal{F})\oplus\mathrm{Gr}^n(\mathcal{F})),q\oplus q|_{\mathrm{Gr}^0(\mathcal{F})\oplus\mathrm{Gr}^n(\mathcal{F})}\right].$$

We introduce the $\mathbb{Z}$-grading of length 2 on $\left(\mathcal{F}\oplus(\mathrm{Gr}^0(\mathcal{F})\oplus\mathrm{Gr}^n(\mathcal{F})),q\oplus q|_{\mathrm{Gr}^0(\mathcal{F})\oplus\mathrm{Gr}^n(\mathcal{F})}\right)$ such that $\mathrm{Gr}^n(\mathcal{F})$ sits in degree zero, $\mathcal{F}$ is in degree one, and $\mathrm{Gr}^0(\mathcal{F})$ is in degree two. There is a compatible differential v given by the inclusion $\mathrm{Gr}^n(\mathcal{F})\to\mathcal{F}$ and the negative of the projection $\mathcal{F}\to\mathrm{Gr}^0(\mathcal{F})$. Using Lemma 2.5 (2), we have

$$\left[\mathcal{F}\oplus(\mathrm{Gr}^0(\mathcal{F})\oplus\mathrm{Gr}^n(\mathcal{F})),q\oplus q|_{\mathrm{Gr}^0(\mathcal{F})\oplus\mathrm{Gr}^n(\mathcal{F})}\right]=[\mathcal{F}',q'],$$

where $\mathcal{F}' = F^1\mathcal{F}/F^n\mathcal{F}$ and q' is the induced ϵ–symmetric duality structure. Note that $(\mathcal{F}', q')$ has an induced decreasing filtration of length $n-2$.

Now we iterate this procedure. If n is odd, then it terminates at $0 \in L_\epsilon(X)$, and if n is even, then we finally obtain $(\mathrm{Gr}^{n/2}(\mathcal{F}), \mathrm{Gr}(q)|_{\mathrm{Gr}^{n/2}(\mathcal{F})})$. □

Let $\mathcal{I} \subset \mathcal{F}$ be an isotropic subsheaf. Then we can consider the locally constant sheaf $\mathcal{F}_\mathcal{I} := \mathcal{I}^\perp/\mathcal{I}$. Furthermore, we let $q_\mathcal{I}\colon \mathcal{F}_\mathcal{I} \to \mathcal{F}_\mathcal{I}^*$ be given by $\mathcal{I}^\perp \xrightarrow{q|_{\mathcal{I}^\perp}} \mathcal{F}^* \to (\mathcal{I}^\perp)^*$. Then $q_\mathcal{I}$ is an ϵ–symmetric duality structure on $\mathcal{F}_\mathcal{I}$.

DEFINITION 2.8. We call $(\mathcal{F}_\mathcal{I}, q_\mathcal{I})$ the isotropic reduction of $(\mathcal{F}, q)$ by $\mathcal{I}$.

LEMMA 2.9. *In* $L_\epsilon(M)$ *we have* $[\mathcal{F}, q] = [\mathcal{F}_\mathcal{I}, q_\mathcal{I}]$.

PROOF. We consider the filtration of length two on $\mathcal{F}$ such that $\mathcal{I} := F^2\mathcal{F}$, $\mathcal{I}^\perp := F^1\mathcal{F}$. Then we can identify $(\mathrm{Gr}^1(\mathcal{F}), \mathrm{Gr}(q)|_{\mathrm{Gr}^1(\mathcal{F})}) \cong (\mathcal{F}_\mathcal{I}, q_\mathcal{I})$. We now apply the Lemmas 2.7 and 2.5 (1). □

2.3. Computation of $L_\epsilon(X)$.

2.3.1. *Definition of* $L_\epsilon(G)$. Let G be any group. We define a $\mathbb{Z}_2$-graded ring $L(G)$. First we define the abelian semigroups $\hat{L}_\epsilon(G)$, $\epsilon \in \mathbb{Z}_2$, which consists of isomorphism classes of tuples (F, q, ρ). Here F is a finite-dimensional real vector space, $q\colon F \to F^*$ is an ϵ–symmetric duality structure, and $\rho\colon G \to \mathrm{Aut}(F, q)$ is a representation of G which is compatible with q, i.e., $q\colon F \to F^*$ is G-equivariant, where G acts on F^* by the adjoint representation ρ^* given by $\rho^*(g) = \rho(g^{-1})^*$. The operation in $\hat{L}_\epsilon(G)$ is induced by the direct sum of representatives.

We obtain $L_\epsilon(G)$ as the quotient of $\hat{L}_\epsilon(G)$ with respect to the equivalence relation generated by *lagrangian reduction.* First we declare that $(\mathcal{F}, q, \rho) \sim 0$ if there exists a G-invariant lagrangian subspace $L \subset F$, i.e., a G-invariant isotropic subspace such that $L^\perp = L$. Then we extend $\sim$ to the minimal equivalence relation on $\hat{L}_\epsilon(G)$ which contains lagrangian reduction and which is compatible with the semigroup structure. Let $[F, q, \rho]$ denote the class in $L_\epsilon(G)$ represented by (F, q, ρ).

Let (F, q, ρ) be a generator of $L_\epsilon(G)$ and $i\colon L \to F$ be the inclusion of a G-invariant isotropic subspace. Then $L^\perp$ is G-invariant and the quotient $L^\perp/L =: F_L$ carries an induced ϵ–symmetric form q_L and a representation ρ_L.

DEFINITION 2.10. We say that (F_L, q_L, ρ_L) is the isotropic reduction of (F, q, ρ) with respect to L.

LEMMA 2.11. *In* $L_\epsilon(G)$ *we have* $[F, q, \rho] = [F_L, q_L, \rho_L]$.

PROOF. We consider $(F \oplus F_L, q \oplus (-q_L), \rho \oplus \rho_L)$. This tuple represents the zero element in $L_\epsilon(G)$, since it contains the invariant lagrangian subspace which is the image of $L^\perp \to F \oplus F_L$, $x \mapsto x \oplus [x]$. Thus $[F, q, \rho] - [F_L, q_L, \rho_L] = 0$. □

The ring structure is given by

$$[F, q, \rho] \bullet [F', q', \rho'] = \frac{\sqrt{\epsilon}\sqrt{\epsilon'}}{\sqrt{\epsilon\epsilon'}}[F \otimes F', q \otimes q', \rho \otimes \rho'].$$

If X is path-connected, then $L_\epsilon(X) \cong L_\epsilon(\pi_1(X,x))$ for any base point $x \in X$. Furthermore, if $f\colon G' \to G$ is a homomorphism of groups, then there is a natural ring homomorphism $f^*\colon L_\epsilon(G) \to L_\epsilon(G')$ given by $f^*[F,q,\rho] = [F,q,\rho\circ f]$.

2.3.2. *Classification of irreducible ϵ–symmetric forms.* Let (F,ρ) be an irreducible representation of G. In the following we classify the invariant ϵ–symmetric duality structures on (F,ρ). We distinguish various cases, and in each case we define a $\mathbb{Z}_2$-graded group $A(F,\rho)$ by the following rule. $A^\epsilon(F,\rho)$ is trivial if (F,ρ) does not admit an ϵ–symmetric form. If it admits one isomorphism class of such forms, then we set $A^\epsilon(F,\rho) := \mathbb{Z}_2$. In the remaining case it admits two isomorphism classes, and we set $A^\epsilon(F,\rho) := \mathbb{Z}$. The group $A^\epsilon(F,\rho)$ can naturally be interpreted as the part of $L_\epsilon(G)$ which is generated by triples with underlying representation of the form $(F\otimes W, \rho\otimes 1)$, where W a finite-dimensional real vector space.

If q,p are two G-equivariant duality structures on F, then $p^{-1}\circ q \in \mathrm{Aut}_G(F)$. Thus there exists $\lambda \in \mathrm{Aut}_G(F)$ such that $p = q^\lambda$, where $q^\lambda(x,y) = q(\lambda x, y)$.

We call (F,ρ)

- *real* if $\mathrm{End}_G(F) \cong \mathbb{R}$,
- *complex* if $\mathrm{End}_G(F) \cong \mathbb{C}$ and
- *quarternionic* if $\mathrm{End}_G(F) \cong \mathbb{H}$

as algebras over $\mathbb{R}$. By Schur's lemma, every non-zero element in $\mathrm{End}_G(F)$ is invertible and hence $\mathrm{End}_G(F)$ is a division algebra over $\mathbb{R}$. By Frobenius' Theorem, $\mathrm{End}_G(F)$ must be one of the above three possibilities.

The real case. In this case $\lambda \in \mathbb{R}^*$. If $\lambda > 0$, then q and p are isomorphic, namely $p(x,y) = q(\sqrt{\lambda}x, \sqrt{\lambda}y)$. If $\lambda < 0$, then p and q are not isomorphic. Thus given a real representation (F,ρ) which admits an ϵ–symmetric duality structure q, then ϵ is determined, and there are two isomorphism classes represented by (F,q,ρ) and $(F,-q,\rho)$.

We define the $\mathbb{Z}_2$-graded groups

$$A(F,\rho) := \begin{cases} \mathbb{Z}\oplus 0 & \text{if } \epsilon = 1, \\ 0\oplus\mathbb{Z} & \text{if } \epsilon = -1. \end{cases}$$

If G is compact, by [**13**, §2.6, Prop.6.5], ϵ must be 1.

The complex case. In this case there is a unique up to sign $I \in \mathrm{Aut}_G(F)$ satisfying $I^2 = -1$. For $X \in \mathrm{End}_G(F)$ we define $X^q := q^{-1}\circ X^* \circ q$. Then $(I^q)^2 = -1$, and therefore we distinguish two subcases:

- Case $\mathbb{C}_+$: $I^q = I$
- Case $\mathbb{C}_-$: $I^q = -I$.

CASE $\mathbb{C}_+$. In this case $q(\mu x, y) = q(x, \mu y)$ for all $\mu \in \mathrm{End}_G(F)$. There exists a root $\sqrt{\lambda} \in \mathrm{Aut}_G(F)$ and we can write $p(x,y) = q(\sqrt{\lambda}x, \sqrt{\lambda}y)$. Thus p and q are isomorphic. We conclude that given (F,ρ) in case $\mathbb{C}_+$ admitting an ϵ–symmetric duality structure q, then ϵ is determined and there is one isomorphism class represented by (F,q,ρ).

We define the $\mathbb{Z}_2$-graded groups

$$A(F,\rho) := \begin{cases} \mathbb{Z}_2 \oplus 0 & \text{if } \epsilon = 1, \\ 0 \oplus \mathbb{Z}_2 & \text{if } \epsilon = -1. \end{cases}$$

CASE $\mathbb{C}_-$. In this case $q(Ix,y) = -q(x,Iy) = -\epsilon q(Iy,x)$. Thus if q is ϵ–symmetric, then q^I is $-\epsilon$–symmetric. If p and q are ϵ–symmetric, then we write $\lambda = a + bI$ and $p = aq + bq^I$ and conclude that $b = 0$. Moreover, q is isomorphic to p exactly if $a > 0$. If (F,ρ) admits an ϵ–symmetric duality structure q, then it also admits an $-\epsilon$–symmetric duality structure. The isomorphism classes are represented by (F,q,ρ), $(F,-q,\rho)$ for ϵ and (F,q^I,ρ) and $(F,-q^I,\rho)$ for $-\epsilon$.

We define the $\mathbb{Z}_2$-graded group

$$A(F,\rho) := \mathbb{Z} \oplus \mathbb{Z}.$$

The quarternionic case. Let $S^2 \subset \mathrm{Im}(\mathbb{H})$ be the unit sphere of complex structures. The $\mathbb{R}$-linear involution $X \mapsto X^q$ acts on $\mathrm{Im}(\mathbb{H})$ and restricts to an involution of S^2. We distinguish the following three cases

- Case $\mathbb{H}_0$: The involution is trivial.
- Case $\mathbb{H}_+$: The involution is non-trivial, but has a fixed point on S^2.
- Case $\mathbb{H}_-$: The involution has no fixed points on S^2.

CASE $\mathbb{H}_-$. In this case $X^q = -X$ for any $X \in \mathrm{Im}(H)$. We write $\lambda = a + bI$ for some $I \in S^2$. Then the same discussion as in the case $\mathbb{C}_-$ shows the following: If (F,ρ) admits an ϵ–symmetric duality structure q in case $\mathbb{H}_-$, then it also admits an $-\epsilon$–symmetric duality structure. The isomorphism classes are represented by (F,q,ρ), $(F,-q,\rho)$ for ϵ and (F,q^I,ρ) and $(F,-q^I,\rho)$ for $-\epsilon$.

We define the $\mathbb{Z}_2$-graded group

$$A(F,\rho) = \mathbb{Z} \oplus \mathbb{Z}.$$

CASE $\mathbb{H}_0$. We can write $\lambda = a + bI$ with $I^q = I$. The same discussion as in case $\mathbb{C}_+$ shows the following: If (F,ρ) admits an ϵ–duality structure q in case $\mathbb{H}_0$, then ϵ is determined. There is one isomorphism class represented by (F,q,ρ).

We define the $\mathbb{Z}_2$-graded groups

$$A(F,\rho) := \begin{cases} \mathbb{Z}_2 \oplus 0 & \text{if } \epsilon = 1, \\ 0 \oplus \mathbb{Z}_2 & \text{if } \epsilon = -1. \end{cases}$$

CASE $\mathbb{H}_+$. In this case we can write $\lambda = a + bI + cJ$, where $I^q = I$ and $J^q = -J$. Writing $p = aq + bq^I + cq^J$, as q^J is a $-\epsilon$ symmetric duality structure, we see that $c = 0$. On the one hand we argue as in the case $\mathbb{C}_+$ that p and q are isomorphic. On the other hand (F,ρ) also admits the $-\epsilon$–symmetric duality structure q^J. If (F,ρ) admits an ϵ–symmetric duality structure q in case $\mathbb{H}_+$, then it also admits an $-\epsilon$–symmetric duality structure. The isomorphism classes are represented by (F,q,ρ) for ϵ and (F,q^J,ρ) for $-\epsilon$.

We define the $\mathbb{Z}_2$-graded groups

$$A(F,\rho) = \mathbb{Z}_2 \oplus \mathbb{Z}_2.$$

2.3.3. *Computation of $L_\epsilon(G)$.* Let $\mathrm{Rep}(G)$ be the set isomorphism classes of finite-dimensional irreducible representations (F,ρ) of G on real vector spaces. For each $(F,\rho) \in \mathrm{Rep}(G)$ and $\epsilon = \pm 1$ we fix one ϵ–symmetric form q if it exists.

THEOREM 2.12. *We have an isomorphism of $\mathbb{Z}_2$-graded groups*

$$L_1(G) \oplus L_{-1}(G) \cong \bigoplus_{(F,\rho)\in \mathrm{Rep}(G)} A(F,\rho).$$

PROOF. Fix $\epsilon \in \mathbb{Z}_2$ and let $\mathrm{Rep}^\epsilon(G) \subset \mathrm{Rep}(G)$ denote the set of classes admitting an ϵ–symmetric form. We consider a generator (V,p,σ) of $L_\epsilon(G)$.

LEMMA 2.13. *If (V,p,σ) has no invariant isotropic subspace, then it is isomorphic to a direct sum $\bigoplus_{(F,\rho)\in \mathrm{Rep}^\epsilon(G)} n(F,\rho)(F,q,\rho)$, where $n(F,\rho) \in \mathbb{Z}$ is non-zero for at most finitely many (F,ρ). Furthermore, $n(F,\rho) \in \{0,1\}$ if $A^\epsilon(F,\rho) = \mathbb{Z}_2$, while in case $A^\epsilon(F,\rho) = \mathbb{Z}$ we use the convention that $-n(F,q,\rho)$ stands for $n(F,-q,\rho)$.*

PROOF. Let $i\colon W \hookrightarrow V$ be a minimal G-invariant subspace. Then there is a G-invariant decomposition $V = W \oplus W^\perp$. In fact, $W \cap W^\perp = 0$, since W was assumed to be minimal and it cannot be isotropic by assumption. Iterating this argument replacing V by $W^\perp$ we obtain the required decomposition into irreducibles.

The multiplicity of (F,ρ) with $A^\epsilon(F,\rho) = \mathbb{Z}_2$ cannot be greater than 1. If there were two summands, then $(F,q,\rho) \oplus (F,q,\rho)$ would admit an invariant isotropic subspace $W = \{x \oplus I(x) \,|\, x \in F\} \subset F \oplus F$, where $I \in \mathrm{Aut}_G(F)$ is an isomorphism such that $q(Ix, Iy) = -q(x,y)$ for any $x, y \in F$.

If $A^\epsilon(F,\rho) = \mathbb{Z}$, then either (F,q,ρ) or $(F,-q,\rho)$ can occur with positive multiplicity. If they occurred both, then the sum $(F,q,\rho) \oplus (F,-q,\rho)$ would admit the invariant isotropic subspace $W := \big\{x \oplus x \;\big|\; x \in F\big\} \subset F \oplus F$. □

LEMMA 2.14. *If (V,p,σ) does not admit an invariant isotropic subspace, then the multiplicities $n(F,\rho)$ are uniquely determined.*

PROOF. Assume that

$$(V,p,\sigma) \cong \bigoplus_{(F,\rho)\in \mathrm{Rep}^\epsilon(G)} n(F,\rho)(F,q,\rho) \cong \bigoplus_{(F,\rho)\in \mathrm{Rep}^\epsilon(G)} n(F,\rho)'(F,q,\rho)$$

are two decompositions. Consider $(F,\rho) \in \mathrm{Rep}^\epsilon(G)$ with $n(F,\rho) \neq 0$ and the inclusion $i\colon \mathrm{sign}(n(F,\rho))(F,q,\rho) \hookrightarrow (V,p,\sigma)$ given by the first decomposition. Then one can check that there is a summand (F',q',ρ') of the second decomposition such that the composition of i with the projection onto this summand is an isomorphism. Therefore we can take off a summand $\mathrm{sign}(n(F,\rho))(F,q,\rho)$ from both decompositions. Repeating this argument finitely many times we obtain the assertion of the lemma. □

LEMMA 2.15. *Isotropic reduction in stages can be combined to a single isotropic reduction.*

PROOF. Fix a generator (F, q, ρ). Given an invariant isotropic subspace $L \subset F$, we form the reduction (F_L, q_L, ρ_L). If $N \subset F_L$ is an invariant isotropic subspace, then we further form $\big((F_L)_N, (q_L)_N, (\rho_L)_N\big)$. The preimage $\tilde{N}$ of N under $L^\perp \to F_L$ is isotropic and G-invariant. There is a natural isomorphism

$$(F_{\tilde{N}}, q_{\tilde{N}}, \rho_{\tilde{N}}) \cong \big((F_L)_N, (q_L)_N, (\rho_L)_N\big).$$

□

So given (F, q, ρ) the possible maximal isotropic reductions are parameterized by maximal invariant isotropic subspaces.

LEMMA 2.16. *Let $L, N \subset F$ be two maximal isotropic invariant subspaces. Then the corresponding reductions (F_L, q_L, ρ_L) and (F_N, q_N, ρ_N) are isomorphic.*

PROOF. First of all $L \cap N$ is an invariant isotropic. After reduction by $L \cap N$ we can assume that $L \cap N = 0$.

We now show that $(L+N)^\perp \cap (L+N) = 0$.

We claim that $N \cap L^\perp = 0$. In fact, $L + (N \cap L^\perp)$ is isotropic and invariant. Since L is maximal, we conclude $L + (N \cap L^\perp) = L$. Thus $N \cap L^\perp \subset N \cap L = 0$.

We have $(L+N)^\perp = L^\perp \cap N^\perp$. Let $l + n \in (L+N) \cap (L+N)^\perp$. From $l \in L^\perp$ we conclude $n \in L^\perp$. By the claim above $n = 0$. Interchanging the roles of L and N, we also conclude $l = 0$.

Thus $(L+N) \oplus (L+N)^\perp = F$ and $L^\perp = L \oplus (L+N)^\perp$. Therefore, we can decompose $(F, q, \rho) = \big(L+N, q_{L+N}, \rho_{L+N}\big) \oplus ((L+N)^\perp, q_{(L+N)^\perp}, \rho_{(L+N)^\perp})$. The second summand is now naturally isomorphic to both, (F_L, q_L, ρ_L) and (F_N, q_N, ρ_N). □

Given a generator (V, p, σ), we have well-defined multiplicities $n_{(V,p,\sigma)}(F, \rho) \in A(F, \rho)$ given by any maximal isotropic reduction of (V, p, σ). One easily checks that these multiplicities are additive and satisfy $n_{(V,-p,\sigma)}(F, \rho) = -n_{(V,p,\sigma)}(F, \rho)$ and $n_{(V,p,\sigma)}(F, \rho) = n_{(V_L, p_L, \sigma_L)}(F, \rho)$ for any isotropic reduction. They therefore define the isomorphism

$$L_\epsilon(G) \to \bigoplus_{(F,\rho) \in \mathrm{Rep}^\epsilon(G)} A^\epsilon(F, \rho).$$

This finishes the proof of the theorem. □

2.4. The natural transformation to K-theory.

2.4.1. *The* `bundle`*-construction.* By Top_{met} we denote the full subcategory of Top of paracompact metrizable topological spaces. Let $K^0(X)$ be the complex K-theory functor. We construct a natural transformation $b \colon L_\epsilon \to K^0$ of functors from Top_{met} to rings.

A locally constant sheaf of finite-dimensional $\mathbb{R}$-modules on X gives rise to a locally trivial real vector bundle `bundle`$(\mathcal{F})$ in a natural way. We will describe `bundle`$(\mathcal{F})$ by providing the local trivializations and the transition maps. Let $x \in X$ and $U \subset X$ be a neighborhood of x such that the restriction $\mathcal{F}|_U$ is isomorphic to the constant sheaf $\underline{\mathcal{F}_x}_U$, where $\mathcal{F}_x$ denotes the stalk of $\mathcal{F}$ at x. Then we have a local trivialization $\mathtt{bundle}(\mathcal{F})|_U \cong U \times \mathcal{F}_x$. Consider another point $x' \in X$ and the corresponding local trivialization $\mathtt{bundle}(\mathcal{F})|_{U'} \cong U' \times \mathcal{F}_{x'}$ of this type such that $U \cap U' \neq \emptyset$. The isomorphism

$$(\underline{\mathcal{F}_x}_U)|_{U\cap U'} \cong \mathcal{F}|_{U'\cap U} \cong (\underline{\mathcal{F}_{x'}}_{U'})|_{U\cap U'}$$

induces an isomorphism $\mathcal{F}_x \overset{\phi_{UU'}}{\cong} \mathcal{F}_{x'}$ which we consider as the (constant) transition map $\phi_{UU'}\colon U \cap U' \to \operatorname{Hom}(\mathcal{F}_x, \mathcal{F}_{x'})$ between the two local trivializations.

The correspondence bundle is functorial with respect to morphisms of sheaves and compatible with the linear operations direct sum, tensor product, and duality on sheaves and vector bundles. Thus applying the bundle construction to $(\mathcal{F}, q)$ we obtain a pair (F, Q) consisting of a finite-dimensional real vector bundle and an isomorphism $Q\colon F \to F^*$. The bundle-construction is also compatible with pull-back, i.e., if $f\colon Y \to X$ is a morphism in Top_{met}, then there is a natural isomorphism $f^*\mathtt{bundle}(\mathcal{F}) \cong \mathtt{bundle}(f^*\mathcal{F})$.

2.4.2. *Metric structures.* Fix $\epsilon \in \mathbb{Z}_2$. Let (F, Q) be a real vector bundle with an isomorphism $Q\colon F \to F^*$ such that $Q^* = \epsilon Q$. Following the language introduced by Lott [**21**] we define the notion of a metric structure.

DEFINITION 2.17. An isomorphism $J\colon F \to F$ is called a metric structure if
(1) $J^* \circ Q$ defines a scalar product on F,
(2) $J^2 = \epsilon \operatorname{id}_F$,
(3) $J^* \circ Q = \epsilon Q \circ J$, i.e., $Q(x, Jy) = \epsilon Q(Jx, y)$.

Since we assume that X is metrizable and paracompact, it admits partitions of unity. This implies that metric structures exist and that the space of all metric structures is contractible.

Given (F, Q) as above, we choose a metric structure J. Let $F_{\mathbb{C}}$ be the complexification of F. Then $z^J := \frac{1}{\sqrt{\epsilon}} J$ is a $\mathbb{Z}_2$-grading of $F_{\mathbb{C}}$, and $F_{\pm} = \{x \in F_{\mathbb{C}}, z^J x = \pm x\}$ are sub-bundles of $F_{\mathbb{C}}$, thus the pair $(F_{\mathbb{C}}, z^J)$ represents an element $F_+ - F_-$ of $K^0(X)$ which does not depend on the choice of J.

DEFINITION 2.18. We define the natural transformation $b\colon L_\epsilon \to K^0$ by composing the latter construction with bundle.

2.5. Push-forward for L_ϵ.

2.5.1. *Definition of* π_*^L. Let $\pi\colon X \to B$ be a locally trivial fibre bundle where the fibre is a closed topological n-dimensional manifold Z. There is an open covering $\mathcal{U} = \{U_\lambda\}_\lambda$ of B such that $\phi_\lambda\colon X|_{U_\lambda} \cong U_\lambda \times Z$. If $U_\lambda \cap U_\mu \neq \emptyset$, then we have an isomorphism

$$U_\lambda \cap U_\mu \times Z \xrightarrow{\phi_\lambda^{-1}} X|_{U_\lambda \cap U_\mu} \xrightarrow{\phi_\mu} U_\lambda \cap U_\mu \times Z,$$

which is of the form $(b, z) \mapsto (b, \phi_{\lambda\mu}(b)(z))$, where $\phi_{\lambda\mu}\colon U_\lambda \cap U_\mu \to \operatorname{Aut}(Z)$ is a continuous family of homeomorphisms of Z.

A fibrewise orientation of the bundle $\pi\colon X \to B$ is a choice of an orientation of Z and of an atlas of local trivializations such that the $\phi_{\lambda\mu}$ are orientation preserving.

Set $\epsilon_n := (-1)^{\frac{n(n+1)}{2}} = (-1)^{[\frac{n+1}{2}]}$, $\sqrt{\epsilon_n} := (\sqrt{-1})^{[\frac{n+1}{2}]}$. To define

$$\pi_*^L\colon L_\epsilon(X) \to L_{\epsilon\epsilon_n}(B),$$

we assume that the bundle comes equipped with a fibrewise orientation. Let $[\mathcal{F}, q] \in L_\epsilon(X)$. Then we construct a representative of $\pi_*^L([\mathcal{F}, q])$ as follows. Note that $\pi(\mathcal{F}) := HR\pi_*\mathcal{F} := \bigoplus_{i=0}^n R^i\pi_*(\mathcal{F})$ is a locally constant sheaf of finite dimensional $\mathbb{R}$-modules. In fact, let $b \in U_\lambda$ and $Z_b = \pi^{-1}(\{b\})$. Then we have

$$\pi(\mathcal{F})\big|_{U_\lambda} \cong \underline{H(Z_b, \mathcal{F}\big|_{Z_b})}_{U_\lambda}.$$

By Poincaré duality over Z_b, we have an isomorphism such that $H^k(Z_b, \mathcal{F}\big|_{Z_b}^*) \cong H^{n-k}(Z_b, \mathcal{F}\big|_{Z_b})^*$ for all $k \in \mathbb{N}$. This isomorphism is preserved by the transition maps so that we obtain an isomorphism $\pi(\mathcal{F}^*) \cong \pi(\mathcal{F})^*$. If we compose this isomorphism with the sum of the isomorphisms $R^i\pi_*(q)\colon R^i\pi_*\mathcal{F} \to R^i\pi_*\mathcal{F}^*$, then we obtain an $\epsilon\epsilon_n$-symmetric duality structure $\pi(q)$ on $\pi(\mathcal{F})$ (cf. Subsections 2.5.2, 5.2.2).

DEFINITION 2.19. We define $\pi_*^L([\mathcal{F}, q]) := \big[\pi(\mathcal{F}), \pi(q)\big]$.

LEMMA 2.20. *π_*^L is well-defined.*

PROOF. By construction $\pi(q)$ is an isomorphism with the correct symmetry properties. Thus our prescription $(\mathcal{F}, q) \mapsto [\pi(\mathcal{F}), \pi(q)]$ provides a homomorphism of semigroups $\hat{\pi}_*^L\colon \hat{L}_\epsilon(X) \to L_{\epsilon\epsilon_n}(B)$. We must show that it factors over $\hat{L}_\epsilon(X) \to L_\epsilon(X)$. Let $\mathcal{L} \subset \mathcal{F}$ be a lagrangian subsheaf. It leads to a compatible (see Definition 2.6) filtration $\big(F^i\mathcal{F}\big)_{i=0,1}$ of length 1 by $F^0\mathcal{F} := \mathcal{F}$, $F^1\mathcal{F} := \mathcal{L}$.

We obtain an induced filtration $\big(F^i\pi(\mathcal{F})\big)_{i=0,1}$ such that $F^1\pi(\mathcal{F}) = \mathrm{im}(HR\pi_*\mathcal{L} \to HR\pi_*\mathcal{F})$ of length 1 which is compatible with $\pi(q)$. Here one has to check that $F^1\pi(\mathcal{F})$ is a lagrangian subsheaf. This can be verified either directly by looking at the long exact cohomology sequence associated to $0 \to \mathcal{L} \to \mathcal{F} \to \mathcal{F}/\mathcal{L} \to 0$, or by invoking the spectral sequence induced by the filtration and the discussion in Subsection 4.4.1 below.

By a combination of Lemmas 2.5 (1) and 2.7, we have $[\pi(\mathcal{F}), \pi(q)] = 0$. □

Note that $(\pi(\mathcal{F}), \pi(q))$ comes with a natural $\mathbb{Z}$-grading of length n. Thus we can apply Lemma 2.5 to reduce to the middle term. In particular, we have $\pi_*^L = 0$ if n is odd.

REMARK 2.21. In [**21**], Lott defines $\pi_*^{L,\mathrm{Lott}}\colon L_\epsilon^{\mathrm{Lott}}(X) \to L_{\epsilon\epsilon_n}^{\mathrm{Lott}}(B)$ which induces π_*^L by passing to quotients.

2.5.2. *Functoriality.* Here we will show that π_*^L is functorial with respect to iterated fibre bundles. Let $\pi_1\colon W \to V$ and $\pi_2\colon V \to S$ be locally trivial fibre bundles where the fibres are closed topological manifolds. We assume that both bundles come equipped with fibrewise orientations. We set $\pi_3 := \pi_2 \circ \pi_1$. We furthermore assume that $\pi_3\colon W \to S$ is a locally trivial fibre bundle with fibrewise orientation with fiber Z. The fibre Z_s ($s \in S$) is itself a fibre bundle. We assume that this bundle structure is preserved by the transition maps between local trivializations of π_3. We assume that the orientation of the fibres Z_s are induced by the orientations of the fibres of π_1 and π_2. We will call this situation an iterated fibre bundle with compatible fibrewise orientations.

In the smooth category, i.e., if W, V, and S are smooth manifolds and if π_1, π_2, and π_3 are smooth maps, to give a locally trivial fibre bundle is the same as to give a proper submersion. Furthermore, to give a fibrewise orientation is the same as to give an orientation of the vertical bundle. The composition of proper submersion is again a proper submersion. Thus if $\pi_1\colon W \to V$ and $\pi_2\colon V \to S$ are locally trivial fibre bundles in the smooth category with fibre X, Y, then automatically $\pi_3\colon W \to S$ is a locally trivial fibre bundle with fiber Z. Using the existence of connections and parallel transport on can produce local trivializations of π_3 preserving the bundle structure of the fibres. The compatible fibrewise orientation is obtained by the orientation of the vertical bundle TZ, which can be identified with the sum of oriented vertical bundles $TX \oplus \pi_1^*TY$.

We now turn back to the general situation.

THEOREM 2.22. *We have equality of homomorphisms $L_\epsilon(W) \to L_{\epsilon\epsilon_n}(S)$,*

$$\pi_{3*}^L = \pi_{2*}^L \circ \pi_{1*}^L.$$

PROOF. Let $(\mathcal{F}, q)$ be a generator of $L_\epsilon(W)$. We show that the equality $[\pi_3(\mathcal{F}), \pi_3(q)] = [\pi_2(\pi_1(\mathcal{F})), \pi_2(\pi_1(q))]$ using the fibrewise Leray-Serre spectral sequence $({}_{LS}E_r, {}_{LS}d_r)$ ($r \geq 2$) associated to the composition of functors π_{1*} and π_{2*} applied to $\mathcal{F}$ (cf. [**18**, Thm. 3.7.3], [**17**, p. 464]). The term ${}_{LS}E_2$ is given by ${}_{LS}E_2^{p,q} = R^p\pi_{2*}(R^q\pi_{1*}\mathcal{F})$. Furthermore, there are decreasing filtrations $(F^iR^k\pi_{3*}\mathcal{F})_i$ on $R^k\pi_{3*}\mathcal{F}$, $k \in \mathbb{N}$, such that $\mathrm{Gr}(R^k\pi_{3*}\mathcal{F}) \cong \sum_{p+q=k} E_\infty^{p,q}$.

One checks that the filtration $\left(F^i\pi_3(\mathcal{F})\right)_i$ is compatible with $\pi_3(q)$. To do so we can restrict to the fibre over some $s \in S$. Let n be the dimension of the base Y_s of the bundle Z_s. The length of the filtration is n. One has to show that

$$\pi_3(q)(x)(y) = 0 \text{ whenever } x \in F^p\pi_3(\mathcal{F})_s,\ y \in F^q\pi_3(\mathcal{F})_s,\ p+q > n. \tag{2.1}$$

If one computes the cohomology $\pi_3(\mathcal{F})_s = H(Z_s, \mathcal{F}|_{Z_s})$ using the chain complexes $\overline{\Omega}(Z_s, \mathcal{F}|_{Z_s})$ which are functorial in sheaves, and on which one can implement Poincaré duality $\int_{Z_s} : \overline{\Omega}(Z_s, \mathcal{F}|_{Z_s}) \wedge \overline{\Omega}(Z_s, \mathcal{F}^*|_{Z_s}) \to \mathbb{R}$ as well as the filtration $\left(F^i\overline{\Omega}(Z_s, \mathcal{F}|_{Z_s})\right)_i$ leading to the Leray-Serre spectral sequence (cf. [**25**]), then one checks on this level that $\int_{Z_s} \omega\wedge\omega' = 0$ if $\omega \in F^p\overline{\Omega}(Z_s, \mathcal{F}|_{Z_s})$, $\omega' \in F^{p'}\overline{\Omega}(Z_s, \mathcal{F}^*|_{Z_s})$, and $p+p' > n$. It follows that the filtration is compatible with the form Q given by

$Q(\omega,\omega') = \frac{\sqrt{\epsilon\epsilon_n}}{\sqrt{\epsilon}\sqrt{\epsilon_n}} \int_{Z_s}((-1)^{\frac{N(N-1)}{2}+N\dim Z} q(\omega)) \wedge \omega'$, $\omega, \omega' \in \overline{\Omega}(Z_s, \mathcal{F}|_{Z_s})$, where N is the $\mathbb{Z}$-grading of $\overline{\Omega}(Z_s, \mathcal{F}|_{Z_s})$. The form Q induces $\pi_3(q)$ in cohomology. This immediately implies (2.1).

To be more explicit in the smooth case we can take for $\overline{\Omega}(Z_s, \mathcal{F}^*|_{Z_s})$ as $\Omega(Z_s, \mathcal{F}|_{Z_s})$ the smooth $\mathcal{F}$-valued forms so that the symbols above aquire their usual meaning. In this case, $F^i\Omega(Z_s, \mathcal{F}|_{Z_s}) = \pi^*_{1|Z_s}\Omega^{\geq i}(Y_s)\Omega(X, \mathcal{F}|_X)$ (also cf. [**24**, §2.1]).

By Lemma 2.7 we have $[\pi_3(\mathcal{F}), \pi_3(q)] = [\mathrm{Gr}(\pi_3(\mathcal{F})), \mathrm{Gr}(\pi_3(q))]$.

Now $\pi_2(\pi_1(\mathcal{F}))$ can be identified with the term ${}_{LS}E_2(\mathcal{F})$ of the spectral sequence and $\pi_2(\pi_1(q))$ is the induced form ${}_{LS}E_2(q)$ (see Subsection 4.4.1 below). The same model as above can be used to check that the n-th term ${}_{LS}E_n(\mathcal{F})$, $n \geq 2$, is a locally constant sheaf on S with an induced form ${}_{LS}E_n(q)$, which carries a compatible $\mathbb{Z}$-grading and a compatible differential. We obtain the $(n+1)$th stage of the spectral sequence by taking cohomology. Thus, by Lemma 2.5 (2). we have $[{}_{LS}E_n(\mathcal{F}), {}_{LS}E_n(q)] = [{}_{LS}E_{n+1}(\mathcal{F}), {}_{LS}E_{n+1}(q)]$. We conclude with

$$\begin{aligned} [\pi_2(\pi_1(\mathcal{F})), \pi_2(\pi_1(q))] &= [{}_{LS}E_2(\mathcal{F}), {}_{LS}E_2(q)] = [{}_{LS}E_\infty(\mathcal{F}), {}_{LS}E_\infty(q)] \\ &= [\mathrm{Gr}(\pi_3(\mathcal{F})), \mathrm{Gr}(\pi_3(q))] = [\pi_3(\mathcal{F}), \pi_3(q)]. \end{aligned}$$

□

REMARK 2.23. Note that this functoriality does not hold in general for $\pi_*^{L,\mathrm{Lott}}$.

2.5.3. *Compatibility and naturality.* In the present subsection we work in the smooth category. Let $\pi\colon M \to B$ be a locally trivial fibre bundle over a compact base B such that the fibres are compact even-dimensional smooth manifolds. We further assume that the fibrewise tangent bundle TM/B is oriented. Then the bundle has a fibrewise orientation. We have the following maps:

- $\pi_*^{\mathrm{sign}}\colon H^*(M,\mathbb{R}) \to H^*(B,\mathbb{R})$ defined by $\pi_*(\omega) = \int_{M/B} \omega \cup \mathbf{L}(TM/B)$, where $\int_{M/B}$ is integration over the fibre and $\mathbf{L}(TM/B)$ denotes the Hirzebruch $\mathbf{L}$-class of the fibrewise tangent bundle.
- $\pi_!^{\mathrm{sign}}\colon K^0(M) \to K^0(B)$ defined by $\pi^{\mathrm{sign}}([E]) = \mathrm{ind}(D_E^{\mathrm{sign}})$, where D_E^{sign} is the fibrewise signature operator twisted by the complex vector bundle $E \to M$ and $\mathrm{ind}(D_E^{\mathrm{sign}}) \in K^0(B)$ denotes the class of the index bundle.
- $\pi_*^L\colon L_\epsilon(M) \to L_{\epsilon\epsilon_n}(B)$ given in Definition 2.19.

Let $\mathbf{ch}\colon K^0 \to H^{\mathrm{ev}}(.,\mathbb{R})$ denote the natural transformation of ring-valued functors given by the Chern character.

THEOREM 2.24. *The following diagram commutes:*

$$\begin{array}{ccccc} L_\epsilon(M) & \xrightarrow{b} & K^0(M) & \xrightarrow{\mathbf{ch}} & H^{\mathrm{ev}}(M,\mathbb{R}) \\ \pi_*^L \downarrow & & \pi_!^{\mathrm{sign}} \downarrow & & \pi_*^{\mathrm{sign}} \downarrow \\ L_{\epsilon\epsilon_n}(B) & \xrightarrow{b} & K^0(B) & \xrightarrow{\mathbf{ch}} & H^{\mathrm{ev}}(B,\mathbb{R}) \end{array}$$

PROOF. Recall that the map b is defined in Definition 2.18. Commutativity of the right square is the assertion of the index theorem for families applied to the family of twisted fibrewise signature operators. The commutativity of the left square is a consequence of Hodge theory by which we can identify $\mathtt{bundle}(\pi(\mathcal{F}))$ with the bundle $\ker(D_{\mathtt{bundle}(\mathcal{F})_{\mathbb{C}}})$. □

The following proposition is an immediate consequence of the definition.

PROPOSITION 2.25. *The push-forward π^L_* is natural with respect to pull-back of fibre bundles, i.e., given $f\colon B' \to B$ we consider the pull back*

$$\begin{array}{ccc} f^*M & \xrightarrow{f_\sharp} & M \\ {\scriptstyle f^*\pi}\downarrow & & \downarrow{\scriptstyle \pi} \\ B' & \xrightarrow{f} & B \end{array},$$

and we then have $(f^*\pi)^L_* \circ f^*_\sharp = f^* \circ \pi^L_*$.

3. The functor $\bar{L}_\epsilon$

3.1. Introduction and summary. This subsection shall be considered as an introduction to both, the present section and Section 5. Our purpose here is to motivate the introduction of $\bar{L}_\epsilon$.

If we compose the natural transformation $b\colon L_\epsilon \to K^0$ with the Chern character $\mathbf{ch}\colon K^0 \to H^{\mathrm{ev}}(.,\mathbb{R})$, then we obtain a natural transformation $\mathbf{ch}\circ b\colon L_\epsilon \to H^{\mathrm{ev}}(.,\mathbb{R})$. In the present section we study the kernel of this map in detail.

Note that L_ϵ is given by a topological construction. Staying in the topological framework we first define a functor $\bar{L}^{\mathbb{R}/\mathbb{Z}}$ from topological spaces to $\mathbb{Z}_2$-graded groups together with a surjective natural transformation to $\ker(\mathbf{ch}\circ b)$. We can consider $\mathbf{ch}\colon K \to H(.,\mathbb{R})$ as a map between classifying spaces. Its homotopy fibre is again a classifying space of a cohomology theory $K_{\mathbb{R}/\mathbb{Z}}$. The functor $\bar{L}^{\mathbb{R}/\mathbb{Z}}$ is defined by a pull-back of $K^{-1}_{\mathbb{R}/\mathbb{Z}} \to K^0$ via $b\colon L_\epsilon \to K^0$. Using the geometric description of $K_{\mathbb{R}/\mathbb{Z}}$ given in [**20**], one could also obtain a geometric description of $\bar{L}^{\mathbb{R}/\mathbb{Z}}$.

As a cohomology theory $K_{\mathbb{R}/\mathbb{Z}}$ admits wrong-way maps for suitably oriented fibre bundles. The topic of secondary index theory [**20**] is to relate these topological secondary index maps with their analytic counterparts. The main constituent of the construction of the secondary analytic index is the η-form of a family of Dirac operators. The well-definedness and functoriality of the wrong-way maps in the geometric picture encode properties of η-forms. As a formal consequence of its definition we have wrong-way maps for $\bar{L}^{\mathbb{R}/\mathbb{Z}}$ with nice functorial properties.

The functor $\bar{L}^{\mathbb{R}/\mathbb{Z}}$ fits into an exact sequence

$$\begin{aligned} 0 \longrightarrow H^{\mathrm{odd}}(X,\mathbb{R})/\mathrm{im}(\mathbf{ch}\colon K^1(X) \to H^{\mathrm{odd}}(X,\mathbb{R})) \longrightarrow \bar{L}^{\mathbb{R}/\mathbb{Z}}(X) \longrightarrow \\ \longrightarrow \ker(\mathbf{ch}\circ b\colon K^0(X) \longrightarrow H^{\mathrm{ev}}(X,\mathbb{R})) \longrightarrow 0. \end{aligned}$$

It turns out that, working in the smooth category, we can define a refinement $\bar{L}_\epsilon \to \bar{L}^{\mathbb{R}/\mathbb{Z}}$ which fits into the sequence

$$0 \longrightarrow H^{\mathrm{odd}}(X,\mathbb{R}) \longrightarrow \bar{L}_\epsilon(X) \longrightarrow \ker(\mathbf{ch}\circ b\colon K^0(X) \longrightarrow H^{\mathrm{ev}}(X,\mathbb{R})) \longrightarrow 0.$$

The definition of $\bar{L}_\epsilon$ is geometric and not (obviously) related to a cohomology theory. So it can be considered as a nontrivial fact that $\bar{L}_\epsilon$ still admits wrong-way maps with nice functorial properties. It is in fact *the main purpose of the present paper to construct these maps and verify their functorial properties.* These results thus encode some finer properties of the η-forms of families of signature operators twisted with flat vector bundles.

In the present section we give the definition of the secondary L-functors and discuss their simplest properties and relations to other functors. The wrong-way maps are introduced in Section 5 after a digression to η-forms and η-invariants.

3.2. Secondary K-theory.

3.2.1. *Definition of* $K^{-1}_{\mathbb{R}/\mathbb{Z}}$. We are going to recall the definition of the 2-periodic cohomology theory $K_{\mathbb{R}/\mathbb{Z}}$ introduced in [**19**], [**20**]. Let BU be the classifying space of complex K-theory. The Chern character (with real coefficients) is induced by a map $\mathbf{ch}\colon BU \to \prod_{n=1}^\infty K(\mathbb{R},2n)$. The homotopy fibre of this map classifies $K_{\mathbb{R}/\mathbb{Z}}$. In particular, for any space X there is a natural exact sequence of $K^0(X)$-modules

$$\dots \longrightarrow K^{-1}(X) \xrightarrow{\mathbf{ch}} H^{\mathrm{odd}}(X,\mathbb{R}) \longrightarrow K^{-1}_{\mathbb{R}/\mathbb{Z}}(X) \xrightarrow{\beta} K^0(X) \xrightarrow{\mathbf{ch}} H^{\mathrm{ev}}(X,\mathbb{R}) \longrightarrow \dots,$$

where $K^0(X)$ acts on cohomology via the Chern character.

3.2.2. *A geometric description.* For a manifold M we recall the definition of $K^{-1}_{\mathbb{R}/\mathbb{Z}}(M)$ in terms of generators and relations as given in [**20**, Defs. 5, 6]. We form the abelian semigroup $\hat{K}^{-1}_{\mathbb{R}/\mathbb{Z}}(M)$ consisting of isomorphism classes of tuples (E,h^E,∇^E,ρ), where $E = E_+ \oplus E_-$ is a $\mathbb{Z}_2$-graded complex vector bundle, h^E is a hermitian metric and $\nabla^E = \nabla^{E_+}\oplus\nabla^{E_-}$ is a metric connection, both being compatible with the grading, and $\rho \in \Omega^{\mathrm{odd}}(M)/\mathrm{im}(d)$ satisfies $d\rho = \mathbf{ch}(\nabla^E) := \mathbf{ch}(\nabla^{E_+}) - \mathbf{ch}(\nabla^{E_-})$. The semigroup operation is induced by the direct sum of generators. On $\hat{K}^{-1}_{\mathbb{R}/\mathbb{Z}}(M)$ we consider the minimal equivalence relation $\sim$ which is compatible with the semigroup structure and such that the following holds:

(1) (Change of connections) We have $(E,h^E,\nabla,\rho) \sim (E,h^{E'},\nabla',\rho')$ if and only if $\rho' = \rho + \tilde{\mathbf{ch}}(\nabla',\nabla)$. (See Section 3.6 for a definition of the transgression Chern form.)
(2) (Trivial elements) If (E,h^E,∇^E) is a $\mathbb{Z}_2$-graded hermitean vector bundle with connection, then $(E\oplus E^{\mathrm{op}}, h^{E\oplus E}, \nabla^E\oplus\nabla^E, 0) \sim 0$, where E^{op} denotes E with the opposite grading.

The group $K^{-1}_{\mathbb{R}/\mathbb{Z}}(M)$ is the quotient of $\hat{K}^{-1}_{\mathbb{R}/\mathbb{Z}}(M)$ by $\sim$. By $[E,h^E,\nabla^E,\rho]$ we denote the class of (E,h^E,∇^E,ρ) in $K^{-1}_{\mathbb{R}/\mathbb{Z}}(M)$.

It was shown by Lott [**20**, §2] that the group given by this geometric definition is naturally isomorphic to the topological defined object of Subsection 3.2.1.

3.3. The functor $\bar{L}^{\mathbb{R}/\mathbb{Z}}$.

3.3.1. *The definition.*

DEFINITION 3.1. We define the functor $X \mapsto \bar{L}^{\mathbb{R}/\mathbb{Z}}(X)$ from Top_{para} to groups by the following pull-back diagram:

$$\begin{array}{ccc} \bar{L}^{\mathbb{R}/\mathbb{Z}}(X) & \longrightarrow & L(X) \\ \downarrow & & \downarrow b \\ K^{-1}_{\mathbb{R}/\mathbb{Z}}(X) & \xrightarrow{\beta} & K^0(X) \end{array}$$

Especially, $\bar{L}^{\mathbb{R}/\mathbb{Z}}(X) = \{(u,v) \in K^{-1}_{\mathbb{R}/\mathbb{Z}}(X) \times L(X) \mid \beta u = bv\}$.

On morphisms the functor $\bar{L}^{\mathbb{R}/\mathbb{Z}}$ only depends on homotopy classes. Note that $K^0(X)$ and $K^{-1}_{\mathbb{R}/\mathbb{Z}}(X)$ are $L(X)$-modules via b. This induces an $L(X)$-module structure on $\bar{L}^{\mathbb{R}/\mathbb{Z}}(X)$. We have the following natural exact sequence of $L(X)$-modules

$$K^{-1}(X) \xrightarrow{\mathbf{ch}} H^{\mathrm{odd}}(X,\mathbb{R}) \longrightarrow \bar{L}^{\mathbb{R}/\mathbb{Z}}(X) \longrightarrow L(X) \xrightarrow{\mathbf{ch}\circ b} H^{\mathrm{ev}}(X,\mathbb{R}).$$

3.3.2. *Secondary push-forwards.* Let $\pi\colon X \to B$ be a smooth locally trivial fibre bundle with closed even n-dimensional fibres and equipped with an orientation of the vertical tangent bundle TX/B. In order to define an index map for $K_{\mathbb{R}/\mathbb{Z}}$ we need the further assumption that π is K-oriented. Thus assume that TX/B has a Spin_c-structure. Then there are maps $\pi_!^{\mathrm{Spin}_c}\colon K^0(X) \to K^0(B)$ and $\pi_!^{\mathrm{Spin}_c,\mathbb{R}/\mathbb{Z}}\colon K^{-1}_{\mathbb{R}/\mathbb{Z}}(X) \to K^{-1}_{\mathbb{R}/\mathbb{Z}}(B)$ (the topological secondary index, cf. e.g. [**20**]), such that the following diagram commutes:

$$\begin{array}{ccccccc} H^{\mathrm{odd}}(X,\mathbb{R}) & \longrightarrow & K^{-1}_{\mathbb{R}/\mathbb{Z}}(X) & \longrightarrow & K^0(X) & \longrightarrow & H^{\mathrm{ev}}(X,\mathbb{R}) \\ \pi_*^{\mathrm{Spin}_c}\downarrow & & \pi_!^{\mathrm{Spin}_c,\mathbb{R}/\mathbb{Z}}\downarrow & & \pi_!^{\mathrm{Spin}_c}\downarrow & & \pi_*^{\mathrm{Spin}_c}\downarrow \\ H^{\mathrm{odd}}(B,\mathbb{R}) & \longrightarrow & K^{-1}_{\mathbb{R}/\mathbb{Z}}(B) & \longrightarrow & K^0(B) & \longrightarrow & H^{\mathrm{ev}}(B,\mathbb{R}) \end{array},$$

where $\pi_*^{\mathrm{Spin}_c}(\omega) = \int_{X/B} \hat{\mathbf{A}}(TX/B) \cup e^{c_1/2} \cup \omega$ and c_1 is the first Chern class determined by the Spin_c-structure. There is an unique element $E_{\mathrm{sign}} \in K_0(X)$ such that $\pi_!^{\mathrm{sign}}(x) = \pi_!^{\mathrm{Spin}_c}(E_{\mathrm{sign}} \bullet x)$. Note that $\mathbf{ch}(E_{\mathrm{sign}}) \cup \hat{\mathbf{A}}(TX/B) \cup e^{c_1/2} = \mathbf{L}(TX/B)$.

DEFINITION 3.2. (1) The secondary signature index map

$$\pi_!^{\mathrm{sign},\mathbb{R}/\mathbb{Z}}\colon K^{-1}_{\mathbb{R}/\mathbb{Z}}(X) \to K^{-1}_{\mathbb{R}/\mathbb{Z}}(B)$$

is defined by

$$\pi_!^{\mathrm{sign},\mathbb{R}/\mathbb{Z}}(x) := \pi_!^{\mathrm{Spin}_c,\mathbb{R}/\mathbb{Z}}(E_{\mathrm{sign}} \bullet x).$$

(2) The push-forward $\pi_*^{\bar{L},\mathbb{R}/\mathbb{Z}}\colon \bar{L}^{\mathbb{R}/\mathbb{Z}}(X) \to \bar{L}^{\mathbb{R}/\mathbb{Z}}(B)$ is defined as the map induced by $\pi_!^{\mathrm{sign},\mathbb{R}/\mathbb{Z}}$ and π_*^L.

COROLLARY 3.3. *The following diagrams commute:*

$$\begin{array}{ccccccc} H^{\mathrm{odd}}(X,\mathbb{R}) & \longrightarrow & K^{-1}_{\mathbb{R}/\mathbb{Z}}(X) & \longrightarrow & K^0(X) & \longrightarrow & H^{\mathrm{ev}}(X,\mathbb{R}) \\ \pi_*^{\mathrm{sign}}\downarrow & & \pi_!^{\mathrm{sign},\mathbb{R}/\mathbb{Z}}\downarrow & & \pi_!^{\mathrm{sign}}\downarrow & & \pi_*^{\mathrm{sign}}\downarrow \\ H^{\mathrm{odd}}(B,\mathbb{R}) & \longrightarrow & K^{-1}_{\mathbb{R}/\mathbb{Z}}(B) & \to & K^0(B) & \longrightarrow & H^{\mathrm{ev}}(B,\mathbb{R}), \end{array}$$

$$\begin{array}{ccccccc} H^{\mathrm{odd}}(X,\mathbb{R}) & \longrightarrow & \bar{L}^{\mathbb{R}/\mathbb{Z}}(X) & \longrightarrow & L(X) & \longrightarrow & H^{\mathrm{ev}}(X,\mathbb{R}) \\ \pi_*^{\mathrm{sign}}\downarrow & & \pi_*^{\bar{L},\mathbb{R}/\mathbb{Z}}\downarrow & & \pi_*^{L}\downarrow & & \pi_*^{\mathrm{sign}}\downarrow \\ H^{\mathrm{odd}}(B,\mathbb{R}) & \longrightarrow & \bar{L}^{\mathbb{R}/\mathbb{Z}}(B) & \to & L(B) & \longrightarrow & H^{\mathrm{ev}}(B,\mathbb{R}). \end{array}$$

All push-forward maps are natural with respect to the pull-back of fibre bundles. Moreover, they are functorial with respect to iterated fibre bundles. In greater detail we have the following: Let $\pi_1\colon W \to V$ and $\pi_2\colon V \to S$ be locally trivial smooth fibre bundles with closed even-dimensional fibres X, Y. Further assume that the vertical bundles TX and TY are oriented and carry Spin_c-structures. Then the composition $\pi_3 = \pi_2 \circ \pi_1\colon W \to S$ is a locally trivial fibre bundle with closed even-dimensional fibres Z and the vertical bundle TZ carries an induced orientation and Spin_c-structure. In this situation the index maps on complex K-theory, $K^{-1}_{\mathbb{R}/\mathbb{Z}}$-theory, L-theory (Theorem 2.22), and in cohomology behave functorially with respect to the iterated fibre bundle. It now follows immediately from the definition that

COROLLARY 3.4. $\pi_{3*}^{\bar{L},\mathbb{R}/\mathbb{Z}} = \pi_{2*}^{\bar{L},\mathbb{R}/\mathbb{Z}} \circ \pi_{1*}^{\bar{L},\mathbb{R}/\mathbb{Z}}$.

3.4. The functor $\bar{L}$.

3.4.1. *Definition of* $\bar{L}$. The functor $X \mapsto \bar{L}^{\mathbb{R}/\mathbb{Z}}(X)$ from Top_{met} to $L(X)$-modules was defined by a purely homotopy-theoretic construction as an extension of the functor

$$X \mapsto \ker(\mathbf{ch} \circ b\colon L(X) \to H^{\mathrm{ev}}(X,\mathbb{R}))$$

by $X \mapsto H^{\mathrm{odd}}(X,\mathbb{R})/\mathbf{ch}(K^{-1}(X))$.

Let Top_{smooth} denote the full subcategory of paracompact metrizable spaces Top_{met} which are homotopy equivalent to smooth manifolds. In the present subsection we use a differential geometric construction to define on Top_{smooth} a functor $X \mapsto \bar{L}(X)$ to graded $L(X)$-modules which extends $X \mapsto \ker(\mathbf{ch} \circ b)$ by $X \mapsto H^{\mathrm{odd}}(X,\mathbb{R})$.

It suffices to define $\bar{L}$ as a homotopy invariant functor on the category of smooth manifolds and smooth maps. Again we start with defining a abelian semigroups $\hat{\bar{L}}_\epsilon(M)$, $\epsilon = \pm 1$, with identity and obtain the group $\bar{L}_\epsilon(M)$ as the quotient of $\hat{\bar{L}}_\epsilon(M)$ by an equivalence relation.

Let $(\mathcal{F}, q)$ be a representative of an element of $\hat{L}_\epsilon(M)$. Then we have a real vector bundle $F := \mathtt{bundle}(\mathcal{F})$ which carries a natural flat connection ∇^F such that $\ker(\nabla^F) = \mathcal{F}$. The form $Q \in \mathrm{Hom}(F, F^*)$ which is induced by q is parallel with respect to ∇^F. Let J be a smooth metric structure on (F, Q). It induces a $\mathbb{Z}_2$-grading $z^J := \frac{1}{\sqrt{\epsilon}} J$ of the complexification $F_\mathbb{C}$. In general, since J is not parallel with respect to ∇^F, this grading is not preserved by the connection $\nabla^{F_\mathbb{C}}$ induced by ∇^F. The even part of $\nabla^{F_\mathbb{C}}$ with respect to z^J is a connection on $F_\mathbb{C}$ which preserves the $\mathbb{Z}_2$-grading. It will be denoted by $\nabla^{F_\mathbb{C},J}$. Then $\nabla^{F_\mathbb{C},J} = \nabla^{F_\mathbb{C}} + \frac{1}{2} J^{-1}(\nabla^F J)$. Let $(\Omega^*(M), d)$ be the real de Rham complex of M. We can use this connection to define a characteristic form which represents the Chern class of $F_\mathbb{C}$.

DEFINITION 3.5. We define $p(\nabla^F, J) \in \Omega^{4*-\epsilon+1}(M)$ by

$$p(\nabla^F, J) := \mathbf{ch}(\nabla^{F_\mathbb{C},J}) := \mathrm{Tr}_s\left[\exp(-(\nabla^{F_\mathbb{C},J})^2/2\pi\mathrm{i})\right] := \mathrm{Tr}\left[z^J \exp(-(\nabla^{F_\mathbb{C},J})^2/2\pi\mathrm{i})\right]. \tag{3.1}$$

An element of $\hat{\hat{L}}_\epsilon(M)$ is an isomorphism class of tuples $(\mathcal{F}, q, J, \rho)$, where $(\mathcal{F}, q)$ is a representative of an element of $\hat{L}_\epsilon(M)$, J is a metric structure on $F := \mathtt{bundle}(\mathcal{F})$, and $\rho \in \Omega^{4*-\epsilon}(M)/\mathrm{im}(d)$ satisfies $d\rho = p(\nabla^F, J)$. The semigroup operation is induced by direct sum of representatives:

$$(\mathcal{F}, q, J, \rho) + (\mathcal{F}', q', J', \rho') := (\mathcal{F} \oplus \mathcal{F}', q \oplus q', J \oplus J', \rho + \rho').$$

Before introducing the equivalence relation we recall the definition of the transgression Chern form. Let $E \to M$ be a $\mathbb{Z}_2$-graded complex vector bundle and let ∇, ∇' be two connections on E preserving the grading. Then we consider the bundle $\tilde{E} := \mathrm{pr}^* E \to [0,1] \times M$ with connection $\tilde{\nabla}$ which is given by $\tilde{\nabla}_{\partial t} := \partial_t$ (t is the coordinate in $[0,1]$) and $\tilde{\nabla}_X = (1-t)\nabla_X + t\nabla'_X$ for $X \in TM$. We decompose $\mathbf{ch}(\tilde{\nabla}) = dt \wedge \gamma + r$, where r does not contain dt and $\gamma \colon [0,1] \to \Omega(M)$ is a smooth family of forms.

DEFINITION 3.6. The transgression Chern form is defined by

$$\tilde{\mathbf{ch}}(E, \nabla', \nabla) := \int_0^1 \gamma(t)dt.$$

It satisfies

$$d\tilde{\mathbf{ch}}(E, \nabla', \nabla) = \mathbf{ch}(\nabla') - \mathbf{ch}(\nabla).$$

We now introduce the equivalence relation which is again generated by *lagrangian reduction.* We consider $(\mathcal{F}, q)$ and a metric structure J on F. Let $\mathcal{L} \subset \mathcal{F}$ be a locally constant lagrangian subsheaf and $L := \mathtt{bundle}(\mathcal{L})$. Then we have a decomposition $F = L \oplus J(L)$. Let $\nabla^\oplus$ denote be the part of $\nabla^{F_\mathbb{C},J}$ which preserves this decomposition.

DEFINITION 3.7. We define

$$\tilde{p}(\mathcal{F}, q, J, \mathcal{L}) := \tilde{\mathbf{ch}}(F_\mathbb{C}, \nabla^\oplus, \nabla^{F_\mathbb{C},J}).$$

We require that the equivalence relation contains the relation

$$(\mathcal{F}, q, J, \rho) \sim \big(0, 0, 0, \rho + \tilde{p}(\mathcal{F}, q, J, \mathcal{L})\big).$$

Note that $(0, 0, 0, \rho+\tilde{p}(\mathcal{F}, q, J, \mathcal{L}))$ is a generator of $\hat{\hat{L}}_\epsilon(M)$, since $\mathbf{ch}(\nabla^\oplus) = \frac{1}{\sqrt{\epsilon}}\operatorname{Tr}[J \exp(-\nabla^{\oplus,2}/2\pi i)] = 0$ and hence

$$d\rho + d\tilde{p}(\mathcal{F}, q, J, \mathcal{L}) = \mathbf{ch}(\nabla^{F_{\mathbb{C}},J}) + \mathbf{ch}(\nabla^\oplus) - \mathbf{ch}(\nabla^{F_{\mathbb{C}},J}) = 0.$$

Then we extend $\sim$ to the minimal equivalence relation on $\hat{\hat{L}}_\epsilon(M)$ which contains lagrangian reduction and which is compatible with the semigroup structure.

DEFINITION 3.8. We define the semigroup $\bar{L}_\epsilon(M) := \hat{\hat{L}}_\epsilon(M)/\sim$.

By $[\mathcal{F}, q, J, \rho]$ we denote the class in $\bar{L}_\epsilon(M)$ represented by $(\mathcal{F}, q, J, \rho)$.

LEMMA 3.9. *$\bar{L}_\epsilon(M)$ is a group.*

PROOF. We have $[\mathcal{F}, q, J, \rho] + [\mathcal{F}, -q, -J, -\rho] = 0$. Indeed, we consider the locally constant lagrangian subsheaf $\mathcal{L} \subset \mathcal{F} \oplus \mathcal{F}$ which is the image of the diagonal embedding $\mathcal{F} \to \mathcal{F} \oplus \mathcal{F} =: \bar{\mathcal{F}}$. Let $\bar{J} := J \oplus (-J)$ be the metric structure on $\bar{F} := \mathtt{bundle}(\bar{\mathcal{F}})$. Then $\nabla^{\bar{F}_{\mathbb{C}},\bar{J}}$ already preserves the decomposition $\bar{F} = L \oplus \bar{J}L$. Hence $\tilde{p}(\bar{\mathcal{F}}, \bar{q}, \bar{J}, \mathcal{L}) = 0$, where $\bar{q} := q \oplus -q$. Therefore, we have $(\bar{\mathcal{F}}, \bar{q}, \bar{J}, 0) \sim 0$. □

3.4.2. *Change of the metric structure.* We consider $(\mathcal{F}, q)$ and two metric structures J, J' on the associated bundle F. We define the transgression form

$$\tilde{p}(\nabla^F, J', J) := \tilde{\mathbf{ch}}(F_{\mathbb{C}}, \nabla^{F_{\mathbb{C}},J'}, \nabla^{F_{\mathbb{C}},J})$$

such that we have

$$d\tilde{p}(\nabla^F, J', J) = p(\nabla^F, J') - p(\nabla^F, J).$$

LEMMA 3.10. *In $\bar{L}_\epsilon(M)$ we have*

$$[\mathcal{F}, q, J', \rho'] - [\mathcal{F}, q, J, \rho] = [0, 0, 0, \rho' - \rho - \tilde{p}(\nabla^F, J', J)].$$

PROOF. Let $\mathcal{M}$ be the space of all metric structures on M and $\mathrm{pr}\colon M \times \mathcal{M} \to M$ be the projection. Let $\tilde{\mathcal{F}} := \mathrm{pr}^*\mathcal{F}$, $\tilde{q} := \mathrm{pr}^* q$. On the associated bundle $\tilde{F}$ we consider the tautological metric structure $\tilde{J}$ which over $M \times \{J'\}$ restricts to J'.

Furthermore, we consider the sheaf $\hat{\mathcal{F}} := \mathcal{F} \oplus \mathcal{F}$ with the form $\hat{q} = q \oplus (-q)$ and the metric structure $\hat{J} = \tilde{J} \oplus (-J)$ on the associated bundle $\hat{F}$.

Finally, we define $\bar{\mathcal{F}} := \mathrm{pr}^*\hat{\mathcal{F}}$ and $\bar{q} = \mathrm{pr}^*\hat{q}$. As metric structure on $\bar{F}$ we take $\bar{J} = \tilde{J} \oplus (-\mathrm{pr}^* J)$. We set $\bar{J}_0 := \mathrm{pr}^* J \oplus (-\mathrm{pr}^* J)$.

Let $\mathcal{L} \subset \hat{\mathcal{F}}$ be the image of the diagonal $\mathcal{F} \to \mathcal{F} \oplus \mathcal{F}$. Furthermore, we set $\bar{\mathcal{L}} = \mathrm{pr}^*\mathcal{L}$. We define the form $\omega := \tilde{p}(\nabla^{\tilde{F}}, \tilde{J}, \mathrm{pr}^* J) + \tilde{p}(\bar{\mathcal{F}}, \bar{q}, \bar{J}, \bar{\mathcal{L}})$ on $M \times \mathcal{M}$. It is closed, since

$$\begin{aligned} d\omega &= p(\nabla^{\tilde{F}}, \tilde{J}) - p(\nabla^{\tilde{F}}, \mathrm{pr}^* J) - \mathbf{ch}(\nabla^{\bar{F}_{\mathbb{C}},\bar{J}}) \\ &= p(\nabla^{\tilde{F}}, \tilde{J}) - p(\nabla^{\tilde{F}}, \mathrm{pr}^* J) - p(\nabla^{\tilde{F}}, \tilde{J}) + p(\nabla^{\tilde{F}}, \mathrm{pr}^* J) = 0. \end{aligned}$$

The metric structure J' induces an embedding $i_{J'}\colon M \to M \times \{J'\} \to M \times \mathcal{M}$. We have

$$\tilde{p}(\nabla^F, J', J) = i_{J'}^* \tilde{p}(\nabla^{\tilde{F}}, \tilde{J}, \mathrm{pr}^* J)$$
$$\tilde{p}(\hat{F}, \hat{q}, \hat{J}, \mathcal{L}) = i_{J'}^* \tilde{p}(\bar{\mathcal{F}}, \bar{q}, \bar{J}, \bar{\mathcal{L}}).$$

Therefore,

$$\tilde{p}(\nabla^F, J', J) + \tilde{p}(\hat{F}, \hat{q}, \hat{J}, \mathcal{L}) = i_{J'}^* \omega.$$

Since ω is closed, $i_{J'}$ is homotopic to i_J, and

$$i_J^* \tilde{p}(\nabla^{\tilde{F}}, \mathrm{pr}^* J, \tilde{J}) = 0 = i_J^* \tilde{p}(\bar{\mathcal{F}}, \bar{q}, \bar{J}, \bar{\mathcal{L}}),$$

we conclude that $\tilde{p}(\nabla^F, J', J) + \tilde{p}(\hat{F}, \hat{q}, \hat{J}, \mathcal{L})$ is exact. In $\bar{L}_\epsilon(M)$ we have the identity

$$\begin{aligned}[][\mathcal{F}, q, J', \rho'] - [\mathcal{F}, q, J, \rho] &= [\hat{\mathcal{F}}, \hat{q}, \hat{J}, \rho' - \rho] \\ &= [0, 0, 0, \rho' - \rho + p(\hat{F}, \hat{q}, \hat{J}, \mathcal{L})] = [0, 0, 0, \rho' - \rho - \tilde{p}(\nabla^F, J', J)].\end{aligned}$$

□

REMARK 3.11. In [**21**, Def. 23], Lott defined a similar functor $\bar{L}^{\mathrm{Lott}}_\epsilon(M)$. It is also obtained as a quotient of $\hat{\bar{L}}_\epsilon(M)$ by an equivalence relation $\sim^{\mathrm{Lott}}$. This relation is the minimal equivalence relation which is compatible with the semigroup operation and which contains $(\mathcal{F}, q, J', \rho') \sim (\mathcal{F}, q, J, \rho) + (0, 0, 0, \rho' - \rho - \tilde{p}(\nabla^F, J', J))$, and "lagrangian reduction" in the special case that the lagrangian subsheaf admits a lagrangian complement. As a consequence of Lemma 3.10 the relation $\sim$ is coarser than $\sim^{\mathrm{Lott}}$ so that we have a natural surjective map $\bar{L}^{\mathrm{Lott}}_\epsilon(M) \to \bar{L}_\epsilon(M)$.

3.4.3. *The module structure.*

DEFINITION 3.12. The graded module structure of $\bar{L}(M) = \bigoplus_{\epsilon \in \mathbb{Z}_2} \bar{L}_\epsilon(M)$ over $L(M)$ is defined by

$$[\mathcal{F}, q, J, \rho] \bullet [\mathcal{E}, p] := \frac{\sqrt{\epsilon}\sqrt{\epsilon'}}{\sqrt{\epsilon\epsilon'}} [\mathcal{F} \otimes \mathcal{E}, q \otimes p, J \otimes J^E, \rho \wedge \mathbf{ch}(\nabla^{E_{\mathbb{C}}, J^E})],$$

where J^E is any metric structure on $\mathtt{bundle}(\mathcal{E})$.

LEMMA 3.13. *The $L(M)$-module structure of $\bar{L}(M)$ is well-defined.*

PROOF. Let J_i^E, $i = 0, 1$ be two choices of metric structures on $(\mathcal{E}, q)$. Then we must show that

$$[\mathcal{F} \otimes \mathcal{E}, q \otimes p, J \otimes J_0^E, \rho \wedge \mathbf{ch}(\nabla^{E_{\mathbb{C}}, J_0^E})] = [\mathcal{F} \otimes \mathcal{E}, q \otimes p, J \otimes J_1^E, \rho \wedge \mathbf{ch}(\nabla^{E_{\mathbb{C}}, J_1^E})].$$

In view of Lemma 3.10 we must show that

$$\rho \wedge \mathbf{ch}(\nabla^{E_{\mathbb{C}}, J_0^E}) - \rho \wedge \mathbf{ch}(\nabla^{E_{\mathbb{C}}, J_1^E}) - \tilde{p}(\nabla^{F \otimes E}, J \otimes J_0^E, J \otimes J_1^E)$$

is exact. We compute

$$\begin{aligned}
&\rho \wedge \mathbf{ch}(\nabla^{E_{\mathbb{C}},J_0^E}) - \rho \wedge \mathbf{ch}(\nabla^{E_{\mathbb{C}},J_1^E}) - \tilde{p}(\nabla^{F\otimes E}, J\otimes J_0^E, J\otimes J_1^E) \\
&\quad = \rho \wedge d\tilde{\mathbf{ch}}(\nabla^{E_{\mathbb{C}},J_0^E}, \nabla^{E_{\mathbb{C}},J_1^E}) - \tilde{\mathbf{ch}}(\nabla^{F_{\mathbb{C}}\otimes E_{\mathbb{C}},J\otimes J_0^E}, \nabla^{F_{\mathbb{C}}\otimes E_{\mathbb{C}},J\otimes J_1^E}) \\
&\quad = -d\big(\rho \wedge \tilde{\mathbf{ch}}(\nabla^{E_{\mathbb{C}},J_0^E}, \nabla^{E_{\mathbb{C}},J_1^E})\big) + d\rho \wedge \tilde{\mathbf{ch}}(\nabla^{E_{\mathbb{C}},J_0^E}, \nabla^{E_{\mathbb{C}},J_1^E}) \\
&\qquad - \mathbf{ch}(\nabla^{F_{\mathbb{C}},J}) \wedge \tilde{\mathbf{ch}}(\nabla^{E_{\mathbb{C}},J_0^E}, \nabla^{E_{\mathbb{C}},J_1^E}) \\
&\quad = -d\big(\rho \wedge \tilde{\mathbf{ch}}(\nabla^{E_{\mathbb{C}},J_0^E}, \nabla^{E_{\mathbb{C}},J_1^E})\big).
\end{aligned}$$

□

3.4.4. *Complexes and η-forms.* We consider a pair $(\mathcal{F}, q)$ together with a $\mathbb{Z}$-grading N of length n and a compatible differential v (cf. Definition 2.4).

DEFINITION 3.14. A metric structure J is called compatible with N if $JN + NJ = nJ$.

Equivalently, one could require that the decomposition $F = \bigoplus_{k=0}^{n} F^k$ of F into eigenspaces of N is orthogonal with respect to the metric induced by J.

The adjoint of v with respect to the metric induced by J is given by $v^{*_J} = -\epsilon J \circ v \circ J$. Set $V = v^{*_J} + v$. By Hodge theory we can canonically identify the cohomology bundle $H = \mathtt{bundle}(\mathcal{H})$ with $\ker(v) \cap \ker(v^{*_J})$. Since the latter is J-invariant we obtain an induced metric structure $J^H := J|_H$ for $(\mathcal{H}, q_{\mathcal{H}})$.

The theory of characteristic classes and forms extends to superconnections (cf. [**27**], [**3**, §1.4]). Since we consider several $\mathbb{Z}_2$-gradings at the same time we will speak of z-superconnection in order to indicate that z is the relevant grading. In particular, given a z-superconnection A we set

$$\mathbf{ch}(A) := \varphi \mathrm{Tr}\Big[\, z \, \exp(-A^2)\Big],$$

where φ multiplies a p-form by $(2\pi i)^{-p/2}$. Furthermore, if A' is a $(-1)^N$-superconnection on F, then the odd part A with respect to $z^J = \frac{1}{\sqrt{\epsilon}} J$ is a z^J-superconnection. In this case we define $p(A', J) := \mathbf{ch}(A)$.

Let $\tilde{M} := (0,\infty) \times M$ and $\mathrm{pr}\colon \tilde{M} \to M$ be the projection. We consider $(\tilde{\mathcal{F}}, \nabla^{\tilde{F}}) = \mathrm{pr}^*(\mathcal{F}, \nabla^F)$ and $\tilde{q} := \mathrm{pr}^* q$ on $\tilde{M}$. It has the $\mathbb{Z}$-grading $\tilde{N} := \mathrm{pr}^* N$ of length n and the compatible differential $\tilde{v} := \mathrm{pr}^* v$. We further consider the compatible metric structure $\tilde{J}$ which restricts to $t^{-N/2+n/4} \circ J \circ t^{N/2-n/4}$ on $\{t\} \times M$. We have a $(-1)^{\tilde{N}}$-superconnection $A' := \nabla^{\tilde{F}} + \tilde{v}$. Let us decompose $p(A', \tilde{J}) = dt \wedge \gamma + r$, where r does not contain dt. Here $\gamma\colon (0,\infty) \to \Omega(M)$ is a smooth family of forms on M. More precisely,

$$\gamma(t) = -(2\pi i)^{-1/2} \frac{1}{4\sqrt{t}} \varphi \, \mathrm{Tr}\Big[z^J V \exp(-(\nabla^{F_{\mathbb{C}},J} + \frac{\sqrt{t}}{2} V)^2)\Big].$$

Remark that $JV = -VJ$. Thus V is odd with respect to the $\mathbb{Z}_2$-grading z^J. By [**21**, Prop. 28 and Prop. 29], (also cf. [**8**, (2.26)]) the following integral exists.

DEFINITION 3.15. [**21**, Def. 32]. The η-form $\tilde{\eta}(\mathcal{F},N,J,v) \in \Omega(M)$ is defined by

$$\tilde{\eta}(\mathcal{F},N,J,v) := -\int_0^\infty \gamma(t)dt.$$

It was shown in [**21**, (216)] (cf. [**7**, Thm. 2.8]) that

$$d\tilde{\eta}(\mathcal{F},N,J,v) = p(\nabla^F,J) - p(\nabla^H,J^H). \tag{3.2}$$

PROPOSITION 3.16. *In* $\bar{L}_\epsilon(M)$ *we have*

$$[\mathcal{F},q,J,\rho] = [\mathcal{H},q_{\mathcal{H}},J^H,\rho - \tilde{\eta}(\mathcal{F},N,J,v)].$$

PROOF. We consider $(\mathcal{G},q_{\mathcal{G}}) := (\mathcal{F}\oplus\mathcal{H},q\oplus -q_{\mathcal{H}})$. It admits a locally constant lagrangian subsheaf $\mathcal{L}$ given by the image of $\ker(v) \hookrightarrow \mathcal{F}\oplus\mathcal{H}$, $x \mapsto x\oplus[x]$. We consider the metric structure $J^G := J \oplus -J^H$. Then we have

$$\begin{aligned}
&[\mathcal{F},q,J,\rho] - [\mathcal{H},q_{\mathcal{H}},J^H,\rho-\tilde{\eta}(\mathcal{F},N,J,v)]\\
&\qquad = [\mathcal{G},q_{\mathcal{G}},J^G,\tilde{\eta}(\mathcal{F},N,J,v)]\\
&\qquad = [0,0,0,\tilde{\eta}(\mathcal{F},N,J,v)+\tilde{p}(\mathcal{G},q_{\mathcal{G}},J^G,\mathcal{L})].
\end{aligned}$$

It remains to show that $\tilde{\eta}(\mathcal{F},N,J,v)+\tilde{p}(\mathcal{G},q_{\mathcal{G}},J^G,\mathcal{L})$ is exact. This will be a consequence of the following general result.

LEMMA 3.17. *Let* **Eta** *be a construction which associates to a tuple* (F,∇^F,Q,N,J,v) *(a real vector bundle with flat connection, parallel* ϵ*-symmetric form, compatible parallel* $\mathbb{Z}$*-grading, a metric structure (not necessarily parallel), and compatible parallel differential) over a manifold* M *a form* $\mathbf{Eta}(F,\nabla^F,Q,N,J,v) \in \Omega(M)$ *such that*
(1) $d\mathbf{Eta}(F,\nabla^F,Q,N,J,v) = p(\nabla^F,J) - p(\nabla^H,J^H)$.
(2) *For a smooth map* $f\colon M' \to M$ *we have*

$$f^*\mathbf{Eta}(F,\nabla^F,Q,N,J,v) = \mathbf{Eta}(f^*F,f^*\nabla^F,f^*Q,f^*N,f^*J,f^*v).$$

(3) $\mathbf{Eta}(F,\nabla^F,Q,N,J,v)$ *depends smoothly on the data* ∇^F,Q,J *(note that we fix* N *and* v*).*
(4) *If* (F,∇^F,Q,N,J,v) *splits, then* $\mathbf{Eta}(F,\nabla^F,Q,N,J,v) = 0$. *Here* (F,∇^F,Q,N,J,v) *splits if and only if the complex* (F,N,v) *is of the form*

$$0\to E^0\oplus H^0 \to E^0\oplus H^1\oplus E^1 \to \ldots\ldots E^{n-2}\oplus H^{n-1}\oplus E^{n-1} \to \oplus E^{n-1}\oplus H^n \to 0$$

with flat vector bundles E^i, H^i*, where the differential* v *is given by the obvious maps. Furthermore,* $E^{n-i-1} = (E^i)^*$*,* $H^i = (H^{n-i})^*$*, and this identification gives (with the suitable signs) the form* Q*. The metric structure* J *shall induce a metric such that this splitting is orthogonal and the identifications* $E^{n-i-1} = (E^i)^*$*,* $H^i = (H^{n-i})^*$ *are isometries.*

Then $\mathbf{Eta}(F,\nabla^F,Q,N,J,v) - \tilde{\eta}(\mathcal{F},N,J,v)$ *is exact.*

PROOF. The proof is very similar to the axiomatic characterizations of analytic torsion forms [**12**, A 1.2] (in the acyclic case) and its extension to the case with cohomology [**24**, Lemma 3.1].

First one shows that the η-form $\tilde{\eta}(\mathcal{F}, N, J, v)$ has the four properties above. Given another construction $\mathbf{Eta}(F, \nabla^F, Q, N, J, v)$ with these properties one considers the difference $\Delta(F, \nabla^F, Q, N, J, v) := \mathbf{Eta}(F, \nabla^F, Q, N, J, v) - \tilde{\eta}(\mathcal{F}, N, J, v)$. Then one applies the argument of [**12**, A1.2] to show that $\Delta(F, \nabla^F, Q, N, J, v)$ is exact. First observe by property (1) that $\Delta(F, \nabla^F, Q, N, J, v)$ is closed. Then the idea is that we can deform ∇, Q, J to approximate the split case. Using the properties (2), (3), and (4), we see that $\Delta(F, \nabla^F, Q, N, J, v)$ can be smoothly deformed to zero without changing its cohomology class. This implies exactness of $\Delta(F, \nabla^F, Q, N, J, v)$. □

We now finish the proof of Lemma 3.16. Given $(F, \nabla^F, Q, N, J, v)$ we can construct $(\mathcal{G}, q_\mathcal{G}, J^G, \mathcal{L})$ as above. Let $\mathbf{Eta}(F, \nabla^F, Q, N, J, v) := -\tilde{p}(\mathcal{G}, q_\mathcal{G}, J^G, \mathcal{L})$. One checks that it satisfies the four conditions of the lemma. We conclude that $\tilde{\eta}(\mathcal{F}, N, J, v) + \tilde{p}(\mathcal{G}, q_\mathcal{G}, J^G, \mathcal{L})$ is exact. □

3.4.5. *Filtration.* Let $(\mathcal{F}, q, J, \rho)$ be given with a compatible filtration $\left(F^i\mathcal{F}\right)_i$ of length n. We assume that we have chosen a metric structure $J^{\mathrm{Gr}(\mathcal{F})}$ on $\big(\mathrm{Gr}(\mathcal{F}), \mathrm{Gr}(q)\big)$ which is compatible with the $\mathbb{Z}$-grading. For $i < n/2$ we consider the following sheaves

$$\mathcal{E}_i := F^i\mathcal{F}/F^{n-i+1}\mathcal{F} \oplus \mathrm{Gr}^{n-i}(\mathcal{F}) \oplus \mathrm{Gr}^i(\mathcal{F}).$$

We introduce forms q_i and metric structures J_i by induction on i.

The sheaf $\mathcal{E}_i$ admits a $\mathbb{Z}$-grading N_i of length 2 such that $\mathcal{E}_i^0 = \mathrm{Gr}^{n-i}(\mathcal{F})$, $\mathcal{E}_i^1 = F^i\mathcal{F}/F^{n-i+1}$, and $\mathcal{E}_i^2 = \mathrm{Gr}^i(\mathcal{F})$. The differential v_i is given by the embedding $\mathrm{Gr}^{n-i}(\mathcal{F}) \hookrightarrow F^i\mathcal{F}/F^{n-i+1}\mathcal{F}$ and the negative of the projection $F^i\mathcal{F}/F^{n-i+1}\mathcal{F} \to \mathrm{Gr}^i(\mathcal{F})$. For $i = 0$ we have an obvious form $q_0 := q_\mathcal{F} \oplus \mathrm{Gr}(q)|_{\mathrm{Gr}^0(\mathcal{F})\oplus\mathrm{Gr}^n(\mathcal{F})}$ and a compatible metric structure $J_0 := J^F \oplus J^{\mathrm{Gr}(F)}|_{\mathrm{Gr}^0(\mathcal{F})\oplus\mathrm{Gr}^n(\mathcal{F})}$ such that v_0 is compatible.

Assume that we already have defined q_i such that the grading and the differential are compatible. We identify $F^{i+1}\mathcal{F}/F^{n-i}\mathcal{F}$ with the cohomology of v_i. Therefore, we have an induced form $q_{F^{i+1}\mathcal{F}/F^{n-i}\mathcal{F}}$. We now define $q_{i+1} := q_{F^{i+1}\mathcal{F}/F^{n-i}\mathcal{F}} \oplus \mathrm{Gr}(q)|_{\mathrm{Gr}^{n-i-1}(\mathcal{F})\oplus\mathrm{Gr}^{i+1}(\mathcal{F})}$. The grading and the differential of $\mathcal{E}_{i+1}$ are again compatible with q_{i+1}.

Assume that we have already defined J_i such that it is compatible with the $\mathbb{Z}$-grading. Then we have an induced metric structure $J^{F^{i+1}\mathcal{F}/F^{n-i}\mathcal{F}}$. We now define $J_{i+1} := J^{F^{i+1}\mathcal{F}/F^{n-i}\mathcal{F}} \oplus J^{\mathrm{Gr}(F)}|_{\mathrm{Gr}^{n-i-1}(\mathcal{F})\oplus\mathrm{Gr}^{i+1}(\mathcal{F})}$. Then J_{i+1} is again compatible with the $\mathbb{Z}$-grading.

DEFINITION 3.18. We define the η-form $\tilde{\eta}(\mathcal{F}, \mathrm{Gr}(\mathcal{F}), J, J^{\mathrm{Gr}(\mathcal{F})}) \in \Omega(M)$ of the filtration by

$$\tilde{\eta}(\mathcal{F}, \mathrm{Gr}(\mathcal{F}), J, J^{\mathrm{Gr}(\mathcal{F})}) := \sum_{i=0}^{[(n-1)/2]} \tilde{\eta}(\mathcal{E}_i, N_i, J_i, v_i) + \tilde{p}\left(\mathrm{Gr}^{n/2} F, J_{n/2}, J^{\mathrm{Gr}(F)}|_{\mathrm{Gr}^{n/2}(\mathcal{F})}\right).$$

Then

$$d_{,}\tilde{\eta}(\mathcal{F}, \mathrm{Gr}(\mathcal{F}), J, J^{\mathrm{Gr}(\mathcal{F})}) = p(\nabla^F, J) - p(\nabla^{\mathrm{Gr}(\mathcal{F})}, J^{\mathrm{Gr}(\mathcal{F})}).$$

LEMMA 3.19. *In $\bar{L}_\epsilon(M)$ we have*

$$[\mathcal{F}, q, J, \rho] = \left[\mathrm{Gr}(\mathcal{F}), \mathrm{Gr}(q), J^{\mathrm{Gr}(\mathcal{F})}, \rho - \tilde{\eta}\big(\mathcal{F}, \mathrm{Gr}(\mathcal{F}), J, J^{\mathrm{Gr}(\mathcal{F})}\big)\right]$$

$$= \begin{cases} \left[\mathrm{Gr}^{[\frac{n}{2}]}(\mathcal{F}), \mathrm{Gr}(q)|_{\mathrm{Gr}^{[\frac{n}{2}]}(\mathcal{F})}, J^{\mathrm{Gr}(\mathcal{F})}|_{\mathrm{Gr}^{[\frac{n}{2}]}(\mathcal{F})}, \rho - \tilde{\eta}\big(\mathcal{F}, \mathrm{Gr}(\mathcal{F}), J, J^{\mathrm{Gr}(\mathcal{F})}\big)\right] \\ \hfill n \equiv 0\,(2), \\ \left[0, 0, 0, \rho - \tilde{\eta}\big(\mathcal{F}, \mathrm{Gr}(\mathcal{F}), J, J^{\mathrm{Gr}(\mathcal{F})}\big)\right] \\ \hfill n \equiv 1\,(2). \end{cases}$$

PROOF. We reduce the length of the filtration by two as in the proof of Lemma 2.7. Note that

$$\left[\mathrm{Gr}^{n-i}(\mathcal{F}) \oplus \mathrm{Gr}^i(\mathcal{F}), \mathrm{Gr}(q)|_{\mathrm{Gr}^{n-i}(\mathcal{F})\oplus\mathrm{Gr}^i(\mathcal{F})}, J^{\mathrm{Gr}(\mathcal{F})}|_{\mathrm{Gr}^{n-i}(\mathcal{F})\oplus\mathrm{Gr}^i(\mathcal{F})}, 0\right] = 0$$

by lagrangian reduction along the lagrangian subsheaf $\mathrm{Gr}^{n-i}(\mathcal{F})$. In fact, in this case

$$\nabla^{(\mathrm{Gr}^{n-i}(\mathcal{F})\oplus\mathrm{Gr}^i(\mathcal{F}))_{\mathbb{C}}, J^{\mathrm{Gr}(\mathcal{F})}|_{\mathrm{Gr}^{n-i}(\mathcal{F})\oplus\mathrm{Gr}^i(\mathcal{F})}} = \nabla^{\oplus}$$

so that by Definition 3.7

$$\tilde{p}\Big(\mathrm{Gr}^{n-i}(\mathcal{F}) \oplus \mathrm{Gr}^i(\mathcal{F}), \mathrm{Gr}(q)|_{\mathrm{Gr}^{n-i}(\mathcal{F})\oplus\mathrm{Gr}^i(\mathcal{F})}, J^{\mathrm{Gr}(\mathcal{F})}|_{\mathrm{Gr}^{n-i}(\mathcal{F})\oplus\mathrm{Gr}^i(\mathcal{F})}, \mathrm{Gr}^{n-i}(\mathcal{F})\Big) = 0.$$

Therefore, we obtain by Lemma 3.16

$$\left[F^i\mathcal{F}/F^{n-i+1}\mathcal{F}, q_{F^i\mathcal{F}/F^{n-i+1}\mathcal{F}}, J^{F^i\mathcal{F}/F^{n-i+1}\mathcal{F}}, \rho - \sum_{j<i} \tilde{\eta}(\mathcal{E}_i, N_i, J_i, v_i)\right]$$

$$= \left[\mathcal{E}_i, q_i, J_i, \rho - \sum_{j<i} \tilde{\eta}(\mathcal{E}_i, N_i, J_i, v_i)\right]$$

$$= \left[F^{i+1}\mathcal{F}/F^{n-i}\mathcal{F}, q_{F^{i+1}\mathcal{F}/F^{n-i}\mathcal{F}}, J^{F^{i+1}\mathcal{F}/F^{n-i}\mathcal{F}}, \rho - \sum_{j<i+1} \tilde{\eta}(\mathcal{E}_i, N_i, J_i, v_i)\right].$$

The assertion of the lemma now follows by induction on i. □

3.4.6. *Isotropic reduction.* Assume that we are given a tuple $(\mathcal{F}, q, J, \rho)$ and an isotropic locally constant subsheaf $\mathcal{I}$, i.e., $\mathcal{I} \subset \mathcal{I}^\perp$. We consider the filtration of length two of $\mathcal{F}$ such that $\mathcal{I} := F^2\mathcal{F}$, $\mathcal{I}^\perp := F^1\mathcal{F}$ so that we can identify $(\mathrm{Gr}^1(\mathcal{F}), \mathrm{Gr}(q)|_{\mathrm{Gr}^1(\mathcal{F})}) \cong (\mathcal{F}_\mathcal{I}, q_\mathcal{I})$. We choose any metric structure $J^{\mathrm{Gr}(\mathcal{F})}$. Let $J^{F_\mathcal{I}} := J^{\mathrm{Gr}(\mathcal{F})}|_{\mathrm{Gr}^1(\mathcal{F})}$. Then we have the form $\tilde{\eta}(\mathcal{F}, \mathrm{Gr}(\mathcal{F}), J, J^{Gr(\mathcal{F})})$. The following assertion is an immediate consequence of Lemma 3.19.

COROLLARY 3.20. *In $\bar{L}_\epsilon(M)$ we have*

$$[\mathcal{F}, q, J, \rho] = \left[\mathcal{F}_\mathcal{I}, q_\mathcal{I}, J^{F_\mathcal{I}}, \rho - \tilde{\eta}(\mathcal{F}, \mathrm{Gr}(\mathcal{F}), J, J^{\mathrm{Gr}(\mathcal{F})})\right].$$

Sometimes we would like to work with easier transgressed characteristic classes than η-forms. It is an interesting observation that one can do isotropic reduction without appealing to η-forms.

Assume again that we are given a tuple $(\mathcal{F}, q, J, \rho)$ and an isotropic locally constant subsheaf $\mathcal{I}$. Then we obtain an orthogonal decomposition $F = I \oplus (I^\perp \ominus I) \oplus (F \ominus I^\perp)$, here we denote by $F \ominus I^\perp$ the orthogonal complement sub-bundle of $I^\perp$ in F with respect to the metric $J^* \circ q$. Note that J induces an isomorphism $I \cong F \ominus I^\perp$. There is a natural identification of bundles $I^\perp \ominus I = F_\mathcal{I}$. The metric structure J restricts to $I^\perp \ominus I$ and therefore defines a metric structure $J^{F_\mathcal{I}}$ on $F_\mathcal{I}$.

We consider the $\mathbb{Z}$-grading on F which is given with respect to the decomposition above by $N = \mathrm{diag}(0, 1, 2)$. The connection ∇^F is upper-triangular with respect to this decomposition. Therefore,

$$\nabla^F_u := u^{-N} \nabla^F u^N \tag{3.3}$$

is regular at $u = 0$. In fact, $\nabla^F_0 = \nabla^I \oplus \nabla^{F_\mathcal{I}} \oplus \nabla^{F/I^\perp}$. The tuple (F, ∇^F_0, J) is thus isomorphic to the tuple obtained as follows: We consider the sum $\mathcal{F}_\mathcal{I} \oplus (\mathcal{I} \oplus \mathcal{F}/\mathcal{I}^\perp)$ with associated bundle $F_\mathcal{I} \oplus (I \oplus I)$, metric structure

$$J^{\mathcal{F}_\mathcal{I}} \oplus \begin{pmatrix} 0 & 1 \\ \epsilon & 0 \end{pmatrix},$$

and connection $\nabla^{\mathcal{F}_\mathcal{I}} \oplus \nabla^I \oplus (\nabla^I)^*$.

We now consider $\tilde{M} = [0,1] \times M$. Let $\mathrm{pr}\colon \tilde{M} \to M$ be the projection, $\tilde{\mathcal{F}} = \mathrm{pr}^*\mathcal{F}$, $\tilde{q} := \mathrm{pr}^* q$. On the corresponding bundle we consider the metric structure $\tilde{J}$ which on $\{u\} \times M$ restricts to $J_u := u^N J u^{-N}$. We decompose $p(\nabla^{\tilde{F}}, \tilde{J}) = du \wedge \gamma + r$, where r does not contain du and where $\gamma\colon [0,1] \to \Omega(M)$ is a smooth family of forms. By the discussion above, after shifting the rescaling from the metric structure to the connection we see that γ is regular at $u = 0$ so that we can make the following definition.

DEFINITION 3.21. We define

$$\tilde{p}(\mathcal{F}, q, J, \mathcal{I}) := -\int_0^1 \gamma(u)\, du.$$

If $\mathcal{I}$ is lagrangian, then it is easy to check that this definition coincides with the former Def. 3.7. Furthermore, note that

$$d\tilde{p}(\mathcal{F},q,J,\mathcal{I}) = p(\nabla^{F_{\mathcal{I}}}, J^{F_{\mathcal{I}}}) - p(\nabla^{F}, J).$$

Next we consider isotropic reduction in stages. Assume that $\mathcal{L}$ is a locally constant lagrangian subsheaf of $\mathcal{F}$ such that $\mathcal{I} \subset \mathcal{L}$. Then we obtain an induced lagrangian subsheaf $\mathcal{L}_{\mathcal{I}} \subset \mathcal{F}_{\mathcal{I}}$.

LEMMA 3.22. *Modulo exact forms we have*

$$\tilde{p}(\mathcal{F},q,J,\mathcal{L}) \equiv \tilde{p}(\mathcal{F}_{\mathcal{I}},q_{\mathcal{I}},J^{\mathcal{F}_{\mathcal{I}}},\mathcal{L}_{\mathcal{I}}) + \tilde{p}(\mathcal{F},q,J,\mathcal{I}).$$

PROOF. We consider the orthogonal decompositions

$$F = I \oplus (L \ominus I) \oplus (I^{\perp} \ominus L) \oplus (F \ominus I^{\perp})$$

and the commuting $\mathbb{Z}$-gradings $M := \mathrm{diag}(0,1,1,2)$ and $K := \mathrm{diag}(0,0,1,1)$. Note that ∇^F is upper triangular with respect to these gradings. Thus, if we define $\nabla_{u,v} := u^{-M}v^{-K}\nabla^F v^K u^M$, then this family of connections extends to $(u,v) \in [0,1] \times [0,1]$. We identify $F \cong I \oplus F_{\mathcal{I}} \oplus I$ such that $F_{\mathcal{I}} \cong I^{\perp} \ominus I$ and the last component is identified with $F \ominus I^{\perp}$ by J. Furthermore, let $\nabla_v^{F_{\mathcal{I}}}$ be constructed as in (3.3) using $\mathcal{L}_{\mathcal{I}} \subset \mathcal{F}_{\mathcal{I}}$ and $J^{\mathcal{F}_{\mathcal{I}}}$. In this identification, we can write

$$\nabla_{0,v} = \nabla^{I} \oplus \nabla_v^{F_{\mathcal{I}}} \oplus (\nabla^{I})^{*}.$$

We decompose $L := I \oplus (L \ominus I)$ and set $N := \mathrm{diag}(0,1)$. Then we define $\nabla_u^L := u^{-N}\nabla^L u^N$. We now identify $F = L \oplus L$ such that the second component goes to $F \ominus L$ via J. Then we can write

$$\nabla_{u,0} = \nabla_u^L \oplus (\nabla_u^L)^{*}.$$

Let us now consider the manifold $\bar{M} := [0,1] \times [0,1] \times M$. Let $\bar{F} := \mathrm{pr}^* F$ and $\bar{q} = \mathrm{pr}^* q$. We consider the metric structure $\bar{J}$ on $\bar{F}$ which restricts on $\{(u,v)\} \times M$ to $u^M v^K J v^{-K} u^{-M}$. This works for $uv \neq 0$. Nevertheless, it follows from the discussion above after shifting the rescaling from the metric structure to the connection that $p(\bar{\mathcal{F}}, \bar{J})$ is a closed smooth form on $\bar{M}$. We decompose $p(\bar{\mathcal{F}}, \bar{J}) = du \wedge dv \wedge \sigma + r$, where r contains at most one of du, dv, and $\sigma \colon [0,1] \times [0,1] \to \Omega(M)$ is a smooth family of forms. It follows from Stokes' theorem and the structure of $\nabla_{0,v}$ and $\nabla_{u,0}$ that

$$d\int_{[0,1]\times[0,1]} \sigma(u,v)dudv = \tilde{p}(\mathcal{F}_{\mathcal{I}},q_{\mathcal{I}},J^{\mathcal{F}_{\mathcal{I}}},\mathcal{L}_{\mathcal{I}}) + \tilde{p}(\mathcal{F},q,J,\mathcal{I}) - \tilde{p}(\mathcal{F},q,J,\mathcal{L}),$$

where the contributions come from the boundary components $\{u = 0\}$, $\{v = 1\}$, and $\{u = 1\}$, respectively. This proves the assertion. □

LEMMA 3.23. *In $\bar{L}_{\epsilon}(M)$ we have $[\mathcal{F},q,J,\rho] = [\mathcal{F}_{\mathcal{I}},q_{\mathcal{I}},J^{F_{\mathcal{I}}},\rho + \tilde{p}(\mathcal{F},q,J,\mathcal{I})]$.*

PROOF. We consider the sheaf $\mathcal{G} := \mathcal{F} \oplus \mathcal{F}_{\mathcal{I}}$ together with the form $q_{\mathcal{G}} = q \oplus (-q_{\mathcal{I}})$. We have the lagrangian subsheaf $\mathcal{L}$ of $\mathcal{G}$ which is the image of the diagonal map $\mathcal{I}^\perp \to \mathcal{F} \oplus \mathcal{F}_\mathcal{I}$. It contains the isotropic subsheaf $\mathcal{I} \oplus 0$. We consider the metric structure $J^G := J \oplus -J^{F_\mathcal{I}}$. Then we have by Lemma 3.22

$$\begin{aligned}
&[\mathcal{F}, q, J, \rho] - [\mathcal{F}_\mathcal{I}, q_\mathcal{I}, J^{F_\mathcal{I}}, \rho + \tilde{p}(\mathcal{F}, q, J, \mathcal{I})] \\
&\quad = [\mathcal{G}, q_\mathcal{G}, J^G, -\tilde{p}(\mathcal{F}, q, J, \mathcal{I})] \\
&\quad = [0, 0, 0, \tilde{p}(\mathcal{G}, q_\mathcal{G}, J^G, \mathcal{L}) - \tilde{p}(\mathcal{F}, q, J, \mathcal{I})] \\
&\quad = [0, 0, 0, \tilde{p}(\mathcal{G}, q_\mathcal{G}, J^G, \mathcal{I}) + \tilde{p}(\mathcal{G}_\mathcal{I}, (q_\mathcal{G})_\mathcal{I}, J^{G_\mathcal{I}}, \mathcal{L}_\mathcal{I}) - \tilde{p}(\mathcal{F}, q, J, \mathcal{I})]
\end{aligned}$$

Now note that $\tilde{p}(\mathcal{G}, q_\mathcal{G}, J^G, \mathcal{I}) = \tilde{p}(\mathcal{F}, q, J, \mathcal{I})$ and $\tilde{p}(\mathcal{G}_\mathcal{I}, (q_\mathcal{G})_\mathcal{I}, J^{G_\mathcal{I}}, \mathcal{L}_\mathcal{I}) = 0$. The reason for the first equality is that ∇^G and J^G preserve the decomposition $G = F \oplus F_\mathcal{I}$ and that $I \subset F \oplus 0$. The second identity can be seen by noting that $\nabla^{G_\mathcal{I}}$ preserves the decomposition $G_\mathcal{I} = L_\mathcal{I} \oplus J^{G_\mathcal{I}} L_\mathcal{I}$. □

3.5. Functorial properties.

3.5.1. *Homotopy invariance of pull-back.* If $f\colon M \to N$ is a smooth map of manifolds, then we obtain an induced map $f^*\colon \bar{L}_\epsilon(N) \to \bar{L}_\epsilon(M)$, which is given on generators by pull-back of structures.

LEMMA 3.24. *f^* only depends on the smooth homotopy class of f.*

PROOF. Consider a generator $(\mathcal{F}, q, J, \rho)$. Let $(f_t)_{t\in[0,1]}$ be a smooth homotopy of maps from M to N. Then we have an isomorphism $\phi\colon (f_0^*\mathcal{F}, f_0^*q) \to (f_1^*\mathcal{F}, f_1^*q)$. Let $\tilde{M} := [0,1] \times M$ and $H\colon \tilde{M} \to N$ be induced by (f_t). We compute

$$\begin{aligned}
f_0^*\rho - f_1^*\rho - \tilde{p}(\nabla^{f_0^*F}, f_0^*J, \phi^* f_1^* J) &\equiv f_0^*\rho - f_1^*\rho - \int_{\tilde{M}/M} p(\nabla^{H^*F}, H^*J) \\
&= -d\int_{\tilde{M}/M} H^*\rho + \int_{\tilde{M}/M} dH^*\rho - \int_{\tilde{M}/M} p(\nabla^{H^*F}, H^*J) \\
&\equiv \int_{\tilde{M}/M} (p(\nabla^{H^*F}, H^*J) - p(\nabla^{H^*F}, H^*J)) = 0,
\end{aligned}$$

where $\equiv$ means equality modulo exact forms. It now follows from Lemma 3.10 that

$$[f_0^*\mathcal{F}, f_0^*q, f_0^*J, f_0^*\rho] = [f_1^*\mathcal{F}, f_1^*q, f_1^*J, f_1^*\rho].$$

□

One can also check that the pull-back f^* is compatible with the $L(M)$ (resp. $L(N)$)-module structures.

3.5.2. *Exact sequences.* We define natural maps $H^{\mathrm{odd}}(M,\mathbb{R}) \to \bar{L}(M)$ and $\bar{L}(M) \to L(M)$ by $[\rho] \mapsto [0,0,0,\rho]$ and $[\mathcal{F}, q, J, \rho] \mapsto [\mathcal{F}, q]$, respectively. We consider $H^{\mathrm{odd}}(M,\mathbb{R})$ as an $L(M)$-module such that every element of $L(M)$ acts trivially. $H^{\mathrm{ev}}(M,\mathbb{R})$ becomes an $L(M)$-module via the ring structure of $H^{\mathrm{ev}}(M,\mathbb{R})$ and the homomorphism $\mathbf{ch} \circ b\colon L(M) \to H^{\mathrm{ev}}(M,\mathbb{R})$.

LEMMA 3.25. *We have the exact sequence of $L(M)$-modules*

$$H^{\mathrm{odd}}(M,\mathbb{R}) \longrightarrow \bar{L}(M) \longrightarrow L(M) \longrightarrow H^{\mathrm{ev}}(M,\mathbb{R}).$$

PROOF. It is obvious from the definition of the maps that the composition of two of them vanishes. Therefore, it remains to check exactness. We have the following commutative diagram

$$\begin{array}{ccccccc} H^{\mathrm{odd}}(M,\mathbb{R}) & \longrightarrow & \bar{L}^{\mathrm{Lott}}(M) & \longrightarrow & L^{\mathrm{Lott}}(M) & \longrightarrow & H^{\mathrm{ev}}(M,\mathbb{R}) \\ \| & & \downarrow & & \downarrow & & \| \\ H^{\mathrm{odd}}(M,\mathbb{R}) & \longrightarrow & \bar{L}(M) & \longrightarrow & L(M) & \longrightarrow & H^{\mathrm{ev}}(M,\mathbb{R}). \end{array}$$

The upper sequence is exact by [**21**, Prop. 21]. The vertical maps are all surjective. It follows by diagram chasing that the lower sequence is exact at $L(M)$.

We now show exactness at $\bar{L}(M)$. Let $(\mathcal{F},q,J,\rho)$ be given such that $[\mathcal{F},q]=0$. Then there exists $(\mathcal{F}_1,q_1)$ which admits a lagrangian subsheaf $\mathcal{L}_1$ such that $(\mathcal{F}\oplus\mathcal{F}_1,q\oplus q_1)$ also admits a lagrangian subsheaf $\mathcal{L}$. We choose a metric structure J_1 and define ρ_1 such that $[\mathcal{F}_1,q_1,J_1,\rho_1]=0$. In fact we must take $\rho_1 := -\tilde{p}(\mathcal{F}_1,q_1,J_1,\mathcal{L}_1)$. Lagrangian reduction by $\mathcal{L}$ shows that $[\mathcal{F}\oplus\mathcal{F}_1,q\oplus q_1,J\oplus J_1,\rho+\rho_1]=[0,0,0,\omega]$ for the form $\omega:=\rho+\rho_1+\tilde{p}(\mathcal{F}\oplus\mathcal{F}_1,q\oplus q_1,J\oplus J_1,\mathcal{L})$. It follows that $[\mathcal{F},q,J,\rho]=[0,0,0,\omega]$ comes from $H^{\mathrm{odd}}(M,\mathbb{R})$. □

3.5.3. *Injectivity of $H^{\mathrm{odd}}(M,\mathbb{R})\to\bar{L}(M)$.*

PROPOSITION 3.26. *The map $H^{\mathrm{odd}}(M,\mathbb{R})\to\bar{L}(M)$ is injective.*

PROOF. Let $\omega\in\Omega^{\mathrm{odd}}(M)$ be a closed form. If $[0,0,0,\omega]=0$ in $\bar{L}(M)$, then, by Definition 3.8, there exists $(\mathcal{F},q,J,\rho)$ together with two lagrangian subsheaves $\mathcal{L}_0$, $\mathcal{L}_1$ such that

$$[\omega]=\left[\tilde{p}(\mathcal{F},q,J,\mathcal{L}_0)-\tilde{p}(\mathcal{F},q,J,\mathcal{L}_1)\right].$$

We claim that $\left[\tilde{p}(\mathcal{F},q,J,\mathcal{L}_0)-\tilde{p}(\mathcal{F},q,J,\mathcal{L}_1)\right]$ is independent of J. To show this we consider the space $\mathcal{M}$ of all metric structures and $\tilde{M}:=M\times\mathcal{M}$. Furthermore, let $(\tilde{\mathcal{F}},\tilde{q},\tilde{\rho}):=\mathrm{pr}^*(\mathcal{F},q,\rho)$. We define the metric structure $\tilde{J}$ on $\tilde{F}$ such that it restricts to J' on $M\times\{J'\}$. Using the two lagrangian subspaces $\tilde{\mathcal{L}}_i:=\mathrm{pr}^*\mathcal{L}_i$ we define the closed form

$$\alpha:=\tilde{p}(\tilde{\mathcal{F}},\tilde{q},\tilde{J},\tilde{\mathcal{L}}_0)-\tilde{p}(\tilde{\mathcal{F}},\tilde{q},\tilde{J},\tilde{\mathcal{L}}_1).$$

The metric structure J provides an embedding $i_J\colon M\to\tilde{M}$, and we have $i_J^*\alpha=\tilde{p}(\mathcal{F},q,J,\mathcal{L}_0)-\tilde{p}(\mathcal{F},q,J,\mathcal{L}_1)$. Since the embeddings I_J and $I_{J'}$ for two metric structures J,J' are homotopic, the classes $[i_J^*\alpha]$ and $[i_{J'}^*\alpha]$ coincide. This proves the claim.

We next claim that we can assume that $\mathcal{L}_0\cap\mathcal{L}_1=\{0\}$. Note that $\mathcal{I}:=\mathcal{L}_0\cap\mathcal{L}_1$ is an isotropic subsheaf. The reduction $(\mathcal{F}_\mathcal{I},q_\mathcal{I})$ admits two induced lagrangian subsheaves $(\mathcal{L}_i)_\mathcal{I}$. We have by Lemma 3.22 that

$$\left[\tilde{p}(\mathcal{F},q,J,\mathcal{L}_0)-\tilde{p}(\mathcal{F},q,J,\mathcal{L}_1)\right]=\left[\tilde{p}(\mathcal{F}_\mathcal{I},q_\mathcal{I},J^{F_\mathcal{I}},(\mathcal{L}_0)_\mathcal{I})-\tilde{p}(\mathcal{F}_\mathcal{I},q_\mathcal{I},J^{F_\mathcal{I}},(\mathcal{L}_1)_\mathcal{I})\right].$$

Thus we can replace $(\mathcal{F}, q, J, \rho)$ by $(\mathcal{F}_{\mathcal{I}}, q_{\mathcal{I}}, J^{F_{\mathcal{I}}}, \rho + p(\mathcal{F}, q, J, \mathcal{I}))$ and $\mathcal{L}_i$ by $(\mathcal{L}_i)_{\mathcal{I}}$.

Let us now assume that $L_0 \cap L_1 = \{0\}$. Then (cf. e.g. [**26**, Prop. 2.50]) we can choose J such that $JL_i = L_{1-i}$, $i = 0, 1$. With this choice $\tilde{p}(\mathcal{F}, q, J, \mathcal{L}_0) - \tilde{p}(\mathcal{F}, q, J, \mathcal{L}_1) = 0$ so that $[\omega] = 0$. □

3.5.4. *The transformation to $K^{-1}_{\mathbb{R}/\mathbb{Z}}$-theory.* We construct a natural transformation $\bar{\gamma}\colon \bar{L} \to K^{-1}_{\mathbb{R}/\mathbb{Z}}$. Let M be a manifold. Note that $K^{-1}_{\mathbb{R}/\mathbb{Z}}(M)$ is a module over the ring $K^0(M)$ and, therefore, by $b\colon L(M) \to K^0(M)$, a $L(M)$-module. We in fact construct morphism of $L(M)$-modules $\bar{\gamma}_M : \bar{L}(M) \to K^{-1}_{\mathbb{R}/\mathbb{Z}}(M)$ which depends naturally on M.

To define this morphism on generators $(\mathcal{F}, q, J, \rho)$ we use the geometric definition of $K^{-1}_{\mathbb{R}/\mathbb{Z}}(M)$ in terms of generators and relations which was recalled in Subsection 3.2.2.

We define $\bar{\gamma}_M\colon \bar{L}(M) \to K^{-1}_{\mathbb{R}/\mathbb{Z}}(M)$ by $\bar{\gamma}_M[\mathcal{F}, q, J, \rho] = [F_{\mathbb{C}}, h^{F_{\mathbb{C}}}, \nabla^{F_{\mathbb{C}}, J}, \rho]$, where $h^{F_{\mathbb{C}}}$ is the hermitian extension of the metric $J^* \circ Q$ on F.

LEMMA 3.27. *$\bar{\gamma}_M$ is well-defined and has the properties as stated above.*

PROOF. Since $\bar{\gamma}_M$ is induced by an obvious homomorphism $\hat{\bar{\gamma}}_M\colon \hat{\bar{L}}(M) \to K^{-1}_{\mathbb{R}/\mathbb{Z}}(M)$ of semigroups (which is natural in M) it suffices to show that $\hat{\bar{\gamma}}_M$ is compatible with lagrangian reduction and that it induces a $L(M)$-module isomorphism. The latter property we leave as an exercise.

Let $\mathcal{L} \subset \mathcal{F}$ be a locally constant lagrangian subsheaf. Then $[\mathcal{F}, q, J, \rho] = [0, 0, 0, \rho + \tilde{p}(\mathcal{F}, q, J, \mathcal{L})]$. In $K^{-1}_{\mathbb{R}/\mathbb{Z}}(M)$ we have (using $h^{F_{\mathbb{C}}} = h^L \oplus h^L$, $\nabla^{\oplus} = \nabla^L \oplus \nabla^L$) that

$$\begin{aligned}
[F_{\mathbb{C}}, h^{F_{\mathbb{C}}}, \nabla^{F_{\mathbb{C}}, J}, \rho] &= [F_{\mathbb{C}}, h^{F_{\mathbb{C}}}, \nabla^{\oplus}, \rho + \tilde{\mathbf{ch}}(\nabla^{\oplus}, \nabla^{F_{\mathbb{C}}, J})] \\
&= [L \oplus L, h^L \oplus h^L, \nabla^L \oplus \nabla^L, \rho + \tilde{p}(\mathcal{F}, q, J, \mathcal{L})] \\
&= [0, 0, 0, \rho + \tilde{p}(\mathcal{F}, q, J, \mathcal{L})].
\end{aligned}$$

This shows that γ_M is well-defined. □

As an immediate consequence of the definition we get:

COROLLARY 3.28. *The following diagram commutes:*

$$\begin{array}{ccc}
\bar{L}(M) & \longrightarrow & L(M) \\
\downarrow & & \downarrow \\
K^{-1}_{\mathbb{R}/\mathbb{Z}}(M) & \longrightarrow & K^0(M)
\end{array}$$

It follows from Definition 3.1 of $\bar{L}^{\mathbb{R}/\mathbb{Z}}(M)$ as a pull-back that:

COROLLARY 3.29. *There is a natural surjective map*

$$\bar{L}(M) \to \bar{L}^{\mathbb{R}/\mathbb{Z}}(M).$$

4. Eta homomorphisms

4.1. Introduction and summary. It is an interesting problem to detect non-trivial elements of $\bar{L}_\epsilon(M)$ for a manifold M.

From the homotopy theoretic definition of $K^{-1}_{\mathbb{R}/\mathbb{Z}}$ there is a natural pairing

$$K^{-1}_{\mathbb{R}/\mathbb{Z}}(M) \otimes K_1(M) \to \mathbb{R}/\mathbb{Z},$$

where $K_1(M)$ denotes K-homology of M. In [**20**], Lott gave an analytic description of this pairing. If M is odd-dimensional and the K-homology class is represented by a Dirac operator, then its pairing with a class in $K^{-1}_{\mathbb{R}/\mathbb{Z}}(M)$ given in the geometric picture can be expressed through spectral invariants of the corresponding twisted Dirac operator, in particular through the η-invariant. In the present section we consider the K-homology class given by the signature operator, which leads to a homomorphism

$$\eta^{-1}_{\mathbb{R}/\mathbb{Z}} \colon K^{-1}_{\mathbb{R}/\mathbb{Z}}(M) \to \mathbb{R}/\mathbb{Z}.$$

The latter pulls back to

$$\eta^{\mathbb{R}/\mathbb{Z}} \colon \bar{L}^{\mathbb{R}/\mathbb{Z}}(M) \to \mathbb{R}/\mathbb{Z}.$$

It further induces a homomorphism from $\bar{L}(M)$ to $\mathbb{R}/\mathbb{Z}$.

The main objective of the present section is to refine the homomorphism to an $\mathbb{R}$-valued one. From the analytic definition of $\eta^{-1}_{\mathbb{R}/\mathbb{Z}}$ it is quite obvious that it cannot be lifted to $\mathbb{R}$ since the relevant η-invariants jump by integers if the kernels of the corresponding operators change. But the dimension of the kernel is not an invariant of the data given by the classes $K^{-1}_{\mathbb{R}/\mathbb{Z}}(M)$ and $K_1(M)$.

However, in our special case we work with the signature operator and with flat vector bundles. The kernels are tied to cohomology and thus have stronger invariance properties. Lott has defined a lift $\eta^{\mathrm{Lott}} \colon \bar{L}^{\mathrm{Lott}}(M) \to \mathbb{R}$. Unfortunately, this homomorphism does not factor over $\bar{L}(M)$. But we can analyze this failure in detail. These considerations lead to the definition of the extended groups $\bar{L}^{\mathrm{ex}}_\epsilon(M)$, $\epsilon = \pm 1$, fitting into exact sequences

$$0 \to \mathbb{Z} \to \bar{L}^{\mathrm{ex}}_\epsilon(M) \to \bar{L}_\epsilon(M) \to 0.$$

The homomorphism from $\bar{L}(M) \to \mathbb{R}/\mathbb{Z}$ lifts to a homomorphism $\eta \colon \bar{L}^{\mathrm{ex}}(M) \to \mathbb{R}$.

The extended groups are only defined for closed odd-dimensional oriented manifolds M. Note that they are not functorial with respect to smooth maps (only with respect to homotopy equivalences). But they admit extended secondary index maps.

4.2. $\eta^{-1}_{\mathbb{R}/\mathbb{Z}} \colon K^{-1}_{\mathbb{R}/\mathbb{Z}}(M) \to \mathbb{R}/\mathbb{Z}$.

4.2.1. *The definition.* Let M be an oriented n-dimensional closed manifold. If we choose a Riemannian metric, then we can define the signature operator D^{sign}. Assume now that n is odd. Then D^{sign} is a self-adjoint operator which induces a class $[M^{\mathrm{sign}}] \in K_1(M)$, and which is independent of the choice of the metric.

DEFINITION 4.1. We define $\eta^{-1}_{\mathbb{R}/\mathbb{Z}}\colon K^{-1}_{\mathbb{R}/\mathbb{Z}}(M) \to \mathbb{R}/\mathbb{Z}$ by $\eta^{-1}_{\mathbb{R}/\mathbb{Z}}(x) := \langle x, [M^{\mathrm{sign}}]\rangle$. Furthermore, we define $\eta^{\mathbb{R}/\mathbb{Z}}\colon \bar{L}^{\mathbb{R}/\mathbb{Z}}(M) \to \mathbb{R}/\mathbb{Z}$ as the composition of $\eta^{-1}_{\mathbb{R}/\mathbb{Z}}$ with the natural transformation $\bar{L}^{\mathbb{R}/\mathbb{Z}} \to K^{-1}_{\mathbb{R}/\mathbb{Z}}$.

4.2.2. *Geometric description of* $\eta^{-1}_{\mathbb{R}/\mathbb{Z}}$. We fix a Riemannian metric g^{TM} on M. Let $[E, h^E, \nabla^E, \rho] \in K^{-1}_{\mathbb{R}/\mathbb{Z}}(M)$. Then one can express $\eta^{-1}_{\mathbb{R}/\mathbb{Z}}(x)$ in analytic terms. Let $D^{\mathrm{sign}}_E : \Omega(M,E) \to \Omega(M,E)$ be the odd signature operator of M twisted by the bundle E. Then its η-invariant is given by

$$\eta(D^{\mathrm{sign}}_E) := \frac{1}{\sqrt{\pi}} \int_0^\infty \mathrm{Tr}\left[D^{\mathrm{sign}}_E \mathrm{e}^{-t(D^{\mathrm{sign}}_E)^2}\right] \frac{dt}{t^{1/2}}.$$

Let $L(\nabla^{TM}) \in \Omega(M)$ be the L-form of the tangent bundle TM equipped with the Levi-Civita connection ∇^{TM}. Then we have

$$\eta^{\mathbb{R}/\mathbb{Z}}([E, h^E, \nabla^E, \rho]) := \left[\eta(D^{\mathrm{sign}}_E) - 2\int_M L(\nabla^{TM}) \wedge \rho\right],$$

where $[.]$ on the right-hand side takes the class in $\mathbb{R}/\mathbb{Z}$. Using the local variation formula for the η-invariant we see that this combination indeed is independent of the Riemannian metric and only depends on the class of $[E, h^E, \nabla^E, \rho]$. For details we refer to Lott [**20**, Prop. 3].

Note that this homomorphisms cannot be refined to have values in $\mathbb{R}$, since the dimension of the kernel of D^{sign}_E depends on the choice of the connection for the representative $[E, h^E, \nabla^E, \rho]$.

4.3. $\mathbb{R}$-valued η-homomorphisms. Since the $\bar{L}_\epsilon$-groups involve flat bundles $F = \mathtt{bundle}(\mathcal{F})$, the kernel of the twisted signature operator D^{sign}_F defined in (4.2) is isomorphic to the sheaf cohomology $H^*(M, \mathcal{F})$. In particular, it is independent of the choice of a metric structure. This eventually allows us to define $\mathbb{R}$-valued η-homomorphisms.

4.3.1. η^{Lott}. We start with describing the homomorphism $\eta^{\mathrm{Lott}}\colon \bar{L}^{\mathrm{Lott}}_\epsilon(M) \to \mathbb{R}$ which has been defined by Lott in [**21**, §3.3]. Let M be a closed smooth m-dimensional oriented manifold. The $\mathbb{Z}$-graded vector space of real differential forms $\Omega(M)$ carries the $\epsilon_m = (-1)^{[\frac{m+1}{2}]}$—symmetric duality structure (cf. (4.4))

$$q(\omega, \omega') := \int_M \left((-1)^{\frac{N_M(N_M-1)}{2} + mN_M}\omega\right) \wedge \omega',$$

where N_M denotes the $\mathbb{Z}$-grading on $\Omega(M)$. We fix the convention $\sqrt{\epsilon_m} := i^{[\frac{m+1}{2}]}$.

A choice of a Riemannian metric g^{TM} and of the orientation on M induces the Hodge-$*$ operator which is characterized by

$$\omega \wedge *\omega' = (\omega, \omega')_{g^{TM}} \mathrm{vol}.$$

Here vol is the Riemannian volume element of (M, g^{TM}). We define the N_M-compatible metric structure

$$J^M := * \, (-1)^{\frac{N_M(N_M-1)}{2}+mN_M}.$$

Let now $(\mathcal{F}, q)$ be ϵ–symmetric. By $\Omega(M, F)$ we denote the space of F-valued smooth forms on M, where $F := \texttt{bundle}(\mathcal{F})$. If $\omega \in \Omega(M)$ and $\phi \in C^\infty(M, F)$, then we have a product $\omega \otimes \phi \in \Omega(M, F)$. The space $\Omega(M, F)$ carries the $\epsilon_m\epsilon$–duality structure

$$(4.1)\quad q_{M,F}(\omega\otimes\phi, \omega'\otimes\phi') := \frac{\sqrt{\epsilon\epsilon_m}}{\sqrt{\epsilon}\sqrt{\epsilon_m}}\int_M \left((-1)^{\frac{N_M(N_M-1)}{2}+mN_M}\omega\right)\wedge\omega'\, q(\phi,\phi')_F.$$

It further has a natural $\mathbb{Z}$-grading N of length m. If J^F is a metric structure on F, then $J^{M,F}(\omega\otimes\phi) := \frac{\sqrt{\epsilon\epsilon_m}}{\sqrt{\epsilon}\sqrt{\epsilon_m}} J^M\omega \otimes J^F\phi$ is a N-compatible metric structure on $\Omega(M, F)$ (cf. Definition 3.14). Let d^F be the twisted de Rham differential on $\Omega(M, F)$ induced by the flat connection ∇^F on F. Let $(d^F)^*$ be the adjoint of d^F with respect to this metric on $\Omega(M, F)$. Then we have $(d^F)^* = (-1)^{m+1}\epsilon\epsilon_m J^{M,F} d^F J^{M,F}$.

We now assume that $m \equiv -\epsilon \bmod 4$, then $\epsilon\epsilon_m = 1$. Then we define the self-adjoint operator (cf. [**2**, (4.6)], [**7**, (1.38)])

$$(4.2)\qquad D_F^{\mathrm{sign}} := J^{M,F} d^F + d^F J^{M,F}.$$

DEFINITION 4.2. The homomorphism $\eta^{\mathrm{Lott}}\colon \bar{L}_\epsilon^{\mathrm{Lott}}(M) \to \mathbb{R}$ is defined by

$$\eta^{\mathrm{Lott}}(\mathcal{F}, q, J, \rho) := \eta(D_F^{\mathrm{sign}}) - 2\int_M L(\nabla^{TM}) \wedge \rho.$$

By the argument in [**21**, Prop. 24], η^{Lott} is well-defined and independent of the choice of the Riemannian metric g^{TM} of M. It also follows from the definitions that the following diagram commutes:

$$\begin{array}{ccc} \bar{L}_\epsilon^{\mathrm{Lott}}(M) & \xrightarrow{\eta^{\mathrm{Lott}}} & \mathbb{R} \\ \downarrow & & \downarrow \\ K_{\mathbb{R}/\mathbb{Z}}^{-1}(M) & \xrightarrow{\eta^{\mathbb{R}/\mathbb{Z}}} & \mathbb{R}/\mathbb{Z} \end{array}$$

4.3.2. *Motivation of extended L-groups.* A natural problem is now to lift the composition

$$\bar{L}_\epsilon(M) \to K_{\mathbb{R}/\mathbb{Z}}^{-1}(M) \xrightarrow{\eta^{\mathbb{R}/\mathbb{Z}}} \mathbb{R}/\mathbb{Z}$$

to a $\mathbb{R}$-valued homomorphism "η": $\bar{L}_\epsilon(M) \to \mathbb{R}$. This problem could be solved by factoring η^{Lott} over the quotient $\bar{L}_\epsilon^{\mathrm{Lott}}(M) \to \bar{L}_\epsilon(M)$. Unfortunately, this factorization does not exist in general. For this reason we will define modified L- and

$\bar{L}$-groups which will be denoted by L_ϵ^{ex} and $\bar{L}_\epsilon^{\text{ex}}$. They fit into the diagram

(4.3)

$$\begin{array}{ccccccc}
H^{4*-\epsilon}(M,\mathbb{R}) & \longrightarrow & \bar{L}_\epsilon^{\text{Lott}}(M) & \longrightarrow & L_\epsilon^{\text{Lott}}(M) & \longrightarrow & H^{4*+1-\epsilon}(M,\mathbb{R}) \\
\| & & \downarrow & & \downarrow & & \| \\
H^{4*-\epsilon}(M,\mathbb{R}) & \longrightarrow & \bar{L}_\epsilon^{\text{ex}}(M) & \longrightarrow & L_\epsilon^{\text{ex}}(M) & \longrightarrow & H^{4*+1-\epsilon}(M,\mathbb{R}) \\
\| & & \downarrow & & \downarrow & & \| \\
H^{4*-\epsilon}(M,\mathbb{R}) & \longrightarrow & \bar{L}_\epsilon(M) & \longrightarrow & L_\epsilon(M) & \longrightarrow & H^{4*+1-\epsilon}(M,\mathbb{R}).
\end{array}$$

The advantage of $\bar{L}_\epsilon^{\text{ex}}$ is that we can extend the homomorphism η^{Lott} to an homomorphism $\eta\colon \bar{L}_\epsilon^{\text{ex}}(M) \to \mathbb{R}$. The draw-back of the extended L_ϵ-groups is that they are *not functors* on the category of manifolds.

4.4. Definition of the extended L-groups.

4.4.1. *Some linear algebra.* Let V be a real vector space with a non-degenerate form $r\colon V \to V^*$. We will call (V,r) a duality space.

DEFINITION 4.3. A $\mathbb{Z}$-graded duality space of length $n \in \mathbb{Z}$ is a triple (V,r,N), where (V,r) is a duality space and N is a $\mathbb{Z}$-grading operator such that $r^{-1}\circ N^* \circ r = n - N$.

If (V_i,r_i,N_i), $i = 1,2$, are $\mathbb{Z}$-graded duality spaces of length n_i, then their tensor product is a $\mathbb{Z}$-graded duality space of length n_1+n_2.

Let (V,r,N) be a $\mathbb{Z}$-graded duality space of length n. Assume further that r is *graded symmetric*, i.e., $r(v,w) = (-1)^{|v||w|}r(w,v)$. Then we can define an $\epsilon_n := (-1)^{[\frac{n+1}{2}]}$–duality structure on V by

$$q(v,w) := r\big((-1)^{\frac{N(N-1)}{2}+nN}v, w\big). \tag{4.4}$$

Let $\mathbb{R}[-n]$ be the graded vector space which has $\mathbb{R}$ in degree n and 0 in all remaining degrees. The duality structure r induces a map of graded vector spaces $\hat{r}\colon V\otimes V \to \mathbb{R}[-n]$ by $v\otimes w \to r(v,w)$.

We now consider a differential $d\colon V \to V$ of degree one. It induces a differential on $V\otimes V$ by $d(v\otimes w) = dv\otimes w + (-1)^{|v|}v\otimes dw$, where v is homogeneous of degree $|v|$. On $\mathbb{R}[-n]$ we consider the trivial differential.

DEFINITION 4.4. The differential d is called compatible with r if $\hat{r}\colon V\otimes V \to \mathbb{R}[-n]$ is a map of complexes. Equivalently, $r(dv,w) + (-1)^{|v|}r(v,dw) = 0$ for all v, $w \in V$ with v homogeneous of degree $|v|$.

We now assume that d is compatible with r. Let $H(V)$ denote the cohomology of (V,d). It is a $\mathbb{Z}$-graded vector space. The map $H(\hat{r})\colon H(V\otimes V) \to H(\mathbb{R}[-n]) = \mathbb{R}[-n]$ together with the Künneth formula $H(V)\otimes H(V) \xrightarrow{\cong} H(V\otimes V)$ induces a graded–symmetric duality structure r_H on $H(V)$. In order to see that this pairing is non-degenerate, note that $\operatorname{im}(d)^\perp = \ker(d)$ and hence $\ker(d)^\perp = \operatorname{im}(d)$.

Let (V, r, N) be a $\mathbb{Z}$-graded duality space of length n with compatible differential d, where r is graded symmetric. Then we have a $(-1)^{n+1}\epsilon_n$—symmetric form

$$Q(v, w) := r((-1)^{\frac{N(N-1)}{2}+nN} v, dw) = q(v, dw).$$

In particular, if V is finite-dimensional and $(-1)^{n+1}\epsilon_n = 1$, then we can consider the signature of the quadratic form Q.

DEFINITION 4.5. $\tau(V, r, N, d) := 2\,\mathrm{sign}(Q)$.

Let J be a metric structure on (V, q). Let d^{*_J} denote the adjoint of d with respect to the scalar product $q(., J.)$ induced by J. We have $d^{*_J} = (-1)^{n+1}\epsilon_n JdJ$. Assume that $(-1)^{n+1}\epsilon_n = 1$. Then we can form the self-adjoint operator $D := Jd + (-1)^{n+1}dJ = J(d + d^{*_J})$.

LEMMA 4.6. *We have* $\mathrm{sign}(D) = \frac{1}{2}\left(1 + (-1)^{n+1}\right)\tau(V, r, N, d)$.

PROOF. By Hodge theory we can identify $H(V)$ with $\mathcal{H} := \ker(d) \cap \ker(d^{*_J})$. We decompose V into D-invariant subspaces $V = \mathrm{im}(d) \oplus \mathrm{im}(d^{*_J}) \oplus \mathcal{H}$. The metric structure J maps $\mathrm{im}(d)$ isomorphically to $\mathrm{im}(d^{*_J})$. Furthermore, for $x, y \in \mathrm{im}(d)$ we have $q(DJx, JJy) = (-1)^{n+1}q(JdJx, y) = q(dJx, Jy) = (-1)^{n+1}q(Dx, Jy)$. Let P be the projection to $\mathrm{im}(d^{*_J})$. Then $\mathrm{sign}(D) = \left(1 + (-1)^{n+1}\right)\mathrm{sign}(DP)$. For all x, y we have $q(x, JDPy) = (-1)^{n+1}q(x, dy)$. Thus we have $\mathrm{sign}(DP) = \frac{1}{2}(-1)^{n+1}\,\tau(V, r, N, d)$. □

Let (V, r) be a duality space with grading N of length n and compatible differential d. Furthermore, let $\ldots \subset F^{p+1}V \subset F^pV \subset \ldots$ be a decreasing filtration of the complex (V, d, N) by subcomplexes. Further we assume that there exists n_B such that $r(F^pV, F^qV) = 0$ if $p + q > n_B$. On $V \otimes V$ we consider the induced filtration $F^r(V \otimes V) = \sum_{p+q=r} F^pV \otimes F^qV$. Let $E_u^{p,q}(V)$ be the corresponding spectral sequence (cf. [**17**, §3.5]).

We filter $\mathbb{R}[-n]$ such that $F^s\mathbb{R}[-n] = \mathbb{R}[-n]$ if $s \leq n_B$ and $F^s\mathbb{R}[-n] = 0$ for $s > n_B$. The map $\hat{r}$ is then a map of filtered complexes. We therefore obtain a morphism of associated spectral sequences, which we also denote by $\hat{r}$. The target $E_s^{p,q}(\mathbb{R}[-n])$ is easy to describe. We have $E_s^{p,q}(\mathbb{R}[-n]) = 0$ for all $s \geq 0$, p, q except for $p + q = n$ and $p = n_B$, where $E_s^{p,q} \cong \mathbb{R}$. All differentials vanish.

The Künneth formula for spectral sequences yields a map

$$E_u^{p,q}(V) \otimes E_u^{s,t}(V) \to E_u^{p+s,q+t}(V \otimes V).$$

If we compose this with the morphism of spectral sequences induced by $\hat{r}$ we obtain pairings $r_u \colon E_u^{p,q}(V) \otimes E_u^{s,t}(V) \to \mathbb{R}$ if $p + q + s + t = n$ and $p + s = n_B$. In particular, we obtain a pairing on $E_u(V)$ which is compatible with the $\mathbb{Z}$-grading N_{E_u} (induced by N) of length n and has the same symmetry properties as r.

The differential d_u is compatible with the duality structure r_u on E_u and the duality structure r_{u+1} on E_{u+1} is obtained from r_u by considering E_{u+1} as the cohomology of E_u as described above. If r_0 is non-degenerate, then so are the r_u for all $u \geq 0$.

Assume now that r is graded symmetric and $(-1)^{n+1}\epsilon_n = 1$. Then the form r_u induced on $E_u(V)$ is graded symmetric as well. If $E_u(V)$ is finite-dimensional for $u \geq i$, then we define

DEFINITION 4.7. $\tau_i(E_*) := \sum_{u\geq i} \tau(E_u(V), r_u, N_{E_u}, d_u)$.

4.4.2. *Hypercohomology and spectral sequences.* Let M be a smooth manifold. Let $\mathcal{F}$ be a locally constant sheaf of finite-dimensional real vector spaces and (F, ∇^F) be the associated flat vector bundle. We further assume that $\mathcal{F}$ is $\mathbb{Z}$-graded by N_F and has a differential v (cf. Definition 2.4). We use the same symbols in order to denote the corresponding parallel endomorphisms of F. The cohomology $\mathcal{H}$ of $(\mathcal{F}, v)$ is again a locally constant sheaf of real vector spaces with $\mathbb{Z}$-grading N_H.

There are two spectral sequences which converge to the hypercohomology $\mathbf{H}(M, (\mathcal{F}, N_F, v))$ of this complex, the local-global spectral sequence and the hypercohomology spectral sequence. Let us take the soft resolution of $(\mathcal{F}, N, v)$ given by the twisted de Rham complex $(\Omega(M, F), d^F + v)$ with the total $\mathbb{Z}$-grading $N_{M,F} = N_M + N_F$. Its cohomology is naturally isomorphic to $\mathbf{H}(M, (\mathcal{F}, N_F, v))$. The two spectral sequences are associated to the two natural filtrations of this complex.

We first describe the local-global spectral sequence which is associated to the filtration

$$F^p\Omega(M, F) = \sum_{q\geq p} \Omega^q(M, F).$$

We have ${}_{lg}E_0^{p,q} = \Omega^p(M, F^q)$ and ${}_{lg}d_0 = v\colon {}_{lg}E_0^{p,q} \to {}_{lg}E_0^{p,q+1}$. The first term is given by ${}_{lg}E_1^{p,q} = \Omega^p(M, H^q)$ and ${}_{lg}d_1 = d^H\colon {}_{lg}E_1^{p,q} \to {}_{lg}E_1^{p+1,q}$. We obtain ${}_{lg}E_2^{p,q} = H^p(M, \mathcal{H}^q)$. The local-global spectral sequences is natural and finite-dimensional starting with its second term.

We now describe the hypercohomology spectral sequence which is associated to the filtration

$$F^p\Omega(M, F) = \sum_{q\geq p} \Omega(M, F^q).$$

We have ${}_{hc}E_0^{p,q} = \Omega^q(M, F^p)$ and ${}_{hc}d_0 = d^F\colon {}_{hc}E_0^{p,q} \to {}_{hc}E_0^{p,q+1}$. The first term is given by ${}_{hc}E_1^{p,q} = H^q(M, \mathcal{F}^p)$ and ${}_{hc}d_1 := H(v)\colon {}_{hc}E_1^{p,q} \to {}_{hc}E_1^{p+1,q}$, where $H(v)$ is induced by v. The hypercohomology spectral sequence is natural and finite-dimensional starting with its first term.

Now assume that q is an ϵ–duality structure on $\mathcal{F}$ such that N_F and v are compatible with q (cf. Definition 2.4). Let M be closed and oriented, and set $m := \dim M$. On $\Omega(M, F)$ we define the $\epsilon\epsilon_m$–duality structure $q_{M,F}$ by (4.1). The differential $d_F = d^F + v$ and the total grading $N_{M,F} = N_M + N_F$ are compatible with $q_{M,F}$. As explained in Subsection 4.4.1. we obtain induced pairings on the spectral sequences.

We first consider the local-global spectral sequence. We put $n_B := m$ and obtain pairings

$$_{lg}q_r\colon {}_{lg}E_r^{p,q} \otimes {}_{lg}E_r^{s,t} \to \mathbb{R}$$

for $p+q+s+t = m+n$ and $p+s = m$. On ${}_{lg}E_0^{p,q} \otimes {}_{lg}E_0^{s,t} = \Omega^p(M, F^q) \otimes \Omega^s(M, F^t)$ this pairing is just $q_{M,F}$. In particular, it is non-degenerate. As in Subsection 4.4.1, the $(-1)^{m+1}\epsilon\epsilon_m$–symmetric form $q_{M,F}(x, (d^F + v)y)$ induces the corresponding $(-1)^{m+1}\epsilon\epsilon_m$–symmetric forms on ${}_{lg}E_r$. If $\epsilon\epsilon_m = 1$, and m is odd, then we can define

DEFINITION 4.8. ${}_{lg}\tau_2(\mathcal{F}, q, N_F, v) := \tau_2({}_{lg}E_*)$.

In the case of the hypercohomology spectral sequence we put $n_B = n$. We have parings

$$_{hc}q_r\colon {}_{hc}E_r^{p,q} \otimes {}_{lg}E_r^{s,t} \to \mathbb{R}$$

for $p+q+s+t = m+n$ and $p+s = n$. On ${}_{hc}E_0^{p,q} \otimes {}_{hc}E_0^{s,t} = \Omega^q(M, F^p) \otimes \Omega^t(M, F^s)$ this pairing is again just $q_{M,F}$. In particular, it is non-degenerate. If $\epsilon\epsilon_m = 1$, and m is odd, then we can define

DEFINITION 4.9. ${}_{hc}\tau_1(\mathcal{F}, q, N_F, v) := \tau_1({}_{hc}E_*)$.

Finally, we define

DEFINITION 4.10.

$$\tau(\mathcal{F}, q, N_F, v) := {}_{hc}\tau_1(\mathcal{F}, q, N_F, v) - {}_{lg}\tau_2(\mathcal{F}, q, N_F, v).$$

4.4.3. *Definition of $L_\epsilon^{\mathrm{ex}}(M)$ and $\bar{L}_\epsilon^{\mathrm{ex}}(M)$.* We fix closed odd-dimensional oriented manifold M. We assume that $\epsilon \in \mathbb{Z}_2$ is such that $\epsilon\epsilon_m = 1$, where $\epsilon_m := (-1)^{[\frac{m+1}{2}]} \in \mathbb{Z}_2$ and $m = \dim M$.

Let $(\mathcal{F}, q)$ be ϵ–symmetric and $\mathcal{L}$ be a locally constant lagrangian subsheaf of $\mathcal{F}$. Then we define an integer $\tau(\mathcal{F}, q, \mathcal{L})$ by the following construction.

We consider the sheaf $\mathcal{G} := \mathcal{L} \oplus \mathcal{F} \oplus \mathcal{F}/\mathcal{L}$. It carries a natural ϵ–symmetric duality structure $q_\mathcal{G}$. Note that q induces maps $q_\mathcal{L}\colon \mathcal{L} \to (\mathcal{F}/\mathcal{L})^*$ and $q_{\mathcal{F}/\mathcal{L}}\colon \mathcal{F}/\mathcal{L} \to \mathcal{L}^*$. Then $q_\mathcal{G}\colon \mathcal{G} \to \mathcal{G}^*$ is given by

$$\begin{pmatrix} 0 & 0 & q_{\mathcal{F}/\mathcal{L}} \\ 0 & q & 0 \\ q_\mathcal{L} & 0 & 0 \end{pmatrix}.$$

A compatible $\mathbb{Z}$-grading is given by $N_\mathcal{G} = \mathrm{diag}(0, 1, 2)$. We define a differential on $\mathcal{G}$ by

$$v_\mathcal{G} := \begin{pmatrix} 0 & 0 & 0 \\ i_\mathcal{L} & 0 & 0 \\ 0 & -\mathrm{pr} & 0 \end{pmatrix},$$

where $i_\mathcal{L}\colon \mathcal{L} \to \mathcal{F}$ is the inclusion and $\mathrm{pr}\colon \mathcal{F} \to \mathcal{F}/\mathcal{L}$ is the projection. We introduce the sign to make the differential compatible with $q_\mathcal{G}$. We now define the following integer.

DEFINITION 4.11. $\tau(\mathcal{F}, q, \mathcal{L}) := \tau(\mathcal{G}, q_{\mathcal{G}}, N_{\mathcal{G}}, v_{\mathcal{G}})$.

LEMMA 4.12. *If $\mathcal{L}$ has a lagrangian complement, then we have $\tau(\mathcal{F}, q, \mathcal{L}) = 0$.*

PROOF. In this case the spectral sequences degenerate. □

We now define the group L_ϵ^{ex} (note that $\epsilon\epsilon_m = 1$). First we define the semigroup $\hat{L}_\epsilon^{\mathrm{ex}}(M)$ of isomorphism classes of tuples $(\mathcal{F}, q, z)$, where $(\mathcal{F}, q)$ is locally constant sheaf of real vector spaces with ϵ–symmetric duality structure q and $z \in \mathbb{Z}$. The semigroup operation is given by the sum of the entries. Next we define an equivalence relation $\sim$ generated by lagrangian reduction. Let $(\mathcal{F}, q, z)$ as above and $\mathcal{L} \subset \mathcal{F}$ a locally constant lagrangian subsheaf. Then we require that

$$(\mathcal{F}, q, z) \sim (0, 0, z + \tau(\mathcal{F}, q, \mathcal{L})).$$

We extend $\sim$ to the minimal equivalence relation which contains lagrangian reduction and which is compatible with the sum.

DEFINITION 4.13. We define $L_\epsilon^{\mathrm{ex}}(M) := \hat{L}_\epsilon^{\mathrm{ex}}(M)/\sim$.

The semigroup $L_\epsilon^{\mathrm{ex}}(M)$ is in fact a group. Let $[\mathcal{F}, q, z]$ denote the class represented by $(\mathcal{F}, q, z)$. Its inverse is given by $[\mathcal{F}, -q, -z]$, since by Lemma 4.12 we have

$$\tau(\mathcal{F} \oplus \mathcal{F}, q \oplus -q, \mathcal{F}) = 0.$$

There is a surjective homomorphism $L_\epsilon^{\mathrm{ex}}(M) \to L_\epsilon(M)$ induced by $(\mathcal{F}, q, z) \mapsto (\mathcal{F}, q)$. Furthermore, there is homomorphism $\mathbb{Z} \to L_\epsilon^{\mathrm{ex}}(M)$ induced by $z \mapsto (0, 0, z)$.

LEMMA 4.14. *The following sequence is exact:*

$$0 \longrightarrow \mathbb{Z} \longrightarrow L_\epsilon^{\mathrm{ex}}(M) \longrightarrow L_\epsilon(M) \longrightarrow 0.$$

PROOF. We must show that $\mathbb{Z} \to L_\epsilon^{\mathrm{ex}}(M)$ is injective. This will be a consequence of the fact that $\eta \colon L_\epsilon^{\mathrm{ex}}(M) \to \mathbb{R}$ is well-defined which will be proved below (cf. Lemma 4.19). □

Of course, there should be a purely algebraic proof of Lemma 4.14. Another open problem is to turn $M \to L_\epsilon^{\mathrm{ex}}(M)$ into a functor.

Using Lemma 4.12 we see that $(\mathcal{F}, q) \mapsto (\mathcal{F}, q, 0)$ defines a homomorphism

$$L_\epsilon^{\mathrm{Lott}}(M) \to L_\epsilon^{\mathrm{ex}}(M).$$

DEFINITION 4.15. We define $\bar{L}_\epsilon^{\mathrm{ex}}(M)$ by the following pull-back diagram:

$$\begin{array}{ccc} \bar{L}_\epsilon^{\mathrm{ex}}(M) & \longrightarrow & L_\epsilon^{\mathrm{ex}}(M) \\ \downarrow & & \downarrow \\ \bar{L}_\epsilon(M) & \longrightarrow & L_\epsilon(M) \end{array}$$

An element of $\bar{L}_\epsilon^{\mathrm{ex}}(M)$ can be written as $[\mathcal{F}, q, J, \rho, z]$. The following assertions follow immediately from the definition, Lemma 3.25, and Proposition 3.26.

COROLLARY 4.16. *We have the exact sequences*

$$0 \longrightarrow \mathbb{Z} \longrightarrow \bar{L}_\epsilon^{\mathrm{ex}}(M) \longrightarrow \bar{L}_\epsilon(M) \longrightarrow 0$$

and

$$0 \longrightarrow H^{4*-\epsilon}(M,\mathbb{R}) \longrightarrow \bar{L}_\epsilon^{\mathrm{ex}}(M) \longrightarrow L_\epsilon^{\mathrm{ex}}(M) \longrightarrow H^{4*+1-\epsilon}(M,\mathbb{R}).$$

By $(\mathcal{F},q,J,\rho) \mapsto (\mathcal{F},q,J,\rho,0)$ we define a morphism

$$\bar{L}_\epsilon^{\mathrm{Lott}}(M) \to \bar{L}_\epsilon^{\mathrm{ex}}(M).$$

It is now easy to check that the diagram (4.3) commutes.

4.5. Construction of η: $\bar{L}_\epsilon^{\mathrm{ex}}(M) \to \mathbb{R}$.

4.5.1. *The η-invariant of a complex.* Let $\epsilon \in \mathbb{Z}_2$. We assume that M is a smooth closed oriented manifold of dimension m such that m is odd and $\epsilon\epsilon_m = 1$, equivalently $m \equiv -\epsilon \bmod 4$. Let $(\mathcal{F},q)$ be a locally constant sheaf of finite dimensional real vector spaces with ϵ–symmetric form. Let furthermore N_F be a compatible $\mathbb{Z}$-grading and v be a compatible differential. By $(\mathcal{H},q_{\mathcal{H}})$ we denote the associated cohomology. We choose a Riemannian metric g^{TM}, a compatible metric structure J^F, and let J^H be the induced metric structure. Then we can define the operators D_F^{sign} and D_H^{sign} as in (4.2). The following theorem is the main ingredient in the construction of η. Recall Definitions 3.15 and 4.10 of the η-form $\tilde{\eta}(\mathcal{F},N_F,J^F,v)$ and of the integer $\tau(\mathcal{F},q,N_F,v)$.

THEOREM 4.17. *We have*

$$\eta(D_F^{\mathrm{sign}}) - \eta(D_H^{\mathrm{sign}}) = 2\int_M L(\nabla^{TM}) \wedge \tilde{\eta}(\mathcal{F},N_F,J^F,v) - \tau(\mathcal{F},q,N_F,v).$$

PROOF. A sketch of the proof is given in Subsection 6.2. □

4.5.2. *Definition and well-definedness of η.* Let $[\mathcal{F},q,J,\rho,z] \in \bar{L}_\epsilon^{\mathrm{ex}}(M)$. We choose a Riemannian metric g^{TM}.

DEFINITION 4.18. We define

$$\begin{aligned}\eta(\mathcal{F},q,J,\rho,z) &:= \eta^{\mathrm{Lott}}(\mathcal{F},q,J^F,\rho) - z \\ &= \eta(D_F^{\mathrm{sign}}) - 2\int_M L(\nabla^{TM}) \wedge \rho - z.\end{aligned}$$

LEMMA 4.19. *The map $(\mathcal{F},q,J,\rho,z) \mapsto \eta(\mathcal{F},q,J,\rho,z)$ induces a well-defined homomorphism η: $\bar{L}_\epsilon^{\mathrm{ex}}(M) \to \mathbb{R}$.*

PROOF. It follows from the well-definedness of η^{Lott} [**21**, Prop.24] that η is independent of the choice of the Riemannian metric on M. It therefore suffices to show that $(\mathcal{F},q,J,\rho,z) \sim 0$ implies that $\eta(\mathcal{F},q,J,\rho,z) = 0$. Thus let $\mathcal{L} \subset \mathcal{F}$ be a locally constant lagrangian subsheaf of $\mathcal{F}$ such that $\rho + \tilde{p}(\mathcal{F},q,J,\mathcal{L}) \equiv 0$ and $z + \tau(\mathcal{F},q,\mathcal{L}) = 0$.

We construct the complex $(\mathcal{G}, q_{\mathcal{G}}, N_{\mathcal{G}}, v_{\mathcal{G}})$ as in Subsection 4.4.3. Using the compositions $J_L \colon L \longrightarrow F \xrightarrow{J} F \xrightarrow{\mathrm{pr}} F/L$ and $J_{F/L} \colon F/L \longrightarrow F \ominus L \xrightarrow{J} L$, we define the metric structure

$$J^{\mathcal{G}} := \begin{pmatrix} 0 & 0 & J_L \\ 0 & J & 0 \\ J_{F/L} & 0 & 0 \end{pmatrix}.$$

As in the proof of Proposition 3.16, we see that modulo exact forms

$$\tilde{\eta}(\mathcal{G}, N_{\mathcal{G}}, J^{\mathcal{G}}, v_{\mathcal{G}}) \equiv -\tilde{p}(\mathcal{F}, q, J, \mathcal{L}).$$

By definition we have

$$\tau(\mathcal{F}, q, \mathcal{L}) = \tau(\mathcal{G}, q_{\mathcal{G}}, N_{\mathcal{G}}, v_{\mathcal{G}}).$$

The complex $\mathcal{G}$ is exact. By Theorem 4.17, we have

$$\eta(D^{\mathrm{sign}}_{\mathcal{G}}) = 2\int_M L(\nabla^{TM}) \wedge \tilde{\eta}(\mathcal{G}, N_{\mathcal{G}}, J^{\mathcal{G}}, v_{\mathcal{G}}) - \tau(\mathcal{G}, q_{\mathcal{G}}, N_{\mathcal{G}}, v_{\mathcal{G}}).$$

Thus

$$\begin{aligned}
\eta(\mathcal{F}, q, J, \rho, z) &= \eta(D^{\mathrm{sign}}_{F}) - 2\int_M L(\nabla^{TM}) \wedge \rho - z \\
&= 2\int_M L(\nabla^{TM}) \wedge \tilde{\eta}(\mathcal{G}, N_{\mathcal{G}}, J^{\mathcal{G}}, v_{\mathcal{G}}) - \tau(\mathcal{G}, q_{\mathcal{G}}, N_{\mathcal{G}}, v_{\mathcal{G}}) \\
&\qquad\qquad - 2\int_M L(\nabla^{TM}) \wedge \rho - z \\
&= -2\int_M L(\nabla^{TM}) \wedge (\tilde{p}(\mathcal{F}, q, J, \mathcal{L}) + \rho) - (z + \tau(\mathcal{G}, q_{\mathcal{G}}, N_{\mathcal{G}}, v_{\mathcal{G}})) \\
&= 0.
\end{aligned}$$

□

5. The secondary index map

5.1. Introduction and summary. In this section we consider the secondary index map (i.e., the wrong-way or push-forward map) for $\bar{L}_\epsilon$ associated to fibre bundles. It is constructed by refining the geometric construction of $\pi_!^{\mathrm{sign},\mathbb{R}/\mathbb{Z}}$. This construction naturally involves η-forms for fibre bundles. The proofs of the facts that $\pi_*^{\bar{L}}$ is well-defined, and that it has nice functorial properties, are all based on the study of various adiabatic limits of these η-forms. We start this section with the introduction of the η-form associated to a fibre bundle and the statements of the adiabatic limit results. Then we introduce the secondary index maps and discuss their functorial properties. We give the algebraic parts of the proofs in detail using the corresponding adiabatic limit results.

5.2. Adiabatic limits of eta invariants, the eta form.

5.2.1. *Generalized connections.* Let M be a smooth manifold and $F \to M$ be a smooth real vector bundle. Let ∇ be a connection on F. Using the Leibniz rule we extend ∇ to $\Omega(M, F)$ (cf. [**3**, Def. 1.14]).

DEFINITION 5.1. A generalized connection on F is an operator $A: \Omega(M, F) \to \Omega(M, F)$ of the form $A = \nabla + S$ with $S \in \Omega(M, \mathrm{End}(F))$.

Let $q: F \to F^*$ be a duality structure on F. We extend q to a form $q: \Omega(M, F) \otimes \Omega(M, F) \to \Omega(M)$ by $q(u \otimes x, w \otimes y) = u \wedge w q(x, y)$.

DEFINITION 5.2. We say that a generalized connection A on $\Omega(M, F)$ is compatible with q if $dq(\phi, \psi) = q(A\phi, \psi) + q(\phi, A\psi)$.

5.2.2. *The η-form of a fibre bundle.* Let $\pi: M \to B$ be a smooth locally trivial fibre bundle with closed fibres Z_b, $b \in B$, of dimension n. Set $\epsilon_n = (-1)^{[\frac{n+1}{2}]}$, $\sqrt{\epsilon_n} = (\sqrt{-1})^{[\frac{n+1}{2}]}$. We assume that the vertical bundle $TZ := \ker(d\pi) \subset TM$ is oriented. Let $(\mathcal{F}, q_{\mathcal{F}})$ be a locally constant sheaf of finite dimensional real vector spaces over M with a ϵ–symmetric duality structure $q_{\mathcal{F}}$ and let F be the corresponding flat vector bundle (cf. Subsection 2.4.1).

We consider the infinite-dimensional $\mathbb{Z}$-graded vector bundle $\Omega(Z, F) \to B$ (with grading N_Z by form degree) with fibre $\Omega(Z, F)_b = \Omega(Z_b, F|_{Z_b})$ such that its space of smooth sections on B is $C^\infty(M, \Lambda^*(T^*Z) \otimes F)$. The space $\Omega(Z, F)$ carries an $\epsilon\epsilon_n$–duality structure $q_{Z,F}$ induced by $q_{\mathcal{F}}$ as in (4.1). Then the $\mathbb{Z}$-grading N_Z on $\Omega(Z, F)$ is a compatible $\mathbb{Z}$-grading of length n.

Our next goal is the interpretation of the twisted de Rham differential d^F on $\Omega(M, F)$ induced by the flat connection ∇^F on F as a superconnection (cf. [**12**, §III (a)]). We choose a horizontal distribution $T^H M \subset TM$, i.e., a complement to TZ. This choice induces an identification $\Omega(M, F) \cong \Omega(B, \Omega(Z, F))$. Then d^F can be viewed as a generalized connection (see Def. 5.2) on $\Omega(Z, F)$. For a vector field $X \in C^\infty(B, TB)$ we denote by $X^H \in C^\infty(M, T^H M)$ its horizontal lift. If $\omega \otimes \phi \in C^\infty(B, \Omega(Z, F)) \subset \Omega(M, F)$, then we set

$$\nabla_X^{Z,F} \omega \otimes \phi := L_{X^H} \omega \otimes \phi + \omega \otimes \nabla_{X^H}^F \phi,$$

where L_{X^H} denotes the Lie derivative. In this way we define a connection on $\Omega(Z, F)$. We extend $\nabla^{Z,F}$ to $\Omega(B, \Omega(Z, F))$ by using the Leibniz rule. Now we decompose (cf. [**12**, Prop. 3.4])

$$d^F = d^{Z,F} + \nabla^{Z,F} + i_T,$$

where $d^{Z,F}$ is the fibrewise twisted de Rham differential along the fiber Z and i_T is the insertion of a tensor field $T \in C^\infty(M, \Lambda^2(T^H M)^* \otimes TZ)$. To be precise, $\Lambda^2(T^H M)^*$ is considered here as a subspace of $\Lambda^2(T^*M)$ and i_T is interpreted as an element of $C^\infty(B, \Lambda^2(T^*B) \otimes \mathrm{End}(\Omega(Z, F)))$. It turns out that d^F is a $(-1)^{N_Z}$-superconnection of N_Z-degree one (cf. Subsection 3.4.4) which is flat because of $d^F \circ d^F = 0$.

We can form the (finite-dimensional) cohomology $H(Z,F)$ of $d^{Z,F}$ which comes equipped with a flat connection $\nabla^{H(Z,F)}$ (cf. [**12**, §III (f)]) and a parallel $\epsilon\epsilon_n$-duality structure $q_{H(Z,F)}$. The sheaf of parallel sections of $H(Z,F)$ is naturally isomorphic to $\mathcal{H}(Z,F) := \pi(\mathcal{F})$ so that $q_{H(Z,F)}$ corresponds to the form $q_{\mathcal{H}(Z,F)} := \pi(q_{Z,F})$ on $\mathcal{H}(Z,F)$ (cf. Subsection 2.5.1 for notation).

We now choose a vertical Riemannian metric g^{TZ} on TZ and a metric structure J^F on F. The vertical metric and orientation on TZ together induce a Hodge $*$-operator $*\colon \Omega(Z) \to \Omega(Z)$ as in Subsection 4.3.1. We define a metric structure on $\Omega(Z,F)$ by

$$J^{Z,F} := \frac{\sqrt{\epsilon\epsilon_n}}{\sqrt{\epsilon}\sqrt{\epsilon_n}}(*(-1)^{\frac{N_Z(N_Z-1)}{2}+nN_Z}) \otimes J^F. \tag{5.1}$$

This metric structure is compatible with the $\mathbb{Z}$-grading N_Z. On the cohomology $H(Z,F)$ we obtain a metric structure $J^{H(Z,F)}$ induced by $J^{Z,F}$. Let $(d^F)^*$, $(d^{Z,F})^*$, $(i_T)^*$, $(\nabla^{Z,F})^*$ be the adjoints of d^F, $d^{Z,F}$, i_T, $\nabla^{Z,F}$, respectively, with respect to the scalar product $q_{Z,F}(\cdot, J^{Z,F}\cdot)$ on $\Omega(Z,F)$ defined by $J^{Z,F}$.

• If n is even, by an easy computation we see that d^F is compatible with $q_{Z,F}$ in the sense of Def. 5.2. This in particular implies that the differential $d^{Z,F}$ is compatible with $q_{Z,F}$. Though $\Omega(Z,F)$ is infinite-dimensional, the theory of characteristic classes and forms extends to certain nice superconnections. In particular, $p(d^F, J^{Z,F}) \in \Omega(B)$ is well-defined. We consider the grading $z^{J^{Z,F}} := \frac{1}{\sqrt{\epsilon\epsilon_n}} J^{Z,F}$. Then the odd part of d^F with respect to this grading is given by $A = \frac{1}{2}\left(d^F + (d^F)^*\right)$. It is a $z^{J^{Z,F}}$-superconnection and we have

$$p(d^F, J^{Z,F}) := \varphi \operatorname{Tr}\left[z^{J^{Z,F}} \exp(-A^2)\right], \tag{5.2}$$

here φ multiplies a p-form by $(2\pi i)^{-p/2}$.

We now introduce the rescaling. Let $\tilde{B} := (0,\infty) \times B$ and $\mathrm{pr}\colon \tilde{B} \to B$ be the projection. We consider the bundle $\tilde{M} := \mathrm{pr}^* M$ over $\tilde{B}$ together with the canonical projection $\mathrm{Pr}\colon \tilde{M} \to M$. We define $(\tilde{\mathcal{F}}, \tilde{q}_{\mathcal{F}}, T^H\tilde{M}, \tilde{J}^F) := \mathrm{Pr}^*(\mathcal{F}, q_{\mathcal{F}}, T^H M, J^F)$. We obtain the $\mathbb{Z}$-graded $\epsilon\epsilon_n$-duality bundle $\Omega(Z,\tilde{F})$ over $\tilde{B}$ with $(-1)^{N_Z}$-superconnection $A' := \tilde{d}^F$, which is the twisted de Rham differential on $\Omega(\tilde{M},\tilde{F})$ induced by ∇^F. We fix the vertical metric $\tilde{g}^{TZ}$ which restricts to $t^{-1}g^{TZ}$ over $\{t\} \times M$. It induces the metric structure $\tilde{J}$ on $\Omega(Z,\tilde{F})$. The form $p(A',\tilde{J}) \in \Omega(\tilde{B})$ is now well-defined as in (5.2). Let us decompose $p(A',\tilde{J}) = dt \wedge \gamma + r$, where r does not contain dt. Here $\gamma\colon (0,\infty) \to \Omega(B)$ is a smooth family of forms on B. More precisely, for $t > 0$ we set

$$C_t = \frac{1}{2}\left(t^{N_Z/2} d^F t^{-N_Z/2} + t^{-N_Z/2}(d^F)^* t^{N_Z/2}\right). \tag{5.3}$$

Then

$$\gamma(t) = -(2\pi i)^{-1/2}\varphi \operatorname{Tr}\left[z^{J^{Z,F}} (\frac{\partial}{\partial t} C_t) \exp(-C_t^2)\right]. \tag{5.4}$$

By [**21**, Prop. 31], we have $\gamma(t) = O(t^{-3/2})$ as $t \to \infty$ and $\gamma(t) = O(1)$ as $t \to 0$ so that we can define ([**21**, Def. 33], [**8**, Def. 4.33])

DEFINITION 5.3. $\tilde{\eta}(\Omega(Z,F), N_Z, J^{Z,F}, d^F) := -\int_0^\infty \gamma(t)dt$.

Note that g^{TZ} and T^HM induce a canonical connection ∇^{TZ} on TZ (cf. [**5**, Thm. 1.9], [**3**, Prop. 10.2]). Let $L(\nabla^{TZ}) \in \Omega(M)$ be the L-form of TZ as in Subsection 4.2.2. We have

$$d\tilde{\eta}(\Omega(Z,F), N_Z, J^{Z,F}, d^F) = \int_Z L(\nabla^{TZ}) \wedge p(\nabla^F, J^F) - p(\nabla^{H(Z,F)}, J^{H(Z,F)}). \tag{5.5}$$

• If n is odd, then A, $A^{(1)} = \frac{1}{2}(\nabla^{Z,F} + (\nabla^{Z,F})^*)$ commute with $J^{Z,F}$. To stay in the superconnection formalism we introduce an extra odd variable σ such that $\sigma^2 = 1$. We multiply all components of A with even form degree of $\Lambda(T^*B)$ by this variable and denote the result still A. This modified A is then again a superconnection (cf. [**9**, §II (f)]).

If B_0, B_1 are trace class in $\mathrm{End}(\Omega(Z,F))$, $\omega_0, \omega_1 \in \Lambda(T^*B)$, we put

$$\mathrm{Tr}'_\sigma\left[\omega_0 B_0 + \omega_1 B_1 \sigma\right] := \omega_0 \mathrm{Tr}\left[B_0\right], \quad \mathrm{Tr}_\sigma\left[\omega_0 B_0 + \omega_1 B_1 \sigma\right] := \omega_1 \mathrm{Tr}\left[B_1\right]. \tag{5.6}$$

By [**9**, Thm. 2.10], the form

$$p(A', J) = (2i)^{1/2} \varphi \mathrm{Tr}_\sigma[z^{J^{Z,F}} \exp(-A^2)] \tag{5.7}$$

is a closed odd form. Let us decompose $p(A', \tilde{J}) = dt \wedge \gamma + r$, where r does not contain dt. Then

$$\gamma(t) = -\frac{1}{\sqrt{\pi}} \varphi \mathrm{Tr}_\sigma\left[z^{J^{Z,F}} (\frac{\partial}{\partial t} C_t) \exp(-C_t^2)\right]. \tag{5.8}$$

By the same argument as in [**10**, Thm. 2.11], we have $\gamma(t) = O(1)$ as $t \to 0$ as well as $\gamma(t) = O(t^{-3/2})$ as $t \to \infty$ as in [**3**, Thm. 9.23]. So we can define (cf. [**8**, Def. 4.93]):

DEFINITION 5.4. $\tilde{\eta}\big(\Omega(Z,F), N_Z, J^{Z,F}, d^F\big) := -\int_0^\infty \gamma(t)dt$.

Then the degree zero part $\tilde{\eta}^{(0)}$ of $\tilde{\eta}$ is $\frac{1}{2}\eta(D^{\mathrm{sign}}_{Z,F})$ for the fibrewise operator in (4.2). By the argument in [**9**, Thm. 2.10], as in [**8**, Thm. 4.95] we have

$$d\tilde{\eta}(\Omega(Z,F), N_Z, J^{Z,F}, d^F) = \int_Z L(\nabla^{TZ}) \wedge p(\nabla^F, J^F). \tag{5.9}$$

Since the fibres Z are odd-dimensional, we must specify our sign conventions when integrating differential form along the fibres Z. If $\alpha \in \Omega(B)$, β a section of $\Lambda(T^*Z)$ on M with compact support, then $\int_Z \pi^*(\alpha) \wedge \beta = \alpha \int_Z \beta$. This sign convention is compatible with the sign convention of Tr in Subsection 3.4.4 and (5.7).

REMARK 5.5. Assume that n is odd. In [**21**, (233)], Lott defined also an even form $p^{\mathrm{Lott}}(A', J^W(t))$. Actually, $p^{\mathrm{Lott}}(A', J^W(t)) = 0$. In fact, we only need to show $p^{\mathrm{Lott}}(A', J^W(1)) = 0$. Set $\bar{A} = A^{(1)} + z^{J^{Z,F}}(A - A^{(1)})$. If we use our notation,

$$\begin{aligned} p^{\mathrm{Lott}}(A', J^W(1)) &= \sqrt{\epsilon_n}\,\varphi\,\mathrm{Tr}'_\sigma\big[z^{J^{Z,F}}(-1)^{N_Z}\exp(-A^2)\big] \\ &= \sqrt{\epsilon_n}\,\varphi\,\mathrm{Tr}'_\sigma\big[z^{J^{Z,F}}(-1)^{N_Z}\exp(-\bar{A}^2)\big]. \end{aligned}$$

Now $\bar{A}$ preserves the parity of the N_Z-grading, while $z^{J^{Z,F}}$ changes the parity of the N_Z-grading. Thus the above trace must be zero. It also follows from this observation that the eta form in [**21**, (238)] is zero if n is odd.

We now describe how $\tilde{\eta}\big(\Omega(Z,F), N_Z, J^{Z,F}, d^F\big)$ depends on its arguments. Let (T^HM, g^{TZ}, J^F), (T'^HM, g'^{TZ}, J'^F) be two triples of geometric data. We will mark the objects associated to the second triple by $'$.

Let $\tilde{L}(TZ, \nabla^{TZ}, \nabla'^{TZ})$ be the transgression of the L-form such that

$$d\tilde{L}(TZ, \nabla^{TZ}, \nabla'^{TZ}) = L(\nabla^{TZ}) - L(\nabla'^{TZ}).$$

THEOREM 5.6. *Modulo exact forms on B, we have*

$$\begin{aligned} &\tilde{\eta}\big(\Omega(Z,F), N_Z, J^{Z,F}, d^F\big) - \tilde{\eta}\big(\Omega(Z,F), N_Z, J'^{Z,F}, d^F\big) \\ &= \int_Z \tilde{L}(TZ, \nabla^{TZ}, \nabla'^{TZ}) \wedge p(\nabla^F, J^F) + \int_Z L(TZ, \nabla'^{TZ}) \wedge \tilde{p}(\nabla^F, J^F, J'^F) \\ &\qquad - \tilde{p}(\nabla^{H(Z,F)}, J^{H(Z,F)}, J'^{H(Z,F)}). \end{aligned}$$

PROOF. Note that if n is odd, $J^{H(Z,F)}$ changes the parity of the N_Z-grading on $H(Z,F)$, thus $p(\nabla^{H(Z,F)}, J^{H(Z,F)})$, $\tilde{p}(\nabla^{H(Z,F)}, \ldots)$ are zero. Now this equation is a formal consequence of (5.5) and (5.9). □

5.2.3. *Adiabatic limits and the formula of Dai.* We keep the notation which was introduced in Subsection 5.2.2. Furthermore, we assume that $\dim Z := n$ is even and that B is closed, oriented, and of odd dimension n_B such that $\epsilon\epsilon_{n+n_B} = 1$. We choose a Riemannian metric g^{TB} on the base B. Using the horizontal distribution T^HM, we define the family of Riemannian metrics $g_T^{TM} := \pi^* g^{TB} \oplus \frac{1}{T^2} g^{TZ}$, $T > 0$. The orientations of B and TZ induce an orientation of M. Let $D_T^F := D_F^{\mathrm{sign}}$ be defined as in (4.2) using the metric g_T^{TM} and let $\eta(D_T^F)$ be its η-invariant. On the base B we consider the twisted signature operator $D^{H(Z,F)} := D^{\mathrm{sign}}_{H(Z,F)}$ and its η-invariant $\eta(D^{H(Z,F)})$ with respect to $g^{TB}, J^{H(Z,F)}$.

There is a decreasing filtration of the cohomological group $H(M, \mathcal{F})$ and a Leray spectral sequence $({}_{LS}E_r, {}_{LS}d_r)$ $(r \geq 2)$ with ${}_{LS}E_2^{p,q} = H^p(B, \mathcal{H}^q(Z,F))$ converging to $\mathrm{Gr}H^*(M, \mathcal{F})$. In the present smooth model this spectral sequence is associated to the filtration of $\Omega(M,F)$ given by $F^p\Omega(M,F) = \oplus_{q \geq p} \pi^*\Omega^q(B, \Omega(Z,F))$ (cf. [**17**, §3.5]). Then ${}_{LS}E_0^{p,q} = \Omega^p(B, \Omega^q(Z,F))$. The induced 1-symmetric pairing ${}_{LS}E_0^{p,q} \otimes {}_{LS}E_0^{s,t} \to \mathbb{R}$, $p + q + s + t = \dim M$, $p + s = N_B = \dim B$, is given by (4.1). In particular, it is non-degenerate. This spectral sequence with duality structure gives rise to the integer ${}_{LS}\tau_2(Z, F) := \tau_2({}_{LS}E_*)$ as in Definition 4.8.

Let ∇^{TB} be the Levi-Civita connection on TB. Let $L(\nabla^{TB})$ be the corresponding L-form on TB. We can now state the main result of [**16**].

THEOREM 5.7 (Dai [**16**, Thm. 0.3]).

$$\lim_{T\to\infty} \eta(D_T^F) =$$
$$\eta(D^{H(Z,F)}) + 2\int_B L(\nabla^{TB}) \wedge \tilde{\eta}(\Omega(Z,F), N_Z, J^{Z,F}, d^F) + {}_{LS,F}\tau_2(Z,F).$$

5.2.4. *The η-form and complexes.* We make the same assumptions as in Subsection 5.2.2. In addition we assume that $\mathcal{F}$ admits a $\mathbb{Z}$-grading N_F of length n_F and a compatible differential v. Let $\mathcal{H}$ be the cohomology of v. Let H be the flat vector bundle corresponding to $\mathcal{H}$.

We choose a metric structure J^F on F such that it is compatible with N_F. Then we obtain an induced metric structure J^H on H.

We modify the construction of the eta form in Subsection 5.2.2 by replacing d^F by $\mathbf{d} := d^F + v$. This differential is now a flat $(-1)^{\mathbf{N}}$-superconnection, where $\mathbf{N} := N_Z + N_F$ is the total grading. Here $\mathbf{N}$ is a grading of length $n + n_F$ which is compatible with the form $q_{Z,F}$.

The cohomology of the zero part $d^{Z,F}+v$ of $\mathbf{d}$ is the fibrewise hypercohomology $\mathcal{H}H$ of the complex $(\mathcal{F}, v)$ along the fibre Z. By HH we denote the corresponding bundle which acquires an metric structure J^{HH} induced by $J^{Z,F}$.

We now introduce the rescaling. Let again $\tilde{B} := (0,\infty) \times B$ and pr: $\tilde{B} \to B$ be the projection. We consider the bundle $\tilde{M} := \mathrm{pr}^*M$ over $\tilde{B}$ together with the canonical projection Pr: $\tilde{M} \to M$. We define $(\tilde{\mathcal{F}}, \tilde{q}_{\mathcal{F}}, T^H\tilde{M}, \tilde{v}, \tilde{N}_F) := \mathrm{Pr}^*(\mathcal{F}, q_{\mathcal{F}}, T^H M, v, N_F)$. We obtain the (by $\tilde{N} := \tilde{N}_Z + \tilde{N}_F$) $\mathbb{Z}$-graded $\epsilon\epsilon_n$-duality bundle $\Omega(Z, \tilde{F})$ over $\tilde{B}$ with the $(-1)^{\tilde{N}}$-superconnection $A' := \tilde{\mathbf{d}} = d^{\tilde{F}} + \tilde{v}$. We fix the vertical metric $\tilde{g}^{TZ}$ which restricts to $t^{-1}g^{TZ}$ over $\{t\} \times M$. Furthermore, we consider the metric structure $\tilde{J}^F$ which restricts to $t^{-\tilde{N}_F/2+n_F/4}J^F t^{\tilde{N}_F/2-n_F/4} = J^F t^{\tilde{N}_F - n_F/2}$ (cf. Definition 3.14) on $\{t\} \times M$. These choices induce the metric structure $\tilde{J}$ on $\Omega(Z, \tilde{F})$. The form $p(A', \tilde{J}) \in \Omega(\tilde{B})$ is now well-defined as in Subsection 5.2.2. Let us decompose $p(A', \tilde{J}) = dt \wedge \gamma + r$, where r does not contain dt. Here γ: $(0,\infty) \to \Omega(B)$ is a smooth family of forms on B.

LEMMA 5.8. *We have $\gamma(t) = O(t^{-3/2})$ as $t \to \infty$ and $\gamma = O(1)$ as $t \to 0$. The form*

$$\tilde{\eta}(\Omega(Z,F), \mathbf{N}, J^{Z,F}, \mathbf{d}) := -\int_0^\infty \gamma(t)dt$$

satisfies

$$d\tilde{\eta}(\Omega(Z,F),\mathbf{N},J^{Z,F},\mathbf{d}) = \begin{cases} \int_Z L(\nabla^{TZ})\wedge p(\nabla^F,J^F) - p(\nabla^{HH},J^{HH}) & \text{if } n \text{ is even,} \\ \int_Z L(\nabla^{TZ})\wedge p(\nabla^F,J^F) & \text{if } n \text{ is odd.} \end{cases}$$

PROOF. This lemma can be shown using the techniques of the proof of [**12**, Thm. 3.21], [**9**, Thm. 2.10]. □

There are two spectral sequences of locally constant sheaves of finite-dimensional real vector spaces which converge to the hypercohomology $\mathcal{H}H$ of $(\mathcal{F},v)$ along the fibre Z, namely the fibrewise hypercohomology spectral sequence $({}_{hc}\mathcal{E}_r, {}_{hc}d_r)$ and the fibrewise local-global spectral sequence $({}_{lg}\mathcal{E}_r, {}_{lg}d_r)$. Both are obtained from filtrations of $\Omega(Z,F)$ which are compatible with the duality. Therefore, we obtain $\epsilon\epsilon_n$ duality structures $q_{{}_{hc}E_r}$ and $q_{{}_{lg}E_r}$ induced by $q_{Z,F}$. Let $N_{{}_{hc}E_r}$, $N_{{}_{gl}E_r}$ be the $\mathbb{Z}$-gradings on ${}_{hc}E_r, {}_{gl}E_r$ induced by $\mathbf{N}$. We further obtain corresponding metric structures $J^{{}_{hc}E_r}$ and $J^{{}_{lg}E_r}$. Let ${}_{hc}d_r^*$ be the adjoint of ${}_{hc}d_r$ with respect to $J^{{}_{hc}E_r}$. We identify ${}_{hc}\mathrm{Gr}(\mathcal{H}H) \cong {}_{hc}\mathcal{E}_\infty$ and ${}_{lg}\mathrm{Gr}(\mathcal{H}H) \cong {}_{lg}\mathcal{E}_\infty$ (cf. Subsection 4.4.2).

To consider the case of even and odd n in parallel we adopt the following convention. If n is even, then the eta form of a complex or a filtration was defined in Subsections 3.4.4 and 3.4.5, respectively.

If n is odd, then we proceed as in (5.7). Note that ${}_{hc}d_r + {}_{hc}d_r^*$ commutes with $J^{{}_{hc}E_r}$. We form ${}_{hc}A := \nabla^{{}_{hc}E_r, J^{{}_{hc}E_r}} + ({}_{hc}d_r + {}_{hc}d_r^*)\sigma$. By [**7**, Prop. 2.12], [**27**, §5], the form

$$p(\nabla^{{}_{hc}E_r} + {}_{hc}d_r, J^{{}_{hc}E_r}) := (2i)^{1/2}\varphi\,\mathrm{Tr}_\sigma\left[z^{J^{{}_{hc}E_r}}\exp(-{}_{hc}A^2)\right] \tag{5.10}$$

is an exact odd form. To define the η-form $\tilde{\eta}({}_{hc}\mathcal{E}_r, N_{{}_{hc}E_r}, J^{{}_{hc}E_r}, {}_{hc}d_r)$ we now employ the usual rescaling induced by the $\mathbb{Z}$-grading $N_{{}_{hc}E_r}$. As in [**8**, Thm. 2.43], [**7**, Thm. 2.17], it is a closed form on B, and its cohomology class is $\frac{1}{2}[\mathbf{ch}({}_{hc}E_{r,>0}) - \mathbf{ch}({}_{hc}E_{r,<0})]$, where ${}_{hc}E_{r,>0}$ and ${}_{hc}E_{r,<0}$ are the sub-bundles of ${}_{hc}E_r$ associated to positive or negative eigenvalues of $z^{J^{{}_{hc}E_r}}({}_{hc}d_r + {}_{hc}d_r^*)$. In a similar manner we define $\tilde{\eta}({}_{lg}\mathcal{E}_r, N_{{}_{lg}\mathcal{E}_r}, J^{{}_{lg}\mathcal{E}_r}, {}_{lg}d_r)$ associated to the local-global spectral sequence. We do not change the definition of the η-forms of the filtrations.

The following theorems are the main ingredients of the proof of well-definedness of the secondary index map.

THEOREM 5.9. *Modulo exact forms on B we have*

$$\tilde{\eta}(\Omega(Z,F),\mathbf{N},J^{Z,F},\mathbf{d}) \equiv \tilde{\eta}(\Omega(Z,F),N_Z,J^{Z,F},d^F) + \sum_{r\geq 1}\tilde{\eta}({}_{hc}\mathcal{E}_r, N_{{}_{hc}E_r}, J^{{}_{hc}E_r}, {}_{hc}d_r) - \tilde{\eta}(\mathcal{H}H, {}_{hc}\mathrm{Gr}(\mathcal{H}H), J^{HH}, J^{{}_{hc}\mathrm{Gr}(HH)}).$$

THEOREM 5.10. *Modulo exact forms on B we have*

$$\tilde{\eta}(\Omega(Z,F),\mathbf{N},J^{Z,F},\mathbf{d}) \equiv \tilde{\eta}(\Omega(Z,H),N_Z,J^{Z,H},d^H) + \sum_{r\geq 2}\tilde{\eta}({}_{lg}\mathcal{E}_r,N_{{}_{lg}\mathcal{E}_r},J^{{}_{lg}\mathcal{E}_r},{}_{lg}d_r) - \tilde{\eta}(\mathcal{H}H,{}_{lg}\mathrm{Gr}(\mathcal{H}H),J^{HH},J^{{}_{lg}\mathrm{Gr}(HH)}) + \int_Z L(\nabla^{TZ})\wedge\tilde{\eta}(F,N_F,J^F,v).$$

The proof of both results will be sketched in Section 6. Note that if n_F is odd, then $\tilde{\eta}(\Omega(Z,F),N_Z,J^{Z,F},d^F) = 0$ and $\tilde{\eta}(\Omega(Z,H),N_Z,J^{Z,H},d^H) = 0$, as $J^{Z,F}$ changes the parity of N_F-grading.

5.2.5. *The η-form of an iterated fibre bundle.* Let W,V,S be smooth manifolds. Let $\pi_1\colon W\to V$, $\pi_2\colon V\to S$ be smooth fibrations with compact fibres X and Y, respectively. Then $\pi_3=\pi_2\circ\pi_1\colon W\to S$ is a smooth fibration with compact fiber Z of dimension n. Let $n_0=\dim X, m_0=\dim Y$. Let TX,TY,TZ be the corresponding vertical tangent bundles. We assume that TX,TY,TZ are compatibly oriented such that we have an iterated fibre bundle with compatible orientation as in Subsection 2.5.2.

Let $\mathcal{F}$ be a locally constant sheaf of finite dimensional real vector spaces over W with a ϵ–symmetric duality structure $q_{\mathcal{F}}$. We choose a metric structure J^F on $F:=\mathtt{bundle}(\mathcal{F})$.

By taking fibrewise cohomology we obtain locally constant sheaves of finite-dimensional real vector spaces $\mathcal{H}(X,\mathcal{F})$ over V, and $\mathcal{H}(Y,\mathcal{H}(X,\mathcal{F}))$ and $\mathcal{H}(Z,\mathcal{F})$ over S. These come with induced duality structures $q_{\mathcal{H}(X,\mathcal{F})}, q_{\mathcal{H}(Z,\mathcal{F})}, q_{\mathcal{H}(Y,\mathcal{H}(X,\mathcal{F}))}$.

We choose horizontal sub-bundles T_1^HW, T_2^HV, T_3^HW, i.e., complements in TW, TV, TW to TX, TY, TZ. Furthermore, we choose vertical metrics g^{TZ}, g^{TX}, g^{TY} on TZ, TX, TY. Then we obtain induced metric structures $J^{H(X,\mathcal{F})}$, $J^{H(Z,\mathcal{F})}$, $J^{H(Y,\mathcal{H}(X,\mathcal{F}))}$.

Let ∇^{TX}, ∇^{TY}, ∇^{TZ} be the connections on TX, TY, TZ which are induced by the choice of horizontal subspaces and vertical metrics (cf. [**5**, Def. 1.6], [**3**, Prop. 10.2]). We get a further connection ${}^0\nabla^{TZ} := \pi_1^*\nabla^{TY}\oplus\nabla^{TX}$ on $TZ = TX\oplus(T_1^HW\cap TZ)$. Let $L(\nabla^{TX})$, $L(\nabla^{TY})$, $L(\nabla^{TZ})$, $L({}^0\nabla^{TZ})$ be the associated L-forms as in Subsection 5.2.2. By $\tilde{L}(TZ,\nabla^{TZ},{}^0\nabla^{TZ})$ we denote the transgression L-form such that

$$d\widetilde{L}(TZ,\nabla^{TZ},{}^0\nabla^{TZ}) = L(\nabla^{TZ}) - L({}^0\nabla^{TZ}). \tag{5.11}$$

Since the fibre Z has the structure of a fibre bundle, we have a fibrewise Leray spectral sequence $({}_{LS}\mathcal{E}_r, {}_{LS}d_r)$ $(r\geq 2)$ of locally constant sheaves on S (cf. also [**24**, Prop. 2.1]). All terms have $\epsilon\epsilon_n$–duality structures $q_{{}_{LS}\mathcal{E}_r}$ induced by $q_{Z,F}$ on $\Omega(Z,F)$. The $\mathbb{Z}$-gradings $N_{{}_{LS}\mathcal{E}_r}$ (induced by N_Z) and the differentials ${}_{LS}d_r$ are compatible with the duality. In particular, we have ${}_{LS}\mathcal{E}_2 = H(Y,\mathcal{H}(X,\mathcal{F}))$. Hence, we have an induced compatible metric structure $J^{{}_{LS}\mathcal{E}_2} := J^{H(Y,\mathcal{H}(X,\mathcal{F}))}$. Since the other terms are obtained by taking cohomology successively, we get induced metric structures $J^{_{LS}\mathcal{E}_r}$ on ${}_{LS}\mathcal{E}_r$ for all $r \geq 2$. Thus we can define $\tilde{\eta}({}_{LS}\mathcal{E}_r, N_{{}_{LS}\mathcal{E}_r}, J^{_{LS}E_r}, {}_{LS}d_r)$ (with the usual modifications if one or two of the dimensions n_0, m_0 are odd, see the remark below).

There is a filtration on $\mathcal{H}(Z,\mathcal{F})$ and a natural isomorphism $\mathrm{Gr}\mathcal{H}(Z,\mathcal{F}) \cong {}_{LS}\mathcal{E}_\infty$. Therefore, we have the form $\tilde{\eta}(\mathcal{H}(Z,\mathcal{F}), {}_{LS}\mathcal{E}_\infty, J^{H(Z,\mathcal{F})}, J^{_{LS}E_\infty})$ as in Definition 3.18.

The following theorem provides the relation of the η-forms of the total fibration π_3 with the η-forms of the partial fibrations π_1 and π_2. It is the main ingredient in the proof of the fact that the secondary index map is functorial with respect to iterated fibre bundles.

THEOREM 5.11. *The following identity holds modulo exact forms on S:*

$$\begin{aligned}\tilde{\eta}(\Omega(Z,F), N_Z, J^{Z,F}, d^F) \equiv \int_Y & L(\nabla^{TY}) \wedge \tilde{\eta}(\Omega(X,F), N_X, J^{X,F}, d^F) \\ & + \tilde{\eta}(\Omega(Y, H(X,\mathcal{F})), N_Y, J^{Y,H(X,\mathcal{F}))}, d^{H(X,\mathcal{F})}) \\ & - \tilde{\eta}(\mathcal{H}(Z,\mathcal{F}), {}_{LS}\mathcal{E}_\infty, J^{H(Z,\mathcal{F})}, J^{_{LS}E_\infty}) \\ & + \sum_{r=2}^{\infty} \tilde{\eta}({}_{LS}\mathcal{E}_r, N_{{}_{LS}\mathcal{E}_r}, J^{_{LS}E_r}, {}_{LS}d_r) \\ & + \int_Z \tilde{L}(TZ, \nabla^{TZ}, {}^0\nabla^{TZ}) \wedge p(\nabla^F, J^F).\end{aligned}$$

PROOF. The proof of this theorem will be sketched in Subsection 6.5. The formula above arrises from a detailed investigation of the adiabatic limit of $\tilde{\eta}(\Omega(Z,F)$, N_Z, $J^{Z,F}, d^F)$. □

REMARK 5.12. Theorem 5.11 has four cases. Assume first that n is even. Then $\tilde{\eta}(\Omega(Z,F),\ldots)$ and $\tilde{\eta}({}_{LS}\mathcal{E}_r,\ldots)$ are defined as in Definitions 5.3, 3.15. If n_0 is even, then $\tilde{\eta}(\Omega(Y,H(X,\mathcal{F})),\ldots)$, $\tilde{\eta}(\Omega(X,F),\ldots)$ are defined as in Definition 5.3. If n_0 is odd, then $\tilde{\eta}(\Omega(Y,H(X,\mathcal{F})),\ldots)$, $\tilde{\eta}(\Omega(X,F),\ldots)$ are defined as in Definition 5.4 and $\tilde{\eta}(\Omega(Y,H(X,\mathcal{F})),\ldots)$ is zero by the remark following Theorem 5.10.

Assume now that n is odd. Then $\tilde{\eta}(\Omega(Z,F),\ldots)$ and $\tilde{\eta}({}_{LS}\mathcal{E}_r,\ldots)$ are defined as in Definition 5.4 and Subsection 5.2.4. If n_0 is odd, then $\tilde{\eta}(\Omega(Y,H(X,\mathcal{F})),\ldots)$ and $\tilde{\eta}(\Omega(X,F),\ldots)$ are defined as in Definitions 5.3 and 5.4 and $\tilde{\eta}(\Omega(Y,H(X,\mathcal{F})),\cdot)$ is zero by the remark following Theorem 5.10. If n_0 is even, then $\tilde{\eta}(\Omega(X,F),\ldots)$ is defined as in Definition 5.3 and $\tilde{\eta}(\Omega(Y,H(X,\mathcal{F})),\ldots)$ is defined as in Definition 5.4.

5.3. Definition of the secondary index maps.

5.3.1. *The map* $\pi_*^{\bar{L},\mathrm{Lott}}$. Let $M \to B$ be a smooth locally trivial fibre bundle with even-dimensional closed fibres Z over a compact base B such that the vertical bundle TZ is oriented. We set $n := \dim Z$, $\epsilon_n = (-1)^{[\frac{n+1}{2}]}$.

In [**21**, §3.6], Lott constructed a secondary index map

$$\pi_*^{\bar{L},\mathrm{Lott}} \colon \bar{L}_\epsilon^{\mathrm{Lott}}(M) \to \bar{L}_{\epsilon\epsilon_n}^{\mathrm{Lott}}(B)$$

which fits into the commutative diagram

$$\begin{array}{ccccccc} H^{4*-\epsilon}(M,\mathbb{R}) & \longrightarrow & \bar{L}_\epsilon^{\mathrm{Lott}}(M) & \longrightarrow & \bar{L}_\epsilon^{\mathbb{R}/\mathbb{Z}}(M) & \longrightarrow & K_{\mathbb{R}/\mathbb{Z}}^{-1}(M) \\ \pi_*^{\mathrm{sign}}\downarrow & & \pi_*^{\bar{L},\mathrm{Lott}}\downarrow & & \pi_*^{\bar{L},\mathbb{R}/\mathbb{Z}}\downarrow & & \pi_*^{\mathrm{sign},\mathbb{R}/\mathbb{Z}}\downarrow \\ H^{4*-\epsilon\epsilon_n}(B,\mathbb{R}) & \longrightarrow & \bar{L}_{\epsilon\epsilon_n}^{\mathrm{Lott}}(B) & \longrightarrow & \bar{L}_{\epsilon\epsilon_n}^{\mathbb{R}/\mathbb{Z}}(B) & \longrightarrow & K_{\mathbb{R}/\mathbb{Z}}^{-1}(B). \end{array}$$

We recall the definition of $\pi_*^{\bar{L},\mathrm{Lott}}$. We choose a metric g^{TZ} and a horizontal distribution T^HM. This fixes a canonical connection ∇^{TZ} on the vertical bundle TZ.

DEFINITION 5.13. Given a class $\left[\mathcal{F}, q_{\mathcal{F}}, J^F, \rho\right]_{\mathrm{Lott}} \in \bar{L}_\epsilon^{\mathrm{Lott}}(M)$, its secondary index is defined by

$$\pi_*^{\bar{L},\mathrm{Lott}}[\mathcal{F}, q_F, J^F, \rho]_{\mathrm{Lott}} :=$$
$$\left[\mathcal{H}(Z,\mathcal{F}), q_{\mathcal{H}(Z,\mathcal{F})}, J^{H(Z,\mathcal{F})}, \int_Z L(\nabla^{TZ}) \wedge \rho - \tilde{\eta}(\Omega(Z,F), N_Z, J^{Z,F}, d^F)\right]_{\mathrm{Lott}}.$$

By [**21**, Prop. 32], $\pi_*^{\bar{L},\mathrm{Lott}}$ is well-defined. In particular, it is independent of the additional choices.

5.3.2. *Construction of* $\boldsymbol{\pi_*^{\bar{L}}}$. We can now define the secondary analytic index map

$$\pi_*^{\bar{L}} \colon \bar{L}_\epsilon(M) \to \bar{L}_{\epsilon\epsilon_n}(B).$$

Recall that we have surjective homomorphisms

$$\bar{L}_\epsilon^{\mathrm{Lott}}(M) \to \bar{L}_\epsilon(M), \quad \bar{L}_{\epsilon\epsilon_n}^{\mathrm{Lott}}(B) \to \bar{L}_{\epsilon\epsilon_n}(B).$$

DEFINITION 5.14. We define $\pi_*^{\bar{L}}$ by the condition that the following diagram commutes:

$$\begin{array}{ccc} \bar{L}_\epsilon^{\mathrm{Lott}}(M) & \xrightarrow{\pi_*^{\bar{L},\mathrm{Lott}}} & \bar{L}_{\epsilon\epsilon_n}^{\mathrm{Lott}}(B) \\ \downarrow & & \downarrow \\ \bar{L}_\epsilon(M) & \xrightarrow{\pi_*^{\bar{L}}} & \bar{L}_{\epsilon\epsilon_n}(B). \end{array}$$

THEOREM 5.15. *The index map* $\pi_*^{\bar{L},\mathrm{Lott}}$ *induces a well-defined secondary index map*

$$\pi_*^{\bar{L}} \colon \bar{L}_\epsilon(M) \to \bar{L}_{\epsilon\epsilon_n}(B).$$

PROOF. Let $[\mathcal{F},q_{\mathcal{F}},J^F,\rho]\in \bar{L}_\epsilon(M)$ and $\mathcal{L}\subset\mathcal{F}$ be a locally constant lagrangian subsheaf. We must show that

$$\Big[\mathcal{H}(Z,\mathcal{F}),q_{\mathcal{H}(Z,\mathcal{F})},J^{H(Z,\mathcal{F})},\int_Z L(\nabla^{TZ})\wedge\rho-\tilde{\eta}(\Omega(Z,F),N_Z,J^{Z,F},d^F)\Big] = \Big[0,0,0,\int_Z L(\nabla^{TZ})\wedge(\rho+\tilde{p}(\mathcal{F},q_{\mathcal{F}},J^F,\mathcal{L}))\Big]. \tag{5.12}$$

We repeat the construction of $\mathcal{G}$, q_{cG}, J^{cG}, $v_{\mathcal{G}}$, etc. from Subsection 4.5.2 to write

$$-\tilde{p}(\mathcal{F},q_{\mathcal{F}},J^F,\mathcal{L})=\tilde{\eta}(\mathcal{G},N_{\mathcal{G}},J^{\mathcal{G}},v_{\mathcal{G}}). \tag{5.13}$$

We consider the sum $\mathcal{G}^0\oplus\mathcal{G}^2$ together with the corresponding form and metric structure. Note that

$$\big[\mathcal{H}(Z,\mathcal{G}^0)\oplus\mathcal{H}(Z,\mathcal{G}^2),q_{\mathcal{H}(Z,\mathcal{G}^0)\oplus\mathcal{H}(Z,\mathcal{G}^2)},J^{H(Z,\mathcal{G}^0)\oplus H(Z,\mathcal{G}^2)},0\big]=0.$$

Moreover, we have $\tilde{\eta}(\Omega(Z,G^0\oplus G^2),N_Z,J^{G^0\oplus G^2},d^{Z,G^0\oplus G^2})=0$. Thus

$$\begin{aligned}
&\Big[\mathcal{H}(Z,\mathcal{F}),q_{\mathcal{H}(Z,\mathcal{F})},J^{H(Z,\mathcal{F})},\int_Z L(\nabla^{TZ})\wedge\rho-\tilde{\eta}(\Omega(Z,F),N_Z,J^{Z,F},d^F)\Big]\\
&=\Big[\mathcal{H}(Z,\mathcal{F}),q_{\mathcal{H}(Z,\mathcal{F})},J^{H(Z,\mathcal{F})},\int_Z L(\nabla^{TZ})\wedge\rho-\tilde{\eta}(\Omega(Z,F),N_Z,J^{Z,F},d^F)\Big]\\
&\quad+\big[\mathcal{H}(Z,\mathcal{G}^0)\oplus\mathcal{H}(Z,\mathcal{G}^2),q_{\mathcal{H}(Z,\mathcal{G}^0)\oplus\mathcal{H}(Z,\mathcal{G}^2)},J^{H(Z,\mathcal{G}^0)\oplus H(Z,\mathcal{G}^2)},0\big]\\
&=\Big[\mathcal{H}(Z,\mathcal{G}),q_{\mathcal{H}(Z,\mathcal{G})},J^{H(Z,\mathcal{G})},\int_Z L(\nabla^{TZ})\wedge\rho-\tilde{\eta}(\Omega(Z,G),N_Z,J^{Z,G},d^G)\Big].
\end{aligned}$$

Next observe that we have $\mathcal{H}(Z,\mathcal{G})={}_{hc}\mathcal{E}_1$. This relation respects the other structures. Let $\mathcal{H}H$ be the fibrewise hypercohomology of the complex $\mathcal{G}$ along the fibre Z. Using Lemma 3.16 repeatedly and finally Lemma 3.19, we get

$$\begin{aligned}
&\Big[\mathcal{H}(Z,\mathcal{F}),q_{\mathcal{H}(Z,\mathcal{F})},J^{H(Z,\mathcal{F})},\int_Z L(\nabla^{TZ})\wedge\rho-\tilde{\eta}(\Omega(Z,F),N_Z,J^{Z,F},d^F)\Big]\\
&=\Big[{}_{hc}\mathcal{E}_2,q_{{}_{hc}\mathcal{E}_2},J^{{}_{hc}E_2},\int_Z L(\nabla^{TZ})\wedge\rho-\tilde{\eta}(\Omega(Z,G),N_Z,J^{Z,G},d^G)\\
&\qquad-\tilde{\eta}({}_{hc}\mathcal{E}_1,N_{{}_{hc}E_1},J^{{}_{hc}E_1},{}_{hc}d_1)\Big]\\
&=\Big[{}_{hc}\mathcal{E}_\infty,q_{{}_{hc}\mathcal{E}_\infty},J^{{}_{hc}E_\infty},\int_Z L(\nabla^{TZ})\wedge\rho-\tilde{\eta}(\Omega(Z,G),N_Z,J^{Z,G},d^G)\\
&\qquad-\sum_{r\geq1}\tilde{\eta}({}_{hc}\mathcal{E}_r,N_{{}_{hc}E_r},J^{{}_{hc}E_r},{}_{hc}d_r)\Big]\\
&=\Big[\mathcal{H}H,q_{\mathcal{H}H},J^{HH},\int_Z L(\nabla^{TZ})\wedge\rho-\tilde{\eta}(\Omega(Z,G),N_Z,J^{Z,G},d^G)\\
&\qquad-\sum_{r\geq1}\tilde{\eta}({}_{hc}\mathcal{E}_r,N_{{}_{hc}E_r},J^{{}_{hc}E_r},{}_{hc}d_r)+\tilde{\eta}(\mathcal{H}H,{}_{hc}\mathrm{Gr}(\mathcal{H}H),J^{HH},J^{\mathrm{Gr}(HH)})\Big].
\end{aligned}$$

(Note that in fact $\mathcal{H}H = 0$.) We now apply first Theorem 5.9 and then Theorem 5.10 to conclude that

$$\Big[\mathcal{H}(Z,\mathcal{F}), q_{\mathcal{H}(Z,\mathcal{F})}, J^{H(Z,\mathcal{F})}, \int_Z L(\nabla^{TZ}) \wedge \rho - \tilde{\eta}(\Omega(Z,F), N_Z, J^{Z,F}, d^F)\Big]$$
$$= \Big[\mathcal{H}H, q_{\mathcal{H}H}, J^{\mathcal{H}H}, \int_Z L(\nabla^{TZ}) \wedge \rho - \tilde{\eta}(\Omega(Z,G), \mathbf{N}, J^{Z,G}, \mathbf{d})\Big]$$
$$= \Big[\mathcal{H}H, q_{\mathcal{H}H}, J^{HH}, \int_Z L(\nabla^{TZ}) \wedge (\rho - \tilde{\eta}(\mathcal{G}, N_G, J^G, v_{\mathcal{G}}))$$
$$- \sum_{r\geq 2} \tilde{\eta}({}_{lg}\mathcal{E}_r, N_{{}_{lg}E_r}, J^{{}_{lg}E_r}, {}_{lg}d_r) + \tilde{\eta}(\mathcal{H}H, {}_{lg}\mathrm{Gr}(\mathcal{H}H), J^{HH}, J^{{}_{lg}\mathrm{Gr}(HH)})\Big].$$

Remark that in this case, ${}_{lg}E_r = 0$ for $r \geq 2$ and $\mathcal{H}H = 0$, as $\mathcal{G}$ is a short exact sequence. By (5.13) and the above equation, we get (5.12). The proof of Theorem 5.15 is complete. □

5.4. Functorial properties.

5.4.1. *Compatibility and naturality.* The following two propositions are immediate consequences of the definition of the secondary index map.

PROPOSITION 5.16. *The following diagram commutes:*

$$\begin{array}{ccccc} H^{4*-\epsilon}(M,\mathbb{R}) & \longrightarrow & \bar{L}_\epsilon(M) & \longrightarrow & L_\epsilon(M) \\ \pi_*^{\mathrm{sign}}\big\downarrow & & \pi_*^{\bar{L}}\big\downarrow & & \pi_*^{L}\big\downarrow \\ H^{4*-\epsilon\epsilon_n}(B,\mathbb{R}) & \longrightarrow & \bar{L}_{\epsilon\epsilon_n}(B) & \longrightarrow & L_{\epsilon\epsilon_n}(B). \end{array}$$

PROPOSITION 5.17. *The secondary index map is natural with respect to pull-back of fibre bundles, i.e., given $f\colon B' \to B$ we consider the pull back*

$$\begin{array}{ccc} f^*M & \xrightarrow{f_\sharp} & M \\ f^*\pi\big\downarrow & & \pi\big\downarrow \\ B' & \xrightarrow{f} & B, \end{array}$$

where $f^*M \to B'$ has the induced fibrewise orientation, and we have $(f^*\pi)_*^{\bar{L}} \circ f_\sharp^* = f^* \circ \pi_*^{\bar{L}}$.

5.4.2. *Functoriality.* We adopt the notation of Subsection 5.2.5. In particular, we have smooth fibre bundles $\pi_1\colon W \to V$, $\pi_2\colon V \to S$ with closed fibres X, Y. The composition $\pi_3 = \pi_2 \circ \pi_1\colon W \to S$ is a fibre bundle with closed fibre Z. Let $n = \dim Z, n_0 = \dim X, m_0 = \dim Y$ such that $n = n_0 + m_0$. We assume that n, n_0 are even. Furthermore, we assume that the relative tangent bundles TX, TY, TZ are compatibly oriented.

Note that $\epsilon_n = \epsilon_{n_0}\epsilon_{m_0}$. We have well-defined secondary index maps

$$\begin{aligned}\pi^{\bar{L}}_{3,*} &: \bar{L}_\epsilon(W) \to \bar{L}_{\epsilon\epsilon_n}(S),\\ \pi^{\bar{L}}_{1,*} &: \bar{L}_\epsilon(W) \to \bar{L}_{\epsilon\epsilon_{n_0}}(V),\\ \pi^{\bar{L}}_{2,*} &: \bar{L}_{\epsilon\epsilon_{n_0}}(V) \to \bar{L}_{\epsilon\epsilon_n}(S).\end{aligned}$$

THEOREM 5.18. *We have the equality of homomorphisms* $\bar{L}_\epsilon(W) \to \bar{L}_{\epsilon\epsilon_n}(S)$:

$$\pi^{\bar{L}}_{3,*} = \pi^{\bar{L}}_{2,*} \circ \pi^{\bar{L}}_{1,*}.$$

PROOF. Let $(\mathcal{F}, q_{\mathcal{F}}, J^F, \rho)$ represent some element of $\bar{L}_\epsilon(W)$. Then the element $\pi^{\bar{L}}_{3,*}[\mathcal{F}, q_{\mathcal{F}}, J^F, \rho]$ is represented by

$$\Big[\mathcal{H}(Z,\mathcal{F}), q_{\mathcal{H}(Z,\mathcal{F})}, J^{H(Z,\mathcal{F})}, \int_Z L(\nabla^{TZ}) \wedge \rho - \tilde{\eta}(\Omega(Z,F), N_Z, J^{Z,F}, d^F)\Big].$$

There is a filtration of $\mathcal{H}(Z,\mathcal{F})$ such that the corresponding graded sheaf is the limit ${}_{LS}\mathcal{E}_\infty$ of the fibrewise Leray-Serre spectral sequence. Using Lemma 3.19, we get

$$\begin{aligned}&\pi^{\bar{L}}_{3,*}[\mathcal{F}, q_{\mathcal{F}}, J^F, \rho]\\ &= \Big[\mathcal{H}(Z,\mathcal{F}), q_{\mathcal{H}(Z,\mathcal{F})}, J^{H(Z,\mathcal{F})}, \int_Z L(\nabla^{TZ}) \wedge \rho - \tilde{\eta}(\Omega(Z,F), N_Z, J^{Z,F}, d^F)\Big]\\ &= \Big[{}_{LS}\mathcal{E}_\infty, q_{{}_{LS}\mathcal{E}_\infty}, J^{{}_{LS}E_\infty}, \int_Z L(\nabla^{TZ}) \wedge \rho - \tilde{\eta}(\Omega(Z,F), N_Z, J^{Z,F}, d^F)\\ &\qquad - \tilde{\eta}(\mathcal{H}(Z,\mathcal{F}), {}_{LS}\mathcal{E}_\infty, J^{H(Z,\mathcal{F})}, J^{{}_{LS}E_\infty})\Big].\end{aligned}$$

Note that $\mathcal{H}(Y, \mathcal{H}(X,\mathcal{F})) = {}_{LS}\mathcal{E}_2$ with all induced structures. Next we use Lemma 3.16 several times to get

$$\begin{aligned}\pi^{\bar{L}}_{3,*}[\mathcal{F}, q_{\mathcal{F}}, J^F, \rho] = \Big[&\mathcal{H}(Y,\mathcal{H}(X,\mathcal{F})), q_{\mathcal{H}(Y,\mathcal{H}(X,\mathcal{F}))}, J^{\mathcal{H}(Y,\mathcal{H}(X,\mathcal{F}))},\\ &\int_Z L(\nabla^{TZ}) \wedge \rho - \tilde{\eta}(\Omega(Z,F), N_Z, J^{Z,F}, d^F)\\ - \tilde{\eta}(\mathcal{H}(Z,F), {}_{LS}\mathcal{E}_\infty, J^{H(Z,\mathcal{F})}, J^{{}_{LS}E_\infty}) &+ \sum_{r=2}^{\infty} \tilde{\eta}({}_{LS}\mathcal{E}_r, N_{{}_{LS}E_r}, J^{{}_{LS}E_r}, {}_{LS}d_r)\Big].\end{aligned}$$

Now we start with the other side. We have

$$\begin{aligned}&\pi^{\bar{L}}_{1,*}[\mathcal{F}, q_{\mathcal{F}}, J^F, \rho]\\ &= \Big[\mathcal{H}(X,\mathcal{F}), q_{\mathcal{H}(X,\mathcal{F})}, J^{H(X,\mathcal{F})}, \int_X L(\nabla^{TX}) \wedge \rho - \tilde{\eta}(\Omega(X,F), N_X, J^{X,F}, d^F)\Big].\end{aligned}$$

Furthermore, we get

$$
\begin{aligned}
&(\pi^{\bar{L}}_{2,*} \circ \pi^{\bar{L}}_{1,*})[\mathcal{F}, q_{\mathcal{F}}, J^F, \rho] \\
&\qquad = \Big[\mathcal{H}(Y, \mathcal{H}(X, \mathcal{F})), q_{\mathcal{H}(Y,\mathcal{H}(X,\mathcal{F}))}, J^{H(Y,\mathcal{H}(X,\mathcal{F}))}, \\
&\qquad + \int_Y L(\nabla^{TY}) \wedge \Big(\int_X L(\nabla^{TX}) \wedge \rho - \tilde{\eta}(\Omega(X,F), N_X, J^{X,F}, d^F)\Big) \\
&\qquad\qquad - \tilde{\eta}(\Omega(Y, H(X,\mathcal{F})), N_Y, J^{Y,H(X,\mathcal{F})}, d^{H(X,\mathcal{F})})\Big].
\end{aligned}
$$

Thus the assertion of the theorem is proved if we show that

$$
\begin{aligned}
(5.14)\quad &\int_Z L(\nabla^{TZ}) \wedge \rho - \tilde{\eta}(\Omega(Z,F), N_Z, J^{Z,F}, d^F) \\
&- \tilde{\eta}\big(\mathcal{H}(Z,\mathcal{F}), {}_{LS}\mathcal{E}_\infty, J^{H(Z,\mathcal{F})}, J^{{}_{LS}E_\infty}\big) + \sum_{r=2}^{\infty} \tilde{\eta}\big({}_{LS}\mathcal{E}_r, N_{{}_{LS}E_r}, J^{{}_{LS}E_r}, {}_{LS}d_r\big) \\
&\equiv \int_Y L(\nabla^{TY}) \wedge \Big(\int_X L(\nabla^{TX}) \wedge \rho - \tilde{\eta}(\Omega(X,F), N_X, J^{X,F}, d^F)\Big) \\
&\qquad - \tilde{\eta}\big(\Omega(Y, H(X,\mathcal{F})), N_Y, J^{Y,H(X,\mathcal{F})}, d^{H(X,\mathcal{F})}\big)
\end{aligned}
$$

modulo exact forms. Note that we have

$$
\begin{aligned}
&\int_Y L(\nabla^{TY}) \wedge \int_X L(\nabla^{TX}) \wedge \rho \\
&\qquad = \int_Z \pi_2^* L(\nabla^{TY}) \wedge L(\nabla^{TX}) \wedge \rho \\
&\qquad = \int_Z \Big(L(\nabla^{TZ}) \wedge \rho - d\tilde{L}(TZ, \nabla^{TZ}, {}^0\nabla^{TZ}) \wedge \rho\Big) \\
&\qquad \equiv \int_Z \Big(L(\nabla^{TZ}) \wedge \rho - \tilde{L}(TZ, \nabla^{TZ}, {}^0\nabla^{TZ}) \wedge d\rho\Big) \\
&\qquad \equiv \int_Z \Big(L(\nabla^{TZ}) \wedge \rho - \tilde{L}(TZ, \nabla^{TZ}, {}^0\nabla^{TZ}) \wedge p(\nabla^F, J^F)\Big).
\end{aligned}
$$

Using this identity, we see that (5.14) is equivalent to

$$
\begin{aligned}
&- \tilde{\eta}(\Omega(Z,F), N_Z, J^{Z,F}, d^F) - \tilde{\eta}\big(\mathcal{H}(Z,\mathcal{F}), {}_{LS}\mathcal{E}_\infty, J^{H(Z,\mathcal{F})}, J^{{}_{LS}E_\infty}\big) \\
&\qquad + \sum_{r=2}^{\infty} \tilde{\eta}\big({}_{LS}\mathcal{E}_r, N_{{}_{LS}E_r}, J^{{}_{LS}E_r}, {}_{LS}d_r\big) \\
&\quad \equiv - \int_Z \tilde{L}(TZ, \nabla^{TZ}, {}^0\nabla^{TZ}) \wedge p(\nabla^F, J^F) \\
&\qquad - \int_Y L(\nabla^{TY}) \wedge \tilde{\eta}\big(\Omega(X,F), N_X, J^{X,F}, d^F\big) \\
&\qquad\qquad - \tilde{\eta}\big(\Omega(Y, H(X,\mathcal{F})), N_Y, J^{Y,H(X,\mathcal{F})}, d^{Y,H(X,\mathcal{F})}\big).
\end{aligned}
$$

The latter relation, however, is exactly the assertion of Theorem 5.11. □

5.5. The index map for L_ϵ^{ex} and $\bar{L}_\epsilon^{\mathrm{ex}}$.

5.5.1. *Definition.* Let $\pi\colon M \to B$ be a fibre bundle with even-dimensional closed fibres Z of dimension n. We assume that the relative tangent bundle TZ is oriented. Furthermore, we assume that B is closed, oriented, and of dimension m such that $\epsilon\epsilon_n\epsilon_m = 1$ and m is odd. Let $(\mathcal{F}, q_{\mathcal{F}})$ be a locally constant sheaf of finite-dimensional real vector spaces on M with an ϵ–symmetric duality structure $q_{\mathcal{F}}$. Recall from Subsection 5.2.3 that in this situation we have the Leray spectral sequence $({}_{LS}E_r, {}_{LS}d_r)$ $(r \geq 2)$ of finite dimensional vector spaces which carries induced duality structures $q_{{}_{LS}E_r}$. It gives rise to the integer

$$ {}_{LS}\tau_2(Z,F) := \tau_2({}_{LS}E_*) := \sum_{r\geq 2} \tau\big({}_{LS}E_r, q_{{}_{LS}E_r}, N_{{}_{LS}E_r}, {}_{LS}d_r\big). $$

DEFINITION 5.19. (1) We introduce the extended primary index map $\pi_*^{L^{\mathrm{ex}}}$: $L_\epsilon^{\mathrm{ex}}(M) \to L_{\epsilon\epsilon_n}^{\mathrm{ex}}(B)$ by

$$ \pi_*^{L^{\mathrm{ex}}}[\mathcal{F}, q_{\mathcal{F}}, z] := \Big[\mathcal{H}(Z,\mathcal{F}), q_{\mathcal{H}(Z,\mathcal{F})}, z - {}_{LS}\tau_2(Z,F)\Big]. $$

(2) We introduce the extended secondary index map $\pi_*^{\bar{L}^{\mathrm{ex}}} : \bar{L}_\epsilon^{\mathrm{ex}}(M) \to \bar{L}_{\epsilon\epsilon_n}^{\mathrm{ex}}(B)$ by

$$ \pi^{\bar{L}^{\mathrm{ex}}}[\mathcal{F}, q_{\mathcal{F}}, J^F, \rho, z] := \Big[\mathcal{H}(Z,\mathcal{F}), q_{\mathcal{H}(Z,\mathcal{F})}, J^{H(Z,\mathcal{F})}, \int_Z L(\nabla^{TZ}) \wedge \rho - \tilde{\eta}\big(\Omega(Z,F), N_Z, J^{Z,F}, d^F\big), z - {}_{LS}\tau_2(Z,F)\Big]. $$

THEOREM 5.20. $\pi_*^{L^{\mathrm{ex}}}$ *and* $\pi_*^{\bar{L}^{\mathrm{ex}}}$ *are well-defined.*

PROOF. Let $\hat{\pi}_*^{L^{\mathrm{ex}}} : \hat{L}_\epsilon^{\mathrm{ex}}(M) \to \hat{L}_{\epsilon\epsilon_n}^{\mathrm{ex}}(B)$ and $\hat{\pi}_*^{\bar{L}^{\mathrm{ex}}} : \hat{\bar{L}}_\epsilon^{\mathrm{ex}}(M) \to \hat{\bar{L}}_{\epsilon\epsilon_n}^{\mathrm{ex}}(B)$ be given by the above formulas.

LEMMA 5.21. *We have*

$$ \eta\big(\hat{\pi}_*^{\bar{L}^{\mathrm{ex}}}(\mathcal{F}, q_{\mathcal{F}}, J^F, \rho, z)\big) = \eta\big(\mathcal{F}, q_{\mathcal{F}}, J^F, \rho, z\big). $$

PROOF. This is just a reformulation of Theorem 5.7. In fact, since the η-homomorphism of Definition 4.18 is independent of the metric, we can perform the adiabatic limit and obtain (using that in this limit $L(\nabla_T^{TM}) \to \pi^* L(\nabla^{TB}) \wedge$

$L(\nabla^{TZ})$ (cf. [**24**, Thm. 5.1]))

$$\begin{aligned}
\eta(\mathcal{F}, q_{\mathcal{F}}, J^F, \rho, z) &= \eta(D_F^{\mathrm{sign}}) - 2\int_Z L(\nabla^{TZ}) \wedge \rho - z \\
&= \eta(D^{\mathrm{sign}}_{H(Z,\mathcal{F})}) + 2\int_Z L(\nabla^{TB}) \wedge \tilde{\eta}(\Omega(Z,F), N_Z, J^{Z,F}, d^F) \\
&\qquad + {}_{LS}\mathcal{T}_2(Z,F) - 2\int_B L(\nabla^{TB}) \wedge \int_Z L(\nabla^{TZ}) \wedge \rho - z \\
&= \eta(\mathcal{H}(Z,\mathcal{F}), q_{\mathcal{H}(Z,\mathcal{F})}, J^{H(Z,\mathcal{F})}, \\
&\qquad \int_Z L(\nabla^{TZ}) \wedge \rho - \tilde{\eta}(\Omega(Z,F), N_Z, J^{Z,F}, d^F), z - {}_{LS}\mathcal{T}_2(Z,F)) \\
&= \eta\big(\hat{\pi}_*^{\bar{L}^{\mathrm{ex}}}(\mathcal{F}, q_{\mathcal{F}}, J^F, \rho, z)\big).
\end{aligned}$$

□

LEMMA 5.22. *$\pi_*^{\bar{L}^{\mathrm{ex}}}$ is well-defined.*

PROOF. Assume that $(\mathcal{F}, q_F, J, \rho, z) \in \hat{\bar{L}}^{\mathrm{ex}}_\epsilon(M)$ satisfies $(\mathcal{F}, q_F, J, \rho, z) \sim 0$. Then we have $[\mathcal{F}, q_F, J, \rho] = 0$ in $\bar{L}_\epsilon(M)$. Since $\pi_*^{\bar{L}}$ is well-defined, we have $\pi_*^{\bar{L}}([\mathcal{F}, q_F, J, \rho]) = 0$. By an inspection of the definitions we further observe that $\hat{\pi}_*^{\bar{L}^{\mathrm{ex}}}(\mathcal{F}, q_F, J, \rho, z) \sim (0,0,0,0,u)$ for some $u \in \mathbb{Z}$. We must show that $u = 0$. In fact, since the η-homomorphism is well-defined, we can compute (using also Lemma 5.21)

$$0 = \eta(\mathcal{F}, q_F, J, \rho, z) = \hat{\pi}_*^{\bar{L}^{\mathrm{ex}}}(\mathcal{F}, q_F, J, \rho, z) = \eta(0,0,0,0,u) = -u.$$

□

LEMMA 5.23. *$\pi_*^{L^{\mathrm{ex}}}$ is well-defined.*

PROOF. Let $(\mathcal{F}, q_{\mathcal{F}}, z) \in \hat{L}^{\mathrm{ex}}_\epsilon(M)$ satisfy $(\mathcal{F}, q, z) \sim 0$. Then we can find a metric structure J^F and a form ρ such that $(\mathcal{F}, q_{\mathcal{F}}, J^F, \rho, z) \sim 0$ in $\hat{\bar{L}}^{\mathrm{ex}}_\epsilon(M)$. It follows from Lemma 5.22 that

$$\hat{\pi}_*^{\bar{L}^{\mathrm{ex}}}(\mathcal{F}, q_{\mathcal{F}}, J^F, \rho, z) \sim 0.$$

This implies $\hat{\pi}_*^{L^{\mathrm{ex}}}(\mathcal{F}, q, z) \sim 0$. □

REMARK 5.24. The assertion of Lemma 5.23 should have a purely algebraic proof. In particular, such a proof should be independent of analytic results about η-invariants and η-forms. We were not able to find such an argument.

The proof of Theorem 5.20 is now finished. □

Let us state as a corollary the following consequence of Lemma 5.21.

COROLLARY 5.25. *The following diagram commutes:*

$$\begin{array}{ccc} \bar{L}^{\mathrm{ex}}_{\epsilon}(M) & \xrightarrow{\eta} & \mathbb{R} \\ {\scriptstyle \pi_*^{\bar{L}^{\mathrm{ex}}}}\downarrow & & \| \\ \bar{L}^{\mathrm{ex}}_{\epsilon\epsilon_n}(B) & \xrightarrow{\eta} & \mathbb{R}. \end{array}$$

5.5.2. *Functoriality.* We adopt the notation and assumptions of Subsection 5.4.2. In addition we assume that S is compact, the dimension n_S is odd, and $\epsilon\epsilon_{n+n_S} = 1$. We have well-defined extended secondary index maps

$$\pi_{3,*}^{\bar{L}^{\mathrm{ex}}} : \bar{L}^{\mathrm{ex}}_{\epsilon}(W) \to \bar{L}^{\mathrm{ex}}_{\epsilon\epsilon_n}(S)$$
$$\pi_{1,*}^{\bar{L}^{\mathrm{ex}}} : \bar{L}^{\mathrm{ex}}_{\epsilon}(W) \to \bar{L}^{\mathrm{ex}}_{\epsilon\epsilon_{n_0}}(V)$$
$$\pi_{2,*}^{\bar{L}^{\mathrm{ex}}} : \bar{L}^{\mathrm{ex}}_{\epsilon\epsilon_{n_0}}(V) \to \bar{L}^{\mathrm{ex}}_{\epsilon\epsilon_n}(S).$$

THEOREM 5.26. *We have the equality of homomorphisms $\bar{L}_\epsilon(W) \to \bar{L}_{\epsilon\epsilon_n}(S)$:*

$$\pi_{3,*}^{\bar{L}^{\mathrm{ex}}} = \pi_{2,*}^{\bar{L}^{\mathrm{ex}}} \circ \pi_{1,*}^{\bar{L}^{\mathrm{ex}}}.$$

PROOF. Let $[\mathcal{F}, q_{\mathcal{F}}, J^F, \rho, z] \in \bar{L}_\epsilon(W)$ be given. Then we have in view of Theorems 2.22 and 5.18

$$\pi_{3,*}^{\bar{L}}([\mathcal{F}, q_{\mathcal{F}}, J^F, \rho]) - (\pi_{2,*}^{\bar{L}} \circ \pi_{1,*}^{\bar{L}})([\mathcal{F}, q_{\mathcal{F}}, J^F, \rho]) = 0.$$

Thus

$$\pi_{3,*}^{\bar{L}^{\mathrm{ex}}}([\mathcal{F}, q_{\mathcal{F}}, J^F, \rho, z]) - (\pi_{2,*}^{\bar{L}^{\mathrm{ex}}} \circ \pi_{1,*}^{\bar{L}^{\mathrm{ex}}})([\mathcal{F}, q_{\mathcal{F}}, J^F, \rho, z]) = [0,0,0,0,u]$$

for some $u \in \mathbb{Z}$. We must show that $u = 0$. We again use the η-homomorphism and Lemma 5.25:

$$\begin{aligned} -u &= \eta(0,0,0,0,u) \\ &= \eta\big(\pi_{3,*}^{\bar{L}^{\mathrm{ex}}}([\mathcal{F}, q_{\mathcal{F}}, J^F, \rho, z])\big) - \eta\big((\pi_{2,*}^{\bar{L}^{\mathrm{ex}}} \circ \pi_{1,*}^{\bar{L}^{\mathrm{ex}}})([\mathcal{F}, q_{\mathcal{F}}, J^F, \rho, z])\big) \\ &= \eta\big([\mathcal{F}, q_{\mathcal{F}}, J^F, \rho, z]\big) - \eta\big(\pi_{1,*}^{\bar{L}^{\mathrm{ex}}}([\mathcal{F}, q_{\mathcal{F}}, J^F, \rho, z])\big) \\ &= \eta\big([\mathcal{F}, q_{\mathcal{F}}, J^F, \rho, z]\big) - \eta\big([\mathcal{F}, q_{\mathcal{F}}, J^F, \rho, z]\big) \\ &= 0. \end{aligned}$$

□

REMARK 5.27. We conjecture that we also have

$$\pi_{3,*}^{L^{\mathrm{ex}}} = \pi_{2,*}^{L^{\mathrm{ex}}} \circ \pi_{1,*}^{L^{\mathrm{ex}}}.$$

There should be a purely algebraic-topological proof of this identity. This together with Theorem 2.22 would again imply Theorem 5.26. Unfortunately, we were not able to find such a proof. The difficulties are very similar to the problems in an algebraic approach to Lemma 5.23.

In fact, we could conclude the well-definedness and the functoriality of $\pi^{\bar{L}}$ from a combination of these hypothetical algebraic-topological results with Dai's formula for adiabatic limits of η-invariants and the well-definedness of the η-homomorphisms. In this case we would obtain an independent verification of the adiabatic limit results for η-forms.

6. Adiabatic limits - sketches of proofs

6.1. Remarks. This section contains the proofs of the assertions about adiabatic limits of η–forms and invariants which were formulated earlier. The general techniques were developed mainly in the work of Bismut and coworkers, but also by Dai and others. Unfortunately, the details were worked out in specific cases which are similar to the situations of the present paper, but not exactly the same. In order to show the results needed in the present paper one can use the methods after adaptation.

We decided to choose for the present section a coarser level of detailedness of our arguments. While Subsection 6.2 is still rather detailed, in the remaining subsections we just stated the main intermediate results with references to the literature, where proofs of similar results in slightly different situations can be found, which can be adapted to the present cases. It is not by coincidence that the formulations of these intermediate results in the last three subsections almost agree.

6.2. The proof of Theorem 4.17. Let $\epsilon \in \mathbb{Z}_2$. We assume that M is a smooth closed oriented Riemannian manifold of odd dimension m such that $\epsilon\epsilon_m = 1$, i.e., $m \equiv -\epsilon \bmod 4$. Let $(\mathcal{F}, q_{\mathcal{F}})$ be a locally constant sheaf of finite-dimensional real vector spaces with ϵ–symmetric form. Furthermore, let N_F be a compatible $\mathbb{Z}$-grading of length n_F and v be a compatible differential on $\mathcal{F}$. By $(\mathcal{H}, q_{\mathcal{H}})$ we denote the associated cohomology. We choose a compatible metric structure J^F on F and let J^H be the induced metric structure on the cohomology H. Then we can define the operators D_F^{sign} and D_H^{sign} as in (4.2). In the present section we sketch the proof of the following formula.

THEOREM 6.1 (Theorem 4.17). *We have*

$$\eta(D_F^{\mathrm{sign}}) - \eta(D_H^{\mathrm{sign}}) = 2\int_M L(\nabla^{TM}) \wedge \tilde{\eta}(\mathcal{F}, N_F, J^F, v) - \tau(\mathcal{F}, q_{\mathcal{F}}, N_F, v).$$

PROOF. We are going give a detailed sketch of the proof. The methods have been developed in connection with similar questions about analytic torsion (forms) and for the study of adiabatic limits of η-invariants. The proof of Theorem 4.17 is achieved by adapting these methods correspondingly to the present situation.

We abbreviate $D := D_F^{\mathrm{sign}}$. Let $J^{M,F}$ be the metric structure on $\Omega(M, F)$ induced by J^F and the Riemannian metric g^{TM} as in Subsection 4.3.1. We proceed with v in the same manner as we did for d^F to define D. We introduce $W := J^{M,F}v + vJ^{M,F}$ and set $\mathcal{D} := D + W$. Note that $\mathcal{D}$ is not a compatible Dirac operator. Therefore, we must define its η-invariant by zeta function regularization. Thus, we define the η-invariant $\eta(\mathcal{D})$ as the value of the function

$$s \mapsto \frac{1}{\Gamma(s+\frac{1}{2})} \int_0^\infty \operatorname{Tr} \mathcal{D} e^{-t\mathcal{D}^2} t^{s-\frac{1}{2}} dt$$

at $s = 0$, where the integral converges for $\operatorname{Re}(s) \gg 0$ and has a meromorphic continuation which is regular at $s = 0$. In fact, we will see below (Proposition 6.4) that the integral converges locally uniformly for $\operatorname{Re}(s) > -\frac{1}{2}$.

For $T > 0$, we define the rescaled metric structure

$$J^F(T) := T^{-N_F/2+n_F/4} J^F T^{N_F/2-n_F/4}.$$

Let $J^{M,F}(T)$ be the metric structure on $\Omega(M,F)$ induced by $J^F(T)$ and g^{TM} and define $\mathcal{D}(T)$ using $J^{M,F}(T)$. Then we have

$$\mathcal{D}_T := T^{N_F/2}\mathcal{D}(T)T^{-N_F/2} = D + T^{1/2}W.$$

Note that $\mathcal{D}_0 = D$ is well-defined. Furthermore, we have $\eta(\mathcal{D}_T) = \eta(\mathcal{D}(T))$. Below we show that $\eta(\mathcal{D}_T)$ is independent of $T \in (0,\infty)$. We obtain the proof of Theorem 4.17 by considering the limits $T \to 0$ and $T \to \infty$.

PROPOSITION 6.2. *$\eta(\mathcal{D}_T)$ is constant for $T \in (0,\infty)$.*

PROOF. By Hodge theory the kernel of $\mathcal{D}(T)$ for $T \in (0,\infty)$ can be identified with the cohomology of the total complex $(\Omega(M,F), d^F + v)$, i.e., with the hypercohomology $HH(M,F)$. The kernel of $\mathcal{D}_T$ is isomorphic to the kernel of $\mathcal{D}(T)$ and thus has constant dimension. It follows that $\eta(\mathcal{D}_T)$ is a smooth function of T, and its derivative is given by the coefficient $-\frac{2}{\sqrt{\pi}} b_{-1/2}$, where $b_{-1/2}$ is a coefficient of the asymptotic expansion

$$\operatorname{Tr} \frac{\partial \mathcal{D}_T}{\partial T} e^{-t\mathcal{D}_T^2} \underset{t\to 0}{\sim} \sum_{i \in \frac{1}{2}\mathbb{Z}} b_i t^i.$$

We now show that we can apply [**8**, Lemma 2.11] which states that $b_{-1/2} = 0$. Let $\mathbf{D}$ be the restriction of $D^M := J^M d + dJ^M$ to $\Omega(M)^{\mathrm{ev}}$, where J^M denotes the metric structure on $\Omega(M)$ induced by the Riemannian metric g^{TM}. By $\mathbf{D}^F$ we denote the twist with F, or, what is the same, the restriction of D to $\Omega^{\mathrm{ev}}(M,F)$ (the superscript refers to the form degree). We have an isomorphism $z^{J^{M,F}} : \Omega^{\mathrm{odd}}(M,F) \cong \Omega^{\mathrm{ev}}(M,F)$ such that we can identify $\Omega(M,F) \cong \Omega^{\mathrm{ev}}(M,F) \oplus \Omega^{\mathrm{ev}}(M,F)$. In this identification D and W correspond to

$$\begin{pmatrix} \mathbf{D}^F & 0 \\ 0 & \mathbf{D}^F \end{pmatrix}, \quad \begin{pmatrix} 0 & W_- \\ W_+ & 0 \end{pmatrix}.$$

Up to the sign, $\mathcal{D}$ has the form [**8**, (2.5)]. The role of that sign in the argument of [**8**, Lemma 2.11] is to assure that the anti-commutator of the Dirac operator and the potential is a zeroth-order operator. Since in our situation $DW + WD$ is still of zero order, the argument of [**8**] applies. □

Let ${}_{hc}\tau_1 := {}_{hc}\tau_1(\mathcal{F}, q_{\mathcal{F}}, N_F, v)$ be as in Paragraph 4.4.2 associated with the hypercohomology spectral sequence.

PROPOSITION 6.3. *For $T \in (0,\infty)$ we have $\eta(\mathcal{D}_T) - \eta(\mathcal{D}_0) = {}_{hc}\tau_1$.*

PROOF. $\eta(\mathcal{D}_T)-\eta(\mathcal{D}_0)$ is the difference of the numbers of eigenvalues (counted with multiplicity) of $\mathcal{D}_T$ which become positive and negative when T moves from 0 to positive values.

The eigenvalues of $\mathcal{D}_T$ which tend to zero as $T \to 0$ can be described in terms of the hypercohomology spectral sequence $({}_{hc}E_r, {}_{hc}d_r)$. We employ the method developed in [**4**, §VI]. In particular, we realize the spaces ${}_{hc}E_r$ using Hodge theory to obtain natural metric structures $J^{{}_{hc}E_r}$ induced by $J^{Z,F}$.

Fix $r \geq 1$ and $\varepsilon > 0$ sufficiently small. We can find $a > 0$ such that $\pm a T^{r/2}$ is not in the spectrum of $\mathcal{D}_T$ for $T \in (0,\varepsilon)$. Let $P^a_{r,T} := E_{T^{-r/2}\mathcal{D}_T}(-a,a)$ be the spectral projection, i.e., the orthogonal projection from $\Omega(M,F)$ on the direct sum of the eigenspaces of $T^{-r/2}\mathcal{D}_T$ associated to eigenvalues lying in $(-a,a)$. Then as $T \to 0$ the spectrum of $T^{-r/2}\,\mathcal{D}_T\,P^a_{r,T}$ converges to the spectrum of ${}_{hc}D_r := {}_{hc}d_r J^{{}_{hc}E_r} + J^{{}_{hc}E_r}{}_{hc}d_r$ (cf. [**4**, (6.55)]). This implies (using Definition 4.7 and Lemma 4.6 for the second equality) that

$$\eta(\mathcal{D}_T) - \eta(\mathcal{D}_0) = \sum_{r\geq 1} \operatorname{sign}({}_{hc}D_r) = {}_{hc}\tau_1.$$

□

Recall the definition of the families of forms $\gamma(t)$ in $\Omega(M)$ which enter the Definition 3.15 of $\tilde{\eta}(\mathcal{F}, N_F, J^F, v)$. Recall also that $L(\nabla^{TM})$ is defined in Subsection 4.2.2.

PROPOSITION 6.4. *There is some $N \in \mathbb{N}$ such that*

$$\operatorname{Tr} \mathcal{D}_T \mathrm{e}^{-t\mathcal{D}_T^2} = 2\sqrt{\pi}\int_M L(\nabla^{TM}) \wedge \gamma(Tt) + t^{1/2} O(1 + T^N t^N).$$

PROOF. This follows from [**8**, (3.1)] and some local index theory calculation. In fact, our operator is (locally) the spin Dirac operator twisted with the tensor product of the spinor bundle and F. This explains the appearance of the L-form above and the matrix structure of D, W in the proof of Proposition 6.2. □

It follows from Proposition 6.4 and the fact that $\gamma(t) = O(1)$ as $t \to 0$ (see [**21**, Prop. 29]) that

$$\eta(\mathcal{D}_T) = \frac{1}{\sqrt{\pi}} \int_0^\infty \operatorname{Tr} \mathcal{D}_T \mathrm{e}^{-t\mathcal{D}_T^2} \frac{dt}{t^{1/2}}.$$

If we choose $\alpha > 0$ so small that $1 - \alpha > \frac{1/2+N}{1+N}$, then we have

$$\lim_{T\to\infty} \int_0^{T^{-1+\alpha}} T^{1/2} t^{1/2} O(1 + T^N t^N) \frac{dt}{t^{1/2}} = 0$$

and thus after some transformation

COROLLARY 6.5.

$$\lim_{T\to\infty}\frac{1}{\sqrt{\pi}}\int_0^{T^{-1+\alpha}}\operatorname{Tr}\mathcal{D}_T e^{-t\mathcal{D}_T^2}\frac{dt}{t^{1/2}} = 2\int_M L(\nabla^{TM})\wedge\tilde{\eta}(\mathcal{F},N_F,J^F,v).$$

LEMMA 6.6. *Given $N\in\mathbb{N}$, there are functions $a_i(z)$ bounded for $z\in[1,\infty)$ and $r_N(t,T)$ bounded for $T\in[1,\infty)$, $t\geq T^{-1}$ such that*

$$T^{-1/2}\operatorname{Tr}\mathcal{D}_T e^{-t\mathcal{D}_T^2} = \sum_{i=-n}^{N-1} a_{i/2}(Tt)t^{i/2} + t^{N/2}r_N(t,T).$$

PROOF. This is essentially [**16**, Thm. 1.7] (cf. [**8**, (4.81)]). In [**16**], the situation is more complicated, since it corresponds to an infinite-dimensional bundle F and unbounded differential v, namely a fibrewise de Rham complex. But the proof given in [**16**, §3] can be applied in our situation with many simplifications. First we write $\mathcal{D}_T = T^{1/2}(T^{-1/2}D+W)$ so that the parameters x and t in [**16**] correspond to $T^{-1/2}$ and Tt in the present paper.

The localization part [**16**, §3.1] is just the usual finite propagation speed argument. Now we construct the rough parametrix as in [**16**, §3.2]. Then we verify that the proof of [**16**, Lemma 3.4] goes through in our situation. Note that formula [**16**, (3.5)] simplifies a lot in the present situation. The remaining argument using Duhamel's principle can be taken without change. □

We employ the suggestive notation $\mathcal{D}_\infty := D_H^{\mathrm{sign}} = J^{M,H}d^H + d^H J^{M,H}$ and establish the estimate which corresponds to [**16**, (1.14)].

PROPOSITION 6.7. *Given $\alpha\in(0,1)$, there exists constants $C>0$, $N\in\mathbb{N}$ such that, for $T\geq 1$, $t\geq T^{-1+\alpha}$, we have*

$$\left|\operatorname{Tr}\mathcal{D}_T e^{-t\mathcal{D}_T^2} - \operatorname{Tr}\mathcal{D}_\infty e^{-t\mathcal{D}_\infty^2}\right| \leq \frac{C}{T^{1/2}\min(1,t^N)}.$$

PROOF. We decompose $\Lambda^*T^*M\otimes F := E_0\oplus E_1$, where $E_0 := \ker W$ and $E_1 := E_0^\perp$. Let Q be the projection onto E_0 and $Q^\perp = 1-Q$. With respect to this decomposition we write

$$\mathcal{D}_T := \begin{pmatrix}\mathcal{D}_{T,1} & \mathcal{D}_{T,2}\\ \mathcal{D}_{T,3} & \mathcal{D}_{T,4}\end{pmatrix}.$$

$\mathcal{D}_{T,1}$ is independent of T and can be identified with $\mathcal{D}_\infty$. We extend $\mathcal{D}_\infty$ by zero to $C^\infty(M,E_1)$.

Let $\sigma(A)\subset\mathbb{C}$ denote the spectrum of the operator A. Let

$$U_T := \left\{\lambda\in\mathbb{C}\;\middle|\;|\lambda|\leq c_1\sqrt{T},\ \inf_{\mu\in\sigma(\mathcal{D}_\infty)}|\lambda-\mu|\geq c_2\right\},$$

where $c_1, c_2>0$ are sufficiently small and will be fixed below. Following [**11**, §9], we first show the following lemma:

LEMMA 6.8. *For $p \geq \dim(M)+1$, there exist constants $C > 0$, $T_1 \geq 1$ such that, for $T \geq T_1$ and $\lambda \in U_T$,*

$$\left\|(\lambda - \mathcal{D}_T)^{-p} - (\lambda - \mathcal{D}_\infty)^{-p}Q\right\|_1 \leq \frac{C}{\sqrt{T}}(1+|\lambda|)^p, \tag{6.1}$$

where $\|.\|_1$ denotes the norm on trace class operators.

PROOF. In some sense, this is a special case of [**11**, §9]. Instead of writing all details we explain how the the arguments given in [**11**] can be employed in the present case.

The operator $\mathcal{D}_{T,1} = QDQ$ is as in [**11**, §9 (b)], but much simpler. Furthermore, $\mathcal{D}_{T,2} = -Q[D,Q] \in C^\infty(M, \mathrm{Hom}(E_0^\perp, E_0))$, $\mathcal{D}_{T,3} = Q^\perp[D,Q] \in C^\infty(M, \mathrm{Hom}(E_0, E_0^\perp))$ are independent of T and bounded. Therefore, we have a stronger version of the estimate [**11**, (9.38)].

If $\phi \in C^\infty(M, E_1)$, then using elliptic regularity, the fact that $DW + WD$ is bounded, and that $W|_{E_1}$ is injective, we obtain the estimate [**11**, (9.48)]:

$$\begin{aligned} \|\mathcal{D}_T\phi\|^2 &= \langle \mathcal{D}_T\phi, \mathcal{D}_T\phi\rangle \\ &= \langle \phi, \left(D^2 + T^{1/2}(DW+WD) + TW^2\right)\phi\rangle \\ &= \|D\phi\|^2 + T^{1/2}\langle\phi, (DW+WD)\phi\rangle + T\|W\phi\|^2 \\ &\geq d_1(\|\phi\|^2_{W^1(M,E)} + (T-d_2)\|\phi\|^2), \end{aligned} \tag{6.2}$$

where $d_1, d_2 > 0$ are independent of $T \geq 1$ and ϕ, $\|.\|$ denotes the L^2-norm, and $W^1(M, E_1)$ is the L^2-based Sobolev space of order one. We now conclude [**11**, (9.104)]: There are constants $T_0 \geq 1$ and $d_3 > 0$ such that, for $T \geq T_0$, $\phi \in C^\infty(M, E_1)$,

$$\|\mathcal{D}_{T,4}\phi\| \geq d_3(\|\phi\|_{W^1(M,E_1)} + \sqrt{T}\|\phi\|). \tag{6.3}$$

Thus there exists a constant C_1 such that, for $T \geq T_0$, $|\lambda| \leq \frac{d_3}{2}\sqrt{T}$, we have [**11**, (9.106)],

$$\begin{aligned} \|(\lambda - \mathcal{D}_{T,4})^{-1}\phi\| &\leq \frac{C_1}{\sqrt{T}}\|\phi\|, \\ \|(\lambda - \mathcal{D}_{T,4})^{-1}\phi\|_{W^1(M,E_1)} &\leq C_1\|\phi\|, \end{aligned}$$

for all $\phi \in C^\infty(M, E_1)$. The following estimates are proved exactly as in [**11**, Prop. 9.18]. For $p \geq \dim(M)+1$, there is a constant $C_2 > 0$ such that, for $T \geq T_0$, $|\lambda| \leq \frac{d_3}{2}\sqrt{T}$,

$$\begin{aligned} \|(\lambda - \mathcal{D}_{T,4})^{-1}\|_\infty &\leq \frac{C_2}{\sqrt{T}}, \\ \|(\lambda - \mathcal{D}_{T,4})^{-1}\|_p &\leq C_2, \\ \|\mathcal{D}_{T,2}(\lambda - \mathcal{D}_{T,4})^{-1}\|_\infty &\leq \frac{C_2}{\sqrt{T}}, \end{aligned}$$

where $\|.\|_p$ denotes the norm of the pth Schatten class.

We now fix c_2 such that $\sigma(\mathcal{D}_\infty) \cap (-2c_2, 2c_2) \subset \{0\}$. Furthermore, we assume that at least $2c_1 \le d_3$. Following [**11**, (9.120)], for $\lambda \in U_T$ and $T \ge T_0$, we define

$$M_T(\lambda) := \lambda - \mathcal{D}_{T,1} - \mathcal{D}_{T,2}(\lambda - \mathcal{D}_{T,4})^{-1}\mathcal{D}_{T,3}.$$

We have the estimates which correspond to [**11**, (9.124)]: If c_1 is sufficiently small, then there are constants $C_3 > 0$, $T_1 \ge T_0$ such that, for $T \ge T_1$, $\lambda \in U_T$, we have

$$\begin{aligned} \|M_T(\lambda)^{-1}\|_\infty &\le C_3, \\ \|\mathcal{D}_{T,3}M_T(\lambda)^{-1}\|_\infty &\le C_3, \\ \|M_T(\lambda)^{-1}\|_p &\le C_3(1+|\lambda|), \\ \|M_T(\lambda)^{-p} - (\lambda - \mathcal{D}_{1,T})^{-p}\|_1 &\le \frac{C_3}{\sqrt{T}}(1+|\lambda|)^p. \end{aligned}$$

Note that in our case the operator defined in [**11**, (9.125)] is trivial, and this gives the better power p instead of $p+1$ in the last estimate above. Now we follow the proof of [**11**, Thm. 9.23] to finish the proof of (6.1). □

Let $\delta \subset \mathbb{C}$ be the circle of radius c_2 centered at the origin and oriented counterclockwise. For $T \ge T_1$, we have $\sigma(\mathcal{D}_T) \cap \delta = \emptyset$. Let P_T and P_∞ be the spectral projections of $\mathcal{D}_T$ and $\mathcal{D}_\infty$ corresponding to the interval $(-c_2/2, c_2/2)$. Recall that Q is the projection onto E_0. Then, for $T \ge T_1$, we have

$$P_T - P_\infty Q = \frac{1}{2\pi \mathrm{i}} \int_\delta \lambda^{p-1}\left((\lambda - \mathcal{D}_T)^{-p} - (\lambda - \mathcal{D}_\infty)^{-p}Q\right) d\lambda.$$

We conclude from (6.1) that $\|P_T - P_\infty Q\|_1 \le C_4/\sqrt{T}$ for some C_4 independent of T. Since $\mathcal{D}_T P_\infty Q = 0$ and $P_\infty \mathcal{D}_\infty = 0$, there exists a constant C_5 such that, for $T \ge T_1$,

$$\|\mathcal{D}_T P_T - \mathcal{D}_\infty P_\infty\|_1 \le \|P_T \mathcal{D}_T (P_T - P_\infty Q)\|_1 \le \frac{C_5}{\sqrt{T}}. \tag{6.4}$$

We conclude that there is a constant C_6 such that, for $T \ge T_1$,

$$\left|\mathrm{Tr}\, P_T \mathcal{D}_T \mathrm{e}^{-t\mathcal{D}_T^2} - \mathrm{Tr}\, P_\infty \mathcal{D}_\infty \mathrm{e}^{-t\mathcal{D}_\infty^2}\right| \le \frac{C_6}{T^{1/2}}. \tag{6.5}$$

Let Δ be the oriented path in $\mathbb{C}$ which goes parallel to the real axis from $-\infty - \mathrm{i}c_2$ to $c_2 - \mathrm{i}c_2$, then parallel to the imaginary axis to $c_2 + \mathrm{i}c_2$, and then parallel to the real axis to $-\infty + \mathrm{i}c_2$, and which goes parallel to the real axis from $\infty + \mathrm{i}c_2$ to $c_2 + \mathrm{i}c_2$, then parallel to the imaginary axis to $c_2 - \mathrm{i}c_2$, and then parallel to the real axis to $\infty - \mathrm{i}c_2$.

Let $h_p(\lambda)$ be holomorphic on $\mathbb{C} \setminus \mathrm{i}\mathbb{R}$ such that $h_p^{(p-1)}(\lambda) = (p-1)!\,\lambda \mathrm{e}^{-\lambda^2}$. So up to a constant it is the function f_{p-1} as defined in [**11**, (9.165)]. In particular, we have the estimate [**11**, (9.169)],

$$|h_p(\lambda)| \le C_7\, \mathrm{e}^{-c_4|\lambda|^2}, \quad \lambda \in \Delta,$$

where $c_4 > 0$ and C_7 are independent of λ. We have

$$\begin{aligned}(1-P_T)\,\mathcal{D}_T\mathrm{e}^{-t\mathcal{D}_T^2} &- (1-P_\infty)\,\mathcal{D}_\infty\mathrm{e}^{-t\mathcal{D}_\infty^2}\\ &= \frac{1}{2\pi\mathrm{i}}\int_\Delta \lambda\mathrm{e}^{-t\lambda^2}\left((\lambda-\mathcal{D}_T)^{-1}-(\lambda-\mathcal{D}_\infty)^{-1}Q\right)d\lambda\\ &= \frac{1}{2\pi\mathrm{i}}\int_\Delta \frac{h_p(t^{1/2}\lambda)}{t^{(p+1)/2}}\left((\lambda-\mathcal{D}_T)^{-p}-(\lambda-\mathcal{D}_\infty)^{-p}Q\right)d\lambda.\end{aligned}$$

We split the integral into the parts $\Delta_1(T) := \Delta\cap\{\lambda\in U_T\} = \Delta\cap\{|\lambda|\le c_1\sqrt{T}\}$ and $\Delta_2(T) := \Delta\cap\{|\lambda|\ge c_1\sqrt{T}\}$. Let $p := \dim(M)+1$. By Lemma 6.8, there are constants C_8, C_9 such that, for $T \ge T_1$, $t > 0$,

$$\begin{aligned}(6.6)\quad &\left\|\frac{1}{2\pi\mathrm{i}}\int_{\Delta_1(T)} \frac{h_p(t^{1/2}\lambda)}{t^{(p+1)/2}}\left((\lambda-\mathcal{D}_T)^{-p}-(\lambda-\mathcal{D}_\infty)^{-p}Q\right)d\lambda\right\|_1\\ &\qquad\le \frac{C_8}{T^{1/2}t^{(p+1)/2}}\int_{\Delta_1(T)} \mathrm{e}^{-c_4t|\lambda|^2}(1+|\lambda|)^p\,d\lambda \le \frac{C_9}{T^{1/2}t^{p+1}}.\end{aligned}$$

It follows from (6.2) that there is a constant C_{10} such that, for all $T \ge T_1$, $\lambda \in \Delta_2(T)$,

$$\|(\lambda-\mathcal{D}_T)^{-p}\|_1 \le C_{10},\quad \|(\lambda-\mathcal{D}_\infty)^{-p}Q\|_1 \le C_{10}.$$

We obtain constants C_{11}, C_{12}, C_{13} such that, for all $T \ge T_1$ and $t \ge T^{-1+\alpha}$,

$$\begin{aligned}(6.7)\quad &\left\|\frac{1}{2\pi\mathrm{i}}\int_{\Delta_2(T)} \frac{h_p(t^{1/2}\lambda)}{t^{(p+1)/2}}\left((\lambda-\mathcal{D}_T)^{-p}-(\lambda-\mathcal{D}_\infty)^{-p}Q\right)d\lambda\right\|_1\\ &\qquad\le \frac{C_{11}}{t^{(p+1)/2}}\int_{\Delta_2(T)} \mathrm{e}^{-c_4t|\lambda^2|}\,d\lambda \le \frac{C_{12}}{t^{(p+2)/2}}\mathrm{e}^{-c_4Tt/2} \le \frac{C_{13}}{\sqrt{T}t^{(p+2)/2}}.\end{aligned}$$

We combine (6.6) and (6.7) to obtain

$$(6.8)\qquad \left\|(1-P_T)\,\mathcal{D}_T\mathrm{e}^{-t\mathcal{D}_T^2} - (1-P_\infty)\,\mathcal{D}_\infty\mathrm{e}^{-t\mathcal{D}_\infty^2}\right\|_1 \le \frac{C_{14}}{\sqrt{T}\min(t^{p+1},1)}$$

for $T \ge T_1$ and $t \ge T^{-1+\alpha}$, where C_{14} is independent of T, t. Together with (6.5) this implies the assertion of the proposition. □

The estimates in Proposition 6.4, Lemma 6.6, and Proposition 6.7 are all the ingredients which are necessary to perform the proof of [**16**, Prop. 1.8]. We conclude

COROLLARY 6.9. *For $\beta > 0$ sufficiently small, we have*

$$\lim_{T\to\infty}\int_{T^{-1+\alpha}}^{T^{-\beta}} \operatorname{Tr}\mathcal{D}_T\mathrm{e}^{-t\mathcal{D}_T^2}\,\frac{dt}{t^{1/2}} = 0.$$

Fix $1-\alpha > \beta > 0$ such that Corollary 6.9 holds. Note that $\operatorname{Tr}\mathcal{D}_\infty\mathrm{e}^{-t\mathcal{D}_\infty^2} = O(1)$ as $t \to 0$ (cf. [**9**, Thm. 2.4]). The following is an immediate consequence of Proposition 6.7:

COROLLARY 6.10.

$$\lim_{T\to\infty}\int_{T^{-\beta}}^{1}\operatorname{Tr}\mathcal{D}_T e^{-t\mathcal{D}_T^2}\frac{dt}{t^{1/2}}=\int_0^1\operatorname{Tr}\mathcal{D}_\infty e^{-t\mathcal{D}_\infty^2}\frac{dt}{t^{1/2}}.$$

It follows from (6.8) that $\|(1-P_T)\mathcal{D}_T e^{-\mathcal{D}_T^2}\|_1$ is uniformly bounded for $T\geq 1$. Since $P_T\mathcal{D}_T\geq c_2$ for all $T\geq T_1$, we have a constant C such that $\left\|(1-P_T)\,e^{-(t-1)\mathcal{D}_T^2}\right\|_1\leq Ce^{-tc_2}$ for all $T\geq T_1$, $t\geq 1$. We conclude that there is a constant C_1 such that, for all $T\geq T_1$,

$$\left\|(1-P_T)\,\mathcal{D}_T e^{-t\mathcal{D}_T^2}\right\|_1\leq C_1 e^{-tc_2}.$$

Using (6.8) and Lebesgue's theorem about majorized convergence, we obtain

COROLLARY 6.11.

$$\lim_{T\to\infty}\int_1^\infty\operatorname{Tr}(1-P_T)\mathcal{D}_T e^{-t\mathcal{D}_T^2}\frac{dt}{t^{1/2}}=\int_1^\infty\operatorname{Tr}(1-P_\infty)\mathcal{D}_\infty e^{-t\mathcal{D}_\infty^2}\frac{dt}{t^{1/2}}.$$

Note that P_T and P_∞ are finite-dimensional projections. Since $\mathcal{D}_\infty P_\infty=0$ it follows from (6.4) that the spectrum of $P_T\mathcal{D}_T$ converges to zero as $T\to\infty$. We have

$$\begin{aligned}\lim_{T\to\infty}\frac{1}{\sqrt{\pi}}\int_1^\infty\operatorname{Tr}P_T\mathcal{D}_T e^{-t\mathcal{D}_T^2}\frac{dt}{t^{1/2}}&=\lim_{T\to\infty}\sum_{\mu\in\sigma(\mathcal{D}_TP_T)}\frac{1}{\sqrt{\pi}}\int_1^\infty\mu e^{-t\mu^2}\frac{dt}{t^{1/2}}\\&=\lim_{T\to\infty}\sum_{\mu\in\sigma(\mathcal{D}_TP_T)}\frac{1}{\sqrt{\pi}}\int_{\mu^2}^\infty\operatorname{sign}(\mu)\,e^{-t}\frac{dt}{t^{1/2}}=\lim_{T\to\infty}\sum_{\mu\in\sigma(\mathcal{D}_TP_T)}\operatorname{sign}(\mu).\end{aligned}$$

Let ${}_{lg}\tau_2:={}_{lg}\tau_2(\mathcal{F},q,N,v)$ be associated to the local-global spectral sequence as in Definition 4.8.

PROPOSITION 6.12. *We have*

$$\lim_{T\to\infty}\sum_{\mu\in\sigma(\mathcal{D}_TP_T)}\operatorname{sign}(\mu)={}_{lg}\tau_2.$$

PROOF. The eigenvalues of $\mathcal{D}_T$ which tend to zero as $T\to\infty$ can be described in terms of the local-global spectral sequence $({}_{lg}E_r,{}_{lg}d_r)$. We apply the method of [**4**, §VI] to $\mathcal{D}_T=T^{1/2}W+D$. In particular, we realize the spaces ${}_{lg}E_r$ using Hodge theory to obtain natural metric structures $J^{{}_{lg}E_r}$ (induced by $J^{M,F}$).

Fix $r\geq 2$. We can find $a>0$, $T_1\geq 1$ such that $\pm aT^{-(r-1)/2}$ is not in the spectrum of $\mathcal{D}_T$ for $T\geq T_1$. Let $P^a_{r,T}:=E_{T^{(r-1)/2}\mathcal{D}_T}(-a,a)$ be the spectral projection. Then as $T\to 0$ the spectrum of $T^{(r-1)/2}\mathcal{D}_TP^a_{r,T}$ converges to the spectrum of ${}_{lg}D_r:={}_{lg}d_rJ^{{}_{lg}E_r}+J^{{}_{lg}E_r}{}_{lg}d_r$. This implies

$$\lim_{T\to\infty}\sum_{\mu\in\sigma(\mathcal{D}_TP_T)}\operatorname{sign}(\mu)=\sum_{r\geq 2}\operatorname{sign}({}_{lg}D_r)={}_{lg}\tau_2$$

and hence the proposition. □

Combining Propositions 6.2, 6.3, 6.5, 6.9, 6.10, 6.11, and 6.12 we finish the proof of Theorem 4.17. □

6.3. Proof of Theorem 5.9. We consider a family of superconnections depending on two parameters $(T,u)\in(0,\infty)\times(0,\infty)$. The parameter u is the usual rescaling parameter associated to the total grading of $\Omega(Z,F)$. The parameter T is associated to the form degree. If T becomes large, then the de Rham differential is scaled large compared to the differential v of the complex. To stay in our formalism we are going to work over the base $\hat{B}:=(0,\infty)\times(0,\infty)\times B$. Let $\mathrm{pr}\colon \hat{B}\to B$ denote the projection and define $\hat{\pi}\colon \hat{M}:=\mathrm{pr}^*M\to\hat{B}$ with fiber $\hat{Z}$. Let $\mathrm{pr}_M\colon \hat{M}\to M$ be the projection. Let $T\hat{Z}$ be the relative tangent bundle of $\hat{\pi}$, i.e., $T\hat{Z}=\mathrm{pr}_M^*TZ$. This bundle is equipped with the vertical metric $\hat{g}^{TZ}$ on $T\hat{Z}$ which restricts to $(Tu)^{-2}g^{TZ}$ on the fibre over $(T,u)\times\{b\}\in\hat{B}$. We define the metric structure $\hat{J}^F$ on pr_M^*F such that it restricts to $u^{-N_F+n_F/2}J^F u^{N_F-n_F/2}$ over $(T,u)\times M$. The metric $\hat{g}^{TZ}$ and $\hat{J}^F$ together induce the metric structure $\hat{J}^{Z,F}$ on $\Omega(\hat{Z},\mathrm{pr}_M^*F)$ (cf. (5.1)). Let $d^{\mathrm{pr}_M^*F}$ be the twisted de Rham differential on $\Omega(\hat{M},\mathrm{pr}_M^*F)$. We define $\hat{\mathbf{d}}=d^{\mathrm{pr}_M^*F}+\mathrm{pr}_M^*v$ and consider the form $p(\hat{\mathbf{d}},\hat{J}^{Z,F})$ on $\hat{B}$ defined in (5.2) if n is even (resp. in (5.7) if n is odd).

DEFINITION 6.13. We define $\beta:=du\wedge\beta^u+dT\wedge\beta^T$ to be the part of $p(\hat{\mathbf{d}},\hat{J}^{Z,F})\in\Omega(\hat{B})$ of degree one with respect to the coordinates (T,u), with functions $\beta^u,\beta^T\colon(0,\infty)\times(0,\infty)\to\Omega(B)$.

The following corollary is an immediate consequence of the fact that $p(\hat{\mathbf{d}},\hat{J}^{Z,F})$ is closed. Let $d=d_{T,u}+d^B$ be the decomposition of the de Rham differential on $(0,\infty)\times(0,\infty)\times B$.

COROLLARY 6.14. *There exists a smooth family* $\alpha\colon(0,\infty)\times(0,\infty)\to\Omega(B)$ *such that*

$$d_{T,u}\beta=dT\wedge du\,d^B\alpha. \tag{6.9}$$

The following theorems can be shown by adapting the methods of the corresponding references to our present situation. Actually, the following results are analogues of some related properties of Bismut–Lott's real analytic torsion forms, see especially [**24**, Thms. 4.3-4.9], where we work in a much more complicated situation (instead of the finite-dimensional flat vector bundle here we have an infinite-dimensional vector bundle there). See also Subsection 6.5 for more details.

Let $\check{B}:=(0,\infty)\times B$. Let $N_{H(Z,\mathcal{F})}$ be the $\mathbb{Z}$-grading on $H(Z,\mathcal{F})$ induced by N_F. We consider the metric structure $\check{J}^{H(Z,\mathcal{F})}$ on $\check{H}(Z,\mathcal{F}):=\mathrm{pr}^*H(Z,\mathcal{F})$ which restricts to $u^{-N_{H(Z,\mathcal{F})}+n_{H(Z,\mathcal{F})}/2}J^{H(Z,\mathcal{F})}u^{N_{H(Z,\mathcal{F})}-n_{H(Z,\mathcal{F})}/2}$ over $\{u\}\times B$. We consider the flat $(-1)^{N_{H(Z,\mathcal{F})}}$-superconnection $B':=\nabla^{H(Z,\mathcal{F})}+v_{H(Z,\mathcal{F})}$, where the differential $v_{H(Z,\mathcal{F})}$ on $H(Z,\mathcal{F})$ is induced by v.

Let $p(d^{H(Z,\mathcal{F})},\check{J}^{H(Z,\mathcal{F})})$ be the form on $\check{B}$ defined in Subsection 3.4.4 if n is even (resp. in (5.10) if n is odd). Let $\gamma\colon(0,\infty)\to\Omega(B)$ be such that

$$p\big(d^{H(Z,\mathcal{F})},\check{J}^{H(Z,\mathcal{F})}\big)=du\wedge\gamma+r,$$

where the remainder r does not contain du.

THEOREM 6.15. (1) *For any $u > 0$,*

$$\lim_{T\to\infty} \beta^u(T,u) = \gamma(u). \tag{6.10}$$

(2) *For $0 < u_1 < u_2$ fixed, there exists $C > 0$ such that, for $u \in [u_1, u_2]$, $T \geq 1$, we have*

$$|\beta^u(T,u)| \leq C. \tag{6.11}$$

(3) *We have the following identity:*

$$\lim_{T\to\infty} \int_1^\infty \beta^u(T,u)du = \int_1^\infty \gamma(u)du - \sum_{r\geq 2} \tilde{\eta}\big({}_{hc}\mathcal{E}_r, N_{{}_{hc}E_r}, J^{{}_{hc}E_r}, {}_{hc}d_r\big). \tag{6.12}$$

THEOREM 6.16. (1) *There exists a smooth family $\sigma\colon (0,\infty) \to \Omega(B)$ such that, for $T \geq 1$, we have*

$$\lim_{u\to\infty} \beta^T(T,u) = \sigma(T).$$

(2) *There exist constants $C > 0$, $\delta > 0$ such that, for $T \geq 1$, we have the following estimate:*

$$|\sigma(T)| \leq \frac{C}{T^{1+\delta}}. \tag{6.13}$$

(3) *Modulo exact forms on B we have the following identity:*

$$\int_1^\infty \sigma(T)dT = -\tilde{\eta}\big(\mathcal{H}H, {}_{hc}\mathrm{Gr}(\mathcal{H}H), J^{HH}, J^{{}_{hc}\mathrm{Gr}(HH)}\big). \tag{6.14}$$

Note that (6.14) follows from Lemma 3.17 as in the proof of [**24**, Thm. 4.5].

Let $\breve{M} = (0,\infty)\times M$, $\mathrm{Pr}\colon \breve{M} \to M$ be the canonical projection. We consider the vertical metric $\breve{g}^{TZ}$ on $\breve{M}$ for the fibration $\breve{M} \to \breve{B}$ which restricts to $u^{-2}\mathrm{Pr}^* g^{TZ}$ on the fibre over $(u,b) \in (0,\infty) \times B$. $\mathrm{Pr}^* J^F$,$\breve{g}^{TZ}$ induce a metric structure $\breve{J}^{Z,F}$ on $\mathrm{Pr}^*\,\Omega(Z,F)$ on $\breve{B}$. We define the smooth family $\theta\colon (0,\infty) \to \Omega(B)$ such that (cf. Subsection 5.2.2)

$$p\big(d^{\mathrm{Pr}^* F}, \breve{J}^{Z,F}\big) = du \wedge \theta + r$$

and r does not contain du. Note that by Definitions 5.3 and 5.4

$$\int_0^\infty \theta(u)\, du = -\tilde{\eta}\big(\Omega(Z,F), N_Z, J^{Z,F}, d^F\big).$$

THEOREM 6.17. (1) *For any $u > 0$, there exist $C > 0$,$\delta > 0$ such that, for $T \geq 1$,*

$$\big|\beta^T(T,u)\big| \leq \frac{C}{T^{\delta+1}}. \tag{6.15}$$

(2) *For any $T > 0$, we have*

$$\lim_{\varepsilon\to 0} \varepsilon^{-1}\beta^T(T\varepsilon^{-1},\varepsilon) = \theta(T). \tag{6.16}$$

(3) *There exists $C > 0$ such that, for $\varepsilon \in (0,1]$, $\varepsilon \leq T \leq 1$,*

$$\varepsilon^{-1}\left|\beta^T(T\varepsilon^{-1},\varepsilon)\right| \leq C. \tag{6.17}$$

(4) *There exist $\delta \in (0,1]$, $C > 0$ such that, for $\varepsilon \in (0,1]$, $T \geq 1$,*

$$\varepsilon^{-1}\left|\beta^T(T\varepsilon^{-1},\varepsilon)\right| \leq \frac{C}{T^{1+\delta}}. \tag{6.18}$$

PROOF OF THEOREM 5.9. We now finish the proof of Theorem 5.9. For $0 < \varepsilon < A$ and $1 < T_0$, we consider the rectangle $(T,u) \in R := [1,T_0] \times [\varepsilon, A]$. By Corollary 6.14 we have $\int_{\partial R}\beta = d^B\int_R \alpha$. Thus

$$\int_\varepsilon^A \beta^u(T_0,u)du - \int_1^{T_0}\beta^T(T,A)dT - \int_\varepsilon^A \beta^u(1,u)du + \int_1^{T_0}\beta^T(T,\varepsilon)dT$$
$$= I_1 + I_2 + I_3 + I_4$$

is an exact form on B. We take the limits $A \to \infty$, $T_0 \to \infty$, and then $\varepsilon \to 0$ in the indicated order ([**4**, §4 (c)], [**22**, §4 (c)]). Let I_j^k, $j = 1,\dots 4$, $k = 1,2,3$ denote the value of the part I_j after the kth limit. Note that by [**28**, §22, Thm. 17], if α_k is a family of smooth exact forms on B which converges uniformly on any compact set $K \subset B$ to a smooth form α, then α is exact. Thus modulo exact forms on B, $\sum_{j=1}^4 I_j^3 \equiv 0$. We obtain by the definition of $\tilde{\eta}\big(\Omega(Z,F), q, J^{Z,F}, \mathbf{d}\big)$ that

$$I_3^3 = \tilde{\eta}\big(\Omega(Z,F), N_{Z,F}, J^{Z,F}, \mathbf{d}\big).$$

Furthermore, by Theorem 6.16 and in particular (6.14) we get

$$I_2^2 = I_2^3 = \tilde{\eta}\big(\mathcal{H}H, {}_{hc}\mathrm{Gr}(\mathcal{H}H), J^{HH}, J^{{}_{hc}\mathrm{Gr}(HH)}\big).$$

Now $\gamma(u) = O(1)$ as $u \to 0$, and by Definition 3.15, (5.10),

$$\int_0^\infty \gamma(u)\,du = -\tilde{\eta}\big({}_{hc}\mathcal{E}_1, N_{{}_{hc}E_1}, J^{{}_{hc}E_1}, {}_{hc}d_1\big).$$

From (6.10), (6.11), and (6.12) we conclude that

$$I_1^3 = -\sum_{r\geq 1}\tilde{\eta}\big({}_{hc}\mathcal{E}_r, N_{{}_{hc}E_r}, J^{{}_{hc}E_r}, {}_{hc}d_r\big).$$

Finally, using Theorem 6.17, we get

$$I_4^3 = -\tilde{\eta}\big(\Omega(Z,F), N_Z, J^{Z,F}, d^F\big).$$

These four equations imply Theorem 5.9. □

6.4. Proof of Theorem 5.10. We consider again two parameters (T,u). Here u is the rescaling parameter corresponding to the total grading of $\Omega(Z,F)$ and T is associated to the grading of F. For large T, the rescaled differential v becomes large in comparison with the de Rham differential.

Let $\hat{B} := (0,\infty)\times(0,\infty)\times B$. Let $\mathrm{pr}\colon \hat{B}\to B$ denote the projection and define $\hat{M} := \mathrm{pr}^*M \to \hat{B}$ with fiber $\hat{Z}$. This bundle is equipped with the vertical metric $\hat{g}^{TZ}$ on $T\hat{Z}$ which restricts to $u^{-2}g^{TZ}$ on the fibre over $(T,u)\times\{b\}\in\hat{B}$. We define the metric structure $\hat{J}^F$ such that it restricts to $(Tu)^{-N_F+n_F/2}J^F(Tu)^{N_F-n_F/2}$ over $(T,u)\times M$. The metric $\hat{g}^{TZ}$ and $\hat{J}^F$ together induce the metric structure $\hat{J}^{Z,F}$ on $\Omega(\hat{Z},\mathrm{pr}_M^*F)$ (cf. (5.1)).

As in Section 6.3, let $\hat{\mathbf{d}} = d^{\mathrm{pr}_M^*F}+\mathrm{pr}_M^*v$ be the pull-back of $\mathbf{d}$. Let $p(\hat{\mathbf{d}},\hat{J}^{Z,F})$ be the form on $\hat{B}$ defined in (5.2) if n is even (resp. in (5.7) if n is odd).

DEFINITION 6.18. We define $\beta = du\wedge\beta^u+dT\wedge\beta^T$ to be the part of $p(\hat{\mathbf{d}},\hat{J}^{Z,F})$ of degree one with respect to the coordinates (T,u), with functions $\beta^u,\beta^T : (0,\infty)\times(0,\infty)\to\Omega(B)$.

The following corollary is an immediate consequence of the fact that $p(\hat{\mathbf{d}},\hat{J}^{Z,F})$ is closed.

COROLLARY 6.19. *There exists a smooth family* $\alpha\colon (0,\infty)\times(0,\infty)\to\Omega(B)$ *such that*

$$d_{T,u}\beta = dT\wedge du\, d^B\alpha. \tag{6.19}$$

Let $\check{B} := (0,\infty)\times B$ and $\mathrm{pr}_1 : \check{B}\to B$ be the projection. We consider the bundle $\check{M} := \mathrm{pr}_1^*M\to\check{B}$ with fiber $\check{Z}$. Let $\mathrm{Pr}\colon \check{M}\to M$ be the induced map and $\check{H} = \mathrm{Pr}^*H$ be the pull-back of the flat cohomology bundle H of (F,v) on M. We consider the metric structure $\check{J}^H$ on $\check{H}$ which restricts to $u^{-N_H+n_H/2}(\mathrm{Pr}^*J^H)u^{N_H-n_H/2}$ over $\{u\}\times M$. Furthermore, we consider the vertical metric $\check{g}^{TZ}$ on $T\check{Z}$ which restricts to $u^{-2}\mathrm{Pr}^*g^{TZ}$ on the fibre over $(u,b)\in\check{B}$. They induce the metric structure $\check{J}^{Z,H}$ on $\mathrm{pr}_1^*\,\Omega(Z,H)$ as in (5.1). Let N_H be the $\mathbb{Z}$-grading on H induced by N_F. The total $\mathbb{Z}$-grading on $\Omega(Z,H)$ is $N_{Z,H}=N_Z+N_H$. Let $\check{d}^H$ be the twisted de Rham differential on $\Omega(\check{M},\check{H})$. Then $\check{d}^H$ is a flat $(-1)^{\check{N}_{Z,H}}$-superconnection on $\mathrm{pr}_1^*\,(\Omega(Z,H))$; here $\check{N}_{Z,H}=\mathrm{pr}_1^*N_{Z,H}$. We define the family $\gamma\colon (0,\infty)\to\Omega(B)$ such that

$$p(\check{d}^H,\check{J}^{Z,H}) = du\wedge\gamma + r,$$

where r does not contain du. By Definitions 5.3 and 5.4,

$$\begin{aligned}\int_0^\infty \gamma(u)\,du &= -\tilde{\eta}({}_{hc}\mathcal{E}_1, N_{{}_{hc}E_1}, J^{{}_{hc}E_1}, {}_{hc}d_1)\\ &= -\tilde{\eta}\big(\Omega(Z,H),N_Z,J^{Z,H},d^H\big).\end{aligned} \tag{6.20}$$

THEOREM 6.20. (1) *For any* $u>0$, *we have*

$$\lim_{T\to\infty}\beta^u(T,u) = \gamma(u). \tag{6.21}$$

(2) *For $0 < u_1 < u_2$ fixed, there exists $C > 0$ such that, for $u \in [u_1, u_2]$, $T \geq 1$,*

$$|\beta^u(T,u)| \leq C. \tag{6.22}$$

(3) *We have the following identity:*

$$\lim_{T\to\infty} \int_1^\infty \beta^u(T,u)\,du = \int_1^\infty \gamma(u)\,du - \sum_{r\geq 2} \tilde\eta\big({}_{\iota g}\mathcal{E}_r, N_{{}_{\iota g}E_r}, J^{{}_{\iota g}E_r}, {}_{\iota g}d_r\big). \tag{6.23}$$

THEOREM 6.21. (1) *There exists a smooth family $\sigma\colon (0,\infty) \to \Omega(B)$ such that, for $T \geq 1$, we have*

$$\lim_{u\to\infty} \beta^T(T,u) = \sigma(T).$$

(2) *There exist constants $C > 0$, $\delta > 0$ such that, for $T \geq 1$, we have the following estimate:*

$$|\sigma(T)| \leq \frac{C}{T^{1+\delta}}. \tag{6.24}$$

(3) *We have the following equality modulo exact forms on B:*

$$\int_1^\infty \sigma(T)\,dT = -\tilde\eta\big(\mathcal{H}H, {}_{\iota g}\mathrm{Gr}(\mathcal{H}H), J^{HH}, J^{{}_{\iota g}\mathrm{Gr}(HH)}\big). \tag{6.25}$$

We consider the metric structure $\check{J}^F$ on Pr^*F which restricts to $u^{-N_F+n_F/2}J^F u^{N_F-n_F/2}$ over $\{u\}\times M$. We consider $\nabla^F + v$ as $(-1)^{N_F}$-superconnection on F. We define the smooth family $\theta\colon (0,\infty)\to\Omega(M)$ such that

$$p\big(\mathrm{Pr}^*(\nabla^F + v), \check{J}^F\big) = du \wedge \theta + r,$$

where r does not contain du. Note that by Definition 3.15

$$\int_0^\infty \theta(u)\,du = -\tilde\eta\big(\mathcal{F}, N_F, J^F, v\big).$$

.

THEOREM 6.22. (1) *For any $u > 0$, there exist $C > 0$, $\delta > 0$ such that, for $T \geq 1$, we have*

$$\big|\beta^T(T,u)\big| \leq \frac{C}{T^{\delta+1}}. \tag{6.26}$$

(2) *For any $T > 0$, we have*

$$\lim_{\varepsilon\to 0} \varepsilon^{-1}\beta^T(T\varepsilon^{-1},\varepsilon) = \int_Z L(\nabla^{TZ}) \wedge \theta(T). \tag{6.27}$$

(3) *There exists $C > 0$ such that, for $\varepsilon \in (0,1]$, $\varepsilon \leq T \leq 1$, we have*

$$\varepsilon^{-1}\big|\beta^T(T\varepsilon^{-1},\varepsilon)\big| \leq C. \tag{6.28}$$

(4) *There exist $\delta \in (0,1]$, $C > 0$ such that, for $\varepsilon \in (0,1]$, $T \geq 1$, we have*

$$\varepsilon^{-1}\big|\beta^T(T\varepsilon^{-1},\varepsilon)\big| \leq \frac{C}{T^{1+\delta}}. \tag{6.29}$$

Proof of Theorem 5.10. We now finish the proof of Theorem 5.10. For $0 < \varepsilon < A$ and $1 < T_0$, we consider the rectangle $(T, u) \in R := [1, T_0] \times [\varepsilon, A]$. By Corollary 6.19 we have $\int_{\partial R} \beta = d^B \int_R \alpha$. Thus

$$\int_\varepsilon^A \beta^u(T_0, u)\, du - \int_1^{T_0} \beta^T(T, A)\, dT - \int_\varepsilon^A \beta^u(1, u) du + \int_1^{T_0} \beta^T(T, \varepsilon)\, dT$$
$$= I_1 + I_2 + I_3 + I_4$$

is an exact form. We take the limits $A \to \infty$, $T_0 \to \infty$, and then $\varepsilon \to 0$ in the indicated order. Let I_j^k, $j = 1, \dots 4$, $k = 1, 2, 3$ denote the value of the part I_j after the kth limit. Then modulo exact forms on B, $\sum_{j=1}^4 I_j^3 \equiv 0$. We obtain by the definition of $\tilde{\eta}(\Omega(Z, F), N_{Z,F}, J^{Z,F}, \mathbf{d})$ that

$$I_3^3 = \tilde{\eta}(\Omega(Z, F), N_{Z,F}, J^{Z,F}, \mathbf{d}).$$

Furthermore, by Theorem 6.21 and in particular (6.25) we get

$$I_2^2 = I_2^3 = \tilde{\eta}(\mathcal{H}H, {}_{hc}\mathrm{Gr}(\mathcal{H}H), J^{HH}, J^{{}_{hc}\mathrm{Gr}(HH)}).$$

From (6.20), (6.21), (6.22), and (6.23), we conclude that

$$I_1^3 = -\tilde{\eta}(\Omega(Z, H), N_Z, J^{Z,H}, d^Z) - \sum_{r \geq 2} \tilde{\eta}({}_{lg}\mathcal{E}_r, N_{{}_{lg}E_r}, J^{{}_{lg}E_r}, {}_{lg}d_r).$$

Finally, using Theorem 6.22, we get

$$I_4^3 = -\int_Z L(\nabla^{TZ}) \wedge \tilde{\eta}(\mathcal{F}, N_F, J^F, v).$$

These four equations imply the theorem. □

6.5. Proof of Theorem 5.11. By the variation formula for eta forms (Theorem 5.6) it suffices to prove Theorem 5.11 for a particular choice of $T_1^H W$, $T_2^H V$, $T_3^H W$, and g^{TZ}, g^{TX}, g^{TY}. We will suppose that

$$T_3^H W \subset T_1^H W, \qquad (6.30)$$
$$g^{TZ} = g^{TX} \oplus \pi_1^* g^{TY}.$$

We consider a family of superconnections depending on two parameters $(T, u) \in (0, \infty) \times (0, \infty)$. The parameter u is the usual rescaling parameter associated to the total grading of $\Omega(Z, F)$. In the present case the fibre Z is the total space of a fibre bundle $Z \to Y$ with fibre X. The parameter T is introduced to perform an adiabatic limit in this fibration. For large T the vertical part corresponding to $d^{X,F}$ of the differential d^F is scaled to become large with respect to the horizontal part.

Let us now fit this idea into the formalism. We consider the space $\hat{S} := (0, \infty) \times (0, \infty) \times S$. Let $\mathrm{pr}\colon \hat{S} \to S$ denote the projection and define $\hat{W} := \mathrm{pr}^* W \to \hat{S}$ with fiber $\hat{Z}$. Let $\mathrm{pr}_W\colon \hat{W} \to W$ be the canonical projection. We consider the decomposition of the vertical bundle $TZ = TX \oplus T^H Z$, where $T^H Z := T_1^H W \cap TZ$.

Then $T^H Z \cong \pi_1^* TY$. We define the metric $\hat{g}^{TZ}$on $\hat{Z} = \mathrm{pr}_W^*(TX \oplus \pi_1^* TY)$ such that it restricts to $u^{-2}(T^{-2} g^{TX} \oplus \pi_1^* g^{TY})$ over $(T, u) \times S$. The metric $\hat{g}^{TZ}$ and $\mathrm{pr}_W^* J^F$ together induce the metric structure $\hat{J}^{Z,F}$ on $\mathrm{pr}^* \Omega(Z, F)$ as in (5.1).

Let $d^{\mathrm{pr}_W^* F}$ be the twisted de Rham differential on $\Omega(\hat{W}, \mathrm{pr}_W^* F)$. Further let $p\big(d^{\mathrm{pr}_W^* F}, \hat{J}^{Z,F}\big)$ be the form on $\hat{S}$ defined by (5.2) if n is even (resp. by (5.7) if n is odd).

DEFINITION 6.23. We define $\beta = du \wedge \beta^u + dT \wedge \beta^T$ to be the part of $p\big(d^{\mathrm{pr}_W^* F}, \hat{J}^{Z,F}\big)$ of degree one with respect to the coordinates (T, u), with functions $\beta^u, \beta^T\colon (0, \infty) \times (0, \infty) \to \Omega(S)$.

The following fact is an immediate consequence of the fact that $p\big(\hat{\mathbf{d}}, \hat{J}^{Z,F}\big)$ is closed. Let $d = d_{T,u} + d^S$ be the decomposition of the de Rham differential on $(0, \infty) \times (0, \infty) \times S$.

COROLLARY 6.24. *There exists a smooth family* $\alpha\colon (0, \infty) \times (0, \infty) \to \Omega(S)$ *such that*

$$d_{T,u} \beta = dT \wedge du\, d^S \alpha. \tag{6.31}$$

To compare easily to [**24**, §4–§9], in the following we will write down explicitly β^u, β^T. For α_1, α_2 two differential forms on S, we denote $\{\alpha_1 + du\, \alpha_2\}^{du} = \alpha_2$, and $\{\alpha_1 + dT\, \alpha_2\}^{dT} = \alpha_2$.

Let $J_T^{Z,F}$, $*_T$ be the metric structure and the Hodge star operator on $\Omega(Z, F)$ with respect to the metrics $T^{-2} g^{TX} \oplus \pi_1^* g^{TY}$ and J^F (cf. Subsection 4.3.1). Then we have $J_T^{Z,F} = T^{-N_X + \dim X/2} J^{Z,F} T^{N_X - \dim X/2}$. Denote $z^{J_T^{Z,F}} = \sqrt{\epsilon \epsilon_n}^{\,-1} J_T^{Z,F}$. Let $d^{Z,F}$, $\nabla^{Z,F}$, i_T be the operators defined in Subsection 5.2.2. Let $(d^F)_T^*$, $(d^{Z,F})_T^*$, $(i_T)_T^*$, $(\nabla^{Z,F})_T^*$ be the formal adjoints of d^F, $d^{Z,F}$, i_T, $\nabla^{Z,F}$, respectively, with respect to the metric structure $J_T^{Z,F}$ on $\Omega(Z, F)$. Then

$$\begin{aligned} *_T^{-1} \frac{\partial *_T}{\partial T} &= \frac{1}{T} (2N_X - \dim X), \\ (d^F)_T^* &= T^{-2N_X} (d^F)_1^* T^{2N_X}. \end{aligned} \tag{6.32}$$

For $u > 0$, we set

$$C'_{3,u^2,T} = u^{N_Z} d^F u^{-N_Z}, \qquad C''_{3,u^2,T} = u^{-N_Z} (d^F)_T^* u^{N_Z},$$

$$C_{3,u^2,T} = \frac{1}{2}(C'_{3,u^2,T} + C''_{3,u^2,T}), \qquad D_{3,u^2,T} = \frac{1}{2}(C''_{3,u^2,T} - C'_{3,u^2,T}).$$

We denote by $[A, B] = AB - BA$ the commutator. For $T \geq 1$, set $A_{u,T} = T^{N_X} C_{3,u^2,T} T^{-N_X}$. Then from the above equations we get

$$\big[T^{N_X} D_{3,u^2,T} T^{-N_X}, N_X\big] = T \frac{\partial}{\partial T} A_{u,T}.$$

• If n is even, then

$$\beta^u = (2\pi i)^{-1/2}\varphi\,\mathrm{Tr}\Big[z^{J_T^{Z,F}}\big(\frac{\partial}{\partial u}C_{3,u^2,T}\big)\exp(-C^2_{3,u^2,T})\Big], \tag{6.33}$$
$$\beta^T = (2\pi i)^{-1/2}\varphi\,\mathrm{Tr}\Big[z^{J_T^{Z,F}}\big[\frac{1}{2}D_{3,u^2,T}, *_T^{-1}\frac{\partial *_T}{\partial T}\big]\exp(-C^2_{3,u^2,T})\Big].$$

Thus we get

$$\beta^u = (2\pi i)^{-1/2}\varphi\Big\{\mathrm{Tr}\Big[z^{J^{Z,F}}\exp(-A^2_{u,T}+du\,(\frac{\partial}{\partial u}A_{u,T}))\Big]\Big\}^{du}, \tag{6.34}$$
$$\beta^T = (2\pi i)^{-1/2}\varphi\Big\{\mathrm{Tr}\Big[z^{J^{Z,F}}\exp(-A^2_{u,T}+dT\,(\frac{\partial}{\partial T}A_{u,T}))\Big]\Big\}^{dT}.$$

• If n is odd, we introduce an extra odd variable σ as in Subsection 5.2.2. Then

$$\beta^u = \frac{1}{\sqrt{\pi}}\varphi\mathrm{Tr}_\sigma\Big[z^{J_T^{Z,F}}\big(\frac{\partial}{\partial u}C_{3,u^2,T}\big)\exp(-C^2_{3,u^2,T})\Big], \tag{6.35}$$
$$\beta^T = \frac{1}{\sqrt{\pi}}\varphi\mathrm{Tr}_\sigma\Big[z^{J_T^{Z,F}}\big[\frac{1}{2}D_{3,u^2,T}, *_T^{-1}\frac{\partial *_T}{\partial T}\big]\exp(-C^2_{3,u^2,T})\Big].$$

Remark that $J_T^{Z,F}$ commute with $(\nabla^{Z,F}+(\nabla^{Z,F})^*_T$, $d^{Z,F}+(d^{Z,F})^*_T$, $i_T+(i_T)^*_T$. Let $A^{(i)}_{u,T}$ be the part of $A_{u,T}$ of degree i in $\Lambda(T^*S)$. Then $A^{(1)}_{u,T}=\frac{1}{2}(\nabla^{Z,F}+(\nabla^{Z,F})^*_T)$. Set $\bar{A}_{u,T}=A^{(1)}_{u,T}+z^{J^{Z,F}}(A^{(0)}_{u,T}+A^{(2)}_{u,T})$. Then

$$\beta^u = \frac{1}{\sqrt{\pi}}\varphi\Big\{\mathrm{Tr}_\sigma\Big[\exp(-\bar{A}^2_{u,T}+du(\frac{\partial}{\partial u}\bar{A}_{u,T}))\Big]\Big\}^{du}, \tag{6.36}$$
$$\beta^T = \frac{1}{\sqrt{\pi}}\varphi\Big\{\mathrm{Tr}_\sigma\Big[\exp(-\bar{A}^2_{u,T}+dT(\frac{\partial}{\partial T}\bar{A}_{u,T}))\Big]\Big\}^{dT}.$$

The following theorems can be shown by adapting the the method [**24**] to our present situation.

Let $\check{S} := (0,\infty)\times S$ and $\mathrm{pr}_1\colon \check{S}\to S$ be the projection. We consider the bundle $\check{V} := \mathrm{pr}^*V\to\check{S}$. Let $\mathrm{pr}_V\colon\check{V}\to V$ be the induced map and $H(\check{X},\mathcal{F}) = \mathrm{pr}_V^*H(X,\mathcal{F})$ be the pull-back of the flat cohomology bundle $H(X,\mathcal{F})$ of $(\Omega(X,F)$, $d^{X,F})$. We consider the metric structure $\check{J}^{H(X,\mathcal{F})}$ which restricts to

$$u^{-N_{H(X,\mathcal{F})}+n_{H(X,\mathcal{F})}/2}\mathrm{pr}_V^*J^{H(X,\mathcal{F})}u^{N_{H(X,\mathcal{F})}-n_{H(X,\mathcal{F})}/2}$$

over $\{u\}\times V$. Furthermore, we consider the vertical metric $\check{g}^{TY}$, which restricts to $\frac{1}{u^2}\mathrm{pr}_V^*g^{TY}$ on $\{u\}\times V$. They induce the metric structure $\check{J}^{Y,H(X,\mathcal{F})}$ on $\mathrm{pr}_1^*\Omega(Y,H(X,\mathcal{F}))$ as in (5.1). Let $\check{d}^{H(X,\mathcal{F})}$ be the twisted de Rham differential on $\Omega(\check{V},\mathrm{pr}_V^*H(X,\mathcal{F}))$. Note that the $\mathbb{Z}$-grading operators N_Z, N_X, N_Y act naturally on $\Omega(Y,H(X,\mathcal{F}))$. Then $\check{d}^{H(X,\mathcal{F})}$ is a flat $(-1)^{\check{N}_Z}$-superconnection on $\mathrm{pr}_1^*\Omega(Y,H(X,\mathcal{F}))$. We define the family $\gamma\colon(0,\infty)\to\Omega(S)$ such that

$$p\big(\check{d}^{H(X,\mathcal{F})},\check{J}^{Y,H(X,\mathcal{F})}\big) = du\wedge\gamma + r,$$

where r does not contain du. Note that $p(\check{d}^{H(X,\mathcal{F})}, \check{J}^{Y,H(X,\mathcal{F})}) = 0$ if $\dim X$ is odd. Furthermore, note that by Definitions 5.3 and 5.4

$$(6.37)\quad \int_0^\infty \gamma(u)\,du = -\tilde{\eta}\big({}_{hc}\mathcal{E}_1, N_{{}_{hc}E_1}, J^{{}_{hc}E_1}, {}_{hc}d_1\big)$$
$$= -\tilde{\eta}\big(\Omega(Y, H(X,\mathcal{F})), N_Y, J^{Y,H(X,\mathcal{F})}, d^{H(X,\mathcal{F})}\big).$$

THEOREM 6.25. (1) *For any $u > 0$, we have*

$$(6.38)\quad \lim_{T\to\infty} \beta^u(T,u) = \gamma(u).$$

(2) *For $0 < u_1 < u_2$ fixed, there exists $C > 0$ such that, for $u \in [u_1, u_2]$, $T \geq 1$, we have*

$$(6.39)\quad |\beta^u(T,u)| \leq C.$$

(3) *We have the following identity*:

$$(6.40)\quad \lim_{T\to\infty} \int_1^\infty \beta^u(T,u)\,du = \int_1^\infty \gamma(u)\,du - \sum_{r\geq 2} \tilde{\eta}\big({}_{LS}\mathcal{E}_r, N_{{}_{LS}E_r}, J^{{}_{LS}E_r}, {}_{LS}d_r\big).$$

THEOREM 6.26. (1) *There exists a smooth family $\sigma\colon (0,\infty) \to \Omega(S)$ such that, for $T \geq 1$, we have*

$$\lim_{u\to\infty} \beta^T(T,u) = \sigma(T).$$

(2) *There exist constants $C > 0$, $\delta > 0$ such that, for $T \geq 1$, we have the following estimate*:

$$(6.41)\quad |\sigma(T)| \leq \frac{C}{T^{1+\delta}}$$

(3) *We have the following identity*:

$$(6.42)\quad \int_1^\infty \sigma(T)\,dT = -\tilde{\eta}\big(\mathcal{H}(Z,\mathcal{F}), {}_{LS}\mathrm{Gr}(\mathcal{H}(Z,\mathcal{F})), J^{H(Z,\mathcal{F})}, J^{{}_{LS}\mathrm{Gr}(H(Z,\mathcal{F}))}\big).$$

We consider the fibration $\check{W} = (0,\infty) \times W \to \check{V}$ equipped with the vertical metric $\check{g}^{TX}$ which restricts to $u^{-2}g^{TX}$ on $\{u\} \times W$. Let $\mathrm{Pr}_W\colon \check{W} \to W$ be the projection. $\mathrm{Pr}_W^* J^F$ and $\check{g}^{TX}$ induce a metric structure $\check{J}^{X,F}$ on $\mathrm{pr}_V^*\Omega(X,F)$ as in (5.1). Let $d^{\mathrm{Pr}_W^* F}$ be the twisted de Rham differential on $\Omega(\check{W}, \mathrm{Pr}_W^* F)$. We define the smooth family $\theta\colon (0,\infty) \to \Omega(V)$ such that

$$p(d^{\mathrm{Pr}_W^* F}, \check{J}^{X,F}) = du \wedge \theta + r,$$

where r does not contain du. Note that by Definitions 5.3 and 5.4

$$\int_0^\infty \theta(u)\,du = -\tilde{\eta}\big(\Omega(X,F), N_X, J^{X,F}, d^F\big).$$

The adiabatic family of metrics $g^{TX} \oplus \frac{1}{T^2}\pi_1^* g^{TY}$ on $\mathrm{Pr}_W^* TZ$ on $\check{W}$ induces a family of connections $\widetilde{\nabla}^{TZ}$ of $\mathrm{Pr}_W^* TZ$ which after restriction to $\{T\} \times W$ is the connection ∇_T^{TZ} on TZ with respect to g_T^{TZ}, $T_3^H W$, π_3 defined by [**5**, Def. 1.6].

By [**24**, Thm. 5.1], for large T, ∇_T^{TZ} converges to ${}^0\nabla^{TZ} + A_{3,\infty}$, where $A_{3,\infty} \in T^*W \otimes \mathrm{End}(T^H Z, TX)$. We define the smooth family $\lambda\colon (0,\infty) \to \Omega(W)$ such that

$$L(\widetilde{\nabla}^{TZ}) = dT \wedge \lambda + r$$

and r does not contain dT. Now, by the same argument as in [**24**, § 5.7], we have the following proposition for the transgression L class.

PROPOSITION 6.27. *When $T \to \infty$, then we have $\lambda(T) = O(T^{-2})$. Modulo exact forms on W we have*

$$\tilde{L}(TZ, \nabla^{TZ}, {}^0\nabla^{TZ}) \equiv -\int_1^\infty \lambda(T)\, dT. \tag{6.43}$$

THEOREM 6.28. (1) *For any $u > 0$, there exist $C > 0$, $\delta > 0$ such that, for $T \geq 1$, we have*

$$\left|\beta^T(T,u)\right| \leq \frac{C}{T^{\delta+1}}. \tag{6.44}$$

(2) *For any $T > 0$, we have*

$$\lim_{\varepsilon \to 0} \varepsilon^{-1}\beta^T(T\varepsilon^{-1}, \varepsilon) = \int_Y L(\nabla^{TY}) \wedge \theta(T). \tag{6.45}$$

(3) *There exists $C > 0$ such that, for $\varepsilon \in (0,1]$, $\varepsilon \leq T \leq 1$,*

$$\varepsilon^{-1}\left|\beta^T(T\varepsilon^{-1}, \varepsilon) - \int_Z p(\nabla^F, J^F) \wedge \lambda(T\varepsilon^{-1})\right| \leq C. \tag{6.46}$$

(4) *There exist $\delta \in (0,1]$, $C > 0$ such that, for $\varepsilon \in (0,1]$, $T \geq 1$,*

$$\varepsilon^{-1}\left|\beta^T(T\varepsilon^{-1}, \varepsilon)\right| \leq \frac{C}{T^{1+\delta}}. \tag{6.47}$$

PROOF OF THEOREM 5.11. We now finish the proof of Theorem 5.11. For $0 < \varepsilon < A$ and $1 < T_0$, we consider the rectangle $(T,u) \in R := [1, T_0] \times [\varepsilon, A]$. By Corollary 6.24 we have $\int_{\partial R} \beta = d^S \int_R \alpha$. Thus

$$\int_\varepsilon^A \beta^u(T_0, u)\, du - \int_1^{T_0} \beta^T(T, A) dT - \int_\varepsilon^A \beta^u(1,u) du + \int_1^{T_0} \beta^T(T,\varepsilon) dT$$
$$= I_1 + I_2 + I_3 + I_4$$

is an exact form. We take the limits $A \to \infty$, $T_0 \to \infty$, and then $\varepsilon \to 0$ in the indicated order. Let I_j^k, $j = 1, \ldots 4$, $k = 1,2,3$ again denote the value of the part I_j after the kth limit. By [**28**, §22, Thm. 17], $d\Omega(S)$ is closed under uniformly convergence on compact sets of S. Thus modulo exact forms on S, $\sum_{j=1}^4 I_j^3 \equiv 0$. We obtain from the definition of $\tilde{\eta}\big(\Omega(Z,F), N_Z, J^{Z,F}, d^F\big)$ that

$$I_3^3 = \tilde{\eta}\big(\Omega(Z,F), N_Z, J^{Z,F}, d^F\big).$$

Furthermore, by Theorem 6.26 and in particular (6.42) we get

$$I_2^2 = I_2^3 = \tilde{\eta}\big(\mathcal{H}H, {}_{hc}\mathrm{Gr}(\mathcal{H}H), J^{HH}, J^{{}_{hc}\mathrm{Gr}(HH)}\big).$$

From (6.37), (6.38), (6.39), and (6.40), we conclude that

$$I_1^3 = -\tilde{\eta}\big(\Omega(Y, H(X,\mathcal{F})), N_Y, J^{Y,H(X,\mathcal{F})}, d^{H(X,\mathcal{F})}\big) - \sum_{r\geq 2} \tilde{\eta}\big({}_{lg}\mathcal{E}_r, N_{{}_{lg}E_r}, J^{{}_{lg}E_r}, {}_{lg}d_r\big).$$

Finally, using Theorem 6.28, we get

$$I_4^3 = -\int_Y L(\nabla^{TY}) \wedge \tilde{\eta}(\Omega(X,F), N_X, J^{X,F}, d^F) - \int_Z \tilde{L}(TZ, \nabla^{TZ}, {}^0\nabla^{TZ}) \wedge p(\nabla^F, J^F)$$

as follows: Convergence of the integrals below is granted by (6.44). We write

$$\int_1^\infty \beta^T(T,\epsilon)\, dT = \int_\epsilon^\infty \epsilon^{-1}\beta^T(T\epsilon^{-1},\epsilon)\, dT.$$

Using Proposition 6.27, (6.45), and (6.47), we get

$$\lim_{\epsilon\to 0}\int_1^\infty \epsilon^{-1}\beta^T(T\epsilon^{-1},\epsilon)\, dT = \int_Y L(\nabla^{TY}) \wedge \int_1^\infty \theta(T)\, dT,$$

$$\lim_{\epsilon\to 0}\int_\epsilon^1 \epsilon^{-1}\left[\beta^T(T\epsilon^{-1},\epsilon) - \int_Z p(\nabla^F, J^F)\wedge\lambda(T\epsilon^{-1})\right] dT = \int_Y L(\nabla^{TY}) \wedge \int_0^1 \theta(T)\, dT.$$

The remaining part of the integral yields by (6.43)

$$\begin{aligned}\lim_{\epsilon\to 0}\int_\epsilon^1 \epsilon^{-1}\int_Z p(\nabla^F, J^F)\wedge\lambda(T\epsilon^{-1})dT &= \int_Z p(\nabla^F, J^F)\wedge\int_1^\infty \lambda(T)dT \\ &= -\int_Z \tilde{L}(TZ, \nabla^{TZ}, {}^0\nabla^{TZ}) \wedge p(\nabla^F, J^F).\end{aligned}$$

These four equations for I_k^3, $k = 1,\dots,4$, imply Theorem 5.11. □

References

[1] M. F. Atiyah, V. K. Patodi, and I. M. Singer, *Spectral asymmetry and Riemannian geometry. I*, Math. Proc. Cambridge Philos. Soc. **77** (1975), 43–69.

[2] M. F. Atiyah, V. K. Patodi, and I. M. Singer, *Spectral asymmetry and Riemannian geometry. III*, Math. Proc. Cambridge Philos. Soc. **79** (1976), 71–99.

[3] N. Berline, E. Getzler, and M. Vergne, *Heat kernels and the Dirac operator*, Grundl. Math. Wiss., Vol. 298, Springer, Berlin-Heidelberg-New York, 1992.

[4] A. Berthomieu and J.-M. Bismut, *Quillen metrics and higher analytic torsion forms*, J. Reine Angew. Math., **457** (1994), 85–184.

[5] J.-M. Bismut, *The index theorem for families of Dirac operators: two heat equation proofs*, Invent.Math., **83** (1986), 91–151.

[6] J.-M. Bismut, *Families of immersions, and higher analytic torsion,* Astérisque, Vol. 244, 1997.
[7] J.-M. Bismut. *Local index theory, eta invariants, and holomorphic torsion: a survey,* Surv. Differ. Geom., Vol. III, Int. Press, Boston, MA, 1998, 1–76.
[8] J.-M. Bismut and J. Cheeger, *η-invariants and their adiabatic limits,* J. Amer. Math. Soc., **2** (1989), 33–70.
[9] J.-M. Bismut and D. Freed. *The analysis of elliptic families. II. Dirac operators, eta invariants, and the holonomy theorem,* Comm. Math. Phys. **107** (1986), 103–163.
[10] J.-M. Bismut, H. Gillet, and C. Soulé, *Analytic torsion and holomorphic determinant bundles. II,* Comm. Math. Phys. **115** (1988), 79–126.
[11] J.-M. Bismut and G. Lebeau, *Complex immersions and Quillen metrics,* Publ. Math. Inst. Hautes Études Sci., Vol. 74, 1991.
[12] J.-M. Bismut and J. Lott, *Flat vector bundles, direct images, and higher real analytic torsion,* J. Amer. Math. Soc., **8** (1995), 291–363.
[13] T. Bröcker and T. tom Dieck, *Representations of compact Lie groups,* Grad. Texts in Math., Vol. 98, Springer, 1985.
[14] U. Bunke, *On the functoriality of Lott's secondary analytic index,* K-Theory **25** (2002), 51–58.
[15] U. Bunke and X. Ma, *Index and secondary index theory for flat bundles with duality,* arXiv:math.DG/0106068
[16] X. Dai, *Adiabatic limits, non-multiplicativity of signature, and Leray spectral sequence,* J. Amer. Math .Soc., **4** (1991), 265–321.
[17] P. Griffiths and J. Harris, *Principles of algebraic geometry,* Wiley, New-York, 1978.
[18] A. Grothendieck, *Sur quelques points d'algèbre homologique,* Tohoku Math. J. **9**, 1957, 119–221.
[19] M. Karoubi, *Homologie cyclique et K-théorie,* Astérisque, Vol. 149, 1987.
[20] J. Lott, *$\mathbb{R}/\mathbb{Z}$-index theory,* Comm. Anal. Geom., **2** (1994), 279–311.
[21] J. Lott, *Secondary analytic indices,* In: N. Schappacher and A. Reznikov (Eds.), Regulars in Analysis, Geometry, and Number Theory, Progr. Math., Vol. 171., Birkhäuser Boston, 2000, 231–293.
[22] X. Ma, *Formes de torsion analytique et familles de submersions I,* Bull. Soc. Math. France, **127** (1999), 541–621.
[23] X. Ma, *Formes de torsion analytique et familles de submersions II,* Asian J. Math., **4** (2000), 633–668.
[24] X. Ma, *Functoriality of real analytic torsion forms,* Israel J. Math., **131** (2002), 1–50.
[25] J. McCleary, *User's guide to spectral sequences,* Math. Lect. Ser., Vol. 12, Publish or Perish, 1985.
[26] D. McDuff and D. Salamon, *Introduction to symplectic topology,* 2nd edition, Oxford Math. Monogr., 1998.
[27] D. Quillen, *Superconnections and the Chern character,* Topology, **24** (1985), 89–95.
[28] G. de Rham, *Variétés différentiables,* Hermann, Paris, 1973.

Universität Göttingen, Bunsenstr. 3-5, 37073 Göttingen, Germany
E-mail address: bunke@uni-math.gwdg.de

CNRS UMR 7640, Centre de Mathématiques, École Polytechnique, 91128 Palaiseau Cedex, France
E-mail address: ma@math.polytechnique.fr

Operator Theory:
Advances and Applications, Vol. 151, 342–429

Toeplitz operators, and ellipticity of boundary value problems with global projection conditions

B.–W. Schulze

Abstract. Ellipticity of (pseudo-)differential operators A on a compact manifold X with boundary (or with edge) Y is connected with boundary (or edge) conditions of trace and potential type, formulated in terms of global projections on Y together with an additional symbolic structure. This gives rise to operator block matrices $\mathcal{A}$ with A in the upper left corner. We study an algebra of such operators, where ellipticity is equivalent to the Fredholm property in suitable scales of spaces: Sobolev spaces on X plus closed subspaces of Sobolev spaces on Y which are the range of corresponding pseudo-differential projections. Moreover, we express parametrices of elliptic elements within our algebra and discuss spectral boundary value problems for differential operators.

Introduction

Ellipticity of differential (and pseudo-differential) operators on a manifold X with boundary $Y = \partial X$ is connected with a specific control of data near the boundary. More generally, ellipticity on a manifold with edges or with higher (say, polyhedral) singularities includes conditions on the lower-dimensional strata (e.g., edges, corners, etc.) of the configuration.

Such conditions may occur as trace or potential operators, linked to the singularities.

It is a common point of view to interpret A together with the trace and potential conditions as an operator block matrix $\mathcal{A}$ (with A as upper left corner) and to construct an algebra of such block matrices that contains the parametrices of elliptic elements, see Vishik and Eskin **[52]**, Eskin **[12]**, Boutet de Monvel **[7]**, Rempel and Schulze **[31]**, **[32]**, or Schulze **[39]**. In many known cases this is a transparent and satisfying concept. The operators $\mathcal{A}$ then have a principal symbolic hierarchy $\sigma(\mathcal{A})$, and in simplest cases, ellipticity is invertibility of all components of $\sigma(\mathcal{A})$, where parametrices belong to $\sigma^{-1}(\mathcal{A})$.

2000 *Mathematics Subject Classification.* Primary 35J70, 58J05; Secondary 35S05.

To illustrate phenomena we first consider the case of boundary value problems for elliptic differential operators A on a compact C^∞ manifold X with boundary Y with Shapiro-Lopatinskij elliptic conditions (also called SL-ellipticity in this paper).

By a boundary value problem in standard form we understand an equation $Au = f$, together with boundary conditions $Tu = g$, where $A : C^\infty(X, E) \to C^\infty(X, F)$ for $E, F \in \mathrm{Vect}(X)$, is a given differential operator of order μ and $T = {}^{\mathrm{t}}(T_1, \ldots, T_N)$ a column vector of trace operators $T_j : C^\infty(X, E) \to C^\infty(Y, G_j)$, for $G_j \in \mathrm{Vect}(Y)$, $j = 1, \ldots, N$. Here $\mathrm{Vect}(\cdot)$ denotes the set of all smooth complex vector bundles on the space in the brackets. The operators T_j are assumed to be given in the form $T_j = \mathrm{r}' B_j$, where r' is the operator of restriction to the boundary, and $B_j : C^\infty(X, E) \to C^\infty(X, \widetilde{G}_j)$ are differential operators of order μ_j, with $\widetilde{G}_j \in \mathrm{Vect}(X)$, $G_j := \widetilde{G}_j|_Y$, $j = 1, \ldots, N$. The column matrix operator

$$(0.1) \qquad \mathcal{A} := \begin{pmatrix} A \\ T \end{pmatrix} : C^\infty(X, E) \to \begin{array}{c} C^\infty(X, F) \\ \oplus \\ \bigoplus_{j=1}^N C^\infty(Y, G_j) \end{array}$$

then extends to continuous operators

$$(0.2) \qquad \mathcal{A} : H^s(X, E) \to \begin{array}{c} H^{s-\mu}(X, F) \\ \oplus \\ \bigoplus_{j=1}^N H^{s-\mu_j-\frac{1}{2}}(Y, G_j) \end{array}$$

between the respective Sobolev spaces of distributional sections for all real $s > \max(\mu_j + \frac{1}{2})$. The principal symbolic hierarchy in this case consists of two components, namely

$$(0.3) \qquad \sigma(\mathcal{A}) = (\sigma_\psi(\mathcal{A}), \sigma_\partial(\mathcal{A})),$$

where $\sigma_\psi(\mathcal{A}) := \sigma_\psi(A)$ is the standard homogeneous principal symbol of A of order μ, also called the principal interior symbol, which is a bundle morphism

$$(0.4) \qquad \sigma_\psi(A) : \pi_X^* E \to \pi_X^* F,$$

$\pi_X : T^*X \setminus 0 \to X$, and $\sigma_\partial(\mathcal{A})$ is the principal boundary symbol of $\mathcal{A}$. To give a definition of the principal boundary symbol we first look at A. Let us fix a collar neighbourhood $\cong Y \times [0, 1)$ of the boundary in the local splitting of variables $x = (y, t) \in \Omega \times [0, 1)$, $\Omega \subseteq \mathbb{R}^{n-1}$ open, with covariables $\xi = (\eta, \tau)$. (For convenience, transition diffeomorphisms near the boundary are assumed to be independent of t.) We then have a family of continuous operators

$$(0.5) \qquad \sigma_\partial(A)(y, \eta) := \sigma_\psi(A)(y, 0, \eta, D_t) : \mathbb{C}^k \otimes H^s(\mathbb{R}_+) \to \mathbb{C}^l \otimes H^{s-\mu}(\mathbb{R}_+)$$

for $(y, \eta) \in T^*\Omega \setminus 0$, acting in Sobolev spaces $H^s(\mathbb{R}_+) := H^s(\mathbb{R})|_{\mathbb{R}_+}$ in normal direction to the boundary; k and l are the fibre dimensions of E and F, respectively. Using the invariance of (0.5) under transition maps we obtain a bundle morphism

$$(0.6) \qquad \sigma_\partial(A) : \pi_Y^* E' \otimes H^s(\mathbb{R}_+) \to \pi_Y^* F' \otimes H^s(\mathbb{R}_+),$$

$\pi_Y : T^*Y \setminus 0 \to 0$, where prime indicates the restriction of the corresponding bundle to the boundary. In a similar manner we can proceed for the trace operators $T_j = \mathrm{r}' B_j$ and obtain

$$\sigma_\partial(T_j)(y,\eta) := \mathrm{r}'\sigma_\psi(B_j)(y,0,\eta,D_t) : \mathbb{C}^k \otimes H^s(\mathbb{R}_+) \to \mathbb{C}^{l_j}$$

for $(y,\eta) \in T^*\Omega \setminus 0$, where l_j is the fibre dimension of G_j. Globally, we have again bundle morphisms

$$\sigma_\partial(T_j) : \pi_Y^* E' \otimes H^s(\mathbb{R}_+) \to \pi_Y^* G_j,$$

$j = 1, \dots, N$, and we set altogether

$$\sigma_\partial(\mathcal{A}) := \begin{pmatrix} \sigma_\partial(A) \\ \sigma_\partial(T) \end{pmatrix} : \pi_Y^* E' \otimes H^s(\mathbb{R}_+) \to \pi_Y^* \begin{pmatrix} F' \otimes H^{s-\mu}(\mathbb{R}_+) \\ \oplus \\ G \end{pmatrix} \tag{0.7}$$

for $\sigma_\partial(T) := {}^{\mathrm{t}}(\sigma_\partial(T_1), \dots, \sigma_\partial(T_N))$ and $G := \bigoplus_{j=1}^N G_j$.

A boundary value problem $\mathcal{A}$ is called SL-elliptic, if (0.4) is an isomorphism, and if also (0.7) is an isomorphism for any sufficiently large real s. If we talk about the operator A alone, the first condition will also be referred to as σ_ψ-ellipticity of A.

The choice of s is unessential for the condition of SL-ellipticity. Moreover, (0.7) is an isomorphism if and only if

$$\sigma_\partial(\mathcal{A}) : \pi_Y^* E' \otimes \mathcal{S}(\overline{\mathbb{R}}_+) \to \pi_Y^* \begin{pmatrix} F' \otimes \mathcal{S}(\overline{\mathbb{R}}_+) \\ \oplus \\ G \end{pmatrix} \tag{0.8}$$

is an isomorphism, $\mathcal{S}(\overline{\mathbb{R}}_+) := \mathcal{S}(\mathbb{R})|_{\overline{\mathbb{R}}_+}$. Simple examples for SL-elliptic boundary value problems are the Dirichlet or the Neumann problem for Laplace operators.

Let us now observe the specific homogeneity of boundary symbols. For $u(t) \in H^s(\mathbb{R}_+)$ we set

$$(\kappa_\lambda u)(t) := \lambda^{\frac{1}{2}} u(\lambda t), \qquad \lambda \in \mathbb{R}_+. \tag{0.9}$$

In this way we obtain a strongly continuous group $\{\kappa_\lambda\}_{\lambda \in \mathbb{R}_+}$ of isomorphisms on the space $H^s(\mathbb{R}_+)$ (here, if H is a Hilbert space, a group $\{\kappa_\lambda\}_{\lambda \in \mathbb{R}_+}$ of isomorphisms $\kappa_\lambda : H \to H$, $\lambda \in \mathbb{R}_+$, with $\kappa_\lambda \kappa_\varrho = \kappa_{\lambda\varrho}$ for all $\lambda, \varrho \in \mathbb{R}_+$, is called strongly continuous, if $\kappa_\lambda h \in C(\mathbb{R}_+, H)$ for every $h \in H$).

It can easily be verified that

$$\sigma_\partial(A)(y,\lambda\eta) = \lambda^\mu \kappa_\lambda \sigma_\partial(A)(y,\eta)\kappa_\lambda^{-1} \tag{0.10}$$

for all $\lambda \in \mathbb{R}_+$. Similarly, we have

$$\sigma_\partial(T_j)(y,\lambda\eta) = \mathrm{r}'\lambda^{\mu_j}\kappa_\lambda \sigma_\partial(B_j)(y,\eta)\kappa_\lambda^{-1} = \lambda^{\mu_j+\frac{1}{2}}\sigma_\partial(T_j)(y,\eta)\kappa_\lambda^{-1} \tag{0.11}$$

for all $\lambda \in \mathbb{R}_+$, where we employ the relation $\mathrm{r}' \circ \kappa_\lambda = \lambda^{\frac{1}{2}}\mathrm{r}'$ for the corresponding operators on functions in t on the half-axis. This gives us

$$\sigma_\partial(\mathcal{A})(y,\lambda\eta) = \mathrm{diag}(\lambda^\mu\kappa_\lambda, \lambda_1^{\mu_1+\frac{1}{2}}, \dots, \lambda_N^{\mu_N+\frac{1}{2}})\sigma_\partial(\mathcal{A})(y,\eta)\kappa_\lambda^{-1} \tag{0.12}$$

as the homogeneity of boundary symbols.

On the manifold X we fix a Riemannian metric that is equal to the product metric from $Y \times [0,1)$ in a collar neighbourhood of Y. We then have the absolute values of covectors on X and Y in an invariant way.

It is often convenient to unify the orders of trace operators by composing them from the left by classical pseudo-differential operators R_j on Y of order $\mu - \mu_j - \frac{1}{2}$ with homogeneous principal symbol $|\eta|^{\mu-\mu_j-\frac{1}{2}} \operatorname{id}_{\pi_Y^* G_j}$, such that

$$R_j : H^{s-\mu_j-\frac{1}{2}}(Y, G_j) \to H^{s-\mu}(Y, G_j)$$

are isomorphisms for all s. Such a choice of R_j is always possible; we then talk about a corresponding reduction of orders on the boundary.

Set $\sigma_\partial(R_j T_j)(y,\eta) := |\eta|^{\mu-\mu_j-\frac{1}{2}} \sigma_\partial(T_j)(y,\eta)$. We now replace the former T by ${}^t(R_1T_1, \dots, R_NT_N)$ and denote the new boundary value problem again by $\mathcal{A} = \binom{A}{T}$. This gives us continuous operators

$$\mathcal{A} : H^s(X,E) \to \begin{array}{c} H^{s-\mu}(X,F) \\ \oplus \\ H^{s-\mu}(Y,G) \end{array} \tag{0.13}$$

for all sufficiently large s.

For the boundary symbol $\sigma_\partial(\mathcal{A})$ we have

$$\sigma_\partial(\mathcal{A})(y,\lambda\eta) = \lambda^\mu \begin{pmatrix} \kappa_\lambda & 0 \\ 0 & 1 \end{pmatrix} \sigma_\partial(\mathcal{A})(y,\eta)\kappa_\lambda^{-1} \tag{0.14}$$

for all $\lambda \in \mathbb{R}_+$, $(y,\eta) \in T^*Y \setminus 0$, where 1 is the identity map in $\pi_Y^* G$.

There is now the following natural question. Given a σ_ψ-elliptic differential operator $A : H^s(X,E) \to H^{s-\mu}(X,F)$ on X, does there always exist an SL-elliptic boundary value problem $\mathcal{A} = \binom{A}{T}$ with a suitable trace operator T? The answer is negative, 'unfortunately' for many interesting geometric operators, e.g., Dirac operators in even dimensions, in particular, for the Cauchy-Riemann operator in the complex plane. For the existence there is a well known condition of Atiyah and Bott [**3**] that we want to recall here. First note that when A is σ_ψ-elliptic, the boundary symbol

$$\sigma_\partial(A) : \pi_Y^* E' \otimes H^s(\mathbb{R}_+) \to \pi_Y^* F' \otimes H^{s-\mu}(\mathbb{R}_+) \tag{0.15}$$

is a family of Fredholm operators (surjective in the case of differential operators). Let S^*Y be the unit cosphere bundle induced by T^*Y with the canonical projection $\pi_1 : S^*Y \to Y$. The restriction of (0.15) to S^*Y gives us a family of Fredholm operators parametrised by the compact set S^*Y. As such it represents an index element in the K-group of S^*Y, namely

$$\operatorname{ind}_{S^*Y} \sigma_\partial(A) \in K(S^*Y), \tag{0.16}$$

cf. [**2**] and Section 2.1 below. In the present case, $L_+ := \ker \sigma_\partial(A)$ is a vector bundle on $T^*Y \setminus 0$, and we have $\operatorname{ind}_{S^*Y} \sigma_\partial(A) = [L_+|_{S^*Y}]$, where $[\dots]$ denotes the element in the K-group on S^*Y, represented by the bundle in the brackets. Now if there is a SL-elliptic boundary value problem $\mathcal{A}$ for A, i.e., the corresponding boundary symbol (0.7) is an isomorphism, we have necessarily $\ker_{S^*Y}\sigma_\partial(A) := \bigcup_{(y,\eta)\in S^*Y}\ker\sigma_\partial(A)(y,\eta) \cong \pi_1^* G$, i.e.,

$$\operatorname{ind}_{S^*Y}\sigma_\partial(A) \in \pi_1^* K(Y), \tag{0.17}$$

where $\pi_1^* : K(Y) \to K(S^*Y)$ is the pull-back of K-groups under the projection $\pi_1 : S^*Y \to Y$. In other words, (0.17) is a necessary condition for the existence of a SL-elliptic operator $\mathcal{A}$ for a σ_ψ-elliptic A.

As noted before, the condition (0.17) may be violated. We also talk about a topological obstruction for the existence of SL-elliptic boundary value problems. The criterion of Atiyah and Bott says that (0.17) is also sufficient for the existence of a SL-elliptic $\mathcal{A}$ (in a slightly more general context, connected with a stabilisation that we shall see more explicitly below in connection with pseudo-differential boundary value problems, cf. Section 2.1 below.)

Boundary value problems for the case with non-vanishing obstruction for $\sigma_\psi(A)$ have been studied by many authors before, in particular, by Calderón [**9**], and Seeley [**49**] (more references will be given below), and many relations to other classical areas have been discovered, in particular, to the Riemann-Hilbert problem (and its various generalisations), Toeplitz operators, index theory, spectral theory, and geometric analysis.

Among the essential observations there is the aspect that Fredholm operators of the form (0.13) (which are adequate for the SL-elliptic case) have to be replaced by Fredholm operators

$$\mathcal{A} : H^s(X,E) \to \begin{array}{c} H^{s-\mu}(X,F) \\ \oplus \\ P^{s-\mu}(Y,\mathbb{L}) \end{array} \tag{0.18}$$

for a new scale of closed subspaces $P^s(Y,\mathbb{L})$ of $H^s(Y,J)$, $s \in \mathbb{R}$, which are defined as the image under a projection $P_+ : H^s(Y,J) \to H^s(Y,J)$ for a suitable element $J \in \operatorname{Vect}(Y)$, where $P_+ \in L^0_{\mathrm{cl}}(Y;J,J)$ is a classical pseudo-differential operator of order zero (acting between distributional sections in J), cf. Schulze, Sternin and Shatalov [**46**]. Here $\mathbb{L}$ denotes the triple (P_+, J, L), where $L \in \operatorname{Vect}(T^*Y \setminus 0)$ is the image of $\pi_Y^* J$ under the projection $p_+ : \pi_Y^* J \to \pi_Y^* J$, the homogeneous principal symbol of P_+; the symbolic calculus of classical pseudo-differential operators gives us $p_+^2 = p_+$ as a consequence of $P_+^2 = P_+$, such that L is a subbundle of $\pi_Y^* J$. A well known theorem says that for every subbundle $L \subset \pi_Y^* J$ and every choice of a projection $p_+ : \pi_Y^* J \to L$ that is homogeneous in η of order zero there is an associated projection $P_+ \in L^0_{\mathrm{cl}}(Y;J,J)$ with p_+ as homogeneous principal symbol (we will give a proof in Section 1.1 below).

In the present exposition we shall extend an approach of the author [**40**], where operators of the form (0.18) have been completed to an algebra of block matrix operators

$$(0.19)\qquad \mathcal{A} = \begin{pmatrix} A & K \\ T & Q \end{pmatrix} : \begin{matrix} H^s(X,E) \\ \oplus \\ P^s(Y,\mathbb{L}_-) \end{matrix} \rightarrow \begin{matrix} H^{s-\mu}(X,F) \\ \oplus \\ P^{s-\mu}(Y,\mathbb{L}_+) \end{matrix}$$

by conditions of trace and potential type T and K, respectively, connected with suitable triples $\mathbb{L}_\pm = (P_\pm, J_\pm, L_\pm)$ of a similar meaning as before. The operator Q is a Toeplitz operator on Y, see the terminology in Section 1.1 below. Starting point will be an algebra $\mathcal{B}(X)$ of operators of the form

$$(0.20)\qquad \widetilde{\mathcal{A}} = \begin{pmatrix} A & \widetilde{K} \\ \widetilde{T} & \widetilde{Q} \end{pmatrix} : \begin{matrix} H^s(X,E) \\ \oplus \\ H^s(Y,J_-) \end{matrix} \rightarrow \begin{matrix} H^{s-\mu}(X,F) \\ \oplus \\ H^{s-\mu}(Y,J_+) \end{matrix}$$

that are (classical) pseudo-differential boundary value problems on X with the transmission property at Y, cf. Boutet de Monvel [**7**] or Rempel and Schulze [**31**]. The operators (0.19) will be compositions

$$(0.21)\qquad \mathcal{A} = \begin{pmatrix} A & \widetilde{K}R_- \\ P_+\widetilde{T} & P_+\widetilde{Q}R_- \end{pmatrix},$$

where $R_- : P^{s-\mu}(Y,\mathbb{L}_-) \rightarrow H^s(Y,J_-)$ is the canonical embedding, and $P_+ : H^{s-\mu}(Y,J_+) \rightarrow P^{s-\mu}(Y,\mathbb{L}_+)$ is one of the chosen projections. Operators of the form (0.21) will be called boundary value problems for A with global projection conditions. In Section 1.2 we give a brief overview on the algebra $\mathcal{B}(X)$ including its symbolic structure. $\mathcal{B}(X)$ will be a subalgebra of the larger algebra of operators (0.21). In fact, for the case $P_\pm = \mathrm{id}$, $L_\pm = J_\pm$ we just recover the case (0.20).

Elliptic operators A on a manifold with boundary from the point of view of global projection conditions are of interest on their own right. On the other hand, manifolds with boundary may be viewed as particular manifolds with edges, where the boundary is the edge and the inner normal the model cone of a wedge. In general, wedges with non-trivial model cones (i.e., when they have a base of dimension > 0) locally describe manifolds with edges. In such a case ellipticity is also connected with edge operators of trace and potential type, and there is again a topological obstruction for the existence of SL-elliptic conditions, cf. [**35**, Section 3.3.5, Proposition 10]. Similar phenomena may be expected on manifolds with higher (say, polyhedral) singularities with a hierarchy of obstructions for SL-elliptic conditions on the lower-dimensional strata. In all those cases it makes sense to construct operators with global projection data when the corresponding obstructions do not vanish. A joint paper of the author with Seiler [**44**] treats the case of manifolds with egdes. A substructure with special features is the case of boundary value problems without the transmission property, contained in the author's joint paper with Seiler [**45**].

A large variety of examples of edge-degenerate differential operators with non-vanishing obstruction is constructed in Nazaikinskij, Schulze, Sternin and Shatalov

[27]. In another article [28] the authors study the K-theoretic nature of the topological obstruction in general.

In the present paper we will not discuss the aspect of edge operators in detail; this is another fascinating branch of the development of operators with global projection conditions and their index theory. Let us conclude this introduction by references for the case of smooth manifolds with boundary, namely Atiyah, Patodi and Singer [4], Melrose [26], Booß-Bavnbek and Wojciechowski [6], Grubb and Seeley [17], Schulze, Sternin, and Shatalov [46], Schulze and Tarkhanov [47].

Acknowledgement: The author thanks T. Krainer and N. Tarkhanov from Potsdam as well as V. Nazaikinskii and B. Sternin from Moscow for valuable remarks on the manuscript.

1. Elements of the classical calculus of boundary value problems

1.1. Pseudo-differential and Toeplitz operators on a closed manifold.

Let M be a closed, compact C^∞ manifold, $m = \dim M$, with a fixed Riemannian metric and an associated measure dx. Complex smooth vector bundles on M are assumed to be equipped with Hermitian metrics. For every $E, F \in \mathrm{Vect}(M)$ we then have the space $L^2(M, E)$ of square integrable sections in E with a corresponding scalar product. Moreover, we have the scale $H^s(M, E)$, $s \in \mathbb{R}$, of distributional sections in E of Sobolev smoothness s, where we identify $H^0(M, E)$ with $L^2(M, E)$.

Let $E \boxtimes F \in \mathrm{Vect}(M \times M)$ be the external tenser product of bundles $E, F \in \mathrm{Vect}(M)$, i.e., $E \boxtimes F := (p_1^* E) \otimes (p_2^* F)$ with the projections $p_j : (x_1, x_2) \to x_j$, $M \times M \to M$, to the j-th component, $j = 1, 2$. We identify the space $C^\infty(F \boxtimes E^*)$ (where E^* is the dual bundle of E) with the space $L^{-\infty}(M; E, F)$ of all operators $K : C^\infty(M, E) \to C^\infty(M, F)$ which have kernels $c(x, \tilde{x})$ in $C^\infty(F \boxtimes E^*)$, i.e.,

$$Ku(x) = \int_M (c(x, \tilde{x}), u(\tilde{x}))_E \, d\tilde{x}, \tag{1.1}$$

where $(\cdot, \cdot)_E$ means the fibrewise pairing between E^* and E.

We now recall some definitions and results from the standard pseudo-differential calculus on a manifold.

DEFINITION 1.1. (i) The space $S^\mu(U \times \mathbb{R}^n)$ for $\mu \in \mathbb{R}$, and $U \subseteq \mathbb{R}^m$ open, is defined to be the subspace of all $a(x, \xi) \in C^\infty(U \times \mathbb{R}^n)$ such that

$$|D_x^\alpha D_\xi^\beta a(x, \xi)| \le c \langle \xi \rangle^{\mu - |\beta|} \tag{1.2}$$

for all $\alpha \in \mathbb{N}^m$, $\beta \in \mathbb{N}^n$ and $(x, \xi) \in K \times \mathbb{R}^n$ for all $K \Subset U$, with constants $c = c(\alpha, \beta, K) > 0$; here $\langle \xi \rangle := (1 + |\xi|^2)^{1/2}$. The elements of $S^\mu(U \times \mathbb{R}^n)$ are called symbols of order μ.

(ii) $S^{(\mu)}(U \times (\mathbb{R}^n \setminus \{0\}))$ denotes the space of all $a_{(\mu)}(x, \xi) \in C^\infty(U \times (\mathbb{R}^n \setminus \{0\}))$ such that $a_{(\mu)}(x, \lambda\xi) = \lambda^\mu a_{(\mu)}(x, \xi)$ for all $\lambda \in \mathbb{R}_+$, $(x, \xi) \in U \times (\mathbb{R}^n \setminus \{0\})$.

(iii) An excision function $\chi(\xi)$ in $\mathbb{R}^n$ is any element of $C^\infty(\mathbb{R}^n)$ such that $\chi(\xi) = 0$ for $0 \le |\xi| \le c_0$, $\chi(\xi) = 1$ for $c_1 \le |\xi|$ with certain constants $0 < c_0 < c_1$.

(iv) A symbol $a(x,\xi) \in S^\mu(U \times \mathbb{R}^n)$ is said to be classical, if there are elements $a_{(\mu-j)}(x,\xi) \in S^{(\mu-j)}(U \times (\mathbb{R}^n \setminus \{0\}))$, $j \in \mathbb{N}$, such that

$$a(x,\xi) - \sum_{j=0}^{N} \chi(\xi) a_{(\mu-j)}(x,\xi) \in S^{\mu-(N+1)}(U \times \mathbb{R}^n)$$

for every $N \in \mathbb{N}$, where $\chi(\xi)$ is any excision function. Let $S^\mu_{\mathrm{cl}}(U \times \mathbb{R}^n)$ denote the space of all classical symbols of order μ.

Let $\Omega \subseteq \mathbb{R}^n$ be open, and set

$$L^\mu_{(\mathrm{cl})}(\Omega) := \{\mathrm{Op}(a) : \ a(x,x',\xi) \in S^\mu_{(\mathrm{cl})}(\Omega \times \Omega \times \mathbb{R}^n)\},$$

where $\mathrm{Op}(a)u(x) := \iint e^{i(x-x')\xi} a(x,x',\xi)u(x')\,dx'đ\xi$, $đ\xi = (2\pi)^{-n}\,d\xi$, subscript '(cl)' means that we talk about classical or non-classical symbols and operators, respectively. As usual, $\mathrm{Op}(a)u(x)$ makes sense in the oscillatory integral sense, first as a continuous map: $\mathrm{Op}(a) : C_0^\infty(\Omega) \to C^\infty(\Omega)$, and then extended to larger function and distribution spaces.

For $A \in L^\mu_{\mathrm{cl}}(\Omega)$ we set

$$\sigma_\psi(A)(x,\xi) = a_{(\mu)}(x,x',\xi)|_{x'=x},$$

(cf. the notation in Definition 1.1 (iv)) called the homogeneous principal symbol of order μ of the operator A.

Let $L^\mu_{(\mathrm{cl})}(\Omega;\mathbb{C}^j,\mathbb{C}^k)$ denote the space of all $k \times j$ matrices of elements in $L^\mu_{(\mathrm{cl})}(\Omega)$, and let E and F be vector bundles on M of fibre dimension j and k, respectively, with trivialisations $\Omega \times \mathbb{C}^j$ and $\Omega \times \mathbb{C}^k$ for open sets $\Omega \subseteq \mathbb{R}^n$, with transition maps

$$(1.3) \qquad \eta_{\widetilde{\Omega}\Omega} : \Omega \times \mathbb{C}^j \to \widetilde{\Omega} \times \mathbb{C}^j, \qquad \vartheta_{\widetilde{\Omega}\Omega} : \Omega \times \mathbb{C}^k \to \widetilde{\Omega} \times \mathbb{C}^k$$

for open $\Omega, \widetilde{\Omega} \subset \mathbb{R}^n$, with an underlying coordinate diffeomorphism $\kappa : \Omega \to \widetilde{\Omega}$. We then have a push-forward

$$L^\mu_{(\mathrm{cl})}(\Omega;\mathbb{C}^j,\mathbb{C}^k) \to L^\mu_{(\mathrm{cl})}(\widetilde{\Omega};\mathbb{C}^j,\mathbb{C}^k)$$

of operators $A : C_0^\infty(\Omega,\mathbb{C}^j) \longrightarrow C^\infty(\Omega,\mathbb{C}^k)$ to operators $\widetilde{A} : C_0^\infty(\widetilde{\Omega},\mathbb{C}^j) \longrightarrow C^\infty(\widetilde{\Omega},\mathbb{C}^k)$ when we set $\widetilde{A} := T_{\widetilde{\Omega}\Omega} A H^{-1}_{\widetilde{\Omega}\Omega}$, where $H_{\widetilde{\Omega}\Omega} : C_0^\infty(\Omega,\mathbb{C}^j) \to C_0^\infty(\widetilde{\Omega},\mathbb{C}^j)$ and $T_{\widetilde{\Omega}\Omega} : C^\infty(\Omega,\mathbb{C}^k) \to C^\infty(\widetilde{\Omega},\mathbb{C}^k)$ are the isomorphisms induced by (1.3). To every chart $\chi : D \to \Omega$ on M we thus obtain the spaces $L^\mu_{(\mathrm{cl})}(D;E|_D,F|_D)$ of pseudo-differential operators on D, acting between sections in the bundles E and

F. Set

$$L^{\mu}_{(\mathrm{cl})}(M;E,F) := \Big\{ \sum_{j=1}^{N} \varphi_j A_j \psi_j + C : \ A_j \in L^{\mu}_{(\mathrm{cl})}(D_j; E|_{D_j}, F|_{D_j}), \\ C \in L^{-\infty}(M;E,F) \Big\}, \tag{1.4}$$

where $\{D_1, \ldots, D_N\}$ is an open covering of M by coordinate neighbourhoods (say, diffeomorphic to balls), $\{\varphi_1, \ldots, \varphi_N\}$ a subordinate partition of unity, and $\{\psi_1, \ldots, \psi_N\}$ a system of functions $\psi_j \in C_0^\infty(D_j)$ that are equal to 1 on $\operatorname{supp} \varphi_j$ for all j.

REMARK 1.2. $L^{\mu}_{(\mathrm{cl})}(M;E,F)$ is a Fréchet space in a natural way.

Let $A \in L^{\mu}_{\mathrm{cl}}(M;E,F)$, and let $(A_{pq})_{1\le p\le j, 1\le q\le k}$ be a local representative of A on Ω. Then the system of matrices $(\sigma_\psi(A_{pq}))_{1\le p\le j,1\le q\le k}$ has an invariant meaning as a bundle morphism

$$\sigma_\psi(A) : \pi_M^* E \to \pi_M^* F, \tag{1.5}$$

called the homogeneous principal symbol of A; here, $\pi_M : T^*M \setminus 0 \to M$ is the canonical projection of the cotangent bundle of M (minus the zero section) to M.

Let $S^{(\mu)}(T^*M \setminus 0; E, F)$ denote the set of all morphisms $p_{(\mu)} : \pi_M^* E \to \pi_M^* F$ such that $p_{(\mu)}(x, \lambda\xi) = \lambda^\mu p_{(\mu)}(x,\xi)$ for all $(x,\xi) \in T^*M \setminus 0$, $\lambda \in \mathbb{R}_+$.

PROPOSITION 1.3. *The principal symbolic map*

$$\sigma_\psi : L^{\mu}_{\mathrm{cl}}(M;E,F) \to S^{(\mu)}(T^*M \setminus 0; E, F) \tag{1.6}$$

is surjective, and there is a linear map

$$\mathrm{op} : S^{(\mu)}(T^*M \setminus 0; E, F) \to L^{\mu}_{\mathrm{cl}}(M;E,F) \tag{1.7}$$

such that $\sigma_\psi \circ \mathrm{op} = \mathrm{id}$.

A choice of (1.7) directly follows from the existence of local operators in $L^{\mu}_{\mathrm{cl}}(\Omega; \mathbb{C}^j, \mathbb{C}^k)$ associated with principal symbols that correspond to local representations of a given element in $S^{(\mu)}(T^*M \setminus 0; E, F)$; a subsequent globalisation, according to the expressions in (1.4), then gives us (1.6).

THEOREM 1.4. *Let M be a closed, compact C^∞ manifold.*

(i) *Every $A \in L^{\mu}(M;E,F)$ for $E, F \in \mathrm{Vect}(M)$ induces continuous operators*

$$A : H^s(M,E) \to H^{s-\mu}(M,F) \tag{1.8}$$

for all $s \in \mathbb{R}$.

(ii) *$A \in L^{\mu}_{\mathrm{cl}}(M;E,F)$ and $\sigma_\psi(A) = 0$ implies $A \in L^{\mu-1}_{\mathrm{cl}}(M;E,F)$, and hence the operator (1.8) is compact.*

(iii) *$A \in L^{\mu}_{(\mathrm{cl})}(M; E_0, F)$, $B \in L^{\nu}_{(\mathrm{cl})}(M; E, E_0)$ for $E, E_0, F \in \mathrm{Vect}(M)$ implies $AB \in L^{\mu+\nu}_{(\mathrm{cl})}(M; E, F)$. In the classical case we have*

$$\sigma_{\psi}(AB) = \sigma_{\psi}(A)\sigma_{\psi}(B). \tag{1.9}$$

(iv) *Let $A \in L^{\mu}_{(\mathrm{cl})}(M; E, F)$, and let A^* denote the formal adjoint of A (defined by $(Au, v)_{L^2(M,F)} = (u, A^*v)_{L^2(M,E)}$ for all $u \in C^{\infty}(M, E)$, $v \in C^{\infty}(M, F)$). Then we have $A^* \in L^{\mu}_{(\mathrm{cl})}(M; F, E)$ and, in the classical case,*

$$\sigma_{\psi}(A^*) = \sigma_{\psi}(A)^*, \tag{1.10}$$

(the adjoint on the right hand side refers to the Hermitian metrics in the bundles.)

In the sequel we mainly concentrate on classical pseudo-differential operators.

Definition 1.5. Let $A \in L^{\mu}_{\mathrm{cl}}(M; E, F)$, $\mu \in \mathbb{R}$, $E, F \in \mathrm{Vect}(M)$.

(i) The operator A is said to be elliptic (of order μ), if $\sigma_{\psi}(A) : \pi_M^* E \to \pi_M^* F$ is an isomorphism.

(ii) An operator $P \in L^{-\mu}_{\mathrm{cl}}(M; F, E)$ is called a parametrix of A, if P satisfies the following relations:

$$C_l := I - PA \in L^{-\infty}(M; E, E), \qquad C_r := I - AP \in L^{-\infty}(M; F, F), \tag{1.11}$$

where I denotes the corresponding identity operator.

Notice that when P is a parametrix of A, we have $\sigma_{\psi}(P) = \sigma_{\psi}(A)^{-1}$. Moreover, if A is elliptic, so is the formal adjoint A^*.

We call an operator $A \in L^{\mu}_{\mathrm{cl}}(M; E, F)$ underdetermined (overdetermined) elliptic, if $\sigma_{\psi}(A) : \pi_M^* E \to \pi_M^* F$ is injective (surjective).

Remark 1.6. (i) $A \in L^{\mu}_{\mathrm{cl}}(M; E, F)$ is underdetermined elliptic if and only if $A^* \in L^{\mu}_{\mathrm{cl}}(M; F, E)$ is overdetermined elliptic.

(ii) If A is underdetermined (overdetermined) elliptic, then the operator $A^*A \in L^{2\mu}_{\mathrm{cl}}(M; E, E)$ ($AA^* \in L^{2\mu}_{\mathrm{cl}}(M; F, F)$) is elliptic.

Theorem 1.7. *Let $A \in L^{\mu}_{\mathrm{cl}}(M; E, F)$, $\mu \in \mathbb{R}$, $E, F \in \mathrm{Vect}(M)$.*

(i) *The operator A is elliptic (of order μ), if and only if*

$$A : H^s(M, E) \to H^{s-\mu}(M, F) \tag{1.12}$$

is a Fredholm operator for an $s = s_0 \in \mathbb{R}$.

(ii) *If A is elliptic, (1.12) is a Fredholm operator for all $s \in \mathbb{R}$, and* $\dim \ker A$ *and* $\dim \operatorname{coker} A$ *are independent of s. We have $V := \ker A \subset C^{\infty}(M, E)$, and there is a finite-dimensional subspace $W \subset C^{\infty}(M, F)$ such that*

$$W + \operatorname{im} A = H^{s-\mu}(M, F)$$

and $W \cap \operatorname{im} A = \{0\}$ for every $s \in \mathbb{R}$.

(iii) *An elliptic operator* A *has a parametrix* $P \in L_{\mathrm{cl}}^{-\mu}(M; F, E)$, *cf. Definition* 1.5 (ii), *and* P *can be chosen in such a way that the remainders in the relation* (1.11) *are projections*

$$C_l : H^s(M, E) \to V, \qquad C_r : H^{s-\mu}(M, F) \to W$$

to subspaces V *and* W *as in* (ii) *for all* $s \in \mathbb{R}$.

As a consequence of Theorem 1.7 (ii) we see that the index of A

$$\operatorname{ind} A := \dim \ker A - \dim \operatorname{coker} A$$

is independent of s.

Theorem 1.7 (iii) yields elliptic regularity of solutions $u \in H^{-\infty}(M, E)$ to an elliptic equation $Au = f \in H^s(M, F)$, $s \in \mathbb{R}$, namely $u \in H^{s+\mu}(M, E)$. In fact, using a parametrix P of A as a left parametrix, we obtain $PAu = Pf \in H^{s+\mu}(M, E)$, but $C_l = I - PA$ implies $PAu = u - C_l u$ where $C_l u \in H^{\infty}(M, F)$, and it follows that $u = Pf + C_l u \in H^{s+\mu}(M, F)$.

The latter consideration gives us, in particular, $V = \ker A \subset C^{\infty}(M, E)$. Moreover, the relation $\ker A^* + \operatorname{im} A = H^{s-\mu}(M, F)$ (which is a direct decomposition, i.e., $\ker A^* \cap \operatorname{im} A = \{0\}$) allows us to set $W = \ker A^* \subset C^{\infty}(M, F)$, cf. Theorem 1.7 (ii).

The assertions of Theorem 1.7 (ii), (iii) have an abstract background which will be useful in other situations below.

We consider operators A between Hilbert spaces that belong to scales $\{H_i^s\}_{s \in \mathbb{R}}$, where first

$$A : H_1^{\infty} \to H_2^{\infty} \tag{1.13}$$

is continuous, $H_i^{\infty} := \bigcap_{s \in \mathbb{R}} H_i^s$. To every A there is an order $\mu \in \mathbb{R}$ and a constant $c > 0$ such that (1.13) extends to continuous operators

$$A_s : H_1^s \to H_2^{s-\mu} \tag{1.14}$$

for all $s > c$ (the aspect with the constant $c = c(A)$ is not relevant for Theorems 1.4 and 1.7, but in boundary value problems it will play a role). If it is clear from the context which s is considered for the operator, we also write A instead of A_s.

We then assume the following properties:

(i) There are continuous embeddings $H_i^{s'} \hookrightarrow H_i^s$ for $s' \geq s$ that are compact for $s' > s$.

(ii) The space H_i^{∞} is dense in H_i^s for every $s \in \mathbb{R}$, $i = 1, 2$.

(iii) If $V \subset H_i^{\infty}$, $i = 1, 2$, is a finite-dimensional subspace and $C_V : H_i^0 \to V$ a projection, then C_V induces continuous operators $C_V : H_i^s \to V$ for all $s > c$ for some $c \in \mathbb{R}$.

(iv) (1.13) extends to a Fredholm operator (1.14) for all $s > c$, and there is a continuous operator

$$P : H_2^{\infty} \to H_1^{\infty}$$

that extends to a parametrix

$$P_{s-\mu} : H_2^{s-\mu} \to H_1^s$$

of A_s for every $s > c$, i.e., the remainders

$$C_{s,l} := I - P_{s-\mu}A_s, \qquad C_{s-\mu,r} := I - A_s P_{s-\mu}$$

induce continuous operators

$$C_{s,l} : H_1^s \to H_1^\infty, \qquad C_{s-\mu,r} : H_2^{s-\mu} \to H_2^\infty$$

for all $s > c$.

REMARK 1.8. Under the abovementioned conditions the dimensions of $\ker A_s$ and $\operatorname{coker} A_s$ of the Fredholm operator $A : H_1^s \to H_2^{s-\mu}$ are independent of $s > c$, we have $V := \ker A_s \subset H_1^\infty$, and there is a finite-dimensional subspace $W \subset H_2^\infty$ such that $W + \operatorname{im} A_s = H_2^{s-\mu}$, $W \cap \operatorname{im} A_s = \{0\}$ for every $s > c$. Moreover, the parametrix P can be chosen in such a way that the remainders

$$C_l = I - PA, \qquad C_r = I - AP$$

are projections to V and W, respectively.

For references below we now prepare some well known material on parameter-dependent pseudo-differential operators.
First we have the Fréchet space $L^{-\infty}(M; E, F)$ of smoothing operators on M, and we set

$$L^{-\infty}(M; E, F; \mathbb{R}^l) = \mathcal{S}(\mathbb{R}^l, L^{-\infty}(M; E, F))$$

which is the space of smoothing parameter-dependent operators (between sections of $E, F \in \operatorname{Vect}(M)$). The space

$$L^\mu_{(\mathrm{cl})}(M; E, F; \mathbb{R}^l)$$

of parameter-dependent pseudo-differential operators of order μ is defined in a similar manner as (1.4), now for

$$C(\lambda) \in L^{-\infty}(M; E, F; \mathbb{R}^l) \quad \text{and} \quad A_j(\lambda) \in L^\mu_{(\mathrm{cl})}(D_j; E|_{D_j}, F|_{D_j}; \mathbb{R}^l),$$

where the latter spaces are operator pull backs under charts $\chi_j : D_j \to \Omega$ of operators on Ω, with matrices of amplitude functions $a_j(x, x', \xi, \lambda) \in S^\mu_{(\mathrm{cl})}(\Omega \times \Omega \times \mathbb{R}^{n+l}_{\xi,\lambda})$.

THEOREM 1.9. *For every $A(\lambda) \in L^\mu(M; E, F; \mathbb{R}^l)$ and $\nu \geq \mu$ we have*

$$\|A(\lambda)\|_{\mathcal{L}(H^s(X,E), H^{s-\nu}(X,F))} \leq c \begin{cases} (1 + |\lambda|)^\mu & \text{for} \quad \nu \geq 0, \\ (1 + |\lambda|)^{\mu-\nu} & \text{for} \quad \nu \leq 0 \end{cases},$$

for constants $c = c_{s,\nu} > 0$.

For classical parameter-dependent operators $A(\lambda) \in L^{\mu}_{\mathrm{cl}}(M; E, F; \mathbb{R}^l)$ we have a parameter-dependent homogeneous principal symbol (1.5), now for the projection

$$\pi_M : (T^*M \times \mathbb{R}^l) \setminus 0 \to M \qquad (0 \text{ means } (\xi, \lambda) = 0).$$

In this case, $\sigma_\psi(A)(x, \xi, \lambda)$ is defined in terms of the homogeneous principal components of local amplitude functions in $(\xi, \lambda) \neq 0$ (at $x' = x$).

DEFINITION 1.10. An $A(\lambda) \in L^{\mu}_{\mathrm{cl}}(M; E, F; \mathbb{R}^l)$ is called parameter-dependent elliptic (of order μ), if

$$\sigma_\psi(A) : \pi_M^* E \to \pi_M^* F$$

for $\pi_M : (T^*M \times \mathbb{R}^l) \setminus 0 \to M$, is an isomorphism.

THEOREM 1.11. *Let $A(\lambda) \in L^{\mu}_{\mathrm{cl}}(M; E, F; \mathbb{R}^l)$ be parameter-dependent elliptic (of order μ).*

(i) *Then*

$$A(\lambda) : H^s(M, E) \to H^{s-\mu}(M, F) \tag{1.15}$$

is a family of Fredholm operators of index 0. Moreover, there is a constant $C > 0$ such that the operators (1.15) are isomorphisms for all $|\lambda| \geq C$. This holds for all $s \in \mathbb{R}$.

(ii) *$A(\lambda)$ has a parameter-dependent parametrix $P(\lambda) \in L^{-\mu}_{\mathrm{cl}}(M; F, E; \mathbb{R}^l)$, i.e.,*

$$I - P(\lambda)A(\lambda) \in L^{-\infty}(M; E, E; \mathbb{R}^l),$$
$$I - A(\lambda)P(\lambda) \in L^{-\infty}(M; F, F; \mathbb{R}^l).$$

THEOREM 1.12. *For every $\mu \in \mathbb{R}$ there exists an element*

$$R_E^{\mu}(\lambda) \in L^{\mu}_{\mathrm{cl}}(M; E, E; \mathbb{R}^l)$$

that induces isomorphisms

$$R_E^{\mu}(\lambda) : H^s(X, E) \to H^{s-\mu}(X, E)$$

for all $s \in \mathbb{R}$ and $\lambda \in \mathbb{R}^l$.

REMARK 1.13. If $R_E^s(\lambda) \in L^{\mu}_{\mathrm{cl}}(M; E, E; \mathbb{R}^l)$ is an order reducing family of order $s \in \mathbb{R}$ as in Theorem 1.12, then

$$\|R_E^s(\lambda)u\|_{L^2(M,E)}$$

is a parameter-dependent family of norms on the space $H^s(M, E)$, equivalent to each other for different $\lambda \in \mathbb{R}^l$. We shall employ such families in Section 3.3 bolow for constructing Sobolev spaces on an infinite cylinder $M \times \mathbb{R}$.

We now turn to an extension of the concept of classical pseudo-differential operators $L^{\mu}_{\mathrm{cl}}(M; E, F)$ to so called Toeplitz operators, acting between closed subspaces of Sobolev spaces on M. A crucial technical point is the following result:

THEOREM 1.14. *Let* $p : \pi_M^* J \to \pi_M^* J$, $J \in \mathrm{Vect}(M)$, *be a projection, i.e.,* $p^2 = p$, *where* $p(x, \lambda\xi) = p(x, \xi)$ *for all* $(x, \xi) \in T^*M \setminus 0$, $\lambda \in \mathbb{R}_+$. *Then there exists an element* $P \in L^0_{\mathrm{cl}}(M; J, J)$ *such that* $P^2 = P$, *and* $\sigma_\psi(P) = p$.

Moreover, if p *satisfies the condition* $p = p^*$, *there is a choice of the associated pseudo-differential projection* $P \in L^0_{\mathrm{cl}}(M; J, J)$ *such that* $P = P^*$.

COROLLARY 1.15. *To every* $J \in \mathrm{Vect}(M)$ *and every subbundle* $L \subset \pi_M^* J$ *there is an element* $P \in L^0_{\mathrm{cl}}(M; J, J)$ *such that* $P^2 = P$, *where* $\sigma_\psi(P) : \pi_M^* J \to \pi_M^* J$ *is a projection to* L.

The proof of Theorem 1.14 is based on the following general construction. Let H be a (complex) Hilbert space, $\mathcal{L}(H)$ the space of linear continuous operators, $\mathcal{K}(H)$ the subspace of compact operators in H. We then have the Calkin algebra $\mathcal{L}(H)/\mathcal{K}(H)$ and the corresponding canonical map $\pi : \mathcal{L}(H) \to \mathcal{L}(H)/\mathcal{K}(H)$.

LEMMA 1.16. *Let* $p \in \mathcal{L}(H)/\mathcal{K}(H)$ *be an element with* $p^2 = p$, *and choose any* $Q \in \mathcal{L}(H)$ *such that* $\pi Q = p$. *Then the spectrum* $\sigma_{\mathcal{L}(H)}(Q)$ *of* Q *has the property that*

$$\sigma_{\mathcal{L}(H)}(Q) \cap (\mathbb{C} \setminus (\{0\} \cup \{1\}))$$

is discrete.

PROOF. First observe that $p^2 = p$ implies $\sigma_{\mathcal{L}(H)/\mathcal{K}(H}(p) \subseteq \{0\} \cup \{1\}$. In fact, for $\lambda \in \mathbb{C}\setminus(\{0\} \cup \{1\}) =: U$ there exists the inverse $(\lambda e - p)^{-1} = p(\lambda - 1)^{-1} + (e - p)\lambda^{-1}$, where $e \in \mathcal{L}(H)/\mathcal{K}(H)$ is the identity, $e = \pi I$ for the identity $I \in \mathcal{L}(H)$. Now $U \ni \lambda \to \lambda I - Q \in \mathcal{L}(H)$ is a holomorphic Fredholm family in U, and $\lambda I - Q$ is invertible in $\mathcal{L}(H)$ for $|\lambda| > \|Q\|_{\mathcal{L}(H)}$. A well known invertibility result on holomorphic Fredholm families (cf. [**35**, Section 2.2.5] for a proof) implies that $\lambda I - Q$ is invertible for all $\lambda \in U \setminus D$ for a certain discrete subset D (i.e., D is countable and $D \cap K$ finite for every compact subset $K \subset U$). □

PROOF OF THEOREM 1.14. From Lemma 1.16 we see that there exists a $0 < \delta < 1$ such that the circle $C_\delta := \{\lambda \in \mathbb{C} : |\lambda - 1| = \delta\}$ does not intersect $\sigma_{\mathcal{L}(H)}(Q)$. Setting

$$P := \frac{1}{2\pi i} \int_{C_\delta} (\lambda I - Q)^{-1}\, d\lambda \tag{1.16}$$

we obtain $P^2 = P$ and $P \in L^0_{\mathrm{cl}}(M; J, J)$ as a consequence of the holomorphic functional calculus for $L^0_{\mathrm{cl}}(M; J, J)$. Moreover, we have

$$\begin{aligned} \sigma_\psi(P) &= \frac{1}{2\pi i} \int_{C_\delta} (\lambda e - p)^{-1}\, d\lambda \\ &= \left\{\frac{1}{2\pi i} \int_{C_\delta} \frac{1}{\lambda - 1}\, d\lambda\right\} p + \left\{\frac{1}{2\pi i} \int_{C_\delta} \frac{1}{\lambda}\, d\lambda\right\} (e - p). \end{aligned}$$

The second summand on the right hand side vanishes, while the first one is equal to p by the Residue theorem.

To prove the second part of Theorem 1.14 we suppose $p = p^*$. Then, if $P_1 = P_1^2 \in L^0_{\text{cl}}(M; J, J)$ is any choice with $\sigma_\psi(P_1) = p$, also $Q := P_1^* P_1 \in L^0_{\text{cl}}(M; J, J)$ satisfies $\sigma_\psi(Q) = p^* p = p^2 = p$. For Q we have $Q = Q^* \geq 0$. Let η be the spectral measure of Q. Then the projection $P \in L^0_{\text{cl}}(M; J, J)$ defined by the formula (1.16) is equal to the spectral projection

$$\eta(B_\delta(1) \cap \sigma_{\mathcal{L}(L^2(M,J))}(Q)) \quad \text{for } B_\delta = \{\lambda \in \mathbb{C} : \; |\lambda - 1| < \delta\}.$$

In particular, we have $P = P^* = P^2$, and $\sigma_\psi(P) = p$ as before. □

REMARK 1.17. The above construction of projections has a more general functional analytic background. If Ψ is a Fréchet operator algebra with a given ideal $\mathcal{I}$, there is a lifting of idempotent elements of $\Psi/\mathcal{I}$ to idempotent elements in Ψ, provided some natural assumptions on the operator algebra are satisfied, cf. Gramsch [**15**]. In particular, for $\Psi = L^0_{\text{cl}}(M; J, J)$ and $\mathcal{I} = L^{-1}_{\text{cl}}(M; J, J)$ the space $\Psi/\mathcal{I}$ is isomorphic to the space of homogeneous symbols of order zero. The general theory gives a characterisation of the space of all idempotent elements $P \in L^0_{\text{cl}}(M; J, J)$ which belong to the connected component of a given idempotent $P_1 \in L^0_{\text{cl}}(M; J, J)$ and have the same homogeneous principal symbol as P_1. The result says that all those P have the form GP_1G^{-1}, where G varies over the connected component of the identity in the group $\{I + K \in \Psi^{-1} : \; K \in L^{-1}_{\text{cl}}(M; J, J)\}$, where Ψ^{-1} denotes the group of invertible elements of $L^0_{\text{cl}}(M; J, J)$.

Concerning more material and other observations on pseudo-differential projections, see also the papers of Birman and Solomjak [**5**] and Solomjak [**51**].

PROPOSITION 1.18. *Let H be a Hilbert space, and let $P, Q \in \mathcal{L}(H)$ be projections such that $P - Q$ is a compact operator. Then the restrictions of P to $\operatorname{im} Q$ and of Q to $\operatorname{im} P$ are Fredholm operators*

$$P_Q : \operatorname{im} Q \to \operatorname{im} P, \qquad Q_P : \operatorname{im} P \to \operatorname{im} Q$$

between the respective closed subspaces of H, and Q_P is a parametrix (i.e., a Fredholm inverse) of P_Q.

PROOF. The operator Q acts as the identity on $\operatorname{im} Q$. Therefore, we have

$$Q_P P_Q - 1 = Q_P P_Q - Q^2 = Q_P(P_Q - Q_P) : \operatorname{im} Q \to \operatorname{im} Q,$$

i.e., $Q_P P_Q - 1$ is a compact operator in $\operatorname{im} Q$. It follows that Q_P is a Fredholm inverse of P_Q, and we see that both P_Q and Q_P are Fredholm operators. □

Let

$$\operatorname{ind}(P, Q) \tag{1.17}$$

denote the index of $P_Q : \operatorname{im} Q \to \operatorname{im} P$. We then have

$$\operatorname{ind}(P, Q) = -\operatorname{ind}(Q, P). \tag{1.18}$$

For every $L \in \operatorname{Vect}(T^*M \setminus 0)$ there is a $J \in \operatorname{Vect}(M)$ and a projection $p : \pi_Y^* J \to L$ with the property $p(x, \lambda, \xi) = p(x, \xi)$ for all $(x, \xi) \in T^*M \setminus 0$, $\lambda \in \mathbb{R}_+$.

It suffices to set $J = \mathbb{C}^N$ (the trivial bundle with fibre $\mathbb{C}^N$) for sufficiently large N.

Set

$$\mathcal{P}(M) := \{\mathbb{L} := (P, J, L) : \; P \in L^0_{\mathrm{cl}}(M; J, J), P^2 = P, \\ J \in \mathrm{Vect}(M), L = \mathrm{im}\, \sigma_\psi(P)\}. \tag{1.19}$$

Incidentally, the elements of $\mathcal{P}(M)$ will be called projection data on M.

EXAMPLE 1.19. (i) For every $J \in \mathrm{Vect}(M)$ we have

$$(\mathrm{id}, J, \pi_M^* J) \in \mathcal{P}(M).$$

(ii) For $\mathbb{L} := (P, J, L)$, $\mathbb{M} := (Q, G, M) \in \mathcal{P}(M)$ we have

$$\mathbb{L} \oplus \mathbb{M} := (P \oplus Q, J \oplus G, L \oplus M) \in \mathcal{P}(M);$$

the direct sum for bundles is defined as usual. We then have $L^0_{\mathrm{cl}}(M, J \oplus G) = L^0_{\mathrm{cl}}(M, J) \oplus L^0_{\mathrm{cl}}(M, G)$, and $P \oplus Q$ in the latter space is again a pseudo-differential projection.

(iii) For every $\mathbb{L} := (P, J, L) \in \mathcal{P}(M)$ we have a complementary element $\mathbb{L}^\perp \in \mathcal{P}(M)$ in the sense that $L \oplus L^\perp = (\mathrm{id}, J, \pi_M^* J)$. In fact, it suffices to write $\pi_M^* J$ in the form $L \oplus L^\perp$ for a subbundle $L^\perp$ of $\pi_M^* J$ such that $L^\perp = (1 - P, J, L^\perp)$.

(iv) For $\mathbb{L} = (P, J, L)$ we can form an adjoint $\mathbb{L}^* := (P^*, J, L^*)$ by defining $P^* \in L^0_{\mathrm{cl}}(M; J, J)$ to be the formal adjoint of P, cf. Theorem 1.2 (iv), which is again a projection, and $L^* := \sigma_\psi(P^*)\pi_Y^* J$.

Every element $\mathbb{L} = (P, J, L) \in \mathcal{P}(M)$ gives rise to continuous operators

$$P : H^s(M, J) \to H^s(M, J),$$

$s \in \mathbb{R}$, and we set

$$P^s(M, \mathbb{L}) := P H^s(M, J) \tag{1.20}$$

which is a closed subspace of $H^s(M, J)$. In fact, $P^s(M, \mathbb{L})$ is equal to the kernel of $I - P : H^s(M, J) \to H^s(M, J)$.

PROPOSITION 1.20. *We have continuous embeddings*

$$P^{s'}(M, \mathbb{L}) \hookrightarrow P^s(M, \mathbb{L}) \tag{1.21}$$

for all $s' \geq s$ that are compact for $s' > s$.

PROOF. The inclusion in (1.21) is clear; the compactness follows from the fact that (1.21) may be written as a composition PC with the compact embedding $C : P^{s'}(M, \mathbb{L}) \to H^s(M, J)$ and the continuous projection $P : H^s(M, J) \to P^s(M, \mathbb{L})$. □

PROPOSITION 1.21. (i) *The space $P^\infty(M, \mathbb{L}) = \bigcap_{s \in \mathbb{R}} P^s(M, \mathbb{L})$ is dense in $P^s(M, \mathbb{L})$ for every $s \in \mathbb{R}$.*

(ii) *Let $P^0(M;\mathbb{L})$ be equipped with the scalar product from $H^0(M,J) = L^2(M,J)$, let $V \subset P^\infty(M,\mathbb{L})$ be a finite-dimensional subspace. Then the orthogonal projection $C_V : P^0(M,\mathbb{L}) \to V$ induces continuous operators $C_V : P^s(M,\mathbb{L}) \to V$ for all $s \in \mathbb{R}$, and C_V is compact as an operator $P^s(M,\mathbb{L}) \to P^s(M,\mathbb{L})$ for every $s \in \mathbb{R}$.*

PROOF. (i) Every $u \in P^s(M,\mathbb{L})$ can be written as $u = Pv$ for a $v \in H^s(M,J)$. Since $C^\infty(M,J)$ is dense in $H^s(M,J)$, there is a sequence $(v_n)_{n\in\mathbb{N}} \subset C^\infty(M,J)$ such that $v = \lim_{n\to\infty} v_n$ in $H^s(M,J)$. From the continuity of P we get $u = \lim_{n\to\infty} u_n$ for $u_n := Pv_n \in P^\infty(M,\mathbb{L})$, and hence $P^\infty(M,\mathbb{L})$ is dense in $P^s(M,\mathbb{L})$ for every s.

(ii) $V \subset C^\infty(M,J)$ is a finite-dimensional subspace. Hence, every projection $\widetilde{C}_V : H^s(M,J) \to V$ represents a compact operator $\widetilde{C}_V : H^s(M,J) \to H^s(M,J)$ for every $s \in \mathbb{R}$. The projection $C_V : P^s(M,\mathbb{L}) \to P^s(M,\mathbb{L})$ can be extended to a projection $\widetilde{C}_V := C_N P : H^s(M,J) \to H^s(M,J)$ that is a compact operator. Thus, also $\approx{C}_V := P\widetilde{C}_V : H^s(M,J) \to P^s(M,\mathbb{L})$ is compact. The operator C_N itself can be written as a composition $C_V = \approx{C}_V\, R$, where $R : P^s(M,\mathbb{L}) \to H^s(M,J)$ is the canonical embedding. Hence, also C_V is a compact operator. □

DEFINITION 1.22. Let $\mathbb{L}_+ := (P_+, J_+, L_+)$, $\mathbb{L}_- := (P_-, J_-, L_-) \in \mathcal{P}(M)$, and let $R_- : P^s(M,\mathbb{L}_-) \hookrightarrow H^s(M,J_-)$ denote the canonical embedding (given for every $s \in \mathbb{R}$). Then the composition

$$A := P_+\widetilde{A}R_- \tag{1.22}$$

for $\widetilde{A} \in L^\mu_{\mathrm{cl}}(M;J_-,J_+)$ is called a Toeplitz operator on M of order μ. Let $T^\mu(M;\mathbb{L}_-,\mathbb{L}_+)$ denote the set of all Toeplitz operators (1.22) associated with $\mathbb{L}_\pm \in \mathcal{P}(M)$. Set

$$T^{-\infty}(M;\mathbb{L}_-,\mathbb{L}_+) := \{P_+\widetilde{A}R_- : \widetilde{A} \in L^{-\infty}(M;J_-,J_+)\}.$$

REMARK 1.23. The space $T^\mu(M;\mathbb{L}_-,\mathbb{L}_+)$ can be identified with the quotient space $L^\mu_{\mathrm{cl}}(M;J_-,J_+)/\sim$, where $\widetilde{A}_1 \sim \widetilde{A}_2$ means $P_+\widetilde{A}_1P_- = P_+\widetilde{A}_2P_-$.

REMARK 1.24. Let $\widetilde{A} \in L^\mu_{\mathrm{cl}}(M;J_-,J_+)$ and form

$$\approx{A} := P_+\widetilde{A}P_- \in L^\mu_{\mathrm{cl}}(M;J_-,J_+).$$

Then we have $P_+\widetilde{A}R_- = P_+ \approx{A}\, R_-$.

REMARK 1.25. Let $\widetilde{A} \in L^\mu_{\mathrm{cl}}(M;J_-,J_+)$, and choose elements $\mathbb{L}_-, \mathbb{L}_-^\perp$ and $\mathbb{L}_+, \mathbb{L}_+^\perp$ in $\mathcal{P}(M)$ that are complementary in the abovementioned sense. Then, writing $\mathbb{L}_\pm = (P_\pm, J_\pm, L_\pm)$ and $\mathbb{L}_\pm^\perp = (P_\pm^\perp, J_\pm, L_\pm^\perp)$, we can decompose $\widetilde{A}$ into a block matrix operator

$$\begin{pmatrix} P_-\widetilde{A}R_+ & P_+\widetilde{A}R_-^\perp \\ P_+^\perp\widetilde{A}R_- & P_+^\perp\widetilde{A}R_-^\perp \end{pmatrix} : \begin{matrix} P^s(M,\mathbb{L}_-) \\ \oplus \\ P^s(M,\mathbb{L}_-^\perp) \end{matrix} \to \begin{matrix} P^{s-\mu}(M,\mathbb{L}_+) \\ \oplus \\ P^{s-\mu}(M,\mathbb{L}_+^\perp) \end{matrix},$$

where $R_- : P^s(Y, \mathbb{L}_-) \to H^s(Y, J_-)$ and $R_-^\perp : P^s(Y, \mathbb{L}_-^\perp) \to H^s(Y, J_-)$ are the corresponding embedding operators.

Applying that for an operator of the form

$$\widetilde{A} := P_+ \widetilde{A}_1 P_- + P_+^\perp \widetilde{A}_2 P_-^\perp \in L_{\mathrm{cl}}^\mu(M; J_-, J_+)$$

for any $\widetilde{A}_1, \widetilde{A}_2 \in L_{\mathrm{cl}}^\mu(M; J_-, J_+)$ we obtain $\widetilde{A}$ in the form

$$\operatorname{diag}(P_+ \widetilde{A}_1 P_-, P_+^\perp \widetilde{A}_2 P_-^\perp) \in L_{\mathrm{cl}}^\mu(M; J_-, J_+). \tag{1.23}$$

Toeplitz operators have been studied in the literature in many variants, see, for instance, Boutet de Monvel [**8**]. Note that

$$P_+ \widetilde{A} R_- \in \mathcal{T}^{-\infty}(M; \mathbb{L}_-, \mathbb{L}_+) \quad \Leftrightarrow \quad P_+ \widetilde{A} P_- \in L^{-\infty}(M; \mathbb{L}_-, \mathbb{L}_+). \tag{1.24}$$

For every $A \in \mathcal{T}^\mu(M; \mathbb{L}_-, \mathbb{L}_+)$ written in the form (1.22) we have a bundle morphism

$$\sigma_\psi(A) : L_- \to L_+ \tag{1.25}$$

defined by the composition $\sigma_\psi(A)(x,\xi) := p_+(x,\xi)\sigma_\psi(\widetilde{A})(x,\xi)r_-(x,\xi)$, where $p_+(x,\xi) = \sigma_\psi(P_+)(x,\xi)$, while $r_-(x,\xi) : L_{-,(x,\xi)} \to (\pi_M^* J_-)(x,\xi)$ is the canonical embedding of fibres on (x,ξ). We then have $\sigma_\psi(A)(x, \lambda\xi) = \lambda^\mu \sigma_\psi(A)(x,\xi)$ for all $(x,\xi) \in T^*M \setminus 0$, $\lambda \in \mathbb{R}_+$.

Let $S^{(\mu)}(T^*M \setminus 0; L_-, L_+)$ for $L_\pm \in \operatorname{Vect}(T^*M \setminus 0)$ denote the set of all bundle morphisms $p_{(\mu)} : L_- \to L_+$ such that $p_{(\mu)}(x, \lambda\xi) = \lambda^\mu p_{(\mu)}(x,\xi)$ for all $(x,\xi) \in T^*M \setminus 0$; $\lambda \in \mathbb{R}_+$ (we hope this notation does not cause confusion in connection with $S^{(\mu)}(T^*M \setminus 0; E, F)$ for $E, F \in \operatorname{Vect}(M)$ which is an abbreviation of $S^{(\mu)}(T^*M \setminus 0; \pi_M^* E, \pi_M^* F)$).

PROPOSITION 1.26. *The principal symbolic map*

$$\sigma_\psi : \mathcal{T}^\mu(M; \mathbb{L}_-, \mathbb{L}_+) \to S^{(\mu)}(T^*M \setminus 0; L_-, L_+) \tag{1.26}$$

is surjective, and there is a linear map

$$\operatorname{op} : S^{(\mu)}(T^*M \setminus 0; L_-, L_+) \to \mathcal{T}^\mu(M; \mathbb{L}_-, \mathbb{L}_+) \tag{1.27}$$

such that $\sigma_\psi \circ \operatorname{op} = \operatorname{id}$.

PROOF. The map (1.26) directly follows from (1.6) when we represent elements of $\mathcal{T}^\mu(M; \mathbb{L}_-, \mathbb{L}_+)$ as $P_+ \widetilde{A} P_-$ and write $\sigma_\psi(P_+ \widetilde{A} R_-)$ as $\sigma_\psi(P_+ \widetilde{A} P_-)|_{L_-}$, interpreted as a morphism (1.25). To see that (1.26) is surjective we choose a morphism $a_{(\mu)} : \pi_M^* J_- \to \pi_M^* J_+$ which restricts to a given element $t_{(\mu)} \in S^{(\mu)}(T^*M \setminus 0; L_-, L_+)$ (this is always possible), then form an operator $\widetilde{A} := \operatorname{op}(a_{(\mu)}) \in L_{\mathrm{cl}}^\mu(M; J_-, J_+)$ such that $\sigma_\psi(\widetilde{A}) = a_{(\mu)}$, cf. formula (1.7), and finally set $A := P_+ \widetilde{A} R_-$; then $\sigma_\psi(A) = t_{(\mu)}$. □

PROPOSITION 1.27. *Every* $A \in \mathcal{T}^\mu(M; \mathbb{L}_-, \mathbb{L}_+)$ *induces continuous operators*

$$A : P^s(M, \mathbb{L}_-) \to P^{s-\mu}(M, \mathbb{L}_+) \tag{1.28}$$

for all $s \in \mathbb{R}$.

This is an immediate consequence of the definition and of Theorem 1.4 (i).

PROPOSITION 1.28. *Let $A \in T^{\mu}(M; \mathbb{L}_-, \mathbb{L}_+)$ and suppose $\sigma_\psi(A) = 0$. Then $A \in T^{\mu-1}(M; \mathbb{L}_-, \mathbb{L}_+)$, and the operator* (1.28) *is compact for every $s \in \mathbb{R}$.*

PROOF. We have $P_+\widetilde{A}P_- \in L^{\mu}_{\mathrm{cl}}(M; J, J)$ and $\sigma_\psi(P_+\widetilde{A}P_-) = 0$. Hence $P_+\widetilde{A}P_- : H^s(M, J_-) \to H^{s-\mu}(M, J_+)$ is a compact operator, cf. Theorem 1.4 (ii). Thus $P_+(P_+\widetilde{A}P_-) = P_+\widetilde{A}P_- : H^s(M, J_-) \to P^{s-\mu}(M, \mathbb{L}_+)$ is also compact. Finally, since $R_- : P^s(M, \mathbb{L}_-) \to H^s(M, J_-)$ is continuous, also the operator $P_+\widetilde{A}P_-R_- = P_+\widetilde{A}R_- : P^s(M, \mathbb{L}_-) \to P^{s-\mu}(M, \mathbb{L}_+)$ is compact. These conclusions hold for all $s \in \mathbb{R}$. □

THEOREM 1.29. *$A \in T^{\mu}(M; \mathbb{L}_0, \mathbb{L}_+)$, $B \in T^{\nu}(M; \mathbb{L}_-, \mathbb{L}_0)$ for $\mathbb{L}_-, \mathbb{L}_0, \mathbb{L}_+ \in \mathcal{P}(M)$ implies $AB \in T^{\mu+\nu}(M; \mathbb{L}_-, \mathbb{L}_+)$, and we have*

$$\sigma_\psi(AB) = \sigma_\psi(A)\sigma_\psi(B).$$

PROOF. Writing $\mathbb{L}_\pm = (P_\pm, J_\pm, L_\pm)$, $\mathbb{L}_0 = (P_0, J_0, L_0)$, we have $A = P_+\widetilde{A}R_0$, $B = P_0\widetilde{B}R_-$ for certain $\widetilde{A} \in L^{\mu}_{\mathrm{cl}}(M; J_0, J_+)$, $\widetilde{B} \in L^{\nu}_{\mathrm{cl}}(M; J_-, J_0)$, with obvious meaning of R_0, R_-. Then $AB = P_+\widetilde{A}R_0P_0\widetilde{B}R_- = P_-\widetilde{A}P_0\widetilde{B}R_-$. Since $\widetilde{A}P_0\widetilde{B} \in L^{\mu+\nu}_{\mathrm{cl}}(M; J_-, J_+)$, cf. Theorem 1.4 (iii), we obtain $AB \in T^{\mu+\nu}(M; \mathbb{L}_-, \mathbb{L}_+)$. The symbolic rule is a consequence of relation (1.9). □

Given an operator $A \in T^{\mu}(M; \mathbb{L}_-, \mathbb{L}_+)$, $A = P_+\widetilde{A}R_-$, we define the formal adjoint

$$A^* := P_-^*\widetilde{A}^*R_+^*,$$

where $\widetilde{A}^*, P_-^*$ are the formal adjoints in the sense of Theorem 1.4 (iv), and $R_+^* : P^s(M; \mathbb{L}_+^*) \to H^s(M, J_+)$ is the canonical embedding.

THEOREM 1.30. *$A \in T^{\mu}(M; \mathbb{L}_-, \mathbb{L}_+)$ entails*

$$A^* \in T^{\mu}(M; \mathbb{L}_+^*, \mathbb{L}_-^*)$$

and

$$\sigma_\psi(A^*) = \sigma_\psi(A)^*,$$

where the adjoint on the right hand side refers to the Hermitian metrics in the bundles $L_+^* := \sigma_\psi(P_+^*)\pi_Y^*J_+$ and $L_-^* := \sigma_\psi(P_-^*)\pi_Y^*J_-$, induced by $\pi_Y^*J_+$ and $\pi_Y^*J_-$, respectively.

PROOF. It suffices to apply Theorem 1.4 (iv) and Definition 1.22. □

DEFINITION 1.31. Let $A \in T^{\mu}(M; \mathbb{L}_-, \mathbb{L}_+)$, $\mu \in \mathbb{R}$, $\mathbb{L}_\pm = (P_\pm, J_\pm, L_\pm) \in \mathcal{P}(M)$.

(i) The operator A is said to be elliptic (of order μ), if $\sigma_\psi(A) : L_- \to L_+$ is an isomorphism.

(ii) An operator $B \in T^{-\mu}(M; \mathbb{L}_+, \mathbb{L}_-)$ is called a parametrix of A, if B satisfies the following relations:

$$C_l := I - BA \in T^{-\infty}(M; \mathbb{L}_-, \mathbb{L}_-), \qquad C_r := I - AB \in T^{-\infty}(M; \mathbb{L}_+, \mathbb{L}_+), \tag{1.29}$$

where I denotes the corresponding identity operators.

REMARK 1.32. For every $\mathbb{L} := (P, J, L) \in \mathcal{P}(M)$ and every $\mu \in \mathbb{R}$ there exists an elliptic operator $R^\mu_{\mathbb{L}} \in T^\mu(M; \mathbb{L}, \mathbb{L})$.

In fact, let $a_{(\mu)} \in S^{(\mu)}(T^*M \setminus 0; J, J)$ be the unique element that restricts to the identity map on $\pi_1^* J$, with $\pi_1 : S^*M \to M$ being the canonical projection of the unit cosphere bundle S^*M to M. Set $\widetilde{A} := \mathrm{op}(a_{(\mu)})$, cf. the formula (1.7). Then $P\widetilde{A}R$ for the embedding $R : P^s(M, \mathbb{L}) \to H^s(M, J)$ is elliptic because $\sigma_\psi(P\widetilde{A}R) : L \to L$ is an isomorphism.

THEOREM 1.33. *Let $A \in T^\mu(M; \mathbb{L}_-, \mathbb{L}_+)$, $\mu \in \mathbb{R}$, $\mathbb{L}_\pm = (P_\pm, J_\pm, L_\pm) \in \mathcal{P}(M)$.*

(i) *The operator A is elliptic (of order μ) if and only if*

$$A : P^s(M, \mathbb{L}_-) \to P^{s-\mu}(M, \mathbb{L}_+) \tag{1.30}$$

is a Fredholm operator for some $s = s_0 \in \mathbb{R}$.

(ii) *If $\mathcal{A}$ is elliptic,* (1.30) *is Fredholm for all $s \in \mathbb{R}$, and* $\dim\ker \mathcal{A}$ *and* $\dim\operatorname{coker} \mathcal{A}$ *are independent of s.*

(iii) *An elliptic operator $A \in T^\mu(M; \mathbb{L}_-, \mathbb{L}_+)$ has a parametrix B in the space $T^{-\mu}(M; \mathbb{L}_+, \mathbb{L}_-)$, and B can be chosen in such a way that the remainders in the relation* (1.29) *are projections*

$$C_l : P^s(M, \mathbb{L}_-) \to V, \qquad C_l : P^{s-\mu}(M, \mathbb{L}_+) \to W$$

for all $s \in \mathbb{R}$, for $V :=$ $\ker A \subset P^\infty(M, \mathbb{L}_-)$ and a finite-dimensional subspace $W \subset P^\infty(M, \mathbb{L}_+)$ with the property $W + \operatorname{im} A = P^{s-\mu}(M, \mathbb{L}_+)$ and $W \cap \operatorname{im} A = \{0\}$ for every $s \in \mathbb{R}$.

The proof of Theorem 1.33 will be given below. First observe that for arbitrary $\mathbb{L}_\pm = (P_\pm, J_\pm, L_\pm) \in \mathcal{P}(M)$ there exist elements $\mathbb{M}_\pm = (Q_\pm, \mathbb{C}^m, L_\pm) \in \mathcal{P}(M)$ such that

$$T^\mu(M; \mathbb{L}_-, \mathbb{L}_+) = T^\mu(M; \mathbb{M}_-, \mathbb{M}_+). \tag{1.31}$$

In fact, for m sufficiently large, $J_\pm$ may be represented as subbundles of the trivial bundle $\mathbb{C}^m$, with complementary bundles $J_\pm^\perp$ and projections $\pi_M^* \mathbb{C}^m \to \pi_M^* J_\pm$ along $\pi_M^* J_\pm^\perp$ and associated pseudo-differential projections

$$\widetilde{P}_\pm \in L^0_{\mathrm{cl}}(M; \mathbb{C}^m; \mathbb{C}^m), \quad \widetilde{P}_\pm : H^s(M, \mathbb{C}^m) \to H^s(M, J_\pm).$$

Moreover, the compositions of projections

$$\sigma_\psi(P_\pm)\sigma_\psi(\widetilde{P}_\pm) := \sigma_\psi(Q_\pm) : \pi_M^* \mathbb{C}^m \to \pi_M^* J_\pm \to L_\pm$$

are again projections, and $Q_\pm := P_\pm \widetilde{P}_\pm \in L^0_{\mathrm{cl}}(M; \mathbb{C}^m, \mathbb{C}^m)$ are associated pseudo-differential projections, $Q_\pm : H^s(M, \mathbb{C}^m) \to \mathcal{P}^s(M, \mathbb{L}_\pm)$. Relation (1.31) is then obvious when we use that every $\widetilde{A} \in L^\mu_{\rm cl}(M; J_-, J_+)$ can be identified with an element $\widetilde{\widetilde{A}} \in L^\mu_{\rm cl}(M; \mathbb{C}^m, \mathbb{C}^m)$ by $\widetilde{\widetilde{A}} = \widetilde{A}\widetilde{P}_-$.

PROPOSITION 1.34. *Let $A \in T^\mu(M; \mathbb{L}_-, \mathbb{L}_+)$ be an elliptic operator, and let us represent A as an element $A \in T^\mu(M; \mathbb{M}_-, \mathbb{M}_+)$ for $\mathbb{M}_\pm := (Q_\pm, \mathbb{C}^m, L_\pm)$ for a sufficiently large m. Then there exists an elliptic operator $A^\perp \in T^\mu(M; \mathbb{M}^\perp_-, \mathbb{M}^\perp_+)$ for suitable $\mathbb{M}^\perp_\pm \in \mathcal{P}(M)$ such that $A \oplus A^\perp \in L^\mu(M; \mathbb{C}^m, \mathbb{C}^m)$ is elliptic in the sense of Definition* 1.5 (i).

PROOF. Let us write the bundles $L_\pm$ as subbundles of the trivial bundle $\mathbb{C}^m$ on $T^*M \setminus 0$ in such a way that $L_{-,(x,\xi)} \cap L_{+,(x,\xi)} = \{0\}$ for every $(x,\xi) \in T^*M \setminus 0$. We then have a $G \in \mathrm{Vect}(T^*M \setminus 0)$ such that $L_- \oplus L_+ \oplus G = \mathbb{C}^m$. Let us define an isomorphism

$$(L_- \oplus L_+ \oplus G)|_{S^*M} \to (L_+ \oplus L_- \oplus G)|_{S^*M} \tag{1.32}$$

by $\mathrm{diag}(\sigma_\psi(A)|_{S^*M}, \sigma_\psi^{-1}(A)|_{S^*M}, \mathrm{id}_{G|_{S^*M}})$, set $L^\perp_- := L_+ \oplus G$, $L^\perp_+ := L_- \oplus G$, and define $\sigma_\psi(A^\perp) \in S^{(\mu)}(T^*M \setminus 0; L^\perp_-, L^\perp_+)$ to be the unique element such that $\sigma_\psi(A^\perp) = \mathrm{diag}(\sigma_\psi^{-1}(A)|_{S^*M}, \mathrm{id}_{G|_{S^*M}})$. Moreover, set

$$\mathbb{M}^\perp_\pm := (Q^\perp_\pm, \mathbb{C}^m, L^\perp_\pm).$$

By definition, the operator A has the form $A = Q_+\widetilde{A}_1R_-$ for some element $\widetilde{A}_1 \in L^\mu_{\rm cl}(M; \mathbb{C}^m, \mathbb{C}^m)$, where $R_- : P^s(M, \mathbb{M}_-) \to H^s(M, \mathbb{C}^m)$ is the canonical embedding. Moreover, let $\sigma_\psi(\widetilde{A}_2) \in S^{(\mu)}(T^*M \setminus 0; \mathbb{C}^m, \mathbb{C}^m)$ denote the unique element such that $\sigma_\psi(\widetilde{A}_2)|_{S^*M}$ coincides with (1.32). Then, applying (1.7) to $\sigma_\psi(\widetilde{A}_2)$, we obtain an associated elliptic operator $\widetilde{A}_2 \in L^\mu_{\rm cl}(M; \mathbb{C}^m, \mathbb{C}^m)$ with $R^\perp_- : P^s(M, \mathbb{M}^\perp_-) \to H^s(M, \mathbb{C}^m)$ being the canonical embedding, where $\sigma_\psi(A^\perp)$ coincides with the abovementioned symbol. The operator

$$\widetilde{A} := A \oplus A^\perp = \mathrm{diag}(Q_+\widetilde{A}_1R_-, Q^\perp_+\widetilde{A}_2R^\perp_-)$$

can be identified with

$$Q_+\widetilde{A}_1P_- + Q^\perp_+\widetilde{A}_2P^\perp_- : H^s(M, \mathbb{C}^m) \to H^{s-\mu}(M, \mathbb{C}^m),$$

i.e., as a sum of standard pseudo-differential operators, cf. the formula (1.23), and hence belongs to $L^\mu_{\rm cl}(M; \mathbb{C}^m, \mathbb{C}^m)$, and $\widetilde{A}$ is elliptic, since $\sigma_\psi(\widetilde{A}) = \sigma_\psi(\widetilde{A}_2)$. □

PROPOSITION 1.35. *Let $A_j \in T^{\mu-j}(M; \mathbb{L}_-, \mathbb{L}_+)$, $j \in \mathbb{N}$, be an arbitrary sequence. Then there exists an $A \in T^\mu(M; \mathbb{L}_-, \mathbb{L}_+)$ such that*

$$A - \sum_{j=0}^{N} A_j \in T^{\mu-(N+1)}(M; \mathbb{L}_-, \mathbb{L}_+) \tag{1.33}$$

for every $N \in \mathbb{N}$, and A is unique mod $T^{-\infty}(M; \mathbb{L}_-, \mathbb{L}_+)$.

PROOF. By definition, every A_j has the form $A_j = P_+\widetilde{A}_j R_-$ for suitable $\widetilde{A}_j \in L^{\mu-j}_{\mathrm{cl}}(M;J_-,J_+)$. A well known result on standard pseudo-differential operators says that there is an $\widetilde{A} \in L^{\mu}_{\mathrm{cl}}(M;J_-,J_+)$ such that $\tilde{A} - \sum_{j=0}^{N}\tilde{A}_j \in L^{\mu-(N+1)}_{\mathrm{cl}}(M;J_-,J_+)$ for every $N \in \mathbb{N}$, and $\widetilde{A}$ is unique mod $L^{-\infty}(M;J_-,J_+)$. We set $A := P_+\widetilde{A}R_-$; then the relation (1.33) holds. Moreover, if an element $A' \in T^{\mu}(M;\mathbb{L}_-,\mathbb{L}_+)$ satisfies (1.33) with A' in place of A, we have $A - A' = P_+(\widetilde{A}-\widetilde{A}')R_-$, where $P_+(\widetilde{A}-\widetilde{A}')P_- \in \bigcap_{j\in\mathbb{N}} L^{\mu-j}_{\mathrm{cl}}(M;J_-,J_+) = L^{-\infty}(M;J_-,J_+)$, and hence, $A - A' \in T^{-\infty}(M;\mathbb{L}_-,\mathbb{L}_+)$, by virtue of (1.24). □

PROOF OF THEOREM 1.33. First assume that $A \in T^{\mu}(M;\mathbb{L}_-,\mathbb{L}_+)$ is elliptic. According to the relation (1.31), without loss of generality we may assume $\mathbb{L}_\pm = (P_\pm,\mathbb{C}^m,L_\pm)$ for some sufficiently large m. Applying Proposition 1.34 we find an elliptic element $A^\perp \in T^{\mu}(M;\mathbb{L}_+^\perp,\mathbb{L}_-^\perp)$ such that $\widetilde{A} = A \oplus A^\perp$ is elliptic in $L^{\mu}_{\mathrm{cl}}(M;\mathbb{C}^m,\mathbb{C}^m)$. We now choose a parametrix $\widetilde{P} \in L^{-\mu}_{\mathrm{cl}}(M;\mathbb{C}^m,\mathbb{C}^m)$ in the sense of Theorem 1.7 (iii) and set $B_0 := P_-\widetilde{P}R_+ \in T^{-\mu}(M;\mathbb{L}_+,\mathbb{L}_-)$. Writing $A = P_-\widetilde{A}R_-$ we obtain

$$\begin{aligned} C_l^0 &:= I - B_0A = I - P_-\widetilde{P}R_+P_+\widetilde{A}R_- \\ &= I - P_-\widetilde{P}P_+\widetilde{A}R_- \end{aligned}$$

which yields $\sigma_\psi(C_l^0) = 0$, i.e., $C_l^0 \in T^{-1}(M;\mathbb{L}_-,\mathbb{L}_-)$. In a similar manner it follows that $C_r^0 := I - AB_0 \in T^{-1}(M;\mathbb{L}_+,\mathbb{L}_+)$.

By virtue of Proposition 1.28 the operators $C_l^0 \in \mathcal{L}(P^s(M,\mathbb{L}_-))$ and $C_r^0 \in \mathcal{L}(P^{s-\mu}(M,\mathbb{L}_+))$ are compact. Thus (1.30) is a Fredholm operator for every $s \in \mathbb{R}$. Applying a formal Neumann series argument we find a $K \in T^{-1}(M;\mathbb{L}_-,\mathbb{L}_-)$ such that $K \sim -\sum_{j=1}^{\infty}(C_l^0)^j$, cf. Proposition 1.35.

For $B_l := (I-K)B_0 \in T^{-\mu}(M;\mathbb{L}_+,\mathbb{L}_-)$ we then obtain $C_l := I - B_lA \in T^{-\infty}(M;\mathbb{L}_-,\mathbb{L}_-)$. In an analogous manner we find a $B_r \in T^{-\mu}(M;\mathbb{L}_+,\mathbb{L}_-)$ such that $C_r := I - AB_r \in T^{-\infty}(M;\mathbb{L}_-,\mathbb{L}_-)$. A standard algebraic argument shows that B_l is a two-sided parametrix of A. In fact, $B_lA = I - C_l^1$ implies $AB_lA = A - AC_l$ and $AB_lAB_r = AB_l(I-C_r) = AB_r - AC_l^1B_r$; then Theorem 1.29 and relation $AB_r^1 = I \bmod T^{-\infty}(M;\mathbb{L}_+,\mathbb{L}_+)$ yield $AB_l^1 = I \bmod T^{-\infty}(M;\mathbb{L}_+,\mathbb{L}_+)$. Propositions 1.20 and 1.21 together with the first part of the proof allow us to apply Remark 1.8 to the present situation. This gives us the assertions of Theorem 1.33 (ii), (iii).

It remains to show the second part of Theorem 1.33 (i), namely that the Fredholm property of (1.30) for an $s = s_0 \in \mathbb{R}$ implies the ellipticity of A. Without loss of generality we may assume $s_0 = \mu = 0$. In fact, Remark 1.24 gives us elliptic operators

$$R_{\mathbb{L}_-}^{-s_0} : P^0(M,\mathbb{L}_-) \to P^{s_0}(M,\mathbb{L}_-),\, R_{\mathbb{L}_+}^{s_0-\mu} : P^{s_0-\mu}(M,\mathbb{L}_+) \to P^0(M,\mathbb{L}_+).$$

These are Fredholm operators by the first part of the proof of Theorem 1.33. Thus, also $A_0 : R_{\mathbb{L}_+}^{s_0-\mu}AR_{\mathbb{L}_-}^{-s_0} : P^0(M,\mathbb{L}_-) \to P^0(M,\mathbb{L}_+)$ is Fredholm. If we show the ellipticity of A_0, we obtain at once the ellipticity of A, because ellipticity remains preserved under compositions. In other words, we may consider the case $A := A_0$,

$$A : P^0(M, \mathbb{L}_-) \to P^0(M, \mathbb{L}_+). \tag{1.34}$$

Moreover, let

$$E : P^0(M, \mathbb{L}_-^\perp) \to P^0(M, \mathbb{L}_-^\perp), \tag{1.35}$$

be the identity operator, cf. the notation in Remark 1.25.

We have

$$L^2(M, J_-) = P^0(M, \mathbb{L}_-) \oplus P^0(M, \mathbb{L}_-^\perp), \tag{1.36}$$

and there are continuous embeddings

$$R_+ : P^0(M, \mathbb{L}_+) \to L^2(M, J_+), \qquad R_-^\perp : P^0(M, \mathbb{L}_-^\perp) \to L^2(M, J_-). \tag{1.37}$$

Then we can pass to the operator

$$B := \begin{pmatrix} R_+ & 0 \\ 0 & R_-^\perp \end{pmatrix} \begin{pmatrix} A & 0 \\ 0 & E \end{pmatrix} : L^2(M, J_-) \to L^2(M, J_+ \oplus J_-)$$

which is an element of $L^0_{\mathrm{cl}}(M; J_-, J_+ \oplus J_-)$. By assumption, the operator (1.34) is Fredholm. In particular, there is an operator $Q : P^0(M, \mathbb{L}_+) \to P^0(M, \mathbb{L}_-)$ such that $I - QA : P^0(M, \mathbb{L}_-) \to P^0(M, \mathbb{L}_-)$ is compact.

Let $S : L^2(M, J_+ \oplus J_-) \to P^0(M, \mathbb{L}_+) \oplus P^0(M, \mathbb{L}_-^\perp)$ denote a projection. Then $T := \mathrm{diag}(Q, E) \circ S$ has the property that $I - TB =: K$ is compact in $L^2(M, J_-)$. Since $I - K$ is a Fredholm operator, we have $\dim \ker(I - K) < \infty$, and then $\dim \ker B < \infty$, since $Bu = 0$ implies $TBu = 0$ which yields $(I - K)u = 0$, i.e., $\ker B \subseteq \ker(I - K)$.

The operator $\widetilde{B} := B^*B : L^2(M, J_-) \to L^2(M, J_-)$ belongs to $L^0_{\mathrm{cl}}(M, J_-)$ and is self-adjoint and Fredholm.

From Theorem 1.6 (i) we know that $\widetilde{B}$ is elliptic. It follows that $\sigma_\psi(B)$ is injective. Hence, also $\sigma_\psi(A)$ is injective. By passing to adjoint operators, in an analogous manner we can show that $\sigma_\psi(A)$ is also surjective.

This completes the proof of Theorem 1.33. □

We now discuss the question to what extent the specific choice of projections $P_\pm$ in an elliptic operator $A \in T^\mu(M; \mathbb{L}_-, \mathbb{L}_+)$ may affect the index. This is an aspect on more general properties of Fredholm operators and projections in Hilbert spaces, see, for instance, [**45**].

Let H_+ and H_- be Hilbert spaces, and let $P_+, Q_+ \in \mathcal{L}(H_+)$ and $P_-, Q_- \in \mathcal{L}(H_-)$ be continuous projections, such that $P_\pm - Q_\pm$ are compact operators in $H_\pm$.

THEOREM 1.36. *Let $\widetilde{A} \in \mathcal{L}(H_-, H_+)$ such that*

$$A := P_+ \widetilde{A} : \operatorname{im} P_- \to \operatorname{im} P_+$$

is a Fredholm operator. Then, also

$$B := Q_+\widetilde{A} : \operatorname{im} Q_- \to \operatorname{im} Q_+$$

is a Fredholm operator, and we have

$$\operatorname{ind} A - \operatorname{ind} B = \operatorname{ind}(P_-, Q_-) - \operatorname{ind}(P_+, Q_+), \tag{1.38}$$

cf. the notation (1.17).

PROOF. From Proposition 1.18 we know that

$$P_- : \operatorname{im} Q_- \to \operatorname{im} P_- \quad \text{and} \quad Q_+ : \operatorname{im} P_+ \to \operatorname{im} Q_+$$

are Fredholm operators, and Q_- is a Fredholm inverse of P_-. Thus $Q_-P_- : \operatorname{im} Q_- \to \operatorname{im} Q_-$ is a Fredholm operator of index zero. An analogous observation holds for the projections Q_+, P_+. Thus, the operator

$$D : \operatorname{im} Q_- \xrightarrow{P_-} \operatorname{im} P_- \xrightarrow{A} \operatorname{im} P_+ \xrightarrow{Q_+} \operatorname{im} Q_+$$

is Fredholm of index

$$\operatorname{ind} D = \operatorname{ind} A + \operatorname{ind}(P_-, Q_-) - \operatorname{ind}(P_+, Q_+), \tag{1.39}$$

cf. the relation (1.18). We have

$$\begin{aligned} D &= (Q_+P_+)B(Q_-P_-) \\ &- Q_+[P_+, Q_+]\widetilde{A}Q_-P_- + Q_+P_+\widetilde{A}(1 - Q_-)P_-, \end{aligned}$$

where $[P_+, Q_+]$ is the commutator in H_+ which is compact, since

$$\begin{aligned} [P_+, Q_+] &= P_+Q_+ - Q_+P_+ \\ &= (P_+ - Q_+)(1 - Q_+ - P_+). \end{aligned}$$

Moreover, $(1 - Q_-)P_- = (P_- - Q_-)P_- : H_- \to H_-$ is compact. Hence the operator $(Q_+P_+)B(Q_-P_-) - D$ is compact, i.e., $(Q_+P_+)B(Q_-P_-)$ is Fredholm, and we have

$$\operatorname{ind} D = \operatorname{ind}(Q_+P_+)B(Q_-P_-). \tag{1.40}$$

By virtue of Proposition 1.18 the operators $Q_-P_- : \operatorname{im} Q_- \to \operatorname{im} Q_-$ and $Q_+P_+ : \operatorname{im} Q_+ \to \operatorname{im} Q_+$ are Fredholm and of index zero. Therefore, we have $\operatorname{ind} B = \operatorname{ind}(Q_+P_+)B(Q_-P_-)$, and the assertion is a consequence of the relations (1.40) and (1.39). □

COROLLARY 1.37. *Consider elliptic operators $A \in T^\mu(M; \mathbb{L}_-, \mathbb{L}_+)$ for $\mathbb{L}_\pm := (P_\pm, J_\pm, L_\pm)$ and $B \in T^\mu(M; \mathbb{N}_-, \mathbb{N}_+)$ for $\mathbb{N}_\pm := (Q_\pm, J_\pm, L_\pm)$, and assume that $\sigma_\psi(A) = \sigma_\psi(B)$. Then the Fredholm indices of A and B as operators*

$$A : P^s(M, \mathbb{L}_-) \to P^{s-\mu}(M, \mathbb{L}_+), \qquad B : P^s(M, \mathbb{N}_-) \to P^{s-\mu}(M, \mathbb{N}_+)$$

are related by the formula (1.38).

REMARK 1.38. Given projection data $\mathbb{L} := (P, J, L)$ and $\mathbb{M} := (Q, J, L) \in \mathcal{P}(M)$ (with the same J, L but different projections), the operators

$$(1.41) \qquad P : P^s(M, \mathbb{M}) \to P^s(M, \mathbb{L}), \qquad Q : P^s(M, \mathbb{L}) \to P^s(M, \mathbb{N})$$

are Fredholm, and Q is a Fredholm inverse of P (and vice versa), cf. Proposition 1.18. We have

$$P \in T^0(M; \mathbb{M}, \mathbb{L}), \qquad Q \in T^0(M; \mathbb{L}, \mathbb{M}),$$

P and Q are elliptic of order zero, and hence, in particular, the Fredholm indices (as well as the dimensions of kernel and cokernel) of (1.41) are independent of s.

REMARK 1.39. Let $A_0, A_1 \in T^\mu(M; \mathbb{L}_-, \mathbb{L}_+)$ be elliptic, and assume that the principal symbols

$$\sigma_\psi(A_i) : L_- \to L_+$$

coincide for $i = 0, 1$. Then we have

$$\operatorname{ind} A_0 = \operatorname{ind} A_1.$$

In fact, Proposition 1.28 gives us $\sigma_\psi(A_0 - A_1) = 0$, i.e.,

$$A_0 - A_1 \in T^{\mu-1}(M; \mathbb{L}_-, \mathbb{L}_+),$$

and hence A_0 is equal to A_1 modulo a compact operator.

Let us now assume that

$$\mathbb{L}_\pm(t) := (P_\pm(t), J, L_\pm(t)), \quad 0 \le t \le 1,$$

is a family of elements in $\mathcal{P}(M)$, where

$$P_\pm(\cdot) \in C([0,1], L^0_{\mathrm{cl}}(M; J, J))$$

are families of projections, such that

$$L_\pm(t) = \sigma_\psi(P_\pm(t))\pi_M^* J_\pm$$

are families of subbundles in $\pi_M^* J_\pm$. Let us assume that

$$a_{(\mu)}(t) : L_-(t) \to L_+(t)$$

is a continuous family of isomorphisms, smooth in $(x, \xi) \in T^*M \setminus 0$ and homogeneous of degree μ. We can complete $a_{(\mu)}(t)$ to a continuous family of morphisms

$$\widetilde{a}_{(\mu)} : \pi_M^* J_- \to \pi_M^* J_+$$

such that $a_{(\mu)}(t)$ can be identified with $\sigma_\psi(P_+(t))\widetilde{a}_{(\mu)}(t)\sigma_\psi(P_-(t))$ for every t. Let us set $\widetilde{A}_t := \operatorname{op}(\widetilde{a}_{(\mu)}(t))$, cf. Proposition 1.3, which gives us an element of $C([0,1], L^\mu_{\mathrm{cl}}(M; J_-, J_+))$.

We then obtain a family

$$A_t := P_+(t)\widetilde{A}(t)R_-(t) \in T^\mu(M; \mathbb{L}_-(t), \mathbb{L}_+(t))$$

with $R_-(t) : P_-(t)H^s(M, J_-) \to H^s(M, J_-)$ being the canonical embeddings. The operators A_t are elliptic for all $t \in [0, 1]$.

THEOREM 1.40. *Under the conditions on* $A_t \in T^\mu(M; \mathbb{L}_-(t), \mathbb{L}_+(t))$, $0 \leq t \leq 1$, *mentioned above we have*

$$\operatorname{ind} A_0 = \operatorname{ind} A_1,$$

where the index of A_t *refers to the Fredholm operator*

$$A_t : P_-(t)H^s(M, J_-) \to P_+(t)H^{s-\mu}(M, J_+).$$

Theorem 1.40 has a more general functional analytic background. The following considerations up to Remark 1.43 have been contributed to this paper by Thomas Krainer from the University of Potsdam.

THEOREM 1.41. *Let* H *and* $\widetilde{H}$ *be Hilbert spaces, and consider families of operators* $(A_t)_{0\leq t\leq 1} \in C([0,1], \mathcal{L}(H, \widetilde{H}))$,

$$(P_t)_{0\leq t\leq 1} \in C([0,1], \mathcal{L}(\widetilde{H})) \quad \text{and} \quad (Q_t)_{0\leq t\leq 1} \in C([0,1], \mathcal{L}(H)).$$

Assume $P_t^2 = P_t$, $Q_t^2 = Q_t$ *for all* $t \in [0,1]$. *Moreover, let*

$$P_t A_t Q_t : \operatorname{im} Q_t \to \operatorname{im} P_t$$

be a Fredholm operator for every $t \in [0,1]$. *Then we have*

$$\operatorname{ind}(P_0 A_0 Q_0 : \operatorname{im} Q_0 \to \operatorname{im} P_0) = \operatorname{ind}(P_1 A_1 Q_1 : \operatorname{im} Q_1 \to \operatorname{im} P_1).$$

To prove this theorem we first show another result. Consider the set

$$\Pi(\widetilde{H}) \times \mathcal{L}(H, \widetilde{H}) \times \Pi(H), \tag{1.42}$$

where $\Pi(H)$ for a Hilbert space H denotes the set of all $P \in \mathcal{L}(H)$ such that $P^2 = P$. Let $\Phi_k(H, \widetilde{H})$ denote the set of all triples (P, A, Q) in (1.42) such that

$$PAQ : \operatorname{im} Q \to \operatorname{im} P$$

is a Fredholm operator of index k.

PROPOSITION 1.42. *For every* $k \in \mathbb{Z}$ *the set* $\Phi_k(H, \widetilde{H})$ *is open in* (1.42).

PROOF. As is well known, the set of Fredholm operators of index k between Hilbert spaces L and $\widetilde{L}$ is open in $\mathcal{L}(L, \widetilde{L})$. Applying this to $L := \operatorname{im} Q$ and $\widetilde{L} := \operatorname{im} P$, it follows that for a triple $(P, A, Q) \in \Phi_k(H, \widetilde{H})$ there exists an $\varepsilon_0 > 0$ such that $(P, A + K, Q) \in \Phi_k(H, \widetilde{H})$ for every $K \in \mathcal{L}(H, \widetilde{H})$, $\|K\| < \varepsilon_0$.

We now prove that for every

$$(P, A, Q) \in \Phi_k(H, \widetilde{H}) \tag{1.43}$$

there exist constants $\alpha > 0$, $\varepsilon > 0$, $\beta > 0$, such that

$$(M, B, N) \in \Pi(\widetilde{H}) \times \mathcal{L}(H, \widetilde{H}) \times \Pi(H)$$

and

$$\|M - P\| < \alpha, \ \|B - A\| < \varepsilon, \ \|N - Q\| < \beta$$

imply

$$(M, B, N) \in \Phi_k(H, \widetilde{H}). \tag{1.44}$$

Let $G \in \mathcal{L}(H), \widetilde{G} \in \mathcal{L}(\widetilde{H})$ be invertible elements, such that

$$\|\widetilde{G} - I\| < \delta_1, \qquad \|G - I\| < \delta_2 \tag{1.45}$$

for sufficiently small $\delta_1, \delta_2 < 1$. Set

$$M := \widetilde{G}P\widetilde{G}^{-1}, \qquad N := GQG^{-1}.$$

We prove that relation (1.44) holds for ε and δ_1, δ_2 so small that

$$\frac{1+\delta_2}{1-\delta_1}\varepsilon + \frac{\delta_1+\delta_2}{1-\delta_1}\|A\| < \varepsilon_0 \tag{1.46}$$

holds. From Neumann series arguments we obtain

$$\|\widetilde{G}^{-1}\| < \frac{1}{1-\delta_1}, \qquad \|\widetilde{G}^{-1} - I\| < \frac{\delta_1}{1-\delta_1} \tag{1.47}$$

and

$$\|G\| < 1 + \delta_2. \tag{1.48}$$

We now reformulate the operator MBN as follows:

$$MBN = \widetilde{G}P(A+K)QG^{-1} = \widetilde{G}\{PKQ + PAQ\}G^{-1} \tag{1.49}$$

for

$$K := \widetilde{G}^{-1}(B-A)G + (\widetilde{G}^{-1} - I)A + A(G-I) + (\widetilde{G}^{-1} - I)A(G-I).$$

Let us verify that $\|K\| < \varepsilon_0$. In fact, using (1.45), (1.46), (1.47), (1.48), we obtain

$$\begin{aligned} \|K\| \ &\leq \ \|\widetilde{G}^{-1}\| \, \|B-A\| \, \|G\| \\ &+ \ \|\widetilde{G}^{-1} - I\| \, \|A\| + \|A\| \, \|G-I\| \\ &+ \ \|\widetilde{G}^{-1} - I\| \, \|A\| \, \|G-I\| \\ &\leq \ \frac{1}{1-\delta_1}\varepsilon(1+\delta_2) + \frac{\delta_1}{1-\delta_1}\|A\| + \delta_2\|A\| + \frac{\delta_1}{1-\delta_1}\delta_2\|A\| < \varepsilon_0. \end{aligned}$$

Thus, from the first part of the proof it follows that $(P, A+K, Q) \in \Phi_k(H, \widetilde{H})$. Moreover, (1.49) together with the isomorphisms

$$\widetilde{G} : \operatorname{im} P \to \operatorname{im} M, \qquad G : \operatorname{im} N \to \operatorname{im} Q$$

gives us the relation (1.49).

Let us finally consider the map

$$s : \Pi(\widetilde{H}) \to \mathcal{L}(\widetilde{H}),$$

defined by $s(M) := MP + (I-M)(I-P)$. The map s is continuous, and we have $s(P) = I$. Let us choose $\widetilde{\delta}_1 > 0$ in such a way that $\|s(M) - s(P)\| < \delta_1$ holds when $\|M - P\| < \widetilde{\delta}_1$. Since $\delta_1 < 1$ is very small, we have the inverse $s(M)^{-1} \in \mathcal{L}(\widetilde{H})$ for $\|M - P\| < \widetilde{\delta}_1$. We then obtain $Ms(M) = MP = s(MP)$, i.e., $M = \widetilde{G}P\widetilde{G}^{-1}$ for $\widetilde{G} := s(M)$. In a similar manner it follows that $\|N - Q\| < \widetilde{\delta}_2$ for a suitable small $\widetilde{\delta}_2 > 0$ implies $N = GQG^{-1}$ for an invertible $G \in \mathcal{L}(H)$, $\|G - I\| < \delta_2$. This completes the proof of Proposition 1.42. □

PROOF OF THEOREM 1.41. By virtue of Proposition 1.42 the map $[0,1] \to \mathbb{Z}$ defined by $t \to \operatorname{ind} P_t A_t Q_t$ is continuous and hence constant.

□

REMARK 1.43. Theorem 1.41 can be generalised to the case when $[0,1]$ is replaced by any connected topological space X. It follows that the index of Fredholm operators $P_x A_x Q_x : \operatorname{im} Q_x \to \operatorname{im} P_x$ which are continuously parametrised by $x \in X$, is constant.

1.2. Operators with the transmission property at the boundary. In this section we prepare some necessary material on symbols and operators with the transmission property at the boundary and establish a connection with operators on the boundary with operator-valued symbols.

Let $\Omega \subseteq \mathbb{R}^q \ni y$ be open, and let $U := \Omega \times \mathbb{R} \ni x = (y,t), \xi = (\eta, \tau)$.

Set

$$H^s_{\mathrm{loc}(y)}(U) := \{u \in H^s_{\mathrm{loc}}(U) \quad : \quad \varphi u \in H^s(\mathbb{R}^{q+1}) \text{ for every } \varphi(y) \in C_0^\infty(\Omega)\},$$

and define

$$H^s_{\mathrm{comp}(y)}(U) := \{u \in H^s_{\mathrm{loc}}(U) \quad : \quad \psi u \in H^s_{\mathrm{comp}}(\mathbb{R}^{n+1}) \text{ for every } \psi(t) \in C_0^\infty(\mathbb{R})\}.$$

Moreover, set

$$\begin{aligned} H^s_{\mathrm{loc}(y)}(\Omega \times \mathbb{R}_\pm) &:= H^s_{\mathrm{loc}(y)}(\Omega \times \mathbb{R})|_{\Omega \times \mathbb{R}_\pm}, \\ H^s_{\mathrm{comp}(y)}(\Omega \times \mathbb{R}_\pm) &:= H^s_{\mathrm{comp}(y)}(\Omega \times \mathbb{R})|_{\Omega \times \mathbb{R}_\pm}. \end{aligned}$$

DEFINITION 1.44. A symbol $a(x,\xi) \in S^\mu_{\mathrm{cl}}(U \times \mathbb{R}^{q+1})$ for $\mu \in \mathbb{Z}$ is said to have the transmission property at $t = 0$, if the homogeneous components $a_{(\mu-j)}(x,\xi)$ satisfy the following relations:

$$D_t^k D_\eta^\alpha \{a_{(\mu-j)}(y,t,\eta,\tau) - (-1)^{\mu-j} a_{(\mu-j)}(y,t,-\eta,-\tau)\} = 0$$

for $y \in \Omega, t = 0$, $\eta = 0$, $\tau \in \mathbb{R} \setminus \{0\}$, for all $k \in \mathbb{N}$, $\alpha \in \mathbb{N}^q$, and for all $j \in \mathbb{N}$.

Let $S^\mu_{\mathrm{cl}}(U \times \mathbb{R}^{q+1})_{\mathrm{tr}}$ denote the space of all symbols $a(x,\xi) \in S^\mu_{\mathrm{cl}}(U \times \mathbb{R}^{q+1})$ with the transmission property at $t = 0$. Moreover, set

$$S^\mu_{\mathrm{cl}}(\Omega \times \overline{\mathbb{R}}_\pm \times \mathbb{R}^{q+1})_{\mathrm{tr}} := \{a|_{\Omega \times \overline{\mathbb{R}}_\pm \times \mathbb{R}^{q+1}} : a \in S^\mu_{\mathrm{cl}}(U \times \mathbb{R}^{q+1})_{\mathrm{tr}}\}.$$

REMARK 1.45. $S^\mu_{\mathrm{cl}}(U \times \mathbb{R}^{q+1})_{\mathrm{tr}}$ is a closed subspace of $S^\mu_{\mathrm{cl}}(U \times \mathbb{R}^{q+1})$; analogously, $S^\mu_{\mathrm{cl}}(\Omega \times \overline{\mathbb{R}}_\pm \times \mathbb{R}^{q+1})_{\mathrm{tr}}$ is closed in $S^\mu_{\mathrm{cl}}(\Omega \times \overline{\mathbb{R}}_\pm \times \mathbb{R}^{q+1})$.

To illustrate the structure of the transmission property we want to have a look at the one-dimensional case, say on $\overline{\mathbb{R}}_+ \times \mathbb{R}$ (the case $\overline{\mathbb{R}}_- \times \mathbb{R}$ is analogous; the transmission property is an invariant condition with respect to the reflection map $t \longrightarrow -t$).

DEFINITION 1.46. Let $a(t,\tau) \in S^{\mu}_{\mathrm{cl}}(\overline{\mathbb{R}}_+ \times \mathbb{R})$, $\mu \in \mathbb{Z}$, and write the component $a_{(\mu-j)}(t,\tau)$ of $a(t,\tau)$ of homogeneity $\mu - j$ in $\tau \neq 0$ in the form

$$a_{(\mu-j)}(t,\tau) = \{\Theta^+(\tau)a^+(t) + \Theta^-(\tau)a_j^-(t)\}\tau^{\mu-j},$$

$j \in \mathbb{N}$; here $\Theta^{\pm}$ is the characteristic function of $\mathbb{R}_{\pm} \ni \tau$. The symbol $a(t,\tau)$ is said to have the transmission property at $t = 0$, written $a(t,\tau) \in S^{\mu}_{\mathrm{cl}}(\overline{\mathbb{R}}_+ \times \mathbb{R})_{\mathrm{tr}}$, if

$$D_t^k a_j^+(t)|_{t=0} = D_t^k a_j^-(t)|_{t=0} \quad \text{for all} \quad j,k \in \mathbb{N}. \tag{1.50}$$

REMARK 1.47. Definition 1.46 can also be written as follows.
A symbol $a(t,\tau) \in S^{\mu}_{\mathrm{cl}}(\overline{\mathbb{R}}_+ \times \mathbb{R})$, $\mu \in \mathbb{R}$, belongs to $S^{\mu}_{\mathrm{cl}}(\overline{\mathbb{R}}_+ \times \mathbb{R})_{\mathrm{tr}}$ if and only if the coefficients $a_j^{\pm}(t) \in C^{\infty}(\overline{\mathbb{R}}_+)$ in the expansion

$$a(t,\tau) \sim \sum_{j=0}^{\infty} a_j^{\pm}(t)\tau^{\mu-j} \quad \text{for} \quad \tau \longrightarrow \pm\infty$$

satisfy the condition (1.50).

PROPOSITION 1.48. *Let* $a(y,t,\eta,\tau) \in S^{\mu}_{\mathrm{cl}}(\Omega \times \overline{\mathbb{R}}_+ \times \mathbb{R}^{q+1})$, *and set*

$$a_{y,\eta}(t,\tau) := a(y,t,\eta,\tau)$$

for every fixed $(y,\eta) \in \Omega \times \mathbb{R}^q$; *then we have* $a_{y,\eta}(t,\tau) \in S^{\mu}_{\mathrm{cl}}(\overline{\mathbb{R}}_+ \times \mathbb{R})$. *The following conditions are equivalent:*

(i) $a(y,t,\eta,\tau) \in S^{\mu}_{\mathrm{cl}}(\Omega \times \overline{\mathbb{R}}_+ \times \mathbb{R}^{q+1})_{\mathrm{tr}}$,
(ii) $a_{y,\eta}(t,\tau) \in S^{\mu}_{\mathrm{cl}}(\overline{\mathbb{R}}_+ \times \mathbb{R})_{\mathrm{tr}}$ *for every* $(y,\eta) \in \Omega \times \mathbb{R}^q$.

PROOF. Let us first assume the condition (i). Then the symbol $a_{y,\eta}(t,\tau)$ belongs to the space $S^{\mu}_{\mathrm{cl}}(\overline{\mathbb{R}}_+ \times \mathbb{R})_{\mathrm{tr}}$ in the sense of Definition 1.46 if and only if $(\chi(\eta,\tau)a_{(\mu-j)}(y,t,\eta,\tau))_{y,\eta}$ has this property for any excision function χ, for all $j \in \mathbb{N}$. Since χ is equal to 1 for $|\eta,\tau| >$ const for some constant > 0, it suffices to show that the symbol $(a_{(\mu-j)}(y,t,\eta,\tau))_{y,\eta}$ for $\eta \neq 0$ has this property. Let $\nu := \mu - j$ for any $j \in \mathbb{N}$. Let us consider, for simplicity, the case of symbols which are independent of (t,y); the general case is easy as well and left to the reader.

By assumption, we have

$$D_\eta^\alpha a_{(\nu)}(\eta,\tau)|_{\eta=0,\tau\neq 0} = D_\eta^\alpha((-1)^\nu a_{(\nu)}(-\eta,-\tau))|_{\eta=0,\tau\neq 0} \tag{1.51}$$

or, equivalently,

$$(D_\eta^\alpha a_{(\nu)})(0,1) = (-1)^{\nu+|\alpha|}(D_\eta^\alpha a_{(\nu)})(0,-1) \tag{1.52}$$

for all $\alpha \in \mathbb{N}^q$. Set $\rho := |\tau|^{-1}$, and consider the expansion of $a_{(\nu)}(\eta,\tau)$ for $\tau \longrightarrow \pm\infty$. We have

$$\begin{aligned}
a_{(\nu)}(\eta,\tau) &= a_{(\nu)}(\rho^{-1}\rho\eta,\rho^{-1}) = \rho^{-\nu}a_{(\nu)}(\rho\eta,1)\\
&\sim \rho^{-\nu}\sum_{\alpha}\frac{\rho^{|\alpha|}\eta^{\alpha}}{\alpha!}(D_{\eta}^{\alpha}a_{(\nu)})(0,1)\\
&= \sum_{k=0}^{\infty}\Big\{\sum_{|\alpha|=k}\frac{\eta^{\alpha}}{\alpha!}(D_{\eta}^{\alpha}a_{(\nu)})(0,1)\Big\}\tau^{\nu-k} \qquad \text{for} \qquad \tau>0,
\end{aligned}$$

and

$$\begin{aligned}
a_{(\nu)}(\eta,\tau) &= a_{(\nu)}(\rho^{-1}\rho\eta,-\rho^{-1}) = \rho^{-\nu}a_{(\nu)}(\rho\eta,-1)\\
&\sim \rho^{-\nu}\sum_{\alpha}\frac{\rho^{|\alpha|}\eta^{\alpha}}{\alpha!}(D_{\eta}^{\alpha}a_{(\nu)})(0,-1)\\
&= \sum_{k=0}^{\infty}\Big\{\sum_{|\alpha|=k}(-1)^{\nu+|\alpha|}\frac{\eta^{\alpha}}{\alpha!}(D_{\eta}^{\alpha}a_{(\nu)})(0,-1)\Big\}\tau^{\nu-k} \qquad \text{for} \qquad \tau<0.
\end{aligned}$$

By virtue of (1.52) we then obtain (i) $\Longrightarrow$ (ii). On the other hand, (ii) gives us immediately the relation (1.51), i.e., we have also proved (ii) $\Longrightarrow$ (i). □

Let $\mathrm{e}^{\pm}$ be the extension operators of functions on $\mathbb{R}_{\pm}$ by zero to the opposite sides (applied to elements in $H^{s}_{\mathrm{loc}(y)}(\Omega\times\mathbb{R}_{\pm})$ for $s>-\frac{1}{2}$). Moreover, let $\mathrm{r}^{\pm}$ denote the operator of restriction to $\Omega\times\mathbb{R}_{\pm}$ (applied to distributions in $\Omega\times\mathbb{R}$).

Set

$$\mathrm{Op}^{+}(a)u(x) := \mathrm{r}^{+}\,\mathrm{Op}(\widetilde{a})\mathrm{e}^{+}u(x) \tag{1.53}$$

for $a\in S^{\mu}_{\mathrm{cl}}(\Omega\times\overline{\mathbb{R}}_{+}\times\mathbb{R}^{q+1})_{\mathrm{tr}}$, $\widetilde{a}\in S^{\mu}_{\mathrm{cl}}(\Omega\times\mathbb{R}\times\mathbb{R}^{q+1})_{\mathrm{tr}}$, such that $\widetilde{a}|_{\Omega\times\overline{\mathbb{R}}_{+}\times\mathbb{R}^{q+1}}=a$. Clearly, (1.53) is then independent of the specific choice of $\widetilde{a}$. We form (1.53) first for $u\in C_0^{\infty}(\Omega\times\mathbb{R}_{+})$ and then for $u\in H^{s}_{\mathrm{comp}(y)}(\Omega\times\mathbb{R}_{+})$, $s>-\frac{1}{2}$, cf. Theorem 1.56 below. Similarly, we consider operator families on the half-axis

$$\mathrm{op}^{+}(a)(y,\eta) := \mathrm{r}^{+}\,\mathrm{op}(\widetilde{a})(y,\eta)\mathrm{e}^{+}, \tag{1.54}$$

where $\mathrm{op}(\widetilde{a})u(t) := \iint e^{i(t-t')\tau}\widetilde{a}(y,t,\eta,\tau)u(t')dt'đ\tau$, $\widetilde{a}\in S^{\mu}_{\mathrm{cl}}(\Omega\times\mathbb{R}\times\mathbb{R}^{q+1})_{\mathrm{tr}}$.

$\mathrm{op}^{+}(a)(y,\eta)$ will be regarded as an operator-valued symbol.

Let H be a Hilbert space, and let $\{\kappa_{\lambda}\}_{\lambda\in\mathbb{R}_{+}}$ be a strongly continuous group of isomorphisms $\kappa_{\lambda}: H\to H$, i.e., $\kappa_{\lambda}h\in C(\mathbb{R}_{+},H)$ for every $h\in H$, and $\kappa_{\lambda}\kappa_{\rho}=\kappa_{\lambda\rho}$ for all $\lambda,\rho\in\mathbb{R}_{+}$. To have a simple notation, we say that H is endowed with a group action. As is known, there are constants $c>0$ and M such that

$$\|\kappa_{\lambda}\|_{\mathcal{L}(H)} \le c(\max(\lambda,\lambda^{-1}))^{M} \qquad \text{for all} \quad \lambda\in\mathbb{R}_{+}. \tag{1.55}$$

More generally, if F is a Fréchet space, written as a projective limit of Hilbert spaces $F=\varprojlim_{k\in\mathbb{N}}H^{k}$ with continuous embeddings $H^{k+1}\hookrightarrow H^{k}$ for all k, and if H^{0} is endowed with a group action $\{\kappa_{\lambda}\}_{\lambda\in\mathbb{R}_{+}}$, such that $\{\kappa_{\lambda}|_{H^{k}}\}_{\lambda\in\mathbb{R}_{+}}$ is a group action on H^{k} for every k, we say that F is endowed with a group action.

EXAMPLE 1.49. (i) The space $H^s(\mathbb{R}_+)$ $(:= H^s(\mathbb{R})|_{\mathbb{R}_+})$ is endowed with the group action $(\kappa_\lambda u)(t) = \lambda^{\frac{1}{2}} u(\lambda t), \lambda \in \mathbb{R}_+$. The same is true of the spaces $\langle t\rangle^{-k} H^s(\mathbb{R}_+) := \{\langle t\rangle^{-k} u : u \in H^s(\mathbb{R}_+)\}$.

(ii) The Fréchet space $\mathcal{S}(\overline{\mathbb{R}}_+)$ $(:= \mathcal{S}(\mathbb{R})|_{\overline{\mathbb{R}}_+})$, written as

$$\mathcal{S}(\overline{\mathbb{R}}_+) = \varprojlim_{k\in\mathbb{N}} \langle t\rangle^{-k} H^k(\mathbb{R}_+),$$

is endowed with the group action defined as in (i).

DEFINITION 1.50. Let $\{H, \{\kappa_\lambda\}_{\lambda\in\mathbb{R}_+}\}$ and $\{\widetilde{H}, \{\widetilde{\kappa}_\lambda\}_{\lambda\in\mathbb{R}_+}\}$ be Hilbert spaces with group actions, moreover, let $\Omega \subseteq \mathbb{R}^p$ be open, and $\mu \in \mathbb{R}$.

(i) The space $S^\mu(\Omega \times \mathbb{R}^q; H, \widetilde{H})$ is defined to be the set of all $a(y,\eta) \in C^\infty(\Omega \times \mathbb{R}^q, \mathcal{L}(H, \widetilde{H}))$ such that

$$\|\widetilde{\kappa}^{-1}_{\langle\eta\rangle}\{D^\alpha_y D^\beta_\eta a(y,\eta)\}\kappa_{\langle\eta\rangle}\|_{\mathcal{L}(H,\widetilde{H})} \le c\langle\eta\rangle^{\mu-|\beta|}$$

for all $\alpha \in \mathbb{N}^p$, $\beta \in \mathbb{N}^q$ and $(y,\eta) \in K \times \mathbb{R}^q$ for all $K \subset\subset \Omega$, with constants $c = c(\alpha, \beta, K) > 0$. The elements of $S^\mu(\Omega \times \mathbb{R}^q; H, \widetilde{H})$ are called operator-valued symbols of order μ (associated with the given group actions).

(ii) $S^{(\mu)}(\Omega \times (\mathbb{R}^q \setminus \{0\}); H, \widetilde{H})$ denotes the space of all $a_{(\mu)}(y,\eta) \in C^\infty(\Omega \times (\mathbb{R}^q \setminus \{0\}), \mathcal{L}(H, \widetilde{H}))$ such that $a_{(\mu)}(y, \lambda\eta) = \lambda^\mu \widetilde{\kappa}_\lambda a_{(\mu)}(y,\eta)\kappa_\lambda^{-1}$ for all $\lambda \in \mathbb{R}_+$, $(y,\eta) \in \Omega \times (\mathbb{R}^q \setminus \{0\})$.

(iii) A symbol $a(y,\eta) \in S^\mu(\Omega \times \mathbb{R}^q; H, \widetilde{H})$ is said to be classical, if there are elements $a_{(\mu-j)}(y,\eta) \in S^{(\mu-j)}(\Omega \times (\mathbb{R}^q \setminus \{0\}); H, \widetilde{H})$, $j \in \mathbb{N}$, such that

$$a(y,\eta) - \sum_{j=0}^{N} \chi(\eta) a_{(\mu-j)}(y,\eta) \in S^{\mu-(N+1)}(\Omega \times \mathbb{R}^q; H, \widetilde{H})$$

for every $N \in \mathbb{N}$, where $\chi(\eta)$ is any excision function. Let $S^\mu_{\mathrm{cl}}(\Omega \times \mathbb{R}^q; H, \widetilde{H})$ denote the space of all classical symbols of order μ.

Let us extend this notation to the case of Fréchet spaces with group actions as follows: If $\widetilde{F} := \varprojlim_{j\in\mathbb{N}} \widetilde{H}^j$ is a Fréchet space with group action $\{\widetilde{\kappa}_\lambda\}_{\lambda\in\mathbb{R}_+}$, we set

$$S^\mu_{(\mathrm{cl})}(\Omega \times \mathbb{R}^q; H, \widetilde{F}) := \bigcap_{j\in\mathbb{N}} S^\mu_{(\mathrm{cl})}(\Omega \times \mathbb{R}^q; H, \widetilde{H}^j). \tag{1.56}$$

Moreover, if also $F = \varprojlim_{k\in\mathbb{N}} H^k$ is a Fréchet space endowed with a group action $\{\kappa_\lambda\}_{\lambda\in\mathbb{R}_+}$, we first fix a function $r : \mathbb{N} \to \mathbb{N}$ and set

$$S^\mu_{(\mathrm{cl})}(\Omega \times \mathbb{R}^q; F, \widetilde{F})_r := \bigcap_{j\in\mathbb{N}} S^\mu_{(\mathrm{cl})}(\Omega \times \mathbb{R}^q; H^{r(j)}, \widetilde{H}^j). \tag{1.57}$$

Then we define $S^{\mu}_{(\mathrm{cl})}(\Omega \times \mathbb{R}^q; F, \widetilde{F})$ to be the union of (1.57) over all r. Similarly, we have the spaces $S^{(\mu)}(\Omega \times (\mathbb{R}^q \setminus \{0\}); F, \widetilde{F})$.

THEOREM 1.51. *Let $\Omega \subseteq \mathbb{R}^q$ be an open set, and fix a symbol $a(y,t,\eta,\tau) \in S^{\mu}_{\mathrm{cl}}(\Omega \times \overline{\mathbb{R}}_+)_{\mathrm{tr}}$, $\mu \in \mathbb{Z}$, that is independent of t for $t > c$ for a constant $c > 0$.*

(i) We have

$$\mathrm{op}^+(a)(y,\eta) \in S^{\mu}(\Omega \times \mathbb{R}^q; H^s(\mathbb{R}_+), H^{s-\mu}(\mathbb{R}_+)) \tag{1.58}$$

for every real $s > -\frac{1}{2}$, and

$$\mathrm{op}^+(a)(y,\eta) \in S^{\mu}(\Omega \times \mathbb{R}^q; \mathcal{S}(\overline{\mathbb{R}}_+), \mathcal{S}(\overline{\mathbb{R}}_+)). \tag{1.59}$$

The operator-valued symbol $\mathrm{op}^+(a)(y,\eta)$ is classical, when the symbol a is independent of t.

(ii) Let $a_{(\mu-j)}(y,t,\eta,\tau) \in S^{(\mu-j)}(\Omega \times \overline{\mathbb{R}}_+ \times (\mathbb{R}^{q+1} \setminus \{0\}))$ be the homogeneous component of $a(y,t,\eta,\tau)$ of order $\mu - j$. Then we have

$$\mathrm{op}^+(a_{(\mu-j)}|_{t=0})(y,\eta) \in S^{(\mu-j)}(\Omega \times (\mathbb{R}^q \setminus \{0\}); H^s(\mathbb{R}_+), H^{s-\mu+j}(\mathbb{R}_+))$$

for every $s > -\frac{1}{2}$, and

$$\mathrm{op}^+(a_{(\mu-j)}|_{t=0})(y,\eta) \in S^{(\mu-j)}(\Omega \times (\mathbb{R}^q \setminus \{0\}); \mathcal{S}(\overline{\mathbb{R}}_+), \mathcal{S}(\overline{\mathbb{R}}_+)).$$

The technicalities to prove Theorem 1.51 can be found in [**39**].

We now turn to pseudo-differential operators that are associated with symbols $a(y,y',\eta) \in S^{\mu}_{(\mathrm{cl})}(\Omega \times \Omega \times \mathbb{R}^q; H, \widetilde{H})$ (for simplicity, generalities will be formulated for the case of Hilbert spaces $H, \widetilde{H}$ with group actions; the case with Fréchet spaces is analogous and will tacitly be used below).

Let

$$L^{\mu}_{(\mathrm{cl})}(\Omega; H, \widetilde{H}) := \{\mathrm{Op}(a) : a(y,y',\eta) \in S^{\mu}_{(\mathrm{cl})}(\Omega \times \Omega \times \mathbb{R}^q; H, \widetilde{H})\}, \tag{1.60}$$

where $\mathrm{Op}(a)u(y) := \iint e^{i(y-y')\eta} a(y,y',\eta)u(y')dy'đ\eta$, $đ\eta = (2\pi)^{-q}d\eta$.

DEFINITION 1.52. Let $\{H, \{\kappa_\lambda\}_{\lambda \in \mathbb{R}_+}\}$ be a Hilbert space with group action. Then $\mathcal{W}^s(\mathbb{R}^q, H)$ for $s \in \mathbb{R}$ is defined to be the completion of $\mathcal{S}(\mathbb{R}^q, H)$ (or, equivalently, of $C_0^\infty(\mathbb{R}^q, H)$) with respect to the norm

$$\|u\|_{\mathcal{W}^s(\mathbb{R}^q,H)} = \left\{ \int \langle\eta\rangle^{2s} \|\kappa^{-1}_{\langle\eta\rangle} \hat{u}(\eta)\|^2_H d\eta \right\}^{\frac{1}{2}},$$

($\hat{u}(\eta)$ denotes the Fourier transform of u with respect to $y \in \mathbb{R}^q$).

The spaces $\mathcal{W}^s(\mathbb{R}^q, H)$ have been introduced in [**34**] in connection with operators on manifolds with edges, see also [**33**], and their properties are studied, for instance, in [**35**], [**39**], [**21**], [**50**]. Let us summarise some results in the following theorems.

THEOREM 1.53. *Let H be a Hilbert space with group action $\{\kappa_\lambda\}_{\lambda \in \mathbb{R}_+}$, and let $s \in \mathbb{R}$.*

(i) *We have*

$$\mathcal{W}^s(\mathbb{R}^q, H) = \left\{ u \in \mathcal{S}'(\mathbb{R}^q, H) : \langle\eta\rangle^s \kappa^{-1}_{\langle\eta\rangle} \widehat{u}(\eta) \in L^2(\mathbb{R}^q, H) \right\}$$

where $\mathcal{S}'(\mathbb{R}^q, H) := \mathcal{L}(\mathcal{S}(\mathbb{R}^q), H)$. *The space* $\mathcal{W}^s(\mathbb{R}^q, H)$ *is a Hilbert space with the scalar product*

$$(u, v)_{\mathcal{W}^s(\mathbb{R}^q, E)} = \left(\langle\eta\rangle^s \kappa^{-1}_{\langle\eta\rangle} \widehat{u}(\eta), \langle\eta\rangle^s \kappa^{-1}_{\langle\eta\rangle} \widehat{u}(\eta) \right)_{L^2(\mathbb{R}^q, H)}.$$

(ii) *The operator of multiplication* $\mathcal{M}_\varphi$ *by a function* $\varphi \in \mathcal{S}(\mathbb{R}^q)$ *induces a continuous operator*

$$\mathcal{M}_\varphi : \mathcal{W}^s(\mathbb{R}^q, H) \to \mathcal{W}^s(\mathbb{R}^q, H),$$

and $\varphi \to \mathcal{M}_\varphi$ *represents a continuous operator* $\mathcal{S}(\mathbb{R}^q) \to \mathcal{L}(\mathcal{W}^s(\mathbb{R}^q, H))$.

(iii) $(\chi_\lambda u)(y) := \kappa_\lambda \lambda^{q/2} u(\lambda y)$ *for* $\lambda \in \mathbb{R}_+$, $u \in \mathcal{S}(\mathbb{R}^q, H)$, *extends to a group action on* $\mathcal{W}^s(\mathbb{R}^q, H)$, *and we have*

$$\mathcal{W}^s(\mathbb{R}^p, \mathcal{W}^s(\mathbb{R}^q, H)) = \mathcal{W}^s(\mathbb{R}^{p+q}, H), \tag{1.61}$$

where the space on the left of relation (1.61) *refers to* $\{\chi_\lambda\}_{\lambda \in \mathbb{R}_+}$ *on* $\mathcal{W}^s(\mathbb{R}^q, H)$ *and that on the right of* (1.61) *to* $\{\kappa_\lambda\}_{\lambda \in \mathbb{R}_+}$ *on* H.

(iv) *For* $H := H^s(\mathbb{R}^q)$ *with the group action* $(\kappa_\lambda u)(y) := \lambda^{q/2} u(\lambda y)$ *we have*

$$\mathcal{W}^s(\mathbb{R}^p, H^s(\mathbb{R}^q)) = H^s(\mathbb{R}^{p+q}).$$

Let us set for any open $\Omega \subseteq \mathbb{R}^q$, $s \in \mathbb{R}$,

$$\mathcal{W}^s_{\text{loc}}(\Omega, H) := \{ u \in \mathcal{D}'(\Omega, H) \ : \ \varphi u \in \mathcal{W}^s(\mathbb{R}^q, H) \text{ for every } \varphi \in C_0^\infty(\Omega) \} \tag{1.62}$$

and

$$\mathcal{W}^s_{\text{comp}}(\Omega, H) := \{ u \in \mathcal{W}^s_{\text{loc}}(\Omega, H) : \operatorname{supp} u \quad \text{compact} \}. \tag{1.63}$$

The space (1.62) is Fréchet, and (1.63) is an inductive limit of Fréchet spaces.

In particular, we have

$$H^s_{\text{loc}(y)}(\Omega \times \mathbb{R}) = \mathcal{W}^s_{\text{loc}}(\Omega, H^s(\mathbb{R})), H^s_{\text{comp}(y)}(\Omega \times \mathbb{R}) = \mathcal{W}^s_{\text{comp}}(\Omega, H^s(\mathbb{R})) \tag{1.64}$$

for any open set $\Omega \subseteq \mathbb{R}^q$, as well as

$$H^s_{\text{loc}(y)}(\Omega \times \mathbb{R}_\pm) = \mathcal{W}^s_{\text{loc}}(\Omega, H^s(\mathbb{R}_\pm)), H^s_{\text{comp}(y)}(\Omega \times \mathbb{R}_\pm) = \mathcal{W}^s_{\text{comp}}(\Omega, H^s(\mathbb{R}_\pm)). \tag{1.65}$$

THEOREM 1.54. *Let* $a(y, y', \eta) \in S^\mu(\Omega \times \Omega \times \mathbb{R}^q; H, \widetilde{H})$, $\Omega \subseteq \mathbb{R}^q$ *open*, $\mu \in \mathbb{R}$, *with* H *and* $\widetilde{H}$ *being Hilbert spaces with group actions. Then*

$$\operatorname{Op}(a)u(y) := \iint e^{i(y-y')\eta} a(y, y', \eta) u(y') dy' đ\eta$$

induces a continuous operator

$$\operatorname{Op}(a) : C_0^\infty(\Omega, H) \to C^\infty(\Omega, \widetilde{H})$$

which extends to continuous operators

$$\mathrm{Op}(a) : \mathcal{W}^s_{\mathrm{comp}}(\Omega, H) \to \mathcal{W}^{s-\mu}_{\mathrm{loc}}(\Omega, \widetilde{H})$$

for all $s \in \mathbb{R}$. *In particular, for* $a(\eta) \in S^\mu(\mathbb{R}^q; H, \widetilde{H})$ *we obtain continuous operators*

$$\mathrm{Op}(a) : \mathcal{W}^s(\mathbb{R}^q, H) \to \mathcal{W}^{s-\mu}(\mathbb{R}^q, \widetilde{H})$$

for all $s \in \mathbb{R}$.

REMARK 1.55. The notation (1.62) and (1.63) as well as the results of Theorems 1.53, 1.54 extend in a natural way to the case of Fréchet spaces H or $\widetilde{H}$ with group actions.

THEOREM 1.56. *Let* $a(y, t, \eta, \tau) \in S^\mu_{\mathrm{cl}}(\Omega \times \overline{\mathbb{R}}_+)_{\mathrm{tr}}$ *be a symbol which is independent of* t *for* $t > c$ *for some* $c > 0$. *Then* $\mathrm{Op}^+(a) := \mathrm{r}^+ \mathrm{Op}(\tilde{a})\mathrm{e}^+$ *(for any* $\tilde{a} \in S^\mu_{\mathrm{cl}}(\Omega \times \mathbb{R})_{\mathrm{tr}}$ *with* $a = \tilde{a}$ *for* $t \geq 0$*) induces continuous operators*

$$\mathrm{Op}^+(a) : H^s_{\mathrm{comp}(y)}(\Omega \times \mathbb{R}_+) \to H^{s-\mu}_{\mathrm{loc}(y)}(\Omega \times \mathbb{R}_+)$$

for all $s \in \mathbb{R}$, $s > -\frac{1}{2}$.

This result is a consequence of Theorem 1.51 (i), the relations (1.65), and Theorem 1.54.

Let X be a compact C^∞ manifold with boundary Y, and let $2X$ denote the double of X, obtained by gluing together two copies X_+ and X_- of X along their common boundary Y by the identity map; we then identify X with X_+. Moreover, let e^+ denote the operator of extension of functions on $\mathrm{int}\, X_+$ by zero to the opposite side X_-, and let r^+ denote the operator of restriction of distributions on $2X$ to $\mathrm{int}\, X_+$; analogously, we have operators e^- and r^- with respect to the minus-side of $2X$. Let $2X$ be equipped with a Riemannian metric that equals the product metric of $Y \times (-1, +1)$ in a neighbourhood of Y for some Riemannian metric on Y.

Given an $E \in \mathrm{Vect}(X)$ we fix any $\widetilde{E} \in \mathrm{Vect}(2X)$ such that $E = \widetilde{E}|_X$. For E and F in $\mathrm{Vect}(X)$ with fibre dimensions l and k, respectively, we now consider the space $L^\mu_{\mathrm{cl}}(2X; \widetilde{E}, \widetilde{F})$. For every chart $\chi : V \to U$ on $2X$, $U \subseteq \mathbb{R}^n$ open, and trivialisations $\widetilde{E}|_V \cong U \times \mathbb{C}^l$, $\widetilde{F}|_V \cong U \times \mathbb{C}^k$, the push-forward $\chi_* A$ of an operator $A \in L^\mu_{\mathrm{cl}}(2X; \widetilde{E}, \widetilde{F})$ belongs to $L^\mu_{\mathrm{cl}}(U; \mathbb{C}^l, \mathbb{C}^k)$. By notation, the push-forward χ_* also takes into account the chosen trivialisations of $\widetilde{E}|_V$ and $\widetilde{F}|_V$; for simplicity they are not explicity indicated (this should not cause confusion).

Let $V \cap Y \neq \emptyset$, $V := V' \times (-1, +1)$, where V' is a coordinate neighbourhood on the boundary Y, and assume that χ restricts to charts $\chi_\pm : V_\pm \to \Omega \times \overline{\mathbb{R}}_\pm$ on $X_\pm$, $V_\pm := X_\pm \cap V$, and to a chart $\chi' := \chi|_{V'} : V' \to \Omega$ on Y, $\Omega \subseteq \mathbb{R}^{n-1}$.

DEFINITION 1.57. The space $L^\mu_{\mathrm{cl}}(2X; \widetilde{E}; \widetilde{F})_{\mathrm{tr}}$ for $\mu \in \mathbb{Z}$ is defined to be the set of all elements $\widetilde{A} \in L^\mu_{\mathrm{cl}}(2X; \widetilde{E}, \widetilde{F})$ such that for every $\varphi, \psi \in C^\infty_0(\Omega \times \mathbb{R})$ and $e_\xi := e^{ix\xi}$

$$\widetilde{a}(x, \xi) := e_{-\xi}\{\varphi(\chi_* A)\psi\}e_\xi$$

is an $k \times l$ matrix of elements in $S^{\mu}_{\mathrm{cl}}(\Omega \times \mathbb{R})_{\mathrm{tr}}$; here, χ is an arbitrary chart of the described kind.

Moreover, set

$$L^{\mu}_{\mathrm{cl}}(X; E, F)_{\mathrm{tr}} := \{\mathrm{r}^{+}\widetilde{A}\mathrm{e}^{+} : \widetilde{A} \in L^{\mu}_{\mathrm{cl}}(2X; \widetilde{E}, \widetilde{F})_{\mathrm{tr}}\}. \tag{1.66}$$

REMARK 1.58. $L^{\mu}_{\mathrm{cl}}(2X; \widetilde{E}, \widetilde{F})_{\mathrm{tr}}$ is a closed subspace of $L^{\mu}_{\mathrm{cl}}(2X; \widetilde{E}, \widetilde{F})$, cf. Remark 1.2. Moreover, the space

$$\{\widetilde{A} \in L^{\mu}_{\mathrm{cl}}(2X; \widetilde{E}, \widetilde{F})_{\mathrm{tr}} : \mathrm{r}^{+}\widetilde{A}\mathrm{e}^{+} = 0\} \tag{1.67}$$

is closed in $L^{\mu}_{\mathrm{cl}}(2X; \widetilde{E}, \widetilde{F})_{\mathrm{tr}}$, and we have

$$L^{\mu}_{\mathrm{cl}}(X; E, F)_{\mathrm{tr}} = L^{\mu}_{\mathrm{cl}}(2X; \widetilde{E}, \widetilde{F})/\sim, \tag{1.68}$$

where $/\sim$ indicates the quotient space with respect to (1.67). In this way, in the space $L^{\mu}_{\mathrm{cl}}(X; E, F)_{\mathrm{tr}}$ we obtain a natural Fréchet topology.

THEOREM 1.59. *Every $A \in L^{\mu}_{\mathrm{cl}}(X; E, F)_{\mathrm{tr}}$ induces continuous operators*

$$A : H^{s}(X, E) \to H^{s-\mu}(X, F)$$

for all $s \in \mathbb{R}$, $s > -\frac{1}{2}$.

PROOF. Let $\omega, \widetilde{\omega}$ be functions supported in a collar neighbourhood of Y such that $\widetilde{\omega} \equiv 1$ on $\operatorname{supp}\omega$. Then A can be written as $A = \omega A \widetilde{\omega} + (1-\omega)A(1-\widetilde{\omega}) + C$, where C has a smooth kernel up to the boundary. The continuity $C : H^{s}(X, E) \to C^{\infty}(X, F)$ for $s > -\frac{1}{2}$ is then evident, while the continuity of $(1-\omega)A(1-\widetilde{\omega}) : H^{s}(X, E) \to H^{s-\mu}(X, F)$ for all s is a consequence of Theorem 1.4 (i). The continuity of $\omega A \widetilde{\omega} : H^{s}(X, E) \to H^{s-\mu}(X, F)$ for $s > -\frac{1}{2}$ follows from Theorem 1.56 by a simple partition of unity argument and from Theorems 1.53, 1.54. □

Let us look at a special kind of order reducing symbols. Choose a function $\varphi \in \mathcal{S}(\mathbb{R})$ such that $\varphi(0) = 1$ and $\operatorname{supp} F^{-1}\varphi \subset \mathbb{R}_{-}$ (here F is the Fouier transform on the real line). Set

$$r^{\mu}_{-}(\eta, \tau) := \left(\varphi\Big(\frac{\tau}{C\langle\tau\rangle}\Big) - i\tau\right)^{\mu} \tag{1.69}$$

for any $\mu \in \mathbb{R}$ and a constant $C > 0$. We then have $r^{\mu}_{-}(\eta, \tau) \in S^{\mu}_{\mathrm{cl}}(\mathbb{R}^{n})$, and the symbol $r^{\mu}_{-}(\eta, \tau)$ is elliptic for a sufficiently large $C > 0$, cf. Grubb [**16**]. It is well known that

$$\mathrm{r}^{+}\operatorname{Op}(\mathrm{r}^{\mu}_{-})\mathrm{e}^{+}_{s} : H^{s}(\mathbb{R}^{n}_{+}) \to H^{s-\mu}(\mathbb{R}^{n}_{+}) \tag{1.70}$$

is continuous for every $s, \mu \in \mathbb{R}$, with $\mathrm{e}^{+}_{s} : H^{s}(\mathbb{R}^{n}_{+}) \to H^{s}(\mathbb{R}^{n})$ being a continuous extension operator, i.e., $\mathrm{r}^{+}\mathrm{e}^{+}_{s} = \mathrm{id}_{H^{s}(\mathbb{R}^{n}_{+})}$. Moreover, for $s > -\frac{1}{2}$ we may replace e^{+}_{s} by e^{+}, and (1.70) is independent of the choice of the specific extension operator. For $\mu \in \mathbb{Z}$ we have $r^{\mu}_{-}(\eta, \tau) \in S^{\mu}_{\mathrm{cl}}(\mathbb{R}^{n})_{\mathrm{tr}}$. Order reducing operators also may be formulated globally on a compact C^{∞} manifold X with boundary.

THEOREM 1.60. *The space* $L^{\mu}_{\mathrm{cl}}(X;E,E)_{\mathrm{tr}}$ *for* $E \in \mathrm{Vect}(X)$ *and arbitrary* $\mu \in \mathbb{Z}$ *contains an element* $R^{\mu}_{E} := \mathrm{r}^{+}\widetilde{R}^{\mu}_{\widetilde{E}}\mathrm{e}^{+}$ *for some* $\widetilde{R}^{\mu}_{\widetilde{E}} \in L^{\mu}_{\mathrm{cl}}(2X;\widetilde{E},\widetilde{E})_{\mathrm{tr}}$, $E = \widetilde{E}|_{X}$, *such that*

$$R^{\mu}_{E} : H^{s}(X,E) \to H^{s-\mu}(X,F)$$

is an isomorphism for every $s > -\frac{1}{2}$.

This result can be proved in different ways, cf. Boutet de Monvel [**7**], or Grubb [**16**]; the latter paper employs order reducing symbols of the kind (1.69). Such symbols for arbitrary $\mu \in \mathbb{R}$ have been used for an analogous order reducing result for the case without the transmission property in the author's joint paper with Harutjunjan [**20**]. Other constructions in this direction (more general in different ways) may be found in Duduchava and Speck [**10**]. A completely different method to reduce orders in general boundary value problems is given in [**43**], based on ideas of the edge pseudo-differential calculus.

REMARK 1.61. The operators in [**20**] are obtained in a slightly more general form, namely as $\mathrm{r}^{+}\widetilde{R}^{\mu}_{\widetilde{E}}\mathrm{e}^{+}$ for operators $\widetilde{R}^{\mu}_{\widetilde{E}} \in L^{\mu}_{\mathrm{cl}}(M;\widetilde{E},\widetilde{E})$ on any closed compact C^{∞} manifold M which contains $X =: X_{+}$ and $X_{-} := M \setminus \mathrm{int}\, X_{+}$ as submanifolds with common boundary Y, with an evident generalisation of the meaning of operators r^{+}, e^{+} (and $\mathrm{r}^{-}, \mathrm{e}^{-}$ with respect to the minus side). We can also construct an operator $\widetilde{S}^{\mu}_{\widetilde{E}}$ of similar meaning for X_{-}, such that

$$\mathrm{r}^{-}\widetilde{S}^{\mu}_{\widetilde{E}}\mathrm{e}^{-} : H^{s}(X_{-},E|_{X_{-}}) \to H^{s-\mu}(X_{-},E|_{X_{-}})$$

are isomorphisms for $s > -\frac{1}{2}$. The local symbols of $\widetilde{S}^{\mu}_{\widetilde{E}}$ in a collar neighbourhood of Y are just $\mathrm{r}^{\mu}_{+}(y,\eta) := \overline{\mathrm{r}^{\mu}_{+}(y,\eta)}$, the complex conjugate of (1.69).

Given an operator $A \in L^{\mu}_{\mathrm{cl}}(X;E,F)_{\mathrm{tr}}$, $A := \mathrm{r}^{+}\widetilde{A}\mathrm{e}^{+}$ for an $\widetilde{A} \in L^{\mu}_{\mathrm{cl}}(2X;\widetilde{E},\widetilde{F})_{\mathrm{tr}}$, cf. the formula (1.66), we have the homogeneous principal symbol $\sigma_{\psi}(\widetilde{A}) : \pi^{*}_{2X}\widetilde{E} \to \pi^{*}_{2X}\widetilde{F}$, cf. (1.5), and we set $\sigma_{\psi}(A) := \sigma_{\psi}(\widetilde{A})|_{T^{*}X\setminus 0}$,

$$\sigma_{\psi}(A) : \pi^{*}_{X}E \to \pi^{*}_{X}F, \tag{1.71}$$

$\pi_{X} : T^{*}X \setminus 0 \to X$. With (1.71) we associate a family of operators

$$\begin{aligned} \sigma_{\partial}(A)(y,\eta) &:= \mathrm{r}^{+}\sigma_{\psi}(A)(y,0,\eta,D_{t})\mathrm{e}^{+} \\ &:= \mathrm{r}^{+}\,\mathrm{op}(\sigma_{\psi}(A)|_{t=0})(y,\eta)\mathrm{e}^{+} \end{aligned} \tag{1.72}$$

for $(y,\eta) \in T^{*}Y \setminus 0$. This gives us a bundle morphism

$$\sigma_{\partial}(A) : \pi^{*}_{Y}E' \otimes H^{s}(\mathbb{R}_{+}) \to \pi^{*}_{Y}F' \otimes H^{s-\mu}(\mathbb{R}_{+})$$

for every fixed $s \in \mathbb{R}$, $s > -\frac{1}{2}$, $\pi_{Y} : T^{*}Y \setminus 0 \to Y$, where $E' := E|_{Y}$, $F' := F|_{Y}$. Alternatively, we consider $\sigma_{\partial}(A)$ as a family of maps

$$\sigma_{\partial}(A) : \pi^{*}_{Y}E' \otimes \mathcal{S}(\overline{\mathbb{R}}_{+}) \to \pi^{*}_{Y}F' \otimes \mathcal{S}(\overline{\mathbb{R}}_{+}), \tag{1.73}$$

$\mathcal{S}(\overline{\mathbb{R}}_+) := \mathcal{S}(\mathbb{R})|_{\overline{\mathbb{R}}_+}$. We call $\sigma_\psi(A)$ the principal interior symbol, $\sigma_\partial(A)$ the principal boundary symbol of the operator A.

REMARK 1.62. For $(\kappa_\lambda u)(t) := \lambda^{\frac{1}{2}} u(\lambda t)$, $\lambda \in \mathbb{R}_+$, we have

$$\sigma_\partial(A)(y, \lambda\eta) = \lambda^\mu \kappa_\lambda \sigma_\partial(A)(y, \eta) \kappa_\lambda^{-1} \tag{1.74}$$

for all $\lambda \in \mathbb{R}_+$.

We now pass to the algebra of boundary value problems on X with trace and potential conditions, cf. [7]. First, for $E, F \in \mathrm{Vect}(X)$, $J_-, J_+ \in \mathrm{Vect}(Y)$, $\boldsymbol{v} := (E, F; J_-, J_+)$, by $\mathcal{B}^{-\infty,0}(X; \boldsymbol{v})$ we denote the space of all

$$\mathcal{G} := \begin{pmatrix} G & K \\ T & Q \end{pmatrix} : \begin{matrix} C^\infty(X, E) \\ \oplus \\ C^\infty(Y, J_-) \end{matrix} \longrightarrow \begin{matrix} C^\infty(X, F) \\ \oplus \\ C^\infty(Y, J_+) \end{matrix}$$

such that $\mathcal{G}$ and $\mathcal{G}^*$ extend to continuous operators

$$\mathcal{G} : \begin{matrix} H^s(X, E) \\ \oplus \\ H^s(Y, J_-) \end{matrix} \longrightarrow \begin{matrix} C^\infty(X, F) \\ \oplus \\ C^\infty(Y, J_+) \end{matrix}, \qquad \mathcal{G}^* : \begin{matrix} H^s(X, F) \\ \oplus \\ H^s(Y, J_+) \end{matrix} \longrightarrow \begin{matrix} C^\infty(X, E) \\ \oplus \\ C^\infty(Y, J_-) \end{matrix}$$

for all $s \in \mathbb{R}$, $s > -\frac{1}{2}$. Here $\mathcal{G}^*$ is the formal adjoint of $\mathcal{G}$ in the sense

$$(u, \mathcal{G}^* v)_{L^2(X,E) \oplus L^2(Y,J_-)} = (\mathcal{G} u, v)_{L^2(X,F) \oplus L^2(Y,J_+)} \tag{1.75}$$

for all $u \in C^\infty(X, E) \oplus C^\infty(Y, J_-)$, $v \in C^\infty(X, F) \oplus C^\infty(Y, J_+)$; the L^2 - scalar products refer to the chosen Riemannian metrics on X and Y and to the Hermitian metrics in the respective vector bundles.

For every vector bundle $E \in \mathrm{Vect}(X)$ we fix a first order differential operator $T : C^\infty(X, E) \to C^\infty(X, E)$ that is equal to $\partial_t \otimes \mathrm{id}_E$ in a collar neighbourhood of the boundary, with t being the normal variable. Then $\mathcal{B}^{-\infty,d}(X; \boldsymbol{v})$ for $d \in \mathbb{N} \setminus \{0\}$ is defined to be the space of all operators

$$\mathcal{G} = \mathcal{G}_0 + \sum_{j=1}^{d} \mathcal{G}_j \,\mathrm{diag}(T^j, 0) \tag{1.76}$$

for arbitrary $\mathcal{G}_j \in \mathcal{B}^{-\infty,0}(X; \boldsymbol{v})$. The elements of $\mathcal{B}^{-\infty,d}(X; \boldsymbol{v})$ are called smoothing operators of type d.

We now turn to Green symbols that are operator-valued symbols in the sense of Definition 1.50 (iii) for Hilbert spaces

$$H = L^2(\mathbb{R}_+, \mathbb{C}^l) \oplus \mathbb{C}^{j_-}, \qquad \widetilde{H} = L^2(\mathbb{R}_+, \mathbb{C}^k) \oplus \mathbb{C}^{j_+}$$

or Schwartz subspaces, where κ_λ as well as $\widetilde{\kappa}_\lambda$ are defined by $u \oplus c \to \lambda^{\frac{1}{2}} u(\lambda t) \oplus c$, $\lambda \in \mathbb{R}_+$, for a vector-valued function $u(t)$ and a vector c of complex numbers.

DEFINITION 1.63. Let $l, k, j_-, j_+ \in \mathbb{N}$, $\mu \in \mathbb{R}$, and $\Omega \subseteq \mathbb{R}^q$ open. An element

$$g(y, \eta) \in S^\mu_{\mathrm{cl}}(\Omega \times \mathbb{R}^q; L^2(\mathbb{R}_+, \mathbb{C}^l) \oplus \mathbb{C}^{j_-}, L^2(\mathbb{R}_+, \mathbb{C}^k) \oplus \mathbb{C}^{j_+})$$

is called a Green symbol of order μ and type 0, if it has the properties

$$g(y,\eta) \in S^{\mu}_{\mathrm{cl}}(\Omega \times \mathbb{R}^q; L^2(\mathbb{R}_+, \mathbb{C}^l) \oplus \mathbb{C}^{j_-}, \mathcal{S}(\overline{\mathbb{R}}_+, \mathbb{C}^k) \oplus \mathbb{C}^{j_+})$$

and

$$g^*(y,\eta) \in S^{\mu}_{\mathrm{cl}}(\Omega \times \mathbb{R}^q; L^2(\mathbb{R}_+, \mathbb{C}^k) \oplus \mathbb{C}^{j_+}, \mathcal{S}(\overline{\mathbb{R}}_+, \mathbb{C}^l) \oplus \mathbb{C}^{j_-}),$$

where $g^*(y,\eta)$ denotes the pointwise adjoint in the sense

$$(u, g^*(y,\eta)v)_{L^2(\mathbb{R}_+,\mathbb{C}^l)\oplus\mathbb{C}^{j_-}} = (g(y,\eta)u, v)_{L^2(\mathbb{R}_+,\mathbb{C}^k)\oplus\mathbb{C}^{j_+}} \tag{1.77}$$

for all $u \in L^2(\mathbb{R}_+, \mathbb{C}^l) \oplus \mathbb{C}^{j_-}$, $v \in L^2(\mathbb{R}_+, \mathbb{C}^k) \oplus \mathbb{C}^{j_+}$. Let $\mathcal{R}^{\mu,0}_G(\Omega \times \mathbb{R}^q; \boldsymbol{w})$, $\boldsymbol{w} = (l, k; j_-, j_+)$, denote the space of all such symbols.

REMARK 1.64. It can be proved that every $g(y,\eta) \in \mathcal{R}^{\mu,0}_G(\Omega \times \mathbb{R}^q; \boldsymbol{w})$ induces elements

$$g(y,\eta) \in S^{\mu}_{\mathrm{cl}}(\Omega \times \mathbb{R}^q; H^s(\mathbb{R}_+, \mathbb{C}^l) \oplus \mathbb{C}^{j_-}, \mathcal{S}(\overline{\mathbb{R}}_+, \mathbb{C}^k) \oplus \mathbb{C}^{j_+})$$

for all $s \in \mathbb{R}$, $s > -\frac{1}{2}$.

This is a consequence of a kernel characterisation of [**37**, Theorem 3.1], cf. also [**39**, Section 4.2.3].

Note that the differential operator ∂_t^j on $\mathbb{R}_+$ represents an operator-valued symbol

$$\partial_t^j \in S^{j}_{\mathrm{cl}}(\Omega \times \mathbb{R}^q; H^s(\mathbb{R}_+, \mathbb{C}^l), H^{s-j}(\mathbb{R}_+, \mathbb{C}^l))$$

for every $s \in \mathbb{R}$; there is in this case no dependence on $(y,\eta) \in \Omega \times \mathbb{R}^q$.

DEFINITION 1.65. By $\mathcal{R}^{\mu,d}_G(\Omega \times \mathbb{R}^q; \boldsymbol{w})$ for $\mu \in \mathbb{R}$, $d \in \mathbb{N}$, we denote the space of all operator functions

$$g(y,\eta) := g_0(y,\eta) + \sum_{j=1}^{d} g_j(y,\eta)\,\mathrm{diag}(\partial_t^j, 0),$$

for arbitrary $g_j(y,\eta) \in \mathcal{R}^{\mu-j,d}(\Omega \times \mathbb{R}^q, \boldsymbol{w})$. The elements of $\mathcal{R}^{\mu,d}_G(\Omega \times \mathbb{R}^q; \boldsymbol{w})$ are called Green symbols of order μ and type d.

Notice that $g(y,\eta) \in \mathcal{R}^{\mu,d}_G(\Omega \times \mathbb{R}^q; \boldsymbol{w})$ implies

$$g(y,\eta) \in S^{\mu}_{\mathrm{cl}}(\Omega \times \mathbb{R}^q; H^s(\mathbb{R}_+, \mathbb{C}^l) \oplus \mathbb{C}^{j_-}, \mathcal{S}(\overline{\mathbb{R}}_+, \mathbb{C}^k) \oplus \mathbb{C}^{j_+}) \tag{1.78}$$

for every $s \in \mathbb{R}$, $s > d - \frac{1}{2}$.

The nature of Green symbols as operator-valued symbols in the sense of Definition 1.63 has been first observed in [**37**, Theorem 3.1]. A similar characterisation holds for Green operators belonging to boundary value problems without the transmission property, see [**38**]. In this case, the space $\mathcal{S}(\overline{\mathbb{R}}_+)$ is to be replaced by a space of functions with more general conormal asymptotics rather than Taylor asymptotics at $t = 0$ (and Schwartz function behaviour for $t \to \infty$).

THEOREM 1.66. *For every* $g(y,\eta) \in \mathcal{R}_G^{\mu,d}(\Omega \times \mathbb{R}^q; \boldsymbol{w})$, $\boldsymbol{w} = (l,k;j_-,j_+)$, *the associated (so called Green operator)* $\mathcal{G} := \mathrm{Op}(g)$ *induces continuous operators*

$$\begin{aligned}(1.79)\qquad \mathcal{G} : H^s_{\mathrm{comp}(y)}(\Omega \times \mathbb{R}_+, \mathbb{C}^l) \oplus H^s_{\mathrm{comp}}(\Omega, \mathbb{C}^{j_-}) \\ \to H^{s-\mu}_{\mathrm{loc}(y)}(\Omega \times \mathbb{R}_+, \mathbb{C}^k) \oplus H^{s-\mu}_{\mathrm{loc}}(\Omega, \mathbb{C}^{j_+})\end{aligned}$$

for all $s \in \mathbb{R}$, $s > d - \frac{1}{2}$.

PROOF. We have

$$g(y,\eta) \in S^\mu_{\mathrm{cl}}(\Omega \times \mathbb{R}^q; H^s(\mathbb{R}_+, \mathbb{C}^l) \oplus \mathbb{C}^{j_-}, H^{s-\mu}(\mathbb{R}_+, \mathbb{C}^k) \oplus \mathbb{C}^{j_+})$$

(which is more crude than (1.78) but sufficient for the moment), and then

$$\mathcal{G} : \mathcal{W}^s_{\mathrm{comp}}(\Omega, H^s(\mathbb{R}_+, \mathbb{C}^l) \oplus \mathbb{C}^{j_-}) \to \mathcal{W}^{s-\mu}_{\mathrm{loc}}(\Omega, H^{s-\mu}(\mathbb{R}_+, \mathbb{C}^k) \oplus \mathbb{C}^{j_+})$$

is continuous for $s > d - \frac{1}{2}$, cf. Theorem 1.54 (the $\mathcal{W}^s$-spaces refer to the same group actions as in Definition 1.63.) Now we have

$$\mathcal{W}^s_{\mathrm{comp}}(\Omega, H^s(\mathbb{R}_+, \mathbb{C}^l) \oplus \mathbb{C}^{j_-}) = H^s_{\mathrm{comp}(y)}(\Omega, \times \mathbb{R}_+, \mathbb{C}^l) \oplus H^s_{\mathrm{comp}}(\Omega, \mathbb{C}^{j_-})$$

and, similarly, with subscripts 'loc', cf. Theorem 1.53 (iv). □

By definition, the symbol $g(y,\eta)$ has a homogeneous principal component $g_{(\mu)}(y,\eta)$ of order μ, cf. Definition 1.50 (iii). It will be interpreted as the boundary symbol

$$(1.80)\qquad \sigma_\partial(\mathcal{G})(y,\eta) : H^s(\mathbb{R}_+, \mathbb{C}^l) \oplus \mathbb{C}^{j_-} \to \mathcal{S}(\overline{\mathbb{R}}_+, \mathbb{C}^k) \oplus \mathbb{C}^{j_+}$$

of the associated pseudo-differential operator $\mathcal{G} = \mathrm{Op}(g)$. Alternatively, we also write

$$(1.81)\qquad \sigma_\partial(\mathcal{G})(y,\eta) : \mathcal{S}(\overline{\mathbb{R}}_+, \mathbb{C}^l) \oplus \mathbb{C}^{j_-} \to \mathcal{S}(\overline{\mathbb{R}}_+, \mathbb{C}^k) \oplus \mathbb{C}^{j_+}.$$

Now let $E, F \in \mathrm{Vect}(X)$ and $J_-, J_+ \in \mathrm{Vect}(Y)$, and consider trivialisations

$$(1.82)\qquad E|_V \cong \Omega \times \overline{\mathbb{R}}_+ \times \mathbb{C}^l, \qquad F|_V \cong \Omega \times \overline{\mathbb{R}}_+ \times \mathbb{C}^k,$$

$$(1.83)\qquad J_\pm|_V \cong \Omega \times \mathbb{C}^{j_\pm}$$

on a coordinate neighbourhood V on X near the boundary Y, such that $V' := V \cap Y \neq \emptyset$, and let $\chi : V \to \Omega \times \overline{\mathbb{R}}_+$ and $\chi' : V' \to \Omega$ be corresponding charts on X and Y, respectively.

Green operators (1.79), interpreted for a moment as operators

$$\mathcal{G} : C_0^\infty(\Omega \times \overline{\mathbb{R}}_+, \mathbb{C}^l) \oplus C_0^\infty(\Omega, \mathbb{C}^{j_-}) \to C^\infty(\Omega \times \overline{\mathbb{R}}_+, \mathbb{C}^k) \oplus C^\infty(\Omega, \mathbb{C}^{j_+})$$

can be pulled back to X with respect to the mapping (1.82), (1.83) as operators

$$\mathcal{G}_V : C_0^\infty(V, E|_V) \oplus C_0^\infty(V', J_-|_{V'}) \to C^\infty(V, F|_V) \oplus C^\infty(V', J_+|_{V'}).$$

Let us write, for simplicity, $\mathcal{G}_V = (\chi^{-1})_* \mathrm{Op}(g)$; the pull back also refers to the cocycles of transition maps of the involved bundles. Let us now fix a finite system $\{V_j\}_{j=1,\dots,L}$ of such coordinate neighbourhoods on X near Y such that $\{V_1'\}_{j=1,\dots,L}$

form an open covering of Y, and choose functions $\varphi_j, \psi_j \in C_0^\infty(V_j)$, $j = 1, \dots, L$ such that $\sum_{j=1}^{L} \varphi_j \equiv 1$ in a collar neighbourhood of Y and $\psi_j \equiv 1$ on $\operatorname{supp} \varphi_j$ for all j, and set $\varphi'_j := \varphi_j|_{V'}$, $\psi'_j := \psi_j|_{V'}$.

DEFINITION 1.67. (i) The space $\mathcal{B}_G^{\mu,d}(X; \boldsymbol{v})$ for $\boldsymbol{v} := (E, F; J_-, J_+)$, $\mu \in \mathbb{Z}$, $d \in \mathbb{N}$, is defined to be the set of all operators

$$\mathcal{G} := \sum_{j=1}^{L} \operatorname{diag}(\varphi_j, \varphi_j^{'})(\chi_j^{-1})_* \operatorname{Op}(g_j) \operatorname{diag}(\psi_j, \psi_j^{'}) + \mathcal{C}$$

for arbitrary and $\mathcal{C} \in \mathcal{B}^{-\infty,d}(X; \boldsymbol{v})$ and $g_j(y, \eta) \in \mathcal{R}_G^{\mu,d}(\Omega \times \overline{\mathbb{R}}_+; \boldsymbol{w})$, $1 \le j \le L$.

The elements of $\mathcal{B}_G^{\mu,d}(X; \boldsymbol{v})$ are called Green operators on X of order μ and type d. The families of maps (1.80) or (1.81), applied to localisations of Green operators $\mathcal{G}$ in a coordinate neighbourhood V, have an invariant meaning as bundle morphisms

$$\sigma_\partial(\mathcal{G}) : \pi_Y^* \begin{pmatrix} E' \otimes H^s(\mathbb{R}_+) \\ \oplus \\ J_- \end{pmatrix} \to \pi_Y^* \begin{pmatrix} F' \otimes H^{s-\mu}(\mathbb{R}_+) \\ \oplus \\ J_+ \end{pmatrix}, \tag{1.84}$$

$\pi_Y : T^*Y \setminus 0 \to Y$, $s > d - \frac{1}{2}$, (alternatively, we may write $\mathcal{S}(\overline{\mathbb{R}}_+)$ instead of $H^{s-\mu}(\mathbb{R}_+)$ on the right hand side, or $\mathcal{S}(\overline{\mathbb{R}}_+)$ on both sides).

(ii) The space $\mathcal{B}^{\mu,d}(X; \boldsymbol{v})$, $\mu \in \mathbb{Z}$, $d \in \mathbb{N}$, is defined as the set of all operators

$$\mathcal{A} = \begin{pmatrix} A & 0 \\ 0 & 0 \end{pmatrix} + \mathcal{G} \tag{1.85}$$

for arbitrary $A \in L_{\mathrm{cl}}^\mu(X; E, F)_{\mathrm{tr}}$ and $\mathcal{G} \in \mathcal{B}_G^{\mu,d}(X; \boldsymbol{v})$. The elements of $\mathcal{B}^{\mu,d}(X; \boldsymbol{v})$ are called (pseudo-differential) boundary value problems for the operator A, of order μ and type d.

(iii) We set $\sigma_\psi(\mathcal{A}) := \sigma_\psi(A)$, called the (homogeneous principal) interior symbol of $\mathcal{A}$ of order μ, and

$$\sigma_\partial(\mathcal{A}) := \begin{pmatrix} \sigma_\partial(A) & 0 \\ 0 & 0 \end{pmatrix} + \sigma_\partial(\mathcal{G}),$$

called the (homogeneous principal) boundary symbol of $\mathcal{A}$. Set

$$\sigma(\mathcal{A}) := (\sigma_\psi(\mathcal{A}), \sigma_\partial(\mathcal{A})), \tag{1.86}$$

called the principal symbol of the operator $\mathcal{A}$.

The homogeneity of $\sigma_\psi(\mathcal{A})$ is as usual, i.e., $\sigma_\psi(\mathcal{A})(x, \lambda\xi) = \lambda^\mu \sigma_\psi(\mathcal{A})(x, \xi)$ for all $\lambda \in \mathbb{R}_+$, $(x, \xi) \in T^*X \setminus 0$. For $\sigma_\partial(\mathcal{A})$ we have

$$\sigma_\partial(\mathcal{A})(y, \lambda\eta) = \lambda^\mu \operatorname{diag}(\kappa_\lambda, \mathrm{id}) \sigma_\partial(\mathcal{A})(y, \eta) \operatorname{diag}(\kappa_\lambda^{-1}, \mathrm{id}) \tag{1.87}$$

for all $\lambda \in \mathbb{R}_+$, $(y, \eta) \in T^*Y \setminus 0$.

By

$$B_G^{\mu,d}(X;E,F) \quad \text{and} \quad B^{\mu,d}(X;E,F) \tag{1.88}$$

we denote the space of upper left cornes of 2×2 block matrices in $B_G^{\mu,d}(X;\boldsymbol{v})$ and $B^{\mu,d}(X;\boldsymbol{v})$, respectively. Given an element $\mathcal{A} \in B^{\mu,d}(X;\boldsymbol{v})$ in the form $\mathcal{A} = (A_{ij})_{i,j=1,2}$, we also set $A_{11} := \text{ulc}\,\mathcal{A}$. The operator A_{21} is often called a trace and A_{12} a potential operator in $B^{\mu,d}(X;\boldsymbol{v})$. Note that A_{22} belongs to $L_{\text{cl}}^{\mu}(Y;J_-,J_+)$.

REMARK 1.68. The following conditions are equivalent:

(i) $G \in B^{\mu,d}(X;E,F)$ and $\varphi G \psi \in L^{-\infty}(\text{int}\,X;E,F)$ for every $\varphi,\psi \in C_0^{\infty}(\text{int}\,X)$,

(ii) $G \in B_G^{\mu,d}(X;E,F)$.

REMARK 1.69. Every $G \in B_G^{\mu,d}(X;E,F)$ has a unique representation

$$G = G_0 + \sum_{j=0}^{d-1} K_j \circ T_j \tag{1.89}$$

for a $G_0 \in B_G^{\mu,0}(X;E,F)$ and potential operators $K_j \in \mathcal{B}^{\mu-j-\frac{1}{2},0}(X;0,F;E',0)$ and $T_j u = \text{r}' T^j$, cf. the notation of (1.76).

THEOREM 1.70. *Let X be a compact C^∞ manifold with boundary Y.*

(i) *Every $\mathcal{A} \in \mathcal{B}^{\mu,d}(X;\boldsymbol{v})$ for $\boldsymbol{v} := (E,F;J_-,J_+)$ induces continuous operators*

$$\mathcal{A} : \begin{array}{c} H^s(X,E) \\ \oplus \\ H^s(Y,J_-) \end{array} \rightarrow \begin{array}{c} H^{s-\mu}(X,F) \\ \oplus \\ H^{s-\mu}(Y,J_+) \end{array} \tag{1.90}$$

for all $s \in \mathbb{R}$, $s > d - \frac{1}{2}$.

(ii) *$\mathcal{A} \in \mathcal{B}^{\mu,d}(X;\boldsymbol{v})$ and $\sigma(\mathcal{A}) = 0$ imply $\mathcal{A} \in \mathcal{B}^{\mu-1,d}(X;\boldsymbol{v})$, and hence the operator (1.90) is compact.*

(iii) *$\mathcal{A} \in \mathcal{B}^{\mu,d}(X;\boldsymbol{v})$ for $\boldsymbol{v} := (E_0,F;J_0,J_+)$ and $\mathcal{B} \in \mathcal{B}^{\nu,e}(X;\boldsymbol{w})$ for $\boldsymbol{w} := (E,E_0;J_-,J_0)$ implies $\mathcal{AB} \in \mathcal{B}^{\mu+\nu,h}(X;\boldsymbol{v}\circ\boldsymbol{w})$ for $\boldsymbol{v}\circ\boldsymbol{w} := (E,F;J_-,J_+)$ and $h = \max(\nu+d,e)$, and we have*

$$\sigma(\mathcal{AB}) = \sigma(\mathcal{A})\sigma(\mathcal{B})$$

(with componentwise multiplication).

(iv) *The formal adjoint $\mathcal{A}^*$ (cf. formula (1.75)) of an operator $\mathcal{A} \in \mathcal{B}^{0,0}(X;\boldsymbol{v})$ for $\boldsymbol{v} := (E,F;J_-,J_+)$ belongs to $\mathcal{B}^{0,0}(X;\boldsymbol{v}^*)$ for $\boldsymbol{v}^* := (F,E;J_+,J_-)$, and we have*

$$\sigma(\mathcal{A}^*) = \sigma(\mathcal{A})^*$$

(with componentwise adjoint, cf. formulas (1.10) and (1.77)).

THEOREM 1.71. *Let $\boldsymbol{v} := (E, F; J_-, J_+)$ for vector bundles $E, F \in \mathrm{Vect}(X)$, $J_-, J_+ \in \mathrm{Vect}(Y)$, and let $\mathcal{A}_j \in \mathcal{B}^{\mu-j,d}(X; \boldsymbol{v})$, $j \in \mathbb{N}$, be an arbitrary sequence. Then there exists an element $\mathcal{A} \in \mathcal{B}^{\mu,d}(X; \boldsymbol{v})$ such that*

$$\mathcal{A} - \sum_{j=0}^{N} \mathcal{A}_j \in \mathcal{B}^{\mu-(N+1),d}(X; \boldsymbol{v})$$

for every $N \in \mathbb{N}$, and $\mathcal{A}$ is unique $\mod \mathcal{B}^{-\infty,d}(X; \boldsymbol{v})$.

1.3. SL-elliptic boundary value problems.

DEFINITION 1.72. Let $\mathcal{A} \in \mathcal{B}^{\mu,d}(X; \boldsymbol{v})$, $\mu \in \mathbb{Z}$, $d \in \mathbb{N}$, $\boldsymbol{v} := (E, F; J_-, J_+)$ for $E, F \in \mathrm{Vect}(X)$, $J_-, J_+ \in \mathrm{Vect}(Y)$.

(i) The operator $\mathcal{A}$ is said to be SL-elliptic (of order μ), if both

$$\sigma_\psi(\mathcal{A}) : \pi_X^* E \to \pi_X^* F, \tag{1.91}$$

$\pi_X : T^*X \setminus 0 \to X$, and

$$\sigma_\partial(\mathcal{A}) : \pi_Y^* \begin{pmatrix} E' \otimes \mathcal{S}(\overline{\mathbb{R}}_+) \\ \oplus \\ J_- \end{pmatrix} \to \pi_Y^* \begin{pmatrix} F' \otimes \mathcal{S}(\overline{\mathbb{R}}_+) \\ \oplus \\ J_+ \end{pmatrix} \tag{1.92}$$

$\pi_Y : T^*Y \setminus 0 \to Y$, are isomorphisms.

(ii) An operator $\mathcal{P} \in \mathcal{B}^{-\mu,e}(X; \boldsymbol{v}^{-1})$ for some $e \in \mathbb{N}$, $\boldsymbol{v}^{-1} := (F, E; J_+, J_-)$, is called a parametrix of $\mathcal{A}$, if $\mathcal{P}$ satisfies the following relations:

$$\mathcal{C}_l := \mathcal{I} - \mathcal{P}\mathcal{A} \in \mathcal{B}^{-\infty,d_l}(X; \boldsymbol{v}_l), \quad \mathcal{C}_r := \mathcal{I} - \mathcal{A}\mathcal{P} \in \mathcal{B}^{-\infty,d_r}(X; \boldsymbol{v}_r) \tag{1.93}$$

for certain $d_l, d_r \in \mathbb{N}$, $\boldsymbol{v}_l := (E, E; J_-, J_-)$, $\boldsymbol{v}_r := (F, F; J_+, J_+)$, where $\mathcal{I}$ denotes the corresponding identity operator.

REMARK 1.73. If (1.91) is an isomorphism, a bundle morphism (1.92) for $\mathcal{A} \in \mathcal{B}^{\mu,d}(X; \boldsymbol{v})$ is an isomorphism, if and only if

$$\sigma_\partial(\mathcal{A}) : \pi_Y^* \begin{pmatrix} E' \otimes H^s(\mathbb{R}_+) \\ \oplus \\ J_- \end{pmatrix} \to \pi_Y^* \begin{pmatrix} F' \otimes H^{s-\mu}(\mathbb{R}_+) \\ \oplus \\ J_+ \end{pmatrix} \tag{1.94}$$

is an isomorphism for any fixed $s = s_0 \in \mathbb{R}$, $s_0 > \max(\mu, d) - \frac{1}{2}$ (or, equivalently, for all $s > \max(\mu, d) - \frac{1}{2}$).

REMARK 1.74. If $A \in L^\mu_{\mathrm{cl}}(X; E, F)_{\mathrm{tr}}$ is elliptic, i.e., (1.91) an isomorphism, we have

$$E' \cong F'. \tag{1.95}$$

In fact, let us consider the composition $A_0 := AR_E^{-\mu}$, cf. Theorem 1.60. Then we have $A_0 \in B^{\mu,0}(X; E, F)$, and $\sigma_\psi(A_0) : \pi_X^* E \to \pi_X^* F$ also is an isomorphism. Let $\Xi := S^*X|_Y \cup N$ with the conormal interval $\{(y, (\eta, \tau)) : \eta = 0, \ \ -1 \le \tau \le +1\}$.

The transmission property of symbols of order zero has the consequence that

$$\sigma_\psi(A_0)|_{S^*X|_Y} : \pi_2^* E' \to \pi_2^* F'$$

for $\pi_2 : S^*X|_Y \to Y$ extends to an isomorphism $\pi_3^* E' \to \pi_3^* F$, where $\pi_3 : \Xi \to Y$ is the canonical projection of $\Xi := S^*X|_Y \cup N$ to Y.

Since $Y \subset \Xi$, the latter isomorphism then restricts to an isomorphism (1.95).

THEOREM 1.75. *Let $\mathcal{A} \in \mathcal{B}^{\mu,d}(X; \boldsymbol{v})$, $\mu \in \mathbb{Z}$, $d \in \mathbb{N}$, $\boldsymbol{v} := (E, F; J_-, J_+)$.*

(i) *The operator $\mathcal{A}$ is SL-elliptic (of order μ), if and only if (1.90) is a Fredholm operator for an $s = s_0 \in \mathbb{R}$, $s_0 > \max(\mu, d) - \frac{1}{2}$.*

(ii) *If $\mathcal{A}$ is elliptic, (1.90) is a Fredholm operator for all $s > \max(\mu, d) - \frac{1}{2}$, and $\dim \ker \mathcal{A}$ and $\dim \operatorname{coker} \mathcal{A}$ are independent of s.*

(iii) *An elliptic operator $\mathcal{A}$ has a parametrix $\mathcal{P} \in \mathcal{B}^{-\mu,(d-\mu)^+}(X; \boldsymbol{v}^{-1})$ ($\rho^+ := \max(\rho, 0)$ for any $\rho \in \mathbb{R}$) which can be chosen in such a way that the remainders in the relation (1.93) are projections*

$$\mathcal{C}_l : H^s(X, E) \oplus H^s(Y, J_-) \to V,$$
$$\mathcal{C}_r : H^{s-\mu}(X, F) \oplus H^{s-\mu}(Y, J_+) \to W$$

for all $s > \max(\mu, d) - \frac{1}{2}$ and are of type $d_l = \max(\mu, d)$ and $d_r = (d - \mu)^+$, respectively, for $V := \ker \mathcal{A} \subset C^\infty(X, E) \oplus C^\infty(Y; J_-)$, and some finite-dimensional subspace $W \subset C^\infty(X, F) \oplus C^\infty(Y; J_+)$ such that

$$W + \operatorname{im} \mathcal{A} = H^{s-\mu}(X, F) \oplus H^{s-\mu}(Y; J_+)$$

and $W \cap \operatorname{im} \mathcal{A} = \{0\}$ for every $s > \max(\mu, d) - \frac{1}{2}$.

An operator $A \in B^{\mu,d}(X; E, F)$ will be called σ_ψ-elliptic (of order μ), if (1.91) is an isomorphism.

REMARK 1.76. Let $A \in B^{\mu,d}(X; E, F)$ be σ_ψ-elliptic. Then

$$\sigma_\partial(A)(y, \eta) : E'_y \otimes H^s(\mathbb{R}_+) \to F'_y \otimes H^{s-\mu}(\mathbb{R}_+) \tag{1.96}$$

is a family of Fredholm operators for all $s > \max(\mu, d) - \frac{1}{2}$, and $\ker \sigma_\partial(A)(y, \eta)$, $\operatorname{coker} \sigma_\partial(A)(y, \eta)$ are independent of s. For every $G \in B_G^{\mu,d}(X; E, F)$ we have $\operatorname{ind} \sigma_\partial(A)(y, \eta) = \operatorname{ind} \sigma_\partial(A + G)(y, \eta)$ for all $(y, \eta) \in T^*Y \setminus 0$. Moreover, by virtue of the homogeneity (1.87) it follows that $\operatorname{ind} \sigma_\partial(A)(y, \eta) = \operatorname{ind} \sigma_\partial(A)(y, \eta/|\eta|)$.

Thus it makes sense to interpret (1.96) as a family of Fredholm operators, parametrised by $(y, \eta) \in S^*Y$, the unit cosphere bundle. As such there is an index element

$$\operatorname{ind}_{S^*Y} \sigma_\partial(A) \in K(S^*Y), \tag{1.97}$$

cf. Atiyah and Bott [**3**], Boutet de Monvel [**7**].

REMARK 1.77. Let $\mathcal{A} \in \mathcal{B}^{\mu,d}(X; \boldsymbol{v})$ be SL-elliptic. Then, in the notation of Definition 1.72 for $A := \operatorname{ulc} \mathcal{A}$ we have

$$\operatorname{ind}_{S^*Y} \sigma_\partial(A) = [\pi_1^* J_+] - [\pi_1^* J_-], \tag{1.98}$$

where $\pi_1 : S^*Y \to Y$ is the canonical projection. In other words, SL-ellipticity of $\mathcal{A}$ entails the relation

$$\operatorname{ind}_{S^*Y} \sigma_\partial(A) \in \pi_1^* K(Y). \tag{1.99}$$

2. Ellipticity with global projection conditions

2.1. The index obstruction. We first discuss the problem, whether a σ_ψ-elliptic operator $A \in B^{\mu,d}(X; E, F)$ admits SL-elliptic boundary conditions.

THEOREM 2.1. *Let $A \in B^{\mu,d}(X; E, F)$ be σ_ψ-elliptic. Then the following conditions are equivalent:*

(i) *There is an SL-elliptic element $\mathcal{A} \in B^{\mu,d}(X; \boldsymbol{v})$, $\boldsymbol{v} := (E, F; J_-, J_+)$ for certain $J_\pm \in \operatorname{Vect}(Y)$ such that $A = \operatorname{ulc} \mathcal{A}$.*
(ii) *A satisfies the relation* (1.99).

Remark 1.77 shows that (1.99) is necessary for the existence of an SL-elliptic operator $\mathcal{A}$ with A as upper left corner. For the converse direction we first establish a result on general families of Fredholm operators.

In the following consideration we assume M to be a compact topological space. In this connection by $\operatorname{Vect}(M)$ we understand the set of complex vector bundles on M in the continuous category, i.e., with continuous transition maps between local trivialisations. Moreover, let H_1 and H_2 be separable infinite-dimensional Hilbert spaces. For every function

$$a \in C(M, \mathcal{L}(H_1, H_2))$$

with values in the set $\mathcal{F}(H_1, H_2)$ of Fredholm operators between H_1, H_2 there is an index element

$$\operatorname{ind}_M a \in K(M). \tag{2.1}$$

The construction is based on the following observation. There exists a finite-dimensional vector space and an injective linear map $k : W \to H_2$ such that

$$\begin{pmatrix} a(m) & k \end{pmatrix} : \begin{matrix} H_1 \\ \oplus \\ W \end{matrix} \to H_2 \tag{2.2}$$

is surjective for all $m \in M$. Then (2.2) is again a continuous family of Fredholm operators, now surjective. Hence the family of kernels $\{\ker(a(m) \;\; k) : m \in M\}$ represents a finite-dimensional subbundle $\widetilde{V}$ of $M \times (H_1 \oplus W)$. In other words, there is a $V \in \operatorname{Vect}(M)$ and a continuous family of isomorphisms

$$v_m : \ker(a(m) \;\; k) \to V_m \tag{2.3}$$

(with V_m being the fibre of V over m). Let $p_m : H_1 \oplus W \to \ker(a(m) \;\; k)$ denote the orthogonal projection and set $(t(m) \;\; g(m)) := v_m \circ p_m$.

Then

$$\begin{pmatrix} a(m) & k \\ t(m) & g(m) \end{pmatrix} : \begin{matrix} H_1 \\ \oplus \\ W_m \end{matrix} \to \begin{matrix} H_2 \\ \oplus \\ V_m \end{matrix}$$

is a continuous family of isomorphisms, and we define

$$\text{ind}_M\, a := [V] - [W] \in K(M), \tag{2.4}$$

where W is identified with the trivial bundle $M \times \mathbb{C}^{\dim W}$.

REMARK 2.2. Let $a \in C(M, \mathcal{F}(H_1, H_2))$ be as before, and let $V, W \in \text{Vect}(M)$ such that there is a continuous family of isomorphisms

$$\begin{pmatrix} a(m) & k(m) \\ t(m) & g(m) \end{pmatrix} : \begin{matrix} H_1 \\ \oplus \\ W_m \end{matrix} \to \begin{matrix} H_2 \\ \oplus \\ V_m \end{matrix}, \qquad m \in M$$

for suitable operator functions k, t and g. Then we have

$$\text{ind}_M\, a = [V] - [W].$$

In other words, the element $\text{ind}_M\, a$ is independent of the choice of the bundles V, W and of the operator families k, t, q.

REMARK 2.3. Let $c \in C(M, \mathcal{L}(H_1, H_2))$ be a family such that $c(m)$ is a compact operator for every $m \in M$. Then we have

$$\text{ind}_M(a + c) = \text{ind}_M\, a.$$

For purposes below we need the following more specific construction.

PROPOSITION 2.4. *Let $a \in C(M, \mathcal{F}(H_1, H_2))$ be a Fredholm function, and let $L_\pm \in \text{Vect}(M)$ be a fixed choice of vector bundles such that $\text{ind}_M\, a = [L_+] - [L_-]$. Then there exists an element $c \in C(M, \mathcal{L}(H_1, H_2))$ with values in operators of finite rank such that $\widetilde{a} := a + c$ has the following properties:*

(i) $\ker \widetilde{a} \cong L_+$, $\text{coker}\, \widetilde{a} \cong L_-$, *i.e., there are subbundles $\widetilde{L}_+ \subset M \times H_1$, $\widetilde{L}_+ \cong L_+$ and $\widetilde{L}_- \subset M \times H_2$, $\widetilde{L}_- \cong L_-$, such that $\widetilde{L}_{+,m} = \ker \widetilde{a}(m)$, $\widetilde{L}_{-,m} + \text{im}\, \widetilde{a}(m) = H_2$, and $\widetilde{L}_{-,m} + \text{im}\, \widetilde{a}(m) = \{0\}$ for all $m \in M$*

(ii) *There are (continuous) bundle morphisms*

$$\widetilde{k} : L_- \to M \times H_2, \qquad \widetilde{t} : M \times H_1 \to L_+ \tag{2.5}$$

such that

$$\begin{pmatrix} \widetilde{a} & \widetilde{k} \\ \widetilde{t} & 0 \end{pmatrix} : \begin{matrix} H_1 \\ \oplus \\ L_- \end{matrix} \to \begin{matrix} H_2 \\ \oplus \\ L_+ \end{matrix} \tag{2.6}$$

is an isomorphism.

PROOF. As before we first pass to the surjective family (2.2). By assumption, we have

$$\mathrm{ind}_M\, a = [V] - [W] = [L_+] - [L_-]$$

in $K(M)$. We can choose $\dim W$ as large as we want, and we now replace W by $W \oplus W_1$ for another finite-dimensional W_1, and together with k we choose a $k_1 : W_1 \to H_2$, and such that $k^0 := \mathrm{diag}(k, k_1) : W \oplus W_1 \to H_2$ is injective. Let $p : H_2 \to \mathrm{im}\, k^0$ denote the orthogonal projection, and write $a^0 := (1-p)a$. Then, for $W^0 := W \oplus W_1$ the operator family

$$\begin{pmatrix} a^0 & k^0 \end{pmatrix} : \begin{matrix} H_1 \\ \oplus \\ W^0 \end{matrix} \to H_2$$

is surjective, and $\widetilde{V}^0 := \ker_M a^0$ is a subbundle of H_1 isomorphic to a bundle $V^0 \in \mathrm{Vect}(M)$ where $V^0 \cong V \oplus W_1$ (here vector spaces F are identified with the respective trivial bundles $M \times F$). By the formula (2.4) applied to a^0 we obtain

$$\mathrm{ind}_M\, a^0 = [V^0] - [W^0] = [L_+] - [L_-] \tag{2.7}$$

for $V^0 := V \oplus W_1$. We may assume that the given bundles $L_\pm$ are both subbundles of the trivial bundle W_1 for a sufficienty large choice of $\dim W_1$.

There are then complementary bundles $L_\pm^\perp$ of $L_\pm$ in V^0 and W^0, respectively, i.e., we have

$$L_+ \oplus L_+^\perp = V^0, \qquad L_- \oplus L_-^\perp = W^0.$$

Then the relation (2.7) implies $[L_+^\perp] = [L_-^\perp]$ in $K(M)$, i.e., there is an R such that $L_+^\perp \oplus \mathbb{C}^R \cong L_-^\perp \oplus \mathbb{C}^R$. Replacing W_1 by $W_1 \oplus \mathbb{C}^R$ in the construction before and returning to the former notation, we thus obtain $L_+^\perp \cong L_-^\perp$. By construction there are subbundles $\widetilde{L}_-, \widetilde{L}_-^\perp \subset H_2$ such that $\widetilde{L}_- \cong L_-$, $\widetilde{L}_-^\perp \cong L_-^\perp$ wich $\widetilde{L}_- \oplus \widetilde{L}_-^\perp = \mathrm{im}\, k^0$, and subbundles $\widetilde{L}_+$, $\widetilde{L}_+^\perp \subset H_1$ such that $\widetilde{L}_+ \cong L_+$, $\widetilde{L}_+^\perp \cong L_+^\perp$ and $\widetilde{L}_+ \oplus \widetilde{L}_+^\perp = \ker a^0$. Choose any isomorphism $\lambda : \widetilde{L}_+^\perp \to \widetilde{L}_-^\perp$, and let $\pi^\perp : H_1 \to \widetilde{L}_+^\perp$ denote the orthogonal projection, $\iota^\perp : \widetilde{L}_-^\perp \to H_2$ the canonical embedding. Then $q := \iota^\perp \circ \lambda \circ \pi^\perp : H_1 \to H_2$ is a continuous family of operators of finite rank, and $a^0 + q$ satisfies the relations

$$\ker_M (a^0 + q) \cong L_+, \qquad \mathrm{coker}_M (a^0 + q) \cong L_-.$$

Because of $a^0 = (1-p)a$ we may set $\widetilde{a} := a + c$ for $c := -pa + q$. Then $\widetilde{a}$ satisfies the relations of Proposition 2.4 (i). To construct the isomorphism (2.6) it suffices to choose isomorphisms $h : L_- \to \widetilde{L}_-$ and $l : \widetilde{L}_+ \to L_+$, and to set $\widetilde{k} := \iota h$, $\widetilde{t} := l\pi$, where $\iota : \widetilde{L}_- \to H_2$ is the canonical embedding, $\pi : H_1 \to \widetilde{L}_+$ the orthogonal projection. □

REMARK 2.5. Let $A \in B^{\mu,d}(X; E, F)$ be σ_ψ-elliptic, and consider the Fredholm operators (1.96) for any fixed $s > \max(\mu, d) - \frac{1}{2}$. Then there is a subbundle $\widetilde{W} \subset \pi_1^* F' \otimes \mathcal{S}(\overline{\mathbb{R}}_+)$ of finite fibre dimension such that

$$\widetilde{W}_{y,\eta} + \mathrm{im}\, \sigma_\partial(A)(y, \eta) = F'_y \otimes H^{s-\mu}(\mathbb{R}_+)$$

for all $(y,\eta) \in S^*Y$.

Choose a vector bundle $W \in \mathrm{Vect}(S^*Y)$, and let

$$k : W \to \widetilde{W}$$

be an isomorphism. Then

$$(2.8) \qquad (\sigma_\partial(A)(y,\eta) \quad k(y,\eta)) : \begin{pmatrix} E'_y \otimes H^s(\mathbb{R}_+) \\ \oplus \\ W_{y,\eta} \end{pmatrix} \to F'_y \otimes H^{s-\mu}(\mathbb{R}_+)$$

is a surjective family of Fredholm operators. W can be chosen as the pull-back of a bundle on Y with respect to $\pi_1 : S^*Y \to Y$ (in fact, we may assume that it is trivial).

Let

$$p : \pi_1^* F' \otimes H^{s-\mu}(\mathbb{R}_+) \to \widetilde{W}$$

be a projection (orthogonal with respect to the $F'_y \otimes L^2(\mathbb{R}_+)$ - scalar product in the fibres). Then

$$(1-p)\sigma_\partial(A) : \pi_1^* E' \otimes H^s(\mathbb{R}_+) \to \pi_1^* F' \otimes H^{s-\mu}(\mathbb{R}_+)$$

is again a Fredholm family such that

$$\widetilde{V} := \ker_{S^*Y}(1-p)\sigma_\partial(A) \subset \pi_1^* E' \otimes \mathcal{S}(\overline{\mathbb{R}}_+)$$

is a subbundle, and

$$\mathrm{ind}_{S^*Y}(1-p)\sigma_\partial(A) = [V] - [W]$$

for any $V \in \mathrm{Vect}(S^*Y)$ which is isomorphic to $\widetilde{V}$.

PROPOSITION 2.6. *Let $A \in B^{\mu,d}(X; E, F)$ be σ_ψ-elliptic, and $L_\pm \in \mathrm{Vect}(S^*Y)$ such that*

$$(2.9) \qquad \mathrm{ind}_{S^*Y}\, \sigma_\partial(A) = [L_+] - [L_-].$$

Then there exists an element $G \in B_G^{\mu,0}(X; E, F)$ such that

$$(2.10) \qquad \ker_{S^*Y}(\sigma_\partial(A+G)) \cong L_+, \qquad \mathrm{coker}_{S^*Y}(\sigma_\partial(A+G)) \cong L_-.$$

PROOF. The construction of G may be given along the lines of Proposition 2.4 (i). Let us fix some $s \in \mathbb{R}$ such that $s > \max(\mu, d) - \frac{1}{2}$ (the specific choice of s will be unessential), and let

$$a : \pi_1^* E' \otimes H^s(\mathbb{R}_+) \to \pi_1^* F' \otimes H^{s-\mu}(\mathbb{R}_+)$$

for $\pi_1 : S^*Y \to Y$ denote the restriction of $\sigma_\partial(A)$ to S^*Y. According to Remark 2.5 there is a surjective bundle morphism

$$(a \quad k) : \begin{pmatrix} \pi_1^* E' \otimes H^s(\mathbb{R}_+) \\ \oplus \\ W \end{pmatrix} \to \pi_1^* F' \otimes H^{s-\mu}(\mathbb{R}_+)$$

for a $W \in \mathrm{Vect}(S^*Y)$, where $k : W \to \widetilde{W}$ is an isomorphism to some subbundle $\widetilde{W}$ of $\pi_1^* F' \otimes \mathcal{S}(\overline{\mathbb{R}}_+)$. Without loss of generality, we assume W to be trivial. Let

$p : \pi_1^* F' \otimes H^{s-\mu}(\mathbb{R}_+) \to \widetilde{W}$ be a projection that is orthogonal in the fibres with respect to the scalar products of $F'_y \otimes L^2(\mathbb{R}_+)$. By adding, if necessary, another finite-dimensional subbundle to W (and denoting the new bundle again by W) we obtain the following properties: There are subbundles $\widetilde{L}_- \subset \widetilde{W}$ and

$$\widetilde{L}_+ \subset \widetilde{V} := \ker_{S^*Y}((1-p)a) \subset \pi_1^* E' \otimes H^s(\mathbb{R}_+),$$

such that $\widetilde{L}_- \cong L_-$, $\widetilde{L}_+ \cong L_+$. In addition, choosing complements $\widetilde{L}_-^\perp$ in $\widetilde{W}$ and $\widetilde{L}_+^\perp$ in $\widetilde{V}$, we have $\widetilde{L}_-^\perp \cong \widetilde{L}_+^\perp$, provided the fibre dimension of W is sufficiently large. If $\lambda : \widetilde{L}_+^\perp \to \widetilde{L}_-^\perp$ is an isomorphism, and if

$$\iota^\perp : \widetilde{L}_-^\perp \to \pi_1^* F' \otimes H^{s-\mu}(\mathbb{R}_+)$$

is the canonical embedding, and

$$\pi^\perp : \pi_1^* E' \otimes H^s(\mathbb{R}_+) \to \widetilde{L}_+^\perp$$

the orthogonal projection, the operator family $a_0 := (1-p)a + q$ for $q := \iota^\perp \circ \lambda \circ \pi^\perp$ has the property

$$L_+ \cong \ker_{S^*Y} a_0, \quad L_- \cong \operatorname{coker}_{S^*Y} a_0. \tag{2.11}$$

The operator function $g := -pa + q : \pi_1^* E' \otimes H^s(\mathbb{R}_+) \to \pi_1^* F' \otimes H^{s-\mu}(\mathbb{R}_+)$ can be extended by homogeneity μ to a morphism

$$g_{(\mu)} : \pi_Y^* E' \otimes H^s(\mathbb{R}_+) \to \pi_Y^* F' \otimes H^{s-\mu}(\mathbb{R}_+),$$

i.e., $g_{(\mu)}(y, \lambda\eta) = \lambda^\mu \kappa_\lambda g_{(\mu)}(y,\eta)\kappa_\lambda^{-1}$ for all $\lambda \in \mathbb{R}_+$, $(y,\eta) \in T^*Y \setminus 0$, and $g_{(\mu)}|_{S^*Y} = g$. Now we may set

$$G := \sum_{j=1}^{L} \varphi_j (\chi_j^{-1})_* \operatorname{Op}(g_j) \psi_j, \tag{2.12}$$

cf. Definition 1.67, where $g_j(y,\eta)$ are local Green symbols of order μ which have $g_{(\mu)}(y,\eta)$ as homogeneous principal components (in local coordinates it suffices to set $g_j(y,\eta) := \chi(\eta) g_{(\mu)}(y,\eta)$ for any excision function $\chi(\eta)$). Because of $a_0 = \sigma_\partial(A+G)|_{S^*Y}$, the assertion follows from the relations (2.11). □

THEOREM 2.7. *Let $A \in B^{\mu,d}(X; E, F)$ be a σ_ψ-elliptic operator. Then there exist vector bundles*

$$J_\pm \in \operatorname{Vect}(Y) \quad \text{and} \quad L_\pm \in \operatorname{Vect}(T^*Y \setminus 0)$$

and an operator $\mathcal{A} \in \mathcal{B}^{\mu,d}(X; \boldsymbol{v})$ for $\boldsymbol{v} := (E, F; J_-, J_+)$ such that (1.94) *restricts to an isomorphism*

$$\begin{matrix} \pi_Y^* E' \otimes H^s(\mathbb{R}_+) \\ \oplus \\ L_- \end{matrix} \to \begin{matrix} \pi_Y^* F' \otimes H^{s-\mu}(\mathbb{R}_+) \\ \oplus \\ L_+ \end{matrix}. \tag{2.13}$$

PROOF. If $A \in B^{\mu,d}(X;E,F)$ is σ_ψ-elliptic, the boundary symbol $\sigma_\partial(A)$ represents a family of Fredholm operators (1.96), and there is an index element (1.97). Choose any $L_\pm \in \mathrm{Vect}(S^*Y)$ such that

$$\mathrm{ind}_{S^*Y}\,\sigma_\partial(A) = [L_+] - [L_-].$$

For abbreviation, $L_\pm$ will also denote the pull-backs of these bundles to $T^*Y \setminus 0$ under the canonical projection $T^*Y \setminus 0 \to S^*Y$. Applying Proposition 2.6 we find a Green operator $G \in B_G^{\mu,0}(X;E,F)$ such that the relations (2.10) hold.

Choose arbitrary bundle morphisms

$$k_1 : L_- \to \pi_1^* F' \otimes \mathcal{S}(\overline{\mathbb{R}}_+), \quad t_1 : \pi_1^* E' \otimes \mathcal{S}(\overline{\mathbb{R}}_+) \to L_+,$$

such that k_1 represents an isomorphism $L_- \to \widetilde{L}_-$, and t_1 restricts to an isomorphism $\widetilde{L}_+ \to L_+$, cf. the notation in the proof of Proposition 2.6. Then the block matrix

$$\begin{pmatrix} \sigma_\partial(A+G) & k_1 \\ t_1 & 0 \end{pmatrix} : \begin{matrix} \pi_1^* E' \otimes H^s(\mathbb{R}_+) \\ \oplus \\ L_- \end{matrix} \longrightarrow \begin{matrix} \pi_1^* F' \otimes H^{s-\mu}(\mathbb{R}_+) \\ \oplus \\ L_+ \end{matrix} \tag{2.14}$$

is an isomorphism for every $s > \max(\mu, d) - \frac{1}{2}$. Now let $J_\pm \in \mathrm{Vect}(Y)$ be arbitrary bundles such that $L_\pm$ are subbundles of $\pi_1^* J_\pm$ (for $J_\pm$ we may always take trivial bundles of sufficiently large fibre dimension). We then obtain a bundle morphism

$$\begin{pmatrix} \sigma_\partial(A+G) & k_0 \\ t_0 & 0 \end{pmatrix} : \begin{matrix} \pi_1^* E' \otimes H^s(\mathbb{R}_+) \\ \oplus \\ \pi_1^* J_- \end{matrix} \longrightarrow \begin{matrix} \pi_1^* F' \otimes H^{s-\mu}(\mathbb{R}_+) \\ \oplus \\ \pi_1^* J_+ \end{matrix} \tag{2.15}$$

when we set $k_0 := k_1 \circ \pi_-$ for a bundle projection $\pi_- : \pi_1^* J_- \to L_-$ and $t_0 := \iota_+ \circ t_1$ for the canonical embedding $\iota_+ : L_+ \to \pi_1^* J_+$. By construction, (2.15) restricts to the isomorphism (2.14). Next we extend (2.15) by κ_λ-homogeneity μ to a boundary symbol of the form (1.94), which has the form

$$\begin{pmatrix} \sigma_\partial(A+G) & k_{(\mu)} \\ t_{(\mu)} & 0 \end{pmatrix}$$

for unique $k_{(\mu)}(y,\eta)$ and $t_{(\mu)}(y,\eta)$ satisfying $k_{(\mu)}(y,\eta/|\eta|) = k_1(y,\eta/|\eta|)$ and $t_{(\mu)}(y,\eta/|\eta|) = t_1(y,\eta/|\eta|)$, respectively. Similarly to the construction of (2.12) we find potential and trace operators K and T, respectively, such that $\sigma_\partial(K) = k_{(\mu)}$, $\sigma_\partial(T) = t_{(\mu)}$.

Setting

$$\mathcal{A} = \begin{pmatrix} A+G & K \\ T & 0 \end{pmatrix} \tag{2.16}$$

we obtain an element in $\mathcal{B}^{\mu,d}(X;\boldsymbol{v})$ as desired. □

REMARK 2.8. For $\mathrm{ind}_{S^*Y}\,\sigma_\partial(A) \in \pi_1^* K(Y)$ we can carry out the construction in the latter proof with bundles $\pi_1^* J_\pm$ for suitable $J_\pm \in \mathrm{Vect}(Y)$ in place of $L_\pm$. Then (2.13) shows that the operator (2.16) is SL-elliptic.

PROOF OF THEOREM 2.1, (ii) $\Rightarrow$ (i). If the relation (1.99) is satisfied, the family of Fredholm operators (1.96) for any fixed $s > \max(\mu, d) - \frac{1}{2}$ can be completed to a block matrix (1.94) that is an isomorphism. For the construction we first restrict $\sigma_\partial(A)$ to S^*Y (denote it again by $\sigma_\partial(A)$) and set $J_- := \mathbb{C}^{j_-}$. For a sufficiently large j_- there is an injective bundle morphism

$$k_0 : \pi_1^* J_- \to \pi_1^* F' \otimes \mathcal{S}(\overline{\mathbb{R}}_+)$$

such that

$$(\sigma_\partial(A) \quad k_0) : \begin{matrix} \pi_1^* E' \otimes H^s(\mathbb{R}_+) \\ \oplus \\ \pi_1^* J_- \end{matrix} \to \pi_1^* F' \otimes H^{s-\mu}(\mathbb{R}_+)$$

is surjective. Then $\ker_{S^*Y}(\sigma_\partial(A) \quad k_0)$ is a finite-dimensional subbundle of

$$\begin{matrix} \pi_1^* E' \otimes \mathcal{S}(\overline{\mathbb{R}}_+) \\ \oplus \\ \pi_1^* J_- \end{matrix} . \tag{2.17}$$

As we saw by the above constructions, for a sufficiently large choice of j_- the bundle $\ker_{S^*Y}(\sigma_\partial(A) \quad k_0)$ is isomorphic to $\pi_1^* J_+$ for some $J_+ \in \mathrm{Vect}(S^*Y)$.

Now let

$$\kappa_0 : \ker_{S^*Y}(\sigma_\partial(A) \quad k_0) \to \pi_1^* J_+$$

be any isomorphism, and let

$$\pi_0 : \begin{matrix} \pi_1^* E' \otimes H^s(\mathbb{R}_+) \\ \oplus \\ \pi_1^* J_- \end{matrix} \to \ker_{S^*Y}(\sigma_\partial(A) \quad k_0) \tag{2.18}$$

be the orthogonal projection with respect to the $(E'_y \otimes L^2(\mathbb{R}_+)) \oplus J_{-,y}$-scalar products in the fibres, first for $s \geq \max(\mu, d)$ and then extended by continuity to all $s > \max(\mu, d) - \frac{1}{2}$.

Setting $(t_0 \quad q_0) := \kappa_0 \circ \pi_0$ we obtain an isomorphism

$$\begin{pmatrix} \sigma_\partial(A) & k_0 \\ t_0 & q_0 \end{pmatrix} : \quad \pi_1^* \begin{pmatrix} E' \otimes H^s(\mathbb{R}_+) \\ \oplus \\ J_- \end{pmatrix} \to \pi_1^* \begin{pmatrix} F' \otimes H^{s-\mu}(\mathbb{R}_+) \\ \oplus \\ J_+ \end{pmatrix} . \tag{2.19}$$

There is then a unique extension to an isomorphism

$$\sigma_\partial(\mathcal{A}) := \begin{pmatrix} \sigma_\partial(A) & k_{(\mu)} \\ t_{(\mu)} & q_{(\mu)} \end{pmatrix} : \quad \pi_Y^* \begin{pmatrix} E' \otimes H^s(\mathbb{R}_+) \\ \oplus \\ J_- \end{pmatrix} \to \pi_Y^* \begin{pmatrix} F' \otimes H^{s-\mu}(\mathbb{R}_+) \\ \oplus \\ J_+ \end{pmatrix} ,$$

homogeneous in the sense

$$\sigma_\partial(\mathcal{A})(y, \lambda\eta) = \lambda^\mu \operatorname{diag}(1, \kappa_\lambda) \sigma_\partial(\mathcal{A})(y, \eta) \operatorname{diag}(1, \kappa_\lambda^{-1})$$

for all $(y, \eta) \in T^*Y \setminus 0$ and all $\lambda \in \mathbb{R}_+$.

We finally pass to an element

$$\mathcal{A} = \begin{pmatrix} A & K \\ T & Q \end{pmatrix} \in \mathcal{B}^{\mu,d}(X; \boldsymbol{v})$$

for $\boldsymbol{v} = (E, F; J_-, J_+)$ that has $\sigma_\partial(\mathcal{A})$ as homogeneous principal symbol. The construction of the entries T, K and Q in terms of $t_{(\mu)}, k_{(\mu)}$ and $q_{(\mu)}$, respectively, is analogous to that for (2.16). □

2.2. A Toeplitz algebra of boundary value problems. For every $L \in \operatorname{Vect}(T^*Y \setminus 0)$ there exists an element $J \in \operatorname{Vect}(Y)$ such that L is a subbundle of $\pi_Y^* J$. In fact, we may choose J as a trivial bundle $Y \times \mathbb{C}^N$ (also written as $\mathbb{C}^N$) for a sufficiently large N. Let

$$p_{(0)} : \pi_Y^* J \to L$$

be a bundle morphism that is a projection to L, such that $p_{(0)}(y, \lambda\eta) = p_{(0)}(y, \eta)$ for all $\lambda \in \mathbb{R}_+$, $(y, \eta) \in T^*Y \setminus 0$. Then, by Theorem 1.14 there exists a $P \in L^0_{\mathrm{cl}}(Y; J, J)$ such that $P^2 = P$ and $p_{(0)} = \sigma_\psi(P)$. Recall that triples of the form

$$\mathbb{L} := (P, J, L) \tag{2.20}$$

are called global projection data and recall that $\mathcal{P}(Y)$ denotes the set of all such triples, cf. (1.19).

DEFINITION 2.9. Let

$$\mathbb{L}_+ := (P_+, J_+, L_+), \quad \mathbb{L}_- := (P_-, J_-, L_-) \in \mathcal{P}(Y) \tag{2.21}$$

be projection data, and let $R_- : P^s(Y, \mathbb{L}_-) \to H^s(Y, J_-)$ denote the canonical embedding, $s \in \mathbb{R}$. Moreover, let $\boldsymbol{v} := (E, F; J_-, J_+)$, and set $\boldsymbol{l} := (E, F; \mathbb{L}_-, \mathbb{L}_+)$. Then $\mathcal{S}^{\mu,d}(X; \boldsymbol{l})$ for $\mu \in \mathbb{Z}$, $d \in \mathbb{N}$, is defined as the set of all operators

$$\mathcal{A} := \begin{pmatrix} 1 & 0 \\ 0 & P_+ \end{pmatrix} \widetilde{\mathcal{A}} \begin{pmatrix} 1 & 0 \\ 0 & R_- \end{pmatrix} \tag{2.22}$$

for arbitrary $\widetilde{\mathcal{A}} \in \mathcal{B}^{\mu,d}(X; \boldsymbol{v})$.

THEOREM 2.10. *Every $\mathcal{A} \in \mathcal{S}^{\mu,d}(X; \boldsymbol{l})$ (with notation of Definition 2.9) induces continuous operators*

$$\mathcal{A} : \begin{matrix} H^s(X, E) \\ \oplus \\ P^s(Y, \mathbb{L}_-) \end{matrix} \to \begin{matrix} H^{s-\mu}(X, F) \\ \oplus \\ P^{s-\mu}(Y, \mathbb{L}_+) \end{matrix} \tag{2.23}$$

for all $s \in \mathbb{R}$, $s > d - \frac{1}{2}$.

PROOF. It suffices to apply Theorem 1.70 to the operator $\widetilde{\mathcal{A}}$ in (2.22) and to employ the definitions of R_- and P_+, cf. (1.20). □

We now introduce the principal symbolic structure of $\mathcal{S}^{\mu,d}(X;\boldsymbol{l})$. By definition, elements in that space are 2×2 block matrices $\mathcal{A} = (A_{ij})_{i,j=1,2}$ with $A_{11} = \text{ulc } \mathcal{A} \in B^{\mu,d}(X;E,F)$. We then call

$$\sigma_\psi(\mathcal{A}) := \sigma_\psi(\text{ulc } \mathcal{A}) : \pi_X^* E \to \pi_X^* F \tag{2.24}$$

the (homogeneous principal) interior symbol of $\mathcal{A}$. Furthermore, the operator family

(2.25)
$$\sigma_\partial(\mathcal{A}) := \begin{pmatrix} 1 & 0 \\ 0 & p_+ \end{pmatrix} \sigma_\partial(\widetilde{\mathcal{A}}) \begin{pmatrix} 1 & 0 \\ 0 & r_- \end{pmatrix} : \begin{matrix} \pi_Y^* E' \otimes H^s(\mathbb{R}_+) \\ \oplus \\ L_- \end{matrix} \to \begin{matrix} \pi_Y^* F' \otimes H^{s-\mu}(\mathbb{R}_+) \\ \oplus \\ L_+ \end{matrix}$$

is called the (homogeneous principal) boundary symbol of $\mathcal{A}$. Here $p_+(y,\eta)$ is the homogeneous principal symbol of order zero of the projection $P_+ \in L^0_{\text{cl}}(Y;J_+,J_+)$, and $r_- : L_- \to \pi_Y^* J_-$ is the canonical embedding.

Let us set

$$\sigma(\mathcal{A}) := (\sigma_\psi(\mathcal{A}), \sigma_\partial(\mathcal{A})). \tag{2.26}$$

REMARK 2.11. Let $\widetilde{\mathcal{A}} \in \mathcal{B}^{\mu,d}(X;\boldsymbol{v})$ for $\boldsymbol{v} = (E,F;J_-,J_+)$, and form the operator

$$\tilde{\widetilde{\mathcal{A}}} := \begin{pmatrix} 1 & 0 \\ 0 & P_+ \end{pmatrix} \widetilde{\mathcal{A}} \begin{pmatrix} 1 & 0 \\ 0 & P_- \end{pmatrix}$$

(with notation of Definition 2.9). Then we also have $\tilde{\widetilde{\mathcal{A}}} \in \mathcal{B}^{\mu,d}(X;\boldsymbol{v})$, if we interpret P_+ as a map $H^{s-\mu}(Y,J_+) \to H^{s-\mu}(Y,J_+)$ (not as $H^{s-\mu}(Y,J_+) \to P^{s-\mu}(Y,\mathbb{L}_+)$ as in (2.22)), and the operator (2.22) can also be written as

$$\mathcal{A} := \begin{pmatrix} 1 & 0 \\ 0 & P_+ \end{pmatrix} \tilde{\widetilde{\mathcal{A}}} \begin{pmatrix} 1 & 0 \\ 0 & R_- \end{pmatrix} \tag{2.27}$$

(with $P_+ : H^{s-\mu}(Y,J_+) \to P^{s-\mu}(Y,\mathbb{L}_+)$).
Then $\sigma(\tilde{\widetilde{\mathcal{A}}}) = 0$ in the sense of $\mathcal{B}^{\mu,d}(X;\boldsymbol{v})$, cf. (1.86), is equivalent to $\sigma(\mathcal{A}) = 0$ in the sense of $\mathcal{S}^{\mu,d}(X;\boldsymbol{l})$.

THEOREM 2.12. (i) $\mathcal{A} \in \mathcal{S}^{\mu,d}(X;\boldsymbol{l})$ *(cf. Definition 2.9) and* $\sigma(\mathcal{A}) = 0$ *imply* $\mathcal{A} \in \mathcal{S}^{\mu-1,d}(X;\boldsymbol{l})$, *and (2.23) is compact for every* $s > d - \frac{1}{2}$.

(ii) $\mathcal{A} \in \mathcal{S}^{\mu,d}(X;\boldsymbol{l}_0)$ *for* $\boldsymbol{l}_0 := (E_0,F;\mathbb{L}_0,\mathbb{L}_+)$, $\mathcal{B} \in \mathcal{S}^{\nu,e}(X;\boldsymbol{l}_1)$ *for* $\boldsymbol{l}_1 := (E,E_0;\mathbb{L}_-,\mathbb{L}_0)$, $\mathbb{L}_0,\mathbb{L}_\pm \in \mathcal{P}(Y)$, *implies* $\mathcal{A}\mathcal{B} \in \mathcal{S}^{\mu+\nu,h}(X;\boldsymbol{l}_0 \circ \boldsymbol{l}_1)$ *for* $\boldsymbol{l}_0 \circ \boldsymbol{l}_1 := (E,F;\mathbb{L}_-,\mathbb{L}_+)$, *and* $h = \max(\nu + d, e)$, *and we have*

$$\sigma(\mathcal{A}\mathcal{B}) = \sigma(\mathcal{A})\sigma(\mathcal{B})$$

(with componentwise multiplication).

(iii) $\mathcal{A} \in \mathcal{S}^{0,0}(X;\boldsymbol{l})$ *for* $\boldsymbol{l} := (E,F;\mathbb{L}_-,\mathbb{L}_+)$, *implies* $\mathcal{A}^* \in \mathcal{S}^{0,0}(X;\boldsymbol{l}^*)$ *for* $\boldsymbol{l}^* := (F,E;\mathbb{L}_+^*,\mathbb{L}_-^*)$ *in the sense of*

$$(u, \mathcal{A}^* v)_{L^2(X,E)\oplus P^0(Y,\mathbb{L}_-)} = (\mathcal{A}u, v)_{L^2(X,F)\oplus P^0(Y,\mathbb{L}_+)}$$

for all $u \in L^2(X,E) \oplus P^0(Y,\mathbb{L}_-)$, $v \in L^2(X,F) \oplus P^0(Y,\mathbb{L}_+)$, *and we have* $\sigma(\mathcal{A}^*) = \sigma(\mathcal{A})^*$ *(with componentwise adjoint, cf. Theorem* 1.70 *and Theorem* 1.30*).*

PROOF. (i) Let us write $\mathcal{A}$ in the form (2.27) and apply Remark 2.11. Then we have $\sigma(\tilde{\tilde{\mathcal{A}}}) = 0$. From Theorem 1.70 (ii) we then obtain $\tilde{\tilde{\mathcal{A}}} \in \mathcal{B}^{\mu-1,d}(X;\boldsymbol{v})$ which implies $\mathcal{A} \in \mathcal{S}^{\mu-1,d}(X;\boldsymbol{l})$. Moreover, $\tilde{\tilde{\mathcal{A}}}$ is compact and so is $\mathcal{A}$.

(ii) The operators $\mathcal{A}$ and $\mathcal{B}$ can be written in the form

$$\mathcal{A} = \begin{pmatrix} 1 & 0 \\ 0 & P_+ \end{pmatrix} \widetilde{\mathcal{A}} \begin{pmatrix} 1 & 0 \\ 0 & R_0 \end{pmatrix}, \qquad \mathcal{B} = \begin{pmatrix} 1 & 0 \\ 0 & P_0 \end{pmatrix} \widetilde{\mathcal{B}} \begin{pmatrix} 1 & 0 \\ 0 & R_- \end{pmatrix}$$

with P_+ from $\mathbb{L}_+ = (P_+, J_+, L_+)$, P_0 from $\mathbb{L}_0 = (P_0, J_0, L_0)$, $R_- : P^s(Y,\mathbb{L}_-) \longrightarrow H^s(Y,J_-)$, and the corresponding canonical embeddings $R_0 : P^{s-\nu}(Y,\mathbb{L}_0) \longrightarrow H^{s-\nu}(Y,J_0)$, moreover, $\widetilde{\mathcal{A}} \in \mathcal{B}^{\mu,d}(X;\boldsymbol{v}_0)$, $\boldsymbol{v}_0 := (E_0, F; J_0, J_+)$, as well as $\widetilde{\mathcal{B}} \in \mathcal{B}^{\nu,d}(X;\boldsymbol{v}_1)$, $\boldsymbol{v}_1 := (E, E_0; J_-, J_0)$. Then it follows that

$$\mathcal{A}\mathcal{B} = \begin{pmatrix} 1 & 0 \\ 0 & P_+ \end{pmatrix} \widetilde{\mathcal{A}} \begin{pmatrix} 1 & 0 \\ 0 & P_0 \end{pmatrix} \widetilde{\mathcal{B}} \begin{pmatrix} 1 & 0 \\ 0 & R_- \end{pmatrix}.$$

From $\widetilde{\mathcal{E}} := \widetilde{\mathcal{A}} \operatorname{diag}(1, P_0) \widetilde{\mathcal{B}} \in \mathcal{B}^{\mu+\nu,h}(X;\boldsymbol{v}_0 \circ \boldsymbol{v}_1)$, cf. Theorem 1.70, we obtain $\mathcal{A}\mathcal{B} \in \mathcal{S}^{\mu+\nu,h}(X;\boldsymbol{l}_0 \circ \boldsymbol{l}_1)$. In addition, we have

$$\sigma(\widetilde{\mathcal{E}}) = \sigma(\widetilde{\mathcal{A}}) \operatorname{diag}(1, p_0) \sigma(\widetilde{\mathcal{B}})$$

where p_0 is the homogeneous principal symbol of P_0, and hence we get $\sigma_\psi(\mathcal{A}\mathcal{B}) = \sigma_\psi(\mathcal{A})\sigma_\psi(\mathcal{B})$. Moreover, from

$$\begin{aligned} \sigma_\partial(\widetilde{\mathcal{E}}) &= \sigma_\partial(\widetilde{\mathcal{A}}) \operatorname{diag}(1, p_0) \sigma_\partial(\widetilde{\mathcal{B}}) \\ &= \sigma_\partial(\widetilde{\mathcal{A}}) \operatorname{diag}(1, r_0) \operatorname{diag}(1, p_0) \sigma_\partial(\widetilde{\mathcal{B}}) \end{aligned}$$

with $r_0 : L_0 \longrightarrow \pi_Y^* J_0$ being the canonical embedding, it follows that

$$\begin{aligned} \sigma_\partial(\mathcal{A}\mathcal{B}) &= \operatorname{diag}(1, p_+) \sigma_\partial(\widetilde{\mathcal{A}}) \operatorname{diag}(1, r_0) \operatorname{diag}(1, p_0) \sigma_\partial(\widetilde{\mathcal{B}}) \operatorname{diag}(1, r_-) \\ &= \sigma_\partial(\mathcal{A}) \sigma_\partial(\mathcal{B}). \end{aligned}$$

(iii) Writing $\mathcal{A}$ in the form (2.22) for $\widetilde{\mathcal{A}} \in \mathcal{B}^{0,0}(X;\boldsymbol{v})$, $\boldsymbol{v} := (E, F; J_-, J_+)$, we have

$$\mathcal{A}^* = \begin{pmatrix} 1 & 0 \\ 0 & P_-^* \end{pmatrix} \widetilde{\mathcal{A}}^* \begin{pmatrix} 1 & 0 \\ 0 & R_+^* \end{pmatrix}$$

with $\widetilde{\mathcal{A}}^* \in \mathcal{B}^{0,0}(X;\boldsymbol{v}^*)$ as in Theorem 1.70 (iv) and $P_\pm^*$ as in Theorem 1.30, where $R_+^* : P^s(Y, L_+^*) \to H^s(Y, J_+)$ is the canonical embedding. This yields $\mathcal{A}^* \in \mathcal{S}^{0,0,}(X;\boldsymbol{l}^*)$, cf. also the notation of Theorem 1.30. Moreover, we have

$$\sigma_\partial(\mathcal{A}^*) = \operatorname{diag}(1, p_-^*) \sigma_\partial(\widetilde{\mathcal{A}})^* \operatorname{diag}(1, r_+^*),$$

where p_-^* is the homogeneous principal symbol of the projection P_-^* and $r_+^* : L_+^* \hookrightarrow \pi_Y^* J_+$ the canonical embedding. From $\sigma_\partial(\widetilde{\mathcal{A}})^* = \sigma_\partial(\widetilde{\mathcal{A}}^*)$ we then obtain the assertion. □

THEOREM 2.13. *Let $\boldsymbol{l} := (E, F; \mathbb{L}_-, \mathbb{L}_+)$ for $E, F \in \text{Vect}(X)$, $\mathbb{L}_\pm \in \mathcal{P}(Y)$, and let $\mathcal{A}_j \in \mathcal{S}^{\mu-j,d}(X; \boldsymbol{l})$, $j \in \mathbb{N}$, be an arbitrary sequence. Then there exists an element $\mathcal{A} \in \mathcal{S}^{\mu,d}(X; \boldsymbol{l})$ such that*

$$\mathcal{A} - \sum_{j=0}^{N} \mathcal{A}_j \in \mathcal{S}^{\mu-(N+1),d}(X; \boldsymbol{l})$$

for every $N \in \mathbb{N}$, and $\mathcal{A}$ is unique $\mod \mathcal{S}^{-\infty,d}(X; \boldsymbol{l})$.

PROOF. The assertion is an immediate consequence of Theorem 1.71 and Definition 2.9. $\square$

Given $\mathcal{A} \in \mathcal{S}^{\mu,d}(X; \boldsymbol{l})$, $\mathcal{B} \in \mathcal{S}^{\mu,d}(X; \boldsymbol{m})$ for

$$\boldsymbol{l} := (E, F; \mathbb{L}_-, \mathbb{L}_+), \quad \mathbb{L}_\pm := (P_\pm, J_\pm, L_\pm),$$

and

$$\boldsymbol{m} := (V, W; \mathbb{M}_-, \mathbb{M}_+), \quad \mathbb{M}_\pm := (Q_\pm, G_\pm, M_\pm),$$

$\mathbb{L}_\pm, \mathbb{M}_\pm \in P(Y)$, we can form the direct sum

$$\mathcal{A} \oplus \mathcal{B} \in \mathcal{S}^{\mu,d}(X; \boldsymbol{l} \oplus \boldsymbol{m}) \tag{2.28}$$

for

$$\boldsymbol{l} \oplus \boldsymbol{m} := (E \oplus V, F \oplus W; \mathbb{L}_- \oplus \mathbb{M}_-, \mathbb{L}_+ \oplus \mathbb{M}_+),$$

$$\mathbb{L}_\pm \oplus \mathbb{M}_\pm := (P_\pm \oplus Q_\pm, J_\pm \oplus G_\pm, L_\pm \oplus M_\pm).$$

We then have

$$\sigma_\psi(\mathcal{A} \oplus \mathcal{B}) = \sigma_\psi(\mathcal{A}) \oplus \sigma_\psi(\mathcal{B}), \quad \sigma_\partial(\mathcal{A} \oplus \mathcal{B}) = \sigma_\partial(\mathcal{A}) \oplus \sigma_\partial(\mathcal{B})$$

with an evident meaning of '$\oplus$' for the symbolic components.

2.3. Ellipticity, parametrices, and the Fredholm property. Our next objective is to study ellipticity with global projection conditions.

DEFINITION 2.14. An operator $\mathcal{A} \in \mathcal{S}^{\mu,d}(X; \boldsymbol{l})$ for $\boldsymbol{l} := (E, F; \mathbb{L}_-, \mathbb{L}_+)$, $\mathbb{L}_\pm := (P_\pm, J_\pm, L_\pm)$, is called elliptic, if

(i) the interior symbol

$$\sigma_\psi(\mathcal{A}) : \pi_X^* E \to \pi_X^* F \tag{2.29}$$

is an isomorphism, and

(ii) the boundary symbol

$$\sigma_\partial(\mathcal{A}) : \begin{matrix} \pi_Y^* E' \otimes H^s(\mathbb{R}_+) \\ \oplus \\ L_- \end{matrix} \to \begin{matrix} \pi_Y^* F' \otimes H^{s-\mu}(\mathbb{R}_+) \\ \oplus \\ L_+ \end{matrix} \tag{2.30}$$

is an isomorphism for every $s \in \mathbb{R}$, $s > \max(\mu, d) - \frac{1}{2}$.

REMARK 2.15. Condition (ii) in Definition 2.14 holds if and only if it is satisfied for any fixed $s_0 > \max(\mu, d) - \frac{1}{2}$. Moreover, this is equivalent to the bijectivity of

$$\sigma_\partial(\mathcal{A}) : \begin{matrix} \pi_Y^* E' \otimes \mathcal{S}(\overline{\mathbb{R}}_+) \\ \oplus \\ L_- \end{matrix} \to \begin{matrix} \pi_Y^* F' \otimes \mathcal{S}(\overline{\mathbb{R}}_+) \\ \oplus \\ L_+ \end{matrix} . \tag{2.31}$$

THEOREM 2.16. *For every operator $A \in L^\mu_{\mathrm{cl}}(X; E, F)_{\mathrm{tr}}$ (cf. notation (1.66)) such that $\sigma_\psi(A) : \pi_X^* E \to \pi_X^* F$ is an isomorphism there exist projection data $\mathbb{L}_\pm \in \mathcal{P}(Y)$ and an element $\mathcal{A} \in \mathcal{S}^{\mu,0}(X; \boldsymbol{l})$ for $\boldsymbol{l} = (E, F; \mathbb{L}_-, \mathbb{L}_+)$ which is elliptic in the sense of Definition 2.14.*

PROOF. Let us choose elements $L_\pm \in \mathrm{Vect}(S^*Y)$ such that the relation (2.9) holds, and let $G \in B_G^{\mu,0}(X; E, F)$ be an operator as in Proposition 2.6. We can apply the construction of the proof of Theorem 2.7 and denote the operator (2.16) by $\widetilde{\mathcal{A}}$ instead of $\mathcal{A}$. This shows that the operator (2.16) has the asserted properties. □

PROPOSITION 2.17. *For every $\mu \in \mathbb{Z}$, $E \in \mathrm{Vect}(X)$ and $\mathbb{L} \in \mathcal{P}(Y)$ there exists an elliptic element $\mathcal{R}^\mu_{E,\mathbb{L}} \in \mathcal{S}^{\mu,0}(X; \boldsymbol{l})$ for $\boldsymbol{l} := (E, E; \mathbb{L}, \mathbb{L})$ which induces Fredholm operators*

$$\mathcal{R}^\mu_{E,\mathbb{L}} : \begin{matrix} H^s(X, E) \\ \oplus \\ P^s(Y, \mathbb{L}) \end{matrix} \to \begin{matrix} H^{s-\mu}(X, E) \\ \oplus \\ P^{s-\mu}(Y, \mathbb{L}) \end{matrix}$$

for all $s > \max(\mu, 0) - \frac{1}{2}$, such that $(\mathcal{R}^\mu_{E,\mathbb{L}})^{-1} \in \mathcal{S}^{-\mu,(-\mu)^+}(X; \boldsymbol{l})$.

PROOF. It suffices to set

$$\mathcal{R}^\mu_{E,\mathbb{L}} := \begin{pmatrix} R^\mu_E & 0 \\ 0 & R^\mu_{\mathbb{L}} \end{pmatrix}$$

with R^μ_E from Theorem 1.60 and $R^\mu_{\mathbb{L}}$ from Remark 1.32. □

REMARK 2.18. Let R^μ_E be as in Theorem 1.60. Then

$$\sigma_\partial(R^\mu_E) : \pi_Y^* E' \otimes H^s(\mathbb{R}_+) \to \pi_Y^* E' \otimes H^{s-\mu}(\mathbb{R}_+)$$

is an isomorphism, $s > \max(\mu, 0) - \frac{1}{2}$. This implies

$$\mathrm{ind}_{S^*Y}\, \sigma_\partial(R^\mu_E) = 0.$$

Similarly to R^μ_E we can form an operator $S^\mu_E \in L^\mu_{\mathrm{cl}}(X; E, E)_{\mathrm{tr}}$ in terms of the local symbols $\mathrm{r}^\mu_+(\eta, \tau) := \overline{\mathrm{r}^\mu_-(\eta, \tau)}$ (the complex conjugate), cf. Remark 1.61. The operator $S^\mu_E \in B^{\mu,0}(X; E, E)$ can be chosen in such a way that

$$\sigma_\partial(S^\mu_E) : \pi_Y^* E' \otimes H^s(\mathbb{R}_+) \to \pi_Y^* E' \otimes H^{s-\mu}(\mathbb{R}_+)$$

is surjective and $\ker_{S^*Y} \sigma_\partial(S^\mu_E) = \mu[\pi_1^* E']$; this yields

$$\mathrm{ind}_{S^*Y}\, \sigma_\partial(S^\mu_E) = \mu[\pi_1^* E']. \tag{2.32}$$

THEOREM 2.19. *Let $\mathcal{A} \in \mathcal{S}^{\mu,d}(X;\boldsymbol{l})$, $\boldsymbol{l} := (E,F;\mathbb{L}_-,\mathbb{L}_+)$, be elliptic. Then there exists an elliptic operator $\mathcal{B} \in \mathcal{S}^{\mu,0}(X;\boldsymbol{m})$ for $\boldsymbol{m} := (F,E;\mathbb{M}_-,\mathbb{M}_+)$ with suitable projection data $\mathbb{M}_\pm \in \mathcal{P}(Y)$ of the form $\mathbb{M}_\pm := (Q_\pm,\mathbb{C}^N,M_\pm)$ for some $N \in \mathbb{N}$, such that $\mathcal{A}\oplus\mathcal{B} \in \mathcal{B}^{\mu,d}(X;\boldsymbol{v})$, $\boldsymbol{v} := (E\oplus F, F\oplus E;\mathbb{C}^N,\mathbb{C}^N)$, is SL-elliptic.*

PROOF. Choose any $s \in \mathbb{N}$, $s-\mu \geq 0$, and set $A_0 := R_F^{s-\mu} A R_E^{-s} : L^2(X,E) \to L^2(X,F)$, cf. Theorem 1.60. Then we have $A_0 \in B^{0,0}(X;E,F)$ and

$$\operatorname{ind}_{S^*Y}\sigma_\partial(A_0) = \operatorname{ind}_{S^*Y}\sigma_\partial(A) = [L_+] - [L_-].$$

For the L^2-adjoint $A_0^* \in B^{0,0}(X;F,E)$, cf. Theorem 1.70 (iv), we have

$$\operatorname{ind}_{S^*Y}\sigma_\partial(A_0^*) = [L_-] - [L_+]$$

as well as $\operatorname{ind}_{S^*Y}\sigma_\partial(B_1) = [L_-] - [L_+]$ for $B_1 := R_E^{-s+\mu} A_0^* R_F^s \in B^{\mu,e}(X;F,E)$ with some type $e \in \mathbb{N}$. The operator B_1 can be written as $B + G_1$ for a certain $B \in B^{\mu,0}(X;F,E)$ and a Green operator $G_1 \in B_G^{\mu,e}(X;F,E)$, and then

$$\operatorname{ind}_{S^*Y}\sigma_\partial(B) = [L_-] - [L_+],$$

because $\sigma_\partial(G_1)$ is a family of compact operators in the respective Sobolev spaces on $\mathbb{R}_+$.

There are bundles $M_+, M_- \in \operatorname{Vect}(S^*Y)$ such that $M_- \oplus L_- \cong M_+ \oplus L_+ \cong \mathbb{C}^N$, and we obtain

$$\operatorname{ind}_{S^*Y}\sigma_\partial(B) = [M_+] - [M_-].$$

Applying Theorem 2.7 and the proof of Theorem 2.16 we find an elliptic operator $\mathcal{B} \in \mathcal{S}^{\mu,0}(X;\boldsymbol{m})$ for $\boldsymbol{m} = (F,E;\mathbb{M}_-,\mathbb{M}_+)$, $\mathbb{M}_\pm = (Q_\pm,\mathbb{C}^N,M_\pm)$, such that $\ker_{S^*Y}\sigma_\partial(\mathcal{B}) \cong M_+$, $\operatorname{coker}_{S^*Y}\sigma_\partial(\mathcal{B}) \cong M_-$. The operator $\mathcal{B}$ is then as desired. □

DEFINITION 2.20. Let $\mathcal{A} \in \mathcal{S}^{\mu,d}(X;\boldsymbol{l})$ for $\boldsymbol{l} = (E,F;\mathbb{L}_-,\mathbb{L}_+)$, $\mathbb{L}_\pm \in \mathcal{P}(Y)$. An operator $\mathcal{P} \in \mathcal{T}^{-\mu,e}(X;\boldsymbol{l}^{-1})$ for $\boldsymbol{l}^{-1} := (F,E;\mathbb{L}_+,\mathbb{L}_-)$ and some $e \in \mathbb{N}$ is called a parametrix of $\mathcal{A}$, if the operators

$$\mathcal{C}_l := \mathcal{I} - \mathcal{P}\mathcal{A} \quad \text{and} \quad \mathcal{C}_r := \mathcal{I} - \mathcal{A}\mathcal{P} \tag{2.33}$$

belong to $\mathcal{S}^{-\infty,d_l}(X;\boldsymbol{m}_l)$ and $\mathcal{S}^{-\infty,d_r}(X;\boldsymbol{m}_r)$, respectively, for certain $d_l, d_r \in \mathbb{N}$, $\boldsymbol{m}_l := (E,E;\mathbb{L}_-,\mathbb{L}_-)$, and $\boldsymbol{m}_r := (F,F;\mathbb{L}_+,\mathbb{L}_+)$.

THEOREM 2.21. *Let $\mathcal{A} \in \mathcal{S}^{\mu,d}(X;\boldsymbol{l})$, $\mu \in \mathbb{Z}$, $d \in \mathbb{N}$, $\boldsymbol{v} := (E,F;\mathbb{L}_-,\mathbb{L}_+)$ for $E,F \in \operatorname{Vect}(X)$, $\mathbb{L}_\pm \in \mathcal{P}(Y)$.*

(i) *The operator $\mathcal{A}$ is elliptic if and only if*

$$\mathcal{A} : \begin{array}{c} H^s(X,E) \\ \oplus \\ P^s(Y,\mathbb{L}_-) \end{array} \to \begin{array}{c} H^{s-\mu}(X,F) \\ \oplus \\ P^{s-\mu}(Y,\mathbb{L}_+) \end{array} \tag{2.34}$$

is a Fredholm operator for an $s = s_0 \in \mathbb{R}$, $s_0 > \max(\mu,d) - \frac{1}{2}$.

(ii) *If $\mathcal{A}$ is elliptic, (2.34) is a Fredholm operator for all $s > \max(\mu,d) - \frac{1}{2}$, and $\dim\ker\mathcal{A}$ and $\dim\operatorname{coker}\mathcal{A}$ are independent of s.*

(iii) *An elliptic operator* $\mathcal{A}$ *has a parametrix* $\mathcal{P} \in \mathcal{S}^{-\mu,(d-\mu)^+}(X;\boldsymbol{l}^{-1})$ *(in the sense of Definition* 2.20*) for* $d_l = \max(\mu,d), d_r = (d-\mu)^+$*, and* $\mathcal{P}$ *can be chosen in such a way that the remainders in the relation* (2.33) *are projections*

$$\mathcal{C}_l : H^s(X,E) \oplus P^s(Y,\mathbb{L}_-) \to V, \quad \mathcal{C}_r : H^{s-\mu}(X,F) \oplus P^{s-\mu}(Y,\mathbb{L}_+) \to W$$

for all $s > \max(\mu,d) - \frac{1}{2}$*, for* $V := \ker \mathcal{A} \subset C^\infty(X,E) \oplus P^\infty(Y,\mathbb{L}_-)$ *and a finite-dimensional subspace* $W \subset C^\infty(X,F) \oplus P^\infty(Y,\mathbb{L}_+)$ *with the property* $W + \operatorname{im}\mathcal{A} = H^{s-\mu}(X,F) \oplus P^{s-\mu}(Y,\mathbb{L}_+)$*,* $W \cap \operatorname{im}\mathcal{A} = \{0\}$ *for every* $s > \max(\mu,d) - \frac{1}{2}$.

PROOF. Let $\mathcal{A} \in \mathcal{S}^{\mu,d}(X;\boldsymbol{l})$, $\boldsymbol{l} := (E,F;\mathbb{L}_-,\mathbb{L}_+)$, be elliptic. Choose an elliptic operator $\mathcal{B} \in \mathcal{S}^{\mu,0}(X;\boldsymbol{m})$, $\boldsymbol{m} := (F,E;\mathbb{M}_-,\mathbb{M}_+)$ as in Theorem 2.19 such that $\mathcal{A}\oplus\mathcal{B} \in \mathcal{B}^{\mu,d}(X;\boldsymbol{v})$ is SL-elliptic. Applying Theorem 1.75 (iii) we find a parametrix $(\mathcal{A}\oplus\mathcal{B})^{(-1)} \in \mathcal{B}^{-\mu,(d-\mu)^+}(X;\boldsymbol{v}^{-1})$, such that the remainders $\mathcal{I}-(\mathcal{A}\oplus\mathcal{B})^{(-1)}(\mathcal{A}\oplus\mathcal{B})$ and $\mathcal{I}-(\mathcal{A}\oplus\mathcal{B})(\mathcal{A}\oplus\mathcal{B})^{(-1)}$ are of type $d_l = \max(\mu,d)$ and $d_r = (d-\mu)^+$, respectively. For the principal symbolic components of $(\mathcal{A}\oplus\mathcal{B})^{(-1)}$ we have

$$\sigma_\psi((\mathcal{A}\oplus\mathcal{B})^{(-1)}) = \sigma_\psi(\mathcal{A})^{-1} \oplus \sigma_\psi(\mathcal{B})^{-1}$$

and

$$\sigma_\partial((\mathcal{A}\oplus\mathcal{B})^{(-1)}) = \sigma_\partial(\mathcal{A})^{-1} \oplus \sigma_\partial(\mathcal{B})^{-1}.$$

In particular,

$$\sigma_\partial\Big((\mathcal{A}\oplus\mathcal{B})^{(-1)}\Big) : \begin{matrix} \pi_Y^* F' \otimes H^{s-\mu}(\mathbb{R}_+) \\ \oplus \\ L_+ \\ \oplus \\ \pi_Y^* E' \otimes H^{s-\mu}(\mathbb{R}_+) \\ \oplus \\ M_+ \end{matrix} \to \begin{matrix} \pi_Y^* E' \otimes H^{s}(\mathbb{R}_+) \\ \oplus \\ L_- \\ \oplus \\ \pi_Y^* F' \otimes H^{s}(\mathbb{R}_+) \\ \oplus \\ M_- \end{matrix} \tag{2.35}$$

is an isomorphism which induces isomorphisms $\sigma_\partial(\mathcal{A})^{-1}$ and $\sigma_\partial(\mathcal{B})^{-1}$ between the respective bundles separately. In particular, by omitting the third row and column of (2.35) we obtain a morphism

$$\sigma_\partial(\widetilde{\mathcal{P}}_0) : \begin{matrix} \pi_Y^* F' \otimes H^{s-\mu}(\mathbb{R}_+) \\ \oplus \\ \mathbb{C}^N \end{matrix} \to \begin{matrix} \pi_Y^* E' \otimes H^{s}(\mathbb{R}_+) \\ \oplus \\ \mathbb{C}^N \end{matrix} \tag{2.36}$$

which is the boundary symbol of an operator $\widetilde{\mathcal{P}}_0 \in \mathcal{B}^{-\mu,(d-\mu)^+}(X;\boldsymbol{w})$ for $\boldsymbol{w} := (F,E;\mathbb{C}^N,\mathbb{C}^N)$ such that $\sigma_\psi(\widetilde{\mathcal{P}}_0) = \sigma_\psi(\mathcal{A})^{-1}$, and $(\sigma_\partial(\mathcal{P}_0))$ restricts to

$$\sigma_\partial(\mathcal{A})^{-1} : \begin{matrix} \pi_Y^* F' \otimes H^{s-\mu}(\mathbb{R}_+) \\ \oplus \\ L_+ \end{matrix} \to \begin{matrix} \pi_Y^* E' \otimes H^{s}(\mathbb{R}_+) \\ \oplus \\ L_- \end{matrix}.$$

Thus, if we set

$$\mathcal{P}_0 := \begin{pmatrix} 1 & 0 \\ 0 & P_{L_-} \end{pmatrix} \widetilde{\mathcal{P}}_0 \begin{pmatrix} 1 & 0 \\ 0 & R_{L_+} \end{pmatrix}$$

with the projection $P_{L_-} : H^s(Y, \mathbb{C}^N) \to P^s(Y, \mathbb{L}_-)$ and the canonical embedding $R_{L_+} : P^{s-\mu}(Y, \mathbb{L}_+) \to H^s(Y, \mathbb{C}^N)$, we obtain an operator $\mathcal{P}_0 \in \mathcal{S}^{-\mu,(d-\mu)^+}(X; \boldsymbol{l}^{-1})$ such that

$$\sigma_\psi(\mathcal{P}_0) = \sigma_\psi(\mathcal{A})^{-1}, \qquad \sigma_\partial(\mathcal{P}_0) = \sigma_\partial(\mathcal{A})^{-1}.$$

Thus we have $\mathcal{C}_l^0 := \mathcal{I} - \mathcal{P}_0\mathcal{A} \in \mathcal{S}^{-1,d_l}(X; \boldsymbol{m}_l)$, and $\mathcal{C}_r^0 := \mathcal{I} - \mathcal{A}\mathcal{P}_0 \in \mathcal{S}^{-1,d_r}(X; \boldsymbol{m}_r)$, cf. Theorem 2.12 (i), (ii).

Let us form $\mathcal{K} \sim -\sum_{j=1}^\infty (\mathcal{C}_l^0)^j$, cf. Theorem 2.13, and set $\mathcal{P}_l := (\mathcal{I} - \mathcal{K})\mathcal{P}_0$ which belongs to $\mathcal{S}^{-\mu,d_l}(X; \boldsymbol{l}^{-1})$. Then it follows that $\mathcal{I} - \mathcal{P}_l\mathcal{A} \in \mathcal{S}^{-\infty,d_l}(X; \boldsymbol{m}_l)$. In an analogous manner we find an operator $\mathcal{P}_r \in \mathcal{S}^{-\infty,d_r}(X; \boldsymbol{l}^{-1})$ such that $\mathcal{I} - \mathcal{A}\mathcal{P}_r \in \mathcal{S}^{-\infty,d_r}(X; \boldsymbol{m}_r)$. In other words, there is a left parametrix $\mathcal{P}_l$ and a right parametrix $\mathcal{P}_r$ of $\mathcal{A}$, i.e., we may set $\mathcal{P} := \mathcal{P}_l$. From Theorem 2.12 we thus obtain that (2.34) is a Fredholm operator for every $s > \max(\mu, d) - \frac{1}{2}$. The assertions of Theorem 2.21 (ii) and (iii) now follow in a similar manner as the analogous ones of Theorem 1.33 (ii), (iii), again by applying Remark 1.8 to the present situation. Thus, to complete the proof of Theorem 2.21 it remains to show that the Fredholm property of (2.34) for an $s_0 > \max(\mu, d) - \frac{1}{2}$ entails the ellipticity.

If (2.34) is a Fredholm operator for $s = s_0 > \max(\mu, d) - \frac{1}{2}$, also

$$\mathcal{A}_0 := R_{F,\mathbb{L}_+}^{s_0-\mu} \mathcal{A} \Big(R_{E,\mathbb{L}_-}^{s_0} \Big)^{-1} \in \mathcal{S}^{0,0}(X; \boldsymbol{l})$$

is a Fredholm operator

$$\mathcal{A}_0 : \begin{matrix} L^2(X,E) \\ \oplus \\ P^0(Y,\mathbb{L}_-) \end{matrix} \longrightarrow \begin{matrix} L^2(X,F) \\ \oplus \\ P^0(Y,\mathbb{L}_+) \end{matrix}, \tag{2.37}$$

cf. the notation in Proposition 2.17. If we show the ellipticity of $\mathcal{A}_0$, we also obtain the ellipticity of $\mathcal{A}$ itself, because the order reducing operators are elliptic, and compositions of elliptic operators are again elliptic. To simplify notation we write $\mathcal{A} := \mathcal{A}_0$. We now proceed in a similar manner as in the proof of Theorem 1.33 and use the same notation as in (1.35), (1.36), (1.37) (with Y in place of M). Then we can form the operator

$$\mathcal{B} := \begin{pmatrix} 1 & 0 & 0 \\ 0 & R_+ & 0 \\ 0 & 0 & R_-^\perp \end{pmatrix} \begin{pmatrix} \mathcal{A} & 0 \\ 0 & E \end{pmatrix} : \begin{matrix} L^2(X,E) \\ \oplus \\ L^2(Y,J_-) \end{matrix} \longrightarrow \begin{matrix} L^2(X,F) \\ \oplus \\ L^2(Y,J_+ \oplus J_-) \end{matrix}$$

The operator $\mathcal{A}$ is Fredholm as a map (2.37). Hence there is a

$$\mathcal{Q} : \begin{matrix} L^2(X,F) \\ \oplus \\ P^0(Y,\mathbb{L}_+) \end{matrix} \longrightarrow \begin{matrix} L^2(X,E) \\ \oplus \\ P^0(Y,\mathbb{L}_-) \end{matrix}$$

such that

$$\mathcal{I} - \mathcal{Q}\mathcal{A} : \begin{matrix} L^2(X,E) \\ \oplus \\ P^0(Y,\mathbb{L}_-) \end{matrix} \to \begin{matrix} L^2(X,E) \\ \oplus \\ P^0(Y,\mathbb{L}_-) \end{matrix}$$

is compact. Let

$$\mathcal{S} : \begin{matrix} L^2(X,F) \\ \oplus \\ L^2(Y,J_+ \oplus J_-) \end{matrix} \to \begin{matrix} L^2(X,F) \\ \oplus \\ P^0(Y,\mathbb{L}_+) \oplus P^0(Y,\mathbb{L}_-^\perp) \end{matrix}$$

denote a projection. Then

$$\mathcal{T} := \begin{pmatrix} \mathcal{Q} & 0 \\ 0 & E \end{pmatrix} \circ \mathcal{S} : \begin{matrix} L^2(X,F) \\ \oplus \\ L^2(Y,J_+ \oplus J_-) \end{matrix} \to \begin{matrix} L^2(X,E) \\ \oplus \\ L^2(Y,J_-) \end{matrix}$$

has the property that $\mathcal{I} - \mathcal{T}\mathcal{B} =: \mathcal{K}$ is compact in the space $L^2(X,E) \oplus L^2(Y,J_-)$. Since $\mathcal{I} - \mathcal{K}$ is a Fredholm operator in the latter space, it follows that $\dim\ker(\mathcal{I} - \mathcal{K}) < \infty$, and hence $\dim\ker\mathcal{B} < \infty$ since

$$\ker\mathcal{B} \subset \ker(\mathcal{I} - \mathcal{K}).$$

The operator $\mathcal{B}^*\mathcal{B} : L^2(X,E) \oplus L^2(Y,J_-) \to L^2(X,E) \oplus L^2(Y,J_-)$ belongs to $\mathcal{B}^{0,0}(X;\boldsymbol{w})$ for $\boldsymbol{w} := (E,E;J_-,J_-)$ and is Fredholm. From Theorem 1.75 (i) we know that $\mathcal{B}^*\mathcal{B}$ is elliptic. It follows that both $\sigma_\psi(\mathcal{A})$ and $\sigma_\partial(\mathcal{A})$ are injective. By passing to adjoint operators in an analogous manner we can show that $\sigma_\psi(\mathcal{A})$ and $\sigma_\partial(\mathcal{A})$ are also surjective. This completes the proof of Theorem 2.21. □

2.4. Reduction to the boundary. Let

$$\mathcal{A}_i = \begin{pmatrix} D \\ T_i \end{pmatrix} \in \mathcal{S}^{\mu,d}(X;\boldsymbol{l}) \tag{2.38}$$

for $\boldsymbol{l} := (E,F;0,\mathbb{L}_i)$, $\mathbb{L}_i \in \mathcal{P}(Y)$, $i = 1,2$, be two elliptic boundary value problems for the same σ_ψ-elliptic operator $D \in B^{\mu,d}(X;E,F)$ (without loss of generality we assume the type d in $\mathcal{A}_i$ to be independent of i).

We want to reduce the operator $\mathcal{A}_2$ to the boundary by means of $\mathcal{A}_1$. By virtue of Theorem 2.21 (iii) there are parametrices $\mathcal{P}_i \in \mathcal{S}^{-\mu,(d-\mu)^+}(X;\boldsymbol{l}^{-1})$. They have the form of row matrices

$$\mathcal{P}_i =: (G_i \quad K_i), \tag{2.39}$$

where G_i is an analogue of Green's function of the boundary value problem $\mathcal{A}_i$, and K_i is a potential operator. Using

$$\mathcal{A}_1\mathcal{P}_1 = \begin{pmatrix} 1 & 0 \\ 0 & 1 \end{pmatrix} \quad \operatorname{mod} \mathcal{S}^{-\infty,(d-\mu)^+}(X;\boldsymbol{m}_r)$$

for $\boldsymbol{m}_r := (F,F;\mathbb{L}_1,\mathbb{L}_1)$, we obtain

$$\mathcal{A}_2\mathcal{P}_1 = \begin{pmatrix} 1 & 0 \\ T_2G_1 & T_2K_1 \end{pmatrix} \quad \operatorname{mod} \mathcal{S}^{-\infty,(d-\mu)^+}(X;\boldsymbol{m}), \tag{2.40}$$

for $\boldsymbol{m} := (F, F; \mathbb{L}_1, \mathbb{L}_2)$, and we have

$$Q := T_2 K_1 \in \mathcal{T}^0(Y; \mathbb{L}_1, \mathbb{L}_2),$$

cf. Definition 1.22. Because both $\mathcal{A}_2$ and $\mathcal{P}_1$ are Fredholm operators, also

$$Q : P^s(Y, \mathbb{L}_1) \to P^s(Y, \mathbb{L}_2)$$

is Fredholm, and $\operatorname{ind} Q$ is independent of s, cf. Remark 1.38.

The operator Q is called the reduction of $\mathcal{A}_2$ to the boundary by means of $\mathcal{A}_1$. The following result is an analogue of the Agranovich-Dynin index formula.

THEOREM 2.22. *Let $\mathcal{A}_i$, $i = 1, 2$, be two elliptic operators* (2.38). *Then we have*

$$\operatorname{ind} \mathcal{A}_2 - \operatorname{ind} \mathcal{A}_1 = \operatorname{ind} Q.$$

PROOF. The assertion follows from the relation (2.40), using the fact that smoothing remainders are compact operators and $\operatorname{ind} \mathcal{P}_1 = -\operatorname{ind} \mathcal{A}_1$. □

REMARK 2.23. Let $\mathcal{A}_i$, be two elliptic operators (2.38) for $\mathbb{L}_i = (P_i, J, L)$ with the same bundles J, L but different projections P_i, $i = 1, 2$, and assume $T_i = P_i \widetilde{T}$ for the same trace operator $\widetilde{T}$. Then, if $R_1 : P^s(Y, \mathbb{L}_1) \to H^s(Y, J)$ is the canonical embedding, we have (in the notation of (2.40))

$$T_2 K_1 = P_2 R_1$$

modulo a compact operator and hence, using Proposition 1.18 and notation (1.17)

$$\operatorname{ind} \mathcal{A}_2 - \operatorname{ind} \mathcal{A}_1 = \operatorname{ind}(P_1, P_2).$$

This is a consequence of the fact that when $\widetilde{K}$ denotes the potential operator that appears in the parametrix construction for $\mathcal{A}_1$ as in the proof of Theorem 2.21 we have $\sigma_\partial(\widetilde{T})\sigma_\partial(\widetilde{K}) = \mathrm{id}$ as a map $L \to L$.

REMARK 2.24. The procedure to reduce elliptic elements of $\mathcal{S}^{\mu,d}(X, \boldsymbol{l}_i)$ with the same upper left corners to the boundary can be generalised to arbitrary $\boldsymbol{l} = (E, F; \mathbb{L}_{-,i}, \mathbb{L}_{+,i})$, $i = 1, 2$. The algebraic technique is the same as in [**31**, Section 3.2.1.3]. There is then an immediate analogue of Theorem 2.22 for the general case.

3. Transmission operators and Cauchy data spaces

3.1. Transmission operators. Let M be a closed compact C^∞ manifold which is subdivided into compact C^∞ manifolds X_+ and X_- with common C^∞ boundary Y, i.e.,

$$M = X_+ \cup X_-, \quad Y = X_+ \cap X_-.$$

An example is $M = 2X$, the double of X, where two copies of X are glued together along $Y = \partial X$.

Given an elliptic operator $A \in L^\mu_{\mathrm{cl}}(X; V, W)$, $V, W \in \operatorname{Vect}(M)$, we can consider the restrictions

$$A_\pm := A|_{\operatorname{int} X_\pm} \in L^\mu_{\mathrm{cl}}(\operatorname{int} X_\pm; V_\pm, \mathbb{W}_\pm),$$

$V_\pm := V|_{\text{int } X_\pm}$, $W_\pm := W|_{\text{int } X_\pm}$. We want to study the question to what extent the index of the Fredholm operator

$$A : H^s(X, V) \to H^{s-\mu}(X, W)$$

can be compared with Fredholm indices of elliptic boundary value problems for $A_\pm$ on the $\pm$-sides $X_\pm$ with respect to Y.

At first glance, such a problem appears very natural, for instance, when A is an elliptic differential operator. However, in the pseudo-differential case there is a basic analytic problem: The operators $A_\pm$ have not necessarily the transmission property at the boundary Y (which is, in fact, the exception). Moreover, in general there do not exist Shapiro-Lopatinskij-elliptic boundary conditions for $A_\pm$ (also for differential operators), although there are always elliptic projection conditions, cf. Theorem 2.16, when the transmission property is satisfied.

For convenience, we first consider the case $M = 2X$ and assume that A has the transmission property at Y. Let $\mathrm{r}^\pm$ denote the operator of restriction from M to $\operatorname{int} X_\pm$ and $\mathrm{e}^\pm$ the operator of extension by zero from $\operatorname{int} X_\pm$ to M. Moreover, let ε be the reflection map that maps a point $x_+ \in X_+$ to its counterpart $x_- \in X_-$ and conversely; then Y remains fixed. We use ε as a diffeomorphism $\varepsilon : X_+ \to X_-$ as well as $\varepsilon : X_- \to X_+$ (this should not cause confusion).

We then have

$$\mathrm{r}^+ A \mathrm{e}^+ \in B^{\mu,0}(X_+; V_+, W_+), \mathrm{r}^- A \mathrm{e}^- \in B^{\mu,0}(X_-; V_-, W_-). \tag{3.1}$$

and

$$\mathrm{r}^+ A \mathrm{e}^- \varepsilon^*,\ \varepsilon^* \mathrm{r}^- A \mathrm{e}^+ \in B_G^{\mu,0}(X_+; V_+, W_+), \tag{3.2}$$

$$\mathrm{r}^- A \mathrm{e}^+ \varepsilon^*,\ \varepsilon^* \mathrm{r}^+ A \mathrm{e}^- \in B_G^{\mu,0}(X_-; V_-, W_-). \tag{3.3}$$

Let us first assume that $A_0 \in L^0_{\mathrm{cl}}(M; V, W)$ is an elliptic operator. The ellipticity of A_0 is equivalent to the Fredholm property of the operator

$$A_0 : L^2(M, V) \to L^2(M, W),$$

cf. Theorem 1.7 (i), or, equivalently, of

$$\begin{pmatrix} \mathrm{r}^+ A_0 \mathrm{e}^+ & \mathrm{r}^+ A_0 \mathrm{e}_- \\ \mathrm{r}^- A_0 \mathrm{e}^+ & \mathrm{r}^- A_0 \mathrm{e}_- \end{pmatrix} : \begin{matrix} L^2(X_+, V_+) \\ \oplus \\ L^2(X_-, V_-) \end{matrix} \to \begin{matrix} L^2(X_+, V_+) \\ \oplus \\ L^2(X_-, V_-) \end{matrix}.$$

Writing $X := X_+$, $E_1 := V_+$, $E_2 := \varepsilon^* V_-$, $F_1 := W_+$, $F_2 := \varepsilon^* W_-$ this is equivalent to the Fredholm property of

$$\mathbb{A} := \begin{pmatrix} \mathrm{r}^+ A_0 \mathrm{e}^+ & \mathrm{r}^+ A_0 \mathrm{e}^- \varepsilon^* \\ \varepsilon^* \mathrm{r}^- A_0 \mathrm{e}^+ & \varepsilon^* \mathrm{r}^- A_0 \mathrm{e}^- \varepsilon^* \end{pmatrix} : \begin{matrix} L^2(X, E_1) \\ \oplus \\ L^2(X, E_2) \end{matrix} \to \begin{matrix} L^2(X, F_1) \\ \oplus \\ L^2(X, F_2) \end{matrix}. \tag{3.4}$$

By assumption, A_0 has the transmission property at Y; recall that this condition is symmetric with respect to both sides $X_\pm$. Thus we have $\mathbb{A} \in B^{0,0,}(X; E_1 \oplus E_2, F_1 \oplus F_2)$. Because of the Fredholm property of (3.4) the operator $\mathbb{A}$ is elliptic in $B^{0,0}(X;E_1 \oplus E_2, F_1 \oplus F_2)$, cf. Theorem 1.75 (i). In other words, the symbols $\sigma_\psi(\mathbb{A}) : \pi_X^*(E_1 \oplus E_2) \to \pi_X^*(F_1 \oplus F_2)$ and

$$\sigma_\partial(\mathbb{A}) : \pi_Y^*(E' \oplus E') \otimes L^2(\mathbb{R}_+) \to \pi_Y^*(F' \oplus F') \otimes L^2(\mathbb{R}_+) \tag{3.5}$$

are isomorphisms. Here we use that $E' := E_1' \cong E_2'$ and $F' := F_1' \cong F_2'$. Let us write $\mathbb{A} := (\mathbb{A}_{ij})_{i,j=1,2}$. The boundary symbols $\sigma_\partial(\mathbb{A}_{12}) : \pi_Y^* E' \otimes L^2(\mathbb{R}_+) \to \pi_Y^* F' \otimes L^2(\mathbb{R}_+)$ and

$$\sigma_\partial(\mathbb{A}_{21}) : \pi_Y^* E' \otimes L^2(\mathbb{R}_+) \to \pi_Y^* F' \otimes L^2(\mathbb{R}_+)$$

take values in compact operators, because

$$\mathbb{A}_{12} \in B_G^{0,0}(X;E_1,F_2), \quad \mathbb{A}_{21} \in B_G^{0,0}(X;E_2,F_1),$$

cf. the relations (3.2) and (3.3). Since (3.5) is a family of isomorphisms, we therefore have

$$\operatorname{ind}_{S^*Y} \sigma_\partial(\mathbb{A}) = \operatorname{ind}_{S^*Y} \sigma_\partial(\mathbb{A}_{11}) + \operatorname{ind}_{S^*Y} \sigma_\partial(\mathbb{A}_{22}) = 0. \tag{3.6}$$

Let us also consider the boundary symbols $\sigma_{\partial,(\pm)}(\cdot)$ of $\mathrm{r}^\pm A_0 \mathrm{e}^\pm$ with respect to the plus - and the minus - side, i.e., as operators in the space $B^{0,0}(X_+;E_1,F_1)$ and $B^{0,0}(X_-;E_2,F_2)$, respectively.

Let ε_* denote the operator push-forward under the reflection diffeomorphism $\varepsilon : M \to M$. We have

$$\mathbb{A}_{11} = \mathrm{r}^+ A_0 \mathrm{e}^+ \quad \text{and} \quad \mathbb{A}_{22} = \mathrm{r}^+(\varepsilon_* A_0)\mathrm{e}^+$$

and

$$\operatorname{ind}_{S^*Y} \sigma_{\partial,(+)}(\mathrm{r}^+(\varepsilon_* A_0)\mathrm{e}^+) = \operatorname{ind}_{S^*Y} \sigma_{\partial,(-)}(\mathrm{r}^- A_0 \mathrm{e}^-).$$

Together with the relation (3.6) we thus obtain the following result.

Proposition 3.1. *Let $A_0 \in L^0_{\mathrm{cl}}(M;V,W)$ be an operator with the transmission property at Y. Then we have*

$$\operatorname{ind}_{S^*Y} \sigma_{\partial,(+)}(\mathrm{r}^+ A_0 \mathrm{e}^+) + \operatorname{ind}_{S^*Y} \sigma_{\partial,(-)}(\mathrm{r}^- A_0 \mathrm{e}^-) = 0 \tag{3.7}$$

((3.7) is interpreted as a relation in $K(S^*Y)$).

Let us now consider an arbitrary elliptic operator $A \in L^\mu_{\mathrm{cl}}(M;V,W)_{\mathrm{tr}}$. We use the fact that there is an operator $\widetilde{R}^\mu_V \in L^\mu_{\mathrm{cl}}(M;V,V)_{\mathrm{tr}}$ such that

$$R^\mu_{V_+} = \mathrm{r}^+ \widetilde{R}^\mu_V \mathrm{e}^+ : H^\mu(X_+,V_+) \to L^2(X_+,V_+)$$

is an isomorphism with the inverse $\mathrm{r}^+ \widetilde{R}^{-\mu}_V \mathrm{e}^+$, cf. Remark 1.61. Let us set $A_0 := A R^{-\mu}_V$. We then have

$$\mathrm{r}^+ A_0 \mathrm{e}^+ = (\mathrm{r}^+ A \mathrm{e}^+)(\mathrm{r}^+ R^{-\mu}_V \mathrm{e}^+) + G$$

for some $G \in B_G^{0,0}(X_+;V_+,W_+)$. Because of $\operatorname{ind}_{S^*Y} \sigma_\partial(\mathrm{r}^+ R^{-\mu}_V \mathrm{e}^+) = 0$, cf. Remark 2.18, and since $\sigma_\partial(G)(y,\eta)$ is compact for every $(y,\eta) \in S^*Y$, it follows that

$$\operatorname{ind}_{S^*Y} \sigma_{\partial,(+)}(\mathrm{r}^+ A_0 \mathrm{e}^+) = \operatorname{ind}_{S^*Y} \sigma_{\partial,(+)}(\mathrm{r}^+ A \mathrm{e}^+). \tag{3.8}$$

On the other hand, we have

$$\mathrm{r}^- A_0 \mathrm{e}^- = (\mathrm{r}^- A\mathrm{e}^-)(\mathrm{r}^- R_V^{-\mu}\mathrm{e}^-)$$

which implies

$$\begin{aligned}(3.9)\qquad \mathrm{ind}_{S^*Y}\,\sigma_{\partial,(-)}(\mathrm{r}^- A_0\mathrm{e}^-) &= \mathrm{ind}_{S^*Y}\,\sigma_{\partial,(-)}(\mathrm{r}^- A\mathrm{e}^-)\\ &+ \mathrm{ind}_{S^*Y}\,\sigma_{\partial,(-)}(\mathrm{r}^- R_V^{-\mu}\mathrm{e}^-)\\ &= \mathrm{ind}_{S^*Y}\,\sigma_{\partial,(-)}(\mathrm{r}^- A\mathrm{e}^-) - \mu[\pi_1^* E'],\end{aligned}$$

cf. (2.32). Moreover, the relations (3.8) and (3.9) yield

$$\begin{aligned}&\mathrm{ind}_{S^*Y}\,\sigma_{\partial,(+)}(\mathrm{r}^+ A_0\mathrm{e}^+) + \mathrm{ind}_{S^*Y}\,\sigma_{\partial,(-)}(\mathrm{r}^- A_0\mathrm{e}^-)\\ &\quad= \mathrm{ind}_{S^*Y}\,\sigma_{\partial,(+)}(\mathrm{r}^+ A\mathrm{e}^+) + \mathrm{ind}_{S^*Y}\,\sigma_{\partial,(-)}(\mathrm{r}^- A\mathrm{e}^-) - \mu[\pi_1^* E'].\end{aligned}$$

Together with Proposition 3.1 we thus proved the following theorem.

THEOREM 3.2. *For every elliptic operator $A \in L^\mu_{\mathrm{cl}}(M; V, W)_{\mathrm{tr}}$ we have*

$$\mathrm{ind}_{S^*Y}\,\sigma_{\partial,(+)}(\mathrm{r}^+ A\mathrm{e}^+) + \mathrm{ind}_{S^*Y}\,\sigma_{\partial,(-)}(\mathrm{r}^- A\mathrm{e}^-) = \mu[\pi_1^* E']$$

for $E' := V|_Y$.

We want to specify the latter result for the case that $A := D$ is an elliptic differential operator

$$(3.10)\qquad D : H^s(M, V) \to H^{s-\mu}(M, W)$$

of order $\mu \in \mathbb{N}$ on M. In this case we know that

$$(3.11)\qquad \sigma_{\partial,(+)}(D) : \pi_Y^* E' \otimes H^s(\mathbb{R}_+) \to \pi_Y^* F' \otimes H^{s-\mu}(\mathbb{R}_+)$$

and

$$(3.12)\qquad \sigma_{\partial,(-)}(D) : \pi_Y^* E' \otimes H^s(\mathbb{R}_-) \to \pi_Y^* F' \otimes H^{s-\mu}(\mathbb{R}_+)$$

are both surjective, $s - \mu > -\frac{1}{2}$. (Recall, cf. Remark 1.74, that we always have $E' \cong F'$.) The kernels

$$\ker_{S^*Y}\,\sigma_{\partial,(+)}(D), \quad \ker_{S^*Y}\,\sigma_{\partial,(-)}(D)$$

are then subbundles of $\pi_1^* J$ for $J := E' \oplus \ldots \oplus E'$ (μ summands).

PROPOSITION 3.3. *The Cauchy data spaces*

$$\begin{aligned}L_+(y,\eta) &:= \Big\{\Big(D_t^j u|_{t=0}\Big)_{j=0,\ldots,\mu-1} :\\ &\qquad u \in E' \otimes \mathcal{S}(\overline{\mathbb{R}}_+),\quad \sigma_{\partial,(+)}(D)(y,\eta)u = 0\Big\}\end{aligned}$$

for, $D_t := \frac{1}{i}\frac{\partial}{\partial t}$ and

$$\begin{aligned}L_-(y,\eta) &:= \Big\{\Big(D_t^j u|_{t=0}\Big)_{j=0,\ldots,\mu-1} :\\ &\qquad u \in E' \otimes \mathcal{S}(\overline{\mathbb{R}}_-),\quad \sigma_{\partial,(-)}(D)(y,\eta)u = 0\Big\},\end{aligned}$$

$(y,\eta) \in S^*Y$ form complementary subbundles $L_\pm$ of $\oplus_{j=0}^{\mu-1} E'$, i.e. we have

$$\ker_{S^*Y} \sigma_{\partial,(+)}(D) \oplus \ker_{S^*Y} \sigma_{\partial,(-)}(D) = \oplus_{j=1}^{\mu} \pi_1^* E'. \tag{3.13}$$

The proof of this result will be a consequence of Lemma 3.4 below. In the splitting of variables $x = (y,t) \in Y \times (-1,1)$ in a tubular neighbourhood of Y we write the operator D in the form

$$D = \sum_{j=0}^{\mu} a_j(t) D_t^j \tag{3.14}$$

with coefficients $a_j(t) \in C^\infty((-1,1), \mathrm{Diff}^{\mu-j}(Y; E', F'))$, $D_t = \frac{1}{i}\frac{\partial}{\partial t}$. We then have

$$\sigma_{\partial,(\pm)}(D)(y,\eta) = \sum_{j=0}^{\mu} \sigma_\psi(a_j(0))(y,\eta) D_t^j \tag{3.15}$$

on $\mathbb{R}_\pm \ni t$, where $\sigma_\psi(a_j(0))$ is the homogeneous principal symbol of order $\mu - j$ of the operator $a_j(0) \in \mathrm{Diff}^{\mu-j}(Y; E', F')$, $(y,\eta) \in T^*Y \setminus 0$, $j = 0, \ldots, \mu$.

LEMMA 3.4. *Let*

$$A := \sum_{k=0}^{\mu} b_k D_t^k,$$

be an $m \times m$ system of operators on $\mathbb{R}$ with constant coefficients. Assume that

(i) $b_\mu \in GL(m, \mathbb{C})$,
(ii) $\sum_{j=0}^{\mu} b_j \tau^j$ *is invertible for all* $\tau \in \mathbb{R}$.

Then for

$$L_\pm := \left\{ D_t^k u(0))_{k=0,\ldots,\mu-1} \in \mathbb{C}^{m\mu} : u \in \mathcal{S}(\overline{\mathbb{R}}_\pm, \mathbb{C}^m), Au = 0 \right\}$$

we have $L_+ \oplus L_- = \mathbb{C}^{m\mu}$.

PROOF. Without loss of generality we assume $b_\mu = \mathrm{id}_{\mathbb{C}^m}$, otherwise we pass to a new system with coefficients $b_\mu^{-1} b_k$. The equation $Au = 0$ is equivalent to the system

$$D_t u_{\mu-1} + \sum_{k=0}^{\mu-1} b_k u_k = 0,$$

$$D_t u_j - u_{j+1} = 0 \quad \text{for} \quad j = 0, \ldots, \mu - 2,$$

or

$$\Big(D_t - \mathcal{A}\Big) U = 0 \quad \text{for} \quad U := {}^{\mathrm{t}}(u_0, \ldots, u_{\mu-1}),$$

for

$$\mathcal{A} = \begin{pmatrix} 0 & 1 & 0 & \ldots & 0 \\ 0 & 0 & 1 & \ldots & 0 \\ \vdots & \vdots & \vdots & & \vdots \\ 0 & 0 & 0 & \ldots & 1 \\ -b_0 & -b_1 & -b_2 & \ldots & -b_{\mu-1} \end{pmatrix}.$$

We then have

$$L_\pm = \Big\{U|_{t=0} \in \mathbb{C}^{m\mu} : U \in \mathcal{S}(\overline{\mathbb{R}}_\pm, \mathbb{C}^{m\mu}), \Big(D_t - \mathcal{A}\Big)U = 0\Big\}.$$

The matrix $\tau - \mathcal{A}$ is invertible for all $\tau \in \mathbb{R}$ if and only if condition (ii) is satisfied, i.e., if $\operatorname{spec}(\mathcal{A}) \cap \mathbb{R} = \emptyset$. Let $\lambda \in \operatorname{spec}(\mathcal{A})$ and set $S_\rho(\lambda) := \{z \in \mathbb{C} : |z - \lambda| = \rho\}$. We have $\Big(D_t - \mathcal{A}\Big)U = 0$ and $U|_{t=0} =: U_0$ if and only if $U(t) = e^{it\mathcal{A}}U_0$, and

$$e^{it\mathcal{A}}U_0 = \sum_{\lambda \in \operatorname{spec}(\mathcal{A})} \Big(\frac{1}{2\pi i}\int_{S_\rho(\lambda)} e^{it\zeta}(\zeta - \mathcal{A})^{-1} d\zeta\Big) U_0$$

for every sufficiently small $0 < \rho < 1$. Let us set $\operatorname{spec}_\pm(\mathcal{A}) = \{\lambda \in \operatorname{spec}(\mathcal{A}) : \operatorname{Im}\lambda \gtrless 0\}$; then $\operatorname{spec}(\mathcal{A}) = \operatorname{spec}_+(\mathcal{A}) \cup \operatorname{spec}_-(\mathcal{A})$. Since, by assumption, $\operatorname{spec}(\mathcal{A}) \cap \mathbb{R} = \emptyset$, we have

$$\begin{aligned} U(t) &= \sum_{\lambda \in \operatorname{spec}_+(\mathcal{A})} \Big(\frac{1}{2\pi i}\int_{S_\rho(\lambda)} e^{it\zeta}(\zeta - \mathcal{A})^{-1} d\zeta\Big) U_0 \\ &+ \sum_{\lambda \in \operatorname{spec}_-(\mathcal{A})} \Big(\frac{1}{2\pi i}\int_{S_\rho(\lambda)} e^{it\zeta}(\zeta - \mathcal{A})^{-1} d\zeta\Big) U_0 \end{aligned}$$

and

$$U(0) = U_0 = \sum_{\lambda \in \operatorname{spec}_+(\mathcal{A})} \Pi_\lambda U_0 + \sum_{\lambda \in \operatorname{spec}_-(\mathcal{A})} \Pi_\lambda U_0,$$

where $\Pi_\lambda := (2\pi i)^{-1}\int_{S_\rho(\lambda)}(\zeta - \mathcal{A})^{-1} d\zeta$ is the projection to the eigenspace of $\mathcal{A}$ to the eigenvalue λ.

We have

$$\Big(D_t - \mathcal{A}\Big)\Big\{\frac{1}{2\pi i}\int_{S_\rho(\lambda)} e^{it\zeta}(\zeta - \mathcal{A})^{-1} d\zeta\Big\} = 0$$

for every $\lambda \in \operatorname{spec}(\mathcal{A})$ and

$$\frac{1}{2\pi i}\int_{S_\rho(\lambda)} e^{it\zeta}(\zeta - \mathcal{A})^{-1} d\zeta = e^{i\lambda t}p(t)$$

for a suitable polynominal $p(t)$ (with $m\mu \times m\mu$ matrix-valued coefficients) of order $m(\lambda) - 1$ where $m(\lambda)$ is the multiplicity of λ. This gives us

$$L_+ = \Big(\sum_{\lambda \in \operatorname{spec}_+(\mathcal{A})} \Pi_\lambda\Big)\mathbb{C}^{m\mu}, \quad L_- = \Big(\sum_{\lambda \in \operatorname{spec}_-(\mathcal{A})} \Pi_\lambda\Big)\mathbb{C}^{m\mu}$$

and hence $L_+ \oplus L_- = \mathbb{C}^{m\mu}$. □

REMARK 3.5. We also have

$$(D_t - \mathcal{A})\left\{\frac{1}{2\pi i}\int_{S_\rho(\lambda)} e^{it\zeta}(\zeta - \mathcal{A})^{-1}\frac{d\zeta}{\zeta}\right\} = 0 \tag{3.16}$$

for every $\lambda \in \operatorname{spec}(\mathcal{A})$ and $0 < \rho < \operatorname{dist}(\lambda, 0)$, i.e.,

$$\Pi_\lambda = (2\pi i)^{-1} \int_{S_\rho(\lambda)} (\zeta - \mathcal{A})^{-1} d\zeta = (2\pi i)^{-1} \mathcal{A} \int_{S_\rho(\lambda)} (\zeta - \mathcal{A})^{-1} \frac{d\zeta}{\zeta}. \tag{3.17}$$

Thus, if we set $\Gamma_\pm := \{\zeta \in \mathbb{C} : \operatorname{Im} \zeta = \pm\delta\}$ for some sufficiently small $\delta > 0$, $\Gamma_+(\Gamma_-)$ oriented with increasing (decreasing) $\operatorname{Re} \zeta$, it follows that

$$\mathcal{P}_\pm := \sum_{\lambda \in \operatorname{spec}_\pm(\mathcal{A})} \Pi_\lambda = \frac{1}{2\pi i} \mathcal{A} \int_{\Gamma_\pm} (\zeta - \mathcal{A})^{-1} \frac{d\zeta}{\zeta}. \tag{3.18}$$

Moreover, every solution of

$$(D_t - \mathcal{A})U = 0, \quad U|_{t=0} = U_0 \tag{3.19}$$

can be written in the form

$$U(t) = \frac{1}{2\pi i} \mathcal{A} \int_{\Gamma_+} e^{it\zeta} (\zeta - \mathcal{A})^{-1} \frac{d\zeta}{\zeta} U_0 + \frac{1}{2\pi i} \mathcal{A} \int_{\Gamma_-} e^{it\zeta} (\zeta - \mathcal{A})^{-1} \frac{d\zeta}{\zeta} U_0. \tag{3.20}$$

Now let D be as before an elliptic differential operator of order μ, regarded as a map

$$D : H^\mu(M, V) \to L^2(M, W), \tag{3.21}$$

cf. formula (3.10). Choose arbitrary elliptic elements

$$\mathcal{D}_\pm := \begin{pmatrix} D_\pm \\ T_\pm \end{pmatrix} \in \mathcal{S}^{\mu,\mu}(X_\pm; \boldsymbol{l}_\pm),$$

for $D_\pm := \mathrm{r}^\pm D \mathrm{e}^\pm$, $\boldsymbol{l}_+ := (V_+, W_+; 0, \mathbb{L}_+)$, $\boldsymbol{l}_- := (V_-, W_-; 0, \mathbb{L}_-)$ for suitable projection data $\mathbb{L}_\pm = (P_\pm, J, L_\pm)$; according to Theorem 2.16 there always exist such operators $T_\pm$. Recall that trace operators $T_\pm$ have the form $T_\pm = P_\pm \widetilde{T}_\pm$ for suitable trace operators $\widetilde{T}_\pm \in \mathcal{B}^{\mu,\mu}(X_\pm; V_\pm, W_\pm; 0, J)$. In view of the ellipticity we have the Fredholm operators (3.21) and

$$\mathcal{D}_\pm : H^\mu(X_\pm, V_\pm) \to \begin{matrix} L^2(X_\pm, W_\pm) \\ \oplus \\ P^0(Y, \mathbb{L}_\pm) \end{matrix}. \tag{3.22}$$

We now derive a relation between their indices. To this end we consider the following diagram

$$\begin{array}{ccccccccc} 0 & \longrightarrow & H^\mu(M,V) & \stackrel{i}{\longrightarrow} & \mathcal{M}_+ \oplus \mathcal{M}_- & \stackrel{j}{\longrightarrow} & L^2(Y,J) & \longrightarrow & 0 \\ & & \downarrow D & & \mathcal{D} \downarrow \uparrow \mathcal{B} & & \uparrow R & & \\ 0 & \longleftarrow & L^2(M,V) & \stackrel{b}{\longleftarrow} & \mathcal{N}_+ \oplus \mathcal{N}_- & \stackrel{a}{\longleftarrow} & L^2(Y,J) & \longleftarrow & 0 \end{array} \tag{3.23}$$

for $\mathcal{M}_\pm := H^\mu(X_\pm, V_\pm)$, and $\mathcal{N}_\pm := L^2(X_\pm, W_\pm) \oplus P^0(Y, \mathbb{L}_\pm)$. The maps i and j are defined as follows:

$$i(u) := u|_{X_+} \oplus u|_{X_-}, \quad j(u_+ \oplus u_-) := \oplus_{k=0}^{\mu-1} \Delta_{E'}^{-\mu+k+1/2} (\gamma_+^k u_+ - \gamma_-^k u_-)$$

for $\gamma_\pm^k f := \partial_t^k f|_{Y_\pm}$, with ∂_t^k being the derivative in normal direction to Y and $|_{Y_\pm}$ the restriction to Y from the $\pm$ side. The symbol $\Delta_{E'}^\nu$ denotes any element of

$L^{\nu}_{\mathrm{cl}}(Y;E',E')$ that induces isomorphisms $H^s(Y,E') \to H^{s-\nu}(Y,E')$ for all $s \in \mathbb{R}$. The map a is the canonical embedding, where we use

$$H^s(Y,J) = P^s(Y,\mathbb{L}_+) \oplus P^s(Y,\mathbb{L}_-)$$

for $s = 0$. The map b is the canonical projection, using $L^2(M,W) = L^2(X_+,W_+) \oplus L^2(X_-,W_-)$. Moreover, we set

$$\mathcal{D} := \mathcal{D}_+ \oplus \mathcal{D}_- \quad \text{and} \quad \mathcal{B} := \mathcal{B}_+ \oplus \mathcal{B}_-, \tag{3.24}$$

where $\mathcal{B}_\pm \in \mathcal{S}^{-\mu,0}(X_\pm; \boldsymbol{l}_\pm^{-1})$ are parametrices of $\mathcal{D}_\pm$. Finally, we define

$$R := j \circ \mathcal{B} \circ a : L^2(Y,J) \to L^2(Y,J) \tag{3.25}$$

which is an elliptic pseudo-differential operator on Y and, as such, Fredholm. The rows of the diagram (3.23) are exact, and we have

$$D = b \circ \mathcal{D} \circ i.$$

The assumptions of an abstract lemma (see, e.g. Rempel and Schulze [**31**, Section 3.1.1.3]) are satisfied, and we thus obtain the following index formula.

THEOREM 3.6. *With the notation of* (3.25) *we have*

$$\operatorname{ind} D = \operatorname{ind} \mathcal{D}_+ + \operatorname{ind} \mathcal{D}_- + \operatorname{ind} R.$$

3.2. Examples. We now turn to a number of specific observations and examples. Consider the unit circle

$$S^1 := \{w \in \mathbb{C} : |w| = 1\}$$

in the complex plane, and write the space $L^2(S^1)$ (with the scalar product $(u,v)_{L^2(\mathbb{R}_+)} = \int_0^{2\pi} u(\varphi)\overline{v(\varphi)}d\varphi$) as a direct orthogonal sum

$$L^2(S^1) = W^+ \oplus W^-$$

for

$$W^+ := [[w^j : j \in \mathbb{N}\,]], \quad \mathbb{W}^- := [[w^{-j-1} : j \in \mathbb{N}\,]]$$

(where [[...]] denotes closed subspaces spanned by the elements in the brackets.)

THEOREM 3.7. *The orthogonal projections*

$$P_\pm : L^2(S^1) \to W^\pm$$

are elements of $L^0_{\mathrm{cl}}(S^1)$.

PROOF. Let us set

$$V^+ := \{(F_{t\to\tau} \mathrm{e}^+ f_+)(\tau) : f_+(t) \in L^2(\mathbb{R}_+)\},$$
$$V^- := \{(F_{t\to\tau} \mathrm{e}^- f_-)(\tau) : f_-(t) \in L^2(\mathbb{R}_-)\},$$

where $F = F_{t\to\tau}$ is the standard Fourier transform on $\mathbb{R}$. The canonical projections

$$\Pi^\pm : L^2(\mathbb{R}) \to V^\pm,$$

can be written in the form

$$\Pi^\pm = F\Theta^\pm F^{-1}$$

with the characteristic function $\Theta^{\pm}$ of $\mathbb{R}_{\pm}$. If $\chi(t) \in C^{\infty}(\mathbb{R})$ is any excision function in t (i.e., $\chi(t) = 0$ for $|t| < c_0$, $\chi(t) = 1$ for $|t| > c_1$ for certain $0 < c_0 < c_1$) we have

$$\Pi^{+} = F\chi\Theta^{\pm}F^{-1} + F(1-\chi)\Theta^{\pm}F^{-1},$$

where the operators $F(1-\chi)\Theta^{\pm}F^{-1}$ are smoothing. Since $\chi\Theta^{\pm} \in S^{0}_{\mathrm{cl}}(\mathbb{R})$, it follows that

$$\Pi^{\pm} \in L^{0}_{\mathrm{cl}}(\mathbb{R}). \tag{3.26}$$

We now consider the isomorphism

$$T : L^2(S^1) \to L^2(\mathbb{R})$$

by setting $(Tu)(\tau) := 2(1+i\tau)^{-1}u(w(\tau))$ for

$$w(\tau) := \frac{1-i\tau}{1+i\tau}.$$

We then have

$$T(w^j) = 2\frac{(1-i\tau)^j}{(1+i\tau)^{j+1}}, \qquad T(w^{-j-1}) = 2\frac{(1+i\tau)^j}{(1-i\tau)^{j+1}},$$

for $j \in \mathbb{N}$, $TW^{\pm} = V^{\pm}$, and

$$P_{\pm} = T^{-1}\Pi^{\pm}T. \tag{3.27}$$

Using (3.26) we see that the operators $P_{\pm}$ belong to $L^{0}_{\mathrm{cl}}(S^1)$ because of the invariance of pseudo-differential operators under diffeomorphisms. □

REMARK 3.8. Let $a(\tau) \in S^{0}_{\mathrm{cl}}(\mathbb{R})$ be a symbol with constant coefficients. We then have continuous operators

$$\mathrm{op}^{+}(a) := \mathrm{r}^{+}\,\mathrm{op}(a)\mathrm{e}^{+} : L^2(\mathbb{R}_{+}) \to L^2(\mathbb{R}_{+}) \tag{3.28}$$

and

$$\mathrm{op}^{-}(a) := \mathrm{r}^{-}\,\mathrm{op}(a)\mathrm{e}^{-} : L^2(\mathbb{R}_{-}) \to L^2(\mathbb{R}_{-}).$$

The operator (3.28) can equivalently be formulated as

$$\Pi^{\pm}\mathcal{M}_a : V^{\pm} \to V^{\pm},$$

where $\mathcal{M}_a$ is the operator of multiplication by the function a. In particular, together with the following Remark 3.9 we see that boundary symbols (1.54) can be reformulated as ((y,η)-dependent families of) Toeplitz operators on the circle.

REMARK 3.9. An operator of the form

$$A := P_{+}\mathcal{M}_{\widetilde{a}}P_{+} : W^{+} \to W^{+} \tag{3.29}$$

for an $\widetilde{a}(w) \in C^{\infty}(S^1)$ is a Toeplitz operator (in classical notation). $\mathcal{M}_{\widetilde{a}}$ is the operator of multiplication by $\widetilde{a}$ in $L^2(S^1)$. We have (in the terminology of Definition 1.22) a canonical identification of (3.29) with

$$P\widetilde{A}R \in T^{0}(S^1; \mathbb{L}, \mathbb{L}). \tag{3.30}$$

Here, $\widetilde{A} := \mathcal{M}_{\widetilde{a}}$, $P := P_+$, $R : W^+ \to L^2(S^1)$ is the canonical embedding, and $\mathbb{L} = (P, \mathbb{C}, L) \in \mathcal{P}(S^1)$ for a bundle $L \in \mathrm{Vect}(T^*S^1 \setminus 0)$ that is isomorphic to $(S^1 \times \mathbb{R}_+) \times \mathbb{C}$, where $S^1 \times \mathbb{R}_+$ is the plus-component of $T^*S^1 \setminus 0 \cong (S^1 \times \mathbb{R}_+) \cup (S^1 \times \mathbb{R}_-)$, and $P^0(S^1, \mathbb{L}) = W^+$. Ellipticity of (3.29) in the sense of Definition 1.31 (i) is equivalent to $\widetilde{a}(w) \neq 0$ for all $w \in S^1$.

Clearly if we admit arbitrary operators $\widetilde{A} \in L^{\mu}_{\mathrm{cl}}(S^1)$, we obtain much more general operators than (3.30) (also for $\mu = 0$).

3.3. Spectral boundary value problems. We now consider general (non-homogeneous) elliptic boundary value problems for differential operators as a special case of our pseudo-differential calculus, where we have more explicit information (the material of this section is based on the author's joint papers [**46**], [**30**] with Nazaikinskij, Sternin, and Shatalov.) They are a natural generalisation of (homogeneous) boundary value problems, studied (in an L^2 set-up) by Atiyah, Patoti, and Singer [**4**].

Let X be a compact C^∞ manifold with boundary Y, $n = \dim X$, and let A be an elliptic differential operator on X of order μ with smooth coefficients up to the boundary,

$$A : C^\infty(X, E) \to C^\infty(X, F) \tag{3.31}$$

for $E, F \in \mathrm{Vect}(X)$. In a collar neighbourhood of Y in the splitting of variables $x = (y, t) \in Y \times [0, 1)$ the operator A can be written in the form

$$A = \sum_{j=0}^{\mu} A_j(t) D_t^j, \tag{3.32}$$

$D_t := \frac{1}{i}\frac{\partial}{\partial t}$, with coefficients $A_j \in C^\infty([0, 1), \mathrm{Diff}^{\mu-j}(Y))$. The ellipticity of A implies that

$$A_\mu(0) : E' \to F' \tag{3.33}$$

(for $E' := E|_Y$, $F' := F|_Y$) is an isomorphism. The boundary symbol

$$\sigma_\partial(A)(y, \eta) = \sum_{j=0}^{\mu} \sigma_\psi(A_j(0))(y, \eta) D_t^j : H^s(\mathbb{R}_+, E') \to H^{s-\mu}(\mathbb{R}_+, F') \tag{3.34}$$

(with $\sigma_\psi(A_j(0)) : \pi_Y^* E' \to \pi_Y^* F'$ being the homogeneous principal symbol of order $\mu - j$ of the operator $A_j(0) \in \mathrm{Diff}^{\mu-j}(Y; E', F')$) is a surjective family of Fredholm operators, parametrised by $(y, \eta) \in T^*Y \setminus 0$. Thus, as in the general calculus of boundary value problems, there is a kernel bundle

$$L_+ := \ker \sigma_\partial(A) \in \mathrm{Vect}(T^*Y \setminus 0). \tag{3.35}$$

Let us consider the family of differential operators on Y

$$\sigma_c(A)(w) := \sum_{j=0}^{\mu} A_j(0) w^j, \tag{3.36}$$

parametrised by the complex variable $w \in \mathbb{C}$.

REMARK 3.10. We have

$$\sigma_c(A)(\beta + i\gamma) \in L^{\mu}_{\mathrm{cl}}(Y; E', F'; \mathbb{R}_\beta)$$

for every $\gamma \in \mathbb{R}$, and $\sigma_c(A)(\beta + i\gamma)$ is parameter-dependent elliptic with parameter $\beta \in \mathbb{R}$, cf. Agranovich and Vishik [1]. Moreover, there exists a countable set $D \subset \mathbb{C}$ such that

$$D \cap \{w : \ c < \operatorname{Im} w < c'\}$$

is finite for every $c \leq c'$ and

$$\sigma_c(A)(w) : H^s(Y, E') \to H^{s-\mu}(Y, F') \tag{3.37}$$

is an isomorphism for every $w \in \mathbb{C} \setminus D$ and $s \in \mathbb{R}$.

The bijectivity of (3.37) for large $|\operatorname{Re} w|$ is a well known phenomenon of parameter-dependent elliptic operators, holomorphically dependent on w. First, (3.37) is a holomorphic family of Fredholm operators (kernels and cokernels are independent of s), and for large $|\operatorname{Re} w|$ the operators are isomorphisms, cf. Theorem 1.11 (i). Then there is a countable set $D \subset \mathbb{C}$ of non-bijectivity points with the asserted properties, cf. also [**35**, Section 2.2.5].

Let us set

$$B := \sum_{j=0}^{\mu} A_j(0) D_t^j, \tag{3.38}$$

regarded as a differential operator on the infinite cylinder $Y \times \mathbb{R}$. The coefficient $A_j(0) : E' \to F'$ is an isomorphism. For convenience, we set $J := E' = F'$ and assume $A_\mu(0) = 1$ (otherwise we compose (3.38) from the left by $A_\mu^{-1}(0)$). Let us write B in the form

$$B = D_t^\mu + \sum_{j=0}^{\mu-1} b_j D_t^j \tag{3.39}$$

for $b_j := A_j(0) \in \mathrm{Diff}^{\mu-j}(Y)$. The pull back of the bundle J to $Y \times \mathbb{R}$ will be denoted again by J.

Let $H^s(Y \times \mathbb{R}, J)$ denote the (cylindrical) Sobolev space on $Y \times \mathbb{R}$, of smoothness $s \in \mathbb{R}$, defined as the completion of $C_0^\infty(Y \times \mathbb{R}, J)$ with respect to the norm

$$\Big\{ \int \|R^s(\tau)\hat{u}(\tau)\|^2_{L^2(Y,J)} d\tau \Big\}^{1/2},$$

where $R^s(\tau) \in L^s_{\mathrm{cl}}(Y; J, J; \mathbb{R}_\tau)$ is any classical parameter-dependent elliptic pseudo-differential operator of order s on Y, with parameter $\tau \in \mathbb{R}$. Moreover, set

$$H^{s,\gamma}(Y \times \mathbb{R}, J) := \{e^{\gamma t} u(y,t) : u \in H^s(Y \times \mathbb{R}, J)\}$$

for every $\gamma \in \mathbb{R}$. The operator B then defines continuous maps

$$B : H^{s,\gamma}(Y \times \mathbb{R}, J) \to H^{s-\mu,\gamma}(Y \times \mathbb{R}, J) \tag{3.40}$$

for all $s, \gamma \in \mathbb{R}$.

Let us reformulate the equation $Bu = f$ as

$$D_t U - \mathcal{A} U = F \tag{3.41}$$

for $U := {}^{\mathrm{t}}(u_0, \ldots, u_{\mu-1})$, $F := {}^{\mathrm{t}}(0, \ldots, 0, f)$ and $u_j := D_t^j u$, $j = 0, \ldots, \mu - 1$, and

$$\mathcal{A} := \begin{pmatrix} 0 & 1 & 0 & \ldots & 0 \\ 0 & 0 & 1 & \ldots & 0 \\ \vdots & \vdots & \vdots & & \vdots \\ 0 & 0 & 0 & \ldots & 1 \\ -b_0 & -b_1 & -b_2 & \ldots & -b_{\mu-1} \end{pmatrix}. \tag{3.42}$$

Note that $\det(w - \mathcal{A}) = w^{\mu} + \sum_{j=0}^{\mu-1} b_j w^j$.

LEMMA 3.11. *The operator*

$$w - \mathcal{A} : \bigoplus_{j=0}^{\mu-1} H^{s-j}(Y, J) \to \bigoplus_{k=1}^{\mu} H^{s-k}(Y, J) \tag{3.43}$$

is invertible for a $w \in \mathbb{C}$ *if and only if so is*

$$w^{\mu} + \sum_{j=0}^{\mu-1} b_j w^j : H^s(Y, J) \to H^{s-\mu}(Y, J) \tag{3.44}$$

for any $s \in \mathbb{R}$. *In other words, we have* $\operatorname{spec} \mathcal{A} = \{w \in \mathbb{C} : \sigma_c(A)(w)$ *is bijective*$\}$. *Moreover, we have*

$$(w - \mathcal{A})^{-1} = \sigma_c(A)^{-1}(w) \mathcal{Q}(w) \tag{3.45}$$

for a matrix $\mathcal{Q}(w) = (Q_{jk}(w))_{\substack{k=1,\ldots,\mu \\ j=0,\ldots,\mu-1}}$, *of differential operators polynomially dependent on* w, *where* $Q_{jk}(w)$ *is parameter-dependent with parameter* β *(for* $w = \beta + i\gamma$*) of order* $\mu - k + j$ *for* $k = 1, \ldots, \mu$, $j = 0, \ldots, \mu - 1$.

PROOF. The formula (3.45) is elementary, and the proof gives the characterisation of the entries of the matrix Q.

As a consequence of (3.45) we see that the operator (3.43) is invertible if so is (3.44). Conversely, the inverbility of (3.43) entails that of (3.44) because $\sigma_c(A)^{-1}(w) f$ is equal to the first component of $(w - \mathcal{B})^{-1\mathrm{t}}(0, \ldots, 0, f)$. □

REMARK 3.12. The set spec($\mathcal{A}$) is countable, and every strip $\{w \in \mathbb{C} : c < \operatorname{Im} w < c'\}$ only contains finitely many elements of spec($\mathcal{A}$) for arbitrary reals $c < c'$. The operator function $(w - \mathcal{A})^{-1}$ is meromorphic with poles of finite multiplicity at the points of spec($\mathcal{A}$), and the Laurent coefficients of $(w - \mathcal{A})^{-1}$ at $(w - p)^{-(k+1)}$, $k \in \mathbb{N}$, are operators in $L^{-\infty}(Y; J, J)$ of finite rank.

In fact, $\sigma_c(A)(w)$ is parameter-dependent elliptic with parameter $\operatorname{Re} w$ on every line $\operatorname{Im} w = \text{const}$, and invertible for large $|\operatorname{Re} w|$. Then the asserted properties follow from the relation (3.45) together with classical results of Agranovich and Vishik [**1**], cf. also [**39**, Section 1.2.4].

Similarly to the considerations in Section 3.1 we want to formulate a relation between the spectral points of the operator $\mathcal{A}$ (which is the same as the set of non-bijectivity points of (3.37)) and Cauchy data spaces of Sobolev distributions on X at Y. To this end we interpret (3.39) as a continuous operator

$$B : H^s(Y \times \mathbb{R}_+, J) \to H^{s-\mu}(Y \times \mathbb{R}_+, J),$$

where $H^s(Y \times \mathbb{R}_+, J) := H^s(Y \times \mathbb{R}, J)|_{\mathbb{R}_+ \times Y}$, and $H^s(Y \times \mathbb{R}, J)$ is the cylindrical Sobolev space on $Y \times \mathbb{R}$. The latter space is defined as the completion of $C_0^\infty(Y \times \mathbb{R}, J)$ with respect to the norm

$$\left\{ \int \|R^s(\tau)\hat{u}(\tau)\|^2_{L^2(Y,J)} d\tau \right\}^{\frac{1}{2}}$$

for any choice of an order reducing family $R^s(\tau) \in L^s_{\mathrm{cl}}(Y; J, J; \mathbb{R})$ in the sense of Theorem 1.12. An equivalent definition for $s \in \mathbb{N}$ is

$$\begin{aligned} H^s(Y \times \mathbb{R}, J) = \{&u(y,t) \in L^2(Y \times \mathbb{R}, J) : \\ &D_y^\alpha D_t^k u(y,t) \in L^2(Y \times \mathbb{R}, J) \quad \text{for all} \quad |\alpha| + k \leq s\}. \end{aligned}$$

Here D_y^α runs over the set of all differential operators on Y of order $|\alpha|$, acting between sections of J.

For convenience, we assume that the set D of Remark 3.10 does not intersect the real line $\operatorname{Im} w = 0$. Otherwise, we can pass to a translated operator $(D_t - i\gamma)^\mu + \sum_{j=0}^{\mu-1} b_j (D_t - i\gamma)^j$ for a suitable real γ with a corresponding shifted set D_γ in the complex plane which does not intersect the real line.

For any integer $s \geq \mu$ we form the Cauchy data space

$$\mathcal{C}^{s,\mu}(Y, J) := \{(D_t^k u(y,0))_{k=0,\ldots,\mu-1} : u \in H^s(Y \times \mathbb{R}_+)\}.$$

We then have

$$\mathcal{C}^{s,\mu}(Y, J) = \bigoplus_{k=0}^{\mu-1} H^{s-k-\frac{1}{2}}(Y, J), \tag{3.46}$$

and the operator $\mathcal{T}^\mu := {}^t(\mathrm{r}', \mathrm{r}'D_t, \ldots, \mathrm{r}'D_t^{\mu-1})$ with $\mathrm{r}'u := u|_{t=0}$ defines a continuous map

$$\mathcal{T}^\mu : H^s(Y \times \mathbb{R}_+, J) \to \mathcal{C}^{s,\mu}(Y, J). \tag{3.47}$$

We want to formulate results on the solvability of the boundary value problem

$$Bu_+ = f_+ \in H^{s-\mu}(Y \times \mathbb{R}_+, J), \tag{3.48}$$

$$\mathcal{T}^\mu u_+ = g_+ \in \mathcal{C}^{s,\mu}(Y, J). \tag{3.49}$$

It turns out that, in general, not the whole space $\mathcal{C}^{s,\mu}(Y, J)$ of boundary data g_+ on the right hand side of (3.49) is induced by solutions $u_+ \in H^s(Y \times \mathbb{R}_+, J)$, but a subspace which is the image under a suitable pseudo-differential projection $\mathcal{P}_+$.

Similarly to (3.48), (3.49) we can also consider a boundary value problem on the negative half-cylinder

$$Bu_- = f_- \in H^{s-\mu}(Y \times \mathbb{R}_-, J), \tag{3.50}$$

$$\mathcal{T}^\mu u_- = g_- \in \mathcal{C}^{s,\mu}(Y, J). \tag{3.51}$$

The admissible boundary data g_- are then determined by the corresponding complementary projection $\mathcal{P}_-$.

The projections $\mathcal{P}_\pm$ are obtained as follows. First recall that the spectrum $\operatorname{spec}(\mathcal{A}) = D$ (in the notation of Remark 3.10) does not intersect the real line $\operatorname{Im} w = 0$. Therefore, we have

$$\operatorname{spec}(\mathcal{A}) \cap \{w : -c \le \operatorname{Im} w \le c\} = \emptyset \tag{3.52}$$

for some $c > 0$. Let us fix some $0 < \varepsilon < c$, set

$$\Gamma_\pm := \{w = \tau \pm i(c - \varepsilon) : \tau \in \mathbb{R}\},$$

oriented in direction of increasing τ on Γ_+, and descreasing τ on Γ_-, and form

$$\mathcal{P}_\pm = \frac{1}{2\pi i}\, \mathcal{A} \int_{\Gamma_\pm} (w - \mathcal{A})^{-1} \frac{dw}{w}. \tag{3.53}$$

Note that the operator $\mathcal{A}$ can be written in the form

$$\mathcal{A} = \mathcal{R}\mathcal{A}_1\mathcal{R}^{-1} \tag{3.54}$$

for

$$\mathcal{R} = \operatorname{diag}(\mathcal{R}_J^0, \mathcal{R}_J^1, \dots, \mathcal{R}_J^{(\mu-1)}) \tag{3.55}$$

where $\mathcal{R}_J^k : H^s(Y, J) \to H^{s-k}(Y, J)$ is an order reducing operator of order $-k$ in the sense of Theorem 1.12, and $\mathcal{A}_1$ is a system of operators of order 1. Then $\mathcal{P}_\pm$ takes the form

$$\mathcal{P}_\pm = \mathcal{R}\left\{\frac{1}{2\pi i}\mathcal{A}_1 \int_{\Gamma_\pm} (w - \mathcal{A}_1)^{-1} \frac{dw}{w}\right\}\mathcal{R}^{-1}. \tag{3.56}$$

LEMMA 3.13. (i) *The integral* (3.53) *stronly converges in* $\mathcal{C}^{s,\mu}(Y, J)$ *on the dense subset* $\mathcal{C}^{s+1,\mu}(Y, J)$ *for every* $s \in \mathbb{R}$.

(ii) *The operators* $\mathcal{P}_\pm$ *form a matrix* $(\mathcal{P}_{\pm,jk})_{0 \le j \le \mu-1, 0 \le k \le \mu-1}$ *of elements of* $L_{\mathrm{cl}}^{j-k}(Y; J, J)$. *Thus* $\mathcal{P}_\pm$ *extend to continuous operators*

$$\mathcal{P}_\pm : \mathcal{C}^{s,\mu}(Y, J) \to \mathcal{C}^{s,\mu}(Y, J)$$

for all $s \in \mathbb{R}$.

PROOF. (i) From Lemma 3.11 we know that the entries of the matrix $(w - \mathcal{A})^{-1}$ are classical parameter-dependent pseudo-differential operators of order ≤ -1, with parameter $\tau \in \mathbb{R}$ (when we identify $\mathbb{R}$ with Γ_+ or Γ_- via $\tau \to w = \tau \pm i(c - \varepsilon)$). In fact, $\sigma_c(A)^{-1}(w)$ is parameter-dependent of order $-\mu$, and the orders of the entries of $Q(w)$ are $\le \mu - 1$, cf. formula (3.45). By virtue of Theorem

1.9 the $\mathcal{L}(H^{s+1}(Y,J), H^{s+1}(Y,J))$ - norm of every entry of $(w-\mathcal{A})^{-1}$ can be estimated by $c_s(1+|w|)^{-1}$, $w \in \Gamma_\pm$, for a constant $c_s > 0$.

This gives us immediately assertion (i).

(ii) Let us write $\mathcal{P}_\pm$ in the form (3.56). Then it suffices to observe that

$$\frac{1}{2\pi i}\int_{\Gamma_\pm}(w-\mathcal{A}_1)^{-1}\frac{dw}{w}$$

is a matrix of classical pseudo-differential operators of order -1. The technique of the proof is similar to [**48**]. □

Let us write the operator (3.42) in the form $\mathcal{A} = (\mathcal{A}_{kj})_{1\le k\le\mu, 0\le j\le\mu-1}$,

$$\mathcal{A}: \bigoplus_{j=0}^{\mu-1} H^{s-j}(Y,J) \to \bigoplus_{k=1}^{\mu} H^{s-k}(Y,J),$$

and interpret the orders $\operatorname{ord}\mathcal{A}_{kj} = k-j$ in the Douglis-Nirenberg sense, with homogeneous principal symbols $\sigma_\psi(\mathcal{A}_{kj})(y,\eta)$ of order $k-j$. Setting

$$\sigma_\psi(\mathcal{A})(y,\eta) = (\sigma_\psi(\mathcal{A}_{kj})(y,\eta))_{1\le k\le\mu, 0\le j\le\mu-1}$$

we conclude that $\sigma_\psi(\mathcal{A})(y,\eta)$ equals

$$\begin{pmatrix} 0 & 1 & 0 & \dots & 0 \\ 0 & 0 & 1 & \dots & 0 \\ \vdots & \vdots & \vdots & & \vdots \\ 0 & 0 & 0 & \dots & 1 \\ -\sigma_\psi(b_0)(y,\eta) & -\sigma_\psi(b_1)(y,\eta) & -\sigma_\psi(b_2)(y,\eta) & \dots & -\sigma_\psi(b_{\mu-1})(y,\eta) \end{pmatrix}.$$

Moreover, set

$$L_\pm(y,\eta) := \Big\{u(0) : u(t) \in \mathcal{S}(\overline{\mathbb{R}}_\pm, E_y), \Big(D_t - \sigma_\psi(\mathcal{A})(y,\eta)\Big)u = 0\Big\},$$

$E = \oplus_{k=0}^{\mu-1} J$.

THEOREM 3.14. (i) *The operators $\mathcal{P}_\pm$ are complementary projections, i.e., $\mathcal{P}_\pm^2 = \mathcal{P}_\pm$, $\mathcal{P}_+ + \mathcal{P}_- = 1$, and have the property*

$$\mathcal{P}_\pm\mathcal{A} = \mathcal{A}\mathcal{P}_\pm. \tag{3.57}$$

(ii) *The homogeneous principal symbols $\sigma_\psi(\mathcal{P}_{\pm,jk})$ of order $j-k$ of $\mathcal{P}_{\pm,jk} \in L_{\mathrm{cl}}^{j-k}(Y;E,E)$ (cf. Lemma 3.13 (ii)) form projections*

$$\Big(\sigma_\psi(\mathcal{P}_{\pm,jk})(y,\eta)\Big)_{0\le j\le\mu-1, 0\le k\le\mu-1} =: \sigma_\psi(\mathcal{P}_\pm)(y,\eta) : E_y \to L_\pm(y,\eta)$$

along $L_\mp(y,\eta)$.

(iii) *The operator functions $(w-\mathcal{A})^{-1}\mathcal{P}_\pm$ are holomorphic in* $\operatorname{spec}(\mathcal{A}) \cap \{w : \operatorname{Im} w \gtrless 0\}$.

PROOF. (i) First note that the relation (3.57) is evident. Moreover, to verify that $\mathcal{P}_+ + \mathcal{P}_- = 1$, it suffices to observe that

$$\mathcal{P}_+ + \mathcal{P}_- = \mathcal{A}\frac{1}{2\pi i}\int_{\Gamma_0}(w-\mathcal{A})^{-1}\frac{dw}{w}$$

for a small circle Γ_0 clockwise surrounding the origin; by Cauchy's residue theorem the integral is equal to $2\pi i\mathcal{A}^{-1}$.

Let us now calculate $\mathcal{P}_+^2$ (the consideration for $\mathcal{P}_-$ is analogous and left to the reader). Set $\Gamma'_+ := \Gamma_+ + i\varepsilon$ for some sufficiently small $\varepsilon > 0$. Then, using the relation (3.57) and the resolvent identity we have

$$\begin{aligned}
\mathcal{P}_+^2 &= -\left(\frac{\mathcal{A}}{2\pi}\right)^2\int_{\Gamma'_+}\left\{\int_{\Gamma_+}\frac{1}{w'w}(w'-\mathcal{A})^{-1}(w-\mathcal{A})^{-1}dw\right\}dw' \\
&= -\left(\frac{\mathcal{A}}{2\pi}\right)^2\int_{\Gamma'_+}\left\{\int_{\Gamma_+}\frac{1}{w'w(w-w')}\left[(w'-\mathcal{A})^{-1}-(w-\mathcal{A})^{-1}\right]dw\right\}dw' \\
&= I_0 + I_1
\end{aligned}$$

for

$$\begin{aligned}
I_0 &= -\left(\frac{\mathcal{A}}{2\pi}\right)^2\int_{\Gamma'_+}\left\{\int_{\Gamma_+}(w-\mathcal{A})^{-1}\frac{dw}{w(w-w')}\right\}\frac{dw'}{w'} \\
&= -\left(\frac{\mathcal{A}}{2\pi}\right)^2\int_{\Gamma_+}(w-\mathcal{A})^{-1}\left\{\int_{\Gamma'_+}\frac{dw'}{w'(w-w')}\right\}\frac{dw}{w}
\end{aligned}$$

which vanishes, since the inner integral is zero, and

$$\begin{aligned}
I_1 &= -\left(\frac{\mathcal{A}}{2\pi}\right)^2\int_{\Gamma'_+}(w'-\mathcal{A})^{-1}\left\{\int_{\Gamma_+}\frac{dw}{w(w-w')}\right\}\frac{dw'}{w'} \\
&= -\left(\frac{\mathcal{A}}{2\pi}\right)^2 2\pi i\int_{\Gamma'_+}(w'-\mathcal{A})^{-1}\frac{dw'}{(w')^2} \\
&= \frac{\mathcal{A}}{2\pi i}\left\{\int_{\Gamma_+}(w'-\mathcal{A})^{-1}\frac{dw'}{w'}+\int_{\Gamma_+}\frac{dw'}{(w')^2}\right\} = \mathcal{P}_+.
\end{aligned}$$

In the first line of the latter relation we inserted $\int_{\Gamma_+}\frac{dw}{w(w-w')} = 2\pi i(w')^{-1}$, and in the second line $\mathcal{A} = \mathcal{A} - w' + w'$.

(ii) is simple after the considerations in the proof of Lemma 3.4.

(iii) Let $\operatorname{Im} w > 0$. Then we have

$$\begin{aligned}
(w-\mathcal{A})^{-1}\mathcal{P}_+ &= \frac{\mathcal{A}}{2\pi i}\int_{\Gamma_+}(w'-\mathcal{A})^{-1}(w-\mathcal{A})^{-1}\frac{dw'}{w'} \\
&= \frac{\mathcal{A}}{2\pi i}\int_{\Gamma_+}(w'-\mathcal{A})^{-1}\frac{dw'}{w'(w-w')} - \frac{\mathcal{A}(w-\mathcal{A})^{-1}}{2\pi i}\int_{\Gamma_+}\frac{dw'}{w'(w-w')}.
\end{aligned}$$

The second integral on the right hand side vanishes, and the first one is holomorphic for $\operatorname{Im} w > 0$. □

In the following we employ the Fourier-Laplace transform

$$(Fu)(\zeta) = \widetilde{u}(\zeta) = \int_0^\infty e^{-it\zeta} u(t)dt,$$

$\operatorname{Im}\zeta < 0$, with the inverse

$$(F^{-1}\widetilde{u})(t) = \frac{1}{2\pi}\int_{-\infty+i\delta}^{\infty+i\delta} e^{ity}\widetilde{u}(y)dy, \quad t > 0$$

for $\delta < 0, y = \operatorname{Re}\zeta$.

Let us define the maps

$$E : \mathbb{C} \to \mathbb{C}^\mu, \quad Ez := (0, \dots, 0, z)$$

and

$$Q : \mathbb{C}^\mu \to \mathbb{C}, \quad Q(z_0, \dots, z_{\mu-1}) := z_0.$$

THEOREM 3.15. *The boundary value problem*

$$Bu = f \in H^{s-\mu}(Y \times \mathbb{R}_+, J) \tag{3.58}$$

$$\mathcal{P}_+\mathcal{T}^\mu u = g \in \mathcal{P}_+\mathcal{C}^{s,\mu}(Y, J) \tag{3.59}$$

has a unique solution $u \in H^s(Y \times \mathbb{R}_+, J)$, *given by*

$$u = QF^{-1}\left\{(w-\mathcal{A})^{-1}\left[E(\widetilde{f}(w)) - ig - \frac{1}{2\pi}\int_{\Gamma_-}(z-\mathcal{A})^{-1}E(\widetilde{f}(z))dz\right]\right\}. \tag{3.60}$$

The corresponding map $\mathcal{R} : (f, g) \to u$ *is continuous as an operator*

$$\mathcal{R} : H^{s-\mu}(Y \times \mathbb{R}_+, J) \oplus \mathcal{P}_+\mathcal{C}^{s,\mu}(Y, J) \to H^s(Y \times \mathbb{R}_+, J). \tag{3.61}$$

PROOF. We show that $\mathcal{R}$ is a right inverse of the operator ${}^t(B, P_+\mathcal{T}^\mu)$; the consideration for the left inverse is left to the reader. Concerning the continuity (3.61) we refer to [**30**].

It is convenient to pass from B to the operator $\mathcal{D} := D_t - \mathcal{A}$ with $\mathcal{A}$ being given by (3.42) and the vector function U as in (3.41). Then we can omit the mappings Q and E which only single out the first components of U and replace $\mathcal{T}^\mu$ by r', the restriction to $t = 0$.

Let us set

$$\mathcal{R} = (\mathcal{S} \quad \mathcal{K})$$

for

$$\begin{aligned}\mathcal{S}f &:= F^{-1}(w-\mathcal{A})^{-1}\left\{\widetilde{f}(w) - \frac{1}{2\pi}\int_{\Gamma_-}(z-\mathcal{A})^{-1}\widetilde{f}(z)dz\right\}\\ \mathcal{K}g &:= -i\lim_{t\to+0}F^{-1}(w-\mathcal{A})g.\end{aligned}$$

We then show the relations

$$\mathcal{D}\mathcal{S}f = f, \quad \mathcal{D}\mathcal{K}g = 0, \tag{3.62}$$

$$\mathcal{P}_+ \lim_{t\to+0} \mathcal{S}f = 0, \quad \mathcal{P}_+\mathcal{K}g = g \tag{3.63}$$

for f and g belonging to the respective spaces.

We have

$$\begin{aligned} \mathcal{D}\mathcal{S}f &= \{F^{-1}(w-\mathcal{A})F\}\{F^{-1}(w-\mathcal{A})\tilde{f}(w)\} \\ &- \frac{1}{2\pi}\mathcal{D}F^{-1}(w-\mathcal{A})^{-1}\int_{\Gamma_-}(z-\mathcal{A})^{-1}\tilde{f}(z)dz. \end{aligned} \tag{3.64}$$

The first summand on the right hand side of (3.64) is equal to f. So it remains to observe that

$$\mathcal{D}F^{-1}(w-\mathcal{A})^{-1}h = 0,$$

$h = -\frac{1}{2\pi}\int_{\Gamma_-}(z-\mathcal{A})^{-1}\tilde{f}(z)dz$ (the latter vector is independent of w). For similar reasons it follows that $\mathcal{D}\mathcal{K}g = 0$.

For $P_+\mathcal{K}$ we have

$$\begin{aligned} \mathcal{P}_+\mathcal{K}g &= -i\lim_{t\to+0}\mathcal{P}_+F^{-1}(w-\mathcal{A})^{-1}g \\ &= \lim_{t\to+0}\frac{1}{2\pi i}\int_{\Gamma_-}e^{iwt}\mathcal{P}_+(w-\mathcal{A})^{-1}g\,dw. \end{aligned} \tag{3.65}$$

For $t>0$ the exponent in e^{iwt} has a negative real part in the upper complex w half-plane. Therefore, we can deform the contour of integration to a curve 'surrounding' $\mathrm{spec}_+(\mathcal{A}) := \{\lambda \in \mathrm{spec}(\mathcal{A}) : \mathrm{Im}\,\lambda > 0\}$ (which is a countable set which intersects every strip $\{w \in \mathbb{C} : c < \mathrm{Im}\,w < c'\}$ in a finite set for arbitrary $c<c'$, cf. Remark 3.12. An elementary consideration shows that the limes on the right hand side of (3.65) is just $\mathcal{P}_+g$. By assumption we have $\mathcal{P}_+g = g$, so it follows the second relation of (3.63). In remains to check the first equation of (3.63):

$$\mathcal{P}_+\lim_{t\to+0}F^{-1}(w-\mathcal{A})^{-1}\left\{\tilde{f}(w) - \frac{1}{2\pi}\int_{\Gamma_-}(z-\mathcal{A})^{-1}\tilde{f}(z)dz\right\}$$

$$= F^{-1}\mathcal{P}_+(w-\mathcal{A})^{-1}\tilde{f}(w)|_{t=0} - \frac{1}{2\pi}\int_{\Gamma_-}\mathcal{P}_+(z-\mathcal{A})^{-1}\tilde{f}(z)dz = 0.$$

□

Let us now consider the operator (3.31) on a compact C^∞ manifold with boundary Y, where we fix a collar neighbourhood $\cong Y\times[0,1) \ni (x,t)$ of Y. As in the cylindrical situation we then have for the Cauchy data spaces $\mathcal{C}^{s,\mu}(Y,E')$ for any integer $s \geq \mu$, $E' = E|_Y$, a continuous operator

$$\mathcal{T}^\mu : H^s(X,E) \to \mathcal{C}^{s,\mu}(Y,E'),$$

and a pseudo-differential projection

$$\mathcal{P}_+ : \mathcal{C}^{s,\mu}(Y,E') \to \mathcal{C}^{s,\mu}(Y,E').$$

Concerning the non-bijectivity points of the induced operator family (3.37) we make the same assumptions as before, i.e., that there are no such points on the line $\operatorname{Re} w = 0$.

THEOREM 3.16. *The spectral boundary value problem*

$$Au = f \in H^{s-\mu}(X, F), \tag{3.66}$$

$$\mathcal{P}_+ \mathcal{T}^\mu u = g \in \mathcal{P}_+ \mathcal{C}^{s,\mu}(Y, E') \tag{3.67}$$

defines a Fredholm operator

$$\begin{pmatrix} A \\ \mathcal{P}_+ \mathcal{T}^\mu \end{pmatrix} : H^s(X, E) \to \begin{matrix} H^{s-\mu}(X, F) \\ \oplus \\ \mathcal{P}_+ \mathcal{C}^{s,\mu}(Y, E') \end{matrix} \tag{3.68}$$

for every $s \geq \mu$.

More general boundary value problems with a modification of the trace operators $\mathcal{P}_+ \mathcal{T}^\mu$ are studied in [**46**]. Boundary value problems of type (3.68) as well as those of [**46**] are special cases of elliptic problems of Section 2.3, up to a slight modification of the orders of the trace operators which may consist of components of different order. In fact, similarly as (3.56) we can form

$$\mathcal{P}_\pm = \mathcal{R}_0 \frac{1}{2\pi i} \int_{\Gamma_\pm} \mathcal{A}_0 (w - \mathcal{A}_0)^{-1} \frac{dw}{w} \mathcal{R}_0^{-1}$$

with a diagonal matrix $\mathcal{R}_0 = \operatorname{diag}\left(\mathcal{R}_{E'}^{k+\frac{1}{2}}\right)_{k=0,\ldots,\mu-1}$ and $\mathcal{A}_0 = \mathcal{R}_0^{-1} \mathcal{A} \mathcal{R}_0$.

Set

$$P_\pm = \frac{1}{2\pi i} \int_{\Gamma_\pm} \mathcal{A}_0 (w - \mathcal{A}_0)^{-1} \frac{dw}{w}$$

which are pseudo-differential projections

$$P_\pm : \bigoplus_{0 \leq k \leq \mu-1} H^s(Y, E') \to \bigoplus_{0 \leq k \leq \mu-1} H^s(Y, E')$$

of the class $L^0_{\mathrm{cl}}\left(Y; \bigoplus_{0 \leq k \leq \mu-1} E', \bigoplus_{0 \leq k \leq \mu-1} E'\right)$. The operator $\mathcal{R}_0$

$$\mathcal{R}_0 : \bigoplus_{0 \leq k \leq \mu-1} H^s(Y, E') \to \mathcal{C}^{s,\mu}(Y, E'),$$

and (3.68) can be transformed to an equivalent boundary value problem

$$\begin{pmatrix} A \\ T \end{pmatrix} : H^s(X, E) \to \begin{matrix} H^{s-\mu}(X, F) \\ \oplus \\ P_+ H^s\left(Y, \oplus_{0 \leq k \leq \mu-1} E'\right) \end{matrix} \tag{3.69}$$

for $T = \mathcal{R}_0^{-1} \mathcal{P}_+ \mathcal{T}^\mu$.

Setting

$$\mathbb{L}_+ = \left(P_+, \bigoplus_{0 \leq k \leq \mu-1} E', \quad \operatorname{im} \sigma_\psi \bigoplus_{0 \leq k \leq \mu-1} E' \right)$$

we have

$$P_+H^s\left(Y, \bigoplus_{0\le k\le\mu-1} E'\right) = P^s(Y,\mathbb{L}_+),$$

and the operator (3.69) is elliptic in the sense of Definition 2.14. Thus, Theorem 3.16 is an immediate consequence of Theorem 2.21 (i).

3.4. Calderón-Seeley projections. In the present section we discuss another way of constructing pseudo-differential projections to Cauchy data spaces of solutions to elliptic differential equations, cf. also the author's joint paper [**47**] with Tarkhanov.

Let

$$A : C^\infty(X,E) \to C^\infty(X,F)$$

be a differential operator on a compact C^∞ manifold X with boundary Y; $E, F \in \mathrm{Vect}(X)$. Moreover, let M be a C^∞ manifold (countable at infinity with a Riemannian metric) containing X as a submanifold with boundary. Let $\widetilde{E}, \widetilde{F} \in \mathrm{Vect}(M)$ such that $E = \widetilde{E}|_X, F = \widetilde{F}|_X$, and let

$$\widetilde{A} : C^\infty(M,\widetilde{E}) \to C^\infty(M,\widetilde{F})$$

be a differential operator with $A = \widetilde{A}|_X$.

Let $\Omega := \mathrm{int}\, X$, and interpret the characteristic function χ_Ω as an operator of multiplication

$$\chi_\Omega : H^s_{\mathrm{loc}}(M,\widetilde{E}) \to \mathcal{D}'(M,\widetilde{E})$$

for $s \ge \mu = \mathrm{ord}\, \widetilde{A}$. Then the distributional kernel of the commutator

$$[\widetilde{A},\chi_\Omega] = \widetilde{A}\chi_\Omega - \chi_\Omega\widetilde{A}$$

(as a map $C_0^\infty(M,\widetilde{E}) \to \mathcal{D}'(M,\widetilde{F})$) is supported on $Y \times Y$. Moreover, $f \in H^s_{\mathrm{loc}}(M,\widetilde{E})$, $s \ge \mu$, and $\mathcal{T}^\mu f = 0$ imply $[\widetilde{A},\chi_\Omega]f = 0$.

We now assume that there is an operator $\widetilde{P} \in L^{-\mu}_{\mathrm{cl}}(M;\widetilde{F},\widetilde{E})$ such that

$$\widetilde{A}\widetilde{P}f = f, \qquad \widetilde{P}\widetilde{A}u = u \tag{3.70}$$

for all (distributional sections in the respective bundles) f and u, supported in an ε-neighbourhood of X in M for all $0 < \varepsilon < \varepsilon_0$ for some $\varepsilon_0 > 0$.

In the following we employ the operators e^+ and r^+ in the same meaning as in Section 3.1. Set

$$A := \mathrm{r}^+\widetilde{A}\mathrm{e}^+, \qquad P := \mathrm{r}^+\widetilde{P}\mathrm{e}^+.$$

PROPOSITION 3.17. *Let $\widetilde{A}$ be an elliptic differential operator with the above-mentioned properties. Then*

$$G := 1 - PA$$

(as an operator on $H^s(X,E)$, $s \ge \mu = \mathrm{ord}\,\widetilde{A}$) belongs to the space $B_G^{0,\mu}(X;E,F)$ (cf. the notation (1.88)) and satisfies

$$G^2 = G \tag{3.71}$$

PROOF. By virtue of Theorem 1.70 we have $G \in \mathcal{B}^{0,\mu}(X;E,F)$. We then obtain $G \in \mathcal{B}_G^{0,\mu}(X;E,F)$ from Remark 1.68.

Let us now verify (3.71). First we have the relation

$$AP = 1 \tag{3.72}$$

as a consequence of (3.70) and of the fact that $\widetilde{A}$ is a differential operator. Thus $P : H^{s-\mu}(X,F) \to H^s(X,E)$ is a right inverse of A. This yields a projection PA, and G is just the complementary projection. □

REMARK 3.18. Let $s \geq \mu$, and set

$$\ker_s A = \{u \in H^s(X,E) : Au = 0\}.$$

Then

$$G : H^s(X,E) \to \ker_s A \tag{3.73}$$

is a projection to $\ker_s A$.

In fact, $u \in \ker_s A$ gives us $Gu = u$, and for arbitrary $f \in H^s(X,E)$ we have $AGf = A(1 - PA)f = Af - (AP)Af = 0$.

PROPOSITION 3.19. *The operator* (3.73) *can be written in the form*

$$G = \sum_{j=0}^{\mu-1} K_j \circ T_j \tag{3.74}$$

with (unique) potential operators $K_j \in \mathcal{B}^{0,-j-\frac{1}{2}}(X;0,F;E',0)$ (*cf. Remark* 1.69). *Thus* $f \in H^s(X,E)$ *and* $\mathcal{T}^\mu f = 0$ *entails* $Gf = 0$.

PROOF. We have $G = 1 - PA \in \mathcal{B}_G^{0,\mu}(X;E,F)$, and Remark 1.69 gives us a representation (1.89). Since $PA\varphi = \varphi$ for every $\varphi \in C_0^\infty(\Omega,E)$, the operator G_0 vanishes. This yields the relation (3.74) which implies the second assertion. □

THEOREM 3.20. *Let* $\widetilde{A}$ *satisfy the abovementioned assumptions. Moreover, let* $\mathcal{T}^{-\mu}$ *be any right inverse of the map*

$$\mathcal{T}^\mu : H^s(X,E) \to \mathcal{C}^{s,\mu}(Y,E'), \tag{3.75}$$

cf. (3.46), (3.47) *for* $E' = J$.

Then

$$\Pi := \mathcal{T}^\mu G \mathcal{T}^{-\mu} : \mathcal{C}^{s,\mu}(Y,E') \to \mathcal{C}^{s,\mu}(Y,E')$$

is a projection to the Cauchy data space of solutions of $Au = 0$, $u \in H^s(X,E)$ (*called the Calderón-Seeley projection*).

PROOF. Let $g = \mathcal{T}^\mu u$ for some $u \in \ker_s A$. Then we have $\mathcal{T}^\mu(u - \mathcal{T}^{-\mu}g) = 0$ and hence, by Proposition 3.19, $G(u - \mathcal{T}^{-\mu}g) = 0$, which entails $G\mathcal{T}^{-\mu}g = Gu = u$ (by Remark 3.18) and $\Pi g = \mathcal{T}^\mu u = g$. For arbitrary $h \in \mathcal{C}^{s,\mu}(Y,E')$ we have $\mathcal{T}^{-\mu}h \in H^s(X,E)$ and $G\mathcal{T}^{-\mu}h \in \ker_s A$ by Remark 3.18. This yields $\Pi h \in \mathcal{T}^\mu \ker_s A$. □

REMARK 3.21. (i) The operator Π is independent of the choice of $\mathcal{T}^{-\mu}$. In fact, if $\mathcal{T}_1^{-\mu}$ is another right inverse of $\mathcal{T}^{\mu}$ we have

$$\mathcal{T}^{\mu}G\mathcal{T}^{-\mu} - \mathcal{T}^{\mu}G\mathcal{T}_1^{-\mu} = \mathcal{T}^{\mu}G(\mathcal{T}^{-\mu} - \mathcal{T}_1^{-\mu}) = 0$$

because $\mathcal{T}^{\mu}(\mathcal{T}^{-\mu} - \mathcal{T}_1^{-\mu}) = 0$, cf. the second assertion of Proposition 3.19.

(ii) The operator Π is also independent of the specific choice of $\widetilde{P}$, because for another $\widetilde{P}_1$ the associated projection $G_1 = 1 - P_1 A$ has the property that $G_1\mathcal{T}^{-\mu}g \in \ker_s A$ has the Cauchy data g; the same is true of $G\mathcal{T}^{-\mu}g$, i.e., $\mathcal{T}^{\mu}(G - G_1)\mathcal{T}^{-\mu} = 0$.

REMARK 3.22. Assume that the operator A has the unique continuation property of solutions. Then

$$\mathcal{T}^{\mu} : \ker_s A \to \mathcal{C}^{s,\mu}(Y, E')$$

is injective, i.e., there is a unique solution $u \in \ker_s A$ for every $g \in \Pi\mathcal{C}^{s,\mu}(Y, E')$ such that $\mathcal{T}^{\mu}u = g$.

4. Remarks on the edge calculus with global projection data

4.1. Boundary value problems without the transmission property. In this section we want to make some remarks on the role of the transmission property of boundary value problems in connection with Chapter 2 . Pseudo-differential operators that appear as parametrices of elliptic differential operators have always the transmission property, cf. Definition 1.44. On the other hand, there are interesting cases of pseudo-differential operators on a manifold with boundary, where the transmission property is not satisfied, for instance, for operators obtained by a reduction to the boundary of some mixed elliptic problem.

A classical example of a mixed problem is the Zaremba problem for the Laplace operator with jumping boundary conditions, with Dirichlet conditions on one part, Neumann conditions on the other part of the boundary.

In mixed problems the boundary is subdivided into (smooth) submanifolds with an interface of codimension 1 as the common boundary. The solvability can be discussed in terms of boundary value problems (or transmission problems) on the boundary with respect to the interface. Typical operators in this context have $|\eta|$ (the absolute value of the covariable η on the boundary) as the principal symbol; those fail to have the transmission property at the interface. In general, operators with principal symbols $|\eta|^{\mu}$, $\mu \in \mathbb{R}$, have the transmission property with respect to any interface of codimension 1 if and only if μ is an even integer.

This shows, in particular, in a situation as at the beginning of Section 3.1, that the transmission property of the operators $A_{\pm}$ only holds in exceptional cases (although always in the case of differential operators).

The general program to construct an operator algebra of boundary value problems for the case without the transmission property and with global projection conditions at the boundary is carried out in Schulze and Seiler [**45**]. This paper also contains an analogue of Theorem 2.21 for elliptic operators without the transmission property.

The pseudo-differential analysis of boundary value problems and parametrices for the case of Shapiro-Lopatinskij ellipticity are studied in Rempel and Schulze [**32**], and further in [**38**] and [**43**]. Another (earlier) approach of Vishik and Eskin [**52**], [**53**], [**12**], is not organised in terms of operator algebras with complete and smooth symbolic structures.

The algebra property for boundary value problems with or without the transmission property fits into the concept of edge problems with a specific kind of operator-valued symbols, see also Section 4.2 below. This point of view has been developed in [**38**], cf. also [**33**]. Let us also note that mixed and crack problems have been systematically investigated in a new monograph jointly with Kapanadze [**22**] and in the author's joint papers with Harutjunjan [**18**], [**19**]. More details in the context of parabolic mixed and transmission problems may be found in Krainer, Schulze and Zhou [**24**].

4.2. Edge problems. Boundary value problems for differential (and pseudo-differential) operators on a smooth manifold with boundary have much in common with problems for operators on a manifold with edges. This is based on the fact that the 'half-space' $\Omega \times \overline{\mathbb{R}}_+$ for an open set $\Omega \subseteq \mathbb{R}^q$ can be regarded as a wedge with model cone $\overline{\mathbb{R}}_+$ and edge Ω. To illustrate this, let $A = \sum_{|\alpha|\le\mu} a_\alpha(\tilde{x}) D^\alpha_{\tilde{x}}$ be a differential operator in $\Omega \times \mathbb{R}_+$ with coefficients $a_\alpha(\tilde{x}) \in C^\infty(\Omega \times \overline{\mathbb{R}}_+)$. Inserting $\tilde{x} = (y, t)$ for $y \in \Omega$, $t \in \overline{\mathbb{R}}_+$, the operator A takes the form

$$A = t^{-\mu} \sum_{|\beta|+j\le\mu} a_{\beta j}(y,t)(tD_y)^\alpha \left(-t\frac{\partial}{\partial t}\right)^j \tag{4.1}$$

with coefficients $a_{\beta j}(y,t) \in C^\infty(\Omega \times \overline{\mathbb{R}}_+)$ for all j, β.

A differential operator A in $\Omega \times \mathbb{R}_+$ of the form (4.1) will be called edge-degenerate. Clearly, the class of such operators is much larger than that induced by operators with smooth coefficients.

One may ask, whether the calculus of boundary value problems with the transmission property at $t = 0$ in the sense of Section 1.3 has an analogue for edge-degenerate operators, with an adequate substitute of the standard Sobolev spaces, a corresponding generalisation of trace and potential boundary conditions at $t = 0$, Shapiro-Lopatinskj ellipticity, etc. Such a calculus is possible, indeed, and for the wedge $\Omega \times \overline{\mathbb{R}}_+$, or, more generally, a C^∞ manifold X with boundary, it is just a special case of a corresponding calculus on a manifold W with edge Y. Geometrically, such a W is a space such that both $W \setminus Y$ and Y are C^∞ manifolds ($\dim Y = q$, $\dim W \setminus Y = n + 1 + q$), and every point $y \in Y$ has a neighbourhood V which can be represented (via a 'singular chart') in the form $\Omega \times N^\Delta$ for an open set $\Omega \subseteq \mathbb{R}^q$ and a model cone

$$N^\Delta := (N \times \overline{\mathbb{R}}_+)/(N \times \{0\})$$

with base N which is a closed compact C^∞ manifold. The nature of transition maps and other details may be found, for instance, in [**42**].

Note that the class of edge-degenerate operators of order 2 contains the Laplace-Betrami operators (as well as other geometric operators) when $W \setminus Y$ in the local splitting of variables into (y, t, x) is equipped with a 'wedge metric', e.g.,

$$dt^2 + t^2 g_N(t, y) + dy^2,$$

where $g_N(t, y)$ is a family of Riemannian metrics on N smootly depending on (t, y) up to $t = 0$.

An example is $W = X$ for a C^∞ manifold X with boundary ∂X. In this case we have $Y = \partial X$, and N is a single point, i.e., $N^\Delta = \overline{\mathbb{R}}_+$.

The pseudo-differential algebra of edge problems (say, for the case that W is compact) consists of 2×2 block matrix operators

$$\mathcal{A} : \begin{array}{c} \mathcal{W}^{s,\gamma}(W, E) \\ \oplus \\ H^{s-\frac{n+1}{2}}(Y, J_-) \end{array} \to \begin{array}{c} \mathcal{W}^{s-\mu,\gamma-\mu}(W, F) \\ \oplus \\ H^{s-\mu-\frac{n+1}{2}}(Y, J_+) \end{array} \tag{4.2}$$

of order $\mu \in \mathbb{R}$. Compared with (1.90) the shift of smoothness by $(n+1)/2$ on the edge is not essential for the calculus; a reduction of order on Y allows us unify the orders of the spaces on Y to $H^s(Y, J_-)$ and $H^{s-\mu}(Y, J_+)$ respectively. The spaces $\mathcal{W}^{s,\gamma}(W)$ (for the trivial bundle E on W with fibre $\mathbb{C}$) are contained in $H^s_{\mathrm{loc}}(W \setminus Y)$. Near Y in the variables $(y, x, t) \in \mathbb{R}^q \times N \times \mathbb{R}_+$ the spaces $\mathcal{W}^{s,\gamma}(W)$ are of the form

$$\mathcal{W}^s(\mathbb{R}^q, \mathcal{K}^{s,\gamma}(N^\wedge)), \tag{4.3}$$

cf. Definition 1.52, for a certain scale of weighted Sobolev spaces $\mathcal{K}^{s,\gamma}(N^\wedge)$ on the infinite stretched cone $N^\wedge = N \times \mathbb{R}_+$, cf. [**35**], [**39**], or [**11**]. The group action on the space $\mathcal{K}^{s,\gamma}(N^\wedge)$ is given by $\kappa_\lambda : u(x,t) \to \lambda^{\frac{n+1}{2}} u(x, \lambda t)$, $\lambda \in \mathbb{R}_+$. An easy modification of this construction for the case of arbitrary $E \in \mathrm{Vect}(W)$ then gives us the spaces $\mathcal{W}^{s,\gamma}(W, E)$. The operators (4.2) form a so called edge-algebra that contains all edge-degenerate pseudo-differential operators in the upper left corners, cf. [**34**], [**39**], [**11**].

If $W = X$ is a C^∞ manifold X with boundary, the edge-algebra contains all pseudo-differential boundary value problems without the transmission property as a subalgebra, cf. [**38**] or [**43**]. Similarly to the calculus in Sections 1.2 and 1.3, the 2×2 block matrix structure reflects additional trace and potential conditions with respect to the boundary which satisfy an analogue of the Shapiro-Lopatinskij condition in the elliptic case.

In the general case, on a manifold W with edge Y, we have corresponding edge conditions of trace and potential type, again with an analogue of the Shapiro-Lopatinskij condition as an ellipticity condition for the so called principal edge symbol $\sigma_\wedge(\mathcal{A})$ which is the second component of the principal symbolic hierarchy

$$\sigma(\mathcal{A}) = (\sigma_\psi(\mathcal{A}), \sigma_\wedge(\mathcal{A}))$$

in the edge algebra (the first component $\sigma_\psi(\mathcal{A})$ is the edge-degenerate principal interior symbol of $\mathcal{A}$). Ellipticity of $\mathcal{A}$ in the edge algebra requires the bijecticity of both components.

The edge symbol $\sigma_\wedge(\mathcal{A})$ is an analogue of the boundary symbol $\sigma_\partial(\mathcal{A})$ from the situation of boundary value problems. In the edge algebra it consists of a family of operators

$$\sigma_\wedge(\mathcal{A})(y,\eta) : \begin{array}{c} E'_y \otimes \mathcal{K}^{s,\gamma}(N^\wedge) \\ \oplus \\ J_- \end{array} \rightarrow \begin{array}{c} F'_y \otimes \mathcal{K}^{s-\mu,\gamma-\mu}(N^\wedge) \\ \oplus \\ J_+ \end{array} \tag{4.4}$$

parametrised by $(y,\eta) \in T^*Y \setminus 0$; here $E' = E|_Y$ $F' = F|_Y$. Similarly to the relation (1.87) we have homogeneity of order μ in the sense

$$\sigma_\wedge(\mathcal{A})(y,\lambda\eta) = \lambda^\mu \operatorname{diag}(\kappa_\lambda, \lambda^{\frac{n+1}{2}} \operatorname{id})\sigma_\wedge(\mathcal{A})(y,\eta) \operatorname{diag}(\kappa_\lambda^{-1}, \lambda^{-\frac{n+1}{2}} \operatorname{id}) \tag{4.5}$$

for all $\lambda \in \mathbb{R}_+$, $(y,\eta) \in T^*Y \setminus 0$.

The Shapiro-Lopatinskij condition is just the bijectivity of (4.4) for any $s \in \mathbb{R}$ (it entails the bijectivity for all $s \in \mathbb{R}$; the weight $\gamma \in \mathbb{R}$ is kept fixed). We also talk about $\sigma_\wedge$-ellipticity of the corresponding operator $\mathcal{A}$.

Now, for a σ_ψ-elliptic operator

$$A : \mathcal{W}^{s,\gamma}(W,E) \rightarrow \mathcal{W}^{s-\mu,\gamma-\mu}(W,F) \tag{4.6}$$

we can ask the existence of a 2×2 block matrix operator $\mathcal{A}$ in the edge algebra, containing A in the upper left corner, with suitable bundles $J_\pm \in \operatorname{Vect}(Y)$, such that $\mathcal{A}$ is $\sigma_\wedge$-elliptic.

The answer is similar as in the case of boundary value problems, cf. the Introduction. First, because of the relation (4.5), it suffices to look at $(y,\eta) \in S^*Y$. Then

$$\sigma_\wedge(\mathcal{A})(y,\eta) : E'_y \otimes \mathcal{K}^{s,\gamma}(N^\wedge) \rightarrow F'_y \otimes \mathcal{K}^{s-\mu,\gamma-\mu}(N^\wedge) \tag{4.7}$$

is necessarily a family of Fredholm operators, parametrised by the compact space S^*Y. An analogue of the Atiyah-Bott condition (0.17) in the present case is then

$$\operatorname{ind}_{S^*Y} \sigma_\wedge(A) \in \pi_1^* K(Y), \tag{4.8}$$

$\pi_1 : S^*Y \rightarrow Y$. We then have the following theorem.

THEOREM 4.1 ([**35**]). *Let* (4.6) *be a* σ_ψ*-elliptic operator for which* (4.7) *is a family of Fredholm operators. There is then a* $\sigma_\wedge$*-elliptic operator* $\mathcal{A}$ *in the edge algebra containing* A *in the upper left corner if and only if condition* (4.8) *is fulfilled.*

THEOREM 4.2 ([**39**]). (i) *Every* $(\sigma_\psi, \sigma_\wedge)$*-elliptic operator* $\mathcal{A}$ *in the edge algebra has a parametrix within the calculus.*

(ii) *The ellipticity of* $\mathcal{A}$ *entails the Fredholm property of the operator* (4.2) *for every* $s \in \mathbb{R}$.

Index formulas for elliptic operators in the edge algebra have been constructed in Fedosov, Schulze, and Tarkhanov [**13**], and Nazaikinskij, Savin, Schulze, and Sternin [**29**].

Nazaikiskij, Savin Schulze, and Sternin [**27**], [**28**] constructed large classes of σ_ψ-elliptic operators on manifolds with edge for which the condition (4.8) is not satisfied.

It is now again an interesting problem to extend the edge algebra to an edge-Toeplitz algebra such that every σ_ψ-elliptic operator A admits an elliptic problem with global projection conditions on the edge Y.

Such an algebra was constructed in Schulze and Seiler [**44**]. All essential elements of the calculus of Section 1.3 have a natural analogue in the edge-Toeplitz algebra.

4.3. Analysis on manifolds with singularities. Ellipticity of (pseudo-) differential operators and parametrix constructions within a calculus with symbolic structures are an interesting program also on manifolds with higher (say 'polyhedral') singularities. Locally, such spaces can be generated by iteratively forming cones and wedges, starting from smooth compact base manifolds. The corresponding analysis refers to specific geometric properties, for instance, whether edges and corners are regular or cuspidal. Let us make here a few remarks on the case of regular polyhedra with their system of lower dimensional edges. An example is a cube in $\mathbb{R}^3$ with its two-dimensional faces, one-dimensional edges, and corner points. Outside the corner points close to the one-dimensional edges the configuration is a manifold with edges (with boundary), outside corners and one-dimensional edges it is a C^∞ manifolds with boundary. Thus, a calculus in the cube should contain both edge and boundary conditions.

Operator algebras on manifolds with 'higher' singularities have been constructed and investigated in many variants, cf. [**36**], [**14**], [**41**], [**42**], [**23**], [**25**]. The operators are connected with trace and potential operators on the lower-dimensional strata, and ellipticity of corresponding higher edge/corner problems contains a hierarchy of Shapiro-Lopatinskij conditions.

There is then also a hierarchy of topological obstructions for such edge conditions, and it would be necessary to study Toeplitz extensions of the corresponding operator algebras, in order not to rule out ('most of the') interesting operators.

In this connection, the basic questions are still open, and the analysis of elliptic operators and their index theory on stratified spaces is an awarding task for future activities.

References

[1] M.S. Agranovich and M.I. Vishik, *Elliptic problems with parameter and parabolic problems of general type*, Uspekhi Mat. Nauk **19** (1964), no. 3, 53–161.

[2] M.F. Atiyah, *K-theory*, Harvard University, Cambridge Mass., 1965.

[3] M.F. Atiyah and R. Bott, *The index problem for manifolds with boundary*, Coll. Differential Analysis, Tata Institute Bombay, Oxford University Press, Oxford, 1964, pp. 175–186.

[4] M.F. Atiyah, V. Patodi, and I.M. Singer, *Spectral asymmetry and Riemannian geometry I, II, III*, Math. Proc. Cambridge Philos. Soc. **77,78,79** (1975, 1976, 1976), 43–69, 405–432, 315–330.

[5] M.S. Birman and M.Z. Solomjak, *On the subspaces admitting a pseudodifferential projection*, Vestnik LGU (1982), no. 1, 18–25.

[6] B. Booß-Bavnbek and K. Wojciechowski, *Elliptic boundary problems for Dirac operators*, Birkhäuser, Boston-Basel-Berlin, 1993.

[7] L. Boutet de Monvel, *Boundary problems for pseudo-differential operators*, Acta Math. **126** (1971), 11–51.

[8] ______, *On the index of Toeplitz operators of several complex variables*, Invent. Math. **50** (1979), 249–272.

[9] A.P. Calderón, *Boundary value problems for elliptic equations*, Outlines of the Joint Soviet-American Symposium on PDE's, Novosibirsk, 1963, pp. 303–304.

[10] R. Duduchava and F.-O. Speck, *Pseudodifferential operators on compact manifolds with Lipschitz boundary*, Math. Nachr. **160** (1993), 149–191.

[11] Ju. V. Egorov and B.-W. Schulze, *Pseudo-differential operators, singularities, applications*, Operator Theory, Advances and Applications, vol. 93, Birkhäuser Verlag, Basel, 1997.

[12] G.I. Eskin, *Boundary value problems for elliptic pseudodifferential equations*, Math. Monographs, vol. 52, Amer. Math. Soc., Providence, Rhode Island, 1980, Transl. of Nauka, Moskva, 1973.

[13] B.V. Fedosov, B.-W. Schulze, and N.N. Tarkhanov, *On the index of elliptic operators on a wedge*, J. Funct. Anal. **157** (1998), 164–209.

[14] ______, *Analytic index formulas for elliptic corner operators*, Ann. Inst. Fourier **52** (2002), no. 3, 899–982.

[15] B. Gramsch, *Relative Inversion in der Störungstheorie von Operatoren und Ψ-Algebren*, Math. Ann. **269** (1984), 27–71.

[16] G. Grubb, *Pseudo-differential boundary value problems in L_p spaces*, Comm. Partial Differential Equations **15** (1990), 289–340.

[17] G. Grubb and R.T. Seeley, *Weakly parametric pseudodifferential operators and Atiyah-Patodi-Singer boundary problems*, Inventiones Math. **121** (1995), 481–529.

[18] G. Harutjunjan and B.-W. Schulze, *Mixed problems and edge calculus: symbolic structures*, Preprint 2001/20, Institut für Mathematik, Potsdam, 2001.

[19] ______, *Parametrices of mixed elliptic problems*, Preprint 2002/08, Institut für Mathematik, FU, Berlin, 2002, Math. Nachr. (to appear).

[20] ______, *Reduction of orders in boundary value problems without the transmission property*, Preprint 2002/03, Institut für Mathematik, Potsdam, 2002, Journal of the Math. Soc. of Japan (to appear).

[21] T. Hirschmann, *Functional analysis in cone and edge Sobolev spaces*, Ann. Global Anal. Geom. **8** (1990), no. 2, 167–192.

[22] D. Kapanadze and B.-W. Schulze, *Crack theory and edge singularities*, Kluwer Academic Publ., Dordrecht, 2003.

[23] T. Krainer and B.-W. Schulze, *Long-time asymptotics with geometric singularities in the spatial variables*, Preprint 2000/17, Institut für Mathematik, Potsdam, 2000, Israel Math. Conference Proceedings, Vol. **16**, 2003.

[24] T. Krainer, B.-W. Schulze, and Zhou Xiaofang, *Mixed parabolic boundary value problems*, (in preparation).

[25] L. Maniccia and B.-W. Schulze, *An algebra of meromorphic corner symbols*, Bull. des Sciences Math. **127** (2003), no. 1, 55–99.

[26] R.B. Melrose, *The Atiyah-Patodi-Singer index theorem*, Research Notes in Mathematics, A.K. Peters, Wellesley, 1993.

[27] V. Nazaikinskij, A. Savin, B.-W. Schulze, and B. Ju. Sternin, *Elliptic theory on manifolds with nonisolated singularities: II. Products in elliptic theory on manifolds with edges*, Preprint 2002/15, Institut für Mathematik, Potsdam, 2002.

[28] ———, *Elliptic theory on manifolds with nonisolated singularities: IV. Obstructions to elliptic problems on manifolds with edges*, Preprint 2002/24, Institut für Mathematik, Potsdam, 2002.

[29] ———, *Elliptic theory on manifolds with nonisolated singularities: V. Index formulas for elliptic problems on manifolds with edges.*, Preprint 2003/, Institut für Mathematik, Potsdam, 2003.

[30] V. Nazaikinskij, B.-W. Schulze, B. Ju. Sternin, and V. Shatalov, *Spectral boundary value problems and elliptic equations on singular manifolds*, Differentsial'nye Uravneniya **34** (1998), no. 5, 695–708, (Russian).

[31] S. Rempel and B.-W. Schulze, *Index theory of elliptic boundary problems*, Akademie-Verlag, Berlin, 1982.

[32] ———, *Parametrices and boundary symbolic calculus for elliptic boundary problems without transmission property*, Math. Nachr. **105** (1982), 45–149.

[33] ———, *Asymptotics for elliptic mixed boundary problems (pseudo-differential and Mellin operators in spaces with conical singularity)*, Math. Res., vol. 50, Akademie-Verlag, Berlin, 1989.

[34] B.-W. Schulze, *Pseudo-differential operators on manifolds with edges*, Symposium "Partial Differential Equations", Holzhau 1988, Teubner-Texte zur Mathematik, vol. 112, Teubner, Leipzig, 1989, pp. 259–287.

[35] ———, *Pseudo-differential operators on manifolds with singularities*, North-Holland, Amsterdam, 1991.

[36] ———, *The Mellin pseudo-differential calculus on manifolds with corners*, Symposium: "Analysis in Domains and on Manifolds with Singularities", Breitenbrunn 1990, Teubner-Texte zur Mathematik, vol. 131, Teubner, Leipzig, 1992, pp. 208–289.

[37] ———, *The variable discrete asymptotics of solutions of singular boundary value problems*, Operator Theory: Advances and Applications, vol. 57, Birkhäuser Verlag, Basel, 1992, pp. 271–279.

[38] ———, *Pseudo-differential boundary value problems, conical singularities, and asymptotics*, Akademie Verlag, Berlin, 1994.

[39] ———, *Boundary value problems and singular pseudo-differential operators*, J. Wiley, Chichester, 1998.

[40] ———, *An algebra of boundary value problems not requiring Shapiro-Lopatinskij conditions*, J. Funct. Anal. **179** (2001), 374–408.

[41] ———, *Operator algebras with symbol hierarchies on manifolds with singularities*, Advances in Partial Differential Equations (Approaches to Singular Analysis) (J. Gil, D. Grieser, and Lesch M., eds.), Oper. Theory Adv. Appl., Birkhäuser Verlag, Basel, 2001, pp. 167–207.

[42] ———, *Operators with symbol hierarchies and iterated asymptotics*, Publications of RIMS, Kyoto University **38** (2002), no. 4, 735–802.

[43] B.-W. Schulze and J. Seiler, *The edge algebra structure of boundary value problems*, Annals of Global Analysis and Geometry **22** (2002), 197–265.

[44] ———, *Edge operators with conditions of Toeplitz type*, Preprint 2002/13, Institut für Mathematik, Potsdam, 2002.

[45] ———, *Pseudodifferential boundary value problems with global projection conditions*, Preprint 2002/04, Institut für Mathematik, Potsdam, 2002, J. Funct. Anal. (to appear).

[46] B.-W. Schulze, B.Ju. Sternin, and V.E. Shatalov, *On general boundary value problems for elliptic equations*, Sbornik Mathematics **189** (1998), no. 10, 1573–1586.

[47] B.-W. Schulze and N. Tarkhanov, *Fredholm problems for elliptic operators*, Russian J. of Math. Phys. **8** (2001), no. 3, 346–355.

[48] R. Seeley, *Complex powers of an elliptic operator*, Proc. Sympos. Pure Math., vol. 10, 1967, pp. 288–307.

[49] ———, *Topics in pseudo-differential operators*, C.I.M.E. Conference on pseudo-differential operators, Stresa 1968 (Cremonese, Rome), 1969, pp. 167–305.

[50] J. Seiler, *Continuity of edge and corner pseudo-differential operators*, Math. Nachr. **205** (1999), 163–182.

[51] M.Z. Solomyak, *The Calderón projection. (Russ.)*, Operator Theory and Function Theory **1** (1983), 47–55, (University of Leningrad).

[52] M.I. Vishik and G.I. Eskin, *Convolution equations in a bounded region*, Uspekhi Mat. Nauk **20** (1965), no. 3, 89–152.

[53] ———, *Convolution equations in bounded domains in spaces with weighted norms*, Mat. Sb. **69** (1966), no. 1, 65–110.

Institut für Mathematik, Universität Potsdam, 60 15 53 Potsdam
E-mail address: schulze@math.uni-potsdam.de

Operator Theory:
Advances and Applications, Vol. 151, 430–471

On the tangential oblique derivative problem – methods, results, open problems

Peter Popivanov

ABSTRACT. An introduction to the oblique derivative problem for second order elliptic operators is presented. The main ideas and methods are illustrated by means of model examples. The investigation involves subelliptic pseudodifferential operators in Sobolev and Hölder spaces, microlocal analysis, and Fredholm theory. At the end, two open problems are proposed.

Introduction

This paper can be considered as an introduction to the oblique derivative problem for second order elliptic partial differential equations. We consider the boundary value problem (bvp)

$$Lu = F \text{ in } \Omega, \quad \frac{\partial u}{\partial l} = g \text{ on } \partial\Omega, \tag{0.1}$$

where L is a second order elliptic differential operator and l is a smooth real-valued nondegenerate vector field tangential to some part E of the boundary $\partial\Omega$. The boundary value problem (0.1) was considered for the first time by H. Poincaré when studying the tides in the oceans. The vector fields appearing in these investigations are always tangent to the celestial bodies surfaces at the points of some curves or more massive sets. Another application of (0.1) is found in the theory of strong Markov processes with continuous paths in the interior of Ω, such as Brownian motion in the case of diffusion and absorption phenomena on $\partial\Omega$, see [**P-Pl**], [**Pa**].

From the mathematical point of view the problem (0.1) is of great interest because it is probably the first example of a bvp violating the famous Shapiro-Lopatinskij condition. As it is well known, this condition is satisfied at each point in which l is not tangential to $\partial\Omega$, that is, when E is empty. Assuming that $E = \varnothing$, the bvp (0.1) turns out to be of Fredholm type and therefore it has finite dimensional kernel and cokernel. Moreover, (0.1) is solvable for given F and g if and only if they satisfy finitely many orthogonality conditions. The corresponding

---

2000 *Mathematics Subject Classification.* Primary: 35J25; Secondary: 35J70, 35A27, 35B65.

theory is valid in Sobolev spaces H^s, Hölder spaces C^α, and Besov spaces. Observe that in this case there is no loss of regularity of the solution u, i.e., if $F \in H^s(\Omega)$ and $g \in H^{s+1/2}(\partial\Omega)$, then $u \in H^{s+2}(\Omega)$.

The qualitative behavior of the bvp (0.1) changes drastically when l becomes tangential to $\partial\Omega$. At first, it is assumed that E is a smooth submanifold of $\partial\Omega$ of codimension 1 and that $l|_E$ is transversal to E. Let $\mathbf{n}(x)$ be the unit outward normal to $\partial\Omega$ at the point $x \in E$. The following three cases may occur:

(i) $\langle l(x), \mathbf{n}(x)\rangle \geq 0$ for every $x \in E$,
(ii) $\langle l(x), \mathbf{n}(x)\rangle$ changes its sign on $\partial\Omega$ through E from minus to plus in the direction of the vector field $l(x)$,
(iii) $\langle l(x), \mathbf{n}(x)\rangle$ changes its sign on $\partial\Omega$ through E from plus to minus in the direction of $l(x)$.

In the case (i) the vector field l is called a field of *neutral type*; in the cases (ii) and (iii) it is called a field of *emergent* and *submergent* type, respectively. In these two cases a new effect of loss of regularity of the solution u appears, contrary to the nontangential case. The bvp (0.1) with a neutral type vector field l on $\partial\Omega$ is the only one of Fredholm type. In the case of an emergent vector field l, the bvp (0.1) has an infinite dimensional kernel. In order to guarantee the Fredholm property of the problem, the values of u should be prescribed on the set of tangency E. Even under this additional condition we will have loss of regularity of the solution u. The most difficult situation appears when l is of submergent type. In fact, in this case, the bvp (0.1) has an infinite dimensional cokernel as it is not even locally solvable in a neighborhood of each point $x \in E$. This strange effect can be explained by the discontinuity of u (with finite jump) on E. In other words, a classical solution u of (0.1) does not exist for arbitrary F and g. It is more convenient to look for a solution u in the class $C^2(\overline{\Omega} \setminus E)$ allowing the existence of a finite smooth jump $\{u\}_E$ of u on E. As it was proved in [**Mz-Pa**], for any sufficiently smooth function (F, g) from a subspace in $L_2(\Omega) \times L_2(\partial\Omega)$ of finite codimension, there exists a solution $u \in C^2(\overline{\Omega} \setminus E)$ that satisfies (0.1) on $\overline{\Omega} \setminus E$. This solution is unique modulo a finite dimensional subspace of $C^2(\overline{\Omega} \setminus E)$. Moreover, the restriction $u|_{\partial\Omega}$ has a finite smooth jump $\{u\}_E$ that can be determined by (F, g).

The main tools in studying the tangential oblique derivative problem are the energy estimates in appropriate functional spaces. In some papers, the central results are proved avoiding the pseudodifferential approach. Nevertheless, the investigation of the tangential oblique derivative problem 'in general', and the eventual construction of a parametrix, are rather complicated tasks and use modern methods of harmonic analysis and pseudodifferential operators (ΨDO) including the study of subelliptic operators, the Calderón projector on the boundary $\partial\Omega$, Fourier integral operators (FIO) with real and complex valued phase functions on Sobolev and Hölder spaces, canonical transformations, etc.

The present paper is an extended version of the lectures given by the author at the EPAG conference held in Potsdam, Germany, in August 2001. The author supposes some initial knowledge of the reader in the area of pseudodifferential operators on Sobolev spaces, and some small experience in microlocal analysis.

This paper is divided into two parts. The first section deals with the investigation of some model examples in the theory of pseudodifferential boundary value problems. A small introduction in the theory of subelliptic ΨDO is proposed, and a detailed proof of the subellipticity with sharp loss of regularity in Sobolev spaces is given. Again, by means of simple examples, the cases of emergent and submergent vector fields are considered, and the corresponding modified bvp are stated and studied. Calderón projectors on the boundary and canonical transformations for ΨDO on Sobolev spaces are discussed, too. A few words are mentioned about the oblique derivative problem in Hölder spaces and the difficulties arising there.

The second section of this article may be considered as a short survey on the general theory of the tangential oblique derivative problem. Once the examples of modified bvp are stated and studied in Section 1, the reader might take for granted similar results formulated in a more general situation. Certainly, the qualitative properties of the solutions of (0.1) given in Section 1 and Section 2 are the same. While the proofs in Section 1 are more or less elementary, the rather complicated proofs in the 'general case' can be found in many papers or in some monographs. We particularly recommend the very well written book by Paneah [**Pa**] that contains many deep and interesting results on the subject.

At the end of our exposition, we propose two open problems that are of some interest for the further development of the tangential bvp in Hölder classes and for fully nonlinear second order elliptic differential operators.

Acknowledgment. The author would like to thank Prof. B.-W. Schulze, Prof. T. Krainer, and Prof. J. Gil for inviting him to participate in the EPAG conference in Potsdam, and for the opportunity to give two lectures on the famous Poincaré problem: The tangential oblique derivative problem. Special acknowledgment is due to Prof. Gil and to Prof. Krainer for their precise editorial work.

1. Subelliptic operators and their application to the tangential oblique derivative problem

1.1. Subelliptic estimates. In this section we will study the oblique derivative problem for some simple examples in order to show the main difficulties and tools in proving the famous subelliptic estimates. We are not going to develop the whole theory but rather illustrate the main results and give an idea of how the machinery of Calderón projectors, pseudodifferential operators, and Fourier integral operators works in this situation. We will concentrate on the Sobolev case, but we shall also formulate some results concerning the subelliptic estimates in the Hölder spaces.

Let us consider in

$$\mathbb{R}_+^{n+1} = \{(x, x_{n+1}) \in \mathbb{R}^{n+1} : x = (x_1, \dots, x_n) \in \mathbb{R}^n, x_{n+1} > 0\}$$

the boundary value problem

$$(1.1)\qquad \begin{cases} \Delta u = 0 & \text{in } \mathbb{R}^{n+1}_+, \\ \frac{\partial u}{\partial l} = \frac{\partial u}{\partial x_1} - \varphi(x_1)\frac{\partial u}{\partial x_{n+1}} = g(x) & \text{on } \{x_{n+1} = 0\}, \end{cases}$$

where $\varphi \in C^\infty(\mathbb{R}^1)$, $g \in \mathcal{E}'(\mathbb{R}^n)$. Let $D_j = \frac{1}{i}\frac{\partial}{\partial x_j}$ for $1 \le j \le n+1$. In the case $n = 1$, the operator $D_1 - i\varphi(x_1)|D_1|$ arises on the boundary. Obviously, it is an elliptic ΨDO on the real line. We will concentrate on the degenerate problem (1.1), i.e., we will assume that the real-valued function φ vanishes at some points.

Denote by $\hat{u}(\xi, x_{n+1})$ the partial Fourier transform of $u(x, x_{n+1})$ with respect to x. Then we have $\frac{\partial^2 \hat{u}}{\partial x_{n+1}^2} - |\xi|^2 \hat{u} = 0$, and therefore

$$\hat{u} = \hat{c}(\xi)e^{-x_{n+1}|\xi|} + \hat{d}(\xi)e^{x_{n+1}|\xi|}.$$

Having in mind that $\hat{u}$ should be a tempered distribution with respect to x, we get $\hat{u} = \hat{c}(\xi)e^{-x_{n+1}|\xi|}$ and $u(x,0) = c(x)$. Thus,

$$(1.2)\qquad g(x) = \int_{\mathbb{R}^n} e^{ix\xi}(i\xi_1 + \varphi(x_1)|\xi|)\hat{c}(\xi)\, \bar{d}\xi, \quad \text{where } \bar{d}\xi = (2\pi)^{-n}d\xi.$$

We point out that $-\frac{\partial}{\partial x_{n+1}}$ is the outward normal vector field to the boundary $\{x_{n+1} = 0\}$ of (1.1), and the symbol of the operator in (1.2) is $i\xi_1 + \varphi(x_1)|\xi|$.

Let $\varphi(x_1) = a x_1^k$ with $a \neq 0$, $k \in \mathbb{N}$. There are four cases to be studied in this situation, see Figure 1 and Figure 2.

DEFINITION 1.1. A classical ΨDO P of order m is called subelliptic if there exists some $0 \le \delta < 1$ such that for each compact set $K \subset \Omega$ and for each real s, one can find a positive constant $C_{K,s}$ for which one has

$$(1.3)\qquad \|u\|_s \le C_{K,s}\big(\|Pu\|_{s-m+\delta} + \|u\|_{s-1}\big)$$

for all $u \in C_0^\infty(K)$.

Here $\|\cdot\|_s$ stands for the norm in the standard Sobolev space H^s. One can easily prove the following properties of a subelliptic ΨDO P (cf. [**E1**], [**H2**]):

(1) The estimate (1.3) holds if and only if it holds for some fixed s_0.
(2) If $Pu \in H^s_{\mathrm{loc}}(\Omega)$ and $u \in \mathcal{D}'(\Omega)$, then $u \in H^{s+m-\delta}_{\mathrm{loc}}(\Omega)$. Because of this property the number δ is called *loss of regularity* of P.
(3) The adjoint operator P^* of each subelliptic operator is locally solvable about each point $x_o \in \Omega$.
(4) To study the estimate (1.3), it is sufficient to consider the special case $m = 1$, $s = 1 - \delta$. In fact, if $p^o(x,\xi)$ is the principal symbol of the operator P, $\mathrm{ord}_\xi\, p^o = m$, then we can set $p_1^o = p^o|\xi|^{1-m}$ and consider an operator P_1 having $p_1^o(x,\xi)$ as principal symbol. It follows that P is subelliptic with loss of regularity δ if and only if P_1 is subelliptic with loss of regularity δ. Note that $\mathrm{ord}_\xi\, p_1^o = 1$.

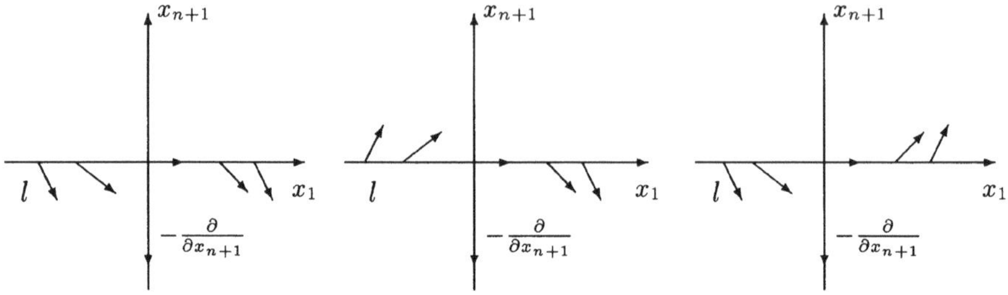

FIGURE 1. Structure of the vector field l near the set of tangency.

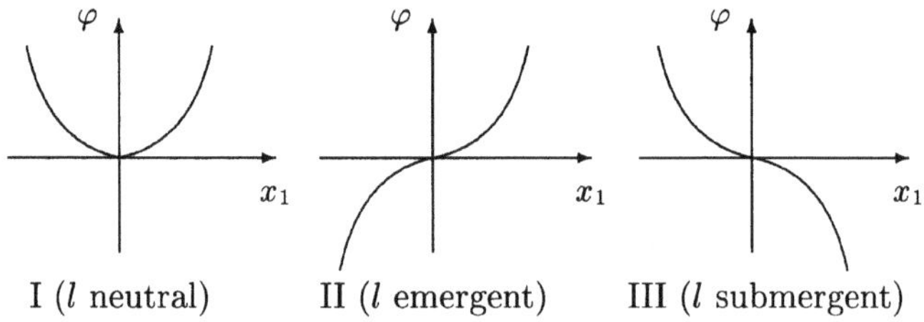

FIGURE 2. Local behavior of φ near 0.

(5) Without loss of generality we can assume that in a conical neighborhood of the point (x_o, ξ^o), $|\xi^o| > 0$, $p_1^o(x_o, \xi^o) = 0$ the principal symbol $p_1^o = i\xi_1 + a(x, \xi)$, $\operatorname{ord}_\xi a = 1$, $a \in C^\infty$ is real valued, $\xi_1^o = 0$, $a(x_o, \xi^o) = 0$. To prove this fact we use two important results. At first, if P_1 is a subelliptic ΨDO, then $p_1^o(x_o, \xi^o) = 0$ implies $\nabla_{x,\xi}\, p_1^o(x_o, \xi^o) \neq 0$, and $\nabla_{x,\xi}\, p_1^o(x_o, \xi^o)$ is not colinear to $(\xi^o, 0)$. Assume for the sake of simplicity that $\frac{\partial}{\partial \xi_1} \operatorname{Re} p_1^o(x_o, \xi^o) \neq 0$. The application of homogeneous canonical transformations [**E1**] leads to the microlocal form $p_1^o = i\xi_1 + a(x, \xi)$.

Next we will formulate two theorems that play important role in the investigation of subelliptic operators. Detailed proofs are given in [**E1**] and [**E2**].

THEOREM 1.2. *Assume that $P(x, D)$ is a classical first order subelliptic ΨDO. Then, for each $k \in \mathbb{N}$ and for each $K \subset \Omega$, there exists a constant $C(K) > 0$ such hat the a priori estimate*

$$\|\psi(y)\|_0 \leq C\Big\{\lambda^\delta \left\|T_k^{(x,\xi)} p^o(x + y\lambda^{\delta-1}, \xi + D\lambda^{-\delta})\psi(y)\right\|_0 + \lambda^{\delta-(k+1)(1-\delta)} \sum_{|\alpha+\beta| \leq k+1} \|y^\beta D^\alpha \psi\|_0\, \lambda^{|\alpha|(1-2\delta)}\Big\} \tag{1.4}$$

holds for each $x \in K$, $\xi \in \mathbb{R}^n$ *with* $|\xi| = 1$, $\lambda \geq 1$, $\psi \in C_0^\infty(\mathbb{R}^n)$. *Here,*

$$T_k^{(x,\xi)} f(y,\eta) = \sum_{|\alpha+\beta|\leq k} \frac{1}{\alpha!\beta!} f_{(\beta)}^{(\alpha)}(x,\xi)(y-x)^\beta(\eta-\xi)^\alpha,$$

where $f_{(\beta)}^{(\alpha)}(x,\xi) = \partial_\xi^\alpha \partial_x^\beta f(x,\xi)$ *for* $f \in C^\infty(T^*(\Omega)\setminus 0)$.

THEOREM 1.3. *Let* $P(x,D)$ *be a first order classical* ΨDO *with principal symbol* p^o *and assume that for each* $K \subset \Omega$, *there exist two constants* $C(K) > 0$, $N = N(K) > 0$, *and a function* $\varepsilon(\lambda) > 0$ *with* $\varepsilon(\lambda) \to 0$ *as* $\lambda \to 0$, *such that the following a priori estimate*

$$\|\psi\|_0 \leq C\Big\{\lambda^\delta \big\|T_k^{(x,\xi)} p^o(x + y\lambda^{\delta-1}, \xi + D\lambda^{-\delta})\psi(y)\big\|_0 + \varepsilon(\lambda) \sum_{|\alpha+\beta|\leq N} \|y^\beta D^\alpha \psi\|_0\, \lambda^{|\alpha|(1-2\delta)}\Big\} \tag{1.5}$$

holds for each $(x,\xi) \in T^*(K)$, $|\xi| = 1$, $\lambda \geq 1$, $\psi \in C_0^\infty(\mathbb{R}^n)$, *where* $\frac{k}{k+1} \leq \delta < \frac{k+1}{k+2}$. *Then one can find a constant* $C_k > 0$ *such that*

$$\|u\|_{1-\delta} \leq C_k\big(\|Pu\|_0 + \|u\|_0\big) \tag{1.6}$$

for all $u \in C_0^\infty(K)$.

Theorem 1.2 enables us to find an interesting necessary condition for the subellipticity of large classes of ΨDO.

THEOREM 1.4. *Consider a first order* ΨDO $P(x,D)$ *with principal symbol* $p^o = i\xi_1 + a(x,\xi)$ *in a conical neighborhood of the characteristic point* (x_o, ξ^o), $|\xi^o| = 1$ (*i.e.* $\xi_1^o = 0$, $a(x_o,\xi^o) = 0$). *Assume that the real-valued function* $a(x,\xi)$, $\operatorname{ord}_\xi a = 1$, *satisfies the following conditions:* $\partial_{x_1}^j a(x_o,\xi^o) = 0$ *for* $1 \leq j \leq k-1$, $\partial_{x_1}^k a(x_o,\xi^o) \neq 0$ *for some* $k \in \mathbb{N}$, *and* $\partial_{x_1}^j \nabla_{x',\xi'} a(x_o,\xi^o) = 0$ *for* $0 \leq j < \frac{l}{2}$, *where* l *is an integer such that* $0 \leq l \leq k$, $x' = (x_2,\dots,x_n)$, *and* $\xi' = (\xi_2,\dots,\xi_n)$. *Then the a priori estimate*

$$\|u\|_{1-\delta} \leq C(\|Pu\|_0 + \|u\|_0), \quad u \in C_0^\infty(K), \tag{1.7}$$

is violated for $\delta < \frac{l}{l+1}$ *if* $x_o \in K$.

REMARK 1.5. Theorem 1.4 is also valid without the assumption

$$\partial_{x_1}^j \nabla_{x',\xi'} a(x_o,\xi^o) = 0 \ \text{ for } 0 \leq j < \frac{l}{2}.$$

PROOF OF THEOREM 1.4. Obviously, $l \geq 1$. We apply Theorem 1.3 with $k = l-1$ and we get the following estimate

$$\|\psi(y)\|_0 \leq C\Big\{\lambda^\delta \|T_{l-1}^{(x_o,\xi^o)} p^o(x_o + y\lambda^{\delta-1}, \xi^o + D\lambda^{-\delta})\psi(y)\|_0 + \lambda^{\delta - l(1-\delta)} \sum_{|\alpha+\beta|\leq l} \|y^\beta D^\alpha \psi\|_0\, \lambda^{|\alpha|(1-2\delta)}\Big\} \tag{1.8}$$

for all $\psi \in C_0^\infty(\mathbb{R}^n)$. Evidently, $0 < \delta < \frac{l}{l+1}$ implies $\delta - l(1-\delta) < 0$. Having in mind the fact that $1 \leq \cdots < l-1 < l \Rightarrow \frac{1}{2} \leq \cdots < \frac{l-1}{l} < \frac{l}{l+1}$, we can assume without loss of generality that $\frac{l-1}{l} \leq \delta$. We will consider several cases.

Case (a): $l = 1$ and so $0 < \delta < \frac{1}{2}$. Then (1.8) takes the form

$$\|\psi\|_0 \leq C\Big\{\lambda^\delta \|p^o(x_o,\xi^o)\psi\|_0 + \sum_{|\alpha+\beta|\leq 1} \|y^\beta D^\alpha \psi\|_0\, \lambda^{(|\alpha|-1)(1-2\delta)}\Big\}.$$

Now, letting $\lambda \to \infty$ we get

$$\|\psi\|_0 \leq C \sum_{|\alpha|=1} \|D^\alpha \psi\|_0 \quad \text{for all } \psi \in C_0^\infty(\mathbb{R}^n).$$

For $\varepsilon > 0$ sufficiently small ($\varepsilon \to 0$), the previous inequality is violated for the function $\psi_\varepsilon(y) = \psi(\varepsilon y)$, where $\psi \in C_0^\infty(\mathbb{R}^n)$.

Case (b): $l \geq 2$ and so $\delta \geq \frac{l-1}{l} \geq \frac{1}{2}$. The change of variables $y_1 = z_1$, $y' = z'\lambda^{1/2-\delta}$ in (1.8) leads to the a priori estimate:

$$\|\psi(z)\|_0 \leq C\Big\{\lambda^\delta \Big\|\partial_1\psi(z)\lambda^{-\delta} + \sum_{1\leq|\alpha+\beta|\leq l-1} \frac{1}{\alpha!\beta!} a^{(\alpha)}_{(\beta)}(x_o,\xi^o) z^\beta D^\alpha \psi \lambda^{\gamma_1}\Big\|_0$$
$$+ \sum_{|\alpha+\beta|\leq l} \|z^\beta D^\alpha\psi\|_0\, \lambda^{\gamma_2}\Big\},$$

where

$$\gamma_1 = (\delta-1)\beta_1 - \frac{1}{2}|\alpha'| - \alpha_1\delta - \frac{1}{2}|\beta'|,$$
$$\gamma_2 = \delta - l(1-\delta) + \alpha_1(1-2\delta) + (\frac{1}{2}-\delta)|\alpha'+\beta'|,$$

that is,

$$\|\psi(z)\|_0 \leq C\Big\{\Big\|\partial_1\psi(z) + \sum_{1\leq|\alpha+\beta|\leq l-1} \frac{1}{\alpha!\beta!} a^{(\alpha)}_{(\beta)}(x_o,\xi^o) z^\beta D^\alpha \psi(z) \lambda^{\delta+\gamma_1}\Big\|_0$$
$$+ \sum_{|\alpha+\beta|\leq l} \|z^\beta D^\alpha\psi\|_0\, \lambda^{\gamma_2}\Big\} \tag{1.9}$$

for all $\psi \in C_0^\infty(\mathbb{R}^n)$, where $\alpha = (\alpha_1, \alpha')$, $\beta = (\beta_1, \beta')$. Evidently, the last term on the right-hand side of (1.9) tends to 0 as $\lambda \to \infty$ since $\delta - l(1-\delta) < 0$ and $\delta \geq \frac{1}{2}$. Consider now the first term. We are looking for the negative powers of λ. Put $\chi = \frac{\partial}{\partial\xi_1} a(x_o,\xi^o)$. Thus, $\nabla_{x,\xi}\, a(x_o,\xi^o) = \chi(0,\ldots,0;1,0,\ldots,0)$ and we conclude that if $|\alpha+\beta| = 1$, then on the corresponding right hand-side the term $(1-i\chi)\partial_1\psi(z)$ will participate. Thus $2 \leq |\alpha+\beta| \leq l-1$. It is obvious that λ will have a negative powers if:

- $|\alpha'| + |\beta'| \geq 2$;
- $\alpha_1 = 1$ and $|\alpha'| + |\beta| > 0$;

- $\alpha_1 = 0$, $|\alpha' + \beta'| = 1$, and $\delta - \frac{1}{2} - (1-\delta)\beta_1 < 0$.

Again we will consider several cases.

1) Assume that $|\alpha' + \beta'| = 1$, which implies $|\alpha_1 + \beta_1| \geq 1$. Nonnegative powers of λ can appear if $\alpha_1 = 0$, $\beta_1 \geq 1$, and $\delta - \frac{1}{2} - (1-\delta)\beta_1 \geq 0$, that is, if $1 \leq \beta_1 \leq \frac{\delta - 1/2}{1-\delta}$. But $\delta < \frac{l}{l+1} \Rightarrow \frac{\delta-1/2}{1-\delta} < \frac{l-1}{2}$, and therefore the coefficients in front of these powers of λ are vanishing. In fact, $\partial_{x_1}^{\beta_1} \nabla_{x',\xi'}\, a(x_o, \xi^o) = 0$ for $\beta_1 \leq \frac{l-1}{2}$.
2) Suppose now that $|\alpha' + \beta'| = 0$ so that $|\alpha_1 + \beta_1| \geq 2$. Nonnegative powers of λ can appear if $\alpha_1 = 0$, $\beta_1 \geq 2$, $\beta_2 \leq l-1$, and $\delta - (1-\delta)\beta_1 \geq 0$. But $\delta < \frac{l}{l+1} \Rightarrow \beta_1 \leq \frac{\delta}{(1-\delta)} < l \leq k$, and $\partial_{x_1}^{\beta_1} a(x_o, \xi^o) = 0$ for $0 \leq \beta_1 \leq k-1$.

Letting $\lambda \to \infty$ in (1.9), we conclude that

$$\|\psi\|_0 \leq C\|(1 - i\chi)\partial_1 \psi\|_0 \quad \text{for all } \psi \in C_0^\infty(\mathbb{R}^n).$$

Since this inequality is violated for some $\psi \in C_0^\infty(\mathbb{R}^n)$ (as we proved it above), our Theorem 1.4 is verified. □

Conclusion. *The ΨDO with symbol* $p_1^o = i\xi_1 + a x_1^k |\xi|$, $a \in \mathbb{R}^1$, $a \neq 0$, *and* $k \in \mathbb{N}$, *is not subelliptic with loss of regularity* $0 < \delta < \frac{k}{k+1}$.

Our next step is to find out all subelliptic operators among the ΨDO having the principal symbol

$$p_1^o = i\xi_1 + a x_1^k |\xi| \quad \text{with } a \neq 0. \tag{1.10}$$

PROPOSITION 1.6. *An operator P with principal symbol* (1.10) *is subelliptic if and only if k is even, or if k is odd and $a < 0$. In both cases, the sharp loss of regularity of P is* $\delta = \frac{k}{k+1}$.

PROOF. We will show that for each power $0 < \gamma < \frac{1}{(k+1)^2}$, $\varepsilon(\lambda) = \lambda^{-\gamma}$, the following estimate holds:

$$\begin{aligned} \|\psi\|_0 \leq C\Big\{ & \lambda^{\frac{k}{k+1}} \Big\| T_k^{(x,\xi)} p^o(x + y\lambda^{-\frac{1}{k+1}}, \xi + D\lambda^{-\frac{k}{k+1}})\psi(y)\Big\|_0 \\ & + \lambda^{-\gamma} \sum_{|\alpha+\beta| \leq N} \|y^\beta D^\alpha \psi\|_0 \, \lambda^{-|\alpha|\frac{k-1}{k+1}} \Big\} \end{aligned}$$

for all $(x,\xi) \in T^*(K)$, $|\xi| = 1$, $\lambda \geq 1$, and $\psi \in C_0^\infty(\mathbb{R}^n)$. The statement of the proposition will then follow from Theorem 1.3.

One obtains immediately that

$$\begin{aligned} & \lambda^{\frac{k}{k+1}} T_k^{(x,\xi)} p^o(x + y\lambda^{-\frac{1}{k+1}}, \xi + D\lambda^{-\frac{k}{k+1}})\psi(y) \\ & \qquad = \sum_{|\alpha+\beta| \leq k} \frac{1}{\alpha!\beta!} p^{o(\alpha)}_{\ (\beta)}(x,\xi) y^\beta D^\alpha \psi(y) \lambda^{\frac{k-|\beta|-k|\alpha|}{k+1}}. \end{aligned}$$

We will assume that $\|\psi\|_0 = 1$ and $\|y^\beta D^\alpha \psi\|_0 \leq C^{-1}\lambda^{\gamma+|\alpha|\frac{k-1}{k+1}}$, $|\alpha+\beta| \leq k$; otherwise the a priori estimate is trivially fulfilled. To prove the estimate, two separate cases will be considered, namely

$$\text{(I)}\quad |x_1| \geq M\lambda^{\gamma-\frac{1}{k+1}}, \quad 0<\gamma<\frac{1}{k+1},$$
$$\text{(II)}\quad |x_1| \leq M\lambda^{\gamma-\frac{1}{k+1}},$$

where the constant $M \gg 1$ will be determined later on.

In case (I) we have $\lambda^{\frac{k}{k+1}} T_k p^o\,\psi = A_1\psi + B_1\psi$, where

$$A_1\psi = \partial_1\psi + i\xi_1\lambda^{\frac{k}{k+1}}\psi + ax_1^k\lambda^{\frac{k}{k+1}}\psi,$$
$$B_1\psi = a\sum_{1\leq|\alpha+\beta_1|\leq k}\frac{1}{\alpha!\beta_1!}\partial_1^{\beta_1}(x_1^k)\frac{\partial^\alpha|\xi|}{\partial\xi^\alpha}\bigg|_{|\xi|=1} y_1^{\beta_1}D^\alpha\psi\,\lambda^{\frac{k-k|\alpha|-\beta_1}{k+1}}.$$

In order to estimate $\|A_1\psi\|_0$ from below, we make the change $\psi = \psi_1\,e^{-i\xi_1 y_1\lambda^{\frac{k}{k+1}}}$, so $\|\psi\|_0 = \|\psi_1\|_0$, and we obtain

$$\|A_1\psi\|_0 = \|\partial_1\psi_1(y) + ax_1^k\lambda^{\frac{k}{k+1}}\psi_1(y)\|_0 \geq |a||x_1|^k\lambda^{\frac{k}{k+1}}.$$

In fact, a simple integration by parts and the identity $\operatorname{Re}\int\overline{\psi}_1\partial_1\psi\,dy = 0$ prove the previous inequality. On the other hand, there exist constants $C_1, C_2 > 0$ independent of M, ψ, a such that

$$\begin{aligned}\|B_1\psi\|_0 &\leq C_1|a|\sum_{1\leq|\alpha+\beta_1|\leq k}|x_1|^{k-\beta_1}\|y_1^{\beta_1}D^\alpha\psi\|_0\,\lambda^{\frac{k-k|\alpha|-\beta_1}{k+1}}\\ &\leq C_2|a|\sum_{1\leq|\alpha+\beta_1|\leq k}|x_1|^{k-\beta_1}\lambda^{\gamma+\frac{k-|\alpha|-\beta_1}{k+1}}\\ &\leq C_2|a||x_1^k|\,\lambda^{\frac{k}{k+1}}\sum_{1\leq|\alpha+\beta_1|\leq k}M^{-\beta_1}\lambda^{\gamma(1-\beta_1)-\frac{|\alpha|}{k+1}}\\ &\leq C_2|a||x_1^k|\,\lambda^{\frac{k}{k+1}}\Big(\frac{1}{M}+o(1)\Big).\end{aligned}$$

The function $o(1)\to 0$ for $\lambda\to\infty$ uniformly with respect to M. Therefore, taking $C_2M^{-1}\leq\frac{1}{2}$ we get

$$\|A_1\psi + B_1\psi\|_0 \geq \frac{1}{3}|a|\,\lambda^{k\gamma}.$$

In case (II) we now have $\lambda^{\frac{k}{k+1}} T_k p^o\,\psi = A_2\psi + B_2\psi$, where

$$A_2\psi = \lambda^{\frac{k}{k+1}} i\xi_1\psi + \partial_1\psi + a(y_1 + x_1\lambda^{\frac{1}{k+1}})^k,$$
$$B_2\psi = a\sum_{|\alpha+\beta_1|\leq k,|\alpha|\geq 1}\frac{1}{\alpha!\beta!}\partial_1^{\beta_1}(x_1^k)\frac{\partial^\alpha|\xi|}{\partial\xi^\alpha}\bigg|_{|\xi|=1} y_1^{\beta_1}D^\alpha\psi\,\lambda^{\frac{k-k|\alpha|-\beta_1}{k+1}}.$$

In fact,

$$\sum_{\alpha=0,0\le\beta_1\le k}\frac{1}{\beta_1!}y_1^{\beta_1}\partial_1^{\beta_1}(x_1^k)\,\lambda^{\frac{k-\beta_1}{k+1}}=(y_1+x_1\lambda^{\frac{1}{k+1}})^k.$$

There are no difficulties to verify that

$$\begin{aligned}\|B_2\psi\|_0&\le C_3\sum_{|\alpha|\ge1,|\alpha+\beta|\le k}|x_1|^{k-\beta_1}\|y_1^{\beta_1}D^\alpha\psi\|_0\,\lambda^{\frac{k-\beta_1-k|\alpha|}{k+1}}\\&\le C_3M^k\sum_{|\alpha|\ge1,|\alpha+\beta_1|\le k}\lambda^{\gamma(k+1-\beta_1)-\frac{|\alpha|}{k+1}}\\&\le C_3M^k\lambda^{\gamma(k+1)-\frac{1}{k+1}},\end{aligned}$$

and $\lambda^{\gamma(k+1)-\frac{1}{k+1}}\to0$ as $\lambda\to+\infty$.

In order to apply Theorem 1.3 we must show that

$$\|\psi(y)\|_0\le C\big\|\partial_1\psi(y)+\psi(y)[i\xi_1\lambda^{\frac{k}{k+1}}+a(y_1+x_1\lambda^{\frac{1}{k+1}})^k]\big\|_0 \tag{1.11}$$

for all $\psi\in C_0^\infty(\mathbb{R}^n)$, where $|\xi_1|\le1$ and $C=\text{const}>0$. To this end, we successively make the following changes:

$$y_1+x_1\lambda^{\frac{1}{k+1}}=z_1,\quad \psi(z)=\psi_1(z)e^{-iz_1\xi_1\lambda^{\frac{k}{k+1}}},$$

which reduce (1.11) to the a priori estimate

$$\|\psi_1(z)\|_0\le C\|\partial_1\psi_1(z)+az^k\psi_1(z)\|_0,$$

for all $\psi\in C_0^\infty(\mathbb{R}^n)$, $C=\text{const}>0$. This multidimensional estimate is an easy consequence of the one-dimensional estimate (1.12) that will be proved below. □

LEMMA 1.7. *The a priori estimate*

$$c_0\|\psi\|_0\le\|\psi'(y)+ay^k\psi(y)\|_0,\quad c_0=\text{const}>0,\ a\in\mathbb{R}^1\setminus0, \tag{1.12}$$

holds for each $\psi\in C_0^\infty(\mathbb{R}^1)$ *if and only if*

(i) *k is even; or*
(ii) *k is odd and $a<0$.*

PROOF. (i) Let k be even. The change $y\to-y$ reduces the case $a<0$ to the case $a>0$. Having in mind the fact that $\frac{d}{dy}+ay^k$ is a linear operator with real-valued coefficients, we may assume that $\psi(y)$ is a real-valued function. Then,

$$\int_{-\infty}^y\psi(x)(\psi'(x)+ax^k\psi(x))\,dx=\frac12\psi^2(y)+a\int_{-\infty}^y x^k\psi^2(x)\,dx.\quad(a>0)$$

So,

$$\int_{-\infty}^y\psi(x)(\psi'(x)+ax^k\psi(x))\,dx\ge\max\Big(\frac12\psi^2(y),\,a\int_{-\infty}^y x^k\psi^2(x)\,dx\Big),$$

and evidently,

$$(\psi' + ax^k\psi, \psi)_{L^2(\mathbb{R})} \geq \max\Big(\frac{1}{2}\max_y \psi^2(y),\ a\int_{-\infty}^{\infty} x^k\psi^2(x)\,dx\Big).$$

We rewrite the previous inequality in the following form

$$2(\psi' + ax^k\psi, \psi)_{L^2(\mathbb{R})} \geq \frac{1}{2}\max_y \psi^2(y) + a\int_{-\infty}^{\infty} x^k\psi^k(x)\,dx. \tag{1.13}$$

If $|x| \geq \frac{1}{2}$, then

$$\int_{-\infty}^{\infty} ax^k\psi^2(x)\,dx \geq a\Big(\frac{1}{2}\Big)^k \int_{|x|\geq\frac{1}{2}} \psi^2(x)\,dx,$$

while for $|x| \leq \frac{1}{2}$, we have

$$\int_{-1/2}^{1/2} \psi^2(x)\,dx \leq \max_x \psi^2(x).$$

Applying Cauchy's inequality to the left-hand side of (1.13) and using the previous two inequalities, we get (1.12).

(ii) Let k be odd and $a < 0$. Then,

$$\int_{-\infty}^{\infty} (\psi'(x) + ax^k\psi(x))^2\,dx = \int_{-\infty}^{\infty} \big[\psi'^2 + a^2x^{2k}\psi^2 - kax^{k-1}\psi^2\big]\,dx$$

with $-ka > 0$ and $k-1$ even. Consider now the cut-off function $\zeta \in C_0^\infty(\mathbb{R}^1)$, $0 \leq \zeta \leq 1$, and $\zeta \equiv 1$ near 0. Thus $\psi = \zeta\psi + (1-\zeta)\psi$. According to the Poincaré's inequality:

$$\|\zeta\psi\|_0^2 \leq C\|\partial_1(\zeta\psi)\|_0^2 \leq \widetilde{C}\int_{-\infty}^{\infty} \big[(\partial_1\psi)^2 + x^{2k}\psi^2\big]\,dx,$$

since $\partial_1\zeta \equiv 0$ near $x = 0$, and consequently, $|\partial_1\zeta| \leq \operatorname{const} x^{2k}$ for all $x \in \mathbb{R}^1$. On the other hand,

$$\|(1-\zeta)\psi\|_0 \leq \widetilde{C}'\int_{-\infty}^{\infty} x^{2k}\psi^2\,dx.$$

Finally, let k be odd and $a > 0$. As $C_0^\infty(\mathbb{R})$ is dense in the Schwartz space $S(\mathbb{R}^1)$, we have that the estimate (1.12) holds true for each $\psi \in S(\mathbb{R})$. The function $\psi = e^{-\frac{ax^{k+1}}{k+1}} \in S(\mathbb{R}^1) \cap \ker(\frac{d}{dx} + ax^k)$ violates (1.12). A detailed discussion for the case $a > 0$ and k odd is proposed below. □

To summarize our previous discussion, we now formulate the main result characterizing the subelliptic operators.

Let $p^o = a + ib$, $\operatorname{ord}_\xi p^o = m$, be the principal symbol of some classical ΨDO of order m. We assume a and b to be real-valued functions. We define

$$\operatorname{Char} p^o = \big\{(x,\xi) \in T^*(\Omega)\setminus 0 : p_m^o(x,\xi) = 0\big\}.$$

The Hamiltonian vector field of the function a is defined as follows:

$$-H_a = \sum_{j=1}^{n} \Big(\frac{\partial a}{\partial x_j}(x,\xi)\frac{\partial}{\partial \xi_j} - \frac{\partial a}{\partial \xi_j}(x,\xi)\frac{\partial}{\partial x_j} \Big).$$

The integral curve $\mathcal{L}$ of H_a passing through the characteristic point (x_o, ξ^o) is called a *zero bicharacteristic* of p^o. Thus,

$$\mathcal{L}: \quad \dot{x} = \nabla_\xi a, \ \ \dot{\xi} = -\nabla_x a; \quad x(0) = x_o, \ \ \xi(0) = \xi^o.$$

Obviously, $a|_{\mathcal{L}} \equiv 0$ and $H_a b(x_o, \xi^o) = \frac{d}{dt} a(x(t), \xi(t))|_{t=0}$.

THEOREM 1.8 ([**E1**]). *A classical ΨDO $P(x, D)$ with principal symbol p^o_m is subelliptic with sharp loss of regularity $0 < \delta < 1$ if and only if, for each $(x, \xi) \in T^*(K) \setminus 0$, there exist $j(x, \xi) \in \mathbb{Z}_+$ and $z \in \mathbb{C}^1 \setminus \{0\}$ such that*

$$H^l_{\mathrm{Re}(zp^o)} \operatorname{Im}(zp^o)(x,\xi) = 0 \ \ \text{for} \ \ 0 \le l \le j-1,$$
$$H^j_{\mathrm{Re}(zp^o)} \operatorname{Im}(zp^o)(x,\xi) \neq 0.$$

If $j(x, \xi)$ is odd, then $H^j_{\mathrm{Re}(zp^o)} \operatorname{Im}(zp^o)(x, \xi) > 0$, and moreover,

$$\sup_{(x,\xi)\in T^*(K)\setminus 0} j(x,\xi) = \frac{\delta}{1-\delta}.$$

REMARK 1.9. If $p^o(x, \xi) \neq 0$, we put $j(x, \xi) = 0$. It is evident that the loss of regularity is $\delta = \frac{k}{k+1}$ for some $k \in \mathbb{N}$.

Assume that the function $\varphi(x_1)$ from (1.1) is flat at 0, i.e., $\varphi^{(l)}(0) = 0$ for all $l \ge 0$, and assume that $\varphi(x_1)$ conserves its sign. Consider the symbol (1.10) with φ instead of ax_1^k. Then subelliptic estimates with loss of regularity $\delta < 1$ are impossible.

In the paper [**Mz-Pa**] the regularity of solutions of corresponding ΨDO is studied in appropriate weighted Sobolev spaces. In many cases, these regularity results are optimal.

DEFINITION 1.10. A pseudodifferential operator $P(x, D)$ is said to be *locally solvable* at the point $x_o \in \Omega$, if there exists a neighborhood $\omega \ni x_o$ such that for each $f \in C_0^\infty(\omega)$, one can find a compactly supported distribution $u \in \mathcal{E}'(\Omega)$ for which $Pu = f$ in ω.

DEFINITION 1.11. A pseudodifferential operator $P(x, D)$ is called *hypoelliptic*, if for each open set $\omega \subset \Omega$ and for each distribution $u \in \mathcal{E}'(\Omega)$, the assumption $Pu \in C^\infty(\omega)$ implies $u \in C^\infty(\omega)$.

The following theorem on local nonsolvability is due to Nirenberg and Treves [**N-Tre**], and Egorov [**E1**].

THEOREM 1.12. *Consider a classical ΨDO $P(x,D)$ with principal symbol $p^o(x,\xi)$. Assume that $p(x^o,\xi^o) = 0$, $|\xi^o| > 0$, and suppose that there exists an odd integer $k \in \mathbb{N}$ for which*

$$H^l_{\mathrm{Re}(zp^o)} \operatorname{Im}(zp^o)(x,\xi) = 0 \quad for \quad 0 \le l \le k-1,$$
$$H^k_{\mathrm{Re}(zp^o)} \operatorname{Im}(zp^o)(x_o,\xi^o) > 0$$

for some $z \in \mathbb{C}^1 \setminus \{0\}$. Then, P is locally nonsolvable at the point x_o.

1.2. Local solvability and microlocal analysis. Keeping in mind the main theorems on subelliptic operators, we now consider the model boundary value problem (1.1) in the emergent case from the point of view of microlocal analysis. The study of a ΨDO P with symbol $i\xi_1 + \varphi(x_1)|\xi| = i(\xi_1 - i\varphi(x_1)|\xi|)$ could be simplified by applying the implicit function theorem on the closed cone $\Gamma = \{|\xi_1| \le C|\xi'| : \xi' = (\xi_2,\dots,\xi_n)\}$. In fact, $p^o = \xi_1 - i\varphi(x_1)|\xi|$ is elliptic outside of Γ. So we have that in Γ the variable ξ_1 satisfies $\xi_1^2 = -\varphi^2(x_1)|\xi|^2$ which implies $\xi_1 = \pm i\varphi(x_1)|\xi'|/\sqrt{1+\varphi^2(x_1)}$, and so

$$\xi_1 - i\varphi(x_1)|\xi| = \left(\xi_1 - i\frac{\varphi(x_1)|\xi'|}{\sqrt{1+\varphi^2(x_1)}}\right)h$$

with $\operatorname{ord}_\xi h = 0$ and $h|_\Gamma \neq 0$. In this way, we have reduced the study of our initial ΨDO with principal symbol p^o to the investigation of a ΨDO with principal symbol $q^o = \xi_1 - i\psi(x_1)|\xi'|$ where $\psi(x_1) = ax_1^k + O(|x_1|^{k+1})$ as $x_1 \to 0$.

We now study the kernel of the operator $Q = D_1 - i\psi(x_1)|D'|$ in the cone Γ, when k is odd and $a > 0$. One can easily see that $\frac{\partial \hat{u}}{\partial x_1} + \psi(x_1)|\xi'|\hat{u} = 0$ implies

$$u(x_1,x',0) = \int_{\mathbb{R}^{n-1}}\int_{\mathbb{R}^{n-1}} e^{i(x'-y')\xi'} e^{-B(x_1)|\xi'|} v(y')\,dy'\,d\!\!{}^-\xi',$$

where $v(y') = u(0,y',0)$ and $B(x_1) = \int_0^{x_1}\psi(s)\,ds$. Consider the kernel K of that FIO with complex-valued phase:

$$Kv(x) = \int_{\mathbb{R}^{n-1}}\int_{\mathbb{R}^{n-1}} e^{i(x'-y')\xi' - B(x_1)|\xi'|} v(y')\,dy' d\!\!{}^-\xi', \quad x \in \mathbb{R}^n.$$

Then $K : C_0^\infty(\mathbb{R}^{n-1}) \to C^\infty(\mathbb{R}^n)$ and $K : \mathcal{E}'(\mathbb{R}^{n-1}) \to \mathcal{D}'(\mathbb{R}^n)$ are continuous, and $QKv = 0$. Denote by $\tilde{\gamma}$ the mapping $\gamma(x') = (0,x')$ for all $x' \in \mathbb{R}^{n-1}$. Certainly, its adjoint $\tilde{\gamma}^*$ is the restriction to the hyperplane $x_1 = 0$. Obviously,

$$\tilde{\gamma}^* Kv(x) = Kv|_{x_1=0} = \int_{\mathbb{R}^{n-1}} e^{ix'\xi'}\hat{v}(\xi')\,d\!\!{}^-\xi' = v(x') \in \mathcal{E}'.$$

The phase function of K is $\Phi(x_1,x',y';\xi_1,\xi',\eta') = (x'-y')\xi' + iB(x_1)|\xi'|$, $|\xi'| > 0$, $B \ge 0$. According to Theorem 8.1.9 from [**H2**, Vol. I]

$$WF(K) = \{(x_1,x',y';\xi_1,\xi',\eta');\ (\xi_1,\xi',\eta') = \Phi'_{x,y'},\ \Phi'_{\xi,\eta'} = 0\},\ \text{i.e.,}$$
$$WF(K) = \{(0,x',x';0,\xi',-\xi'),\ \xi' \neq 0\}.$$

Evidently, $WF'(K)|_{\mathbb{R}^{n-1}_{y'}} = \{(y', \eta') : (x, y', 0, -\eta') \in WF(K)\} = \varnothing$, and therefore Theorem 8.2.13 from [**H2**, Vol. I] gives us:

$$(1.14) \qquad \begin{aligned} WF(Kv) &= WF'(K).WF(v), \text{ where} \\ WF'(K) &= \{(0, x', x'; 0, \xi', \xi'),\ \xi' \neq 0\}. \end{aligned}$$

In other words,

$$WF(Kv) \subset \big\{(x, \xi);\ (x, y'; \xi, \eta') \in WF'(K) \text{ for some } (y', \eta') \in WF(v)\big\}.$$

Assume that $F \neq \varnothing$ is a closed conic subset of $\{(x_1 = 0, x'; \xi_1 = 0, \xi'),\ \xi' \neq 0\}$ and $WF(v) = F$. The relation (1.14) implies

$$WF(Kv) = \{(x_1 = 0, x'; \xi_1 = 0, \xi');\ (x', \xi') \in F\}.$$

The last fact shows that the operator Q is not hypoelliptic and consequently is non-subelliptic ($QKv = 0$, $\operatorname{sing\,supp} Kv \neq \varnothing$). Moreover, $\dim \ker Q = \infty$. The adjoint operator P^* of P has the principal symbol $-i\xi_1 + \varphi(x_1)|\xi| = -i(\xi_1 + i\varphi(x_1)|\xi|)$, where $\varphi(x_1) = a x_1^k$ with k odd, $a > 0$. As we proved in Proposition 1.6, P^* and the corresponding operator $Q^* = D_1 + i\psi(x_1)|D'|$ are both hypoelliptic. Then by Theorem 1.12 we can conclude that P^* and Q^* are both locally nonsolvable.

Consider now the equation $Q^* w = f,\ w \in \mathcal{E}'(\mathbb{R}^n)$. Then,

$$K^* Q^* w = K^* f \Rightarrow (QK)^* w = K^* f \Rightarrow K^* f = 0.$$

We explain this way the local nonsolvability of the operator Q^* with the existence of infinitely many compatibility conditions imposed on the right-hand side f, namely $K^* f = 0$. In order to avoid that $\ker Q$ be infinite dimensional, we prescribe an extra condition on the hyperplane $\{x_1 = 0\}$:

$$(1.15) \qquad \begin{cases} Qw = g \in \mathcal{E}'(\mathbb{R}^n) \\ w|_{x_1 = 0} = 0. \end{cases}$$

Then,

$$w(x) = \int_{\mathbb{R}^{n-1}} \int_0^{x_1} e^{i(x'-y')\xi' - B(x_1, x_1')|\xi'|} g(x_1', y')\, dx_1' dy' \bar{d}\xi'.$$

The last FIO with $B(x_1, x_1') = \int_{x_1'}^{x_1} \psi(s)\, ds$ is of subelliptic type with loss of regularity $\frac{k}{k+1}$. We omit the proof.

To complete the study of the boundary value problem (1.1), we will estimate the norm $\|u\|_{H^s(\mathbb{R}^{n+1}_+)}$ assuming that its trace $c(x) = u(x, 0)$ on the boundary satisfies a subelliptic ΨDO. Thus,

$$\begin{aligned}\|u\|^2_{L^2(\mathbb{R}^{n+1}_+)} &= \int_{\mathbb{R}^n_\xi}\int_0^\infty |\hat{u}(\xi, x_{n+1})|^2\, d\xi dx_{n+1}\\ &= \int_{\mathbb{R}^n} |\hat{c}(\xi)|^2 \left\{\int_0^\infty e^{-2x_{n+1}|\xi|}\, dx_{n+1}\right\} d\xi\\ &\le C\int_{\mathbb{R}^n_\xi} |\xi|^{-1}|\hat{c}(\xi)|^2 d\xi\\ &\le C\|c\|^2_{H^{-1/2}(\mathbb{R}^n)} \le C_1\left(\|g\|^2_{H^{-1/2-\frac{1}{k+1}}(\mathbb{R}^n)} + \|c\|^2_{H^{-1}(\mathbb{R}^n)}\right).\end{aligned}$$

There are no technical difficulties in verifying the estimate

$$\|u\|_{H^s(\mathbb{R}^{n+1}_+)} \le C_s\left(\|g\|_{H^{s-1/2-\frac{1}{k+1}}(\mathbb{R}^n)} + \|u\|_{H^{s-1}(\mathbb{R}^{n+1}_+)}\right) \text{ for } s > \frac{3}{2},$$

which follows from the Sobolev trace theorem $\|c\|_{H^{s-1/2}(\mathbb{R}^n)} \le C\|u\|_{H^s(\mathbb{R}^{n+1}_+)}$ for $s > \frac{1}{2}$. So we conclude that, if $g \in H^{s-1/2-\frac{1}{k+1}}(\mathbb{R}^n)$ and $u \in H^{s_o}(\mathbb{R}^{n+1}_+)$ for some fixed $s_o > \frac{3}{2}$, $s > s_o$, then the solution u of (1.1) belongs to the space $H^s(x_{n+1} \ge 0) \cap C^\infty(x_{n+1} > 0)$ (regularity result).

In the case when Q is nonhypoelliptic one can similarly get the a priori estimate

$$\|u\|_{H^s(\mathbb{R}^{n+1}_+)} \le C_s\Big(\|g\|_{H^{s-1/2-\frac{1}{k+1}}(\mathbb{R}^n)} + \|u_0\|_{H^{s-\frac{1}{2(k+1)}}(\mathbb{R}^{n-1})} + \|u\|_{H^{s-1}(\mathbb{R}^{n+1}_+)}\Big) \text{ for } s > \frac{3}{2},$$

where u_0 is the extra condition prescribed on $\{x_{n+1} = x_1 = 0\}$. In conclusion, we have an estimate for the solution u of the following modified tangential oblique derivative problem with an emergent vector field l:

$$\text{(1.16)}\qquad \begin{cases}\Delta u = 0 & \text{in } x_{n+1} > 0,\\ \frac{\partial u}{\partial l} = g(x) & \text{on } x_{n+1} = 0\\ u|_{x_1=0} = u_0(x'), & x' = (x_2, \dots, x_{n-1}),\ u_0 \in \mathcal{E}'(\mathbb{R}^{n-1}).\end{cases}$$

We point out that u_0 is given on the set of tangency of the vector field l to the boundary $\{x_1 = x_{n+1} = 0\}$.

Combining the previous considerations we come to the conclusion that the Cauchy problem: $Q(x_1, D)u = g$, $\gamma u = v(x')$, where γ is the trace operator on $\{x_1 = 0\}$ and $x = (x_1, x') \in \mathbb{R}^n$, has the solution

$$\hat{u}(x_1, \xi') = e^{-B(x_1)|\xi'|}\hat{v}(\xi') + \int_0^{x_1} e^{-B(x_1, x_1')|\xi'|}\hat{g}(x_1', \xi')\, dx_1',$$

where $q^o = i\xi_1 + \psi(x_1)|\xi'|$, $\psi(x_1) = ax_1^k + O(x_1^{k+1})$ as $x_1 \to 0$, k odd, $a > 0$, and $B(x_1, x_1') = \int_{x_1'}^{x_1} \psi(s)\, ds$. Certainly, $\hat{u}$ is the unique solution of the ODE

$$\partial_{x_1}\hat{u} + \psi(x_1)|\xi'|\hat{u} = \hat{g}(x_1, \xi'), \quad \hat{u}(0, \xi') = \hat{v}(\xi').$$

Moreover, $B(x_1, x_1') \geq 0$ since either $0 \leq x_1' \leq x_1$ or $0 \geq x_1' \geq x_1$. Thus,

$$u(x_1, x') = Kv(x) + Eg$$

with

$$Kv(x) = \int_{\mathbb{R}^{n-1}} \int_{\mathbb{R}^{n-1}} e^{i(x'-y')\xi' - B(x_1)|\xi'|} v(y')\, dy' \bar{d}\xi',$$

$$Eg(x) = \int_0^{x_1} \int_{\mathbb{R}^{n-1}} \int_{\mathbb{R}^{n-1}} e^{i(x'-y')\xi' - B(x_1, x_1')|\xi'|} g(x_1', y')\, dx_1' dy' \bar{d}\xi'.$$

One can easily see that

$$\begin{pmatrix} Q \\ \gamma \end{pmatrix} (E, K) = \begin{pmatrix} Id & 0 \\ 0 & Id \end{pmatrix}.$$

In fact, $QE = Id$, $QK \equiv 0$, $\gamma E = 0$, $\gamma K = Id$. Similar calculations show that

$$(E, K) \begin{pmatrix} Q \\ \gamma \end{pmatrix} = EQ + K\gamma \equiv Id.$$

The proof of the identity $EQu + K\gamma u = u$ is left to the reader. The trace operator γ maps $C_0^\infty(\mathbb{R}^n) \to C_0^\infty(\mathbb{R}^{n-1})$, and therefore its adjoint γ^* is given by the formula $\gamma^* v(x) = v(x') \otimes \delta(x_1)$ for all $v \in C_0^\infty(\mathbb{R}^{n-1})$. To clarify the main idea, our considerations will be rather formal. Put

$$M : u \mapsto \begin{pmatrix} Qu \\ \gamma u \end{pmatrix} : C_0^\infty(\mathbb{R}^n) \to C_0^\infty(\mathbb{R}^{n-1}) \oplus C_0^\infty(\mathbb{R}^{n-1}_{x'}).$$

We equip all these spaces with L_2 norms and scalar product $\langle \cdot, \cdot \rangle$, and we look for the adjoint operator M^*. Let $(u, v) \in C_0^\infty(\mathbb{R}^n) \oplus C_0^\infty(\mathbb{R}^{n-1})$, $w \in C_0^\infty(\mathbb{R}^n)$. Then,

$$\begin{aligned} \langle M^*(u, v), w \rangle &= \langle (u, v), Mw \rangle = \langle (u, v), (Qw, \gamma w) \rangle = \langle u, Qw \rangle + \langle v, \gamma w \rangle \\ &= \langle Q^* u, w \rangle + \langle \gamma^* v, w \rangle = \langle Q^* u + v \otimes \delta(x_1), w \rangle, \end{aligned}$$

that is, $M^*(u(x), v(x')) = Q^* u + v(x') \otimes \delta(x_1)$. Since (E, K) is a left inverse of M, we get $M^*(E, K)^* = Id$, and so

$$(Q^* \ \delta(x_1)) \begin{pmatrix} E^* \\ K^* \end{pmatrix} = Id.$$

Therefore, the operator M^* is locally solvable near 0 and the solution is uniquely determined.

Conclusion. *Consider the operator*

$$Q^* = -\partial_{x_1} + \psi(x_1)|D'|,$$

where $\psi(x_1) = a x_1^k + O(|x_1|^{k+1})$ *with* $a > 0$ *and* k *odd. This operator is locally nonsolvable near the origin. However, the following modified problem*

$$Q^* u(x) + v(x') \otimes \delta(x_1) = g(x)$$

is locally solvable near 0 *for any given function* g*, and the solution* (u, v) *is uniquely determined by* g.

REMARK 1.13. Let $Q^* = -\partial_{x_1} + \psi(x_1)|D'|$, $\psi(x_1) = ax_1^k$ with $a > 0$ and k odd, and let $\theta \subset \mathbb{R}^n$ be a neighborhood of $\{x_1 = 0\}$. Then one can find a constant $C > 0$ such that

$$\|\{u\}_{x_1=0}\|_{L^2(\mathbb{R}^{n-1})} + \|u\|_{L^2(\mathbb{R}^n)} \leq C\|Q^*u + \{u\}_{x_1=0} \otimes \delta(x_1)\|_{L^2(\mathbb{R}^n)}$$

for each function $u \in H^1(\mathbb{R}^n \setminus \{x_1 = 0\})$ having compact support in θ. As usual, $\{u\}_{x_1=0}$ denotes the jump of u on $\{x_1 = 0\}$, i.e.,

$$\{u\}_{x_1=0} = u(+0, x') - u(-0, x') \in H^{1/2}(\mathbb{R}^{n-1}).$$

Note that by the Sobolev trace theorem, $u(\pm 0, x') \in H^{1/2}(\mathbb{R}^{n-1})$.

We will prove the a priori estimate stated above in the 1-dimensional case. The proof in $\mathbb{R}^n$ for $n \geq 2$ follows immediately from $\mathbb{R}^1$. Consider $L = \partial_t + \lambda t^k$ with $\lambda < 0$ and k odd. Let $u_\pm \in C_0^1(\mathbb{R}^1_\pm)$, where $\mathbb{R}^1_\pm = \{t : t \geq (\leq) 0\}$. Then, for some constant $c_1 > 0$:

$$|u_\pm(\pm 0)|^2 + \|u_\pm\|^2_{L^2(\mathbb{R}_\pm)} \leq c_1\|Lu_\pm\|_{L^2(\mathbb{R}_\pm)}.$$

In fact, a simple integration by part shows that for $u \in C_0^1(\mathbb{R}_+)$, we have

$$\begin{aligned}
&\mathrm{Re}(Lu, \partial_t u - u)_{L^2(\mathbb{R}_+)} \\
&\geq \|u_t\|^2_{L^2(\mathbb{R}_+)} - \lambda \| |t|^{\frac{k}{2}} u\|_{L^2(\mathbb{R}_+)} + \frac{1}{2}|u(+0)|^2 \\
&\geq \frac{1}{2}|u(+0)|^2 + c_2\Big(\|u\|^2_{L^2(\mathbb{R}_+)} + \|u_t\|^2_{L^2(\mathbb{R}_+)} + (-\lambda)\| |t|^{\frac{k}{2}} u\|_{L^2(\mathbb{R}_+)}\Big),
\end{aligned}$$

where c_2 is a constant independent of λ. The Cauchy-Schwartz inequality, applied to the left-hand side, gives us the inequality we are looking for. The well-known jump formula (see [**Pa**]),

$$\partial_t u = \partial_t u|_{t \neq 0} + \{u\}_{t=0} \otimes \delta_t$$

for $u \in C_0^1(\mathbb{R}_+) \cap C_0^1(\mathbb{R}_-)$ with $\partial_t u \in \mathcal{D}'(\mathbb{R}^1)$, completes the proof of our a priori estimate:

$$|\{u\}_{t=0}| + \|u\|_{L^2(\mathbb{R})} \leq C\|Lu - \{u\}_{t=0} \otimes \delta_t\|_{L^2(\mathbb{R})}.$$

Summary. We now summarize the local properties of the tangential oblique derivative problem for Laplace operator (1.1).

Case I: l is a neutral vector field (see Figure 1). Then the bvp (1.1) is locally solvable, i.e., for each $g \in H^{s-1/2-\frac{1}{k+1}}(\mathbb{R}^n)$, $s > 1/2$, there exists a solution $u \in H^s(\mathbb{R}^{n+1}_+)$ of (1.1). Moreover, $g \in C^\infty(\mathbb{R}^n)$ implies $u \in C^\infty(\overline{\mathbb{R}^{n+1}_+})$. In fact, the trace $c(x) = u(x, 0)$ on the boundary satisfies a subelliptic ΨDO with principal symbol $p^o = i\xi_1 + ax_1^k|\xi|$ with k even, $k \neq 0$.

Case II: l is a vector field of emergent type. Then the bvp (1.1) has a kernel of infinite dimensions. In order to avoid this difficulty, we modify (1.1) by putting an extra condition u_0 on the tangency set $\{x_1 = x_{n+1} = 0\}$. The modified problem (1.16) is locally solvable, and for each $g \in H^{s-1/2-\frac{1}{k+1}}(\mathbb{R}^n)$, $u_0 \in H^{s-\frac{1}{2(k+1)}}(\mathbb{R}^{n-1})$,

$s > 3/2$, there exists a solution $u \in H^s(\mathbb{R}^{n+1}_+)$ of (1.16). There are at most finitely many solutions. In fact, the trace $c(x) = u(x,0)$ satisfies a non-subelliptic equation with principal symbol of the form $i\xi_1 + \psi(x_1)|\xi'|$ with $\psi(x_1) = ax_1^k + O(|x_1|^{k+1})$ as $x_1 \to 0$, $a > 0$, and k odd. Then we consider the following Cauchy problem: $Q(x,D)c = g$, $c|_{x_1=0} = u_0(x')$ which locally has a unique solution c.

Case III: l is a vector field of submergent type. The bvp (1.1) is in general locally nonsolvable. In fact, the trace $c(x)$ is the solution of a locally nonsolvable subelliptic ΨDO on the boundary with principal symbol $\xi_1 + iax_1^k|\xi|$ with $a > 0$ and k odd. It is interesting to note that the boundary value problem under consideration remains hypoelliptic, i.e., if $u \in H^{s_o}_{\text{loc}}(\overline{\mathbb{R}^{n+1}_+})$, $s_o > 3/2$, is a solution of (1.1) for $g \in C^\infty(\mathbb{R}^n)$, then $u \in C^\infty(\overline{\mathbb{R}^{n+1}_+})$.

In the case of a submergent vector field l, we can replace the tangential oblique derivative problem (1.1) with the following modified problem:

$$\text{(1.1M)}\qquad \begin{aligned} &\Delta u = 0 \quad \text{in } x_{n+1} > 0 \\ &\frac{\partial u}{\partial l} - v(x') \otimes \delta(x_1) = g(x) \quad \text{on } \{x_{n+1} = 0\}. \end{aligned}$$

Not only u is an unknown function but v as well. We can expect the existence (even locally) of a solution $(u(x), v(x'))$ of the bvp (1.1M).

REMARK 1.14. Assume that we are looking for a solution $u \in C^\infty(x_{n+1} > 0)$ of the bvp (1.1), the submergent case, such that u belongs to $C^2(\overline{\mathbb{R}^{n+1}_+} \setminus \{x_1 = 0\})$ and has a finite jump discontinuity on $\{x_1 = x_{n+1} = 0\}$. From the Schwartz distribution theory it is well known that on $\{x_{n+1} = 0\}$,

$$\frac{\partial u}{\partial x_1} = \left\{\frac{\partial u}{\partial x_1}\right\}_{x_1 \neq 0} + \{u\}_{x_1=0} \otimes \delta(x_1),$$

where $\{u\}_{x_1=0} = u(+0,x',0) - u(-0,x',0)$. Therefore, the bvp (1.1) takes the form

$$\begin{aligned} &\Delta u = 0 \quad \text{in } x_{n+1} > 0 \\ &\frac{\partial u}{\partial l} - \{u\}_{x_1=0} \otimes \delta(x_1) = g(x') \quad \text{on} \quad x_{n+1} = 0, \end{aligned}$$

in other words, it is of type (1.1M) with $v = \{u\}_{x_1=0}$. We can expect (possibly under additional assumptions) that the bvp (1.1M) is not only solvable but that $v(x') = \{u\}_{x_1=0}$. Therefore, we can hope that 'in general' the solution u of (1.1) will have a jump discontinuity on $\{x_1 = x_{n+1} = 0\}$ which is uniquely determined by g. The solution u of (1.1M) could be a classical one if $v(x') = 0$, which implies the overdetermination of the bvp (1.1) in the submergent case. The lack of classical solutions for a 'massive' set of right-hand sides g is due to the last observation.

1.3. Canonical transformations. Subellipticity is invariant under homogeneous canonical transformations and under multiplication by zero order elliptic ΨDO, see [**E1**], [**H2**, Vol. IV]. For the sake of completeness, we will remind the reader of the corresponding definitions and theorems.

A *symplectic form* σ on a smooth manifold M, $\dim M = 2n$, is by definition a closed nondegenerate differential two-form on M. Assume that $M = T^*(\Omega)$ is the cotangent bundle over the smooth manifold Ω, $\dim \Omega = n$. Then, M can be equipped with a symplectic form in a standard way. In fact, in local coordinates $(x, \xi) \in T^*(\Omega)$, a symplectic form is given by:

$$\sigma = d\xi \wedge dx = \sum_{j=1}^{n} d\xi_j \wedge dx_j.$$

This definition is actually independent of the choice of local coordinates.

DEFINITION 1.15. A smooth mapping χ from $M_1 \subset T^*(\Omega)$ to $M_2 \subset T^*(\Omega)$ is called *canonical*, if it preserves the symplectic form on $T^*(\Omega)$. In other words, $\chi : M_1 \to M_2$ is canonical if and only if $\chi^*\sigma = \sigma$.

Consider now the transformation defined by the real-valued generating function $S(x, \eta) \in C^\infty$: $dS = \langle \xi, dx\rangle + \langle y, d\eta\rangle$, where $\det S''_{x\eta} \neq 0$. In coordinate form we have

$$\xi_j = \frac{\partial S}{\partial x_j}, \quad y_j = \frac{\partial S}{\partial \eta_j}, \quad 1 \leq j \leq n, \tag{1.17}$$

i.e., locally, $\eta_j = \eta_j(x, \xi) \Rightarrow y_j = y_j(x, \xi)$. We will prove that (1.17) gives a canonical transformation. In fact, the identity

$$d(\langle y, \eta\rangle - S) = \langle y, d\eta\rangle + \langle \eta, dy\rangle - dS = -\langle \xi, dx\rangle + \langle \eta, dy\rangle$$

implies $0 = d^2(\langle y, \eta\rangle - S) = -d\xi \wedge dx + d\eta \wedge dy$, and so $d\xi \wedge dx = d\eta \wedge dy$.

The canonical transformation (1.17) is called *homogeneous* if $S(x, \eta)$ is positively homogeneous of order 1 with respect to η, i.e., $\operatorname{ord}_\eta S = 1$. Then, certainly, $\operatorname{ord}_\xi y = 0$ and $\operatorname{ord}_\xi \eta = 1$. We now define the *Poisson bracket* of two smooth real-valued functions on $T^*(\Omega)$:

$$\{f, g\} = \sum_{j=1}^{n} \Big(\frac{\partial f}{\partial x_j}\frac{\partial g}{\partial \xi_j} - \frac{\partial f}{\partial \xi_j}\frac{\partial g}{\partial x_j}\Big).$$

It is well known that a mapping $\chi : T^*(\Omega) \to T^*(\Omega)$, $\Omega \subset \mathbb{R}^n$, is canonical if and only if it preserves the Poisson bracket of every two functions. The last assertion can be proved by the reader.

DEFINITION 1.16. Consider a smooth real-valued function $S(x, \xi)$ such that $\operatorname{ord}_\xi S(x, \xi) = 1$, and assume that $|\nabla_x S(x, \xi)| \geq C_1|\xi|$ for $|\xi| \geq C_2$, where C_1, C_2 are some positive constants. Then, $S(x, \xi)$ is said to be a *phase function*. We will assume, moreover, that $\det S''_{x,\xi} \neq 0$.

The corresponding FIO with phase function S is defined as follows:

$$\Phi u(x) = \int_{\mathbb{R}^n} a(x, \xi)\hat{u}(\xi)e^{iS(x,\xi)}\, \bar{d}\xi \quad \text{for } u \in C_0^\infty(\mathbb{R}^n),$$

where $a(x, \xi) \in C^\infty(\mathbb{R}^n \times (\mathbb{R}^n \setminus 0))$, $\operatorname{ord}_\xi a = 0$.

The following theorem is a famous result due to Egorov [**E1**].

THEOREM 1.17. *Suppose that $P(x,D)$ is a classical ΨDO of order m, and suppose that the symbols p of P and a of Φ vanish outside some compact $K \subset \mathbb{R}^n_x$. Then there exists a classical ΨDO Q of order m with the following properties:*

(i) *The order of the operator $P\Phi - \Phi Q$ is $-\infty$*

(ii) *The principal symbols p^o of P and q^o of Q are linked by the equality*

$$p^o(x,\xi) = q^o(y,\eta),$$

where ξ, y are given by the homogeneous canonical transformation (1.17).

A simple application of Theorem 1.17 is proposed below.

Let $p^o(x,\xi)$ be a real-valued classical symbol such that $\operatorname{ord}_\xi p^o = 1$, $p^o(x_o,\xi^o) = 0$ for $\xi^o \neq 0$, and $\nabla_\xi p^o(x_o,\xi^o) \neq 0$. Without loss of generality we assume that $\partial_{\xi_1} p^o(x_o,\xi^o) \neq 0$, and consider the following Cauchy problem:

$$\text{(1.18)} \qquad \begin{aligned} &p^o(x, \nabla_x S(x,\eta)) = \eta_1, \\ &S|_{x_1 = x_{j_o}} = \sum_{j=2}^{n} (x_j - x_{j_o}). \end{aligned}$$

In a neighborhood of (x_o,ξ^o), there exists a smooth solution $S(x,\eta)$ of (1.18), and one can check that $S(x,t\eta) = tS(x,\eta)$ for every $t > 0$, $|\nabla_x S(x,\eta)| \geq C|\eta|$ for some constant C, and $\det S''_{x,\eta} \neq 0$.

Conclusion. *In each microlocal investigation near a characteristic point, say (x_o,ξ^o), we can work with a ΨDO having the symbol η_1 instead of p^o. That is, we can work with the operator $D_1 = \frac{1}{i}\frac{\partial}{\partial y}$. Suppose now that the classical complex-valued symbol $r^o = p^o(x,\xi) + iq^o(x,\xi)$ satisfies the conditions $r^o(x_o,\xi^o) = 0$ and $\nabla_\xi p^o(x_o,\xi^o) \neq 0$. Then there exists a homogeneous canonical change with generating function S defined by* (1.18), *which on a conical neighborhood of (x_o,ξ^o), microlocally transforms r^o into $\tilde{r}^o = \eta_1 + i\tilde{q}^o(y,\eta)$ with $\operatorname{ord}_\eta \tilde{q}^o = 1$.*

Since the subelliptic estimates (1.4) *and* (1.5) *are invariant under canonical transformations, in all our microlocal subelliptic investigations we can deal with the symbol $\xi_1 + ia(x,\xi)$ for real-valued $a(x,\xi)$, $\operatorname{ord}_\xi a = 1$, and even with the symbol $\xi_1 + ia(x,\xi')$ where $\xi' = (\xi_2,\dots,\xi_n)$. As we mentioned before, this very important contribution is due to Egorov.*

1.4. The Calderón projector. In this section we will discuss some aspects of the Calderón projector. Let Ω be a bounded domain in $\mathbb{R}^{n+1}$ with smooth boundary $\partial\Omega$, and let Ω lie on one side of $\partial\Omega$. Let $\mathcal{F}$ be a neighborhood of $\partial\Omega$. We identify $\mathcal{F} \cap \overline{\Omega}$ with the product $\partial\Omega \times [0,\bar{\varepsilon}]$, $\bar{\varepsilon} > 0$, by introducing local coordinates $x = (x_1,\dots,x_n)$ in $\partial\Omega \subset \mathbb{R}^n$, and taking a geodesic normal coordinate x_{n+1} with respect to $\partial\Omega$. Thus the points in $\mathcal{F}$ have the form (x, x_{n+1}) with $0 \leq x_{n+1} \leq \bar{\varepsilon}$.

We consider the second order differential operator

$$L = \sum_{i,j=1}^{n+1} a_{ij}(y)\partial_i\partial_j + \sum_{j=1}^{n+1} a_j(y)\partial_j + a_0(y),$$

where $y = (y_1, \ldots, y_n, y_{n+1}) \in \Omega$, $a_{ij}(y) = a_{ji}(y)$, and $\sum_{i,j} a_{ij}(y)\xi_i\xi_j \geq c|\xi|^2$ for some constant $c > 0$ and for all $y \in \overline{\Omega}$, $\xi \in \mathbb{R}^{n+1}$. We assume all the coefficients of L to be C^∞ smooth real-valued functions. Moreover, we will assume that $a_0(y) \leq 0$ everywhere in $\overline{\Omega}$. In the neighborhood $\mathcal{F}$, the operator L can be written as:

$$L = \sum_{j=0}^{2} \varepsilon_j L_{2-j}(x, x_{n+1}; \partial_x)\partial^j_{x_{n+1}},$$

where the L_k are linear differential operators of order k on the boundary $\partial\Omega$, $\varepsilon_0 = 1$, $\varepsilon_1 = 2$, and $\varepsilon_2 = 1$.

Let l be a nonvanishing smooth real-valued vector field on $\partial\Omega$, and let $h \in C^\infty(\partial\Omega)$ be a real-valued function. We will call the boundary value problem

$$\begin{aligned} &Lu = F \quad \text{in } \Omega, \\ &\frac{\partial u}{\partial l} + hu = f \quad \text{on } \partial\Omega, \end{aligned} \tag{1.19}$$

an *oblique derivative problem.*

Denote by τ a smooth real-valued vector field tangential to $\partial\Omega$ and such that $l = \tau + a\mathbf{n}$. As usual, $\mathbf{n}$ stands for the unit exterior normal vector field to $\partial\Omega$. Without loss of generality we assume $|l| = 1$; evidently $a = (l, \mathbf{n}) \in C^\infty(\partial\Omega)$. In our local coordinates $\Omega = \{x_{n+1} > 0\}$, $\partial\Omega = \{x_{n+1} = 0\}$, we have

$$\frac{\partial}{\partial \mathbf{n}} = -\frac{\partial}{\partial x_{n+1}} \quad \text{and} \quad \frac{\partial}{\partial \tau} = \sum_{j=1}^{n} a_j(x)\partial_j.$$

The unique solution of the Dirichlet problem $Lu = F$ in Ω, $u = 0$ on $\partial\Omega$ will be denoted by RF, while the unique solution of the Dirichlet problem $Lu = 0$ in Ω, $u = f$ on $\partial\Omega$ will be denoted by Sf. The solutions are unique due to the maximum principle ($a_0 \leq 0$). Obviously, $Sf|_{\partial\Omega} = f$ implies $\frac{\partial}{\partial\tau}(Sf)|_{\partial\Omega} = \frac{\partial f}{\partial\tau}$.

Let us make now the change $u_1 = u - RF$ in (1.19). In this way, we get

$$\begin{aligned} &Lu_1 = 0 \quad \text{in } \Omega, \\ &\frac{\partial u_1}{\partial l} + hu_1 = f - \gamma\Big(\frac{\partial}{\partial l} + h\Big)RF \quad \text{on } \partial\Omega, \end{aligned} \tag{1.20}$$

where $\gamma : w \in C^\infty(\overline{\Omega}) \to w|_{\partial\Omega}$ is the trace operator.

Put $u_1 = Sv$, where v is an unknown function on $\partial\Omega$, and let

$$f_1 = f - \gamma\Big(\frac{\partial}{\partial l} + h\Big)RF.$$

Using this notation, one obtains immediately that $(\frac{\partial}{\partial\tau} + a\frac{\partial}{\partial\mathbf{n}} + h)Sv = f_1$ on $\partial\Omega$, and so $\frac{\partial v}{\partial\tau} - a\gamma\frac{\partial}{\partial x_{n+1}}Sv + hv = f_1$ on $\partial\Omega$. The operator just obtained,

$$A : v \mapsto \frac{\partial v}{\partial\tau} - a\gamma\frac{\partial}{\partial x_{n+1}}Sv + hv, \quad \text{for } v \in C^\infty(\partial\Omega), \tag{1.21}$$

is a classical ΨDO on $\partial\Omega$.

DEFINITION 1.18. The operator A defined by (1.21) is called a *Calderón operator* of the bvp (1.19).

Conclusion. *The solvability and regularity properties of the oblique derivative problem* (1.19) *are reduced to the investigation of the solvability and regularity of the Calderón operator A on the boundary $\partial\Omega$.*

For simplicity, we will show that A is a ΨDO in the special case when the coefficients of L do not depend on (x, x_{n+1}) and each operator L_j is homogeneous of order j. Having in mind that L is an elliptic operator with symbol

$$-(L_2(\xi) + 2L_1(\xi)\xi_{n+1} + L_0(\xi)\xi_{n+1}^2),$$

we have that $L_0(\xi) = \text{const} \neq 0$, and $L_0(\xi)L_2(\xi) - L_1^2(\xi) > 0$ for $\xi \neq 0$.

A partial Fourier transformation with respect to x in (1.20) leads to the following ordinary differential operator depending on the parameter ξ:

$$L_2(\xi)\hat{u} - 2iL_1(\xi)\partial_{n+1}\hat{u} + L_0\partial_{n+1}^2\hat{u} = 0, \quad \hat{u} = \hat{u}(\xi, x_{n+1}),$$

and therefore,

$$\hat{u}(\xi, x_{n+1}) = \hat{c}(\xi)e^{x_{n+1}\lambda_1(\xi)} + \hat{d}(\xi)e^{x_{n+1}\lambda_2(\xi)},$$

where $\lambda_{1,2} = \frac{-iL_1 \pm \sqrt{L_0L_2 - L_1^2}}{L_0}$, $\operatorname{Re}\lambda_1 < 0$, $\operatorname{Re}\lambda_2 > 0$.

Suppose that $L_0 > 0$. Thus $\lambda_1 = \frac{-iL_1 - \sqrt{L_0L_2 - L_1^2}}{L_0}$, $\operatorname{ord}_\xi \lambda_1(\xi) = 1$, and obviously we must take $\hat{d}(\xi) = 0$, which implies $u(x, 0) = c(x)$ and

$$u(x, x_{n+1}) = \int e^{ix\xi}\hat{c}(\xi)e^{x_{n+1}\lambda_1(\xi)}\, \bar{d}\xi.$$

Then,

$$-a\gamma\frac{\partial}{\partial x_{n+1}}u = -\int e^{ix\xi}a(x)\lambda_1(\xi)\hat{c}(\xi)\, \bar{d}\xi.$$

Since L_1 is a first order differential operator, we have

$$L_1(\partial_x) = \sum_{j=1}^{n} b_j(x)\partial_j = i\sum_{j=1}^{n} b_jD_j, \quad iD_j = \partial_j.$$

We conclude that the symbol of the Calderón operator A is given by

$$i\sum_{j=1}^{n}\Big(a_j(x) - \frac{a(x)b_j(x)}{L_0}\Big)\xi_j + a(x)\frac{\sqrt{L_0L_2 - L_1^2}}{L_0}.$$

Through more complicated calculations, one can show that the same formula for the principal symbol A_o of the Calderón operator also holds in the case of variable coefficients. Indeed,

$$\text{(1.22)} \quad A_o(x,\xi) = i\sum_{j=1}^{n}\Big(a_j(x) - \frac{a(x)b_j(x)}{L_0(x)}\Big)\xi_j + a(x)\frac{\sqrt{L_0(x)L_2(x,\xi) - L_1^2(x,\xi)}}{L_0(x)}$$

with $L_0(x) > 0$, $(x,\xi) \in T^*(\partial\Omega) \setminus 0$.

If $n \geq 2$, it is obvious that $A_o(x,\xi) \neq 0$ for $\xi \neq 0$ if and only if $a(x) = (l, \mathbf{n}) \neq 0$. We will say that (1.19) satisfies the Shapiro-Lopatinskij condition if $A_o(x,\xi) \neq 0$ for every $\xi \neq 0$. Therefore, the oblique derivative problem (1.22) satisfies the Shapiro-Lopatinskij condition if and only if the vector field l is not tangential to $\partial\Omega$ at any point $x \in \partial\Omega$. This condition is violated at each point $x \in \partial\Omega$ at which $l(x)$ becomes tangential to $\partial\Omega$. We assume that $\tau \neq 0$ at such points.

1.5. Hölder regularity. We will complete Section 1 by formulating several results about the solvability of the bvp (1.1) in Hölder classes and about the behavior of subelliptic ΨDO in Hölder spaces. We remind the reader that zero-order classical ΨDO continuously map each Hölder class into itself (cf. [**Mey**]).

We begin by recalling the definition and some properties of the Hölder spaces $\Lambda^\alpha(\mathbb{R}^n)$. They coincide with $C^{k+\beta}(\mathbb{R}^n)$ for $\alpha = k+\beta$, $k \in \mathbb{Z}_+$, $0 < \beta < 1$. However, the definition of Λ^k is quite different.

DEFINITION 1.19 ([**St**]). Let $0 < \alpha < 2$. Then a function f belongs to $\Lambda^\alpha(\mathbb{R}^n)$ if and only if $f \in L^\infty(\mathbb{R}^n)$, and

$$\omega(t) = \|f(x+t) + f(x-t) - 2f(x)\|_{L^\infty(\mathbb{R}^n_x)} \leq A\,|t|^\alpha$$

for some constant $A > 0$.

The corresponding norm is defined by

$$|f|_\alpha = \|f\|_{L^\infty(\mathbb{R}^n)} + \sup_{|t|>0} \frac{1}{|t|^\alpha} \|f(x+t) + f(x-t) - 2f(x)\|_{L^\infty(\mathbb{R}^n)}.$$

It is well known (see [**St**]) that for $\alpha \neq 1$, $\Lambda^\alpha(\mathbb{R}^n)$ coincides with $C^{k+\beta}(\mathbb{R}^n)$ for $k = 0, 1$; and $0 < \beta < 1$. Moreover, the norms in these spaces are equivalent. On the other hand, $\Lambda^1(\mathbb{R}^n) \supsetneqq C^1(\mathbb{R}^n)$, and there exists a function $f \in L^\infty(\mathbb{R}^n) \cap \Lambda^1(\mathbb{R}^n)$ such that $f \notin C^{0,1}$. As usual, $C^{k,1}$ stands for the set of C^k smooth functions having Lipschitz continuous k-th order derivatives. Moreover, for $\alpha > 1$, $f \in \Lambda^\alpha(\mathbb{R}^n)$ if and only if $\nabla f \in \Lambda^{\alpha-1}(\mathbb{R}^n)$. The space $\Lambda^1(\mathbb{R}^n)$ is called *Zygmund space.*

We will consider the bvp (1.1), only in the case of a neutral vector field l. According to the result of Section 1.2, we have the equation

$$\frac{\partial \hat{u}}{\partial x_1} + \psi(x_1)|\xi'|\hat{u} = \hat{g}(x_1, \xi'),$$

where $\psi(x_1) = ax_1^k + O(|x_1|^{k+1})$ as $x_1 \to 0$, $a > 0$, and k even. Setting $B(x_1, x_1') = \int_{x_1'}^{x_1} \psi(s)\, ds$, we get

$$\hat{u}(x_1, \xi', 0) = \int_{-1/2}^{x_1} e^{-B(x_1,x_1')|\xi'|}\hat{g}(x_1', \xi')\, dx_1', \quad \text{so}$$

$$v(x) = u(x,0) = \int_{\mathbb{R}^{n-1}} \int_{-1/2}^{x_1} e^{ix'\xi' - B(x_1,x_1')|\xi'|}\hat{g}(x_1', \xi')\, \bar{d}\xi' dx_1, \tag{1.23}$$

if $\operatorname{supp} g \subset \left\{|x_1| < \frac{1}{2}\right\}$. The term $e^{-|\xi'| \int_0^{x_1} \psi\, ds}$ does not participate in (1.23) because $v(x)$ is a tempered distribution.

The kernel K of the FIO (1.23) is given by the formula

$$K(x_1, x_1'; x', y') = \int_{\mathbb{R}^{n-1}} e^{i(x'-y')\xi' - B(x_1,x_1')|\xi'|}\, \bar{d}\xi', \text{ so}$$

$$v(x) = \int_{\mathbb{R}^n} K(x_1, x_1'; x', y') g(x_1', y')\, dx_1' dy'. \tag{1.23A}$$

A change of variables yields

$$K(x_1, x_1'; x', y') = \int_{|\xi'|=1} \int_0^\infty r^{n-2} e^{i(x'-y')\xi' r - B(x_1,x_1')r}\, dr \bar{d}\sigma_{\xi'},$$

and a standard regularization of K for $x_1' < x_1$ implies

$$K(x_1, x_1'; x', y') = C_n \int_{|\xi'|=1} \frac{1}{\big(i(x'-y')\xi' - B(x_1,x_1')\big)^{n-1}}\, d\sigma_{\xi'}, \tag{1.24}$$

where $C_n = \text{const} \neq 0$. There are no difficulties to verify that

$$|K| \leq C \frac{1}{\big(|x'-y'| + B(x_1,x_1')\big)^{n-1}}$$

for $n \geq 2$ and some constant $C > 0$. After long computations based on this estimate of K, we arrive at the following Lemma.

LEMMA 1.20. *The integral operator defined by* (1.23A) *acts continuously from the space* $\Lambda^\alpha_{\text{comp}}(\mathbb{R}^n)$ *into* $\Lambda^{\alpha+\frac{1}{k+1}}_{\text{comp}}(\mathbb{R}^n)$ *for* $\alpha > 0$.

The proof of this lemma can be found in [**Gu**] (see also [**P-Pl**]).

Going back to the problem (1.1) in the case I (l of neutral type), we can construct a local solution of (1.1) near $x_1 = 0$ having the following properties: If $g \in C^\beta$, $0 < \beta \notin \mathbb{N}$, then $u \in C^{\beta+\frac{1}{(k+1)}}(\{x_{n+1} \geq 0\}) \cap C^\infty(\{x_{n+1} > 0\})$, $\beta + \frac{1}{k+1} \notin \mathbb{N}$. The result is more precise in Zygmund spaces, namely, $g \in \Lambda^\beta$ for $\beta > 0$ implies that $u \in \Lambda^{\beta+\frac{1}{k+1}}(\{x_{n+1} \geq 0\}) \cap C^\infty(\{x_{n+1} > 0\})$.

It is interesting to note that subelliptic estimates in Hölder spaces do not hold in general for differential and pseudodifferential operators that are subelliptic in Sobolev spaces. In the framework of Guan [**Gu**], we will give below an example of a boundary value problem that is subelliptic in Sobolev classes but is not subelliptic in Hölder spaces. We consider the bvp in $\mathbb{R}^{n+1}_+$, $n \geq 2$,

$$\begin{aligned} &\Delta u = 0 \ \text{ in } \{x_{n+1} > 0\}, \\ &\frac{\partial u}{\partial x_1} + (i\alpha + \varphi(x_1))\frac{\partial u}{\partial x_{n+1}} = g \ \text{ on } \{x_{n+1} = 0\}. \end{aligned} \tag{1.25}$$

Here $\alpha \in (0,1)$ is a constant, $\varphi(x_1) = ax_1^k$, $a > 0$, $a \in \mathbb{R}$, $k \geq 2$ even. Certainly, the Shapiro-Lopatinskij condition is violated on the set $\{x_1 = x_{n+1} = 0\}$.

Proceeding in the same way as in the investigation of (1.1), we obtain a ΨDO for the trace $u(x,0) = c(x)$. The principal symbol of the corresponding operator is (up to a factor i):

$$q_1^o = \xi_1 - \alpha|\xi| + i\varphi(x_1)|\xi|,$$

and $\operatorname{Char} q_1^o = \{(x,\xi) : \xi_1 = \frac{\alpha|\xi'|}{\sqrt{1-\alpha^2}},\ x_1 = 0\}$. Since q_1^o is microlocally elliptic outside of the conical surface $\xi_1 = \alpha|\xi|$, we can apply to $q_1^o = 0$ the implicit function theorem in a conical neighborhood Γ of $\{\xi_1 = \alpha|\xi|\}$. Therefore, the study of the bvp (1.25) can be reduced to the investigation of a ΨDO with the symbol

$$p_1 = \xi_1 - |\xi'|\frac{\alpha - i\varphi(x_1)}{\sqrt{1-(\alpha - i\varphi(x_1))^2}}, \qquad \xi' = (\xi_2, \dots, \xi_n).$$

The zero bicharacteristics of $\operatorname{Re} q_1^o$ are given by the formula:

$$\mathcal{L}: \quad \begin{aligned} &x_1(t) = t(1-\alpha^2), \ \ x_j(t) = x_{jo} - \frac{\alpha\xi_j^o}{|\xi^o|}t \ \text{ for } 2 \le j \le n, \\ &\xi(t) = \xi^o, \ \ \xi_1^o = \alpha|\xi^o|, \ \ x_1^o = 0. \end{aligned}$$

Thus $\operatorname{Im} q_1^o|_{\mathcal{L}} = at^k(1-\alpha^2)^k$ and, according to Theorem 1.8, q_1 is a first-order subelliptic ΨDO with sharp loss of regularity $\frac{k}{k+1}$. We will show this result does not hold in Hölder classes. We point out that canonical transformations cannot be applied in Λ^s, $s > 0$.

We write

$$\frac{\alpha - i\varphi(x_1)}{\sqrt{1-(\alpha - i\varphi(x_1))^2}} = a(x_1) + ib(x_1)$$

with real-valued functions $a(x_1)$ and $b(x_1)$. These functions are C^∞ smooth in a neighborhood of the origin. Simple calculations yield:

$$a(x_1) = \frac{\alpha}{\sqrt{1-\alpha^2}} + O(\varphi^2) \ \text{ and } \ b(x_1) = -\frac{\varphi(x_1)}{(1-\alpha^2)^{3/2}} + O(\varphi^2)$$

near $x_1 = 0$. Thus $p_1^o = \xi_1 - |\xi'|d(x_1)$ with $d(x_1) = a(x_1) + ib(x_1)$, and we will study the ΨDO

$$\big(D_1 - |D'|d(x_1)\big)c(x) = g_1(x)$$

for some g_1. Making a partial Fourier transformation with respect to x' in the equation above we get

$$\partial_{x_1}\hat{c}(x_1,\xi') - i|\xi'|d(x_1)\hat{c} = i\hat{g}_1.$$

The change $x_1 \to -x_1$ gives us:

$$\partial_{x_1}\hat{c}(x_1,\xi') + i|\xi'|d(x_1)\hat{c} = -i\hat{g}_1.$$

For simplicity, we have written the same functions $\hat{c}$, d, and $\hat{g}_1$ instead of the corresponding $\hat{c}(-x_1,\xi')$, etc. Then,

$$\hat{c}(x_1,\xi') = -i\int_{-\infty}^{x_1} e^{-i|\xi'|D(x_1,x_1')}\hat{g}_1(x_1',\xi')\,dx_1',$$

where $D(x_1,x_1') = \int_{x_1'}^{x_1} d(\theta)\,d\theta$, $\operatorname{supp} g_1 \subset \big\{|x_1| < \frac{1}{2}\big\}$.

Evidently, $\mathrm{Re}(-iD(x_1, x_1')) = \int_{x_1'}^{x_1} b(\theta)d\theta \leq 0$ near 0, as $x_1' \leq x_1$ and k is even. It follows immediately that

$$c(x) = -i \int_{\mathbb{R}^{n-1}} \int_{-\infty}^{x_1} \int_{\mathbb{R}^{n-1}} e^{i(x'-y')\xi' - i|\xi'|D(x_1, x_1')} g(x_1', y')\, dx_1' dy' \bar{d}\xi'.$$

The FIO just obtained has the kernel

$$K(x_1, x_1'; x', y') = -i \int_{\mathbb{R}^{n-1}} e^{i(x'-y')\xi' - i|\xi'|D(x_1, x_1')}\, \bar{d}\xi'$$

which, as it was shown in (1.24), can be rewritten in the form

$$K(x_1, x_1'; x', y') = C_n \int_{|\xi'|=1} \frac{1}{\left((x'-y')\xi' - D(x_1, x_1')\right)^{n-1}}\, d\sigma_\xi'. \tag{1.26}$$

To clarify the calculations assume $n = 3$, i.e., $n - 1 = 2$. After a polar change in (1.26) one obtains:

$$K(x_1, x_1'; x', y') = C_3' \frac{D(x_1, x_1')}{\left(D^2(x_1, x_1') - |x'-y'|^2\right)^{3/2}},$$

for some constant $C_3' \neq 0$.

Instead of g_1 we consider the functions $g_{1\beta} = \omega(x) x_1^\beta h^{3/2} |h|^{-1-\frac{3}{k+2}}$, where $h(x) = D^2(0, x_1) - |x'|^2 \in C^\infty$ near the origin, $\omega \in C_0^\infty(\mathbb{R}^n)$ with $\omega \equiv 1$ in a small neighborhood of 0, $k > 4$, and $\beta \in [0, 1]$ such that $1/\beta$ is an odd integer. We can see that $g_{1\beta} \in \Lambda_{\mathrm{comp}}^{1/2 - \frac{3}{k+2}}$. Obviously,

$$c_\beta = \int_{-\infty}^{x_1} \int_{\mathbb{R}^2} K(x_1, x_1'; x', y') g_{1\beta}(x_1', y')\, dx_1' dy'.$$

PROPOSITION 1.21. *Suppose that $k > 4$ is an even integer. Then:*

(i) *The trace c_β on $\{x_{n+1} = 0\}$ of the solution u of the bvp (1.25) with right-hand side $g_{1\beta}$ satisfies in a neighborhood of the origin the pointwise estimates*

$$|c_1(x)| \leq \mathrm{const}\, |\ln|x||, \quad |c_\beta(x)| \leq \mathrm{const}\, |x|^{\beta-1} \textit{ for } \beta \in [0, 1).$$

(ii) *The trace c_1 of the solution u of the problem (1.25) with right-hand side g_{11} satisfies in a neighborhood of the origin the lower estimate*

$$|c_1(x_1, 0)| \geq \mathrm{const}\, |\ln|x_1||.$$

This proposition is proved in [**P-Sou**] (see also [**P-Pl**]).

REMARK 1.22. Obviously,

$$D^2(0, x_1) - |x'|^2 = b_0 x_1^2 - |x'|^2 + i b_1 x_1^{k+2} + O(x_1^{2k+2}) \text{ as } x_1 \to 0,$$

where $b_0 > 0$ and $b_1 \neq 0$. Consequently, $\mathrm{sing\,supp}\, g_{1\beta} = \{x_1 = 0\}$ for $0 < \beta < 1$, and $\mathrm{sing\,supp}\, g_{11} = \{0\}$. Having in mind the hypoellipticity of P, we have that $c_1 \in C^\infty(\mathbb{R}^3 \setminus 0)$, while $c_\beta \in C^\infty(\mathbb{R}^3 \setminus \{x_1 = 0\})$ for $0 < \beta < 1$.

Conclusion. *Consider the bvp (1.25) for $n = 3$, $k > 4$, k even, and with boundary data g_{11}. Then the corresponding solution u belongs to $C^\infty(\overline{\mathbb{R}^4_+} \setminus 0)$, and it blows up logarithmically near the origin. The operator P is not subelliptic in the Hölder classes since $Pc \in \Lambda^{\frac{1}{2}-\frac{3}{k+2}}$ implies $c \notin L^\infty$ and $\Lambda^s \subset L^\infty$ for all $s > 0$.*

Concluding remark 1.23. The natural way of proving the existence, uniqueness and Fredholmness of the tangential oblique derivative problem (0.1) is the method of a priori estimates in Sobolev spaces. To do this we localize the problem in inner subdomains and in one-side neighborhoods of the boundary points of Ω by means of a partition of unity. On each neighborhood of a boundary point on which the Shapiro-Lopatinskij condition holds, there are well-known a priori estimates. Therefore, these neighborhoods comprising parts of the set (manifold) E of tangency of the vector field l to the boundary, are the most interesting to us. In fact, the Shapiro-Lopatinskij condition is violated at E. As the order of tangency of l to $\partial\Omega$ is invariant under smooth change of variables, we can concentrate our attention in a local straightening of the boundary and consider the oblique derivative problem in the half space. The main difficulties lay there. After reducing our bvp to the boundary, the famous subelliptic estimates play an important role there. For this reason, in the first section of this paper we decided to present a more detailed analysis (including calculus) of the oblique derivative problem in the half space. Thus, in a natural way, we arrived at a different statement of the problem in question depending on the structure of the boundary vector field around the points of tangency. More precisely, we considered vector fields of neutral, emergent and submergent types. At least in the case of some instructive model examples, we constructed parametrices of the bvp (1.1) and (1.16). We also concluded that, depending on the type of boundary vector field, there were three fundamentally different possible cases: bvp of Fredholm type, bvp having an infinite dimensional kernel, and bvp having an infinite dimensional cokernel. Relatively simple considerations in the latter two cases enabled us to modify the corresponding bvp in order to obtain bvp of Fredholm type again: (1.16), (1.1M).

Since we have proved local existence theorems (though in model examples), it was only natural to estimate appropriate Sobolev norms of the solutions. In this way, we got to the subelliptic estimates of Hörmander-Egorov-Nirenberg-Treves. Their rather technical proofs are carried out through the method of localization (see [**E1**], [**H2**]). Having in mind the purpose of this paper, we preferred to illustrate the localization method mentioned above by means of model examples arising in the tangential oblique derivative problem in the half space (1.1). Through our discussions in Section 1 we hope to have introduced the reader into the theory of the tangential oblique derivative problem for the Laplace equation.

The study of the same problem, but in the case of an arbitrary bounded domain with smooth boundary, is very technical and requires a priori estimates for ΨDO in (weighted) Sobolev spaces and, of course, it needs an appropriate partition of unity. This procedure has been developed in detail in many papers (see also the monographs [**Pa**] and [**E1**]). Once again, depending on the structure of the vector field l at the boundary (neutral, emergent, submergent), we will have a bvp (0.1) of Fredholm type, with an infinite dimensional kernel, or with an infinite dimensional cokernel. The detailed proofs in that situation go beyond the frames of our survey.

We hope that the reader getting acquainted with the methods of study in Section 1, will take for granted the general results presented in Section 2, and will look for their complete proofs in the papers [**Mz-Pa**], [**Mz-Pa1**], [**Mz**], [**Gu**], and in the loc. cit. monographs.

2. Survey on the degenerate oblique derivative problem in Hölder, Sobolev, and Besov spaces

During his investigations on the tides in the oceans the famous French Mathematician Henri Poincaré [**Poi**] considered the following boundary problem

$$Pu = f \text{ in a domain } \Omega \subset \mathbb{R}^n, \quad lu = g \text{ on } \partial\Omega, \tag{2.1}$$

where P is an elliptic second order differential operator, and l is a real-valued non-degenerate vector field on the smooth boundary $\partial\Omega$. The problem (2.1) differs from the classical Neumann problem ($l = \frac{\partial}{\partial \mathbf{n}}$ with $\mathbf{n}$ being the unit outward normal vector field to $\partial\Omega$) as l may be tangential to $\partial\Omega$ at some points.

The investigations from Section 1 show that the oblique derivative problem (2.1) may have an infinite dimensional kernel or cokernel, if l contacts the boundary $\partial\Omega$. At first Bicadze [**Bi**] obtained several interesting results in the case of emergent and submergent vector fields on a ball. A classification of the vector fields l in the tangential case, and some a priori estimates in a special case, were proposed by Borrelly [**Bo**] who had in mind the theory of Brownian motion. We introduced and used this terminology at the beginning of the previous section under the assumption that $\partial\Omega = \{x_{n+1} = 0\}$.

As we mentioned before, the Shapiro-Lopatinskij condition is satisfied in the case when l is not tangential to the boundary $\partial\Omega$, and it is violated at the points where l is tangential to $\partial\Omega$ for $n \geq 3$.

To formulate the first result of this section, we define the elliptic operator

$$P = \sum_{i,j=1}^{n} a_{ij}(x)\partial_{ij}^2 + \sum_{i=1}^{n} b_i(x)\partial_i + c(x)$$

with coefficients in $C^\alpha(\overline{\Omega})$ for $0 < \alpha < 1$, and

$$\sum_{i,j=1}^{n} a_{ij}(x)\xi_i\xi_j \geq c_0|\xi|^2 \text{ for all } \xi \in \mathbb{R}^n,$$

where $c_0 > 0$ is a constant, and $a_{ij}(x) = a_{ji}(x)$. We assume the domain Ω to be bounded with $C^{2+\alpha}$ boundary $\partial\Omega$. We also assume the boundary operator, interpreted as a smooth real-valued vector field $l = \sum_{i=1}^{n} l_i(x)\partial_i$, to have coefficients in $C^{1+\alpha}(\partial\Omega)$, and be such that $l(x)$ is never tangential to $\partial\Omega$, i.e., $\langle l(x), \mathbf{n}(x)\rangle \neq 0$

for every $x \in \partial\Omega$. Then the bvp (2.1) has a finite dimensional kernel, and the following a priori estimate holds for the solution $u \in C^{2+\alpha}(\overline{\Omega})$, see [**G-T**],

$$\|u\|_{C^{2+\alpha}(\overline{\Omega})} \leq C\Big(\|u\|_{C^0(\overline{\Omega})} + \|f\|_{C^\alpha(\overline{\Omega})} + \|g\|_{C^{1+\alpha}(\partial\Omega)}\Big),$$

if $f \in C^\alpha$ and $g \in C^{1+\alpha}$. Moreover, the problem (2.1) is solvable if finitely many orthogonality (complementary) conditions are satisfied by f and g.

Suppose now that the coefficients of P and l, and the boundary $\partial\Omega$ of the bounded domain Ω are all C^∞ smooth, and assume $\langle l(x), \mathbf{n}(x)\rangle \neq 0$ for every $x \in \partial\Omega$. Then the bvp (2.1) is of Fredholm type in the classical Sobolev spaces $H^s(\Omega)$ for $s > 3/2$, and the following a priori estimate is valid, see [**E1**], [**Tre**],

$$\|u\|_{H^s(\Omega)} \leq C\Big(\|u\|_{H^0(\Omega)} + \|f\|_{H^{s-2}(\Omega)} + \|g\|_{H^{s-3/2}(\partial\Omega)}\Big), \tag{2.2}$$

if $f \in H^{s-2}$ and $g \in H^{s-3/2}$. Therefore, the bvp (2.1) admits a finite dimensional kernel contained in $C^\infty(\overline{\Omega})$, a finite dimensional cokernel contained in $C^\infty(\overline{\Omega})$, and (2.1) is solvable if finitely many orthogonality conditions are satisfied by f and g.

REMARK 2.1. We would like to recall the definition of a Fredholm operator. Assume that B_1 and B_2 are two Banach spaces and let $A : B_1 \to B_2$ be a linear continuous operator. The operator A is said to be *Fredholm* if $\dim\ker A < \infty$ and $\dim\operatorname{coker} A < \infty$, where $\operatorname{coker} A = B_2/\operatorname{im} A$. As usual, $\operatorname{im} A$ denotes the range of B_1 under the mapping A. It is well known that for each Fredholm operator A the range $\operatorname{im} A$ is a closed linear subspace of B_2, and $\dim\operatorname{coker} A = \dim\ker A^*$, where $A^* : B_2^* \to B_1^*$ is the adjoint operator of A (see [**R-Sch**]). According to the Fredholm theory, the equation $y = Ax$ is solvable if and only if $y \in (\ker A^*)^\perp$. In other words, a solution x exists if and only if y fulfills finitely many orthogonality conditions. By definition, the index of a Fredholm operator is given by

$$\operatorname{ind} A = \dim\ker A - \dim\operatorname{coker} A.$$

2.1. Tangential vector fields and solvability. We will now introduce the classification of the vector fields l near the set of tangency to the boundary $\partial\Omega$. To this end, we consider

$$l(x) = \tau(x) + a(x)\mathbf{n}(x) \quad \text{for } x \in \partial\Omega,$$

where $\tau(x)$ is a smooth real-valued vector field tangential to $\partial\Omega$, $\mathbf{n}(x)$ is the unit outward normal to $\partial\Omega$, and $a(x) \in C^\infty(\partial\Omega)$ is a real-valued function. Having in mind that $|l(x)| > 0$, we have, by assumption, that $a(x) = 0$ implies $|\tau(x)| > 0$. The Shapiro-Lopatinskij condition holds in the case $n \geq 3$, if and only if, for every $x \in \partial\Omega$, we have $0 \neq \langle l(x), \mathbf{n}(x)\rangle = a(x)$.

Denote by $\gamma_p(t)$ the maximal integral curve of the vector field τ passing through the point $p \in \partial\Omega$ such that $\gamma_p(0) = p$ and $\partial_t\gamma_p(t) = \tau \circ \gamma_p(t)$. Let

$$E = \{x \in \partial\Omega : \langle l(x), \mathbf{n}(x)\rangle = 0\}.$$

Usually, we will suppose that E is a closed smooth submanifold of $\partial\Omega$ of codimension 1 and that $l(x)|_E$ is transversal to E.

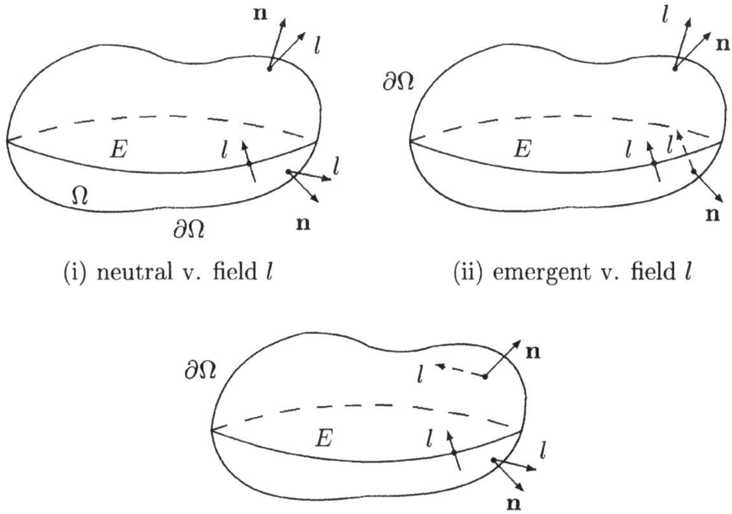

(i) neutral v. field l

(ii) emergent v. field l

(iii) submergent v. field l

FIGURE 3. Structure of the vector field l near E.

There are three possible behaviors of the function $a(x)$ near E:

(i) $a(x)$ conserves its sign on $\partial\Omega$,
(ii) $a(x)$ changes its sign through E from negative to positive in the direction of the vector field $l|_E$,
(iii) $a(x)$ changes its sign through E from positive to negative in the direction of the vector field $l|_E$.

The vector field l in the case (i) is said to be of *neutral* type, in the case (ii) of *emergent* type, and in the case (iii) of *submergent* type, see Figure 3. Certainly, (ii) can be reformulated as: $a(x)$ changes its sign from minus to plus in the positive direction on the integral curve γ. Similarly for (iii).

Egorov and Kondratiev assumed in [**E-K**] that E is a closed smooth submanifold of the boundary of codimension 1 and that $l(x)$ is transversal to E. They proved that in the emergent case (ii), (2.1) becomes a Fredholm problem if an extra condition is prescribed:

$$u(x) = u_0(x) \quad \text{on } E. \tag{2.3}$$

Under the assumptions $u \in H^s(\Omega)$, $s > 0$, $f \in H^s(\Omega)$, $g \in H^{s+1/2}(\partial\Omega)$, and $u_0 \in H^{s+1/2}(E)$, they proved that each solution of the bvp (2.1), (2.3), satisfies the estimate

$$\begin{aligned}\|u\|_{H^{s+1}(\Omega)} \leq C_s\Big(\|u\|_{H^s(\Omega)} + \|f\|_{H^s(\Omega)} \\ + \|g\|_{H^{s+1/2}(\partial\Omega)} + \|u_0\|_{H^{s+1/2}(E)}\Big).\end{aligned} \tag{2.4}$$

The existence theorem is valid under finitely many orthogonality conditions imposed on the triple (f, g, u_0). Existence results cannot be proved in general in the case (iii), but the regularity still holds true. Existence results in the overdetermined case (iii) under additional conditions are not obtained in loc. cit.

In the case (i) of a neutral vector field it was shown in [**E-K**] that the problem (2.1) has finite dimensional kernel and cokernel, and an existence result was proved under finitely many orthogonality conditions satisfied by f and g.

Comparing the a priori estimates (2.2) and (2.4) we see that in the case of the tangential oblique derivative problem, a loss of regularity 1 appears with respect to f and g. By means of the subelliptic theory, we will show that it is possible to improve the regularity that the solution u gains from f and g.

To simplify the problem consider the bvp

$$Pu = 0 \ \text{ in } \Omega, \quad lu = g \ \text{ on } \partial\Omega \tag{2.5}$$

in the cases (i) and (iii). We assume the coefficients and the boundary to be C^∞. The Calderón operator from Section 1.4 enables us to reduce (2.5) to a ΨDO on the boundary and apply Theorem 1.8. The smooth vector field $l = \tau(x) + a(x)\mathbf{n}(x)$ is called a vector field of finite type if there exists an integer k such that

$$\sum_{j=1}^{k} |\tau^j a(x)| > 0 \ \text{ at each point } x \in \partial\Omega \text{ where } a(x) = 0. \tag{2.6}$$

In other words, the function $a(x)$ vanishes at the point $x \in E$ to order less than or equal to k along the integral curve γ of the vector field τ passing through x. The set E is not necessarily a smooth submanifold of $\partial\Omega$. Then, we have the a priori estimate

$$\|u\|_{H^s(\Omega)} \leq C_s\Big(\|u\|_{H^{s-1}(\Omega)} + \|g\|_{H^{s-1/2-\frac{1}{k+1}}(\partial\Omega)}\Big). \tag{2.7}$$

This result is optimal concerning the regularity of u. As we can see, the sharp loss of regularity of the solution u is equal to $\frac{k}{k+1}$, where k is the maximal vanishing order of the function $a(x)$ along the integral curves γ of the vector field τ. Moreover, the subelliptic estimate (2.7), $s > \frac{3}{2}$, holds only in the case of vector fields of neutral and submergent type. This estimate is not valid for vector fields of emergent type.

Interesting results on the subject were proved by Winzell [**W**] (cf. also [**Al**]), but we will briefly discuss subelliptic estimates in Hölder spaces following Guan [**Gu**]. He proved that if P is a subelliptic differential operator of order m in two variables, then it remains subelliptic in Hölder spaces. Thus, if $Pu \in \Lambda^\alpha$ with $\alpha > 0$, and $u \in \mathcal{D}'$, then $u \in \Lambda^{\alpha+m-\frac{k}{k+1}}$. As we mentioned in Section 1, there are subelliptic ΨDO in Sobolev spaces that are not subelliptic in Hölder classes. The main subelliptic Theorem 1.8 remains valid for the following classical ΨDO:

$$L = \frac{\partial}{\partial x_{n+1}} + a(x, x_{n+1})Q \quad \text{in } \mathbb{R}^{n+1},$$

where Q is an elliptic ΨDO of order 1 with a positive principal symbol. We will formulate the corresponding result only in the case (i) (i.e., when the function a conserves its sign), assuming that the condition (2.6) holds. Then, for each right-hand side $f \in \Lambda^{\alpha}_{\text{comp}}$, $\alpha > 0$, one can find $u \in \Lambda_{\text{loc}}^{\alpha+\frac{1}{k+1}}$ such that $Lu = f$. Concerning the regularity of the solution we have that, if $Lu \in \Lambda^{\alpha}_{\text{loc}}$ and $u \in \mathcal{D}'$, then $u \in \Lambda_{\text{loc}}^{\alpha+\frac{1}{k+1}}$.

REMARK 2.2. The assumption

$$\sum_{j=1}^{k} \left| \frac{\partial^j}{\partial x_{n+1}^j} a(x, x_{n+1}) \right| > 0$$

at the points where $a(x, x_{n+1}) = 0$, guarantees (according to Theorem1.8) the subellipticity of L in Sobolev spaces with minimal loss of regularity $\delta = \frac{k}{k+1}$.

Below we formulate the general result of Guan. Let l be a vector field of finite type $k \in \mathbb{N}$, let E be a closed C^∞ smooth submanifold of $\partial\Omega$ of codimension 1, and let τ be transversal to E. In the case (i) of a neutral vector field l, and for a strictly elliptic second order homogeneous linear differential operator L with coefficients in $C^\infty(\overline{\Omega})$, the following bvp is considered:

$$\begin{aligned} Lu &= \sum_{i,j=1}^{n} a_{ij}(x)\partial^2_{ij}u = f(x) \quad \text{in } \Omega \subset \mathbb{R}^n,\ n \geq 3, \\ lu &= g(x) \quad \text{on } \partial\Omega; \end{aligned} \tag{2.8}$$

while in the case (ii) of an emergent vector field l, one studies the bvp

$$\begin{aligned} Lu &= f(x) \quad \text{in } \Omega, \\ lu &= g(x) \quad \text{on } \partial\Omega, \qquad u = 0 \quad \text{on } E. \end{aligned} \tag{2.9}$$

THEOREM 2.3 ([**Gu**]). *Suppose that $f \in C^\alpha(\overline{\Omega})$, $g \in C^\beta(\partial\Omega)$ for $\beta + \frac{1}{k+1} \notin \mathbb{N}$, and assume that the function $u \in C^2(\Omega) \cap C^1(\overline{\Omega})$ satisfies the bvp (2.8) or the bvp (2.9). Then we claim that $u \in C^\theta(\overline{\Omega})$ with $\theta = \min(2 + \alpha, \beta + \frac{1}{k+1})$, and one can find a constant C such that the following subelliptic estimate holds:*

$$\|u\|_{C^\theta(\overline{\Omega})} \leq C\Big(\|u\|_{C^0(\overline{\Omega})} + \|f\|_{C^\alpha(\overline{\Omega})} + \|g\|_{C^\beta(\partial\Omega)}\Big). \tag{2.10}$$

Moreover, a solution of the problem (2.8) (respectively (2.9)) exists, if finitely many orthogonality conditions are satisfied by f and g. This solution belongs to the Hölder class $C^\theta(\overline{\Omega})$.

In [**W**] Winzell proved that, if u solves (2.8) with $Lu \in C^\alpha(\overline{\Omega})$ and $lu \in C^{1+\alpha}(\partial\Omega)$, then $u \in C^{1+\alpha+\frac{1}{k+1}}(\overline{\Omega}) \cap C^{2+\alpha}(\overline{\Omega} \setminus E)$ for $0 < \alpha + \frac{1}{k+1} < 1$. Another interesting result is due to Alimov [**Al**] who studied the bvp (2.9) in the special case when $l = \frac{\partial}{\partial x_n}$, $f = 0$, $\partial\Omega = \{|x'|^2 + |x_n|^m = 1\}$ with $m > 1$, $n \geq 3$, and $E = \partial\Omega \cap \{x_n = 0\}$. He proved that if $\alpha > 0$ and $\alpha + \frac{1}{m} \in \mathbb{N}$, then there exists $g \in C^\alpha(\partial\Omega)$ such that $u \notin C^{\alpha+\frac{1}{m}}(\overline{\Omega})$. We point out that in this case $u \in \Lambda^{\alpha+\frac{1}{m}}(\overline{\Omega})$ as it was shown by Guan [**Gu**].

All the results mentioned above are proved under very strong regularity assumptions (C^∞, $C^{2+\alpha}$) on the coefficients and on the boundary. We now discuss a result on the existence of a solution of the tangential oblique derivative problem in the case (i) under rather weak regularity conditions imposed on the coefficients of the operator, on the boundary vector field l, and on the boundary $\partial\Omega$.

Let Ω be a bounded domain in $\mathbb{R}^n$, $n \geq 3$ with boundary $\partial\Omega$ whose regularity will be specified later on. $W^{k,p}(\Omega)$ denotes the Sobolev space of k times weakly differentiable functions such that $\partial_x^\alpha u \in L_p(\Omega)$ for $|\alpha| \leq k$. As usual, $1 \leq p \leq \infty$. The corresponding norm is defined by

$$(2.11)\qquad \begin{aligned} \|u\|_{W^{k,p}(\Omega)} &= \|u\|_{L^p(\Omega)} + \sum_{|\alpha|=k} \|\partial^\alpha u\|_{L^p(\Omega)} \ \text{ if } 1 \leq p < \infty, \\ \|u\|_{W^{k,\infty}(\Omega)} &= \max_{0\leq l\leq k} \Big(\max_{|\alpha|=l} \text{ess sup}_\Omega |\partial^\alpha u(x)| \Big). \end{aligned}$$

We will deal with $\partial\Omega \in C^{2,1}$, i.e., the second derivatives of the local equations of the boundary are Lipschitz continuous. The regularity properties of the restriction to the boundary $u|_{\partial\Omega}$ of an element $u \in W^{k,p}(\Omega)$ can be described by means of the Besov spaces $W^{s,p}(\partial\Omega)$ with $0 < s \notin \mathbb{N}$. Let $s = k + \theta$ with $k \in \mathbb{N} \cup \{0\}$ and $0 < \theta < 1$. Then the norm in $W^{s,p}(\partial\Omega)$ is defined by

$$(2.12)\qquad \|v\|_{W^{s,p}(\partial\Omega)} = \|v\|_{L^p(\partial\Omega)} + \Big(\sum_{|\alpha|=k} \iint_{\partial\Omega} \frac{|\partial^\alpha v(x) - \partial^\alpha v(y)|^p}{|x-y|^{n-1+\theta p}} \, d\sigma_x d\sigma_y \Big)^{\frac{1}{p}}.$$

More details about these spaces can be found in [**Ad**].

The main result to be used later is the following. Suppose that Ω is a bounded domain in $\mathbb{R}^n$ and $\partial\Omega \in C^{2,1}$. Then, for $1 < p < \infty$, the restriction map (trace)

$$(2.13)\qquad \rho : u \mapsto v = u|_{\partial\Omega}, \quad \rho : W^{2,p}(\Omega) \to W^{2-\frac{1}{p},p}(\partial\Omega),$$

is well defined and surjective.

REMARK 2.4. Let $\partial\Omega \in C^\infty$. Then the trace $\rho : W^{k,p}(\Omega) \to W^{k-\frac{1}{p},p}(\partial\Omega)$ is well defined for all $k > \frac{1}{p}$.

Consider the oblique derivative problem in the neutral case (i):

$$(2.14)\qquad \begin{aligned} Lu &= \sum_{i,j=1}^{n} a_{ij}(x)\partial_{ij}^2 u = f(x) \quad \text{in } \Omega, \\ Bu &= \frac{\partial u}{\partial l} + \sigma(x)u = \varphi(x) \quad \text{on } \partial\Omega. \end{aligned}$$

The following assumptions are imposed on L, B, and $\partial\Omega$:

(2.15a) L is uniformly elliptic, i. e., one can find a constant $\lambda > 0$ such that

$$\lambda|\xi|^2 \leq \sum_{i,j=1}^{n} a_{ij}(x)\xi_i\xi_j \leq \frac{1}{\lambda}|\xi|^2 \quad \text{for every } x \in \overline{\Omega},\ \xi \in \mathbb{R}^n;$$

(2.15b) $a_{ij}(x) = a_{ji}(x)$ for every x, and $a_{ij}(x) \in C^{0,1}(\overline{\Omega}) \equiv W^{1,\infty}(\overline{\Omega})$;

(2.15c) If $l = \sum_{i=1}^{n} l_i(x)\partial_i$, then $l_i,\ \sigma \in C^{1,1}(\partial\Omega)$, $\sigma(x) < 0$,
Ω is a bounded domain with $\partial\Omega \in C^{2,1}$,
$l(x) = \tau(x) - a(x)\mathbf{n}(x)$ and $\langle l(x), \mathbf{n}(x)\rangle \leq 0$ for every $x \in \partial\Omega$.

We also assume that $E = \{x \in \partial\Omega : a(x) = 0\}$ is nonempty. The set E is not required to be a submanifold of $\partial\Omega$. Moreover, the vector field l is tangential to $\partial\Omega$ only on E, and it is oriented inwards Ω on $\partial\Omega \setminus E$.

We are looking for a solution of the bvp (2.14) belonging to the space $W^{2,p}(\Omega)$ for some $1 < p < \infty$. Therefore, the elliptic equation $Lu = f$ is fulfilled almost everywhere (a.e.), and $Bu = \varphi$ holds true on $\partial\Omega$ in the trace sense. In fact, if $u \in W^{2,p}(\Omega)$, then $f \in W^{0,p}(\Omega)$ and $Bu \in W^{1-\frac{1}{p},p}(\partial\Omega)$.

THEOREM 2.5 ([**MaPlSo**]). *Suppose that the bvp* (2.14) *satisfies the conditions* (2.15a), (2.15b), (2.15c), *and the integral curves* γ *of the vector field* l *contained in* E *are nonclosed and of finite length. Assume that* $f \in W^{1,p}(\Omega)$ *and* $\varphi \in W^{2-\frac{1}{p},p}(\partial\Omega)$ *for some* $\frac{n}{2} < p < \infty$. *Then there exists a solution* $u \in W^{2,p}(\Omega)$ *of* (2.14), *and there is a constant* $C > 0$ *such that*

$$(2.16)\qquad \|u\|_{W^{2,p}(\Omega)} \leq C\Big(1 + \|f\|_{W^{1,p}(\Omega)} + \|\varphi\|_{W^{2-\frac{1}{p},p}(\partial\Omega)}\Big).$$

Certainly, C *does not depend on the functions* u, f, φ. *Moreover, if* $p > n$, *then the solution* $u \in W^{2,p}(\Omega)$ *is uniquely determined by* f *and* φ.

REMARK 2.6. In the case (i) of a neutral vector field l, there is a loss of one unit of regularity compared with the nontangential case. In fact, as it is shown in [**MaPlSo**], the strict inequality $\langle l(x), \mathbf{n}(x)\rangle < 0$ on $\partial\Omega$ implies that for each $f \in L^p(\Omega)$, $\varphi \in W^{1-\frac{1}{p},p}(\partial\Omega)$, there exists a unique solution $u \in W^{2,p}(\Omega)$ of (2.14).

According to the Sobolev embedding theorem, the solution u of the bvp (2.14) belongs to the Hölder class $C^{0+2-\frac{n}{p}}(\overline{\Omega})$ if $\frac{n}{2} < p < n$, and it belongs to $C^{1+1-\frac{n}{p}}(\overline{\Omega})$ for $p > n$. Assume that E is a closed smooth submanifold of $\partial\Omega$ of codimension

1, and the vector field $\tau(x)$ is transversal to E. Then the integral curves γ of τ certainly satisfy the requirements of Theorem 2.5.

2.2. Modified oblique derivative problem. We will continue our short survey by paying special attention to the paper of Maz'ya and Paneah [**Mz-Pa**]. They described for the first time the asymptotic behavior of solutions of the tangential oblique derivative problem in a neighborhood of the submanifold μ_2 on which the vector field is of submergent type. They proved that for each pair (f, g) from a subspace of $L_1(\Omega) \times L_2(\partial\Omega)$ of finite codimension N, one can find a function u (with the appropriate regularity) satisfying the bvp (2.1) in $\overline{\Omega} \setminus \mu_2$. The function u is unique modulo an N-dimensional space, and the restriction $u|_{\partial\Omega}$ has a finite smooth jump $\{u\}_{\mu_2}$ on μ_2 which is determined by f and g. This fact leads to a corresponding modification of (2.1). In [**Mz-Pa**], it was shown for the first time that the index of the modified oblique derivative problem is 0. Moreover, under an additional condition, the classical maximum principle was proved for the modified problem.

We should also mention Maljutov [**Ma**], who considered the modified oblique derivative problem and established its unique solvability by using partially probabilistic methods. The appearance of jumps and underdetermined bvp can be found in his paper, too. We will formulate and comment some results of Maz'ya and Paneah [**Mz-Pa**], [**Mz**], [**Mz-Pa1**], [**Pa1**], which form the basis of the very interesting monograph [**Pa**]. For simplicity, we will discuss the homogeneous oblique derivative problem, cf. (2.5).

Let Ω be a bounded domain with C^∞ boundary $M = \partial\Omega$, and assume that Ω lies on one side of M. Consider the vector field $l = \tau + a\nu$, where ν stands for the unit inward normal to M, i.e., $\nu = -\mathbf{n}$. The vector field l may be of emergent and of submergent type simultaneously. The homogeneous oblique derivative problem in Ω can be written as:

$$\begin{aligned} &Pu = 0 \ \text{ in } \Omega, \quad \frac{\partial u}{\partial l} + hu = g \ \text{ on } M, \\ &a = (l, \nu), \ \tau \neq 0 \ \text{ on } E = \{x \in M : a(x) = 0\} \neq \varnothing. \end{aligned} \tag{2.17}$$

Along with the set E we introduce the open subsets $M^\pm \subset M$, and the submanifolds $\mu_{1,2}$ of codimension 1 in M, such that:

(1) $M = M^+ \cup M^- \cup \mu_1 \cup \mu_2$.
(2) $M^+ \cap M^- = \varnothing$.
(3) $\partial M^+ \cap \partial M^- = \mu_1 \cup \mu_2 \equiv \mu_{12}$,
(4) $a \geq 0$ on M^+ and $a \leq 0$ on M^-.
(5) The vector field τ is transversal to $\mu_1 \cup \mu_2$, and it is oriented towards M^- on μ_1 and towards M^+ on μ_2.
(6) The set M does not contain any semitrajectory τ_p, $p \in M$, of the vector field τ.

Denote by $T(M)$ the tangent space of M, and let $\vec{\mu} \in C^\infty(\mu_{12}, T(M))$ be the vector field of unit normals to the submanifold μ_{12}, oriented towards M^+. Put

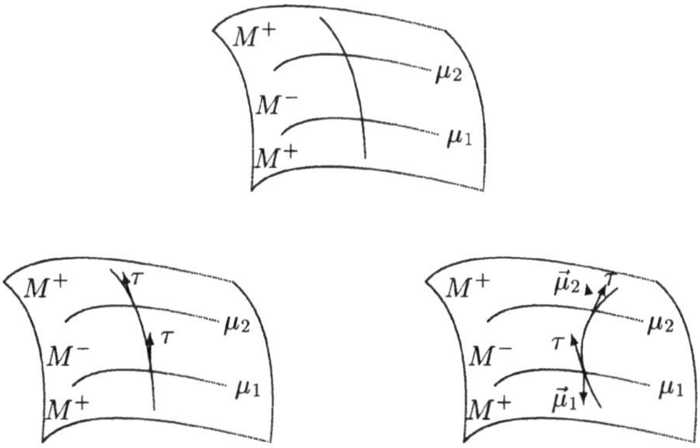

FIGURE 4

$\overrightarrow{\mu}_i = \overrightarrow{\mu}|_{\mu_i}$ for $i = 1, 2$. Evidently, $\mu_1 = \{p \in \partial M^+ \cap \partial M^- : (\tau, \overrightarrow{\mu})(p) < 0\}$. Geometrical illustration is given on Figure 4. The vector field l is of emergent type near μ_1 and of submergent type near μ_2.

REMARK 2.7. Suppose that $u \in H^1(M \setminus \mu_2)$ and let $u_\pm = u|_{M^\pm}$. According to the Sobolev trace result $u_\pm|_{\mu_2} \in H^{1/2}(\mu_2)$. The jump on μ_2 of the function u is defined as follows:

$$\{u\}_{\mu_2} = (u_+)|_{\mu_2} - (u_-)|_{\mu_2} \in H^{1/2}(\mu_2).$$

The well known jump formula from the Schwartz distribution theory (see [**Pa**]) shows that

$$\frac{\partial u}{\partial \tau} = \frac{\partial}{\partial \tau} u|_{M \setminus \mu_2} + (\tau, \overrightarrow{\mu}_2)\{u\}_{\mu_2} \otimes \delta_{\mu_2},$$

where the distribution $\langle \delta_{\mu_2}, \varphi \rangle = \int_{\mu_2} \varphi \, dS$ for every $\varphi \in C_0^\infty(M)$, and $(\tau, \overrightarrow{\mu}_2)$ is the scalar product of the vector fields τ and $\overrightarrow{\mu}_2$.

Having in mind our considerations from Section 1, we will study the following modified Poincaré problem:

$$(2.18) \qquad Pu = 0 \ \text{ in } \Omega, \quad \frac{\partial u}{\partial l} + hu - v \otimes \delta_{\mu_2} = g \ \text{ on } M, \quad u = u_0 \ \text{ on } \mu_1.$$

The functions g and u_0 are given on M and μ_1, respectively, and the functions u and v are sought. Certainly, v is only defined on μ_2. Obviously, the pair (u, v) solves the problem

$$(2.19) \qquad Pu = 0 \ \text{ in } \Omega, \quad \frac{\partial u}{\partial l} + hu = g \ \text{ on } M \setminus \mu_2, \quad u = u_0 \ \text{ on } \mu_1.$$

In other words, the solution u of the modified oblique derivative problem (2.18) satisfies the boundary conditions everywhere on M except, perhaps, on the submanifold μ_2. Thus, this modification is quite reasonable as the bvp (2.17) is locally nonsolvable in each neighborhood of an arbitrary point $p \in \mu_2$. It is shown in [**Mz-Pa**] (see also [**Pa**]) that (2.18) is a Fredholm type problem with index

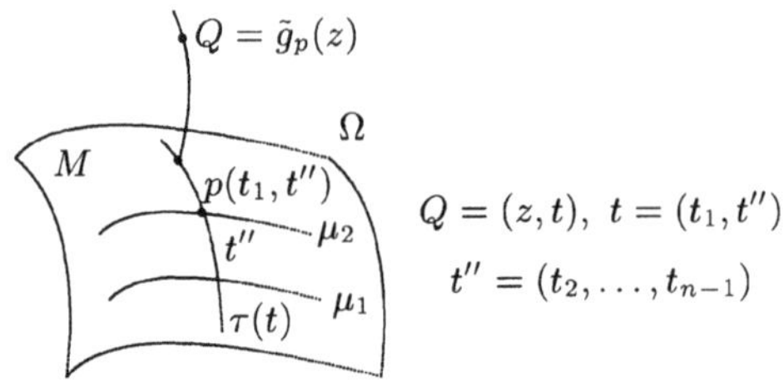

FIGURE 5

zero, and under small restrictions, this problem is uniquely solvable. It is proved, moreover, that if (u, v) is the solution of (2.18), then u is smooth on $\overline{\Omega} \backslash \mu_2$ and the restriction $u|_M = u_M$ has a finite jump on μ_2. Therefore, u satisfies the boundary condition $\frac{\partial u}{\partial l} + hu - (\tau, \overrightarrow{\mu}_2)\{u\}_{\mu_2} \otimes \delta_{\mu_2} = g$. According to the unicity theorem, we conclude that $v = (\tau, \overrightarrow{\mu}_2)\{u\}_{\mu_2} \otimes \delta_{\mu_2}$. In other words, the right-hand side (g, u_0) determines in a unique way the solution u in $\overline{\Omega} \setminus \mu_2$ and the jump of u_M on μ_2.

In order to formulate other interesting results on this subject, we now introduce the corresponding weighted Sobolev spaces on M and Ω.

Let the symbol $\mathcal{F}$ denote a sufficiently small neighborhood of M in Ω, and let $\overrightarrow{g}$ be the vector field of geodesics normal to M, oriented towards Ω on M. The integral curve of $\overrightarrow{g}$ passing through the point $p \in M$ and contained in $\mathcal{F}$ is denoted by $\tilde{g}_p(z)$, $|z| < \varepsilon$. Let $z' = (z_2, \ldots, z_n)$ be the local coordinates of p. To any point $Q = \tilde{g}_p(z)$ we associate the coordinates (z, z'). Let $\theta = \{\tau_p(t_1), p \in \mu_{12}, |t_1| < \varepsilon\} \subset M$ be a neighborhood of μ_{12}. Then, $0 < \varepsilon \ll 1$ implies that θ splits into the union of two disjoint layers θ_1, θ_2 formed by the shifts of μ_1, μ_2 along the vector field τ, $|t_1| < \varepsilon$. Denote by t'' the local coordinates of $p \in \mu_{12}$. Then we define in the layers $\theta_{1,2}$ the coordinates $t = (t_1, t'')$. So we can define neighborhoods $\mathcal{F}_{\theta_i} \subset \Omega$, $i = 1, 2$, equipped with local coordinates (z, t), and such that $M = \{p \in \overline{\Omega} : z = 0\}$ and $\mu_j = \{p \in \mathcal{F}_{\theta_j} : z = t_1 = 0\}$, see Figure 5.

Consider now a smooth partition of unity $\{\psi_j\}_0^2$ on M with supp $\psi_j \subset \theta_j$ and $\psi_j \equiv 1$ near μ_j for $j = 1, 2$.

DEFINITION 2.8. For $k \in \mathbb{N} \cup \{0\}$, define $\mathcal{H}_k(M, \mu_2)$ as the space of functions $u \in L_2(M)$ for which the norm

$$\langle u \rangle_k^2 = \sum_{l=0}^{k} \sum_{|\alpha|=0}^{l} \|(t_1\partial_1)^{l-|\alpha|} \partial_{t''}^{\alpha} \psi_2 u\|_o^2 + \sum_{j=0}^{1} \|\psi_j u\|_k^2$$

is finite. We put $\mathcal{H}_\infty(M, \mu_2) = \bigcap_{k=0}^{\infty} \mathcal{H}_k(M, \mu_2)$. The space $\mathcal{H}_k(M, \mu_1)$ is defined in a similar way.

DEFINITION 2.9. Let Φ and Ψ form a smooth partition of unity on $\overline{\Omega}$ with $\Phi \in C_0^\infty(\mathcal{F}_{\theta_2} \cap \overline{\Omega})$, and $\Phi(z, t) \equiv 1$ for $0 \le z \le \frac{\varepsilon}{4}$, $|t_1| < \frac{\varepsilon}{2}$. For any $m \in \mathbb{N} \cup \{0\}$

and $r \in \mathbb{R}$, define the space $H_{m,r}(\Omega, \mu_2)$ as the completion of $C^\infty(\overline{\Omega})$ in the norm

$$\langle\langle u\rangle\rangle_{m,r} = \sum_{k=1}^{m} \|z^k \Phi u\|_{k+r} + \|t_1^k \Phi u\|_{k+r} + \|\partial_{t''}^{\alpha} \Phi u\|_r + \|\Psi u\|_{m+r}.$$

Evidently, the elements of $H_{m,r}(\Omega, \mu_2)$ have singularities on the submanifold μ_2, and $H_{0,r}(\Omega, \mu_2)$ is the standard Sobolev space $H^r(\Omega)$. As in the previous definition, we put $H_{\infty,r}(\Omega, \mu_2) = \bigcap_{m=0}^{\infty} H_{m,r}(\Omega, \mu_2)$.

The Calderón operator (cf. Section 1) of the bvp (2.17) takes the form:

$$Av = \partial_\tau v + a\Lambda v + hv \quad \text{for } v \in H^1(M),$$

where $\Lambda v = \frac{\partial}{\partial z} Sv|_{z=0}$. Its adjoint becomes $A^* = -\partial_\tau + a\Lambda + h + T_0$, $\operatorname{ord} T_0 = 0$.

THEOREM 2.10 ([**Pa**]). *Let $s \in \mathbb{N} \cup \{0\}$. Then the following assertions hold:*

(a) *The modified oblique derivative problem* (2.18) *has a solution $(u, v) \in H_{s,1/2}(\Omega, \mu_2) \times H^s(\mu_2)$ if and only if $(g, u_0) \in \mathcal{H}_s(M, \mu_2) \times H^s(\mu_1)$ satisfy the orthogonality conditions*

$$(g, w_j)_M + (u_0, (\tau, \overrightarrow{\mu}_1)\{w_j\}_{\mu_1})_{\mu_1} = 0$$

for $1 \leq j \leq N < \infty$. As usual, $\{w_j\}_{\mu_1}$ denotes the jump of the function $w_j \in H^1(M \setminus \mu_1)$ on μ_1, and the functions $w_j \in \mathcal{H}_\infty(M, \mu_1) \cap H^1(M \setminus \mu_1)$ form a basis of the solution space (kernel) of the bvp on M:

$$\begin{cases} A^* w + (\tau, \overrightarrow{\mu}_1)\{w\}_{\mu_1} \otimes \delta_{\mu_1} = 0 & \text{on } M, \\ w = 0 & \text{on } \mu_2. \end{cases}$$

(b) *Let $g = u_0 = 0$. Then* (2.18) *has exactly N linearly independent solutions, and they all belong to $H_{\infty,1/2}(\Omega, \mu_2) \times H^\infty(\mu_2)$.*

(c) *Suppose that $(u, v) \in H_{s,\frac{1}{2}}(\Omega, \mu_2) \times H^s(\mu_2)$ is a solution of* (2.18) *with right-hand side $(g, u_0) \in \mathcal{H}_r(M, \mu_2) \times H^r(\mu_1)$ and $r > s$. Then, the solution (u, v) belongs to $H_{r,1/2}(\Omega, \mu_2) \times H^r(\mu_2)$.*

(d) *Let $s \geq 1$. Then $u|_M \in H^1(M \setminus \mu_2)$ and $v = (\tau, \overrightarrow{\mu}_2)\{u\}_{\mu_2}$.*

Conclusion. *The bvp* (2.18) *is a Fredholm problem with index zero. Its kernel and cokernel consist of $C^\infty(\overline{\Omega} \setminus \mu_2)$ functions, and the restriction $u|_M$ has a finite jump on μ_2. According to* (c)*, we have an increase of regularity (smoothness) for the solution (u, v).*

Suppose now that $\mu_1 \neq \varnothing$ and $\mu_2 \neq \varnothing$. Obviously, the solution (u, v) of the bvp (2.18) satisfies (2.17) if $v = 0$.

THEOREM 2.11 ([**Mz-Pa**], [**Pa**]). *Let $s \geq 1$ and $(g, u_0) \in \mathcal{H}_s(M, \mu_2) \times H^s(\mu_1)$. The bvp*

$$Pu = 0 \ \text{ in } \Omega, \quad \frac{\partial u}{\partial l} + hu = g \ \text{ on } M, \quad u = u_0 \ \text{ on } \mu_1, \tag{2.20}$$

has a solution $u \in H_{s,\frac{3}{2}}(\Omega)$ *if and only if the condition*

$$(g, w_\chi)_M + (u_0, (\tau, \vec{\mu}_1)\{w_\chi\}_{\mu_1})_{\mu_1} = 0$$

is satisfied for each solution $w_\chi \in \mathcal{H}_\infty(M, \mu_1)$ *of the problem*

$$\begin{cases} A^* w_\chi + (\tau, \vec{\mu}_1)\{w_\chi\}_{\mu_1} \otimes \delta_{\mu_1} = 0 \quad \text{on } M, \\ w_\chi = \chi \quad \text{on } \mu_2, \end{cases} \tag{2.21}$$

for an arbitrary function $\chi \in C^\infty(\mu_2)$. *Moreover,* $u \in C^{1+\alpha}(\Omega)$ *for every* $\alpha < 1$.

REMARK 2.12. As the space of functions w_χ satisfying (2.21) is infinite dimensional, we conclude that the classical bvp (2.20) has a cokernel of infinite dimension whenever $\mu_2 \neq \varnothing$. A classical solution u of the bvp (2.20) can exist if and only if infinitely many orthogonality conditions are imposed on g and u_0.

A similar theorem was established by Winzell [**W**] in Hölder spaces:

THEOREM 2.13 ([**W**]). *Given* $\mu_1 \neq \varnothing$ *and* $\mu_2 \neq \varnothing$, *consider the bvp*

$$Pu = 0 \quad \text{in } \Omega, \quad c(x) \leq 0, \quad \frac{\partial u}{\partial l} = g \quad \text{on } M \setminus \mu_2, \quad u = u_0 \quad \text{on } \mu_1.$$

Then, for arbitrary functions $g \in C^{1+\alpha}(M)$ *and* $u_0 \in C^{2+\alpha}(\mu_1)$, $0 < \alpha < 1$, *the above problem has a unique solution* $u \in C^{2+\alpha}_{\text{loc}}(\overline{\Omega} \setminus \mu_2) \cap L^\infty(\Omega)$.

In the previous considerations concerning the bvp (2.17) it was assumed that the vector field τ was transversal to the manifolds μ_1 and μ_2 that separate M^+ from M^-. In this way, the set μ_1 on which (2.17) is underdetermined, and the set μ_2 on which (2.17) is overdetermined, are far from each other. As far as we know, the paper of Maz'ya [**Mz**] is the only one where the above mentioned restriction is overcame. It is worth mentioning that in [**Mz**] the vector field l is allowed to be tangential to a series of submanifolds contained in M, and moreover, l could be simultaneously of both emergent and submergent types along one of these submanifolds.

As before, let Ω be a smooth bounded domain in $\mathbb{R}^n$, $n \geq 3$, $M = \partial\Omega$, and let $M = M_0 \supset M_1 \supset \ldots \supset M_k$, $M_k \neq \varnothing$, be a series of compact C^∞ submanifolds without boundary. The vector field l is tangential to each submanifold M_j only at the points of M_{j+1}, and l is not tangential to M_k. The assumption that M^+ and M^- are nonempty, implies that $\varnothing \neq \mu_1 \cup \mu_2 \subset M_1 \setminus M_2$. The vector field τ fulfills condition (5) formulated after (2.17). Consider now the problem (2.19) with $u_0 = 0$. In [**Mz**] the author defines a weak solution of the bvp (2.19) such that $a^{1/2}u|_M \in H^{1/2}(M)$ and $\rho^{-1}u|_M \in L^2_{\text{loc}}(M \setminus \mu_2)$, where $\rho(x) = \text{dist}(x, \mu_1)$.

THEOREM 2.14 ([**Mz**]). *Suppose that* $P(1) \ll 0$. *Then, for every function* $g \in L^2(M)$ *there exists a unique weak solution of* (2.19).

The previous result can be illustrated by the following example:

$$\Omega = \{x \in \mathbb{R}^n : x_n > 0\}, \ n \geq 3, \quad M = \{x \in \mathbb{R}^n : x_n = 0\},$$
$$l = \partial_1 + x_1\partial_2 + \ldots + x_{n-1}\partial_n.$$

In this situation,

$$M^{\pm} = \left\{x \in \mathbb{R}^n : x_n = 0,\ x_{n-1} \gtrless 0\right\},$$

$$M_j = \{x \in \mathbb{R}^n : x_n = x_{n-1} = \ldots = x_{n-j-1} = 0\} \text{ for } j = 0, \ldots, n-2,$$

$$\mu_{1,2} = \left\{x \in \mathbb{R}^n : x_n = x_{n-1} = 0,\ x_{n-2} \gtrless 0\right\}.$$

3. Open problems

We are going to formulate two open problems concerning the tangential oblique derivative problem. From our point of view, they are interesting problems as they have applications in the theory of probability and in the theory of Brownian motion. On the other hand, their possible solvability will be a contribution to the theory of degenerate boundary value problems for (non)linear second order elliptic operators in Hölder spaces.

PROBLEM 1.

(i) Consider a classical first order ΨDO $P(x, D)$ with the principal symbol $p^o(x, \xi) = \xi_1 - a(x, \xi') + ib(x, \xi')$, where $a(x, \xi')$ and $b(x, \xi')$ are real-valued functions, not identically 0. Assume that $P(x, D)$ is a subelliptic operator in the scale of Sobolev spaces H^s. Under which conditions is $P(x, D)$ a subelliptic operator in the Hölder classes Λ^s, $s > 0$, with the same loss of regularity as in H^s?

(ii) After solving (i) try to develop in the Hölder classes Λ^s, $C^{k,\alpha}$, $0 < \alpha < 1$, the theory of the following bvp:

$$\begin{cases} Lu = F & \text{in } \Omega \subset \mathbb{R}^n,\ n \geq 3, \\ \frac{\partial u}{\partial \tau} + (a + ib)\frac{\partial u}{\partial \mathbf{n}}\Big|_M = g & \text{on } M = \partial\Omega. \end{cases}$$

As usual, Ω is a bounded domain with C^∞ smooth boundary M, $\mathbf{n}$ is the unit outward normal to M, L is a second order elliptic operator, and $a, b \in C^\infty(M)$ are real-valued functions. Moreover, we assume that $a|_E = 0$ and $a|_{(M\setminus E)} \neq 0$, where $E \subset M$ is a smooth submanifold of M of codimension one. The smooth nondegenerate vector field τ is tangential to M and transversal to E.

REMARK 3.1. At the end of of Section 1 we showed an example of a ΨDO which is subelliptic in Sobolev spaces but is not subelliptic in Hölder spaces.

PROBLEM 2. Let Ω be a bounded domain in $\mathbb{R}^n$, $n \geq 3$, $M = \partial\Omega \in C^3$. Consider the bvp

$$\begin{cases} F(x, u, Du, D^2u) = 0 & \text{in } \Omega, \\ \frac{\partial u}{\partial l} + b(x, u) = g(x) & \text{for } x \in M, \end{cases} \tag{3.1}$$

in the cases of a neutral or emergent real smooth vector field $l \in C^2$, $b \in C^2$. When l is of emergent type, the bvp (3.1) is studied with the extra condition $u|_E = u_0$, imposed on the set of tangency E.

Denote by S^n the space of all $n \times n$ symmetric matrices. Then the real-valued function $F \in C^2(\overline{\Omega} \times \mathbb{R} \times \mathbb{R}^n \times S^n)$ defines an elliptic PDE and satisfies the convexity property:

$$\lambda|\xi|^2 \leq \sum_{i,j=1}^{n} \frac{\partial F}{\partial r_{ij}}(x,z,p,r)\xi_i\xi_j \leq \Lambda|\xi|^2 \text{ for every } \xi \in \mathbb{R}^n,$$

where λ and Λ are some positive constants. Moreover,

$$\sum_{i,j=1}^{n} \frac{\partial^2 F}{\partial r_{ij}\partial r_{kl}}\xi_{ij}\xi_{kl} \geq c_0|\xi|^2 \text{ for every } \xi \in \mathbb{R}^{n^2},\ c_0 = \text{const} > 0.$$

(i) Find sufficient conditions for the existence and uniqueness of a classical solution $u \in C^2(\Omega) \cap C^1(\overline{\Omega})$ of the bvp (3.1); with the extra condition $u|_E = u_0$ if l is of emergent type.
(ii) Suppose that $F \in C^{2+\alpha}$, $0 < \alpha < 1$, and $l, b, g, u_0 \in C^{2+\alpha}$. Can we claim that $u \in C^{2+\alpha}(\overline{\Omega})$?

REMARK 3.2. The case of a semilinear PDE was investigated in details in [**P-Pl**]. Under several restrictions, a unique viscosity solution of the bvp (3.1) is constructed in [**P-K**].

References

[Ad] R. Adams, *Sobolev spaces.* Academic Press, New York, 1975.

[Al] Sh. A. Alimov, *A problem involving an inclined derivative.* Diff. Eq. **17** (1982), 1073–1083.

[Bi] A. V. Bicadze, *A homogeneous problem for the slant derivative for harmonic functions in three dimensional regions.* Sov. Math. Dokl. **4** (1963), 156–159.

[Bo] R. L. Borelly, *The singular, second order oblique derivative problem.* J. Math. Mach. **16** (1966), 51–81.

[E1] Y. V. Egorov, *Linear differential equations of principal type.* Contemporary Soviet Math. New York, 1986.

[E2] Y. V. Egorov, *Subelliptic operators.* Soviet Math. Uspehi, **30** (1975), 57–114.

[E-K] Y. V. Egorov and V. Kondratiev, *The oblique derivative problem.* Math. USSR Sbornik **7** (1969), 139–169.

[G-T] D. Gilbarg and N. S. Trudinger, *Elliptic partial differential equations of second order.* 2nd ed., Springer-Verlag, Berlin 1983.

[Gu] P. Guan, *Hölder regularity of subelliptic pseudodifferential operators.* Duke Math. J. **60** (1990), 563–598.

[H1] L. Hörmander, *Pseudodifferential operators and non-elliptic boundary value problems.* Ann. Math. **83** (1966), 129–209.

[H2] L. Hörmander, *The analysis of linear partial differential operators.* vol. I–IV, Springer-Verlag, Berlin, 1983.

[Ma] M. Malyutov, *On the Poincaré boundary value problem.* Trans. Moscow Math. Soc. **20** (1969), 173–204.

[MaPlSo] A. Maugeri, D. K. Palagachev and L. Softova, *Elliptic and paraboloc equations with discontinuous coefficients.* Math. res. ser. **vol. 109**, Wiley- VCH Verlag, Berlin, 2000.

[Mz] V. Maz'ya, *On a degenerating problem with directional derivative.* Math. USSR Sb. **16** (1972), 429–469.

[Mz-Pa] V. Maz'ya and B. P. Paneah, *Degenerate elliptic pseudodifferential operators and oblique derivative problem.* Trans. Moscow Math. Soc. **31** (1974), 247–305.

[Mz-Pa1] V. Maz'ya and B. P. Paneach, *Coercive estimates and regularity of solutions of degenerate elliptic pseudodifferential equations.* Funct. Anal. Appl. **4** (1971), 249–311.

[M-Sj] A. Melin and J. Sjöstrand, *A calculus for Fourier integral operators in domains with boundary and applications to the oblique derivative problem.* CPDE **2** (1977), 857–935.

[Mey] Y. Meyer, *Nouvelles estimations pour les operateurs pseudodifferentieles.* Semin. Goulaouic-Schwartz, Ecole Polytech., Paris, Exposé N X (1977-1978).

[N-Tre] L. Nirenberg and F. Treves, *On local solvability of linear partial differential equations. I: Necessary conditions.* Comm. Pure Appl. Math. **23** (1970), 1–38.

[PPT] D. K. Palagachev, P. R. Popivanov and K. Taira, *A degenerate Neumann problem for quasilinear elliptic integrodifferential operators.* Math. Z. **230** (1999), 679–694.

[Pa] B. P. Paneah, *The oblique derivative problem. The Poincaré problem.* Wiley-VCH, Berlin, 2000.

[Pa1] B. P. Paneach, *On the theory of solvability of the oblique derivative problem.* Math. USSR Sb. **42** (1982), 197–235.

[Poi] H. Poincaré, *Leçons de Mechanique Celeste, Tome III, Théorie de marées.* Gauthiers-Villars, Paris, 1910.

[P-Pl] P. R. Popivanov and D. K. Palagachev, *The degenerate oblique derivative problem for elliptic and parabolic equations.* Akademie-Verlag, Berlin, 1997.

[P-K] P. R. Popivanov and N. D. Kutev, *Viscosity solutions to degenerate oblique derivative problem for fully nonlinear elliptic equations.* C.R. Acad. Sci.-Paris **334** (2002) 661–666.

[P-Sou] P. R. Popivanov and D. M. Souroujon, *Punctual blow up solutions for the tangential oblique derivative problem.* Univ. Potsdam, Inst für Math., Preprint 98/22, 1998.

[R-Sch] S. Rempel and B.-W. Schulze, *Index theory of elliptic boundary problems.* Akademie-Verlag, Berlin, 1982.

[Sch] B.-W. Schulze, *Some classes of non-elliptic boundary problems for elliptic pseudodifferential operators.* Math. Nachr. **95** (1980), 327–350.

[St] E. Stein, *Singular integrals and differentiability properties of functions.* Princeton Univ. Press, Princeton, 1970.

[Ta] K. Taira, *Boundary value problems and Markov Processes.* Lecture notes in Math., **v. 1499**, Springer, 1991.

[Tre] F. Treves, *Introduction to pseudodifferential and Fourier integral operators.* Plenum Press, NY, London, 1980.

[W] B. Winzell, *A boundary value problem with an oblique derivative.* CPDE **6** (1981), 305–328.

Institute of Mathematics and Informatics Bulgarian Academy of Sciences, Acad G. Bonchev str., 1113 Sofia, Bulgaria

E-mail address: popivano@math.bas.bg

Operator Theory:
Advances and Applications, Vol. 151, 472–494

A note on boundary value problems on manifolds with cylindrical ends

Marius Mitrea and Victor Nistor

ABSTRACT. We present an extension of the method of layer potentials to manifolds with boundary and cylindrical ends. To obtain this extension along the classical lines, we have to deal with several technical difficulties due to the non-compactness of the boundary, which prevents us from using the standard characterization of Fredholm and compact (pseudo-)differential operators between Sobolev spaces on compact manifolds.

As an application, we obtain a solution of the non-homogeneous Dirichlet problem in this setting. We also prove the existence of the Dirichlet-to-Neumann map in the class of pseudodifferential operators which are "almost translation invariant at infinity."

Introduction

Starting with the pioneering work of Hodge, de Rham, Kodaira, Spencer, Duff, Kohn, among others (cf. the discussion in the introduction of [**18**]), the method of layer potentials has been widely used to solve boundary value problems on compact manifolds with smooth boundary. See, for instance, [**4, 6, 7**], or [**8**], and the references cited therein. More recently, the method of layer potentials has also lead to a solution of the Dirichlet problem for the Laplace operator on compact manifolds with Lipschitz boundaries in [**16**].

In view of several possible applications, we would like to extend the method of layer potentials to various classes of non-compact manifolds. There are, however, several technical problems that we need to overcome first for such an extension to be possible – at least along the classical lines. The main contribution of this paper is to explain how these difficulties are dealt with in the particular case of manifolds with cylindrical ends, when a number of required results from analysis take a simpler form. We hope that the approach that we outline in this note will

---

2000 *Mathematics Subject Classification.* Primary: 31C12; Secondary: 35J05, 58J40.

Mitrea was partially supported by NSF grants DMS-9870018, DMS-0139801, and a UMC Research Board Grant. Nistor was partially supported by NSF Grants DMS-9971951 and DMS-0200808.

be a paradigm for dealing with some of the issues that arise in the study of elliptic boundary value problems on more general non-compact manifolds with the method of layer potentials.

In fact, let us make it clear now, that, with some additional work, it is also possible to obtain parts of some our results using the classical, ground-breaking paper of Kondratiev [**3, 9, 12, 19**]. Kondratiev deals typically with general boundary value problems and establishes their Fredholm character (or "normal solvability") and asymptotic expansion of the solutions. His methods are elementary, using the gluing of model operators. Our method, although more complicated and less elementary, has the advantages that it provides an explicit formula for the solutions of our elliptic boundary value problems, it describes explicitly the Sobolev spaces that arise on the boundary, and, most importantly, it characterizes the Dirichlet-to-Neumann map. This is important in computational mathematics as well as in some emerging applications of boundary value problems on non-compact manifolds to spectral theory and index theory.

In order to explain some of the technical difficulties encountered in this setting, we need some notation. Let N be a non-compact Riemannian manifold with boundary ∂N and $\Delta_N = d^*d$ be the Laplace operator on scalar functions. A first set of problems consists of defining an elementary solution $E(\cdot,\cdot)$ for Δ_N on N and proving that the associated single and double layer potential integrals converge – issues well-understood when N is compact. A second set of problems has to do with the existence of the non-tangential limits of the aforementioned layer potential integral operators. Finally, even if the non-tangential limits exist and are given by pseudodifferential operators on ∂N, these pseudodifferential operators will not be properly supported in general. Moreover, since ∂N is non-compact, the standard results on the boundedness and compactness of order zero (respectively, negative order) pseudodifferential operators do not (directly) apply.

To overcome the above technical problems, it is natural to make certain additional assumptions on the non-compact manifolds N and ∂N. Thus, in this paper, we have restricted ourselves to the class of manifolds with boundary and cylindrical ends. This is arguably the easiest to understand class of non-compact manifolds.

Let N be a manifold with boundary and cylindrical ends. Such manifolds have a product structure at infinity in a strong sense (that is, including also the metric, see Definition 5.1). To study boundary value problems on N, it is necessary to consider also manifolds with cylindrical ends without boundary. Thus, let in what follows M denote a manifold with cylindrical ends. Typically, M will be either ∂N or a manifold of the same dimension as N containing N as a subdomain.

It is convenient to deal with layer potentials using pseudodifferential operators. We shall consider on M two classes of pseudodifferential operators: $\Psi^m_{\text{inv}}(M)$ and $\Psi^m_{\text{ai}}(M)$, whose distribution kernels form a class large enough to contain those appearing in the boundary layer integrals considered in this paper. These classes are very closely related to the so called b-calculus [**13, 14**], but different.

The first class of operators is the class of order m, classical pseudodifferential operators that are "translation invariant in a neighborhood of infinity" (Definition

1.1). The space $\Psi_{\text{ai}}^{-\infty}(M)$ consists of the closure of $\Psi_{\text{inv}}^{-\infty}(M)$ with respect to a suitable family of semi-norms (see Equations (1.5) and (2.1)). Then we let

$$\Psi_{\text{ai}}^{m}(M) := \Psi_{\text{inv}}^{m}(M) + \Psi_{\text{ai}}^{-\infty}(M). \tag{0.1}$$

An operator $P \in \Psi_{\text{ai}}^{m}(M)$ is called *almost translation invariant in a neighborhood of infinity.*

The reason for introducing this class of operators is that $T^{-1} \in \Psi_{\text{ai}}^{-m}(M)$ for any elliptic operator $T \in \Psi_{\text{ai}}^{m}(M)$ provided that $m \geq 0$ and T is invertible on $L^2(M)$. (Anticipating, this means that the algebras $\Psi_{\text{ai}}^{m}(M)$ are spectrally invariant, Definition 2.3, a property not enjoyed by the b–calculus, but enjoyed by Melrose's c-calculus, [**10**]). This allows us to define our integral kernels – and, implicitly, also the boundary layer integrals – much as in [**16**], namely as follows. First, we embed our manifold with boundary and cylindrical ends N into a manifold with cylindrical ends M (and without boundary). We then prove that for suitable $V \geq 0$, $V \neq 0$, the operator $\Delta_M + V$ is invertible by checking that it is Fredholm of index zero and injective.

Let $\delta_{\partial N}$ the conditional measure on ∂N (defined using the Riemannian metric). If $f \in L^2(\partial N)$, then integration on ∂N against f defines a distribution $f \otimes \delta$ by the relation $\langle f \otimes \delta_{\partial N}, \varphi \rangle = \int_{\partial N} f\varphi$, where φ is an arbitrary test function on N. The single layer potential integral is defined then as

$$\mathcal{S}(f) = (\Delta_M + V)^{-1}(f \otimes \delta_{\partial N}). \tag{0.2}$$

We shall fix in what follows a vector field ∂_ν on M whose restriction to ∂N is the unit outer normal to ∂N. If $f \in L^2(\partial N)$, then we obtain similarly a distribution $f \otimes \delta'_{\partial N}$ by the relation $\langle f \otimes \delta'_{\partial N}, \varphi \rangle = \int_{\partial N} f \partial_\nu \varphi$. Then the double layer potential integral is defined as

$$\mathcal{D}(f) = (\Delta_M + V)^{-1}(f \otimes \delta'_{\partial N}). \tag{0.3}$$

Since we are dealing with non-compact manifolds (M and ∂N), the above integrals are defined by relying on mapping properties of the operators in $\Psi_{\text{ai}}^{m}(M)$.

Next, we show that we can make sense of the restriction to ∂N of the kernel E of $(\Delta_M + V)^{-1}$, thus obtaining $S := [(\Delta_M + V)^{-1}]_{\partial N} \in \Psi_{\text{ai}}^{-1}(\partial N)$. We can then relate the non-tangential limits of the single and double layer potentials of some function f using the operator S.

Similarly, let $K := [(\Delta_M + V)^{-1}\partial_\nu^*]_{\partial N}$, by restricting the kernel of $(\Delta_M + V)^{-1}\partial_\nu^*$ to ∂N. Let $f_\pm$ be the non-tangential pointwise limits of some function f defined on $M \setminus \partial N$, provided that they exist.

Some of the properties of the single and double layer potentials alluded to above are summarized in the following theorem.

THEOREM 0.1. *Given $f \in L^2(\partial N)$, we have*

$$\mathcal{S}(f)_\pm = Sf, \quad \partial_\nu \mathcal{S}(f)_\pm = \left(\pm\tfrac{1}{2}I + K^*\right)f, \quad \mathcal{D}(f)_\pm = \left(\mp\tfrac{1}{2}I + K\right)f, \tag{0.4}$$

where K^* is the adjoint of K.

These theorems are proved by reduction to the compact case [**16**]. As in the classical case of a compact manifold with smooth boundary, we obtain the following result.

THEOREM 0.2. *Let N be a manifold with boundary and cylindrical ends. Then*

$$H^s(N) \ni u \mapsto (\Delta_N u, u|_{\partial N}) \in H^{s-2}(N) \oplus H^{s-1/2}(\partial N)$$

is a continuous bijection, for any $s > 1/2$.

The above theorem for s integer (without an explicit description of the Sobolev spaces on the boundary) can also be obtained from results of Kondratiev [**9, 12, 19**] in the closely related setting of manifolds with conical points and of totally characteristic operators on manifolds with boundary. See [**3**] for a sketch of a derivation of our theorem using Kondratiev's results.

In [**22**], Schrohe and Schulze have generalized the Boutet de Monvel calculus to manifolds with boundary and cylindrical ends. Their results are very general and can probably also be used to prove our Theorem 0.2 above. Our proof, however, is shorter, and also leads to a characterization of the Dirichlet-to-Neumann boundary map, Theorem 5.7. It is worth pointing out that our methods can also handle non-smooth structures (cf. §4) and seem amenable to other basic problems of mathematical physics in non-compact manifolds (such as Maxwell's equations in infinite cylinders). We hope to return to these issues at a later time.

The reader is referred to [**21, 24**], or [**25**] for definitions and background material on pseudodifferential operators. Note that in our paper we work exclusively with manifolds of bounded geometry.

Let us mention here that the proofs of some results are only sketched and some standard proofs are omitted altogether. Moreover, for clarity, we concentrate in this note on the Laplace operator, which is anyway one of the most important cases of the general theory. Full details of the proofs, as well as some generalizations will appear elsewhere.

ACKNOWLEDGMENT. The authors are grateful to B. Ammann, T. Christiansen, J. Gil, R. Lauter, G. Mendoza, B. Monthubert, and the referee for useful discussions.

1. Operators on manifolds with cylindrical ends

We begin by introducing the class of manifolds with cylindrical ends (without boundary) and by reviewing some of the results on the analysis on these manifolds that are needed in this paper.

1.1. Manifolds with cylindrical ends and the Laplace operator. Let M_1 be a compact manifold with boundary $\partial M_1 \neq \emptyset$. We assume that a metric g is given on M_1 and that g_1 is a product metric in a tubular neighborhood $V \cong \partial M_1 \times [0,1)$ of the boundary, namely

$$g_1 = g_\partial + (dx)^2, \tag{1.1}$$

where $x \in [0,1)$ is the second coordinate in $\partial M_1 \times [0,1)$ and g_∂ is a metric on ∂M_1. Let

$$M := M_1 \cup (\partial M_1 \times (-\infty, 0]), \quad \partial M_1 \equiv \partial M_1 \times \{0\}, \tag{1.2}$$

be the union of M_1 and $\partial M_1 \times (-\infty, 0]$ along their boundaries. The above decomposition will be called a *standard decomposition* of M. The resulting manifold M is called a *manifold with cylindrical ends.* Note that a manifold with cylindrical ends is a complete, non-compact, Riemannian manifold without boundary.

Let $M = M_1 \cup (\partial M_1 \times (-\infty, 0])$ be a manifold with cylindrical ends. Let g be the metric on M and assume, as above, that $g = g_\partial + (dx)^2$ on the cylindrical end $\partial M_1 \times (-\infty, 0]$, where $x \in (-\infty, 0]$ and g_∂ is a metric on the boundary of M_1. Let $\Delta_M = d^*d$ be the (scalar) Laplace operator on M, and $\Delta_{\partial M_1}$ be the Laplace operator on ∂M_1, defined using the metric g_∂. Then

$$\Delta_M = -\partial_x^2 + \Delta_{\partial M_1} \tag{1.3}$$

on the cylindrical end $\partial M_1 \times (-\infty, 0]$.

1.2. Operators that are translation invariant in a neighborhood of infinity. Let $M = M_1 \cup (\partial M_1 \times (-\infty, 0])$ be a manifold with cylindrical ends, as above, and let, for any $s \geq 0$,

$$\phi_s \colon \partial M_1 \times (-\infty, 0] \to \partial M_1 \times (-\infty, -s] \tag{1.4}$$

be the isometry given by translation with $-s$ in the x-direction. If $s < 0$, then ϕ_s is defined as the inverse of ϕ_{-s}. We shall write ϕ_s instead of ϕ_s^* for the induced action on functions: $\phi_s(f)(x) = f(\phi_s(x))$, in order to keep the notation simple. The special form of the operator Δ_M obtained at the end of the previous subsection suggests the following definition.

DEFINITION 1.1. A continuous linear map $P \colon \mathcal{C}_c^\infty(M) \to \mathcal{C}^\infty(M)$ will be called *translation invariant in a neighborhood of infinity* if its Schwartz kernel has support in

$$V_\epsilon := \{(x,y) \in M^2;\ \mathrm{dist}(x,y) < \epsilon\},$$

for some $\epsilon > 0$, and there exists $R > 0$ such that $P\phi_s(f) = \phi_s P(f)$, for any $f \in \mathcal{C}_c^\infty(\partial M_1 \times (-\infty, -R))$ and any $s > 0$.

We shall denote by $\Psi^m_{\mathrm{inv}}(M)$ the space of *order m, classical pseudodifferential operators on M that are translation invariant in a neighborhood of infinity.*

Let $\mathcal{D}(T)$ denote the domain of a possibly unbounded operator T. Recall that an unbounded operator $T \colon \mathcal{D}(T) \to X$ defined on a subset of a Banach space Y and with values in another Banach space X is *Fredholm* if T is *Fredholm* as a bounded operator $\mathcal{D}(T) \to X$, where $\mathcal{D}(T)$ is endowed with the graph norm. Equivalently, T is Fredholm if it is closed and has finite dimensional kernel and cokernel. Also, T is called *invertible* if T is invertible as an operator $\mathcal{D}(T) \to X$. For all differential operators considered below, we shall consider the minimal closed extension, that is, the closure of the operators with domain compactly supported smooth functions.

For each $s \in \mathbb{R}$, we shall denote by $H^s(M)$ the domain of $(I+\Delta_M)^{s/2}$, regarded as an unbounded operator on $L^2(M)$, i.e.,

$$H^s(M) := \mathcal{D}((I + \Delta_M)^{s/2}). \tag{1.5}$$

We endow $H^s(M)$ with the norm

$$\|u\|_s = \|(I + \Delta_M)^{s/2}u\|_{L^2}$$

(Below, we shall occasionally write $\|\cdot\|$ instead of $\|\cdot\|_{L^2}$.) Note that $I+\Delta_M \geq I$, so that $\|u\|_s \geq \|u\|$ for $s \geq 0$.

As usual, we shall denote by $H^{-s}(M)$ the dual of $H^s(M)$, via a duality pairing that extends the pairing between functions and distributions. We thus identify $H^{-s}(M)$ with a space of distributions on M.

Let $\sigma_s(P) \in S^s(T^*M)/S^{s-1}(T^*M)$ be the principal symbol of an operator $P \in \Psi^s_{\text{inv}}(M)$. See **[21, 24]**, or **[25]**.

LEMMA 1.2. *Let M be a manifold with cylindrical ends.*

(i) $\Psi^s_{\text{inv}}(M)\Psi^{s'}_{\text{inv}}(M) \subset \Psi^{s+s'}_{\text{inv}}(M)$ *and the principal symbol*

$$\sigma_s \colon \Psi^s_{\text{inv}}(M)/\Psi^{s-1}_{\text{inv}}(M) \to S^s(T^*M)/S^{s-1}(T^*M)$$

induces, for any $s \in \mathbb{R}$, an isomorphism onto the subspace of symbols that are translation invariant in a neighborhood of infinity.

(ii) $P \in \Psi^s_{\text{inv}}(M)$ *extends to a continuous operator* $P\colon H^{s'}(M) \to H^{s'-s}(M)$, *if* $s, s' \in \mathbb{R}$.

Let us now recall a classical and well-known construction. We shall use Melrose's terminology. Any operator $P\colon \mathcal{C}^\infty_c(M) \to \mathcal{C}^\infty(M)$ that is translation invariant in a neighborhood of infinity will be properly supported (that is, $P(\mathcal{C}^\infty_c(M)) \subset \mathcal{C}^\infty_c(M)$) and gives rise to a pseudodifferential operator $\tilde{P}\colon \mathcal{C}^\infty_c(\partial M_1 \times \mathbb{R}) \to \mathcal{C}^\infty_c(\partial M_1 \times \mathbb{R})$ by the formula

$$\tilde{P}(f) = \phi_{-s} P \phi_s(f), \tag{1.6}$$

where ϕ_s is the translation by s on the cylinder $\partial M_1 \times \mathbb{R}$ and s is arbitrary, but large enough so that

$$\operatorname{supp}(P\phi_s(f))\,,\ \operatorname{supp}(\phi_s(f)) \subset \partial M_1 \times (-\infty, 0) \subset M.$$

DEFINITION 1.3. The operator $\tilde{P}$ will be called *the indicial operator associated with P*. The resulting map

$$\Phi\colon \Psi^\infty_{\text{inv}}(M) \ni P \mapsto \tilde{P} \in \Psi^\infty(\partial M_1 \times \mathbb{R})$$

will be called the *indicial morphism.*

Let us notice now that $\partial M_1 \times \mathbb{R}$ is also a manifold with cylindrical ends. The partially defined action of $\mathbb{R}$ on the ends of M extends to a global action of $\mathbb{R}$ on $\partial M_1 \times \mathbb{R}$. We shall denote by $\Psi^\infty_{\text{inv}}(\partial M_1 \times \mathbb{R})^{\mathbb{R}}$ the operators in $\Psi^\infty_{\text{inv}}(\partial M_1 \times \mathbb{R})$ that are translation invariant with respect to the natural action of $\mathbb{R}$ on $\partial M_1 \times \mathbb{R}$.

Let $T \in \Psi^{\infty}_{\mathrm{inv}}(M_1 \times \mathbb{R})^{\mathbb{R}}$ and η be a smooth function on $\mathbb{R} \times \partial M_1$ with support in $(-\infty, -1) \times \partial M_1$, equal to 1 in a neighborhood of infinity. Then

$$s_0(T) := \eta T \eta \tag{1.7}$$

defines an operator in $\Psi^{\infty}_{\mathrm{inv}}(M)$.

LEMMA 1.4. *Let s_0 be as in* Eq. (1.7). *Then $\Phi(s_0(T)) = T$ for all $T \in \Psi^{\infty}_{\mathrm{inv}}(\partial M_1 \times \mathbb{R})^{\mathbb{R}}$. In particular, the range of the indicial morphism Φ of* Definition 1.3 *is $\Psi^{\infty}_{\mathrm{inv}}(\partial M_1 \times \mathbb{R})^{\mathbb{R}}$.*

In order to deal with operators acting on weighted Sobolev spaces, we shall need the following lemma. (See also [**13**].) We shall denote by $[a, b] := ab - ba$ the commutator of two linear maps.

LEMMA 1.5. *Let $P, P_1 \in \Psi^{\infty}_{\mathrm{inv}}(M)$ be arbitrary and $\rho\colon M \to [1, \infty)$ be a smooth function such that $\rho(y, x) = x$ on a neighborhood of infinity in $\partial M_1 \times (-\infty, 0]$. Then*

(i) $\tilde{Q} = \tilde{P}\tilde{P}_1$, *if* $Q = PP_1$.

(ii) $\mathrm{ad}_\rho(P) := [\rho, P] = \rho P - P\rho \in \Psi^{\infty}_{\mathrm{inv}}(M)$.

PROOF. The relation (i) follows by chasing definitions. To prove (ii), we can assume that $M = X \times \mathbb{R}$. Let ϕ_s, $s \in \mathbb{R}$, be translation by s along $\mathbb{R}$. We can assume that P is translation invariant, in the sense that $\phi_s(P) = P$, for any $s > 0$. Then

$$\phi_s([x, P]) = [\phi_s(x), \phi_s](P)] = [x + s, P] = [x, P]. \tag{1.8}$$

Thus $[x, P]$ is also $\mathbb{R}$-invariant.

In general, $\rho = x$ in a neighborhood of the infinity, so the result follows. □

We shall need on several occasions a function ρ as in the above lemma, so we fix one throughout the paper.

The properties of the indicial operators $\tilde{P}$ are conveniently studied in terms of "indicial families." Indeed, by considering the Fourier transform in the $\mathbb{R}$–variable, we obtain by Plancherel's theorem an isometric bijection (that is, a unitary operator)

$$\mathcal{F}\colon L^2(\partial M_1 \times \mathbb{R}) \to L^2(\partial M_1 \times \mathbb{R}), \quad \mathcal{F}(f)(y, \tau) := \frac{1}{\sqrt{2\pi}} \int_{\mathbb{R}} e^{-\imath \tau t} f(y, t)\, dt. \tag{1.9}$$

Hereafter, $\imath := \sqrt{-1}$. The *indicial family* $\hat{P}(\tau)$ is then defined by

$$\hat{P}(\tau)[\mathcal{F}(f)(\,\cdot\,, \tau)] := \mathcal{F}(\tilde{P} f)(\,\cdot\,, \tau). \tag{1.10}$$

2. A spectrally invariant algebra

A serious drawback of the algebra $\Psi^{\infty}_{\mathrm{inv}}(M)$ is that it is not "spectrally invariant," in the sense that the inverse of an elliptic operator $P \in \Psi^{\infty}_{\mathrm{inv}}(M)$ that is invertible on L^2 is not necessarily in this algebra (Definition 2.3 below). In this section we slightly enlarge the algebra $\Psi^{\infty}_{\text{inv}}(M)$ so that it becomes spectrally invariant. This will lead us to an algebra of operators that are "almost translation invariant in a neighborhood of infinity."

2.1. Operators that are almost translation invariant in a neighborhood of infinity. We begin by introducing another algebra of pseudodifferential operators that will be very useful later on. Let ρ be a smooth function as in Lemma 1.5. Recall that $\operatorname{ad}_\rho(T) := [\rho, T]$. Assume $T\colon \mathcal{C}^\infty_c(M) \to \mathcal{C}^\infty(M)$ to be a linear map with the property that

$$\operatorname{ad}_\rho^k(T) := [\rho, [\rho, \dots, [\rho, T]\dots]]$$

extends to a continuous map $\operatorname{ad}_\rho^k(T)\colon H^{-m}(M) \to H^m(M)$, for any $m \in \mathbb{Z}_+$. Let $\|T\|_{k,m}$ denote the norm of the resulting operator $\operatorname{ad}_\rho^k(T)$. Recall the section s_0 defined in Eq. (1.7).

We define $\Psi^{-\infty}_{\text{ai}}(M)$ to be the closure of $\Psi^{-\infty}_{\text{inv}}(M)$ with respect to the countable family of semi-norms

$$T \to \|T\|_{k,m}, \quad \text{and} \quad T \to \|\rho^l(T - s_0(\Phi(T)))\rho^l\|_{0,m}. \tag{2.1}$$

where $k, m, l \in \mathbb{Z}_+$. Then $\Psi^{-\infty}_{\text{ai}}(M)$ is a Fréchet algebra (that is, a Fréchet space endowed with an algebra structure such that the multiplication is continuous).

Finally, we define

$$\Psi^m_{\text{ai}}(M) := \Psi^m_{\text{inv}}(M) + \Psi^{-\infty}_{\text{ai}}(M). \tag{2.2}$$

An element $P \in \Psi^m_{\text{ai}}(M)$ will be called *almost translation invariant in a neighborhood of infinity.*

It is interesting to observe now that we can introduce dependence on ρ at infinity (thus obtaining variants of Melrose's *b*–calculus, see [**13**] and [**10**]). This is done by noticing that for any $P \in \Psi^m_{\text{inv}}(M)$ and any $N \in \mathbb{N}$ there exists a bounded operator $R_N\colon H^{-k}(M) \to H^k(M)$, where $2k \le N - m$, such that

$$\rho^{-a}P\rho^a - \sum_{j=0}^{N-1}(-1)^j\rho^{-j}\binom{a}{j}\operatorname{ad}_\rho^j(P) = \rho^{N/2}R_N\rho^{N/2}.$$

(Above, $\binom{a}{j} := a(a-1)\dots(a-j+1)/j!$ stand for the usual "binomial" coefficients.)

In view of further applications, we also introduce the *weighted Sobolev space* $\rho^a H^s(M)$, which is the space of distributions of the form $\rho^a u$, with $u \in H^s(M)$. We endow $\rho^a H^s(M)$ with the norm

$$\|f\|_{s,a} := \|\rho^{-a}f\|_s.$$

We have then the following classical results about almost translation invariant pseudodifferential operators on the manifold with cylindrical ends M, see [**14**] for a proof in the spirit of our paper, or [**9, 10, 11, 15, 20, 22, 23**]. These results generalize the corresponding even more classical results on pseudodifferential operators on compact manifolds.

THEOREM 2.1. *Let M be a manifold with cylindrical ends and $P \in \Psi_{\mathrm{ai}}^m(M)$. Also, let $\rho > 0$, $\rho(y,x) = x$ on a neighborhood of infinity in $\partial M_1 \times (-\infty, 0]$. Let $s, a \in \mathbb{R}$ be arbitrary, but fixed. Then:*

(i) *P extends to a continuous operator $P\colon \rho^a H^s(M) \to \rho^a H^{s-m}(M)$.*
(ii) *$P\colon \rho^a H^s(M) \to \rho^{a'} H^{s-m'}(M)$ is compact for any $a' < a$ and $m' > m$.*
(iii) *$P\colon \rho^a H^s(M) \to \rho^a H^{s-m}(M)$ is compact $\Leftrightarrow$ $\sigma_m(P) = 0$ and $\tilde{P} = 0$.*
(iv) *$P\colon \rho^a H^s(M) \to \rho^a H^{s-m}(M)$ is Fredholm $\Leftrightarrow$ $\sigma_m(P)$ is invertible and the operator $\tilde{P}\colon H^s(\partial M_1 \times \mathbb{R}) \to H^{s-m}(\partial M_1 \times \mathbb{R})$ is an isomorphism.*

PROOF. This theorem follows for example from the results in [**13**], or the older preprint [**14**]. □

As a consequence of the mapping properties and closeness under composition of the algebra $\Psi_{\mathrm{ai}}^m(M)$, we obtain the following result.

COROLLARY 2.2. *Let $P \in \Psi_{\mathrm{ai}}^m(M)$, $m > 0$, be elliptic. If P is symmetric on $\mathcal{C}_c^\infty(M)$, then it is essentially self-adjoint.*

PROOF. It is enough to prove that the range of $P \pm \imath I$ is dense. We deal only with $P + \imath I$, because the other case is completely similar. Denote the inner product on $L^2(M)$ by $\langle \cdot, \cdot \rangle$. Assume the range of $P + \imath I$ is not dense, then there exists $\eta \in L^2(M)$ such that

$$\langle (P + \imath I)\xi, \eta \rangle = 0$$

for all $\xi \in \mathcal{C}_c^\infty(M)$. Then $(P - \imath I)\eta = 0$ in the sense of distributions. Select $Q \in \Psi_{\mathrm{ai}}^{-m}(M)$ such that $Q(P - \imath I) = I - R$, where $R \in \Psi_{\mathrm{ai}}^{-\infty}(M)$. Then $\eta = R\eta$. Choose $\eta_n \in \mathcal{C}_c^\infty(M)$, $\eta_n \to \eta$ in $L^2(M)$. By the definition of $\Psi_{\mathrm{ai}}^{-\infty}(M)$, we can find operators $R_n \in \Psi_{\mathrm{inv}}^{-\infty}(M)$ such that

$$\|R - R_n\|_{0,m'} := \|(I + \Delta)^{m'/2}(R - R_n)(I + \Delta)^{m'/2}\| \to 0, \tag{2.3}$$

for $m' \geq m$. Then $\xi_n := R_n \eta_n \to R\eta = \eta$, in $L^2(M)$ as well, and $\xi_n \in \mathcal{C}_c^\infty(M)$. Moreover,

$$(P - \imath I)\xi_n = (P - \imath I)R_n \eta_n \to (P - \imath I)R\eta = (P - \imath I)\eta = 0, \tag{2.4}$$

because the operators $(P - \imath I)R_n$ are bounded and converge in norm to $(P - \imath I)R \in \Psi_{\mathrm{ai}}^{-\infty}(M)$. Let $\bar{P}$ be the closure of $P\colon \mathcal{C}_c^\infty(M) \to L^2(M)$. We have thus proved that η is in the domain of $\bar{P}$, which is a contradiction, since it is well known that $\bar{P} - \imath I$ is injective. □

To formulate the following results, it is convenient to use the class of "spectrally invariant" algebras.

DEFINITION 2.3. Let $\mathcal{A}$ be an algebra of bounded operators on some Hilbert space $\mathcal{H}$. We say that $\mathcal{A}$ is *spectrally invariant* if, and only if, $(I + T)^{-1} \in I + \mathcal{A}$, for any $T \in \mathcal{A}$ such that $I + T$ is invertible as an operator on $\mathcal{H}$.

The following theorem is the main result of this section. It states that $\Psi_{\mathrm{ai}}^\infty(M)$ is spectrally invariant in a certain sense.

THEOREM 2.4. *Let $T \in \Psi_{ai}^m(M)$, $m \geq 0$, be such that T is invertible as a (possibly unbounded) operator on $L^2(M)$. If $m > 0$, we assume also that T is elliptic. Then $T^{-1} \in \Psi_{ai}^{-m}(M)$.*

2.2. The "$\Delta + V$ trick". Recall that an operator L, mapping L^2_{loc} into distributions, is said to have the *unique continuation property* if $u = 0$, whenever $Lu = 0$ and u vanishes on some open subset.

PROPOSITION 2.5. *Let $L \in \Psi_{ai}^m(M)$ be non-negative (that is, $L \geq 0$) and satisfy the unique continuation property. Also, let $V \in \mathcal{C}^\infty(M) \cap \Psi_{ai}^0(M)$ (that is, V is translation invariant in a neighborhood of infinity), $V \geq 0$, such that V is strictly positive on some open subset of M. Assume $L + V \colon H^m(M) \to L^2(M)$ is Fredholm. Then $L + V$ is in fact invertible.*

PROOF. The assumptions $L \geq 0$ and $V \geq 0$ imply $L + V \geq 0$, as well. Assume by contradiction that $L + V \colon H^m(M) \to L^2(M)$ is Fredholm but not invertible. This shows that 0 must be an eigenvalue of $L + V$. A simple calculation with inner products, using also the unique continuation property, shows that this is not possible. □

EXAMPLE 2.6. If $T \in \Psi_{ai}^k(M)$ has the unique continuation property then $L := T^*T$ satisfies the hypotheses of the above proposition (with $m = 2k$). In particular, this is the case for $\Delta_M = d_M^* d_M$, since the kernel of d_M consists of only locally constant functions.

The following theorem is crucial for our approach to extending the method of layer potentials to manifolds with cylindrical ends.

THEOREM 2.7. *Let M be a manifold with cylindrical ends and $V \geq 0$ be a smooth function on M that is translation invariant in a neighborhood of infinity and does not vanish at infinity. Denote by Δ_M the Laplace operator on M. Then $\Delta_M + V$ is invertible as an unbounded operator on $L^2(M)$ and $(\Delta_M + V)^{-1} \in \Psi_{ai}^{-2}(M)$.*

PROOF. For starters, Δ_M is non-negative ($\Delta_M \geq 0$) and has the unique continuation property (cf. the previous example). Since the potential V is non-negative, as well as strictly positive on some non-empty open set, our result will follow from Proposition 2.5 as soon as we show that $\Delta_M + V \colon H^2(M) \to L^2(M)$ is Fredholm.

Since Δ_M is elliptic, $P := \Delta_M + V \colon H^2(M) \to L^2(M)$ will be Fredholm if, and only if, $\tilde{P}$ is invertible. In turn, to show that $\tilde{P}$ is invertible it suffices to prove the norm of the inverse of $\hat{P}(\tau) \colon H^2(\partial M_1) \to L^2(\partial M_1)$ is bounded uniformly in $\tau \in \mathbb{R}$.

More specifically, let $V_\infty \in \mathcal{C}^\infty(\partial M_1)$ be the limit at infinity of the function V. (This limit exists because we assumed V to be translation invariant in a neighborhood of infinity.) Denote $\Delta = \Delta_{\partial M_1}$, to simplify notation in what follows. By definition, we have $\hat{P}(\tau) = \Delta + \tau^2 + V_\infty$. Since $V_\infty + \tau^2 \geq 0$ and does not vanish identically for any $\tau \in \mathbb{R}$, by assumption, we obtain as in [**16**] that $\hat{P}(\tau)$ is indeed invertible for any $\tau \in \mathbb{R}$. A standard calculation also shows that the norm of the inverse is uniformly bounded.

This can be seen for example as follows. Assume that P is elliptic of order m. Then there exists $R > 0$ such that $\hat{P}(\tau)$ is invertible as a map $H^m(\partial M_1) \to L^2(\partial M_1)$, for any $|\tau| > R$. Moreover, $\hat{P}(\tau)^{-1}$ depends continuously on τ on its domain of definition. See [**21**], especially Theorem 9.2, for details. □

3. Boundary layer potential integrals

We want to extend the method of boundary layer potential to manifolds with cylindrical ends. We begin by introducing the class of manifolds with boundary that we plan to study in this paper.

3.1. Submanifolds with cylindrical ends. Let $N \subset M$ be a submanifold with boundary of a manifold with cylindrical ends. We want to generalize the method of layer potentials to this non-compact case. We notice that N plays a role in the method of boundary layer potentials mostly through its boundary ∂N. Because of this, we shall formulate some of our results in the slightly more general setting when ∂N is replaced by a suitable submanifold of codimension one. This slightly more general setting actually allows for simpler proofs.

DEFINITION 3.1. Let $M = M_1 \cup (\partial M_1 \times (-\infty, 0])$ be a manifold with cylindrical ends. A submanifold with cylindrical ends of M is a submanifold $Z \subset M$ such that

$$Z \cap (\partial M_1 \times (-\infty, 0]) = Z' \times (-\infty, 0],$$

for some submanifold $Z' \subset \partial M_1$. We shall write then $Z \sim Z' \times (-\infty, 0]$.

We shall fix Z, Z' as above in what follows. Our main interest is of course when $Z = \partial N$, but for certain reasonings, it is useful to allow this slightly greater level of generality.

Let us recall from [**25**, vol. II, Proposition 2.8], that a distribution L on $\mathbb{R}^n \times \mathbb{R}^n$ is the kernel of a classical pseudodifferential operator of order $-j$, $j = 1, 2, \ldots$, if, and only if,

$$L \sim \sum_{l=0}^{\infty} (q_l(x, z) + p_l(x, z) \ln |z|) \tag{3.1}$$

where q_l are smooth functions of x with values distributions in z that are homogeneous of degree $j+l-n$ and smooth for $z \neq 0$, and p_l are polynomials homogeneous of degree $j + l - n$. (The sign "$\sim$" in Eq. (3.1) above means that the difference $L - \sum_{l=0}^{N}(q_l(x, z) + p_l(x, z) \ln |z|)$ is as smooth as we want if N is chosen large enough.)

It is not difficult to check that the converse holds true also for $j = 0$ under some additional conditions, for example when $p_0 = 0$ and $q_0(x, z)$ is odd in z and the associated distribution is defined by a principal value integral. Below, we shall identify M with the diagonal of $M \times M$.

THEOREM 3.2. *Let M be a manifold with cylindrical ends and let $Z \subset M$ be a codimension one submanifold with cylindrical ends, as in* Definition 3.1. *If $P \in \Psi^m_{\rm inv}(M)$, $m < -1$, is given by the kernel $K \in \mathcal{C}^\infty(M^2 \setminus M)$, then the restriction of K to $Z^2 \setminus Z$ extends uniquely to the kernel of an operator $P_Z \in \Psi^{m+1}_{\rm inv}(Z)$. The same result holds true with $\Psi^m_{\rm ai}(M)$ and $\Psi^{m+1}_{\rm ai}(Z)$ replacing $\Psi^m_{\rm inv}(M)$ and $\Psi^{m+1}_{\rm inv}(Z)$.*

Moreover, if $\sigma_m(P)$ is odd, then we can also allow $m = -1$, provided that we define P_Z by using a principal value integral.

PROOF. Let $P \in \Psi^m_{\rm inv}(M)$. Then K is supported in a set of the form

$$V_\epsilon := \{(x, y) \in M^2,\ \mathrm{dist}(x, y) < \epsilon\},$$

by Definition 1.1. Clearly the restriction of K to $Z^2 \setminus Z$ will be supported in $V_\epsilon \cap Z^2$. Moreover, by using the expansion recalled above, Eq. (3.1), we see that $K|_{Z\times Z}$ is the kernel of a unique pseudodifferential operator on Z of order $\leq m+1$. The translation invariance of this operator follows from the definition.

To prove the same result for operators that are almost translation invariant in a neighborhood of infinity, it is enough to do this for order $-\infty$ operators. Because the proof is local in a neighborhood of Z, we can assume that $M = Z \times S^1$, in which case

$$\Psi^{-\infty}_{\rm ai}(M) \simeq \Psi^{-\infty}_{\rm ai}(Z) \hat{\otimes} \mathcal{C}^\infty(S^1). \tag{3.2}$$ □

We need now to investigate the relation between restriction to the submanifold Z of codimension one in M and indicial operators.

PROPOSITION 3.3. *Let $Z \subset M$ be as in* Definition 3.1, *with Z of codimension one, $Z \sim Z' \times (-\infty, 0]$ in a neighborhood of infinity. Let $P \in \Psi^m_{\rm ai}(M)$, $m \leq -1$. Then $\tilde{P}_{Z'\times\mathbb{R}} = \widetilde{P_Z}$ and $[\hat{P}(\tau)]_{Z'} = \widehat{P_Z}(\tau)$.*

3.2. Boundary layer potential integrals. We now proceed to define the boundary layer potential integrals. Let M be a manifold with cylindrical ends and $Z \subset M$ be a submanifold with cylindrical ends *of codimension one.* (Later on we shall restrict ourselves to the case when $Z = \partial N$, where $N \subset M$ is a submanifold with boundary and cylindrical ends. For now though, it is more convenient to consider this more general case.)

Let δ_Z be the surface measure on Z, regarded as a distribution on M. If $f \in L^2(Z)$, then $f \otimes \delta_Z \in H^{-a}(M)$, $a > 1/2$. Similarly, if δ'_Z is the normal derivative of δ_Z, then $f \otimes \delta'_Z \in H^{-a-1}(M)$, $a > 1/2$.

DEFINITION 3.4. Fix a smooth function $V \geq 0$, $V \in \Psi^0_{\rm inv}(M)$, V not identical equal to 0, but vanishing on a neighborhood of Z. Let $f \in L^2(Z)$ and $a > 1/2$. The *single layer potential integral associated to $Z \subset M$ and $\Delta_M + V$* is defined as

$$\mathcal{S}(f) := (\Delta_M + V)^{-1}(f \otimes \delta_Z) \in H^{2-a}(M),$$

and the *double layer potential integral associated to $Z \subset M$ and V* is defined as

$$\mathcal{D}(f) := (\Delta_M + V)^{-1}(f \otimes \delta'_Z) \in H^{1-a}(M).$$

Assume that the normal bundle of Z in M is oriented (so there will be a *positive side* and *negative side* of Z in M). As in [**16**] we shall denote by $f_\pm$ the non-tangential limits of some function defined on $M \setminus Z$, when we approach Z from the positive side (+), respectively from the negative side (−), provided, of course, that these limits exist pointwise almost everywhere. (It is here where we need the normal bundle to Z to be oriented.)

We now begin to follow the strategy of [**16**]. Let

$$S := [(\Delta_M + V)^{-1}]_Z \in \Psi_{\mathrm{ai}}^{-1}(Z). \tag{3.3}$$

We shall fix in what follows a vector field ∂_ν on M that is normal to Z at every point of Z. The principal symbol of the order -1 operator $(\Delta + V)^{-1}\partial_\nu^*$ is odd, so we can also define

$$K := [(\Delta_M + V)^{-1}\partial_\nu^*]_Z \in \Psi_{\mathrm{ai}}^{0}(Z). \tag{3.4}$$

PROPOSITION 3.5. *With the above notation, the operator S of* Eq. (3.3) *is elliptic. Moreover, the zero principal symbol of K vanishes, $\sigma_0(K) = 0$, and hence actually $K \in \Psi_{\mathrm{ai}}^{-1}(Z)$.*

PROOF. First, the fact that S is elliptic follows from a symbol calculation (which is local in nature) analogous to [**18**, (3.42), p. 33]. In fact, similar considerations show that $\sigma_0(K) = 0$ so, in fact, $K \in \Psi_{\mathrm{ai}}^{-1}(Z)$. See also the discussion in [**25**, vol. II, Proposition 11.2, p. 36]. □

THEOREM 3.6. *Let $Z \subset M$ be a codimension one submanifold with cylindrical ends. Assume the normal bundle to Z is oriented. Given $f \in L^2(Z)$, we have*

$$\mathcal{S}(f)_+ = \mathcal{S}(f)_- = Sf$$

as pointwise a.e. limits. Also, using the notation of Eq. (3.4) *above, we have*

$$\partial_\nu \mathcal{S}(f)_\pm = \left(\pm\tfrac{1}{2}I + K^*\right)f,$$

where K^* is the adjoint of K.

PROOF. Let us write $T := (\Delta_M + V)^{-1} = P + R$, where $P \in \Psi_{\mathrm{inv}}^{m}(M)$ (so it is translation invariant in a neighborhood of infinity) and $R \in \Psi_{\mathrm{ai}}^{-\infty}(M)$. The first statement of the proposition, namely

$$[T(f \otimes \delta_Z)]_\pm = T_Z f$$

is clearly linear in $T \in \Psi_{\mathrm{ai}}^{m}(M)$, $m < -1$. It is enough then to prove it for P and R separately.

For $T = (\Delta + V)^{-1}$ replaced by P, this is a local statement, which then follows from [**16**, Proposition 3.8]. For T replaced by R, the proof is local in a tubular neighborhood of Z, so we can assume that $M = Z \times S^1$. Then the result follows again from Eq. (3.2). □

The following theorem is proved in a completely similar way, following the results of [**16**, Proposition 3.8].

THEOREM 3.7. *Let Z be a codimension one submanifold with cylindrical ends of M with oriented normal bundle. Given $f \in L^2(Z)$, we have*

$$\mathcal{D}(f)_\pm = \left(\mp\tfrac{1}{2}I + K\right)f$$

as pointwise a.e. limits.

For further reference, let us discuss now the "trace theorem" for codimension one submanifolds in our setting. See [**2**] for more details and results of this kind for manifolds with a Lie structure at infinity [**1**].

PROPOSITION 3.8. *Let $Z \subset M$ be a submanifolds with cylindrical ends of the manifold with cylindrical ends M. Then the restriction map $\mathcal{C}_c^\infty(M) \to \mathcal{C}_c^\infty(Z)$ extends to a continuous map $H^s(M) \to H^{s-1/2}(Z)$, for any $s > 1/2$.*

4. Layer potentials depending on a parameter

The aim of this section is to investigate the invertibility of layer potential operators which depend on a parameter $\tau \in \mathbb{R}$, via a method initially developed by G. Verchota in [**27**], for the case of the flat-space Laplacian. The novelty here is to derive estimates which are *uniform* with respect to the real parameter τ.

Let M be a smooth, *compact, boundaryless* Riemannian manifold, and fix a reasonably regular subdomain $\Omega \subset M$ (Lipschitz will do). Here, M will play the role of the boundary ∂M_1 of a manifold with boundary M_1 (in our standard notation M_1 is used to define manifolds with cylindrical ends). Anticipating notation introduced in the next section, Ω will play the role of the *exterior* of X.

Set ν for the outward unit conormal to Ω and $d\sigma$ for the surface measure on $\partial\Omega$ (naturally inherited from the metric on M). The departure point is the following Rellich type identity (cf. the discussion in [**16**]):

$$\begin{aligned}(4.1)\quad \int_{\partial\Omega} \langle \nu, w\rangle \big\{|\nabla_{\tan} u|^2 - |\partial_\nu u|^2\big\}\, d\sigma &= 2\mathrm{Re}\int_{\partial\Omega} \langle w_{\tan}, \nabla u\rangle \partial_\nu \bar{u}\, d\sigma + 2\mathrm{Re}\int_\Omega \langle \nabla\bar{u}, w\rangle \Delta_M u\, dx \\ &\quad + \mathrm{Re}\int_\Omega \big\{(\operatorname{div} w)|\nabla u|^2 + 2(\mathcal{L}_w g)(\nabla u, \nabla\bar{u})\big\}\, dx,\end{aligned}$$

which is valid for a (possibly complex-valued) scalar function u and a real-valued vector field w (both sufficiently smooth, otherwise arbitrary) in Ω. Here, $\mathcal{L}_w g$ stands for the Lie derivative of the metric tensor g with respect to the field w, and the subscript 'tan' denotes the tangential component relative to $\partial\Omega$.

Another general identity (itself a simple consequence of the divergence theorem) that is useful here is

$$\int_{\partial\Omega} |u|^2 \langle w, \nu\rangle \, d\sigma = \operatorname{Re} \int_{\Omega} \{2u\langle \nabla \bar{u}, w\rangle + (\operatorname{div} w)|u|^2\} \, dx. \tag{4.2}$$

To proceed, fix a non-negative scalar potential $W \in C^\infty(M)$ and for the remainder of this subsection assume that

$$(\Delta_M + \tau^2 + W)u = 0 \quad \text{in} \quad \Omega, \tag{4.3}$$

where $\tau \in \mathbb{R}$ is an arbitrary parameter (fixed for the moment). Our immediate objective is to show that

$$\int_{\partial\Omega} |\partial_\nu u|^2 \, d\sigma \leq C \int_{\partial\Omega} \{|\nabla_{\tan} u|^2 + (1+\tau^2)\,|u|^2\} \, d\sigma, \tag{4.4}$$

uniformly in τ, and that for each $\varepsilon > 0$ there exists a finite constant $C = C(\Omega, \varepsilon) > 0$ so that

$$\int_{\partial\Omega} \{|\nabla_{\tan} u|^2 + \tau^2 |u|^2\} \, d\sigma \leq C \int_{\partial\Omega} |\partial_\nu u|^2 \, d\sigma + \varepsilon \int_{\partial\Omega} |u|^2 \, d\sigma \tag{4.5}$$

uniformly in the parameter $\tau \in \mathbb{R}$. We shall also need a strengthened version of (4.5) to the effect that

$$W > 0 \text{ in } \Omega \Longrightarrow \int_{\partial\Omega} \{|\nabla_{\tan} u|^2 + (1+\tau^2)\,|u|^2\} \, d\sigma \leq C \int_{\partial\Omega} |\partial_\nu u|^2 \, d\sigma \tag{4.6}$$

uniformly in the parameter $\tau \in \mathbb{R}$.

With an eye on (4.5), let us recall Green's first identity for the function u that we assumed to satisfy Eq. (4.3)

$$\int_{\Omega} \{|\nabla u|^2 + \tau^2 |u|^2 + W|u|^2\} \, dx = \operatorname{Re} \int_{\partial\Omega} \bar{u}\, \partial_\nu u \, d\sigma,$$

which readily yields the energy estimate

$$\int_{\Omega} \{|\nabla u|^2 + \tau^2 |u|^2 + W|u|^2\} \, dx \leq \int_{\partial\Omega} |u|\, |\partial_\nu u| \, d\sigma. \tag{4.7}$$

In turn, this further entails

$$\int_{\Omega} \tau^2 |\nabla u|\, |u| \, dx \leq C|\tau| \int_{\Omega} \{\tau^2 |u|^2 + |\nabla u|^2\} \, dx \leq C|\tau| \int_{\partial\Omega} |u|\, |\partial_\nu u| \, d\sigma, \tag{4.8}$$

uniformly in τ.

Let us now select w to be transversal to $\partial\Omega$, i.e.

$$\operatorname{ess\,inf} \langle w, \nu\rangle > 0 \text{ on } \partial\Omega, \tag{4.9}$$

something which can always be arranged given that $\partial\Omega$ is assumed to be Lipschitz. This, in concert with (4.2), then gives

$$\int_{\partial\Omega} |u|^2 \, d\sigma \leq C \int_{\Omega} \{|u|^2 + |\nabla u|\, |u|\} \, dx. \tag{4.10}$$

Multiplying (4.10) with τ^2 and then invoking (4.7)-(4.8) eventually justifies the estimate

$$\int_{\partial\Omega} \tau^2 |u|^2 \, d\sigma \leq C \int_{\partial\Omega} (1+|\tau|) \, |\partial_\nu u| \, |u| \, d\sigma. \tag{4.11}$$

Next, we make the (elementary) observation that for every $\varepsilon, \delta > 0$ there exists $C = C(\varepsilon, \delta) > 0$ so that

$$(1+|\tau|) \, |\partial_\nu u| \, |u| \leq \delta \tau^2 |u|^2 + C \, |\partial_\nu u|^2 + \varepsilon |u|^2, \tag{4.12}$$

uniformly in τ. When considered in the context of (4.11), the boundary integral produced by the first term in the right side of (4.12) can be absorbed in the left side of (4.11), provided δ is sufficiently small. Thus, with this alteration in mind, (4.11) becomes

$$\int_{\partial\Omega} \tau^2 |u|^2 \, d\sigma \leq C \int_{\partial\Omega} |\partial_\nu u|^2 \, d\sigma + \varepsilon \int_{\partial\Omega} |u|^2 \, d\sigma,$$

which is certainly in the spirit of (4.5). In fact, in order to fully prove the latter estimate, there remains to control the tangential gradient in a similar fashion. To this end, observe that (4.9) and Rellich's identity (4.1) give

$$\int_{\partial\Omega} |\nabla_{\tan} u|^2 \, d\sigma \leq C \int_{\partial\Omega} |\partial_\nu u|^2 \, d\sigma + C \int_\Omega \tau^2 |\nabla u| \, |u| \, dx + C \int_\Omega |\nabla u|^2 \, dx,$$

uniformly in τ. With this at hand, the same type of estimates employed before can be used once again to further bound the solid integrals in terms of (suitable) boundary integrals. The bottom line is that

$$\int_{\partial\Omega} |\nabla_{\tan} u|^2 \, d\sigma \leq C \int_{\partial\Omega} |\partial_\nu u|^2 \, d\sigma + \varepsilon \int_{\partial\Omega} |u|^2 \, d\sigma, \tag{4.13}$$

uniformly in τ, and (4.5) follows.

It is now easy to prove (4.6), having disposed off (4.5). One useful ingredient in this regard is

$$\int_\Omega |u|^2 \, dx \leq C \int_\Omega \{|\nabla u|^2 + W|u|^2\} \, dx, \tag{4.14}$$

itself a version of Poincaré's inequality. When used in conjunction with (4.7) and (4.10), this readily yields

$$\int_{\partial\Omega} |u|^2 \, d\sigma \leq C \int_\Omega \{|\nabla u|^2 + W|u|^2\} \, dx \leq C \int_{\partial\Omega} |u| \, |\partial_\nu u| \, d\sigma \tag{4.15}$$

so that, ultimately,

$$\int_{\partial\Omega} |u|^2 \, d\sigma \leq C \int_{\partial\Omega} |\partial_\nu u|^2 \, d\sigma, \tag{4.16}$$

in the case we are currently considering. In concert with (4.5), this concludes the proof of (4.6). Let us now turn our attention to the estimate (4.4). For starters,

Rellich's identity (4.1) can also be employed, along with the condition (4.9), to produce

$$(4.17)\quad \int_{\partial\Omega} |\partial_\nu u|^2\, d\sigma \leq C \int_{\partial\Omega} |\nabla_{\tan} u|^2\, d\sigma + C \int_\Omega \tau^2 |\nabla u|\,|u|\, dx + C \int_\Omega |\nabla u|^2\, dx,$$

uniformly in τ. Then, much as before,

$$\begin{aligned} \int_\Omega \{\tau^2|\nabla u|\,|u| + |\nabla u|^2\}\, dx &\leq C \int_{\partial\Omega} (1+|\tau|)\, |\partial_\nu u|\,|u|\, d\sigma \\ &\leq \delta \int_{\partial\Omega} |\partial_\nu u|^2\, d\sigma + C \int_{\partial\Omega} (1+\tau^2)\, |u|^2\, d\sigma, \end{aligned} \quad (4.18)$$

where $\delta > 0$ is chosen small and C depends only on Ω and δ. With these two estimates at hand, the endgame in the proof of (4.4) is clear.

After these preliminaries, we can finally address the main theme of this subsection. More concretely, for each $\tau \in \mathbb{R}$, let S_τ, K_τ be, respectively, the single and the double layer potential operators associated with $\Delta_M + \tau^2 + W$ on $\partial\Omega$. From the work in [**16**], it is known that if Ω has a Lipschitz boundary then both

$$S_\tau\colon L^2(\partial\Omega) \to H^1(\partial\Omega) \quad \text{and} \quad \tfrac{1}{2}I + K_\tau\colon L^2(\partial\Omega) \to L^2(\partial\Omega)$$

are invertible operators for each $\tau \in \mathbb{R}$. Our objective is to study how the norms of their inverses depend on the parameter τ. To discuss this issue, for each $\tau \in \mathbb{R}$ and $f \in H^1(\partial\Omega)$, set

$$(4.19)\quad \|f\|_{H^1_\tau(\partial\Omega)} := \|f\|_{H^1(\partial\Omega)} + |\tau|\, \|f\|_{L^2(\partial\Omega)}.$$

Thus, $\mathbb{R} \ni \tau \mapsto \|\cdot\|_{H^1_\tau(\partial\Omega)}$ is a one-parameter family of equivalent norms on the Sobolev space $H^1(\partial\Omega)$. A natural way to come up with this family of seminorms is by studying the Sobolev spaces on $H^1(\Omega \times \mathbb{R})$. The main result of this subsection is as follows.

PROPOSITION 4.1. *Assume that Ω is a fixed, Lipschitz subdomain of M. Then there exits a finite constant $C = C(\partial\Omega) > 0$, depending exclusively on the Lipschitz character of Ω, such that for each $\tau \in \mathbb{R}$, we have*

$$(4.20)\quad \|S_\tau^{-1} f\|_{L^2(\partial\Omega)} \leq C\, \|f\|_{H^1_\tau(\partial\Omega)}$$

uniformly for $f \in H^1(\partial\Omega)$.

Furthermore, if $W > 0$ on a set of positive measure in Ω, then for any $\tau \in \mathbb{R}$ we also have

$$(4.21)\quad \left\|(\tfrac{1}{2}I + K_\tau)^{-1} f\right\|_{L^2(\partial\Omega)} \leq C\, \|f\|_{L^2(\partial\Omega)},$$

uniformly for $f \in L^2(\partial\Omega)$.

PROOF. Consider first (4.21). Let $\Omega_+ := \Omega$, $\Omega_- := M \setminus \bar{\Omega}$. For $f \in L^2(\partial\Omega)$, set $u := \mathcal{S}f$ in $\Omega_\pm$. Thus,

$$(4.22)\quad (u)_+ = (u)_-, \quad (\nabla_{\tan} u)_+ = (\nabla_{\tan} u)_-, \quad (\partial_\nu u)_\pm = \left(\pm\tfrac{1}{2}I + K_\tau^*\right) f.$$

In turn, (4.22), (4.4) and (4.6) allow us to write

$$\begin{aligned}\left\|(-\tfrac{1}{2}I + K_\tau^*)f\right\|_{L^2(\partial\Omega)} &= \|(\partial_\nu u)_-\|_{L^2(\partial\Omega)} \leq C\,\|(u)_-\|_{H^1_\tau(\partial\Omega)} \\ &= C\,\|(u)_+\|_{H^1_\tau(\partial\Omega)} \leq C\,\|(\partial_\nu u)_+\|_{L^2(\partial\Omega)} = C\left\|(\tfrac{1}{2}I + K_\tau^*)f\right\|_{L^2(\partial\Omega)}.\end{aligned}$$

Consequently,

$$\begin{aligned}\|f\|_{L^2(\partial\Omega)} &\leq \left\|(-\tfrac{1}{2}I + K_\tau^*)f\right\|_{L^2(\partial\Omega)} + \left\|(\tfrac{1}{2}I + K_\tau^*)f\right\|_{L^2(\partial\Omega)} \\ &\leq C\left\|(\tfrac{1}{2}I + K_\tau^*)f\right\|_{L^2(\partial\Omega)}\end{aligned} \tag{4.23}$$

for some constant $C = C(\partial\Omega) > 0$ independent of τ. Going further, if $\mathcal{L}(X) := \mathcal{L}(X, X)$, the normed algebra of all bounded operators on a Banach space X, then (4.23) entails

$$\left\|(\tfrac{1}{2}I + K_\tau)^{-1}\right\|_{\mathcal{L}(L^2(\partial\Omega))} = \left\|(\tfrac{1}{2}I + K_\tau^*)^{-1}\right\|_{\mathcal{L}(L^2(\partial\Omega))} \leq C.$$

This takes care of (4.21).

As for (4.20), the argument is rather similar, the main step being the derivation of the estimate

$$\|f\|_{L^2(\partial\Omega)} \leq C\,\|\nabla_{\tan}(S_\tau f)\|_{L^2(\partial\Omega)} + C(1 + |\tau|)\,\|S_\tau f\|_{L^2(\partial\Omega)},$$

out of (4.22) and (4.4), when the latter is written both for Ω_+ and Ω_-. Once again, the crux of the matter is that the intervening constant $C = C(\partial\Omega) > 0$ is independent of τ. The proof is finished. □

See also [**28, 29**].

5. The Dirichlet problem

We now apply the results we have established to solve the inhomogeneous Dirichlet problem on manifolds with boundary and cylindrical ends.

The class of manifolds with boundary and cylindrical ends that we consider have a product structure at infinity (including the boundary and the metric). It is possible to relax these conditions, but for simplicity we do not address this technical question in this paper.

DEFINITION 5.1. Let N be a Riemannian manifold with boundary ∂N. We shall say that N is a *manifold with boundary and cylindrical ends* if there exists an open subset V of N isometric to $(-\infty, 0) \times X$, where X is a compact manifold with boundary, such that $N \setminus V$ is compact.

LEMMA 5.2. *Let N be a Riemannian manifold with boundary ∂N. Then N is a manifold with boundary and cylindrical ends if, and only if, there exists a manifold with cylindrical ends (without boundary) M with a standard decomposition*

$$M = M_1 \cup \big(\partial M_1 \times (-\infty, 0]\big)$$

and containing N such that

$$N \cap \big(\partial M_1 \times (-\infty, 0]\big) = X \times (-\infty, 0],$$

for some compact manifold with boundary $X \subset \partial M_1$.

Let M and N be as above. We define $H^s(N)$ to be the space of restrictions to N of distributions $u \in H^s(M)$ (see also [**2**]). We want to prove that the map

$$H^s(N) \ni u \to (\Delta_N u, u|_{\partial N}) \in H^{s-2}(N) \oplus H^{s-1/2}(\partial N) \tag{5.1}$$

is an isomorphism for $s > 1/2$.

PROPOSITION 5.3. *Assume that the potential V is chosen so that V is not identically zero on $\partial M_1 \setminus \bar{X}$. Then the map $-\frac{1}{2}I + K^*\colon L^2(\partial N) \to L^2(\partial N)$ is injective.*

PROOF. Just follow word for word [**16**, Proposition 4.1]. □

Note that our signs are opposite to those in [**16**] or [**26**], because we use the definition that makes the Laplace operator *positive*.

To prove the Fredholm property of the operators $\frac{1}{2}I + K$ and $\frac{1}{2}I + K^*$, we need to slightly change the corresponding argument in [**16**].

PROPOSITION 5.4. *Retain the same assumptions as in* Proposition 5.3. *Then the operator*

$$-\tfrac{1}{2}I + K\colon L^2(\partial N) \to L^2(\partial N) \tag{5.2}$$

is Fredholm of index zero.

PROOF. The above proposition is known when M is compact ([**16**, Corollary 4.5]). To check that the operator in question is Fredholm, we shall rely on (iv) in Theorem 2.1, which amounts to studying the associated indicial family.

Let $W := V_{\partial M_1}$ and $T = (\Delta_M + V)^{-1}\partial_\nu^*$. Recall that $K := T_{\partial N}$ and that $\partial N \sim \partial X \times (-\infty, 0]$ in a neighborhood of infinity. Then Proposition 3.3 gives

$$\hat{K}(\tau) = \widehat{T_{\partial N}}(\tau) = [\hat{T}(\tau)]_{\partial X} = [(\Delta_{\partial M_1} + \tau^2 + W)^{-1}\partial_\nu^*]_{\partial X} = K_\tau, \tag{5.3}$$

where K_τ is the double layer potential operator associated with the perturbed Laplacian $\Delta_{\partial M_1} + \tau^2 + W$ on ∂X (cf. the discussion in §3.3).

Our result is then true pointwise, because of the result on compact manifolds. We only need to prove that it is uniform in τ. This can be achieved as in the previous section, or by using [**21**, Theorem 9.2]. □

COROLLARY 5.5. *Let V be as before. Then the operator*

$$-\tfrac{1}{2}I + K\colon H^s(\partial N) \to H^s(\partial N)$$

is invertible for each $s \in \mathbb{R}$.

PROOF. To begin with, the case $s = 0$ is easily proved by putting together the above two propositions. In particular, the operator $-\frac{1}{2}I+K\colon H^s(\partial N) \to H^s(\partial N)$, in the statement of this corollary, is injective for each $s \geq 0$. Since the fact that this operator is also surjective is a consequence of the corresponding claim in the case $s = 0$ and the elliptic regularity, the desired conclusion follows for $s \geq 0$. As for the case $s < 0$, a similar reasoning shows that

$$-\tfrac{1}{2}I + K^*\colon H^{-s}(\partial N) \to H^{-s}(\partial N) \tag{5.4}$$

is invertible for each $s < 0$. This and duality then yield the invertibility of $-\frac{1}{2}I + K\colon H^s(\partial N) \to H^s(\partial N)$ for $s < 0$, as wanted. □

Another proof of the above result can be obtained from Theorem 2.4, for the case $m = 0$, the "easy one."

After these preliminaries, we are finally in a position to discuss the following basic result.

THEOREM 5.6. *For any $s > 0$ and any $f \in H^s(\partial N)$, there exists a unique function $u \in H^{s+1/2}(N)$ such that $u|_{\partial N} = f$ and $\Delta_N u = 0$.*

PROOF. Fix a potential V which is not identically zero on $\partial M_1 \setminus \bar{X}$, yet $V = 0$ near $\bar{X}$. In particular, the conclusion in Corollary 5.5 holds. First we claim that

$$\mathcal{D}\colon H^s(\partial N) \to H^{s+1/2}(M), \qquad s \in \mathbb{R}, \tag{5.5}$$

is well-defined and bounded. Indeed, if $s < 0$, then this is a consequence of the implication

$$f \in H^s(\partial N),\ s < 0 \implies f \otimes \delta_{\partial N} \in H^{s-1/2}(\partial N) \tag{5.6}$$

along with the factorization $\mathcal{D}(g) = (\Delta_M + V)^{-1}\partial_\nu^*(g \otimes \delta_{\partial N})$. For $s = 0$, one can employ the techniques of [**18**]. The case $s > 0$ then follows inductively from what we have proved so far with the aid of a commutator identity which essentially reads $\nabla\mathcal{D}f = \mathcal{D}(\nabla_{\tan} f) +$ lower order terms; see (8.19) in [**17**] as well as (6.17) in [**18**].

Having disposed off (5.5) the existence part in the theorem is then easily addressed. Specifically, if $s > 0$, consider $g := (-\frac{1}{2}I + K)^{-1}f \in H^s(\partial N)$ and then set $u := \mathcal{D}(g) \in H^{s+1/2}(N)$ by (5.5). □

We are now ready to prove Theorem 0.2

PROOF. The continuity of the map $\tilde{\Delta}_N$ follows from the continuity of the map $\Delta_N\colon H^s(N) \to H^{s-2}(N)$ and from the continuity of the trace map $H^s(N) \to H^{s-1/2}(\partial N)$.

As before, we fix a potential V which vanishes in a neighborhood of N. Let $g \in H^{s-2}(N)$ be arbitrary. First extend g to a distribution (denoted also g) in $H^{s-2}(M)$, then set $u_1 = (\Delta + V)^{-1}g \in H^s(M)$ and $f_1 = u_1|_{\partial N} \in H^{s-1/2}$. Finally, choose $u_2 \in H^s(N)$ such that $\Delta_N u_2 = 0$ and $u_2|_{\partial N} = f - f_1$. Then $u := u_1 + u_2$ satisfies $\Delta_N u = g$ and $u|_{\partial N} = f$. This proves the surjectivity of $\tilde{\Delta}_N$. The injectivity of this map then follows from the uniqueness part in Theorem 5.6. □

It is likely that some versions of the above two theorems extend to weighted Sobolev spaces. This will likely requires techniques similar to those used in [**5**].

5.1. The Dirichlet-to-Neumann map. Theorem 5.6 allows us to define the Dirichlet-to-Neumann map $\mathcal{N}$

$$\mathcal{N}(f) = (\partial_\nu u)_+$$

for $f \in L^2(\partial N)$ and u solution of $\Delta_N u = 0$, $u_+ := u|_{\partial N} = f$.

THEOREM 5.7. *Let N be a manifold with boundary and cylindrical ends. Then the operator $S\colon H^s(\partial N) \to H^{s+1}(\partial N)$ is invertible for any s and $\left(\frac{1}{2}I + K^*\right) S^{-1} = \mathcal{N}$, the "Dirichlet-to-Neumann map." In particular, $\mathcal{N} \in \Psi^1_{\mathrm{ai}}(\partial N)$.*

PROOF. The operator S is elliptic by Proposition 3.5. For further reference, let us note here that

$$f \in H^s(\partial N) \ \&\ Sf \in H^{s+1}(\partial N) \Longrightarrow f \in H^{s+1}(\partial N), \tag{5.7}$$

by elliptic regularity.

Next, using the notation of Proposition 4.1, we have $\hat{S}(\tau) = S_\tau$. By the results of the same proposition, $\hat{S}(\tau)$ is invertible for any τ, and the norm of the inverse is uniformly bounded (this can be proved also by using the results of [**16**] or [**18**] and the estimates in [**21**]). Consequently, $S\colon H^s(\partial N) \to H^{s+1}(\partial N)$ is Fredholm (cf. Theorem 2.1). Checking that S is injective when $s = 0$ is done much as in the last part of §6 in [**16**]. □

COROLLARY 5.8. *The Cauchy data space*

$$\{(u|_{\partial N}, \partial_\nu u|_{\partial N});\ u \in H^s(N)\}$$

is a closed subspace of $H^{s-1/2}(\partial N) \oplus H^{s-3/2}(\partial N)$ for any $s > 1/2$.

We conclude this section with yet another integral representation formula for the Dirichlet problem.

COROLLARY 5.9. *Retain the usual set of assumptions. Then, for each $s > 0$, the solution to the boundary problem*

$$u \in H^{s+1/2}(N), \quad \Delta_N u = 0, \quad u|_{\partial N} = f \in H^s(\partial N),$$

(*see* Theorem 5.6) *can also be expressed in the form $u = \mathcal{S}(S^{-1}f)$ in N.*

References

[1] B. Ammann, R. Lauter, and V. Nistor, *On the Riemannian geometry of manifolds with a Lie structure at infinity*, preprint, math.DG/0201202.

[2] B. Ammann, A. Ionescu, and V. Nistor, *Sobolev spaces on manifolds with a Lie structure at infinity*, in final preparation.

[3] C. Bacuta, V. Nistor, and L. Zikatanov, *On the rate of convergence of 'high order finite elements' on polygons and domains with cusps*, in final preparation.

[4] G. de Rham, *Differentiable manifolds*, Springer, 1984.

[5] J. Gil and G. Mendoza, *Adjoints of elliptic cone operators*, Amer. J. Math. **125** (2003), 357–408.

[6] W. V. D. Hodge, *The theory and applications of harmonic integrals*, Cambridge Univ. Press, 1941.

[7] K. Kodaira, *Harmonic fields in Riemannian manifolds (Generalized potential theory)*, Ann. of Math. **50** (1949), 587–665.

[8] J. J. Kohn and D. C. Spencer, *Complex Neumann problems*, Ann. of Math. **66** (1957), 89–140.

[9] V. A. Kondrat'ev, *Boundary value problems for elliptic equations in domains with conical or angular points*, Trans. Moscow Math. Soc. **16** (1967), 227–313.

[10] R. Lauter, B. Monthubert, and V. Nistor, *Pseudodifferential analysis on continuous family groupoids*, Doc. Math. **5** (2000), 625–655 (electronic).

[11] R. Lockhart and R. McOwen, *Elliptic differential operators on noncompact manifolds*, Ann. Scuola Norm. Sup. Pisa Cl. Sci (4) **12** (1985), 409–447.

[12] V. Mazya, S. A. Nazarov, and B. A. Plamenevskij, *Aymptotic theory of elliptic boundary value problems in singularly perturbed domains*, Operator Theory Advances Applications, Vol. 111, Birkäuser, Basel-Boston-Berlin, 2000.

[13] R. B. Melrose, *The Atiyah-Patodi-Singer index theorem*, Res. Notes Math., Vol. 4, A. K. Peters, MA, 1993.

[14] R. B. Melrose and G. Mendoza, *Elliptic operators of totally characteristic type*, MSRI Preprint 1983.

[15] R. B. Melrose and V. Nistor, *K-Theory of C^*-algebras of b-pseudodifferential operators*, Geom. Funct. Anal. **8** (1998), 99–122.

[16] M. Mitrea and M. E. Taylor, *Boundary layer methods for Lipschitz domains in Riemannian manifolds*, J. Funct. Anal. **163** (1999), 181–251.

[17] M. Mitrea and M. E. Taylor, *Potential theory on Lipschitz domains in Riemannian manifolds: Sobolev-Besov space results and the Poisson problem*, J. Funct. Anal. **176** (2000), 1–79.

[18] D. Mitrea, M. Mitrea, and M. Taylor, *Layer potentials, the Hodge Laplacian, and global boundary problems in non-smooth Riemannian manifolds*, Mem. Amer. Math. Soc., Vol. 713, 2001.

[19] S. A. Nazarov and B. A. Plamenevsky, *Elliptic problems in domains with piecewise smooth boundaries*, de Gruyter Exp. Math., Vol. 13, de Gruyter, New York, 1994.

[20] E. Schrohe, *Spectral invariance, ellipticity and the Fredholm property for pseudodifferential operators on weighted Sobolev spaces*, Ann. Global Anal. Geom. **10** (1992), 237–254.

[21] M. A. Shubin, *Pseudodifferential operators and spectral theory*, Springer, Berlin, 1987.

[22] E. Schrohe and B.-W. Schulze, *Boundary value problems in Boutet de Monvel's algebra for manifolds with conical singularities II*, Boundary Value Problems, Schrödinger Operators, Deformation Quantization, Math. Top., Vol. 8, Akademie Verlag, Berlin, 1995, 70–205.

[23] B.-W. Schulze, *Boundary value problems and singular pseudo-differential operators*, J. Wiley, Chichester, 1998.

[24] M. E. Taylor, *Pseudodifferential operators*, Princeton Math. Ser., Vol. 34, Princeton Univ. Press, Princeton, NJ, 1981.

[25] M. E. Taylor, *Partial differential equations I–III*, Appl. Math. Sci., Springer, New York, 1995–1997.

[26] M. E. Taylor, *Pseudodifferential operators, paradifferential operators, and layer potentials*, Math. Surveys Monogr., Vol. 81, Amer. Math. Soc., 2000.

[27] G. Verchota, *Layer potentials and boundary value problems for Laplace's equation in Lipschitz domains*, J. Funct. Anal., 59 (1984), 572–611.

[28] G. Verchota and A. Vogel, *A multidirectional Dirichlet problem*, preprint.

[29] G. Verchota and A. Vogel, *A multidirectional Neuman problem in $\mathbb{R}^4$*, preprint.

University of Missouri-Columbia, Dept. of Mathematics, Columbia, MO 65211
URL: www.math.missouri.edu/~marius
E-mail address: marius@math.missouri.edu

Pennsylvania State University, Math. Dept., University Park, PA 16802
URL: www.math.psu.edu/nistor
E-mail address: nistor@math.psu.edu

Operator Theory:
Advances and Applications, Vol. 151, 495–560

Relative elliptic theory

Vladimir Nazaikinskii and Boris Sternin

ABSTRACT. This paper is a review of relative elliptic theory (i.e., elliptic theory in the category of smooth embeddings) closely related to the Sobolev problem, first studied by Sternin in the 1960s. We consider both analytic aspects of the theory (the structure of the algebra of morphisms, ellipticity, Fredholm property) and topological aspects (index formulas and Riemann–Roch theorems). We also study the algebra of Green operators arising as a subalgebra of the algebra of morphisms.

Introduction

1. By relative elliptic theory we mean elliptic theory associated with a pair (M, X), where M is a smooth closed manifold and X a submanifold. Thus, relative elliptic theory is elliptic theory in the category of smooth embeddings.

A trivial (and non-interesting) example of a relative elliptic operator is given by a pair of elliptic operators on the manifolds M and X, respectively. Such an operator can be represented in the form of a diagonal matrix operator

$$\begin{pmatrix} D_M & 0 \\ 0 & D_X \end{pmatrix} : \begin{matrix} H^{s_1}(M) \\ \oplus \\ H^{s_2}(M) \end{matrix} \longrightarrow \begin{matrix} H^{\sigma_1}(X) \\ \oplus \\ H^{\sigma_2}(X) \end{matrix}.$$

with elliptic (pseudodifferential) operators D_M and D_X acting, say, in Sobolev spaces on M and X.

This trivial example, however, suggests that in the general case a relative elliptic operator should apparently be represented by a matrix of the form

$$\begin{pmatrix} D_{MM} & D_{MX} \\ D_{XM} & D_{XX} \end{pmatrix} \tag{0.1}$$

(where the subscripts show, from the right to the left, the direction of action of the corresponding operators), which is not necessarily diagonal. First, we note that *a*

2000 *Mathematics Subject Classification.* Primary: 58J05; Secondary: 58J20, 58J32, 58J40.

The work was supported by the DFG via a project with the Arbeitsgruppe "Partielle Differentialgleichungen und Komplexe Analysis," Institut für Mathematik, Universität Potsdam and by RFBR grants 03-02-16336, 02-01-00118, and 00-01-00161.

posteriori this guess proves to be true. However, it deserves a serious explanation. Indeed, the operators D_{MX} and D_{XM}, representing the off-diagonal entries of this matrix, act from a function space on one manifold to a function space on another manifold and hence, unlike D_{MM} and D_{XX}, are by no means pseudodifferential operators, at least if the term is understood in the classical sense. Thus, what is the nature of these operators?

Let us try to find some candidates for these operators. First, consider the operator D_{XM}. This operator acts from a function space on M into a function space on the submanifold X. The most natural operator of this kind is the restriction operator (or boundary operator) induced by the embedding $i\colon X \hookrightarrow M$. If one requires the set of operators of the form (0.1) be invariant with respect to the passage to the adjoint operator, then the operator D_{MX} must be the adjoint of D_{XM}, that is, the operator of co-restriction, or, as it will be called in what follows, the coboundary operator. Thus, at first glance, our would-be theory is the theory of operators of the form

$$\begin{pmatrix} D_{MM} & i_* \\ i^* & D_{XX} \end{pmatrix}, \tag{0.2}$$

where i^* and i_* are the boundary and coboundary operators and D_{MM} and D_{XX} are pseudodifferential operators on the corresponding manifolds. Unfortunately, operators of the form (0.2) do not form an algebra, which is a serious disadvantage from the viewpoint of elliptic theory. Indeed, the proof of the Fredholm property of an elliptic operator is carried out in the most convenient and natural form by constructing so-called regularizers, that is, almost inverses (inverses modulo compact operators). To apply this method, one must have the structure of an algebra on the set of operators and an effective calculus.

Why do the operators (0.2) not form an algebra?

A straightforward multiplication of two matrices of the form (0.2) shows that if we wish the matrices in question to form an algebra, we must first of all extend the class of (co)boundary operators. For example, instead of the (elementary) restriction operator i^* occurring in (0.2), one should consider a general boundary operator of the form $B'_{XX} i^* B''_{MM}$, and the elementary coboundary operator i_* should be replaced by a general operator of the form $C'_{MM} i_* C''_{XX}$.

However, the main cause is essentially that the upper-left corner of the product of two operators of this type contains an operator of entirely different structure than a pseudodifferential operator. This is not surprising, since the factors occurring in this operator are not pseudodifferential operators.[1] Thus, it seems natural to extend the class of operators in the upper-left corner so that it will include not only pseudodifferential operators, but also operators of the form

$$D'_{MM} i_* D_{XX} i^* D''_{MM}$$

---

[1]Note, however, that the similar entry in the lower-right corner of the product is still a pseudodifferential operator.

(and sums of such operators), which occur in compositions. (It is remarkable that now, after an appropriate closure, the set of operators becomes closed with respect to composition, which in principle solves the problem of constructing a calculus.)

This important class of operators, which does not occur in the original statement, is the class of Green operators. These operators inevitably occur as products of boundary and coboundary operators as well as in the construction of an almost inverse operator in the Sobolev problem. (See point 2 below of this Introduction.) Similar operators (also known as Green operators[2]) arise in the theory of boundary value problems as well as in elliptic theory on manifolds with singularities (e.g., see [**Mon71, Sch91**]). In the present paper, we naturally discuss only Green operators associated with the pair (M, X). Green operators, as well as pseudodifferential operators, form an algebra: The product of two Green operators is itself a Green operator. In contrast with the algebra of pseudodifferential operators, the Green algebra is an algebra without unit: The identity operator is not a Green operator. One can attach the unit in the standard way and consider the inversion problem for operators of the form $1 + T$, where T is a Green operator, in the extended algebra. In other words, one studies the solvability of the equation

$$(1 + T)u = f \tag{0.3}$$

for an unknown function u on M, where f is a given function on M. Equation (0.3) will be called the *Green equation of the second kind.* It is of considerable interest owing to the fact that the operator $1 + T$ occurs in the ellipticity conditions for general operators associated with the pair (M, X).

The algebra of operators of the form "a pseudodifferential operators plus a Green operator" is just one possible extension of the algebra of pseudodifferential operators on M. Such extensions arise in the presence of some additional structures on M (in our case, the additional structure is the embedded submanifold X). Other examples of extensions (not covered in this survey) can be found in [**SS92a, SS92b, SS94**].

However, this is not the whole story. It is desirable to describe the newly introduced operators in known terms. Such a description would certainly be some kind of good luck, since it might well happen that these operators are of absolutely new nature unknown in the literature. Fortunately, this is not the case. These operators can be interpreted as *Fourier integral operators* on special Lagrangian manifolds. In this interpretation, each type of operators (boundary, coboundary, etc.) corresponds to its own Lagrangian manifold. These operators admit also an adequate interpretation in quite different terms in a neighborhood of the submanifold X.

[2]In fact, Green operators arise even in the simplest boundary value problems such as the Dirichlet problem with non-zero right-hand side f for the Laplace operator in a domain. One reduces this problem to a problem with zero right-hand side by subtracting the convolution of f with the Newtonian potential from the solution and then expresses the solution of the new problem as the simple layer potential corresponding to the resulting boundary conditions. The simple layer potential with density equal to the restriction to the boundary of the convolution of the right-hand side with the Newtonian potential is just the simplest Green operator applied to f.

Namely, they can be represented as a special class of pseudodifferential operators acting in sections of infinite-dimensional bundles over X (pseudodifferential operators with operator-valued symbols).

2. Relative elliptic theory allows one to solve an important, interesting problem in the theory of partial differential equations, known as the *Sobolev problem.*

To illustrate how the Sobolev problem arises, we consider simple physical examples. Suppose that we push a soap film spanning a one-dimensional contour with a thin needle. In the case of perfect non-wetting, the film collapses immediately unless we actually only touch it not trying to change its shape. In the case of perfect wetting, the needle passes through the film, whose shape remains unchanged. However, if we push a plate with the same needle, then the plate shape changes. In other words, *one cannot pose "boundary conditions" for the thin film equation at a single point, but such conditions are possible for the plate equation.*

Mathematically, this can be described as follows.

The function $z = z(x, y)$ describing the shape of a thin two-dimensional film spanning a closed one-dimensional contour $\{z = \varphi(x, y) \mid (x, y) \in \Gamma\}$ in the space $\mathbb{R}^3$ satisfies the Dirichlet problem

$$\begin{cases} \Delta z = 0, \\ z\big|_\Gamma = \varphi \end{cases} \tag{0.4}$$

in the domain $D \subset \mathbb{R}^2$ bounded by the plane curve Γ. Let us try to support the film with a thin needle at a point $(x_0, y_0, z_0) \in \mathbb{R}^3$ under the assumption of perfect non-wetting. Then the equation in problem (0.4) must be satisfied everywhere in D except for the point (x_0, y_0):

$$\Delta z \equiv 0 \mod (x_0, y_0). \tag{0.5}$$

(This notation means that the distribution Δu is supported at the point (x_0, y_0).) Next, we must equip problem (0.4) with the additional boundary condition

$$z\big|_{(x_0,y_0)} = z_0. \tag{0.6}$$

By the well-known removable singularity theorem for harmonic functions, every solution of the equation (0.5) in D bounded at the point (x_0, y_0) (the boundedness of $z(x, y)$ follows from (0.6)) is a harmonic function everywhere in D including the point (x_0, y_0) so that the value $z(x_0, y_0)$ is uniquely determined by the data of problem (0.4). Hence problem (0.5)–(0.6) is not solvable unless z_0 is equal to the value of the solution of problem (0.4) at that point.

In the case of perfect wetting, the equation in problem (0.4) still holds in $D \setminus \{(x_0, y_0)\}$, but no additional conditions are imposed at the point (x_0, y_0). By the removable singularity theorem, the problem

$$\begin{cases} \Delta z \equiv 0 \mod (x_0, y_0), \\ z\big|_\Gamma = \varphi \end{cases}$$

is equivalent to problem (0.4), which means that the needle passes through the film without changing the film shape.

The situation is entirely different for the elastic plate problem

$$\begin{cases} \Delta^2 z = 0, \\ z\big|_\Gamma = \varphi, \quad \dfrac{\partial z}{\partial n}\Big|_\Gamma = \psi. \end{cases}$$

Here $\partial/\partial n$ is the outward normal derivative on Γ. If we replace the equation in this problem by the equation

$$\Delta^2 z \equiv 0 \mod (x_0, y_0), \tag{0.7}$$

then an additional condition of the form (0.6) is admissible, since Eq. (0.7) has a non-trivial continuous solution behaving as $r^2 \ln r$ near (x_0, y_0), where r is the distance to that point.

Thus, we see that for equations of sufficiently high order one can pose problems involving "boundary conditions" on manifolds of codimension higher than one. Problems with such conditions are called *Sobolev problems*. The general statement of the Sobolev problem is as follows. Let $X \overset{i}{\hookrightarrow} M$ be a smooth embedding and D an elliptic differential operator on M. Consider the equation

$$Du \equiv f \mod X, \tag{0.8}$$

which means that $Du = f$ everywhere on the manifold M except for the submanifold X. The problem is to equip Eq. (0.8) with well-posed boundary conditions on X. (Well-posedness is understood in the sense that the problem must be Fredholm.) Next, we wish to compute the index of the problem. Note that Sobolev [**Sob37**] was the first to consider such a problem for the polyharmonic equation; he used variational methods to prove the unique solvability of the problem in this case. The general statement and analysis of this problem, as well as the term "Sobolev problem," is due to Sternin [**Ste64**] (see also [**Ste66**]).

Note that Sobolev problems have a number of properties distinguishing them from ordinary elliptic boundary value problems for differential equations.

For example, the number of boundary conditions in these problems depends on the index of the Sobolev space in which the solution is to be sought. In particular, the Sobolev problem is essentially trivial in the space of sufficiently smooth functions, since relation (0.8) turns into an equality everywhere and no boundary conditions on X are needed. Next, the solution of an elliptic Sobolev problem is not an infinitely smooth function in general even if the right-hand side of the equation is infinitely smooth. The solution can have singularities on the submanifold X.

3. Relative elliptic theory associated with the pair (M, X) gives rise to remarkable relationships between elliptic pseudodifferential operators on M and X. These relationships were studied from the topological viewpoint for the first time by Novikov and Sternin [**NS66a, NS66b**] and later by Sternin in the book [**Ste71**].

Associated with the embedding $X \overset{i}{\hookrightarrow} M$ is the natural pullback

$$i^!_{\text{top}}\colon K(T^*M) \longrightarrow K(T^*X)$$

in the topological K-functor, as well as the analytically defined metric trace

$$i^*_{\text{m}}\colon \mathrm{PSD}(M) \longrightarrow \mathrm{PSD}(X)$$

and pullback

$$i^!_a\colon \mathrm{PSD}(M) \longrightarrow \mathrm{PSD}(X),$$

which take pseudodifferential operators on M to pseudodifferential operators on X. (The first of the latter mappings is defined by the metric-dependent restriction of the symbol to the conormal bundle of the submanifold X, and the second is given by the formula

$$i^!_a(D) = i^* D i_*,$$

where i^* and i_* are the boundary and coboundary operators.) Hence there is a natural problem of *comparing the corresponding functors*, i.e., establishing *Riemann–Roch type theorems*. It turns out that such theorems are valid indeed. The simplest theorem compares the metric trace and the topological pullback and states that the diagram

$$\begin{array}{ccc} \mathrm{Ell}(M) & \xrightarrow{\;i^*_{\text{m}}\;} & \mathrm{Ell}(X) \\ \downarrow & & \downarrow \\ K(T^*M) & \xrightarrow[\;i^!_{\text{top}}\;]{} & K(T^*X), \end{array}$$

where the vertical arrows are given by the difference construction for the principal symbol, commutes. Similar theorems comparing the analytic and topological pullbacks are valid in the context of meromorphic families of elliptic operators or for the regularized pullback[3]

$$i^!_{\text{reg}}(D) = \Delta^z i^!_a(\Delta^{-z} D)$$

as $\operatorname{Re} z \longrightarrow \infty$.

However, there is an important class of operators for which the "infinite" regularization is not needed. This is the class of first-order differential operators, which in particular includes all main geometric operators. Nor is the infinite regularization needed in the statement of the theorem comparing the analytic and topological pullbacks for operators whose structure in a neighborhood of X is that of the exterior tensor product [**AS63**] of an elliptic operator on X by a family of elliptic operators in the fibers of the normal bundle. This structure is also typical of geometric operators and permits one to compute the pullback for these operators explicitly.

---

[3]Here Δ stands for an invertible operator whose principal symbol coincides with that of the Laplace operator.

4. Let us briefly outline the exposition of this survey. In the first part, we develop the relative theory of pseudodifferential operators as a theory associated with a smooth embedding $X \hookrightarrow M$. In the framework of this theory, we introduce the notion of ellipticity and establish the theorem on the Fredholm property. The index formula for relative elliptic operators (elliptic morphisms) is proved. By way of example, we develop the theory of Sobolev problems, prove the Fredholm property, and compute the index. Furthermore, we introduce and study the algebras of Green operators associated with the embedding $X \hookrightarrow M$, prove a criterion for the Fredholm property of the Green equation, and apply the result to the analysis of conditions for the Fredholm property of general elliptic morphisms. The second part deals with the topological aspects of elliptic theory in the category of smooth embeddings and establishes the facts described in item 3 above. As an example, we consider the Euler, Dirac, Hirzebruch, and Todd (Cauchy–Riemann) operators.

5. The present paper develops ideas, methods and results contained in [**Ste64, Ste66, Ste67, Ste76, SS96, NS66a, NS66b, Ste71**]. In particular, it gives a new glance on classical elliptic Sobolev problems and relative elliptic theory from the viewpoint of modern theory of differential equations. For example, boundary and coboundary operators, as well as Green operators, are interpreted in terms of Fourier integral operators.

As one can see from the cited literature, the first papers on general Sobolev problems were written about forty years ago (see [**Ste64, Ste66**]). Since then, quite a few remarkable papers on relative elliptic theory have been published. The notion of a coboundary operator, originally introduced in [**Ste67**], is nowadays widely used in general theory of differential equations. In various situations, many authors independently developed constructions close in their spirit to the constructions of relative elliptic theory (e.g., see [**VE65a, VE65b, Esk73, Mon71, Sch91**], etc.).

Conventions and notation. Let us make some remarks concerning conventions and assumptions adopted throughout the text. The dimension of the submanifold X will always be denoted by n and the codimension by ν (i.e., $\dim M = n+\nu$). For convenience, we everywhere assume that the manifolds M and X are equipped with some given volume elements $d\,\mathrm{vol}_M$ and $d\,\mathrm{vol}_X$ so that one need not distinguish between functions and densities. Whenever we deal with Sobolev spaces of sections of a vector bundle[4], we assume that a Hermitian structure is given in the fibers, so the the L^2 inner product of sections is well-defined and one can identify function sections and density sections. All these assumptions do not restrict the generality.

Many of our considerations are carried out in local coordinates on X or M in a neighborhood of some point $x_0 \in X$. We always assume that these coordinates are chosen in the following special way. A *special coordinate system* on X is a coordinate system $(x_1, \dots, x_n)$ such that

$$d\,\mathrm{vol}_X = dx_1 \dots dx_n.$$

[4]Such spaces are considered here only for complex vector bundles

A coordinate system $(x_1, \dots, x_n, t_1, \dots, t_\nu)$ on M near X is called a *special coordinate system* if

(1) the equations of the submanifold X have the form
$$t_1 = \dots = t_\nu = 0;$$
(2) for $t = 0$, the system $(x_1, \dots, x_n)$ is a special coordinate system on X;
(3) one has
$$d\,\mathrm{vol}_M = dx_1 \dots dx_n\, dt_1 \dots dt_n.$$

There always exist coordinate systems with these properties. We usually assume that M and X are Riemannian manifolds with metrics ρ_M and ρ_X, respectively, the embedding is isometric, and the volume elements correspond to the metrics. Then one can construct a special coordinate system on M as follows. Let $\pi\colon NX \longrightarrow X$ be the *normal bundle* of X in M, i.e., the orthogonal complement of the tangent bundle $TX \subset TM\big|_X$. We introduce local coordinates (x, t) on NX, where $x = (x_1, \dots, x_n)$ is a special coordinate system on X in a neighborhood of x_0 and $t = (t_1, \dots, t_\nu)$ are linear coordinates in the fiber N_xX with respect to some orthonormal basis smoothly depending on $x \in X$ in a neighborhood of x_0. The exponential mapping
$$\begin{aligned} \exp\colon NX &\longrightarrow M, \\ \xi &\longmapsto \gamma_\xi(1), \end{aligned}$$
where $\gamma_\xi(\tau)$ is the geodesic on M issuing from the point $\pi(\xi)$ with initial velocity vector $\dot\gamma_\xi(0) = \xi$, diffeomorphically maps some neighborhood V of the zero section of the normal bundle onto a tubular neighborhood U of X in M. (In the following, we identify U with V.) In a neighborhood of a given point $x_0 \in X$, one can take the coordinates (x, t) on M transferred by the exponential mapping from NX. Then the submanifold X is given in these coordinates by the equation $\{t = 0\}$, and
$$\rho_M = \rho_X + \sum_{j=1}^{\nu} dt_j^2, \quad \bigl[d\,\mathrm{vol}_M\bigr]\big|_X = d\,\mathrm{vol}_X\; dt_1 \dots dt_\nu.$$
By multiplying t_1 by a non-zero smooth function equal to 1 for $t = 0$, we ensure that the second equation holds not only on X, but everywhere.

The coordinates dual to x (respectively, (x, t)) in the fiber of the cotangent bundle will be denoted by p (respectively, (p, τ)), $p \in \mathbb{R}^n$, $\tau \in \mathbb{R}^\nu$, and the coordinates (x, p) (respectively, (x, t, p, τ)) will also be referred to as *special coordinates* on the cotangent bundle. For example, the symplectic forms on T^*M and T^*X are given by the expressions
$$\omega_M = dp \wedge dx + d\tau \wedge dt \equiv \sum_{j=1}^{n} dp_j \wedge dx_j + \sum_{j=1}^{\nu} d\tau_j \wedge dt_j,$$
$$\omega_X = dp \wedge dx \equiv \sum_{j=1}^{n} dp_j \wedge dx_j$$

in special coordinates.

To avoid cumbersome formulas, we also make the following simplifying assumptions, which are not important for the validity of the main results.

- All operators considered in the first part of the paper (except for the section concerning Sobolev problems) act in *function spaces* (the generalization to the case of operators acting in *spaces of sections of vector bundles* is standard).
- The manifolds M and X are assumed to be *oriented.* Moreover, we use only special coordinate systems coordinated with the orientation so that the Riemannian volume elements $d\,\mathrm{vol}_M$ and $d\,\mathrm{vol}_X$ on M and X can be treated as ordinary differential forms of maximal order on these manifolds.

ACKNOWLEDGMENT. The paper was written during the authors' stay at the Institute of Mathematics, Potsdam University. The authors are grateful to Professor B.-W. Schulze for his kind hospitality. The authors also wish to thank the referees for many helpful remarks.

Analytic Aspects of Relative Elliptic Theory

Let M be a smooth compact manifold without boundary and $X \hookrightarrow M$ a closed smooth submanifold. In this part, we define and study elliptic morphisms associated with the pair (M, X). These morphisms act in pairs of Sobolev spaces on M and X. The main tool in their construction is the technique of Fourier integral operators (briefly recalled in the Appendix), and our exposition follows the standard semiclassical quantization scheme: We first introduce classical objects associated with the pair (M, X) (Lagrangian manifolds with measure and symbol classes) and then describe the quantization procedure, which produces the operators that are elements of the morphism algebra, and establish its main properties, i.e., formulas for the product of operators and the adjoint operator. Next, we use these properties to state ellipticity conditions for morphisms, prove a theorem on the Fredholm property, and prove an index formula for elliptic morphisms. Finally, the results are applied to the Sobolev problem.

We note that the Lagrangian manifolds used in the construction of the morphism algebra do not satisfy the fundamental condition A.1 (see the Appendix), which is assumed in general in the theory of Fourier integral operators. Namely, three of these manifolds have a non-empty intersection with the zero section of at least one of the cotangent bundles in the Cartesian product of which they lie. Hence the general composition and boundedness theorems (see Subsections A.6 and A.7) are not valid for such Fourier integral operators, and we have to introduce some modifications in the statements of definitions and theorems. Moreover, one encounters three new phenomena.

(1) The operators in question prove not to be continuous in the entire Sobolev scale. For example, the restriction operator

$$i^*\colon H^s(M) \longrightarrow H^{s-\nu/2}(X)$$

is well-defined and continuous only for $s > \nu/2$ and is undefined otherwise.

(2) Products of Fourier integral operators are not always well-defined. For example, the product $i^* D i_*$, where D is a pseudodifferential operator on X, is defined only for $\operatorname{ord} D < -\nu$ (and is meaningless, say, for $D = 1$: the coboundary operator i_* produces the delta function concentrated on X, whose restriction to X is not defined). Hence one has to introduce restrictions on the orders of the operators to be considered.

(3) One has to deal with symbol classes determined by estimates that are more complicated than the estimates (A.3), customarily used in the general theory (see the Appendix). Such symbols inevitably arise from the multiplication of operators even if the symbols of the factors satisfy (A.3). For example, the product $\widehat{\Phi} = D_X i^* D_M$, where D_X and D_M are pseudodifferential operators of the form

$$D_X = a\left(\frac{\partial}{\partial x}\right), \quad D_M = b\left(\frac{\partial}{\partial x}, \frac{\partial}{\partial t}\right)$$

on X and M, respectively, has the symbol $a(p')b(p',\tau')$ on the corresponding Lagrangian manifold. The differentiation of this symbol with respect to p' does not change the rate of growth as $|\tau'|\to\infty$ for bounded p'.

1. Classical objects. Lagrangian formalism

1.1. Lagrangian manifolds with measure. Let

$$i\colon X \hookrightarrow M \tag{1.1}$$

be a C^∞ embedding of a closed smooth manifold X in a closed smooth manifold M. Associated with the pair (M, X) are five remarkable $\mathbb{R}_+$-homogeneous Lagrangian manifolds

$$\begin{aligned} L_M, L_g \subset T^*M \times T^*M, \quad & L_X \subset T^*X \times T^*X, \\ L_b \subset T^*X \times T^*M, \quad & L_c \subset T^*X \times T^*X \end{aligned} \tag{1.2}$$

in Cartesian products of cotangent bundles. Furthermore, these manifolds are equipped with natural measures. All these manifolds can be obtained by the same recipe: One takes the conormal bundle of a certain submanifold in the product of bases of the corresponding cotangent bundles and reverses the sign of the momentum variables in the second factor. (Note that this change of sign automatically transforms the conormal bundle into a Lagrangian manifold with respect to the difference of symplectic forms on the first and second factors.) Let us describe the Lagrangian manifolds (1.2).

1.1.1. *The manifolds L_M and L_X.*

Invariant description. These manifolds are trivial in the sense that they are independent of the embedding and are related to M and X separately. They can be obtained by the above-mentioned construction from the diagonals

$$\begin{aligned} \Delta_M &= \{(\alpha,\beta) \in M \times M \mid \alpha = \beta\} \subset M \times M, \\ \Delta_X &= \{(\alpha,\beta) \in X \times X \mid \alpha = \beta\} \subset X \times X \end{aligned}$$

in $M \times M$ and $X \times X$, respectively. Thus,

$$L_M = \widetilde{N}^*(\Delta_M) \subset T^*M \times T^*M, \quad L_X = \widetilde{N}^*(\Delta_X) \subset T^*X \times T^*X,$$

Here $N^*(Y)$ is the conormal bundle of Y and the tilde on N^* stands for reversing the sign of the cotangent variable in the second factor T^*M (or T^*X). One can readily see that the manifolds L_M and L_X themselves are diagonals in products of cotangent bundles, i.e., the graphs of identity canonical transformations:

$$L_M = \Delta_{T^*M}, \quad L_X = \Delta_{T^*X}. \tag{1.3}$$

The manifolds L_M and L_X bear the natural measures μ_M and μ_X given by the formulas

$$\mu_M = \frac{\pi_j^*\left(\omega_M^{\wedge(n+\nu)}\right)}{(n+\nu)!}, \quad \mu_X = \frac{\pi_j^*\left(\omega_X^{\wedge n}\right)}{n!}, \tag{1.4}$$

where π_j is the projection on the jth factor in the corresponding product of cotangent bundles. (These formulas are the standard formulas defining measures on graphs of canonical transformations [**NOSS81**]. By virtue of the Lagrangian property, the result is independent of the choice of $j = 1, 2$.)

Coordinate description. The manifolds L_M and L_X are given in special coordinate systems on the cotangent bundles by the formulas

$$L_M = \{(x,t,p,\tau,x',p',t',\tau') \mid x = x',\, t = t',\, p = p',\, \tau = \tau'\},$$
$$L_X = \{(x,p,x',p') \mid x = x',\, p = p'\}.$$

(From now on, the coordinates on the second factor in the product of cotangent bundles are indicated by primes.) The canonical coordinates (see Subsection A.2) on L_M and L_X can be taken in the form (x,t,p',τ') and (x,p'), respectively. Then the measures are given by the formulas

$$\mu_M = dp'_1 \wedge dx_1 \wedge \cdots \wedge dp'_n \wedge dx_n \wedge d\tau'_1 \wedge dt_1 \wedge \cdots \wedge d\tau'_\nu \wedge dt_\nu,$$
$$\mu_X = dp'_1 \wedge dx_1 \wedge \cdots \wedge dp'_n \wedge dx_n,$$

and the generating functions of these Lagrangian manifolds (see Subsection A.2) have the form

$$S_M(x,t,p',\tau') = p'x + \tau' t, \quad S_X(x,p') = p'x.$$

1.1.2. *The manifold L_b.*

Invariant description. This manifold can be obtained by applying the above-mentioned construction to the graph of the embedding i:

$$L_b = \widetilde{N}^*(\text{graph } i) \subset T^*X \times T^*M. \tag{1.5}$$

The manifold L_b (as well as all Lagrangian manifolds considered below) is not the graph of any canonical transformation, and so we cannot apply the standard construction (1.4) of the measure. To obtain a natural measure on L_b, consider the mappings

$$\pi_1 \colon L_b \longrightarrow T^*X, \quad \pi_2 \colon L_b \longrightarrow T^*M \tag{1.6}$$

obtained by the restriction to L_b of the natural projections on the first and second factors, respectively, in the Cartesian product $T^*X \times T^*M$. Let $\alpha \in L_b$ be an arbitrary point. The differentials

$$\pi_{1*} \colon T_\alpha L_b \longrightarrow T_{\pi_1(\alpha)} T^*X, \quad \pi_{2*} \colon T_\alpha L_b \longrightarrow T_{\pi_2(\alpha)} T^*M \tag{1.7}$$

possess the following properties:

(1) π_{1*} is an epimorphism;
(2) $\pi_{2*}|_{\ker \pi_{1*}}$ is an isomorphism of the kernel $\ker \pi_{1*}$ onto the subspace in $T_{\pi_2(\alpha)} T^*M$ parallel to the fiber $N^*_{\pi(\pi_1(\alpha))} X$ of N^*X at the point $\pi(\pi_1(\alpha))$ (where $\pi \colon T^*X \longrightarrow X$ is the natural projection).

After natural identifications, we obtain the exact sequence of vector spaces

$$0 \longrightarrow N^*_{\pi(\pi_1(\alpha))}X \longrightarrow T_\alpha L_b \longrightarrow T_{\pi_1(\alpha)}T^*X \longrightarrow 0. \tag{1.8}$$

There are natural volume forms on the fibers of TT^*X (namely, the normalized power

$$d\,\mathrm{vol}_{T^*X} = \frac{1}{n!}(\omega_X)^{\wedge n} \tag{1.9}$$

of the symplectic structure on X) as well as on the fibers of N^*X (namely, the form

$$d\,\mathrm{vol}_{N^*X} = d\tau_1 \wedge \ldots \wedge d\tau_\nu, \tag{1.10}$$

where $\tau_1, \ldots, \tau_\nu$ are the momentum variables dual to the coordinates $t_1, \ldots, t_\nu$ from some special coordinate system (x,t); the form on the right-hand side in (1.10) is independent of the choice of coordinates). We define a volume form μ_b on $T_\alpha L_b$ by arbitrarily splitting the sequence (1.8) and by setting

$$\mu_b = d\,\mathrm{vol}_{T^*X} \wedge d\,\mathrm{vol}_{N^*X}\,. \tag{1.11}$$

(The result is independent of the splitting.)

Thus, we have defined a measure μ_b on the manifold L_b.

Coordinate description. The manifold L_b is given in special coordinate systems on the cotangent bundles by the formulas

$$L_b = \bigl\{(x,p,x',t',p',\tau') \bigm| x = x',\, t' = 0,\, p = p'\bigr\}. \tag{1.12}$$

The canonical coordinates on L_b can be taken conveniently in the form (x,p',τ'). Then the measure is given by the formula

$$\mu_b = dp_1' \wedge dx_1 \wedge \cdots \wedge dp_n' \wedge dx_n \wedge d\tau_1' \wedge \cdots \wedge d\tau_\nu',$$

and the generating functions of L_b has the form

$$S_b(x,p',\tau') = p'x.$$

1.1.3. *The manifold L_c.*

Invariant description. This manifold is defined as

$$L_c = \widetilde{N}^*({}^T\mathrm{graph}\, i) \subset T^*M \times T^*X,$$

where the submanifold ${}^T\mathrm{graph}\, i \subset M \times X$ is obtained as the *transpose* of graph i. (The factors in the Cartesian product are transposed.) Clearly, L_c itself coincides with L_b up to transposition:

$$L_c = {}^T L_b \subset T^*M \times T^*X. \tag{1.13}$$

More formally, consider the transposition isomorphism

$$\begin{aligned} \varkappa\colon T^*X \times T^*M &\longrightarrow T^*M \times T^*X, \\ (\alpha,\beta) &\longmapsto (\beta,\alpha); \end{aligned}$$

then

$$\varkappa(L_b) = L_c. \tag{1.14}$$

We use the isomorphism (1.14) and define a measure on L_c by setting

$$\mu_c = (\varkappa^{-1})^*(\mu_b). \tag{1.15}$$

Coordinate description. The manifold L_c is given in special coordinate systems on the cotangent bundles by the formula

$$L_c = \{(x, t, p, \tau, x', p') \mid x = x',\, t = 0,\, p = p'\}. \tag{1.16}$$

The canonical coordinates on L_c can be taken in the form (x, τ, p'). Then the measure is given by the formula

$$\mu_c = dp'_1 \wedge dx_1 \wedge \cdots \wedge dp'_n \wedge dx_n \wedge d\tau_1 \wedge \cdots \wedge d\tau_\nu,$$

and the generating function of L_b has the form

$$S_b(x, \tau, p') = p'x.$$

1.1.4. *The manifold L_g.*
Invariant description. This Lagrangian manifold is defined as

$$L_g = \widetilde{N}^*(W) \subset T^*M \times T^*M, \tag{1.17}$$

where

$$W = \{(\alpha, \alpha) \in M \times M \mid \alpha \in X\}. \tag{1.18}$$

Let

$$\pi_{1,2}\colon L_g \longrightarrow T^*M|_X \subset T^*M \tag{1.19}$$

be the mappings obtained by the restriction to L_g of the natural projections of $T^*M \times T^*M$ on the first and second factors. Consider the mapping

$$p\colon L_g \longrightarrow T^*X \tag{1.20}$$

given by the composition $p = \widetilde{\pi} \circ \pi_1$, where

$$\widetilde{\pi}\colon T^*M|_X \longrightarrow T^*X \tag{1.21}$$

is the natural projection. (Note that $\widetilde{\pi} \circ \pi_1 = \widetilde{\pi} \circ \pi_2$.) Next, p_* is an epimorphism at an arbitrary point $\alpha \in L_g$, and the mapping

$$(\pi_{1*} \oplus \pi_{2*})|_{\ker p_*}$$

is an isomorphism of $\ker p_*$ onto the direct sum of two copies of the tangent space to a fiber of the conormal bundle of X. Thus, we obtain the exact sequence

$$0 \longrightarrow N^*_{\pi(p(\alpha))}X \oplus N^*_{\pi(p(\alpha))}X \longrightarrow T_\alpha L_g \longrightarrow T_{p(\alpha)}T^*X \longrightarrow 0 \tag{1.22}$$

for an arbitrary point $\alpha \in L_g$ (where $\pi\colon T^* \longrightarrow X$ is the natural projection). This sequence permits one to define a measure μ_g on L_g by the formula

$$\mu_g = d\,\mathrm{vol}_{N^*X} \wedge d\,\mathrm{vol}_{N^*X} \wedge d\,\mathrm{vol}_{T^*X}, \tag{1.23}$$

where the factors on the right-hand side are given by formulas (1.9) and (1.10). (The two factors $d\,\mathrm{vol}_{N^*X}$ correspond to the two copies of $N^*_{\pi(p(\alpha))}X$ in the expansion (1.22).)

Coordinate description. The manifold L_g is given in special coordinate systems on the cotangent bundles by the formulas

$$L_g = \{(x, t, p, \tau, x', t', p', \tau') \mid x = x',\ t = t' = 0,\ p = p'\}. \tag{1.24}$$

The canonical coordinates on L_g can be taken in the form (x, τ, p', τ'). Then the measure is given by the formula

$$\mu_g = dp_1' \wedge dx_1 \wedge \cdots \wedge dp_n' \wedge dx_n \wedge d\tau_1 \wedge \cdots \wedge d\tau_\nu \wedge d\tau_1' \wedge \cdots \wedge d\tau_\nu',$$

and the generating functions of L_g has the form

$$S_b(x, \tau, p', \tau') = p'x.$$

Thus, we have defined the Lagrangian manifolds (1.2), constructed some natural measures on these manifolds, and described canonical coordinates related to special coordinate systems in the cotangent bundles. (In the following, these canonical coordinates on Lagrangian manifolds will often be referred to as *special canonical coordinates.*) The following assertion holds.

PROPOSITION 1.1. *The measures μ_M, μ_X, μ_b, μ_c, and μ_g are quantized.*

(Concerning the notion of quantized measures on Lagrangian manifolds, e.g., see [**NOSS81**].) We omit the proof.

Thus, Maslov's canonical operator, which is used in the construction of Fourier integral operators (see the Appendix), is well-defined on all above-mentioned Lagrangian manifolds with measure.

We should possibly comment on the notation. The manifolds L_X and L_M are the graphs of identity canonical transformations. (The corresponding Fourier integral operators are just pseudodifferential operators.) The subscripts b, c, and g on the manifolds L_b, L_g, and L_c have been chosen in view of the fact that the corresponding Fourier integral operators are known as boundary operators, coboundary operators, and Green operators, respectively.

1.2. Composition and transposition formulas. Now we shall study *compositions* and *transpositions* of the Lagrangian manifolds introduced above. This proves useful, since the composition of Lagrangian manifolds corresponds to the composition of the corresponding Fourier integral operators and the transposition corresponds to the passage to the adjoint operator. The general definition of composition for Lagrangian manifolds is given in the Appendix, Subsection A.3. We have already noted that three of our Lagrangian manifolds, i.e., L_b, L_c, and L_g, fail to satisfy the fundamental condition A.1: Each of these manifolds has a nonempty intersection with the zero section of at least one of the cotangent bundles forming the product where this manifold lies. Hence the composition need not be proper in general (cf. Subsection A.3), and we have to modify the definition.

1.2.1. *Definition of the composition.*

DEFINITION 1.2. Let

$$L_1 \subset T^*M_1 \times T^*M_2, \quad L_2 \subset T^*M_2 \times T^*M_3 \tag{1.25}$$

be two Lagrangian manifolds. Suppose that the following conditions hold.

(1) The composition $L = L_1 \circ L_2$ given by (A.2) is clean and connected (see Subsection A.3).
(2) The projection

$$p\colon \widetilde{L} \longrightarrow L, \tag{1.26}$$

where

$$\widetilde{L} = (L_1 \times L_2) \cap (T^*M_1 \times \Delta_{T^*M_2} \times T^*M_3),$$

is a locally trivial bundle over the manifold L. (The fiber of this bundle over a point $(\alpha, \beta) \in L$ will be denoted by $F_{(\alpha,\beta)}$; the dimension will be denoted by $\dim F_{(\alpha,\beta)} = e$ and is referred to as the *excess* of the composition, just as in the general theory.)

Then we say that the composition $L = L_1 \circ L_2$ is *well-defined.* (Note that L is automatically a Lagrangian manifold.)

Needless to say, if the projection (1.26) is proper, then the local triviality of the bundle (1.26) follows from Thom's theorem, and we return to the standard theory.

1.2.2. *The composition table.* Now we are in a position to state the composition theorem for Lagrangian manifolds.

THEOREM 1.3. *Every meaningful composition of two Lagrangian manifolds from the list* (1.2) *is well-defined. The composition table has the following form* (*the first factor corresponds to a row and the second to a column*):

	L_X	L_M	L_b	L_c	L_g
L_X	L_X	—	L_b	—	—
L_M	—	L_M	—	L_c	L_g
L_b	—	L_b	—	L_X	L_b
L_c	L_c	—	L_g	—	—
L_g	—	L_g	—	L_c	L_g

TABLE 1.1

Here dashes are placeholders for meaningless compositions.

Next, the fiber of the projection (1.26) *is a singleton* ($e = 0$) *for all compositions except for the four compositions in the intersection of the last three rows and the last two columns. For each of these four compositions, the fiber of the projection* (1.26) *is naturally isomorphic to the fiber of the normal bundle* N^*X *over the corresponding point of* X (*i.e.,* $e = \nu$ *in these cases*).

PROOF. The proof goes by a straightforward computation. □

REMARK 1.4. We see that for the four compositions mentioned in the theorem (which will be referred to as *exceptional*), the fiber of the projection (1.26) is not compact. (It is isomorphic to $\mathbb{R}^{\nu}$.) Since the formula for the symbol of the product of Fourier integral operators (see Subsections 1.4 and 2.2) contains integration over the fiber, it is not at all surprising that order restrictions guaranteeing the convergence of the integral arise for such products.

1.2.3. *Transposition.* The behavior of our Lagrangian manifolds and measures under transposition (see (A.7)) is described by the following assertion, which can readily be verified by a straightforward computation.

PROPOSITION 1.5. *One has*

$$^{T}L_X = L_X, \quad ^{T}L_M = L_M, \quad ^{T}L_b = L_c, \quad ^{T}L_c = L_b, \quad ^{T}L_g = L_g.$$

Moreover, the transposition takes the above-defined measure on any of these Lagrangian manifolds to the corresponding measure on the transpose (neglecting the sign).

1.3. Symbols. Now we shall introduce symbol spaces on each of the Lagrangian manifolds (1.2). These symbol spaces will be used in the quantization procedure (i.e., in the construction of the corresponding Fourier integral operators). Just as in the standard theory of pseudodifferential operators (or Fourier integral operators), there are various approaches to the definition of symbol spaces. On the one hand, one can define a symbol as an arbitrary function satisfying certain estimates (which guarantee the continuity of the corresponding operator in appropriate Sobolev spaces). In this case, the notion of principal symbol (which alone is meaningful in classical mechanics) becomes rather complicated and inconvenient: One has to define the principal symbol as an element of a quotient space of the symbol space. On the other hand, one can single out a narrower space of *classical* symbols asymptotic for large values of the momentum variables to some $\mathbb{R}_+$-homogeneous function of these variables. It is the latter function that plays the role of the principal symbol. (Note that it usually does not lie in the original symbol space, since it has a singularity at zero. Thus, one has to use cut-off functions in the quantization procedure.) It is most convenient to combine the two approaches: One first introduces a wider symbol class and proves boundedness and composition theorems and then observes the invariance of the space of operators with classical symbols with respect to composition and studies ellipticity, the Fredholm property, etc. in this narrower class.

In this section, we introduce symbol spaces using the above-mentioned approach. All main statements are given in special coordinates and special canonical coordinates on the Lagrangian manifolds, for we believe that coordinate statements are easier to comprehend than invariant statements in this case. We omit the proof of independence of our definitions of the choice of special local coordinates. Needless to say, one can also give an invariant description of symbol spaces. By way of example, we give such a description for the manifolds L_b and L_g.

Finally, we note that the traditional approach to the definition of classical symbols requires that the symbols have *asymptotic series expansions* in homogeneous functions at infinity. The operators to be constructed here act only in a limited interval of the Sobolev scale, and hence there is no need to require the existence of "long" expansions. Furthermore, we deal with qualitative issues of elliptic theory, where lower-order terms play essentially no role. Therefore, we define classical symbols by imposing the minimal requirement of existence of only one (leading) term of the asymptotic expansion.

Throughout the following, by $\chi(r)$, $r \in \mathbb{R}_+$, we denote a smooth non-negative function such that

$$\chi(r) = \begin{cases} 0, & r \leq 1/2, \\ 1, & r \geq 1. \end{cases} \tag{1.27}$$

1.3.1. *Symbols on L_M and L_X.* These symbols are the symbols of pseudodifferential operators. We briefly recall the well-known spaces of such symbols so as to set the notation and clarify the scheme of the subsequent reasoning by a well-known example. To be definite, we consider the Lagrangian manifold L_X. The notation for L_M is entirely similar.

DEFINITION 1.6. By $S^m(L_X)$, where $m \in \mathbb{R}$, we denote the space of smooth functions a on the manifold L_X such that the following estimates hold in *canonical coordinates*:

$$\left| \frac{\partial^{\alpha+\beta} a(x,p')}{\partial x^\alpha \partial p'^\beta} \right| \leq C_{\alpha\beta}(1+|p'|)^{m-|\beta|}, \quad |\alpha|+|\beta| = 0,1,2,\dots .$$

By $O^m(L_X)$ we denote the space of functions a on the manifold L_X such that a is smooth outside the zero section ($|p'|=0$) and $\mathbb{R}_+$-homogeneous of degree m (with respect to the standard action of the group $\mathbb{R}_+$ in the fibers).

By $S^m_{\mathrm{cl}}(L_X) \subset S^m(L_X)$ we denote the subspace of functions representable in the form

$$a = \chi(|p'|)a_0 + b, \quad a_0 \in O^m(L_X), \quad b \in S^{m-\varepsilon}(L_X) \text{ for some } \varepsilon > 0. \tag{1.28}$$

Elements of the spaces $S^m(L_X)$, $S^m_{\mathrm{cl}}(L_X)$, and $O^m(L_X)$ are called, respectively, *symbols*, *classical symbols*, and *homogeneous (principal) symbols of order* m on L_X. The element $a_0 \in O^m(L_X)$ corresponding to a symbol $a \in S^m_{\mathrm{cl}}(L_X)$ according to (1.28) is called the (homogeneous) *principal part* of a.

One can readily see that the homogeneous principal part of a symbol is uniquely determined.

The definitions of homogeneous principal symbols for the other Lagrangian manifolds look somewhat more complicated, since *these symbols have singularities not only on the zero section.*

1.3.2. *Symbols on L_b and L_c.* First, we introduce symbol spaces on the manifold L_b. The symbol spaces on L_c are obtained from these via the mapping $(\varkappa^{-1})^*$, where $\varkappa$ is the isomorphism (1.14).

Let k and l be given real numbers.

DEFINITION 1.7. By $S_b^{k,l}(L_b)$ we denote the space of smooth functions a on the manifold L_b such that the following estimates hold in canonical coordinates:

$$\left|\frac{\partial^{\alpha+\beta+\gamma}a(x,p',\tau')}{\partial x^\alpha \partial p'^\beta \partial \tau'^\gamma}\right| \le C_{\alpha\beta\gamma}(1+|p'|)^{k-|\beta|}(1+|p'|+|\tau'|)^{l-|\gamma|}, \quad |\alpha|+|\beta|+|\gamma| = 0,1,2,\dots \tag{1.29}$$

By $O_b^{k,l}(L_b)$ we denote the space of functions a on the manifold L_b with the following properties:

(1) a is defined and infinitely smooth outside the set

$$\Gamma = \{p' = 0\} \subset L_b;$$

(2) a is $\mathbb{R}_+$-homogeneous of degree $k+l$ in the variables (p',τ'):

$$a(x,\lambda p',\lambda\tau') = \lambda^{k+l}a(x,p',\tau'), \quad \lambda \in \mathbb{R}_+, \quad p' \neq 0;$$

(3) for $p' \neq 0$, the function a satisfies the estimates

$$\left|\frac{\partial^{\alpha+\beta+\gamma}a(x,p',\tau')}{\partial x^\alpha \partial p'^\beta \partial \tau'^\gamma}\right| \le C_{\alpha\beta\gamma}|p'|^{k-|\beta|}(|p'|+|\tau'|)^{l-|\gamma|}, \quad |\alpha|+|\beta|+|\gamma| = 0,1,2,\dots$$

By $S_{b,\mathrm{cl}}^{k,l}(L_b) \subset S_b^{k,l}(L_b)$ we denote the subspace of functions representable in the form

$$a = \chi(|p'|)a_0 + b, \quad a_0 \in O_b^{k,l}(L_b), \quad b \in S_b^{k-\varepsilon,l}(L_b) \oplus S_b^{k,l-\varepsilon}(L_b) \text{ for some } \varepsilon > 0. \tag{1.30}$$

Elements of the spaces $S_b^{k,l}(L_b)$, $S_{b,\mathrm{cl}}^{k,l}(L_b)$, and $O_b^{k,l}(L_b)$ are called, respectively, *symbols*, *classical symbols*, and *homogeneous (principal) symbols of type* (k,l) on L_b. The element $a_0 \in O_b^{k,l}(L_b)$ corresponding to a symbol $a \in S_{b,\mathrm{cl}}^{k,l}(L_b)$ according to (1.30) is called the (homogeneous) *principal part* of a. The number $r := k+l+\nu/2$ is called the *order* of a (or a_0) and is denoted by

$$\operatorname{ord} a = \operatorname{ord} a_0 = k+l+\nu/2.$$

This is well-defined, as shown by the following almost obvious proposition. (The proof is by straightforward computation and is therefore omitted.)

PROPOSITION 1.8. *Let $a \in O_b^{k,l}(L_b)$. Then $\chi(|p'|)a \in S_{b,\mathrm{cl}}^{k,l}(L_b)$. If $\chi_1(r)$ is another smooth function with property* (1.27), *then*

$$(\chi(|p'|) - \chi_1(|p'|))a \in S_b^{k-1,l}(L_b).$$

It follows from the definition of the spaces $O_b^{k,l}(L_b)$ and $S_{b,\mathrm{cl}}^{k,l}(L_b)$ that for a given $a \in S_{b,\mathrm{cl}}^{k,l}(L_b)$ the function a_0 in (1.30) is determined uniquely and is independent of the choice of the cut-off function χ.

Invariant description. Let us show how one can rewrite the estimates (1.29) in invariant form. To this end, we note that L_b is a vector bundle; we define a function $|\alpha|$, $\alpha \in L_b$, as a norm in the fibers smoothly depending on the point of the base. In a similar way, we introduce a norm in the fibers of T^*X (the one induced by the Riemannian metric on X).

We consider two kinds of vector fields on L_b (see (1.7) for the properties of π_{1*}):

- vector fields annihilated by π_{1*}, where π_1 is the projection on the first factor in the product $T^*X \times T^*M$; such fields will be referred to as *normal to* T^*X;
- vector fields mapped by π_{1*} to well-defined vector fields on T^*X (thus, for each $\beta \in T^*X$ the application of π_{1*} to the vector field at a point $\alpha \in \pi_1^{-1}(\beta)$ gives a result independent of the choice of α); such fields will be referred to as *lifted from* T^*X.

Now we can restate the first part of Definition 1.7 as follows.

DEFINITION 1.9. By $S_b^{k,l}(L_b)$, where $k, l \in \mathbb{R}$, we denote the space of smooth functions a on the manifold L_b such that the following condition holds: Suppose that $\{V_1, \dots, V_j\}$ is an arbitrary (possibly, empty) finite set of smooth vector fields on L_b such that each of the fields is homogeneous of degree 0 or 1 and is either normal to T^*X or lifted from T^*X; then

$$(1.31) \qquad |V_1 \cdots V_j a(\alpha)| \le C(1+|\alpha|)^{l-j_1}(1+|\pi_1(\alpha)|)^{k-j_2},$$

where j_1 is the number of first-degree homogeneous fields normal to T^*X and j_2 is the number of first-degree homogeneous fields lifted from T^*X. (The constant C in (1.31) depends on the choice of the fields $V_1, \dots, V_j$.)

Symbols on L_c. We define symbol spaces on L_c as follows.

DEFINITION 1.10. The spaces

$$(1.32) \quad S_c^{m,k}(L_c) = (\varkappa^{-1})^* S_b^{k,m}(L_b),$$
$$O_c^{m,k}(L_c) = (\varkappa^{-1})^* O_b^{k,m}(L_b), \quad S_{c,\mathrm{cl}}^{m,k}(L_c) = (\varkappa^{-1})^* S_{b,\mathrm{cl}}^{k,m}(L_b),$$

where $\varkappa$ is the diffeomorphism (1.14), are called, respectively, the spaces of *symbols, classical symbols*, and *homogeneous (principal) symbols of type* (m,k) on L_c. The element $a_0 \in O_c^{m,k}(L_c)$ corresponding to a symbol $a \in S_{c,\mathrm{cl}}^{m,k}(L_c)$ according to the formula

$$a = \chi(|p'|)a_0 + b, \quad b \in S_c^{m-\varepsilon,k}(L_c) \oplus S_c^{m,k-\varepsilon}(L_c) \text{ for some } \varepsilon > 0,$$

is called the *(homogeneous) principal part* of a. The number $r := m + k + \nu/2$ is called the *order* of a (or a_0) and is denoted by $\operatorname{ord} a$.

REMARK 1.11. The "correction" $\nu/2$ in the definition of the order of symbols on L_b and L_c ensures that the order of a symbol is equal to the order in the Sobolev scale of the operator obtained by the quantization of the symbol. The correction ν in the definition of the order of symbols on L_g has the same meaning.

1.3.3. *Symbols on L_g.* Let m, k, and l be given real numbers.

DEFINITION 1.12. By $S_g^{m,k,l}(L_g)$ we denote the space of smooth functions a on the manifold L_g such that the following estimates hold in canonical coordinates:

$$\left|\frac{\partial^{\alpha+\beta+\gamma+\delta}a(x,p',\tau,\tau')}{\partial x^\alpha\partial p'^\beta\partial\tau^\gamma\partial\tau'^\delta}\right| \le C_{\alpha\beta\gamma\delta}(1+|p'|)^{k-|\beta|}(1+|p'|+|\tau|)^{m-|\gamma|} \times (1+|p'|+|\tau'|)^{l-|\delta|}, \quad |\alpha|+|\beta|+|\gamma|+|\delta|=0,1,2,\dots \tag{1.33}$$

By $O_g^{m,k,l}(L_g)$ we denote the space of functions a on the manifold L_g with the following properties:

(1) a is defined and infinitely smooth outside the set
$$\Gamma=\{p'=0\}\subset L_g;$$

(2) a is $\mathbb{R}_+$-homogeneous of degree $m+k+l$ in the variables (p',τ,τ'):
$$a(x,\lambda p',\lambda\tau,\lambda\tau')=\lambda^{m+k+l}a(x,p',\tau,\tau'),\quad \lambda\in\mathbb{R}_+,\quad p'\neq 0;$$

(3) For $p\neq 0$, the function a satisfies the estimates
$$\left|\frac{\partial^{\alpha+\beta+\gamma+\delta}a(x,p',\tau,\tau')}{\partial x^\alpha\partial p'^\beta\partial\tau^\gamma\partial\tau'^\delta}\right| \le C_{\alpha\beta\gamma\delta}|p'|^{k-|\beta|}(|p'|+|\tau|)^{m-|\gamma|}(|p'|+|\tau'|)^{l-|\delta|}, \quad |\alpha|+|\beta|+|\gamma|+|\delta|=0,1,2,\dots$$

By $S_{g,\mathrm{cl}}^{m,k,l}(L_g)\subset S_g^{m,k,l}(L_g)$ we denote the subspace of functions representable in the form

$$a=\chi(|p'|)a_0+b,\quad a_0\in O_g^{m,k,l}(L_g), \quad b\in S_g^{m-\varepsilon,k,l}(L_g)\oplus S_g^{m,k-\varepsilon,l}(L_g)\oplus S_g^{m,k,l-\varepsilon}(L_g)\text{ for some }\varepsilon>0. \tag{1.34}$$

Elements of the spaces $S_g^{m,k,l}(L_g)$, $S_{g,\mathrm{cl}}^{m,k,l}(L_g)$, and $O_g^{m,k,l}(L_g)$ are called, respectively, *symbols*, *classical symbols*, and *homogeneous (principal) symbols of type* (m,k,l) on L_g. The element $a_0\in O_g^{m,k,l}(L_g)$ corresponding to a symbol $a\in S_{g,\mathrm{cl}}^{m,k,l}(L_g)$ according to (1.34) is called the *(homogeneous) principal part* of the symbol a. The number $r:=m+k+l+\nu$ is called the *order* of a (or a_0) and is denoted by

$$\operatorname{ord}a=\operatorname{ord}a_0=m+k+l+\nu.$$

Just as in the other cases, one can verify that the principal part of a symbol is well-defined and unique.

Invariant description. Let us show how one can rewrite the estimates (1.33) in invariant form. To this end, we consider the following three kinds of vector fields on L_g (see (1.22)):

- vector fields everywhere tangent to the first copy of N^*X (they will be referred to as *normal fields of the first kind*);
- vector fields everywhere tangent to the second copy of N^*X (they will be referred to as *normal fields of the second kind*);
- vector fields *lifted from* T^*X.

Now we can restate the first part of Definition 1.12 as follows.

DEFINITION 1.13. By $S_g^{m,k,l}(L_g)$, where $m,k,l \in \mathbb{R}$, we denote the space of smooth functions a on the manifold L_g with the following property: If $\{V_1,\dots,V_j\}$ is an arbitrary (possibly, empty) finite set of vector fields on L_g such that each of the fields is homogeneous of degree 0 or 1 and either is normal of the first or second kind or is lifted from T^*X, then the estimate

$$|V_1\cdots V_j a(\alpha)| \le C(1+|\pi_1(\alpha)|)^{m-j_1}(1+|p(\alpha)|)^{k-j_2}(1+|\pi_2(\alpha)|)^{l-j_3} \tag{1.35}$$

holds, where j_1 is the number of first-order homogeneous normal fields of the first kind, j_3 is the number of first-order homogeneous normal fields of the second kind, j_2 is the number of first-order homogeneous normal fields lifted from T^*X, π_1 and π_2 are the projections on the first and second factors, respectively, in the Cartesian product $T^*M \times T^*M$, and p is the projection (1.20).

Thus, we have defined quite a number of symbol spaces. To simplify the notation, in the following we usually omit the manifold argument in the notation of symbol spaces, i.e., write, say, $S_g^{m,k,l}$ instead of $S_g^{m,k,l}(L_g)$, etc., provided this does not lead to confusion.

1.4. The product of symbols (twisted multiplication). Let L_1 and L_2 be two Lagrangian manifolds (1.25) such that the composition $L = L_1 \circ L_2$ is well-defined. Then there is an operation, called the *twisted multiplication* in what follows, that takes pairs of symbols on L_1 and L_2 to symbols on L. (See the Appendix, Subsection A.7.) This operation involves integration over the fibers of the bundle (1.26). In our situation, the fibers need not be compact, and so the twisted product is defined only under certain restrictions on the types of factors. First, let us write out an expression for the twisted product specifying the general assertion in the Appendix.

1.4.1. *The general twisted product formula.* Let the manifold L_1, L_2, and L be equipped with smooth real measures μ_1, μ_2, and μ, respectively. We make some choice of the half-densities $\sqrt{\mu_1}$, $\sqrt{\mu_2}$, $\sqrt{\mu}$. For example, we can adopt the convention that $\sqrt{\mu}$ is represented in local coordinates on L by the arithmetic square root $\sqrt{|\widetilde{\mu}|}$, where $\widetilde{\mu}$ is the density of μ in these coordinates. Then the twisted multiplication

$$\begin{aligned} *\colon C_0^\infty(L_1) \times C_0^\infty(L_2) &\to C_0^\infty(L), \\ (f_1, f_2) &\mapsto f_1 * f_2 \end{aligned} \tag{1.36}$$

can be described as follows. The expression

$$(f_1\sqrt{\mu_1})(\alpha,\beta)(f_2\sqrt{\mu_2})(\beta,\gamma),$$
$$(\alpha,\beta) \in L_1 \subset T^*(M_1 \times M_2), \quad (\beta,\gamma) \in L_2 \subset T^*(M_2 \times M_3),$$

is a half-density on L whose values are densities on the fibers $F_{(\alpha,\gamma)}$ of the bundle (1.26). Hence the expression

$$(f_1 * f_2)(\alpha,\gamma) = \frac{1}{\sqrt{\mu}}\left(\frac{1}{2\pi}\right)^e \int_{F_{(\alpha,\gamma)}} (f_1\sqrt{\mu_1})(\alpha,p)(f_2\sqrt{\mu_2})(\beta,\gamma) \tag{1.37}$$

is a well-defined *function* on L (cf. [**Hör85b**, Theorem 25.2.3]). In local coordinates, this function is represented by an ordinary integral over the fiber variables. The normalizing factor $(1/2\pi)^e$ is introduced so as to ensure that there is no additional factor in the formulas for products of operators.

If the fiber $F_{(\alpha,\gamma)}$ is compact, then the twisted product (1.37) is defined not only for compactly supported, but also for arbitrary smooth functions. In the general case, *we say that the twisted product* (1.37) *is defined for given functions* $f_1 \in C^\infty(L_1)$ *and* $f_2 \in C^\infty(L_2)$ *if the integral on the right-hand side in* (1.37) *is absolutely convergent.*

1.4.2. *A twisted product theorem.* We now describe the twisted products (1.36) in the symbol spaces introduced above for the cases corresponding to the compositions of Lagrangian manifolds described in Theorem 1.3 (see Table 1.1).

First, we consider the case of "trivial" products, where at least one of the factors is the diagonal manifold L_M or L_X. By Theorem 1.3, the fiber of the bundle (1.26) is a singleton so that the twisted products are always defined. The type of the product can readily be expressed via the types of the factors. Leaving aside the case in which both factors are diagonal (and hence one deals with products of pseudodifferential operators), we state the following theorem.

THEOREM 1.14. *The twisted product is well-defined in the following symbol spaces for any values of the indices*:

$$S^m(L_X) * S_b^{k,l} \subset S_b^{m+k,l}, \tag{1.38}$$

$$S^m(L_M) * S_c^{m',k} \subset S_c^{m+m',k}, \quad S^m(L_M) * S_g^{m',k,l} \subset S_g^{m+m',k,l}, \tag{1.39}$$

$$S_b^{k,l} * S^m(L_M) \subset S_b^{k,l+m}, \quad S_g^{m',k,l} * S^m(L_M) \subset S_g^{m',k,l+m}, \tag{1.40}$$

$$S_c^{m',k} * S^m(L_X) \subset S_c^{m',k+m}. \tag{1.41}$$

For any of these compositions, one has

$$\operatorname{ord}(a_1 * a_2) = \operatorname{ord} a_1 + \operatorname{ord} a_2. \tag{1.42}$$

The following theorem describes the twisted products (1.36) corresponding to the compositions in Table 1.1 for the case in which neither of the factors is L_X or L_M.

THEOREM 1.15. *The twisted product of symbols corresponding to the compositions of Lagrangian manifolds in the last three rows and columns of* Table 1.1 *is described by the following table:*

	$S_b^{k_2,l_2}$	$S_c^{m_2,k_2}$	$S_g^{m_2,k_2,l_2}$
$S_b^{k_1,l_1}$	—	$S_X^{k_1+l_1+k_2+m_2+\nu}$	$S_b^{k_1+l_1+k_2+m_2+\nu,l_2}$
$S_c^{m_1,k_1}$	$S_g^{m_1,k_1+k_2,l_2}$	—	—
$S_g^{m_1,k_1,l_1}$	—	$S_c^{m_1,k_1+l_1+k_2+m_2+\nu}$	$S_g^{m_1,k_1+l_1+k_2+m_2+\nu,l_2}$

TABLE 1.2

The twisted product

$$S_c^{m_1,k_1}(L_c) * S_b^{k_2,l_2}(L_b)$$

is defined for arbitrary values of the indices. The remaining four products in Table 1.2 *are defined under the condition*

$$l_1 + m_2 < -\nu. \tag{1.43}$$

Finally, condition (1.42) *holds for each of these products.*

Let us now state the assertion concerning the twisted product of homogeneous principal symbols.

THEOREM 1.16. *Let a_1 and a_2 be classical symbols whose twisted product $a = a_1 * a_2$ is well-defined by one of the two preceding theorems. Next, let a_{10}, a_{20}, and a_0 be the homogeneous principal symbols corresponding to a_1, a_2, and a, respectively. Then the twisted product $a_{10} * a_{20}$ is well-defined and $a_{10} * a_{20} = a_0$.*

Coordinate description. Let us give the expressions for $a = a_1 * a_2$ in special coordinate systems for the cases shown in Table 1.2. (The cases described in Theorem 1.14 are left to the reader.) They have the following form:

$$L_b \circ L_c: \qquad a(x,p') = \left(\frac{1}{2\pi}\right)^{\nu} \int a_1(x,p',\tau)a_2(x,p',\tau)\,d\tau;$$

$$L_c \circ L_b: \qquad a(x,\tau,p',\tau') = a_1(x,\tau,p')a_2(x,p',\tau');$$

$$L_g \circ L_g: \qquad a(x,\tau,p',\tau') = \left(\frac{1}{2\pi}\right)^{\nu} \int a_1(x,\tau,p',\tau'')a_2(x,\tau'',p',\tau')\,d\tau'';$$

$$L_g \circ L_c: \qquad a(x,\tau,p') = \left(\frac{1}{2\pi}\right)^{\nu} \int a_1(x,\tau,p',\tau')a_2(x,\tau',p')\,d\tau';$$

$$L_b \circ L_g: \qquad a(x,p',\tau') = \left(\frac{1}{2\pi}\right)^{\nu} \int a_1(x,p',\tau'')a_2(x,\tau'',p',\tau')\,d\tau''.$$

2. Quantization and the algebra of operator morphisms

Now we proceed to the second stage in the construction of our operator algebra. Namely, we quantize the classical objects introduced in the previous section. The elements of the algebra are 2×2 matrices with operator entries. First, we describe these entries, which will be referred to as the *main operators*, and then we deal with the matrices.

2.1. The main operators.

We have constructed five Lagrangian manifolds with measures in Subsection 1.1 and the corresponding symbol classes in Subsection 1.3. We obtain five types of operators by quantizing these objects, i.e., by considering the corresponding Fourier integral operators.

2.1.1. *Pseudodifferential operators.* The quantization of symbols on L_M and L_X results in pseudodifferential operators on M and X, respectively. This procedure is well-known, and we only mention that pseudodifferential operators themselves will be denoted by upper-case letters (for example, D_M or D_X) and their symbols (or principal symbols, depending on the context) by the corresponding lower-case letters:

$$\sigma(D_M) = d_M, \quad \sigma(D_X) = d_X.$$

2.1.2. *Boundary operators.*

DEFINITION 2.1. The *boundary operator with symbol* $a \in S_b^{k,l}(L_b)$ is the Fourier integral operators $\widehat{\Phi}_b[a]$ on the Lagrangian manifold L_b with measure μ_b and symbol a.

The *boundary operator with principal symbol* $a_0 \in O_b^{k,l}(L_b)$ is the operator $\widehat{\Phi}_b[a_0] \overset{\text{def}}{=} \widehat{\Phi}_b[a]$ with symbol $a = \chi(|p'|)a_0$.

REMARK 2.2. Owing to the ambiguity in the construction of the canonical operator, the operator $\widehat{\Phi}_b[a]$ is determined only modulo operators whose symbols are of lower order than a. (The same remark pertains to all other operators that will be constructed in what follows.)

EXAMPLE 2.3. By taking $a = 1$, we obtain the *elementary boundary operator*

$$i^* \colon C^\infty(M) \longrightarrow C^\infty(X), \tag{2.1}$$

which takes each function $f \in C^\infty(M)$ to the restriction $i^* f \in C^\infty(X)$ of f to X.

THEOREM 2.4. *Let* $a \in S_b^{k,l}(L_b)$. *Then the operator* $\widehat{\Phi}_b[a]$ *is continuous in the spaces*

$$\Phi_b[a] \colon H^s(M) \to H^{s-r}(X), \quad r = \operatorname{ord} a, \tag{2.2}$$

for $s > l + \nu/2$.

Coordinate description. Let us write out the expression for $\Phi_b[a]$ in special coordinates (x, t). The operator $\widehat{\Phi}_b[a]$ is given by the formula

$$
\begin{aligned}
(\widehat{\Phi}_b[a]u)(x) &= \left(\frac{1}{2\pi}\right)^{n+\nu} \int e^{i[p'(x-x')-\tau' t']} a(x,p',\tau')u(x',t')\,dp'd\tau'dx'dt' \\
&= \left(\frac{i}{2\pi}\right)^{\frac{n+\nu}{2}} \int e^{ip'x} a(x,p',\tau')\widetilde{u}(p',\tau')dp'd\tau',
\end{aligned}
\tag{2.3}
$$

where $\widetilde{u}(p', \tau')$ is the Fourier transform of $u(x', t')$.

2.1.3. *Coboundary operators.*

DEFINITION 2.5. The *coboundary operator with symbol* $a \in S_c^{m,k}(L_c)$ is the Fourier integral operator $\widehat{\Phi}_c[a]$ on the Lagrangian manifold L_c with measure μ_c and symbol a.

The *coboundary operator with principal symbol* $a_0 \in O_c^{m,k}(L_c)$ is the operator $\widehat{\Phi}_c[a_0] \stackrel{\text{def}}{=} \widehat{\Phi}_c[a]$ with symbol $a = \chi(|p'|)a_0$.

EXAMPLE 2.6. By taking $a = 1$, we obtain the *elementary coboundary operator*

$$
i_* \colon \mathcal{D}'(X) \longrightarrow \mathcal{D}'(M),
\tag{2.4}
$$

which is the L^2-adjoint of the elementary boundary operator. It can be locally described in the special coordinates (x, t) as follows:

$$
[i_* g](x,t) = g(x) \otimes \delta(t),
\tag{2.5}
$$

where δ is the Dirac delta function. In other words, $i_* g$ is the product of g by the delta function in the transversal directions, or the delta function on X with density g so that one sometimes writes $i_* g = g\delta_X$.

THEOREM 2.7. *Let $a \in S_c^{m,k}(L_c)$. Then the operator $\widehat{\Phi}_c[a]$ is continuous in the spaces*

$$
\widehat{\Phi}_c[a] \colon H^s(X) \to H^{s-r}(M), \quad r = \operatorname{ord} a,
$$

for $s < k$.

REMARK 2.8. The operator $\widehat{\Phi}_c[a]$ is actually continuous in the spaces

$$
\widehat{\Phi}_c[a] \colon H^s(X) \to H^{-m-\nu/2-\varepsilon}(M)
$$

for $s \geq k$. (Here $\varepsilon > 0$ is arbitrary.) This follows from the embedding $H^s(X) \subset H^{k-\varepsilon}(X)$ in conjunction with Theorem 2.7. However, we do not consider these values of s, since the order of the operator $\widehat{\Phi}_c[a]$ is not related to the order of the symbol for these s and composition theorems considered in the next subsection are no longer valid owing to the fact that one cannot neglect lower-order terms in asymptotic expansions of Fourier integrals.

Coordinate description. By analogy with the preceding, we can write out the expression for the coboundary operator in special local coordinates. It has the form

$$(2.6)\qquad (\widehat{\Phi}_c[a]u)(x,t) = \left(\frac{1}{2\pi}\right)^{n+\nu} \int e^{i[p'(x-x')+\tau t]} a(x,\tau,p')u(x')dx'dp'd\tau$$
$$= \left(\frac{1}{2\pi}\right)^{\nu} \left(\frac{i}{2\pi}\right)^{n/2} \int e^{i(p'x+\tau t)} a(x,\tau,p')\widetilde{u}(p')dp'd\tau.$$

2.1.4. *Green operators.*

DEFINITION 2.9. The *Green operator* with symbol $a \in S_g^{m,k,l}(L_g)$ is the Fourier integral operator $\widehat{\Phi}_g[a]$ on the Lagrangian manifold L_g with measure μ_g and symbol a.

The *Green operator with principal symbol* $a_0 \in O_g^{m,k,l}(L_g)$ is the operator $\widehat{\Phi}_g[a_0] \stackrel{\text{def}}{=} \widehat{\Phi}_g[a]$ with symbol $a = \chi(|p'|)a_0$.

The following boundedness theorem is valid for Green operators.

THEOREM 2.10. *Let $a \in S_g^{m,k,l}(L_g)$ be a Green symbol. Then the operator $\widehat{\Phi}_g[a]$ is continuous in the spaces*

$$\widehat{\Phi}_g[a] \colon H^s(M) \to H^{s-r}(M), \qquad r = \operatorname{ord} a,$$

for $\nu/2 + l < s < \nu/2 + l + k$.

REMARK 2.11. It follows from Theorem 2.10 that it makes sense to consider Green symbols only for $k > 0$, since otherwise the set of values of s for which the Green operator is continuous in the above-mentioned spaces is empty.

Coordinate description. The expression for a Green operator in special coordinates has the form

$$(2.7)\quad (\widehat{\Phi}_g[a]u)(x,t) = \left(\frac{1}{2\pi}\right)^{n+2\nu} \int e^{i[p'(x-x')+\tau t-\tau' t']} a(x,\tau,p',\tau')u(x',t')$$
$$= \left(\frac{i}{2\pi}\right)^{\frac{n+\nu}{2}} \left(\frac{1}{2\pi}\right)^{\nu} \int e^{i(p'x+\tau t)} a(x,\tau,p',\tau')\widetilde{u}(p',\tau')dp'd\tau'd\tau.$$

2.1.5. *An interpretation of the main operators as pseudodifferential operators with operator-valued symbols.* In conclusion, we note that, modulo operators with smooth kernel, *boundary, coboundary, and Green operators are concentrated on the submanifold X.* This follows from the form of the corresponding Lagrangian manifolds. Hence the structure of these operators in a neighborhood of X is of main interest. It turns out that in a tubular neighborhood of X these operators can be interpreted as pseudodifferential operators on X with operator-valued symbols acting in function spaces on the fibers of the normal bundle of X. (Recall that a neighborhood of the zero section of the normal bundle is identified via the geodesic exponential map with a tubular neighborhood of X.)

By way of example, let us show this for the case of Green operators.

Indeed, formula (2.7), specifying the operator $\widehat{\Phi}_g[a]$ in special coordinates, can be rewritten in the form

$$\widehat{\Phi}_g[a]u(x,t) = \left(\frac{i}{2\pi}\right)^{n/2} \int e^{ipx} \left[\frac{i^{\nu/2}}{(2\pi)^{3\nu/2}} \int e^{it\tau} a(x,p,\tau,\tau')\widetilde{u}(p,\tau')\, d\tau' d\tau\right] dp,$$

or

$$\widehat{\Phi}_g[a]u(x,t) = H\left(\overset{2}{x}, -i\overset{1}{\frac{\partial}{dx}}\right)u(x,t), \tag{2.8}$$

where $H(x,p)$ is the operator-valued symbol acting on functions of the variables t by the formula

$$H(x,p)v(t) = \left(\frac{i}{2\pi}\right)^{\nu/2} \left(\frac{1}{2\pi}\right)^{\nu} \int e^{it\tau} a(x,p,\tau,\tau')\widetilde{v}(\tau') d\tau' d\tau. \tag{2.9}$$

Actually, the operator $H(x,p)$ is a Green operator with parameters (x,p) acting in a space of functions of t. The symbol of this operator is $a(x,p,\tau,\tau')$ (where x and p are treated as parameters). The passage

$$a(x,p,\tau,\tau') \to H(x,p) \to \widehat{\Phi}_g[a] \tag{2.10}$$

can naturally be interpreted as a two-stage quantization procedure, first along the fibers of the normal bundle of X (which gives an operator-valued symbol over X) and then along X.

2.2. Composition formulas and adjoint operators.

Composition. Here we state the composition theorem for the main operators, viz. pseudodifferential operators on M and X, (co)boundary operators, and Green operators. To unify the statement, we temporarily denote the Fourier integral operator with symbol a on a Lagrangian manifold L from the list (1.2) by $\widehat{\Phi}(L,a)$. (The choice of a measure on L has been described above.)

THEOREM 2.12. *Let L_1 and L_2 be two manifolds from the list* (1.2) *such that the composition $L_1 \circ L_2$ is well-defined and let $\widehat{\Phi}(L_1,a_1)$ and $\widehat{\Phi}(L_2,a_2)$ be Fourier integral operators with symbols in some of the classes introduced above such that the composition*

$$\widehat{\Phi}(L_1,a_1)\widehat{\Phi}(L_2,a_2)$$

is well-defined as a product of bounded operators in accordance with the boundedness theorems. Then the twisted product $a_1 * a_2$ is well-defined, and the composition formula

$$\widehat{\Phi}(L_1,a_1)\widehat{\Phi}(L_2,a_2) \equiv \widehat{\Phi}(L_1 \circ L_2, a_1 * a_2)$$

is valid modulo operators of order $< \operatorname{ord} a_1 + \operatorname{ord} a_2$.

The same assertion remains valid if a_1 and a_2 are the homogeneous principal symbols of the corresponding operators.

The adjoint operator.

THEOREM 2.13. *Let L be a manifold from the list (1.2) and let $\Phi(L,a)$ be a Fourier integral operator with symbol from some of the classes introduced above such that the operator $\widehat{\Phi}(L,a)$ is well-defined as a bounded operator by the boundedness theorems. Then the formally L^2-adjoint operator is given by the formula*

$$(\widehat{\Phi}(L,a))^* = \widehat{\Phi}({}^TL, \overline{(\varphi^*)^{-1}a}) \tag{2.11}$$

modulo operators of order $< \operatorname{ord} a_1 + \operatorname{ord} a_2$. Here the bar over the symbol on the right-hand side stands for complex conjugation and

$$\varphi \colon L \to^T L$$

is the mapping obtained by the restriction to L of the transposition of factors in the Cartesian product of cotangent bundles.

The same assertion remains valid if by a we mean the homogeneous principal symbol of the Fourier integral operator.

2.3. Algebras of operator morphisms. Now we are in a position to construct the operator algebras. They will consist of 2×2 operator matrices of the form

$$\begin{pmatrix} D_M + \widehat{\Phi}_g & \widehat{\Phi}_c \\ \widehat{\Phi}_b & D_X \end{pmatrix}.$$

Since the formal "multiplication" of the corresponding matrices of Lagrangian manifolds gives

$$\begin{pmatrix} L_M + L_g & L_c \\ L_b & L_X \end{pmatrix} \begin{pmatrix} L_M + L_g & L_c \\ L_b & L_X \end{pmatrix} = \begin{pmatrix} L_M + L_g & L_c \\ L_b & L_X \end{pmatrix},$$

one can hope that the set of operator matrices of this form is indeed closed with respect to composition.

Let us proceed to exact statements.

2.3.1. *Definition of the algebras.* Since the range of the Sobolev smoothness exponent s for which Green, boundary, and coboundary operators are defined is not the entire real line, we see that operators of positive order with such entries will not form an algebra. However, for any s and σ one can indicate such symbol spaces for all entries for which the operators

$$\begin{pmatrix} D_M + \widehat{\Phi}_g & \widehat{\Phi}_c \\ \widehat{\Phi}_b & D_X \end{pmatrix} : \begin{matrix} H^s(M) \\ \oplus \\ H^\sigma(X) \end{matrix} \longrightarrow \begin{matrix} H^s(M) \\ \oplus \\ H^\sigma(X) \end{matrix} \tag{2.12}$$

will be continuous and form an algebra. We present the corresponding result for $s = \sigma = 0$. The general case was analyzed in [**SS96**].

Let $m_g, k_g, k_c, k_b \in \mathbb{R}$ be given numbers satisfying the inequalities

$$\begin{gathered} k_g > 0,\ k_c > 0,\ k_b > 0, \\ m_g < -\frac{\nu}{2},\ m_g + k_g > -\frac{\nu}{2}. \end{gathered} \tag{2.13}$$

(Note that system (2.13) is consistent. For example, one can take $m_g = -\nu$, $k_g = \nu$, $k_c = k_b = 1$ as a solution.)

We set

$$l_g = -m_g - k_g - \nu,\ m_c = -k_c - \nu/2,\ l_b = -k_b - \nu/2. \tag{2.14}$$

The following assertion is a straightforward consequence of the boundedness and composition theorems.

THEOREM 2.14. *The set $\mathcal{A}(m_g, k_g, k_c, k_b)$ of operator 2×2 matrices of the form*

$$A = \begin{pmatrix} D_M + \widehat{\Phi}_g[a_g] & \widehat{\Phi}_c[a_c] \\ \widehat{\Phi}_b[a_b] & D_X \end{pmatrix}, \tag{2.15}$$

where D_M and D_X are zero-order pseudodifferential operators on M and X, respectively, and the symbols a_g, a_c, and a_b lie in the spaces

$$a_g \in S_g^{m_g, k_g, l_g}(L_g), \quad a_c \in S_c^{m_c, k_c}(L_c), \quad a_b \in S_b^{k_b, l_b}(L_b),$$

is a linear space of continuous operators in $L^2(M) \oplus L^2(X)$.

The (non-direct) sum

$$\mathcal{A} = \sum_{m_g, k_g, k_c, k_b} \mathcal{A}(m_g, k_g, k_c, k_b)$$

over all m_g, k_g, k_c, k_b satisfying (2.13) *is an algebra.*

REMARK 2.15. From the practical viewpoint, it is of interest to consider operators of non-zero order in Sobolev spaces other than L^2; theoretically, it suffices to study the algebras described in Theorem 2.14, since an operator of the form (2.15) acting from $H^{s_1}(M) \oplus H^{s_2}(X)$ into $H^{\sigma_1}(M) \oplus H^{\sigma_2}(X)$ can always be reduced to an operator in the space $L^2(M) \oplus L^2(X)$: One simply multiplies it on the right and on the left by appropriate diagonal matrices of pseudodifferential operators.

REMARK 2.16. Let $m_g, k_g \in \mathbb{R}$ be given numbers satisfying the inequalities

$$m_g < -\nu/2 < m_g + k_g. \tag{2.16}$$

We set

$$k_b = m_g + k_g + \nu/2 \quad k_c = -m_g - \nu/2. \tag{2.17}$$

Then inequalities (2.13) are satisfied automatically. We write

$$\mathcal{A}(m_g, k_g) = \mathcal{A}(m_g, k_g, k_c, k_b), \tag{2.18}$$

where k_b and k_c are given by formulas (2.17).

A careful computation of the exponents according to Table 1.2 in the composition theorem 1.15 shows that $\mathcal{A}(m_g, k_g) \subset \mathcal{A}$ is a subalgebra.

In what follows we mainly deal with subalgebras of the algebras defined above formed by classical operators (i.e., operators with classical symbols).

2.3.2. *Involution.* Theorem 2.13 implies the following assertion.

THEOREM 2.17. *The algebra $\mathcal{A}$ is an involutive algebra with involution taking each operator to its L^2-adjoint. Moreover,*

$$\mathcal{A}(m_g, k_g, k_c, k_b)^* = \mathcal{A}(m_g^*, k_g^*, k_c^*, k_b^*), \tag{2.19}$$

where

$$m_g^* = -m_g - k_g - \nu, \quad k_g^* = k_g, \quad k_c^* = k_b, \quad k_b^* = k_c. \tag{2.20}$$

In particular, the subalgebras

$$\mathcal{A}(-(k+\nu)/2, k) \subset \mathcal{A} \tag{2.21}$$

are invariant with respect to involution.

2.3.3. *Green subalgebras.* There is an important subalgebra of $\mathcal{A}$ formed by matrices of the form

$$\begin{pmatrix} \widehat{\Phi}_g & 0 \\ 0 & 0 \end{pmatrix}.$$

This subalgebra is naturally isomorphic to the algebra of operators of the form $\widehat{\Phi}_g$ in $L^2(M)$. (As follows from the composition theorem, such operators indeed form an algebra.) More precisely, let us give the following definition. Let numbers $m, k, l \in \mathbb{R}$ satisfy the relations

$$\begin{cases} m + k + l + \nu = 0, \\ l + \nu/2 < 0 < l + \nu/2 + k. \end{cases} \tag{2.22}$$

We set $O^{(m,l)} = O_g^{m,k,l}(L_g)$. (Note that the notation is adequate, since

$$k = k(m, l) = -m - l - \nu \tag{2.23}$$

is uniquely determined by m and l.)

DEFINITION 2.18. The *Green algebra* $\mathcal{A}_g^{(m,l)}$ is the set of continuous linear operators in $L^2(M)$ representable in the form

$$A = \widehat{\Phi}_g[a] + K, \tag{2.24}$$

where $a \in O^{(m,l)}$ and K is a compact operator in $L^2(M)$ ($K \in \mathcal{K}$). Elements of the Green algebra are called *Green operators.*

Green algebras play an important role in the ellipticity conditions (see the next section).

3. Elliptic morphisms and the index formula

In this section, we give the definition of ellipticity for morphisms, prove theorems concerning the Fredholm property, and present an index formula for elliptic morphisms that do not contain Green operators. In the ellipticity conditions for general morphisms, the almost inversion of operators of the form $1 + T$, where $T = \widehat{\Phi}_g[a]$ is a Green operator, plays an important role. We start our exposition with a study of such operators.

3.1. The Green equation.

3.1.1. *Statement of the problem.* Consider the equation

$$u + \widehat{\Phi}_g[a]u = f \in L^2(M) \tag{3.1}$$

for an unknown function $u \in L^2(M)$, where $a \in O^{(m,l)}$ is a given principal symbol. In more detailed notation, Eq. (3.1) has the form

$$u(x,t) + \int K(x,t,x',t')u(x',t')dx'dt' = f(x,t), \tag{3.2}$$

where the kernel $K(x,t,x',t')$ is given by the formula

$$K(x,t,x',t') = (\frac{1}{2\pi})^{n+2\nu} \int e^{i[p(x-x')+\tau t-\tau' t']}(\chi a)(x,p,\tau,\tau')dp d\tau d\tau'. \tag{3.3}$$

Equation (3.1) (or (3.2)) will be called the *Green equation of the second kind*, or simply the *Green equation*. As is often the case in the theory of integral equations, one cannot solve Eq. (3.1) exactly or even write out informative conditions guaranteeing the unique solvability (except for the trivial case in which the norm of the operator $\widehat{\Phi}_g[a]$ is less than one). Thus, we proceed to the analysis of conditions under which Eq. (3.1) is Fredholm. Here one can already obtain non-trivial results, namely, a criterion for the Fredholm property of Eq. (3.1) in terms of the principal symbol a. This criterion is described in the subsequent items.

3.1.2. *An equation for the symbol of an almost inverse.* It is well-known that an equation is Fredholm if and only if the operator specifying the equation has a (two-sided) almost inverse. Let us seek a right-almost inverse of $1 + \widehat{\Phi}_g[a]$ in the form $1 + \widehat{\Phi}_g[b]$, where $b \in O^{(m,l)}$. Then we have

$$(1 + \widehat{\Phi}_g[a])(1 + \widehat{\Phi}_g[b]) = 1 + \widehat{\Phi}_g[a + b + a * b] \tag{3.4}$$

modulo compact operators, and for the symbol b we obtain the equation

$$b + a * b = -a \tag{3.5}$$

Likewise, if we seek a left-almost inverse in the form $1 + \widehat{\Phi}_g[c]$, then for c we obtain the equation

$$c + c * a = -a. \tag{3.6}$$

3.1.3. *Ellipticity and the Fredholm property for the Green equation.* Here we state necessary and sufficient conditions for the Fredholm property of Eq. (3.1). The remaining items deal with the proof of the theorem stated in this item.

DEFINITION 3.1. We say that the Green equation (3.1) (and the operator $1 + \widehat{\Phi}_g[a]$) is *elliptic* if the equation

$$w + a * w = v \tag{3.7}$$

has a solution $w \in O^{(m,l)}$ for every right-hand side $v \in O^{(m,l)}$.

REMARK 3.2. The solvability of Eq. (3.7) with arbitrary right-hand side is *equivalent* to the solvability of Eq. (3.5) (i.e., Eq. (3.7) with special right-hand side $-a$). Indeed, one of the implications is trivial; as to the other, if b is a solution of (3.5), then the solution of (3.7) for arbitrary v is given by the formula

$$w = b + a + v + b * (a + v), \tag{3.8}$$

which can readily be verified by a straightforward substitution into the equation.

THEOREM 3.3. *The Green equation* (3.1) *is Fredholm if and only if it is elliptic.*

COROLLARY 3.4. *The operator* $1 + \widehat{\Phi}_g[a] \in \mathcal{A}_g^{(m,l)}$ *is Fredholm if and only if it has an almost inverse (a regularizer) in the algebra* $\mathcal{A}_g^{(m,l)}$.

Indeed, the symbol of the regularizer is just given by the solution of Eq. (3.5).

REMARK 3.5. Note that the ellipticity condition does not involve Eq. (3.6) or the unique solvability of Eq. (3.5). We shall see from the proof that the solvability of Eq. (3.5) is equivalent to the solvability of Eq. (3.6) and is equivalent to the unique solvability of either one of these equations.

PROOF. The proof splits into several steps described in the forthcoming items. Let us rewrite Eq. (3.7) in coordinates:

$$w(x,p,\tau,\tau') + \left(\frac{1}{2\pi}\right)^{\nu} \int_{\mathbb{R}^{\nu}} a(x,p,\tau,\tau'')w(x,p,\tau'',\tau')d\tau'' = v(x,p,\tau,\tau'). \tag{3.9}$$

It is natural to treat this equation as a family of equations depending on the parameters (x,p,τ'). Let us study the properties of this equation. First, without loss of generality we can assume that $m = l$. Indeed, we set $\widetilde{m} = (m+l)/2$ and $\varkappa = (m-l)/4$. The map

$$a(x,p,\tau,\tau') \longmapsto (p^2+\tau^2)^{-\varkappa} a(x,p,\tau,\tau')(p^2+\tau'^2)^{\varkappa} \tag{3.10}$$

is an isomorphism of $O^{(m,l)}$ onto $O^{(\widetilde{m},\widetilde{m})}$; the form of Eq. (3.9) remains unchanged if we apply the transformation (3.10) to all functions w, a, v occurring in the equation. Thus, from now on we assume that $m = l$. Then it follows from the restrictions imposed on m, l, and k that

$$m < -\nu/2.$$

Since

$$|a(x,p,\tau,\tau')| \le C|p|^k(|p|+|\tau|)^m(|p|+|\tau'|)^m,$$

we see that the kernel in (3.9) is square integrable with respect to the variables τ, τ'' so that for any given (x,p,τ'), $p \neq 0$, this equation is a Fredholm integral equation of the second kind in the space $L^2(\mathbb{R}^{\nu}_{\tau})$. It follows from Fredholm theory that this equation is solvable in $L^2(\mathbb{R}^{\nu}_{\tau})$ for any right-hand side if and only if the homogeneous equation has no non-zero solutions in $L^2(\mathbb{R}^{\nu}_{\tau})$. However, we are interested in solutions of this equation in $O^{(m,l)}$. First, consider the case in which the homogeneous equations (3.9) have no non-zero solutions in $L^2(\mathbb{R}^{\nu}_{\tau})$ for any (x,p), $p \neq 0$.

PROPOSITION 3.6. *In the absence of non-zero solutions of the homogeneous equation, the solution in* $L^2(\mathbb{R}^\nu_\tau)$ *of* Eq. (3.9) *with right-hand side in* $O^{(m,m)}$ *lies in* $O^{(m,m)}$.

PROOF. The desired $\mathbb{R}_+$-homogeneity of the solution follows from the $\mathbb{R}_+$-homogeneity of the kernel and the right-hand side in conjunction with the uniqueness of the solution. Next, the function w is obtained from v by an application of the continuous inverse in $L^2(\mathbb{R}^\nu_\tau)$ to the Fredholm operator of the second kind in Eq. (3.9) so that the differentiability of w (viewed as a function ranging in L^2) with respect to variables (τ, x, p) is obvious. Now it is easy to prove the desired estimates for w and the derivatives of w by induction over the order of the derivative, using the relation obtained from (3.9) by the transposition of the integral term to the right-hand side, which completes the proof. □

Note that Eq. (3.6) is the complex conjugate of the adjoint equation of (3.5), and it follows from Fredholm theory that both equations are solvable or unsolvable simultaneously.

The case in which the homogeneous Fredholm equation has a non-trivial solution is consider in the next item.

3.1.4. *The operator* $1 + \widehat{\Phi}_g[a]$ *is not Fredholm if* Eq. (3.9) *is not uniquely solvable.*

PROPOSITION 3.7. *Suppose that the homogeneous equation*

$$\varphi(\tau) + \left(\frac{1}{2\pi}\right)^\nu \int_{\mathbb{R}^\nu} a(x, p, \tau, \tau'')\varphi(\tau'')d\tau'' = 0 \tag{3.11}$$

has a solution $\varphi \in L^2(\mathbb{R}^\nu)$ *for some values* $(x, p) = (x_0, p_0) \in \mathbb{R}^n \times (\mathbb{R}^n \setminus \{0\})$. *Then the operator* $1 + \widehat{\Phi}_g[a]$ *is not Fredholm.*

PROOF. Without loss of generality, we can assume that $|p_0| = 1$. We shall construct a family of functions $u_\lambda \in L^2(M)$, $\lambda \in \mathbb{R}_+$, with the following properties:

(a) $\|u_\lambda\| = 1$ for all λ;
(b) $\|(1 + \widehat{\Phi}_g[a])u_\lambda\| \to 0$ as $\lambda \to \infty$;
(c) u_λ weakly converges to 0 as $\lambda \to \infty$.

The existence of such a sequence implies that $1 + \widehat{\Phi}_g[a]$ is not Fredholm. Indeed, we argue by contradiction: Let $1 + \widehat{\Phi}_g[a]$ be Fredholm and let R be a regularizer so that

$$R(1 + \widehat{\Phi}_g[a]) = 1 + K, \quad K \in \mathcal{K}.$$

Then

$$u_\lambda = R(1 + \widehat{\Phi}_g[a])u_\lambda - Ku_\lambda \to 0 \tag{3.12}$$

as $\lambda \to \infty$ by (b),(c), and the fact that a compact operator takes each weakly convergent sequence to a strongly convergent sequence. But Eq. (3.12) contradicts (a).

We set

$$u_\lambda(x, t) = c_\lambda e^{i\lambda(x-x_0)p_0} \psi(\sqrt{\lambda}(x - x_0))\breve{\varphi}(\lambda t), \tag{3.13}$$

where c_λ is a constant normalizing u_λ in L^2, $u(x)$ is a smooth compactly supported function, and $\breve{\varphi}(t)$ is the inverse Fourier transform of $\varphi(\tau)$. The function (3.13) satisfies condition (a) by construction. Next, one can readily see that this function weakly converges to zero as $\lambda \to \infty$ (i.e., condition (c) holds): It suffices to show that

$$\int u_\lambda(x,t)\varphi_1(x)\varphi_2(t)\,dx\,dt \to 0$$

as $\lambda \to \infty$ for any smooth compactly supported functions φ_1 and φ_2, but this can readily be proved by integration by parts with respect to x with regard to the fact that $c_\lambda \sim \lambda^{n/2+\nu}$. It remains to verify condition (b). Using the representation (2.8)–(2.9) of the Green operator as a pseudodifferential operator with operator-valued symbol, we obtain

$$\begin{aligned}(1+\widehat{\Phi}_g[a])u_\lambda &\equiv \left(1+H\left(\overset{2}{x}_1, -i\overset{1}{\frac{\partial}{\partial x}}\right)\right)u_\lambda \\ &= c_\lambda e^{i\lambda(x-x_0)p_0}\psi(\sqrt{\lambda}(x-x_0))(1+H(x,\lambda p_0))\breve{\varphi}(\lambda t)+o(1) \\ &= c_\lambda e^{i\lambda(x-x_0)p_0}\psi(\sqrt{\lambda}(x-x_0))(1+H(x,\lambda p_0))\breve{\varphi}(\lambda t)+o(1).\end{aligned}$$

But

$$\begin{aligned}&(1+H_0(x_0,\lambda p_0))\breve{\varphi}(\lambda t) \\ &= \lambda^{-\nu}\mathcal{F}^{-1}_{\tau\to t}\left\{\varphi\left(\frac{\tau}{\lambda}\right)+\left(\frac{1}{2\pi}\right)^{\nu}\int_{\mathbb{R}^\nu} a(x_0,\lambda p_0,\tau,\tau'')\varphi\left(\frac{\tau''}{\lambda}\right)d\tau''\right\} \\ &= \lambda^{-\nu}\mathcal{F}^{-1}_{\tau\to t}\left\{\varphi\left(\frac{\tau}{\lambda}\right)+\left(\frac{1}{2\pi}\right)^{\nu}\int_{\mathbb{R}^\nu} a\left(x_0,p_0,\frac{\tau}{\lambda},\frac{\tau''}{\lambda}\right)\varphi\left(\frac{\tau''}{\lambda}\right)d\left(\frac{\tau''}{\lambda}\right)\right\}=0,\end{aligned}$$

whence it follows that (b) is also satisfied. □

3.1.5. *Completion of the proof.* Now the assertion of the theorem becomes obvious: If the symbol equation is not uniquely solvable at least at one point (x,p), $p\neq 0$, then the operator $1+\widehat{\Phi}_g[a]$ is not Fredholm, which shows that the ellipticity condition is necessary. On the other hand, if the ellipticity condition is satisfied, then one can construct left and right regularizers by solving the corresponding symbol equations, thus proving the Fredholm property. The proof is complete. □

3.2. Ellipticity and the Fredholm property. In this subsection, we establish necessary and sufficient conditions for a given morphism

$$A=\begin{pmatrix} D_M+\widehat{\Phi}_g[g] & \widehat{\Phi}_c[c] \\ \widehat{\Phi}_b[b] & D_X\end{pmatrix}\in\mathcal{A} \tag{3.14}$$

to have an almost inverse (an inverse modulo compact operators) $A^{-1} \in \mathcal{A}$ in the algebra $\mathcal{A}$. These conditions are called *ellipticity conditions*. First, we consider ellipticity conditions for morphisms (3.14) with $\widehat{\Phi}_g[g] = 0$. They are of main interest for two reasons:

- practically interesting problems (Sobolev problems) result only in morphisms that do not contain Green operators;
- in this case, one can obtain ellipticity conditions (and an index formula) in a finite (algebraic) form.

Thus, we consider an operator (3.14) without the Green term, i.e., an operator of the form

$$A = \begin{pmatrix} D_M & \widehat{\Phi}_c[c] \\ \widehat{\Phi}_b[b] & D_X \end{pmatrix}. \tag{3.15}$$

We seek a (right) almost inverse in the form

$$A^{-1} = \begin{pmatrix} \widetilde{D}_M + \widehat{\Phi}_g[\varphi] & \widehat{\Phi}_c[\widetilde{c}] \\ \widehat{\Phi}_b\left[\widetilde{b}\right] & \widetilde{D}_X \end{pmatrix} \tag{3.16}$$

and, multiplying by A on the left, arrive at the system of equations

$$\begin{pmatrix} D_M & \widehat{\Phi}_c[c] \\ \widehat{\Phi}_b[b] & P_X \end{pmatrix} \begin{pmatrix} \widetilde{D}_M + \widehat{\Phi}_g[\varphi] & \widehat{\Phi}_c[\widetilde{c}] \\ \widehat{\Phi}_b\left[\widetilde{b}\right] & \widetilde{D}_X \end{pmatrix} = \begin{pmatrix} 1 & 0 \\ 0 & 1 \end{pmatrix}, \tag{3.17}$$

which must be satisfied modulo compact operators. Let us compute the operator on the left-hand side in (3.17). Modulo compact operators, it has the form

$$\begin{pmatrix} \widetilde{\widetilde{D}}_M + \widehat{\Phi}_g[\psi] & \widehat{\Phi}_c\left[\widetilde{\widetilde{c}}\right] \\ \widehat{\Phi}_b\left[\widetilde{\widetilde{b}}\right] & \widetilde{\widetilde{D}}_X \end{pmatrix},$$

where

$$\widetilde{\widetilde{d}}_M = d_M \widetilde{d}_M, \quad \psi = c * \widetilde{b} + d_M * \varphi \tag{3.18}$$

and the formulas for $\widetilde{\widetilde{c}}$, $\widetilde{\widetilde{b}}$, and $\widetilde{\widetilde{d}}_X$ can readily be written out with the use of the composition law in the algebra $\mathcal{A}$. In (3.18), d_M, $\widetilde{d}_M$, and $\widetilde{\widetilde{d}}_M$ are the principal symbols of the pseudodifferential operators D_M, $\widetilde{D}_M$, and $\widetilde{\widetilde{D}}_M$, respectively. It follows from (3.17) and (3.18) that

$$d_M \widetilde{d}_M = 1.$$

Thus, a necessary condition for the almost invertibility of the element (3.15) in the algebra $\mathcal{A}$ is the ellipticity of the pseudodifferential operator D_M. Assuming that this condition is satisfied, we make some transformations of A that do not affect the almost invertibility.

Namely, we subtract the first row multiplied on the left by $\widehat{\Phi}_b[b]D_M^{-1}$ from the second row. (Here D_M^{-1} is an almost inverse of D_M; we freely use this type of notation, which is unlikely to result in confusion.) As a result, modulo compact operators we obtain the upper-triangular operator matrix

$$A' = \begin{pmatrix} D_M & \widehat{\Phi}_c[c] \\ 0 & D_X - \widehat{\Phi}_b[b] D_M^{-1} \widehat{\Phi}_c[c] \end{pmatrix} \tag{3.19}$$

An upper-triangular matrix is almost invertible if and only if so are its diagonal entries. Thus, we arrive at the following definition.

DEFINITION 3.8. An element (3.15) of the algebra $\mathcal{A}$ is said to be *elliptic* if the following conditions are satisfied:

(i) D_M is an elliptic operator on M;
(ii) the operator

$$\Delta_A = D_X - \widehat{\Phi}_b[b] D_M^{-1} \widehat{\Phi}_c[c] \tag{3.20}$$

(the *non-commutative determinant*) is an elliptic operator on X.

REMARK 3.9. The second condition possibly deserves an explanation. By the composition theorem, the product $\widehat{\Phi}_b[b] D_M^{-1} \widehat{\Phi}_c[c]$ is a pseudodifferential operator on X. Hence so is the non-commutative determinant Δ_A, and we can state the ellipticity condition in finite (algebraic) form at least if the symbols occurring in Δ_A are classical. The symbol of Δ_A has the form

$$\sigma(\Delta_A) = d_X - b * d_M^{-1} * c, \tag{3.21}$$

where b and c are the symbols of $\widehat{\Phi}_b$ and $\widehat{\Phi}_c$, respectively, and the twisted product $*$ in the general form is given by (1.36) and (1.37) and is described in specific cases in the composition theorem 1.15 and the subsequent text. In special local coordinates, the symbol (3.21) can be represented in the form

$$\sigma(\Delta_A)(x,p) = d_x(x,p) + \left(\frac{1}{2\pi}\right)^{\nu} \int b(x,p,\tau) c(x,p,\tau) d_M^{-1}(x,0,p,\tau)\, d\tau.$$

If the symbols $b(x,p,\tau)$ and $c(x,p,\tau)$ (as well as $d_x(x,p)$ and $d_M(x,t,p,\tau)$) are classical, i.e., have homogeneous leading components, then so is the symbol $\sigma(\Delta_A)$, and the ellipticity condition acquires a finite form.

The preceding reasoning implies the following theorem.

THEOREM 3.10. *Let A be an elliptic morphism of the form* (3.15). *Then A is Fredholm.*

Let us now state the ellipticity conditions for general morphisms (3.14) containing a non-zero Green operator in the upper-left corner. We seek an almost inverse in the form (3.16); then the first relation in (3.18) remains the same. Thus, the ellipticity of D_M is still necessary for the almost invertibility of A in the algebra. However, the upper-left corner of A is now

$$A_{11} = D_M + \widehat{\Phi}_g[g], \tag{3.22}$$

and the almost invertibility of A_{11} is not necessary for the almost invertibility of A in the algebra so that the reduction to the upper-triangular case simply does not work.

REMARK 3.11. The special case in which the morphism (3.22) is still almost invertible (one can study this property using the representation

$$A_{11} = D_M\bigl(1 + D_M^{-1}\widehat{\Phi}_g[g]\bigr) = D_M\bigl(1 + \widehat{\Phi}_g[d_M^{-1} * g]\bigr)$$

and then analyzing the Green equation with the operator $1 + \widehat{\Phi}_g[d_M^{-1} * g])$ was considered in [**SS96**]. In this case, the reduction to a triangular form can be carried out with D_M replaced by $D_M + \widehat{\Phi}_g[g]$. In particular, the non-commutative determinant acquires the form

$$\Delta_A = D_X - \widehat{\Phi}_b[b]\bigl(D_M + \widehat{\Phi}_g[g]\bigr)^{-1}\widehat{\Phi}_c[c], \tag{3.23}$$

and sufficient ellipticity conditions are that the operators (3.22) and (3.23) must be elliptic. By the results of the preceding subsection concerning the Green equation, the operator $\bigl(D + \widehat{\Phi}_g[g]\bigr)^{-1}$ again has the form of a pseudodifferential operator plus a Green operator so that the non-commutative determinant (3.23) proves to be a pseudodifferential operator on X by the composition theorems.

We consider the general case without the assumption that the operator (3.22) is almost invertible. How to state the ellipticity conditions then? To this end, note that the almost invertibility of A in the algebra is equivalent to the simultaneous almost invertibility of A^*A and AA^*. Now we note that the almost invertibility of A^*A *implies* the almost invertibility of $(A^*A)_{11}$. (The same is also true of AA^*.) This is a straightforward consequence of the following proposition.

PROPOSITION 3.12. *If*

$$C = \begin{pmatrix} a & b \\ c & d \end{pmatrix}$$

is a non-negative bounded self-adjoint operator matrix ($C = C^* \geq 0$) *and* C *is almost invertible, then* a *is almost invertible.*

PROOF. The proof is based on the following auxiliary lemma:

LEMMA 3.13. *For a bounded operator* $B = B^* \geq 0$, *the following assertions are equivalent:*

(1) B *is not Fredholm;*
(2) *there exists a sequence of vectors* u_j *such that* $\|u_j\| = 1$, $u_j \xrightarrow{\text{weakly}} 0$, *and* $(u_j, Bu_j) \longrightarrow 0$.

Indeed, $(u_j, Bu_j) = \bigl\|\sqrt{B}u_j\bigr\|$, whence we see that the condition $(u_j, Bu_j) \to 0$ is equivalent to $Bu_j \to 0$. Now the implication (2) $\Rightarrow$ (1) is obvious, and to prove (1) $\Rightarrow$ (2) one can take an orthonormal sequence $u_j \in R(f_j(B))$, where

$$f_j(x) = \begin{cases} 1, & x \leq 1/j, \\ 0, & x > 1/j. \end{cases}$$

Such a sequence always exists, since for non-Fredholm B all projections $f_j(B)$ are of infinite rank.

Now it is easy to prove Proposition 3.12. First, it is clear that a is self-adjoint and non-negative. If a is not Fredholm, we find a sequence u_j with the properties stated in the lemma (for $B = a$), and it remains to note that the sequence

$$v_j = {}^t(u_j, 0)$$

can be used to prove that C is not Fredholm, since

$$(v_j, Cv_j) = (u_j, au_j).$$

□

Thus, the operators A^*A and AA^* already fall within the framework of the special case considered in [**SS96**] so that we can state the ellipticity conditions for A in the following form.

DEFINITION 3.14. The morphism (3.14) is said to be *elliptic* if

(1) the operator D_M is elliptic;
(2) the operators $(A^*A)_{11}$ and $(AA^*)_{11}$ are elliptic;
(3) the non-commutative determinants Δ_{A^*A} and Δ_{AA^*} are elliptic.

The above reasoning implies the following theorem.

THEOREM 3.15. *If the morphism* (3.14) *is elliptic, then it is Fredholm.*

REMARK 3.16. We have stated the theorem on the Fredholm property for elements of the algebra $\mathcal{A}$, i.e., morphisms of order zero acting from the space $L^2(X) \oplus L^2(M)$ into itself. Since the multiplication by pseudodifferential operators does not change the structure of the operators in question, the ellipticity conditions and the Fredholm property theorem for operators of non-zero order have the same form. This remark pertains to the index theorem (see the next subsection) as well.

REMARK 3.17. One can prove that the ellipticity conditions in Definition 3.14 are actually redundant and it suffices to require the ellipticity of D_M and the operators in (2) and (3) for only one of the two operators A^*A and AA^*. The situation resembles that for Green equations, where it suffices to require only the solvability of the equation for the symbol of one of the regularizers, right or left.

3.3. The index theorem for elliptic morphisms. Here we state a theorem on the index of morphisms elliptic in the sense of Definition 3.8.

THEOREM 3.18. *Let*

$$A = \begin{pmatrix} D_M & \widehat{\Phi}_c \\ \widehat{\Phi}_b & D_X \end{pmatrix}$$

be an elliptic element of the algebra $\mathcal{A}$ in the sense of Definition 3.8. *Then the index of the operator A is given by the formula*

$$\operatorname{ind} A = \operatorname{ind} D_M + \operatorname{ind} \Delta_A, \tag{3.24}$$

where Δ_A is the non-commutative determinant (3.20).

PROOF. One has

$$\begin{pmatrix} 1 & 0 \\ -\widehat{\Phi}_b D_M^{-1} & 1 \end{pmatrix} A = \begin{pmatrix} D_M & \widehat{\Phi}_c \\ 0 & \Delta_A \end{pmatrix}$$

modulo compact operators. (This identity has actually been used in the proof of the ellipticity conditions.) Since the index of a product of operators is equal to the sum of indices of the factors and the index of a triangular matrix is equal to the sum of indices of its diagonal entries, we arrive at the desired formula (3.24). The proof is complete. □

4. The Sobolev problem

Now we apply the results to Sobolev problems. First, we describe general Sobolev problems and clarify their relation to elliptic morphisms.

4.1. Sobolev problems and elliptic morphisms. Let $X \overset{i}{\hookrightarrow} M$ be an embedding of smooth compact manifolds with $\operatorname{codim} X = \nu$,

$$D\colon H^s(M, E_1) \to H^{s-m}(M, E_2)$$

an elliptic pseudodifferential operator of order m acting in spaces of sections of bundles E_1 and E_2 over M, and

$$B\colon H^s(M, E_1) \to H^{s-b}(M, F)$$

a pseudodifferential operator of order b acting in spaces of sections of the bundles E_1 and F over M. To these data, we assign the problem

$$\begin{cases} Du \equiv f \mod H^{s-m}(M, X, E_2), \\ i^*Bu = g \in H^{s-b-\nu/2}(X, F|_X) \end{cases} \tag{4.1}$$

for an unknown function $u \in H^s(M, E)$. This problem will be called the Sobolev problem. We assume that $s > b + \nu/2$ so that the restriction i^*Bu is well-defined. In (4.1), by $H^{s-m}(M, X, E_2)$ we denote the subspace of $H^{s-m}(X, E_2)$ formed by distributions supported on X.

It was shown in [**Ste76**] (see also [**SS96**]) that problem (4.1) is equivalent to the system

$$\begin{cases} Du + i_{l*}v = f, \\ i^*Bu = g \end{cases} \tag{4.2}$$

for the unknown functions $u \in H^s(M, E)$ and

$$v \in \widetilde{H}^{s-m+\nu/2}(X, \mathcal{J}_X^l(E_2)). \tag{4.3}$$

Here $\mathcal{J}_X^l(E_2)$ is the bundle of lth-order normal jets of the bundle E_2 on X; the elements (4.3) can be represented in an arbitrary special coordinate system (x, t) in the form

$$v = \{v_\alpha\}_{|\alpha| \le l}, \quad v_\alpha \in H^{s-m+\nu/2+|\alpha|}(X, E_2\big|_X), \quad \alpha = (\alpha_1, \dots, \alpha_\nu). \tag{4.4}$$

Next, i_{l*} is the coboundary operator assigning to each element (4.3), (4.4) the sum

$$i_{l*}v = \sum_{|\alpha|=1}^{l} v_\alpha \delta_X^{(\alpha)}$$

consisting of the delta function δ_X on X and its derivatives of order $\leq l$ with coefficients v_α,

$$l = \begin{cases} [m-s-\nu/2] & \text{if } m-s-\nu/2 \text{ is not an integer,} \\ m-s-\nu/2-1 & \text{if } m-s-\nu/2 \text{ is an integer,} \end{cases}$$

and $s < m - \nu/2$.

REMARK 4.1. If $s \geq m - \nu/2$, then the space $H^{s-m}(X, E_2)$ contains no distributions supported on X (a "removable singularity") so that the equation

$$Du = f \mod H^{s-m}(X)$$

is equivalent to the equation

$$Du = f$$

on the entire manifold M for the elliptic operator D. There is no need to equip this operator with any boundary conditions on the submanifold X, and so for $s \geq m - \nu/2$ the Sobolev problem is reduced to an elliptic operator on M.

The operator in system (4.2) is given by a morphism of the special form

$$A = \begin{pmatrix} D & i_{l*} \\ i^*B & 0 \end{pmatrix} \tag{4.5}$$

so that one can apply the general theory of elliptic morphisms to A and find the ellipticity conditions as well as compute the index of an elliptic Sobolev problem.

This will be done in the next subsection. Here we discuss the boundary conditions in problem (4.2) in somewhat more detail. It is of special interest to find conditions under which it is possible to state a Sobolev problem using purely differential boundary conditions. The maximal possible order of the operator B is given by the formula

$$b_{\max} = \begin{cases} [s-\nu/2] & \text{if } m-s-\nu/2 \text{ is not an integer,} \\ s-\nu/2-1 & \text{if } m-s-\nu/2 \text{ is an integer.} \end{cases}$$

The number of boundary conditions in a well-posed Sobolev problem must coincide with the number of the "coboundary conditions" (the dimension of the vector v in (4.3)); this follows from the very form of the ellipticity conditions for morphisms. Hence Sobolev problems with purely differential boundary conditions are possible only for

$$b_{\max} \geq l. \tag{4.6}$$

The equality $b_{\max} = l$ is attained for $s = m/2$.

We divide the real line $\mathbb{R} \ni s$ into the half-open intervals

$$\left[\frac{\nu}{2}+j,\ \frac{\nu}{2}+j+1\right),\quad j \in \mathbb{Z}. \tag{4.7}$$

To each of these intervals there corresponds a certain type of Sobolev problem. If s lies in the interval containing the point $m/2$, then every Sobolev problem with differential conditions is equivalent to the *Sobolev–Dirichlet problem*, in which the values of all derivatives of order less than or equal to

$$l_D = \begin{cases} [(m-\nu)/2] & \text{if } (m-\nu)/2 \text{ is not an integer,} \\ (m-\nu)/2-1 & \text{if } (m-\nu)/2 \text{ is an integer} \end{cases}$$

of the unknown function in transversal directions are specified on X.

As was already mentioned, for $s \geq m - \nu/2$ there are no non-trivial Sobolev problems (a removable singularity).

For s lying to the left of the interval containing the point $m/2$, a well-posed Sobolev problem must necessarily contain *pseudodifferential* boundary conditions.

4.2. The Fredholm property. Let us write out the conditions under which the morphism (4.5) defines a continuous mapping

$$\begin{pmatrix} D & i_{l*} \\ i^*B & 0 \end{pmatrix} : \begin{matrix} H^{s_1}(M) \\ \oplus \\ H^{s_2}(X) \end{matrix} \to \begin{matrix} H^{\sigma_1}(M) \\ \oplus \\ H^{\sigma_2}(X) \end{matrix} \tag{4.8}$$

For our problem, we have

$$a_g = 0, \quad a_c = \{\tau^\alpha\}_{|\alpha|\leq l}, \quad a_b = B(x,p,\tau'). \tag{4.9}$$

Here $B(x,p,\tau')$ is a column consisting of symbols B_j of orders b_j, $j = 1,\dots,J$.

In particular, s_2 and σ_2 in (4.8) are columns[5] rather than numbers:

$$s_2 = \{s_2^\alpha\}_{|\alpha|\leq l}, \quad \sigma_2 = \{\sigma_2^j\}_{j=1,\dots,J},$$

where J is the number of boundary conditions in problem (4.1), i.e., the height of the column a_b.

With regard to this, using the continuity theorems given earlier in this paper, we see that the mapping (4.8) is a continuous operator provided that $\sigma_1 = s_1 - m$,

$$s_2^\alpha = s_1 - m + \frac{\nu}{2} + |\alpha| < 0, \quad \sigma_2^j = s_1 - \frac{\nu}{2} - b_j > 0.$$

Let us now write out the ellipticity condition for the Sobolev problem (4.1) or, equivalently, the morphism (4.2). The first ellipticity condition in Definition 3.8 is equivalent to the ellipticity of the pseudodifferential operator D. Let us write out the second ellipticity condition. One can readily verify that in this case it acquires the form of the ellipticity of the pseudodifferential operator

$$\Delta \overset{\text{def}}{=} -\widehat{\Phi}_b[a_b]D^{-1}\widehat{\Phi}_c[a_c] \tag{4.10}$$

[5]In fact, one should work in the context of Douglis–Nirenberg systems [**DN55**].

on the manifold X. In particular, this means that the operator (4.10) (which is a *matrix operator*) must be represented by a *square matrix*. In other words, the number J of boundary conditions must be equal to the number of coboundary conditions, that is, the number of multi-indices α such that $|\alpha| \le l$.

Thus, the definition of ellipticity of an operator morphism results in the following notion of ellipticity for the Sobolev problem (see [**Ste76**]).

DEFINITION 4.2. The Sobolev problem (4.1) is said to be *elliptic* if

(1) the operator D is an elliptic pseudodifferential operator on the manifold M;

(2) the operator (4.10) is an elliptic pseudodifferential operator on the manifold X.

The following assertion is a straightforward consequence of the theorem on the ellipticity conditions for morphisms.

THEOREM 4.3. *If the Sobolev problem* (4.1) *is elliptic, then it is Fredholm, i.e., has a finite-dimensional kernel and a finite-dimensional cokernel.*

4.3. The Sobolev problem and the adjoint problem. Let us now give an operator interpretation of the Sobolev problem. The problem

$$\begin{cases} Du \equiv f \mod H^{s-m}(M,X;E_2), \\ i^*Bu = g \end{cases}$$

obviously defines an operator

$$S\colon H^s(M,E_1) \to (H^{s-m}(M,E_2)/H^{s-m}(M,X;E_2)) \oplus H^{s-b-\nu/2}(M,F), \tag{4.11}$$

where b is the order of the operator B.

One can also construct the *adjoint* of this operator.

Indeed, if we take into account that $H^{s-m}(M,X;E_2)$ is just the range of the coboundary operator i_{l*}, then the adjoint operator acts in the spaces

$$S^*: \operatorname{Ker}(i_{l*}) \oplus H^{s+b+\nu/2}(M,F) \to H^{-s}(M,E_1),$$

and so the adjoint problem has the form

$$\begin{cases} Du + Bi_*v = f \\ i_l^*u = 0, \end{cases}$$

where i_l^* is the operator taking each function u to its normal jet of order l on the boundary. One can readily see that the almost inverse of the operator (4.11) has the form

$$S^{-1} = \Big(D^{-1}\big(1 - i_{l*}\big(i^*BD^{-1}i_{l*}\big)^{-1}i^*BD^{-1}\big), -D^{-1}i_{l*}\big(i^*BD^{-1}i_{l*}\big)^{-1}\Big). \tag{4.12}$$

4.4. The index formula. Since the Sobolev problem can be reduced to an operator morphism, we obtain the following assertion by using the result of Theorem 3.18.

THEOREM 4.4. *Let*

$$\begin{cases} Du \equiv f \mod H^{s-m}(M, X; E_2), \\ i^* Bu = g \end{cases}$$

be an elliptic Sobolev problem. Then the following index formula holds:

$$\operatorname{ind} S = \operatorname{ind} D + \operatorname{ind} \Delta,$$

where Δ is the pseudodifferential operator

$$\Delta = -i^* B D^{-1} i_{l*}. \tag{4.13}$$

REMARK 4.5. One can readily see that in special coordinates the principal symbol of the pseudodifferential operator (4.13) for $p \neq 0$ is represented by the matrix with entries

$$[\sigma(\Delta)]_{kj} = -\left(\frac{1}{2\pi}\right)^{\nu} \int_{\mathbb{R}^\nu} \frac{\tau^k B_j(x, 0, p, \tau)\, d\tau}{D(x, 0, p, \tau)},$$
$$k = (k_1, \ldots, k_\nu), \quad j = (j_1, \ldots, j_\nu), \quad |k|, |j| \leq l.$$

Thus, we have reduced the computation of the index of the Sobolev problem to the computation of the indices of two pseudodifferential operators on smooth compact manifolds without boundary. The index formula for such operators is well-known (e.g., see [**AS63**]).

TOPOLOGICAL ASPECTS OF RELATIVE ELLIPTIC THEORY

We have already mentioned that relative elliptic theory is an elliptic theory for pairs of the form (M, X), where M is a smooth compact manifold and

$$X \overset{i}{\hookrightarrow} M$$

is a closed submanifold. We have seen that relative elliptic theory involves deep relationships between elliptic pseudodifferential operators on M and X. The aim of this section is to study these relationships from the topological viewpoint. Namely, we describe a number of functors, related to elliptic operators and their symbols, from the category of manifolds with smooth embeddings as morphisms and establish Riemann–Roch type theorems comparing these functors.

5. Preliminaries

Let $\mathcal{K}$ be the category whose objects are smooth compact manifolds M without boundary and each set $\operatorname{Hom}(M_1, M_2)$ of morphisms consists of smooth embeddings $M_1 \hookrightarrow M_2$. This category will be referred to as the *category of (smooth) embeddings*. If X, $M \in \operatorname{Ob}(\mathcal{K})$ and $i\colon X \hookrightarrow M$ is an embedding, then there is a well-defined mapping $i_*\colon TX \longrightarrow TM$, which is also a proper embedding. (The preimage of any compact subset of TM under the mapping i_* is compact.) Using the isomorphisms

$$T^*M \simeq TM, \quad T^*X \simeq TX$$

(depending on the choice of Riemannian metrics on M and X), we obtain a proper embedding $T^*X \longrightarrow T^*M$, which induces a mapping in the K-functor with compact supports, which will be denoted by

$$i^!_{\mathrm{top}}\colon K(T^*M) \longrightarrow K(T^*X). \tag{5.1}$$

The mapping (5.1) is independent of the choice of the Riemannian metrics, and we have a contravariant functor from the category $\mathcal{K}$ into the category $\mathfrak{A}$ of Abelian groups; this functor takes each manifold M to the K-group with compact supports $K(T^*M)$ of the cotangent bundle T^*M and each embedding $i\colon X \hookrightarrow M$ to the morphism (5.1) of the corresponding K-groups, which will naturally be called the *topological pullback*.

In the subsequent sections, we use analytic tools to construct mappings that take each pseudodifferential operator on M to some pseudodifferential operator on the submanifold X, find conditions under which elliptic operators are taken to elliptic operators, and compare the corresponding functors with the K-functor.

6. The metric trace of an elliptic operator

Consider some morphism $X \overset{i}{\hookrightarrow} M$ in the category $\mathcal{K}$. Suppose that the manifolds M and X are equipped with Riemannian metrics ρ_M and ρ_X such that $i^*\rho_M = \rho_X$. Then there is a natural embedding

$$i_m\colon T^*X \longrightarrow T^*M, \tag{6.1}$$

which admits a twofold description, either as the composition of the mappings

$$T^*X \simeq TX \xrightarrow{i_*} TM \simeq T^*M$$

where the isomorphisms are induced by the Riemannian metric, or with the help of the orthogonal expansion

$$T^*M\big|_X = N^*X \oplus T^*X,$$

where N^*X is the conormal bundle of X in M and the orthogonality is understood in the sense of the inner product induced in the fibers of T^*M by the Riemannian metric. One can readily see that the embedding (6.1) is a bundle monomorphism

$$i_m \colon T^*X \longrightarrow T^*M\big|_X,$$

so that its restriction to T_0^*X (where, recall, T_0^*X is T^*X with the zero section deleted) can be interpreted as an embedding (denoted by the same letter)

$$i_m \colon T_0^*X \longrightarrow T_0^*M$$

(or, passing to the quotients with respect to the action of the group $\mathbb{R}_+$ in the fibers, as an embedding $i_m \colon S^*X \longrightarrow S^*M$ of the cosphere bundles). Hence if

$$D \colon H^s(M,E) \longrightarrow H^{s-m}(M,F)$$

is an mth-order pseudodifferential operator on M with principal symbol

$$\sigma(D) \colon \pi_M^*E \longrightarrow \pi_M^*F,$$

where $\pi_M \colon T^*M \longrightarrow M$ is the natural projection, then

$$i_{\mathrm{m}}^*\sigma(D) \colon \pi_X^*\left(E|_X\right) \longrightarrow \pi_X^*\left(F|_X\right) \tag{6.2}$$

is a well-defined mth-order symbol on T_0^*X.

DEFINITION 6.1. The pseudodifferential operator with principal symbol (6.2) is denoted by $i_{\mathrm{m}}^*(D)$ and is called the *metric trace* of the operator D.

REMARK 6.2. Needless to say, the operator $i_{\mathrm{m}}^*(D)$ is uniquely determined only modulo lower-order terms. The term "metric trace" is due to the fact that the embedding (6.1) used in the definition of the restriction of the symbol depends on the metric.

Note that if D is an elliptic operator, i.e., the symbol of D is an isomorphism on T_0^*M, then the same is true of the restriction of the symbol to T_0^*X. It follows that *the metric trace of an elliptic operator is elliptic.* Next, it is obvious that for an appropriate choice of lower-order terms the metric trace commutes with direct sums of operators and the metric trace of a continuous operator family is a continuous family. Since the set of Riemannian metrics on a given manifold is convex (and hence arc-wise connected), we arrive at the following assertion.

PROPOSITION 6.3. *The metric trace is well-defined as a homomorphism*

$$i_{\mathrm{m}}^*\colon \operatorname{Ell}(M) \longrightarrow \operatorname{Ell}(X) \tag{6.3}$$

of Abelian groups of classes of stable homotopy equivalence of elliptic operators on M and X. This homomorphism is independent of the choice of Riemannian metrics on M and X and is functorial: If

$$Y \overset{i_{XY}}{\hookrightarrow} X \overset{i_{MX}}{\hookrightarrow} M$$

is a chain of embeddings and $i_{MY} = i_{MX} \circ i_{XY}$, then

$$i_{MY\,\mathrm{m}}^* = i_{MX\,\mathrm{m}}^* \circ i_{XY\,\mathrm{m}}^*$$

COROLLARY 6.4. *The pair $\operatorname{Ell}_{\mathrm{m}} = (\operatorname{Ell}, \cdot_{\mathrm{m}}^*)$ is a contravariant functor from $\mathcal{K}$ to the category $\mathfrak{A}$ of Abelian groups.*

Now we can state the almost tautological Riemann–Roch theorem comparing the functors K and $\operatorname{Ell}_{\mathrm{m}}$.

THEOREM 6.5. *Let $X \overset{i}{\hookrightarrow} M$ be an embedding of smooth compact manifolds without boundary. Then the diagram*

$$\begin{array}{ccc} \operatorname{Ell}(M) & \xrightarrow{i_{\mathrm{m}}^*} & \operatorname{Ell}(X) \\ {\scriptstyle\chi}\downarrow & & \downarrow{\scriptstyle\chi} \\ K(T^*M) & \xrightarrow[i_{\mathrm{top}}^!]{} & K(T^*X), \end{array} \tag{6.4}$$

where the vertical arrows are given by the difference construction for the principal symbol of an operator, commutes.

Recall that the difference construction assigns the class $[a] \in K(T^*M)$ of the virtual bundle (π_M^*E, π_M^*F, a) with compact support on T^*M to each elliptic symbol $a\colon \pi_M^*E \to \pi_M^*F$ on T_0^*M (e.g., see [**LM98**, p. 210]).

PROOF. This follows from the commutative diagram

$$\begin{array}{ccc} \operatorname{Ell}(M) & \xrightarrow{i_{\mathrm{m}}^*} & \operatorname{Ell}(X) \\ \downarrow & & \downarrow \\ \operatorname{Smbl}(T_0^*M) & \xrightarrow{i_{\mathrm{m}}^*} & \operatorname{Smbl}(T_0^*X) \\ \downarrow & & \downarrow \\ K(T^*M) & \xrightarrow{i_{\mathrm{top}}^!} & K(T^*X), \end{array} \tag{6.5}$$

where the upper square commutes by the construction of i_{m}^* and the lower square commutes since the difference construction is natural. □

7. The pullback of an elliptic operator

The metric trace is not the unique way of assigning a pseudodifferential operator on a submanifold to a pseudodifferential operator on a manifold. In this section, we describe the pullback related to Sobolev problems.

7.1. The pullback of a pseudodifferential operator. We have seen that an important role in the ellipticity condition and the index formula for elliptic morphisms is played by the non-commutative determinant, which is a pseudodifferential operator on X induced by the main operator D_M together with the boundary and coboundary conditions. The simplest version of the non-commutative determinant arises if one considers the elliptic morphism

$$A = \begin{pmatrix} D & i_* \\ i^* & 0 \end{pmatrix}, \tag{7.1}$$

which corresponds to the Sobolev–Dirichlet problem of order 0 for the operator D. The non-commutative determinant is equal to

$$\Delta = -i^* D^{-1} i_*, \tag{7.2}$$

where D^{-1} is the almost inverse of D. Recall that the ellipticity condition for A says that both D and the non-commutative determinant (7.2) must be elliptic. If this condition holds, then A is Fredholm and

$$\operatorname{ind} A = \operatorname{ind} D + \operatorname{ind} \Delta. \tag{7.3}$$

Since the boundary and coboundary conditions in the Sobolev–Dirichlet problem are standard (i.e., depend only on the embedding i), it follows that the correspondence $D \longmapsto \Delta$ commutes with direct sums of operators. Using this correspondence as a model, we define a homomorphism[6]

$$i_a^! \colon \operatorname{PSD}_{<-\nu}(M) \longrightarrow \operatorname{PSD}(X) \tag{7.4}$$

by setting

$$i_a^!(D) = i^* \circ D \circ i_*. \tag{7.5}$$

Here, $\operatorname{PSD}(X)$ is the Abelian semigroup (with respect to direct sum) of pseudodifferential operators on X and $\operatorname{PSD}_{<-\nu} \subset \operatorname{PSD}(M)$ is the sub-semigroup of operators of order $< -\nu$ ($\nu = \operatorname{codim} X$). As follows from the results of the first part, the homomorphism (7.4)–(7.5) is well-defined and, moreover, the principal symbol of the operator (7.5) can be obtained from the principal symbol of D by integrating over the fibers of the conormal bundle. In special local coordinates, this formula reads

$$\sigma\big(i_a^!(D)\big)(x,p) = \left(\frac{1}{2\pi}\right)^{\nu} \int_{\mathbb{R}^\nu} \sigma(D)(x,0,p,\tau)\, d\tau. \tag{7.6}$$

[6]Which differs from the non-commutative determinant in the Sobolev–Dirichlet problem only in sign and the absence of the exponent $^{-1}$ on D.

7.2. The regularized pullback. The pullback thus defined has two disadvantages from the viewpoint of our goals. First, it is well-defined only for operators of sufficiently large negative order (depending on the codimension of the subspace). Secondly, we shall see below that the operator (7.5) need not be elliptic if D is elliptic. It turns out that both disadvantages can be eliminated by the following regularization. Let Δ_M and Δ_X be the Laplace operators on M and X, respectively, with respect to the given Riemannian metrics. We set

$$i_a^!(D)(k) = (1-\Delta_X)^{k-\nu/2} i_a^!\big((1-\Delta_M)^{-k} D\big). \quad k \in \mathbb{R}. \tag{7.7}$$

(If D acts in sections of vector bundles, then one should take Laplacians with coefficients in these bundles.) The operator (7.7) is well-defined for sufficiently large k, namely, for

$$k > \frac{\nu + \operatorname{ord} D}{2}, \tag{7.8}$$

where $\operatorname{ord} D$ is the order of the operator D. The values of k satisfying this inequality will be referred to as *admissible.*

REMARK 7.1. We have included the factor $(1-\Delta_X)^{k-\nu/2}$ in the definition of regularization so as to normalize orders: With this definition, the order of the pullback of an operator coincides with the order of the operator itself. This is not necessary in principle, but useful, as we shall see in Remark 8.6.

THEOREM 7.2. *Let D be an elliptic operator. Then for all sufficiently large k the operator $i_a^!(D)(k)$ defined in* (7.7) *is also elliptic and, moreover, its equivalence class in* $\operatorname{Ell}(X)$ *is independent of k and is uniquely determined by the equivalence class of D in* $\operatorname{Ell}(M)$.

PROOF. Without loss of generality, we can assume (including sufficiently many factors $(1+\Delta)^{-1}$ in D) that D itself has a sufficiently large negative order so that its symbol is bounded and integrable at infinity. The formula for the principal symbol of the operator $i_a^!(D)(k)$ on the sphere $|p| = 1$ (where $|p|$ is the natural norm on the fibers of T^*X) in special coordinates has the form

$$\sigma\big(i_a^!(D)(k)\big) = \left(\frac{1}{2\pi}\right)^{\nu} \int_{\mathbb{R}^\nu} \frac{\sigma(D)(x,0,p,\tau)}{(1+\tau^2)^k}\, d\tau. \tag{7.9}$$

(Here $\tau^2 = \tau_1^2 + \cdots + \tau_\nu^2$.) We set

$$c_{k\nu} = \int_{\mathbb{R}^\nu} \frac{d\tau}{(1+\tau^2)^k}\,. \tag{7.10}$$

Then the sequence $(c_{k\nu}(1+\tau^2)^k)^{-1}$ converges to the delta function as $k \to \infty$, and so (with $\rightrightarrows$ denoting the uniform convergence)

$$c_{k\nu}^{-1}\sigma\big(i_a^!(D)(k)\big)(x,p) \rightrightarrows \left(\frac{1}{2\pi}\right)^{\nu} \sigma(D)(x,0,p,0) \neq 0. \tag{7.11}$$

(The uniform convergence with respect to $(x,p) \in S^*X$ can readily be established.) This proves the desired assertion. □

DEFINITION 7.3. By $i^!_{\mathrm{reg}}(D) \in \mathrm{Ell}(X)$ we denote the class defined in the situation of Theorem 7.2 by the operator $i^!_a(D)(k)$ for sufficiently large k.

REMARK 7.4. One can show that $i^!_a(D)(z)$ is a meromorphic operator-valued function of z. It is of interest to study the residues of this function, but we do not touch upon this topic in the present paper.

7.3. The Riemann–Roch theorem for the regularized pullback. Now we can establish a Riemann–Roch theorem comparing the functors K and $\mathrm{Ell}_{\mathrm{reg}} = (\mathrm{Ell}, \cdot^!_{\mathrm{reg}})$.

THEOREM 7.5. *Let $X \overset{i}{\hookrightarrow} M$ be an embedding of smooth compact manifolds without boundary. Then the diagram*

$$\begin{array}{ccc} \mathrm{Ell}(M) & \xrightarrow{i^!_{\mathrm{reg}}} & \mathrm{Ell}(X) \\ \chi\downarrow & & \downarrow\chi \\ K(T^*M) & \xrightarrow[i^!_{\mathrm{top}}]{} & K(T^*X), \end{array} \tag{7.12}$$

where the vertical arrows are given by the difference construction for the principal symbol of an operator, commutes.

PROOF. It follows from formula (7.11) that $i^!_{\mathrm{reg}} = i^*_{\mathrm{m}}$. It remains to apply Theorem 6.5. □

8. A finitely regularized pullback

The definition of the regularized pullback has the following peculiarity: One does not simply choose a k such that the integral in the definition of the pullback converges (an admissible k); instead, one must take an arbitrary *sufficiently large* k. It is only under this condition that Theorem 7.2 guarantees that a given elliptic operator D produces an elliptic operator, and Theorem 7.5 compares the two functors. The following natural question arises: Is it possible (say, for operators of given order) to use some given *finite* regularization, i.e., choose an admissible k once and for all? Then we would have the additional advantage that the operation $i^!_{\mathrm{reg}}$ would be defined for individual operators rather than only for their equivalence classes in $\mathrm{Ell}(M)$.

The following example shows that the answer is "no". Later in this paper, we shall indicate operator classes for which one can successfully use a finite regularization.

8.1. A counterexample. In the forthcoming argument, we analyze the behavior of symbols over a given point $x \in X$. One can readily obtain specific counterexamples by considering, say, operators with constant coefficients on the tori $T^{m-\nu} \hookrightarrow T^m$.

Thus, let $\mathbb{R}^m = \mathbb{R}^n \oplus \mathbb{R}^\nu$ be the m-dimensional space with coordinates (p, τ), $p \in \mathbb{R}^n$, $\tau \in \mathbb{R}^\nu$. We shall study the pullback (7.9) of a symbol $a(p, \tau)$ of arbitrary given order r for a given admissible k. Without loss of generality, we can assume that k=0 (including the corresponding non-zero factor in a) and $r < -\nu$. For simplicity, we work with scalar symbols, which suffices for constructing a counterexample.

PROPOSITION 8.1. *There exists an elliptic symbol $a(p, \tau)$ of order r such that $i_{\mathrm{m}}^*(a) = 1$ and the pullback $i_a(a)$ is homotopic in the class of elliptic symbols to an arbitrarily chosen elliptic symbol $b(p)$ on the sphere $S^{n-1} \subset \mathbb{R}^n$.*

REMARK 8.2. Prior to proving the proposition, note that it also implies the existence of an elliptic symbol whose pullback under a finite regularization is not elliptic. Indeed, take a symbol b non-homotopic to unity and for the corresponding symbol a consider the family $(1 + p^2 + \tau^2)^{-\alpha} a(p, \tau)$. For $\alpha = 0$, the pullback of this symbol is homotopic to b in the class of elliptic symbols, and as $\alpha \to \infty$, the pullback becomes homotopic to the metric trace, i.e., to unity. Thus, somewhere in between the pullback is necessarily non-elliptic.

However, the existence of an elliptic symbol with non-elliptic pullback can be shown in a much simpler way, by using a second-order differential operator as an example. Namely, let D be the square of the "distorted" Cauchy–Riemann operator,

$$D = \left(\frac{\partial}{\partial x} + i\lambda \frac{\partial}{\partial y} \right)^2,$$

where λ is a non-zero real constant to be chosen later. Then $\sigma(D) = -(p + i\lambda\tau)^2$, and on the sphere $\{|p| = 1\}$, i.e., for $p = \pm 1$, we have

$$\begin{aligned} \sigma\big(i_a^!(D)(k)\big) &= -\left(\frac{1}{2\pi}\right) \int_{-\infty}^{\infty} \frac{d\tau}{(1+\tau^2)^k} - \frac{i\lambda}{\pi} \int_{-\infty}^{\infty} \frac{\tau\, d\tau}{(1+\tau^2)^k} + \frac{\lambda^2}{2\pi} \int_{-\infty}^{\infty} \frac{\tau^2\, d\tau}{(1+\tau^2)^k} \\ &= -\frac{1}{2\pi} \big[c_{k1} - (c_{k-1,1} - c_{k1})\lambda^2 \big]. \end{aligned}$$

(The term of the first order in τ disappears after integration, since the integrand is odd.) For given k, we obtain a pullback with zero principal symbol by setting

$$\lambda = \sqrt{\frac{c_{k1}}{c_{k-1,1} - c_{k1}}}.$$

PROOF OF THE PROPOSITION. We represent the sphere

$$S^{m-1} = \{p^2 + \tau^2 = 1\}$$

as the union

$$S^{m-1} = \bigcup_{\omega \in S^{n-1}} D_\omega$$

of the subsets

$$D_\omega = \big\{ (p, \tau) \in S^{m-1} \;\big|\; p = |p|\, \omega \big\},$$

homeomorphic to the disc $\{|\tau| \leq 1\}$. All these subsets intersect in the common sphere

$$S^{\nu-1} = \partial D_\omega = \{p = 0, |\tau| = 1\}.$$

The formula for the symbol of the pullback on the sphere S^{n-1} acquires the form

$$i_a^!(a)(\omega) = \int_{D_\omega} a(p,\tau)\mu(\tau)\, d\tau, \tag{8.1}$$

where $\mu(\tau)$ is a continuous function in the disk $\{|\tau| \leq 1\}$ that is strictly positive outside the boundary of the disk. (This function, which depends on the order r of the symbol a, is easy to compute, but the specific expression will not be needed here.)

Let $\varepsilon > 0$ be a small positive number. We set

$$\begin{aligned} D_{\omega,\varepsilon} &= D_\omega \cap \{\varepsilon \leq |\tau| \leq 1 - \varepsilon\}, \\ D^+_{\omega,\varepsilon} &= D_{\omega,\varepsilon} \cap \{\tau_1 > 0\}, \\ D^-_{\omega,\varepsilon} &= D_{\omega,\varepsilon} \cap \{\tau_1 < 0\}. \end{aligned}$$

We define a continuous symbol $a(p,\tau)$ on the sphere S^{m-1} by the following conditions, which determine its behavior on each of the sets D_ω:

(1) a is elliptic (i.e., everywhere non-zero on the sphere) and nowhere exceeds 1 in modulus;
(2) $a = 1$ on ∂D_ω;
(3) $\arg a = \arg b(\omega)$ in $D^+_{\omega,\varepsilon}$ for $\omega_1 \geq -\varepsilon/2$;
(4) $a = e^{i \arg b(\omega)}$ in $D^+_{\omega,\varepsilon}$ for $\omega_1 \geq 0$;
(5) $|a| < \varepsilon$ in $D^+_{\omega,\varepsilon}$ for $\omega_1 < -\varepsilon/2$;
(6) $\arg a = \arg b(\omega)$ in $D^-_{\omega,\varepsilon}$ for $\omega_1 \leq \varepsilon/2$;
(7) $a = e^{i \arg b(\omega)}$ in $D^-_{\omega,\varepsilon}$ for $\omega_1 \leq 0$;
(8) $|a| < \varepsilon$ in $D^-_{\omega,\varepsilon}$ for $\omega_1 > \varepsilon/2$.

The existence of a symbol satisfying these conditions is clear, since the sets $\omega_1 \geq -\varepsilon$, $\omega_1 \leq \varepsilon$, etc. are contractible. Condition (2) guarantees that $i^*_{\mathrm{m}}(a) = 1$. Next, for sufficiently small $\varepsilon > 0$ the main contribution to the integral (8.1) is from the domains $D^+_{\omega,\varepsilon}$ for $\omega_1 \geq -\varepsilon$ and $D^-_{\omega,\varepsilon}$ for $\omega_1 \leq \varepsilon$. It follows that the integral is non-zero and $\arg i_a^! a(\omega) \to \arg b(\omega)$ as $\varepsilon \to 0$. Hence, for sufficiently small $\varepsilon > 0$ the symbols $i_a^! a(\omega)$ and $b(\omega)$ are homotopic in the class of elliptic symbols, as desired. □

8.2. The pullback for linear principal symbols. However, there are two cases in which the pullback of an elliptic operator is elliptic even for a finite regularization (7.7) (i.e., for any admissible k). These cases are considered in this and the next subsection.

DEFINITION 8.3. Let M be a Riemannian manifold. We say that D is a pseudodifferential operators with *linear principal symbol* on M if for each point $y \in M$

the restriction of the principal symbol $\sigma(D)$ to the unit sphere $S_y^*M \subset T_y^*M$ coincides with the restriction to the same sphere of some homogeneous linear function on T_y^*M.

REMARK 8.4. This notion of an operator with linear principal symbol depends on the choice of unit spheres, i.e., of the Riemannian metric, with the exception of the operators whose principal symbol is linear on the entire T_y^*M (say, first-order differential operators). The reason for introducing this class of operators is as follows: This class, in contrast with the class of first-order differential operators, is preserved by pullbacks regardless of the normalization. (However, see Remark 8.6 below.) Differential operators are undoubtedly of main interest in specific examples.

THEOREM 8.5. *Let $X \overset{i}{\hookrightarrow} M$ be an embedding of Riemannian manifolds and let D be an elliptic pseudodifferential operator with linear principal symbol on M. Then for each admissible k the pullback $i_{\mathrm{a}}^!(D)(k)$ is an elliptic pseudodifferential operator with linear principal symbol on X homotopic in the class of elliptic operators to the metric trace $i_{\mathrm{m}}^*(D)$ (which is also an elliptic pseudodifferential operator with linear principal symbol on X).*

PROOF. In special coordinates, the principal symbol $\sigma(D)$ has the form

$$\sigma(D) = \big(A(x,t)p + B(x,t)\tau\big)\big(|p|^2 + \tau^2\big)^\alpha, \tag{8.2}$$

where $\alpha = \frac{1}{2}(\operatorname{ord} D - 1)$,

$$A(x,t) = (A_1(x,t), \dots, A_m(x,t)), \quad B(x,t) = (B_1(x,t), \dots, B_\nu(x,t))$$

are some sets of matrices, $Ap = \sum A_j p_j$, and similarly for $B\tau$. On the sphere $\{|p| = 1\}$, the principal symbols of the metric trace and the regularized pullback of D (for admissible values of the regularization parameter k) have the form

$$\sigma\big(i_{\mathrm{m}}^*(D)\big) = i_{\mathrm{m}}^*\sigma(D) = A(x,0)p, \tag{8.3}$$

$$\sigma\big(i_{\mathrm{a}}^!(D)(k)\big) = \left(\frac{1}{2\pi}\right)^\nu c_{k-\alpha,\nu} A(x,0)p. \tag{8.4}$$

(Here, just as in the example in Remark 8.2, the term with $B\tau$ disappears, since the integrand is odd.) The expressions (8.3) and (8.4) coincide up to a non-zero factor; the fact that the metric trace is elliptic has already been discussed. The assertion of the theorem now readily follows. □

REMARK 8.6. One can readily see that with our normalizations (see Remark 7.1) the pullback of a first-order differential operator is (up to lower-order terms) again a first-order differential operator.

By abuse of notation, we denote the pullback of an elliptic operator D with linear principal symbol by $i_{\mathrm{a}}^!(D)$, omitting the regularization parameter k. This is justified, since by the theorem the pullback defines the same element in $\mathrm{Ell}(X)$ for any admissible k. The above theorem implies the following important assertion, which can be treated as the Riemann–Roch theorem for operators with linear principal symbol.

COROLLARY 8.7. *If D is an elliptic operator with linear principal symbol, then*

$$[i_{\mathrm{a}}^{!}(D)] = i_{\mathrm{top}}^{!}\sigma(D) \in K(T^*X), \tag{8.5}$$

where the square brackets stand for the difference construction for the principal symbol.

8.3. Pullbacks and exterior tensor products. Another case in which the ellipticity of the pullback is guaranteed is the case in which the elliptic operator in question has the structure of an exterior tensor product in a neighborhood of X. The exterior tensor product, which takes two elliptic operators A_1 and A_2 on two manifolds to an elliptic operator $A_1 \# A_2$ on the product of these manifolds (or on the bundle, where one of the manifolds is the base and the other is the fiber) plays a fundamental role in elliptic theory. We shall describe this construction in the situation of an embedding $X \overset{i}{\hookrightarrow} M$, which is of interest to us. We identify a tubular neighborhood U of X with a neighborhood of the zero section in the normal bundle NX via the exponential mapping.

Let

$$D_1 \colon C^\infty(X, E_1) \longrightarrow C^\infty(X, F_1) \tag{8.6}$$

be an elliptic differential operator[7] of some order m on X. Next, let

$$D_2 \colon C^\infty(N_x X, E_{2x}) \longrightarrow C^\infty(N_x X, F_{2x}) \tag{8.7}$$

be an elliptic differential operator of the same order m in the fiber N_x smoothly depending on the parameter $x \in X$ and defined at least in U. Here E_{2x} and F_{2x} are the restrictions to N_x of some vector bundles E_2 and F_2 over U. We assume that all bundles in question are Hermitian. The lifts of E_1 and F_1 to U via the natural projection of the normal bundle on the base X will be denoted by the same letters.

DEFINITION 8.8. The *exterior tensor product* of the operator D_1 by the operator family D_2 (or, for brevity, of the operators D_1 and D_2) is the differential operator

$$D_1 \# D_2 \colon C^\infty(U, E_1 \otimes E_2 \oplus F_1 \otimes F_2) \longrightarrow C^\infty(U, F_1 \otimes E_2 \oplus E_1 \otimes F_2) \tag{8.8}$$

in U with principal symbol

$$\sigma(D_1 \# D_2) = \begin{pmatrix} \sigma(D_1) \otimes 1_{\pi^* E_2} & 1_{\pi^* F_1} \otimes \sigma(D_2)^* \\ 1_{\pi^* E_1} \otimes \sigma(D_2) & -\sigma(D_1)^* \otimes 1_{\pi^* F_2} \end{pmatrix}, \tag{8.9}$$

[7]We work only with differential operators to avoid considering non-smooth symbols. The results remain valid for pseudodifferential operators if for operators with non-smooth symbols one uses definitions in the spirit of [**Hör85a**, Chap. 20].

where $\pi\colon T_0^*M \longrightarrow M$ is the natural projection. (In formula (8.9), the symbol $\sigma(D_1)$ is lifted from T^*X to $T^*(NX)$ with the help of the projection

$$T^*(NX) \simeq T(NX) \longrightarrow TX \simeq T^*X,$$

where the isomorphisms are induced by the Hermitian metric.)

THEOREM 8.9. *Let D be an mth-order elliptic operator on M coinciding on X with the exterior tensor product* (8.8):

$$D\big|_X = (D_1 \# D_2)\big|_X.$$

Then for any admissible k the pullback of D on X is elliptic, and

$$i_{\mathrm{a}}^!(D)(k) = i_{\mathrm{m}}^*(D) \in \mathrm{Ell}(X). \tag{8.10}$$

Thus, the pullback is independent of the regularization parameter k up to homotopy equivalence, and so it will be denoted by $i_a^!(D)$ in what follows.

The following theorem can be viewed as a "Riemann–Roch theorem for exterior tensor products".

THEOREM 8.10. *Under the assumptions of the preceding theorem, the formula*

$$[i_{\mathrm{a}}^!(D)] = [D_1]\left([E_{2X}] - [F_{2X}]\right) \in K(T^*X) \tag{8.11}$$

holds, where $[A]$ is the difference construction for the principal symbol of an operator A, $[E]$ is the element generated by a bundle E in the K-group, and E_{2X} and F_{2X} are the restrictions of the bundles E_2 and F_2 to X.

PROOF OF THEOREMS 8.9 AND 8.10. Let us compute the symbols of the metric trace and the pullback of D on S^*X. We have

$$\sigma(i_{\mathrm{m}}^*(D_1 \# D_2)) = \begin{pmatrix} \sigma(D_1)(p) \otimes 1_{\pi^* E_{2X}} & 0 \\ 0 & -\sigma(D_1^*)(p) \otimes 1_{\pi^* F_{2X}} \end{pmatrix}. \tag{8.12}$$

On the other hand,

$$\sigma(i_{\mathrm{a}}^!(D_1 \# D_2))(k) = \begin{pmatrix} c_{k\nu}\sigma(D_1)(p) \otimes 1_{\pi^* E_{2X}} & 1_{\pi^* F_{1X}} \otimes V \\ 1_{\pi^* E_{1X}} \otimes V^* & -c_{k\nu}\sigma(D_1^*)(p) \otimes 1_{\pi^* F_{2X}} \end{pmatrix}, \tag{8.13}$$

where

$$V = \int \frac{\sigma(D_2)(\tau)\, d\tau}{(1+\tau^2)^k}.$$

The symbol (8.13) is obviously elliptic (its matrix is the exterior tensor product of the elliptic symbol $c_{k\nu}\sigma(D_1)(p)$ by the matrix V) and homotopic in the class of elliptic symbols to the matrix (8.12) via the homotopy

$$\begin{pmatrix} [(1-t)c_{k\nu} + t]\sigma(D_1)(p) \otimes 1_{\pi^* E_{2X}} & (1-t)1_{\pi^* F_{1X}} \otimes V \\ (1-t)1_{\pi^* E_{1X}} \otimes V^* & -[(1-t)c_{k\nu} + t]\sigma(D_1^*)(p) \otimes 1_{\pi^* F_{2X}} \end{pmatrix},$$

which proves the assertion of the first theorem. To prove the second theorem, one explicitly computes the element corresponding to the symbol (8.12) in the K-group regarding the fact that $[D_1^*] = -[D_1]$. The proof is complete. □

9. The pullback of geometric operators

In this section, we compute the pullbacks of the Hirzebruch, Euler, Todd (Cauchy–Riemann), and Dirac operators. All these operators are linear and can be represented as exterior tensor products in neighborhoods of submanifolds; thus, they are covered by the assumptions of both Subsections 8.2 and 8.3. Hence their pullbacks are elliptic for any admissible value of the regularization parameter, and the corresponding element of the K-group can be computed according to Theorem 8.10. Geometric operators are a special case of group operators (operators associated with a G-structure of a manifold). We briefly recall this well-known construction mainly to introduce the notation. Details can be found, say, in [**Pal65**].

9.1. Group operators. Let G be a compact Lie group and M a compact manifold without boundary.

DEFINITION 9.1. One says that M is equipped with a *G-structure* if an orientation-preserving isomorphism

$$P \underset{G}{\times} V \simeq TM \tag{9.1}$$

is given, where V is an oriented real G-module of dimension $n = \dim M$ and $P \longrightarrow M$ is a principal G-bundle. (In other words, TM is equipped with the structure of a vector bundle with structure group G associated with the principal bundle P.)

Let M be a manifold with G-structure (9.1), and let $\widetilde{E}$ and $\widetilde{F}$ be finite-dimension complex G-modules. Next, let (with $S(V^*)$ the unit sphere in V^*)

$$\sigma\colon S(V^*) \longrightarrow \operatorname{Hom}(\widetilde{E}, \widetilde{F}) \tag{9.2}$$

be a G-equivariant ($\sigma(g\xi) = g\sigma(\xi)g^{-1}$) mapping such that $\sigma(\xi)$ is invertible for all $\xi \in S(V^*)$. The mapping (9.2) determines an elliptic symbol

$$\sigma_P\colon \pi^*E \longrightarrow \pi^*F \tag{9.3}$$

on the cosphere bundle S^*M, where

$$E = P \underset{G}{\times} \widetilde{E}, \quad F = P \underset{G}{\times} \widetilde{F},$$

and $\pi\colon S^*M \longrightarrow M$ is the natural projection.

DEFINITION 9.2. The operator D on M with symbol $\sigma(D) = \sigma_P$ is called the *group operator* (*G-operator*) on M associated with the principal bundle P and the mapping σ. The method used to construct D from these data is referred to as the *universal construction*.

We also consider group operators with coefficients in an arbitrary vector bundle H. The symbol of such an operator is obtained as the tensor product of the symbol of a group operator by the identity automorphism 1_H.

9.2. The Euler and Hirzebruch operators. Let V be a real vector space of dimension $\dim V = n$ equipped with an inner product. We treat V as an oriented $SO(n)$-module. (To this end, it suffices to choose an orthonormal basis $\varepsilon_1, \dots, \varepsilon_n$ in V.) By

$$\Lambda^{\cdot}_{\mathbb{C}}(V) = \bigoplus_{j=0}^{n} \Lambda^j(V) \otimes \mathbb{C}$$

we denote the complexified direct sum of exterior powers of V. This is also an $SO(n)$-module with respect to the action defined on decomposable vectors by the natural formula

$$g(f_1 \wedge \dots \wedge f_s) \stackrel{\text{def}}{=} g(f_1) \wedge \dots \wedge g(f_s), \quad g \in SO(n), \quad f_j \in V.$$

(The action is extended to the entire $\Lambda^{\cdot}_{\mathbb{C}}(V)$ by linearity.)

We define an $SO(n)$-equivariant mapping

$$\sigma \colon S(V^*) \longrightarrow \operatorname{Hom}(\Lambda^{\cdot}_{\mathbb{C}}(V), \Lambda^{\cdot}_{\mathbb{C}}(V)) \tag{9.4}$$

by the formula

$$\sigma(\xi) = \xi \wedge \cdot + i_\xi, \tag{9.5}$$

where i_ξ is the operator of interior multiplication by the vector ξ, i.e., the substitution of ξ as the first argument into a form (e.g., see [**Stb64**, Chap. 1, Eq. (4.11)]). (We have used the identification $V^* \simeq V$.)

An oriented Riemannian manifold M of dimension n can always be equipped with an $SO(n)$-structure. (The structure group of the tangent bundle is reduced to $SO(n)$.) The universal construction applied to the map (9.4)–(9.5) gives the elliptic operator

$$d + \delta \colon \Gamma(\Lambda^{\cdot}_{\mathbb{C}}(T^*M)) \longrightarrow \Gamma(\Lambda^{\cdot}_{\mathbb{C}}(T^*M)), \tag{9.6}$$

where

$$\Lambda^{\cdot}_{\mathbb{C}}(T^*M) = \bigoplus_{j=0}^{n} \Lambda^j T^*M \otimes_{\mathbb{R}} \mathbb{C} \tag{9.7}$$

is the complexified bundle of exterior forms of all degrees on M and δ is the adjoint of the exterior differentiation operator with respect to the Riemannian metric. The Euler and Hirzebruch operators are obtained from the operator (9.6) by restriction to some subspaces (or, equivalently, by restriction of the symbol (9.4)–(9.5) to some submodules of $\Lambda^{\cdot}_{\mathbb{C}}(V)$).

9.2.1. *The Hirzebruch operator.* Suppose that the module V (and accordingly the manifold M) is even-dimensional: $n = 2N$. Then $\Lambda^{\cdot}_{\mathbb{C}}(V)$ has the involution

$$\alpha = i^{j(j-1)+N} * \colon \Lambda^{\cdot}_{\mathbb{C}}(V) \longrightarrow \Lambda^{\cdot}_{\mathbb{C}}(V), \tag{9.8}$$

which commutes with the group action. Here $*$ is the Hodge operator given by the formula (e.g., see [**Rha55**, Sec. 24])

$$*(\varepsilon_{\beta_1} \wedge \dots \wedge \varepsilon_{\beta_k}) = (-1)^{\sigma(\beta)} \varepsilon_{\beta_{k+1}} \wedge \dots \wedge \varepsilon_{\beta_{2N}}, \tag{9.9}$$

where $\beta = \{\beta_1, \dots, \beta_{2N}\}$ is a permutation of the numbers $1, \dots, 2N$, $\sigma(\beta)$ is the parity of β, and $i^{j(j-1)+N}$ stands for the operator acting by multiplication by the number $i^{j(j-1)+N}$ (where i is the imaginary unit) on the component $\Lambda^j_{\mathbb{C}}(V)$. Thus, we have the direct sum of modules

$$\Lambda^{\cdot}_{\mathbb{C}}(V) = \Lambda_+(V) \oplus \Lambda_-(V), \tag{9.10}$$

where the $\Lambda_\pm(V)$ are the eigenspaces of α corresponding to the eigenvalues ± 1.

The homomorphism (9.4) anti-commutes with α for any ξ:

$$\sigma(\xi)\alpha + \alpha\sigma(\xi) = 0.$$

Hence the restriction

$$\sigma_{\mathcal{H}} = \sigma\big|_{\Lambda_+(V)} : \Lambda_+(V) \longrightarrow \Lambda_-(V) \tag{9.11}$$

is well-defined. An application of the universal construction to $\sigma_{\mathcal{H}}$ gives the operator

$$\mathcal{H}_M = d + \delta \colon \Gamma(\Lambda_+(T^*M)) \longrightarrow \Gamma(\Lambda_-(T^*M)), \tag{9.12}$$

which is called the *Hirzebruch operator* on M. Here

$$\Lambda_\pm(T^*M) \subset \Lambda^{\cdot}_{\mathbb{C}}(T^*M)$$

are the bundles on M associated with the modules $\Lambda_\pm(V)$. The principal symbol of the Hirzebruch operator will be denoted by $\sigma(\mathcal{H}_M)$. The Hirzebruch operator with coefficients in a bundle E over M will be denoted by $\mathcal{H}_M \otimes 1_E$; its principal symbol is $\sigma(\mathcal{H}_M) \otimes 1_{\pi^* E}$.

Now let $X \overset{i}{\hookrightarrow} M$ be an oriented Riemannian submanifold of even dimension $m = 2l$. On X we have a direct sum decomposition of the cotangent bundle:

$$TM|_X = TX \oplus NX. \tag{9.13}$$

Accordingly, one has

$$\Lambda_+(TM|_X) = \big(\Lambda_+(TX) \otimes \Lambda_+(NX)\big) \oplus \big(\Lambda_-(TX) \otimes \Lambda_-(NX)\big), \tag{9.14}$$

$$\Lambda_-(TM_X) = \big(\Lambda_-(TX) \otimes \Lambda_+(NX)\big) \oplus \big(\Lambda_+(TX) \otimes \Lambda_-(NX)\big). \tag{9.15}$$

Moreover, a straightforward computation shows that, according to (9.14)–(9.15), the Hirzebruch operator can be represented on X as the exterior tensor product

$$\mathcal{H}_M\big|_X = (\mathcal{H}_X \# \mathcal{H}_{NX})\big|_X \tag{9.16}$$

of the Hirzebruch operator on X by the family

$$\mathcal{H}_{NX} = \{\mathcal{H}_{N_x X}\}_{x \in X}$$

of Hirzebruch operators in the fibers of the normal bundle NX. By applying Theorem 8.10, we arrive at the following assertion.

THEOREM 9.3. *The pullback on X of the Hirzebruch operator on M satisfies the following relation in $K(T^*X)$:*

$$[i^!_a(\mathcal{H}_M)] = [\mathcal{H}_X]\big([\Lambda_+(NX)] - [\Lambda_-(NX)]\big) \in K(T^*X). \tag{9.17}$$

REMARK 9.4. 1. By Theorem 8.9, the same relation naturally holds for the metric trace of the Hirzebruch operator.

2. Let E be an arbitrary vector bundle over M. Since

$$\left(\mathcal{H}_X \# \mathcal{H}_{NX}\right) \otimes 1_E = \mathcal{H}_X \# \left(\mathcal{H}_{NX} \otimes 1_E\right),$$

where the second factor in the exterior tensor product is the family given by the formula

$$\mathcal{H}_{NX} \otimes 1_E \stackrel{\text{def}}{=} \left\{\mathcal{H}_{N_x X} \otimes 1_E|_{N_x X}\right\}_{x \in X},$$

we arrive at the following formula for the pullback of the Hirzebruch operator with coefficients in E:

$$[i_a^!(\mathcal{H}_M \otimes 1_E)] = [\mathcal{H}_X]\left([\Lambda_+(NX) \otimes E] - [\Lambda_-(NX) \otimes E]\right) \in K(T^*X). \tag{9.18}$$

9.2.2. *The Euler operator.* Now we do not assume that the module V and the manifold M are even-dimensional (even though the results are actually of interest only for the even-dimensional case) and replace (9.10) by the decomposition

$$\Lambda_{\mathbb{C}}^{\cdot}(V) = \Lambda_{\text{odd}}(V) \oplus \Lambda_{\text{even}}(V), \tag{9.19}$$

where $\Lambda_{\text{odd}}(V)$ and $\Lambda_{\text{even}}(V)$ are the subspaces of complex-valued forms of odd and even degrees, respectively.

The restriction of the symbol (9.4)–(9.5) to $\Lambda_{\text{even}}(V)$ acts in the spaces

$$\sigma_{\mathcal{E}} = \sigma\big|_{\Lambda_{\text{even}}(V)} \colon \Lambda_{\text{even}}(V) \longrightarrow \Lambda_{\text{odd}}(V), \tag{9.20}$$

and the universal construction takes $\sigma_{\mathcal{E}}$ to the operator

$$\mathcal{E}_M = d + \delta \colon \Gamma(\Lambda_{\text{even}}(T^*M)) \longrightarrow \Gamma(\Lambda_{\text{odd}}(T^*M)), \tag{9.21}$$

which is called the *Euler operator* on M. Its principal symbol will be denoted by $\sigma(\mathcal{E}_M)$. The Euler operator with coefficients in a vector bundle E over M will be denoted by $\mathcal{E}_M \otimes 1_E$; its principal symbol is $\sigma(\mathcal{E}_M) \otimes 1_{\pi^* E}$.

Now let $X \overset{i}{\hookrightarrow} M$ be an oriented Riemannian manifold of dimension m. The decomposition (9.13) induces the decompositions

$$\Lambda_{\text{even}}(TM|_X) = \left(\Lambda_{\text{even}}(TX) \otimes \Lambda_{\text{even}}(NX)\right) \oplus \left(\Lambda_{\text{odd}}(TX) \otimes \Lambda_{\text{odd}}(NX)\right), \tag{9.22}$$

$$\Lambda_{\text{odd}}(TM_X) = \left(\Lambda_{\text{odd}}(TX) \otimes \Lambda_{\text{even}}(NX)\right) \oplus \left(\Lambda_{\text{even}}(TX) \otimes \Lambda_{\text{odd}}(NX)\right). \tag{9.23}$$

A straightforward computation shows that, in accordance with (9.22)–(9.23), the Euler operator can be represented on X by the exterior tensor product

$$\mathcal{E}_M\big|_X = (\mathcal{E}_X \# \mathcal{E}_{NX})\big|_X \tag{9.24}$$

of the Euler operator on X by the family

$$\mathcal{E}_{NX} = \{\mathcal{E}_{N_x X}\}_{x \in X}$$

of Euler operators in the fibers of the normal bundle NX. Applying Theorem 8.10, we obtain the following assertion.

THEOREM 9.5. *The pullback on X of the Euler operator on M satisfies the following relation in the K-group $K(T^*X)$:*

$$[i_a^!(\mathcal{E}_M)] = [\mathcal{E}_X]\left([\Lambda_{\mathrm{even}}(NX)] - [\Lambda_{\mathrm{odd}}(NX)]\right) \in K(T^*X). \tag{9.25}$$

REMARK 9.6. 1. By Theorem 8.9, the same relation holds for the metric trace of the Euler operator.

2. Let E be an arbitrary vector bundle over M. Arguing as for the Hirzebruch operator, we obtain the following formula for the pullback of the Euler operator with coefficients in E:

$$[i_a^!(\mathcal{E}_M \otimes 1_E)] = [\mathcal{E}_X]\left([\Lambda_{\mathrm{even}}(NX)\otimes E] - [\Lambda_{\mathrm{odd}}(NX)\otimes E]\right) \in K(T^*X). \tag{9.26}$$

9.3. The Todd and Dirac operators. Quite similar considerations give completely analogous results for the pullbacks of the Todd and Dirac operators. Hence, we only give the statements of the corresponding theorems.

9.3.1. *The Todd (Cauchy–Riemann) operator.* Let M be a complex-analytic manifold and $X \overset{i}{\hookrightarrow} M$ be a complex-analytic submanifold. Then the normal bundle NX is a complex vector bundle, and we write

$$\Lambda_{\mathrm{odd}}(NX) = \bigoplus_{j=1 \mod 2} \Lambda^j(NX), \quad \Lambda_{\mathrm{even}}(NX) = \bigoplus_{j=0 \mod 2} \Lambda^j(NX).$$

(Complexification is not needed.)

THEOREM 9.7. *Let $\mathcal{T}_M$ and $\mathcal{T}_X$ be the Todd operators on M and X, respectively, and let E be an arbitrary complex vector bundle over M. Then the following relation holds in the group $K(T^*X)$:*

$$[i_a^!(\mathcal{T}_M \otimes 1_E)] = [\mathcal{T}_X]\left([\Lambda_{\mathrm{even}}(NX)\otimes E] - [\Lambda_{\mathrm{odd}}(NX)\otimes E]\right) \in K(T^*X). \tag{9.27}$$

9.3.2. *The Dirac operator.* Let M be an oriented Riemannian manifold of dimension $2N$ with a spinor structure and let X be an oriented Riemannian submanifold of dimension $2n$ also equipped with a spinor structure which is compatible with that on M.

THEOREM 9.8. *Let $\not{D}_M$ and $\not{D}_X$ be the Dirac operators on M and X, respectively, and let E be an arbitrary complex vector bundle over M. Then the following relation holds in the group $K(T^*X)$:*

$$[i_a^!(\not{D}_M \otimes 1_E)] = [\not{D}_X]\left([\Delta_+(NX)\otimes E] - [\Delta_-(NX)\otimes E]\right) \in K(T^*X), \tag{9.28}$$

where the $\Delta_\pm(V)$ are the half-spin representations of the group $\mathrm{Spin}(2(N-n))$ *corresponding to a* $\mathrm{Spin}(2(N-n))$*-module V.*

Appendix. Fourier Integral Operators

Here we recall some well-known material from the theory of Fourier integral operators and simultaneously introduce the notation used in the main body of the article. For details, the reader is referred to Hörmander's book [**Hör85b**] and also to [**MSS90**]. Our notation is close to that used in the latter book. We do not give any precise statement of theorems from these sources, just because the Fourier integral operators emerging in the theory of Sobolev problems are a special case not covered by the general theorems. Accordingly, the material given here serves only as a motivation for our reasoning. Precise statements of theorems for our special case are given at appropriate places in the main text.

A.1. Homogeneous Lagrangian manifolds. Let X and Y be C^∞ manifolds of dimensions $m = \dim X$ and $n = \dim Y$. The cotangent bundles T^*X and T^*Y bear the canonical 2-forms (symplectic forms) ω_X^2 and ω_Y^2, whose expressions in canonical local coordinates (x, p) and (y, q) on T^*X and T^*Y, respectively, are

$$\omega_X^2 = \sum_{j=1}^{m} dp_j \wedge dx_j \equiv dp \wedge dx, \quad \omega_Y^2 = \sum_{j=1}^{n} dq_j \wedge dy_j \equiv dq \wedge dy.$$

The group $\mathbb{R}_+$ of positive numbers naturally acts on T^*X (and T^*Y) by multiplication in the fibers. Let us equip the product $T^*X \times T^*Y = T^*(X \times Y)$ with the canonical form equal to the difference of the canonical forms on the factors:

$$\omega_{XY}^2 \stackrel{\text{def}}{=} \omega_X^2 - \omega_Y^2.$$

A submanifold

$$L \stackrel{i}{\hookrightarrow} T^*(X \times Y)$$

is called a *homogeneous Lagrangian manifold* if the following conditions hold:

Lagrangian property: $i^*\omega_{XY}^2 \equiv 0$;
maximum dimension: $\dim L = m + n$;
homogeneity: L is $\mathbb{R}_+$-invariant, and

$$L \subset T_0^*(X \times Y) \equiv T^*(X \times Y) \setminus 0,$$

where 0 is the zero section of $T^*(X \times Y)$.

The following additional condition is often imposed on a homogeneous Lagrangian manifold L.

Condition A.1. The manifold L is closed in $T_0^*(X \times Y)$ and is contained in $T_0^*X \times T_0^*Y$.

In particular, it is used in general boundedness and composition theorems for Fourier integral operators (see below). However, the Lagrangian manifolds corresponding to most of the operators considered in this paper (including boundary, coboundary, and Green operators) do not satisfy this condition. That is why they need special treatment and display a variety of new effects.

A.2. Local description of homogeneous Lagrangian manifolds. Let L be a homogeneous Lagrangian manifold in $T^*(X \times Y)$. By the lemma on local coordinates [**Arn67**], the manifold L can be covered by $\mathbb{R}_+$-invariant coordinate charts such that the coordinates in each of the charts are of the form $(x_I, p_{\overline{I}}, y_J, q_{\overline{J}})$, where (x,p) and (y,q) are canonical local coordinates on T^*X and T^*Y, respectively, $I \subset \{1,\dots,m\}$, $J \subset \{1,\dots,n\}$, $\overline{I} = \{1,\dots,m\} \setminus I$, $\overline{J} = \{1,\dots,m\} \setminus J$, $x_I = \{x_j\}_{j\in I}$, etc. These coordinates are referred to as *canonical coordinates*, and the corresponding *canonical chart* on L will be denoted by U_{IJ}. Thus, we have the coordinate map

$$\psi_{IJ} \equiv (x_I, p_{\overline{I}}, y_J, q_{\overline{J}})\colon U_{IJ} \longrightarrow V_{IJ} \subset \mathbb{R}^{m+n}.$$

In the chart U_{IJ}, the manifold L is specified by a uniquely determined *generating function* (*action*) $S_{IJ}(x_I, p_{\overline{I}}, y_J, q_{\overline{J}})$ defined in V_{IJ} and homogeneous of degree one[8] in $(p_{\overline{I}}, q_{\overline{J}})$ such that the equations of L read

$$p_I = \frac{\partial S_{IJ}}{\partial x_I}, \quad x_{\overline{I}} = -\frac{\partial S_{IJ}}{\partial p_{\overline{I}}}, \quad q_J = -\frac{\partial S_{IJ}}{\partial y_J}, \quad y_{\overline{J}} = \frac{\partial S_{IJ}}{\partial q_{\overline{J}}}. \tag{A.1}$$

(The signs in these equations are due to the fact that ω_X^2 and ω_Y^2 occur with opposite signs in ω_{XY}^2.)

A.3. Composition of homogeneous Lagrangian manifolds. Let $L_1 \subset T^*(M_1 \times M_2)$ and $L_2 \subset T^*(M_2 \times M_3)$ be homogeneous Lagrangian manifolds. We can view these manifolds as *relations* in $T^*M_1 \times T^*M_2$ and $T^*M_2 \times T^*M_3$, respectively (that is why they are called "homogeneous canonical relations" in Hörmander's writings), and then we can define the composition $L_1 \circ L_2$ as the product of these relations:

$$L_1 \circ L_2 \overset{\text{def}}{=} \big\{(u,v) \in T^*M_1 \times T^*M_3 \bigm| \exists w \in T^*M_2\colon (u,w) \in L_1,\ (w,v) \in L_2\big\}. \tag{A.2}$$

If the composition (A.2) is *clean* (i.e., $L_1 \times L_2$ intersects $T^*M_1 \times \Delta_{T^*M_2} \times T^*M_3$, where $\Delta_{T^*M_2}$ is the diagonal in $T^*M_2 \times T^*M_2$, in a manifold $\widetilde{L}$ with tangent plane everywhere equal to the intersection of tangent planes of intersecting manifolds), *proper* (i.e., the projection $\widetilde{L} \longrightarrow T^*(M_1 \times M_3)$ is proper), and *connected* (i.e., the set $L_{(u,v)}$ of points of $\widetilde{L}$ projected to the same point $(u,v) \in L_1 \times L_2$ is connected), then the composition $L = L_1 \circ L_2$ is also a homogeneous Lagrangian manifold. The dimension $e = \dim L_{(u,v)}$ of the fiber of the projection $\widetilde{L} \longrightarrow L$ is called the *excess* of the clean intersection.

A.4. Definition of Fourier integral operators. Let L be a homogeneous Lagrangian manifold in $T^*(X \times Y)$. Suppose that L is equipped with a *quantized measure* μ (e.g., see [**NOSS81**] and references therein). Then Maslov's canonical operator $\mathcal{K} \equiv \mathcal{K}_{L,\mu}$ that takes asymptotically homogeneous functions on L to

[8] Since L is homogeneous, it follows that $\overline{I}$ or $\overline{J}$ is not empty.

distributions on $X \times Y$ is well-defined (e.g., see [**NOSS81**] and the appendix in [**MSS90**]; for the original (small parameter) canonical operator, see [**Mas72**]).

A *Fourier integral operator*

$$\widehat{\Phi}\colon C^\infty(Y) \longrightarrow \mathcal{D}'(X)$$

associated with L in $T^*(X \times Y)$ can be defined as follows. One takes a smooth asymptotically homogeneous amplitude function a on L and takes $\widehat{\Phi}$ to be the operator with the (distributional) Schwartz kernel $\mathcal{K}a$. The local description can be obtained in the following manner. Let us cover L with canonical charts. Consider some canonical chart U_{IJ}. Let $a(x_I, p_{\overline{I}}, y_J, q_{\overline{J}})$ be a smooth function supported in V_{IJ} and satisfying the estimates

$$\text{(A.3)}\qquad \left|\frac{\partial^{\alpha+\beta+\gamma+\delta} a(x_I, p_{\overline{I}}, y_J, q_{\overline{J}})}{\partial x_I^\alpha \partial p_{\overline{I}}^\beta \partial y_J^\gamma \partial q_{\overline{J}}^\delta}\right| \le C_{\alpha\beta\gamma\delta}(1 + |p_{\overline{I}}| + |q_{\overline{J}}|)^{r-|\beta|-|\delta|},$$
$$|\alpha| + |\beta| + |\gamma| + |\delta| = 0, 1, 2, \dots$$

for some $r \in \mathbb{R}$. (In practice, one mostly deals with functions $a(x_I, p_{\overline{I}}, y_J, q_{\overline{J}})$ asymptotically homogeneous in $(p_{\overline{I}}, q_{\overline{J}})$.) The local Fourier integral operator with amplitude a associated with the Lagrangian manifold L in the chart U_{IJ} is defined as the integral operator $\widehat{\Phi}(L, a)$ with Schwartz kernel

$$\text{(A.4)}\qquad \begin{aligned} K(x, y) &= \overline{\mathcal{F}}_{p_{\overline{I}} \to x_{\overline{I}}} \mathcal{F}_{q_{\overline{J}} \to y_{\overline{J}}} \{ e^{iS_{IJ}(x_I, p_{\overline{I}}, y_J, q_{\overline{J}})} \widetilde{a}(x_I, p_{\overline{I}}, y_J, q_{\overline{J}}) \} \\ &= \frac{i^{(m-n)/2}}{(2\pi)^{(m+n)/2}} \iint e^{i\left[S_{IJ}(x_I, p_{\overline{I}}, y_J, q_{\overline{J}}) + p_{\overline{I}} x_{\overline{I}} - q_{\overline{J}} y_{\overline{J}}\right]} \widetilde{a}(x_I, p_{\overline{I}}, y_J, q_{\overline{J}})\, dp_{\overline{I}} dq_{\overline{J}}, \end{aligned}$$

where $\widetilde{a} = a\sqrt{\mu_I}$ and μ_I is the density of the measure μ in the coordinates of the chart U_{IJ}. (Here $\mathcal{F}$ is the Fourier transform and $\overline{\mathcal{F}}$ the inverse Fourier transform. Products like $x_I p_I$ are defined as $x_I p_I = \sum_{j \in I} x_j p_j$.) Thus, $\widehat{\Phi}(L, a)$ acts as follows:

$$\text{(A.5)}\qquad \left[\widehat{\Phi}(L, a)\psi\right](x) = \int_Y K(x, y)\psi(y)\, dy.$$

Global Fourier integral operators can be defined as sums of local Fourier integral operators in the charts, with usual precautions taken to ensure all sums to be locally finite etc. (This is, of course, not needed if $L/\mathbb{R}_+$ is compact and hence can be covered by finitely many canonical charts.) The stationary phase method allows one to compare Fourier integral operators associated with charts that have non-empty intersections, and for the case of asymptotically homogeneous amplitudes one can define the notion of principal symbol of a Fourier integral operator, which is globally well-defined as a function on L. If a is asymptotically homogeneous, then the leading part of a is called the *principal symbol* of the Fourier integral operator. However, we do not dwell on this, since in the main text we only deal with local expressions like (A.4).

A.5. Pseudodifferential operators as Fourier integral operators. An important example of Fourier integral operators occurs if one takes the diagonal

$$L_X = \{(x,p,x,p) \mid (x,p) \in T_0^*X\} \subset T^*(X \times X) \tag{A.6}$$

as the homogeneous Lagrangian manifold. The canonical coordinates on L_X can always be taken in the form (x,p') (where primes are used on the variables pertaining to the second factor T^*X in the product $T^*X \times T^*X$), the generating function is $S = p'x$, and one has

$$\left[\widehat{\Phi}(L_X,a)\psi\right](x) = \left(\frac{1}{2\pi}\right)^m \iint e^{ip'(x-x')}a(x,p')\,dp'dx.$$

We see that $\widehat{\Phi}(L_X,a)$ is just the pseudodifferential operator with symbol $a(x,p)$.

Fourier integral operators possess a variety of properties, the main of which are boundedness properties, composition formulas, and the passage to the L^2-adjoint operator. Let us recall these properties.

A.6. Boundedness theorems. Boundedness theorems [**Hör85b**, Theorem 25.3.8] claim that Fourier integral operators associated with a Lagrangian manifold

$$L \overset{i}{\hookrightarrow} T^*(X \times Y)$$

extend to bounded operators between Sobolev spaces $H^s(Y)$ and $H^l(X)$ provided that L satisfies condition A.1; the lower and upper bounds on the order $s-l$ depend on the order of the amplitude as well as on the co-rank of the lift of the canonical form ω_X^2 (or, which is the same, ω_Y^2) to L.

A.7. Composition theorems. Composition theorems for Fourier integral operators proved in [**Hör85b**, Theorem 25.2.3] establish that under mild conditions the product of two Fourier integral operators is again a Fourier integral operator. Namely, let $L_1 \subset T^*(M_1 \times M_2)$ and $L_2 \subset T^*(M_2 \times M_3)$ be Lagrangian manifolds and let $\widehat{\Phi}_i$ be Fourier integral operators associated with L_i, $i = 1,2$, respectively. Next, suppose that the composition $L = L_1 \circ L_2$ (see (A.2)) is clean and that both L_1 and L_2 satisfy condition A.1. Then the product $\widehat{\Phi}_1\widehat{\Phi}_2$ is a Fourier integral operator associated with the Lagrangian manifold L. Moreover, if a_i is the principal symbol of $\widehat{\Phi}_i$ (defined on the Lagrangian manifold L_i), $i = 1,2$, then the principal symbol $a(u,v)$ of $\widehat{\Phi}_1\widehat{\Phi}_2$ at a point $(u,v) \in L$ is equal to the integral of the direct product $a_1 \times a_2 = a_1(u,w)a_2(w,v)$ over the fiber $L_{(u,v)}$ of the bundle $\widetilde{L} \longrightarrow L$ (see Subsection A.3). A special case of this theorem occurs if one of the manifolds L_1 and L_2 is the diagonal (or the identity relation) I. Since $I \circ L = L$ (and likewise $L \circ I = L$) for any Lagrangian manifold for which the composition is well-defined, we see that the product of a Fourier integral operator associated with a Lagrangian manifold L by a pseudodifferential operator is a Fourier integral operator associated with the same manifold L. Moreover, the excess of such compositions is always zero, and so the principal symbol of the product is just the product of the principal symbols of the factors (taken at appropriate points).

A.8. L^2–adjoints of Fourier integral operators. Let $\widehat{\Phi}$ be a Fourier integral operator associated with a homogeneous Lagrangian manifold

$$L \subset T^*(M_1 \times M_2).$$

Then the L^2-adjoint $\widehat{\Phi}^*$ of the operator $\widehat{\Phi}$ is also a Fourier integral operator. It is associated with the Lagrangian manifold

$$ {}^{T}L = \{(v,u) \mid u \in T^*M_1,\ v \in T^*M_2,\ (u,v) \in L\} \subset T^*(M_2 \times M_1) \tag{A.7}$$

obtained from L by transposing the factors.

A.9. Fourier integral operators on sections of vector bundles. All preceding considerations pertain to the case of Fourier integral operators acting in function spaces, but the generalization to the case of operators acting in spaces of sections of vector bundles is straightforward. The (principal) symbol of a Fourier integral operators acting between spaces of sections of vector bundles is a morphism between the lifts of these bundles to the Lagrangian manifold.

References

[Arn67] V. I. Arnold, *On a characteristic class entering into conditions of quantization*, Funktsional. Anal. i Prilozhen. **1** (1967), no. 1, 1–14.

[AS63] M. F. Atiyah and I. M. Singer, *The index of elliptic operators on compact manifolds*, Bull. Amer. Math. Soc. **69** (1963), 422–433.

[DN55] A. Douglis and L. Nirenberg, *Interior estimates for elliptic systems of partial differential equaions*, Comm. Pure Appl. Math. **8** (1955), 503–538.

[Esk73] G. I. Eskin, *Boundary Value Problems for Elliptic Pseudodifferential Equations*, Nauka, Moscow, 1973 [Russian].

[Hör83a] L. Hörmander, *The Analysis of Linear Partial Differential Operators. I*, Springer-Verlag, Berlin–Heidelberg–New York–Tokyo, 1983.

[Hör83b] ———, *The Analysis of Linear Partial Differential Operators. II*, Springer-Verlag, Berlin–Heidelberg–New York–Tokyo, 1983.

[Hör85a] ———, *The Analysis of Linear Partial Differential Operators. III*, Springer-Verlag, Berlin–Heidelberg–New York–Tokyo, 1985.

[Hör85b] ———, *The Analysis of Linear Partial Differential Operators. IV*, Springer-Verlag, Berlin–Heidelberg–New York–Tokyo, 1985.

[LM98] G. Luke and A. S. Mishchenko, *Vector Bundles and Their Applications*, Kluwer Academic Publishers, Dordrecht, 1998.

[Mas72] V. P. Maslov, *Théorie des perturbations et méthod asymptotiques*, Dunod, Paris, 1972 [French transl. from the Russian 1965 edition].

[MSS90] A. Mishchenko, V. Shatalov, and B. Sternin, *Lagrangian manifolds and the Maslov operator*, Springer-Verlag, Berlin–Heidelberg, 1990.

[Mon71] L. Boutet de Monvel, *Boundary problems for pseudodifferential operators*, Acta Math. **126** (1971), 11–51.

[NOSS81] V. E. Nazaikinskii, V. G. Oshmyan, B. Yu. Sternin, and V. E. Shatalov, *Fourier integral operators and the canonical operator*, Uspekhi Mat. Nauk **36** (1981), no. 2, 81–140.

[NS66a] S. P. Novikov and B. Yu. Sternin, *Traces of elliptic operators on submanifolds and K-theory*, Dokl. Akad. Nauk SSSR 1265–1268 **170** (1966), no. 5, 1265–1268.

[NS66b] ———, *Elliptic operators and submanifolds*, Dokl. Akad. Nauk SSSR 525–528 **171** (1966), no. 6, 525–528.

[Pal65] R. S. Palais, *Seminar on the Atiyah–Singer Index Theorem*, Princeton Univ. Press, Princeton, NJ, 1965.

[Rha55] G. de Rham, *Variétès différentiables*, Hermann, Paris, 1955.

[Sch91] B.-W. Schulze, *Pseudodifferential Operators on Manifolds with Singularities*, North–Holland, Amsterdam, 1991.

[Sob37] S. L. Sobolev, *On a boundary value problem for the polyharmonic equation*, Mat. Sb. **2** (1937), no. 3, 467–500.

[Stb64] S. Sternberg, *Lectures on Differential Geometry*, Prentice Hall, Englewood Cliffs, N.J., 1964.

[Ste64] B. Yu. Sternin, *General boundary value problems for elliptic equations in a region whose boundary consists of manifolds of different dimension*, Dokl. Akad. Nauk SSSR **159** (1964), no. 5, 992–994.

[Ste66] ______, *Elliptic and parabolic problems on manifolds with boundary consisting of components of different dimension*, Tr. Mosk. Mat. Obs. **15** (1966), 38–108.

[Ste67] ______, *Elliptic (co)boundary morphisms*, Dokl. Akad. Nauk SSSR **172** (1967), no. 1, 44–47.

[Ste71] ______, *Topological aspects of the Sobolev problem*, Moscow Institute of Electronic Engineering, Moscow, 1971 [Russian].

[Ste76] ______, *Relative elliptic theory and the Sobolev problem*, Dokl. Akad. Nauk SSSR **230** (1976), no. 2, 287–290.

[SS92a] B. Yu. Sternin and V. E. Shatalov, *Fourier–Maslov integral operators associated with a bundle and an extension of the algebra of pseudodifferential operators*, Dokl. Akad. Nauk **323** (1992), no. 1, 25–29.

[SS92b] ______, *A class of non-local elliptic problems*, Dokl. Akad. Nauk **323** (1992), no. 2, 245–249.

[SS94] ______, *Extension of the algebra of pseudodifferential operators and some non-local elliptic problems*, Mat. Sb. **185** (1994), no. 3, 117–159.

[SS96] ______, *Relative elliptic theory and the Sobolev problem*, Mat. Sb. **187** (1996), no. 11, 115–144.

[VE65a] M. I. Vishik and G. I. Eskin, *Convolution equations in a bounded region*, Uspekhi Mat. Nauk **20** (1965), no. 3, 89–152.

[VE65b] ______, *Elliptic convolution equations in a bounded region and their applications*, Uspekhi Mat. Nauk **22** (1965), no. 1, 15–76.

Institute for Problems in Mechanics, Russian Academy of Sciences
Current address: Institut für Mathematik, Universität Potsdam
E-mail address: `nazaik@math.uni-potsdam.de`

Independent University of Moscow
Current address: Institut für Mathematik, Universität Potsdam
E-mail address: `sternin@mail.ru`

Subseries Advances in Partial Differential Equations

Subseries editors:

Bert-Wolfgang Schulze **Institut für Mathematik** **Universität Potsdam** **14415 Potsdam** **Germany**	**Michael Demuth** **Institut für Mathematik** **Technische Universität Clausthal** **38678 Clausthal-Zellerfeld** **Germany**
Sergio Albeverio **Institut für Angewandte Mathematik** **Universität Bonn** **53115 Bonn** **Germany**	**Elmar Schrohe** **Institut für Mathematik** **Universität Hannover** **30060 Hannover** **Germany**

The subseries *Advances in Partial Differential Equations* is intended to report on recent developments in all areas of partial differential equations, in particular, on those belonging to the tradition defined by the series *Operator Theory: Advances and Applications* and new areas. The *Advances* volumes will consist primarily of expository research articles, presenting both an overview of the state of a field and new results, in such areas as microlocal analysis and its applications to asymptotic phenomena, mathematical physics, spectral theory, as well as symplectic geometry, the analysis of singularities of solutions, and geometric analysis. Selected articles of this kind may also appear as a separate monographic volume. A certain number of short communications will also be published.

Proposals for manuscripts may be sent to B.-W. Schulze or to one of the other subseries editors.

Published books in the subseries Advances in Partial Differential Equations:

OT 151: Gil, J.B. / Krainer, T. / Witt, I. (Eds.)
Aspects of Boundary Problems in Analysis and Geometry (2004)
ISBN 3-7643-7069-6

OT 145: Albeverio, S. / Demuth, M. / Schrohe, E. / Schulze, B.-W. (Eds.)
Nonlinear Hyperbolic Equations, Spectral Theory, and Wavelet Transformations (2003)
ISBN 3-7643-2168-7

In the first contribution, the singularities of the solutions of several classes of non-linear partial differential equations are investigated. Applications concern the Monge-Ampère equation, quasi-linear systems arising in fluid mechanics as well as integro-differential equations for media with memory. There follows an article on L_p-L_q decay estimates for Klein-Gordon equations with time-dependent coefficients, explaining, in particular, the influence of the relation between the mass term and the wave propagation speed. The next paper addresses questions of local existence of solutions, blow-up criteria, and C^∞ regularity for quasilinear weakly hyperbolic equations. Spectral theory of semibounded selfadjoint operators is the topic of a further contribution, providing upper and lower bounds for the bottom eigenvalue as well as an upper bound for the second eigenvalue in terms of capacitary estimates. The wavelet transformation and the Gabor transformation are two distinguished tools in signal processing with a wide range of applications. The result presented here concerns the construction of an interpolating family between them.

OT 138: Albeverio, S. / Demuth, M. / Schrohe, E. / Schulze, B.-W. (Eds.)
Parabolicity, Volterra Calculus, and Conical Singularities (2002)
ISBN 3-7643-6906-X

This volume highlights the analysis on noncompact and singular manifolds within the framework of the cone calculus with asymptotics. The first article presents a calculus for pseudodifferential operators with an anisotropic analytic parameter. The subsequent paper develops an algebra of Mellin operators on the infinite space-time cylinder. It is shown how timelike infinity can be treated as a conical singularity. In the third text – the central article of this volume – the authors use these results to obtain precise information on the long-time asymptotics of solutions to parabolic equations and to construct inverses within the calculus. There follows a factorization theorem for meromorphic symbols: It is proven that each of these can be decomposed into a holomorphic invertible part and a smoothing part containing all the meromorphic information. It is expected that this result will be important for applications in the analysis of nonlinear hyperbolic equations. The final article addresses the question of the coordinate invariance of the Mellin calculus with asymptotics.

OT 125: Gil, J.B. / Grieser, D. / Lesch, M. (Eds.)
Approaches to Singular Analysis (2001)
ISBN 3-7643-6518-8

The purpose of this publication is to present, in one book, various approaches to analytic problems that arise in the context of singular spaces.

MIX
Papier aus verantwortungsvollen Quellen
Paper from responsible sources
FSC® C105338
FSC
www.fsc.org

If you have any concerns about our products,
you can contact us on
ProductSafety@springernature.com

In case Publisher is established outside the EU,
the EU authorized representative is:
Springer Nature Customer Service Center GmbH
Europaplatz 3, 69115 Heidelberg, Germany

Printed by Libri Plureos GmbH
in Hamburg, Germany